ESSENTIALS OF COGNITIVE NEUROSCIENCE

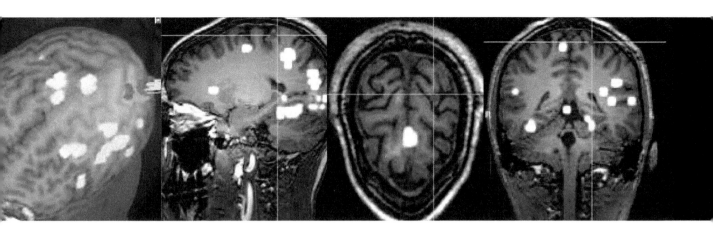

ESSENTIALS OF COGNITIVE NEUROSCIENCE

SECOND EDITION

Bradley R. Postle

WILEY

VP AND EDITORIAL DIRECTOR	Veronica Visentin
EXECUTIVE EDITOR	Glenn Wilson
EDITORIAL ASSISTANT	Jannil Perez
EDITORIAL MANAGER	Judy Howarth
CONTENT MANAGEMENT DIRECTOR	Lisa Wojcik
CONTENT MANAGER	Nichole Urban
SENIOR CONTENT SPECIALIST	Nicole Repasky
PRODUCTION EDITOR	Indirakumari S
COVER PHOTO CREDIT	© Cover image modified from "Rear of Sensory Homunculus" by "Mpj29" is licensed under CC BY-SA;(brain image) Courtesy of Bradley Postle

Founded in 1807, John Wiley & Sons, Inc. has been a valued source of knowledge and understanding for more than 200 years, helping people around the world meet their needs and fulfill their aspirations. Our company is built on a foundation of principles that include responsibility to the communities we serve and where we live and work. In 2008, we launched a Corporate Citizenship Initiative, a global effort to address the environmental, social, economic, and ethical challenges we face in our business. Among the issues we are addressing are carbon impact, paper specifications and procurement, ethical conduct within our business and among our vendors, and community and charitable support. For more information, please visit our website: *www.wiley.com/go/citizenship*.

ISBN: 978-1-119-67616-4 (PBK)
ISBN: 978-1-119-67410-8 (EVALC)

Library of Congress Cataloging-in-Publication Data

Names: Postle, Bradley R., author.
Title: Essentials of cognitive neuroscience / Bradley R. Postle.
Description: Second edition. | Hoboken, NJ : Wiley, [2020] | Includes
   bibliographical references and index.
Identifiers: LCCN 2020001686 (print) | LCCN 2020001687 (ebook) | ISBN
   9781119674153 (paperback) | ISBN 9781119674092 (adobe pdf) | ISBN
   9781119607687 (epub)
Subjects: LCSH: Cognitive neuroscience—Textbooks. | Clinical
   neuropsychology—Textbooks.
Classification: LCC QP360 .P675 2020 (print) | LCC QP360 (ebook) | DDC
   612.8/233—dc23
LC record available at https://lccn.loc.gov/2020001686
LC ebook record available at https://lccn.loc.gov/2020001687

BRIEF CONTENTS

CONTENTS

SECTION II: SENSATION, PERCEPTION, ATTENTION, AND ACTION 87

4 Sensation and Perception of Visual Signals 90

SECTION III: MENTAL REPRESENTATION 241

10 Visual Object Recognition and Knowledge 243

11 Neural Bases of Memory 267

SECTION IV: HIGH-LEVEL COGNITION **363**

PREFACE

Does the world need another cognitive neuroscience textbook? Well, "need" is such a loaded word. What I'll try to do here is summarize my motivation for writing this book and, in the process, convey the pedagogical vision behind choices about style, content, and organization.

For nearly 20 years I have taught cognitive neuroscience to undergraduate and graduate students at the University of Wisconsin–Madison. For undergraduates in the honors program, I offer a seminar that is linked to a large-enrollment lecture course on cognitive psychology. The lecture course is a survey of cognitive psychology, spanning sensation and perception to decision-making, language, and consciousness. The honors seminar, the impetus for this book, is populated by 12–15 students who are concurrently enrolled in the lecture course. In the seminar, which meets only once per week, for only 60 minutes (sic), we discuss the cognitive neuroscience of what's been covered in that week's cognitive psychology lectures. The modal student in this seminar has a background in experimental design and in probability and statistics for experimental psychology; only some have any formal training in biology, and fewer still in neuroscience. In this context, I have found that a textbook that attempts to cover "everything that we know" in the field is not effective – it will either be too superficial in its coverage to support substantive discussions in class or it will be inappropriately lengthy and dense, a cognitive neuroscience analogue of the >1000-page medical-school neuroscience textbook that I lugged around as a graduate student in systems neuroscience. Another approach, which is to assign papers from the primary literature, has proven to be an unsatisfactory alternative, because it results in an inordinate amount of class time being taken up with the remedial work of filling in background that the assigned papers don't cover, and explaining methods so that students can fully understand how the experiments were performed and how their results can be interpreted.

Essentials of Cognitive Neuroscience aspires to be a textbook that will support high-level, substantive engagement with the material in settings where contact hours and/or expectations for outside-of-the-classroom preparation time may be constrained. It introduces and explicates key principles and concepts in cognitive neuroscience in such a way that the reader will be equipped to critically evaluate the ever-growing body of findings that the field is generating. For some students, this knowledge will be needed for subsequent formal study, and for all readers it will be needed to evaluate and interpret reports about cognitive neuroscience research that make their way daily into the news media and popular culture. The book seeks to do so in a style that will give the student a sense of what it's like to be a cognitive neuroscientist: when confronted with a problem, how does one proceed? How does one read and interpret research that's outside of one's subarea of specialization? How do two scientists' advancing mutually incompatible models interrelate? Most importantly, what does it feel like to partake in the wonder and excitement of this most dynamic and fundamental of sciences?

To translate these aspirations into a book, I have made the following pedagogical and organizational choices:

- Background is organized into three preliminary chapters, one on history and key concepts, one a primer on cellular and circuit-level neurobiology, and one on methods. My own preference when teaching is to introduce methods throughout the course of the semester, at the point at which they are relevant for the research question that is being described. Thus, the first edition of this book did not include a third introductory chapter *Methods of cognitive neuroscience*. When we solicited feedback from instructors who had used the first edition, however, we learned not all share this preference. Thus, this edition includes *Chapter 3, Methods of cognitive neuroscience,* and any instructor following the "Postle preference" can do as I will do: omit Chapter 3 as a stand-alone item in syllabus and instruct students to read the relevant section of Chapter 3 each time, in the subsequent chapters, that a method is introduced for the first time.

- Important concepts are introduced, and explained, via experiments that illustrate them. These won't always be the first experiment that is credited with introducing an idea, nor even an experiment performed by the investigator or group that is credited with innovating the idea. Rather, they will be ones that I feel are particularly effective for explicating the concept in question. Further, most of the experiments that make it into this book are considered in a fair amount of depth. In doing so, I seek to model a "best practice" of the cognitive neuroscientist, one that is captured by a quote from the nineteenth-century neuroscientist David Ferrier: "In investigating reports on diseases and injuries of the brain . . . I feel obliged always to go to the fountainhead – dirty and muddy though this frequently turns out." A corollary to this point is that because cognitive neuroscientists learn about and work with key concepts via empirical findings, almost all of the figures in this book are taken directly from the primary scientific literature, just as they appear in the empirical reports or review papers in which they were published. This is also how I teach and reflects my conviction that an important skill for students to develop is how to interpret figures as they appear in scientific publications. Thus, the book contains only a minimum of figures/diagrams/cartoons that were developed expressly for pedagogical purposes.

- Because a book targeting "essentials" can't cover all of the areas of research that fall under the rubric of cognitive neuroscience, each chapter concludes with an annotated list of suggested *Further Reading.* Many of the listed works will provide effective entrées into an area that the book doesn't cover.

- The first edition of this book was accompanied by a companion website that included 6-to-12 (or so)-minute bespoke narrated audiovisual presentations contributed by over 40 experts in different areas of cognitive neuroscience. These have now been spun off from the book and are made publicly available at https://postlab.psych.wisc.edu/cog-neuro-compendium. Through these presentations, students will hear directly from the expert her- or himself about either (a) an important experiment that is described in the book and that I think merits a telling from the perspective of investigator who actually performed the study or (b) an important area of cognitive neuroscience that can't be covered in the book but that the reader really should know about. These narrated videos generate a frisson of "what it's like" to work in this field. I am greatly indebted to my colleagues who contributed their time and expertise to this remarkable collection. (Poignantly, those from Corkin, Eichenbaum, and Lisman capture the wisdom and energy of colleagues from whom we can no longer learn through direct interaction.)

Although conceived and written for an upper-level undergraduate readership, *Essentials of Cognitive Neuroscience* can also be used for graduate teaching. For example, one associate of mine used it as the primary text in a general "Introduction to Cognitive Neuroscience" course in which each week a student presented a chapter to the class and led the ensuing discussion. In my graduate seminar, assigned readings from this book cover the background that every student needs to know before coming the class, and in class we focus more on current controversies and the most recent developments.

Those of us who are lucky enough to make a living grappling with some facet of the mind–brain problem have a responsibility to impart to our students a sense of the wonder, awe, and excitement that permeate our professional lives. I hope this book can help some of us meet this responsibility.

ACKNOWLEDGMENTS

The efforts and/or indulgence of many made this book possible. Going back a few years, then-lab manager Mike Starrett provided unflagging support at all stages of the writing of the first edition, from tracking down obscure, out-of-print volumes to formatting references to proofreading drafts to filming and editing the author's introduction to each chapter. "*MStarr*: We don't know how to do that . . . yet"! In this second edition, Claire Carson played a comparable role in figure production and in assembling and formatting user-friendly pdfs for students in my Fall 2019 classes. Dan Acheson was a wise and resourceful collaborator and was responsible for much of the content that was initially on the book's companion Web page, now accessible at https://postlab.psych.wisc.edu/cog-neuro-compendium/, as well as for much of the instructional support material. Dan was also a critical guide and sounding board during the writing of *Chapter* 19. Additional colleagues (and one spouse) who also provided valued feedback on portions of the book include Sue Corkin, Olivia Gosseries, Mike Koenigs, Bornali Kundu, Maryellen MacDonald, Jay McClelland, Chris Moore, Pat Mulvey, Tim Rogers, and Giulio Tononi. I am fortunate to work in the company of so many smart and generous individuals. Jeff Johnson taught a graduate seminar at North Dakota State University with first drafts of selected chapters of this book's first edition, and feedback from him and his students was very useful at the revision stage.

The indulgence of graduate students and postdocs in my laboratory who tolerated my engagement with this project was much appreciated, as were the substantive contributions of Bornali Kundu, Adam Riggall, Quan Wan, and Qing Yu.

Colleagues who willingly gave of their time and energy to contribute a *Narrated Video* for the companion Web page of the first edition (now accessible at https://postlab.psych.wisc.edu/cog-neuro-compendium/) contributed enormously to the pedagogical value of this enterprise, and I am very grateful to each of them: Dan Acheson, Andy Alexander, Chris Baker, Shaowen Bao, Rasmus Birn, Sue Corkin, Clay Curtis, Mark D'Esposito, Mingzhou Ding, Howard Eichenbaum, Olivia Gosseries, Shawn Green, Peter Hagoort, John-Dylan Haynes, Mike Koenigs, Victor Lamme, John Lisman, Bea Luna, Marcello Massimini, Yuko Munakata, Betsy Murray, Behrad Noodoust, Josef Parvizi, Sean Polyn, Mike Robinson, Tim Rogers, Bas Rokers, Vincenzo Romei, Yuri Saalmann, John Serences, Jon Simons, Heleen Slagter, Leah Somerville, Craig Stark, Mark Stokes, Jack Van Horn, Brad Voytek, and Gagan Wig. Wow! What a list!

I'm also grateful to many additional colleagues who took the time to track down and share high-resolution copies of hard-to-find images appearing as figures in this book.

The team at Wiley, and their associates, provided much-appreciated guidance, assistance, and encouragement at every stage of the process.

A book such as this is necessarily shaped by one's past experience, and I'm grateful to everyone who has guided and influenced my career in cognitive neuroscience: from pre-graduate school mentors, to professors and classmates and lab mates at MIT, to the team in Area 9 (U. Penn instantiation) and the D'Esposito diaspora, to all the colleagues and students with whom I've interacted in the classroom, in the lab, and in the hallway, since coming to Wisconsin. Professional organizations whose members and programming have enriched my career include the Cognitive Neuroscience Society, the International Society for Behavioral Neuroscience, the Memory Disorders

Research Society, the Psychonomic Society, the Society for Human Brain Mapping, and the Society for Neuroscience.

Much of the aforementioned career has been under-written by generous support from the National Institutes of Health and the National Science Foundation, for which I am grateful. These and other funding entities in the United States and around the world, public and private, are necessary and valued contributors to all the knowledge and discoveries that are recounted on the pages that follow.

Finally, I am forever indebted to my loving family, with-out whose support and indulgence this book could not have been written, nor revised.

Madison, WI
October 2019

WALKTHROUGH OF PEDAGOGICAL FEATURES

Key Themes The opening page of each chapter lists the key themes that will be introduced and explored in that chapter, to help the reader navigate the text.

End-of-Chapter Questions The reader can use these to assess how well s/he has understood a chapter's central themes, and as a starting point for further investigations.

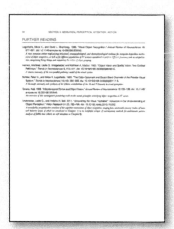

Further Reading An annotated list of readings that either elaborate on a chapter's contents, if one wants to learn more, or offer an entrée into a related topic that the chapter did not cover.

COMPANION WEBSITE

www.wiley.com/go/postle/cognitiveneuroscience2e

This book has a comprehensive companion website that features a number of useful resources for instructors and students alike.

FOR INSTRUCTORS

Sample syllabi

Exam questions

Please note that the resources for instructors are password protected and can only be accessed by instructors who register with the site.

SECTION I
THE NEUROBIOLOGY OF THINKING

INTRODUCTION TO SECTION I
THE NEUROBIOLOGY OF THINKING

Anyone who can read these words, including *you*, experiences conscious awareness. *What explains this? Is the conscious experience of other people "the same" as my own? Is our conscious experience qualitatively different from what may or may not be experienced by nonhuman animals?* These and similar questions have fascinated some of humankind's most celebrated thinkers for as long as humans have been leaving records of their cogitation. They are among the most profound that a human being can ask. Not surprising, then, is the fact that questions that relate either directly or indirectly to the phenomenon consciousness and the related construct of *cognition* (i.e., thinking) have been taken up by many different scholarly disciplines, including philosophy, evolutionary biology, anthropology, economics, linguistics, computer science, and psychology. What distinguishes cognitive neuroscience, the focus of this book, from these and other disciplines, is its grounding in the methods and traditions of neuroscience, and the primacy that it places on understanding the biological bases of mental phenomena. Now it bears noting that cognitive neuroscience doesn't concern itself only with consciousness and conscious awareness. Indeed, I feel confident asserting that the vast majority of articles that have been published in, say, the *Journal of Cognitive Neuroscience* (one of many scientific journals publishing peer-reviewed reports of research in the field) don't even explicitly address the idea of consciousness, much less use the word. Nonetheless, it is also true that it can be difficult to entertain detailed thoughts about human behavior and the principles that govern it without straying into ideas such as those that opened this paragraph. This clear relationship to profound philosophical questions is one of the qualities that differentiates cognitive neuroscience from other physical and biological sciences, including many other domains of neuroscience.

Like other physical and biological scientists, cognitive neuroscientists design and carry out rigorously controlled experiments that yield objective, measurable data, and seek to relate these data to mechanistic models of how a natural system works. And, as suggested above, the system studied by cognitive neuroscientists is one that is also of direct interest to scholars who study it from a very different perspective (e.g., from philosophy, from anthropology, and from cognitive psychology). This overlap turns out to be both a blessing and a curse. It's a blessing in that the cognitive neuroscientist can draw on ideas and observations from a vast array of rich intellectual traditions. And because they are studying questions that are fundamental to the human condition, almost any thinking person that a cognitive neuroscientist encounters is likely to be interested in what they do for a living. (Much easier, in the humble opinion of the author, to be a cognitive neuroscientist attending a social function full of strangers than, say, an economist, or a cell biologist, or a particle physicist.) The curse is found on the flip side of this coin. Because scholars in different fields are often interested in the same object of study (whether it be visual perception, or aesthetics, or antisocial behavior), cognitive neuroscience research can often be characterized by people working in other fields as asking the wrong question, or asking it in the wrong way, or of generating findings that are fundamentally irrelevant for understanding the question at hand. Being aware of this sociological context surrounding cognitive neuroscience is just one piece of knowledge that can be helpful in evaluating the implications and importance of any particular set of facts and ideas that will be described in this book.

COGNITIVE NEUROSCIENCE? OR "HUMAN NEUROSCIENCE"? OR "NEUROSCIENCE-WITH-DIRECT-IMPLICATIONS-FOR-UNDERSTANDING-HUMAN-BEHAVIOR"?

How were the topics to be covered in this book selected? More importantly, when teaching cognitive neuroscience, where does one draw the line that defines the boundaries of the discipline? This is a difficult question that doesn't have a definitive answer. It's really just an accident of history that the confluence of neuroscientific methods and studies of human behavior happened first, or, at least, influentially, with domains of behavior and function that are studied by cognitive psychology (e.g., visual perception, language function, memory), as opposed to, for example, social behavior, personality, emotion, or psychopathology. As a result, one sees the label "cognitive neuroscience" appearing earlier in the literature than, for example, affective neuroscience or, certainly, social cognitive neuroscience. As a consequence, the term "cognitive neuroscience" has come to be used in many contexts, and not always with precisely the same meaning. In one context, "cognitive neuroscience" can be used to refer to the tools and methods used to study the neural bases of human behavior (e.g., brain scans, recording electrical potentials at the scalp, brain stimulation). This might seem like a misnomer, however, if the behavior being studied doesn't directly relate to cognition (e.g., the study of sleep or of the maturation of various brain systems during adolescence). The label "cognitive neuroscience" can also seem like a bad fit for research that spans multiple traditional categories of brain and/or behavior. To pick one real-world example, let's consider the study of how neural systems measured with brain scans relate to performance on an economic decision-making task by high trait-anxiety vs. low trait-anxiety prisoners (all classified as "psychopaths"), and the comparison of these data with data from neurological patients with frank damage to the ventromedial prefrontal cortex. Should this be classified as cognitive neuroscience research? On one level, I hope that the answer is "yes," because we will be considering this research in *Chapter 17*! However, to do so is to not explicitly acknowledge the equally important contributions to this research from clinical psychology, from affective neuroscience, from neuroeconomics, and from neuropsychology.

In view of the above, to capture the interdisciplinary breadth of much of what we'll be considering in this book, might it perhaps have been better to entitle it "Essentials of *Human* Neuroscience"? Here the answer is an unequivocal "no," because to do so would be to exclude the fact that understanding the neural bases of almost all domains of human behavior requires a thorough knowledge of analogous neural functioning in nonhuman animals. This is because, as we shall see in almost every chapter of this book, technical and ethical limitations of what we can measure in humans require us to draw heavily on the results of research performed with nonhuman animals. Hence, the idea, implied in the title of this section, that the most precise title for this book might be "Essentials of Neuroscience with Direct Implications for Understanding Human Behavior." Now I'm no expert in academic publishing, but I'm fairly confident that, were I to have actually proposed this title, my editor would have rejected it. If nothing else, there probably aren't many universities where one can find a course that's called "Introduction to Neuroscience with Direct Implications for Understanding Human Behavior." Nonetheless, this is probably as good a summary, in 10 words or less, of what this book hopes to cover.

And so, with these considerations in mind, we'll stick with the label "cognitive neuroscience." It's not perfect, but if one is comfortable with a reasonably broad definition of cognition as thinking, behaving, and the factors on which these depend, then this label will serve us reasonably well.

CHAPTER 1
INTRODUCTION AND HISTORY

KEY THEMES

- Although the phenomenon of consciousness and the related construct of *cognition* (i.e., thinking) are the focus of many different scholarly disciplines, what distinguishes cognitive neuroscience is its grounding in the methods and traditions of neuroscience, and the primacy that it places on understanding *the neurobiological bases* of mental phenomena.

- There are two levels at which the term "cognitive neuroscience" is used: broadly, it has come to refer to the neuroscientific study of most domains of human behavior; narrowly, it refers to the study of neural bases of *thinking* – what influences it, what it consists of, and how it is controlled.

- The roots of cognitive neuroscience can be traced back to a nineteenth-century debate over two ways of thinking about brain function that both remain relevant today: localization of function vs. mass action.

- Mid-to-late nineteenth-century research, and the vigorous debate that accompanied it, led to models of localization of three functions: motor control (localized to posterior frontal lobes); vision (localized

- to occipital lobes); and speech production (localized to the left posterior inferior frontal gyrus).

- Motor control research introduced the principle of *topographic representation*, that is, adjacent parts of the body can be represented on adjacent parts of the cerebral cortex.

- Studying an aspect of cognition requires careful thought about the validity of the function to be studied; and not all aspects of human behavior can be studied with the same sets of assumptions, or even with the same methods.

- The discipline of cognitive neuroscience could not exist without discoveries yielded by research with nonhuman animals.

- At the dawn of the twentieth century, scientists were studying the brain and behavior from three related, but distinct, perspectives that would eventually give rise to cognitive neuroscience as we know it today: systems neuroscience, behavioral neurology/neuropsychology, and experimental psychology.

CONTENTS

A BRIEF (AND SELECTIVE) HISTORY

Although the term "cognitive neuroscience" as a moniker for a scientific discipline has only been with us for a few decades, the field has roots that extend back thousands of years. Ancient Egyptians, Greeks, and Romans all had ideas about the corporeal bases of human thoughts and emotions, although many of these did not specify a role for the brain. In preparing the bodies of deceased nobles for the afterlife, for example, ancient Egyptians removed and discarded the brain as an early step in the mummification process. The internal organs that were deemed to be important were preserved in urns that were entombed along with the body. In most ancient civilizations for which there are records, up through and including Roman civilization, the heart was believed to be the organ of thought. By the time we get to Enlightenment–era Europe, however, the central importance of "neuro" for cognition was widely accepted. One highly influential (and, more recently, ridiculed) example was that of German anatomists Franz Josef Gall (1758–1828) and Johann Gaspar Spurzheim (1776–1832), who developed a highly detailed scheme, known as phrenology, for how they thought that the shape of different parts of the skull related to one's personality and mental capacities. The underlying premise was that the relative bigness or smallness of various parts of the brain would produce convexities or concavities in the overlying skull. A skilled phrenologist, then, could learn something about an individual by palpating that person's skull. A bulge in the eye socket (inferred from "bulgy eyes") would mean a predilection toward language, whereas an indentation near the left ear would correspond to a relative absence of the trait of "destructiveness" (*Figure 1.1*). One can see how such a scheme, if it had any validity, would have obvious utility for diagnosing maladies of the brain, as well as for assessing personality and aptitude. (Indeed, for a period during the 1800s it was [mis] used in this way quite extensively, particularly in England and in the United States.)

Construct validity in models of cognition

For at least the past 100 years, the psychology and neuroscience communities have viewed virtually all tenets of the phrenological enterprise as being scientifically invalid. For starters, the very "functions" that phrenologists assigned to different parts of the scalp were derived from Gall's intuition rather than from a principled theoretical framework.

Let's consider, for example, conscientiousness. Now it is true, of course, that an organism without a brain cannot exhibit conscientiousness, and, therefore, the phenomenon could not exist without a brain. However, might it not be the case that "conscientiousness" is just a label that we, as denizens of highly organized societies, have given to a certain collection of attributes characteristic of some peoples' behavior and personality? This can be illustrated from two perspectives. The first is a caveat about inferring the existence of a discrete neural correspondence to every describable aspect of behavior. For example, when a student sends me a thank-you note for having written a letter of recommendation for her, I will consider her to have displayed conscientiousness. But could it not be that this reflects the fact that she was conditioned during her upbringing to seek positive reinforcement from her parents ("You're such a good girl for writing those thank-you notes!"), and that her note to me is "merely" the product of an association that she has formed between writing a thank-you note and this reinforcement? Were this the case, it wouldn't make sense to think of conscientiousness as a discrete mental faculty. (And if there's no such faculty, then there's no "thing" to localize to a part of the brain.) The second perspective is that, as it turns out, there is a branch of psychology in which conscientiousness is a valid construct, but it is as one of the so-called "big-5" personality traits. But because each of these five traits (openness-to-experiences, conscientiousness, extraversion, agreeableness, and neuroticism) is a statistical composite derived from multiple measurements (such as responses to many-more-than-five questions on personality assessment questionnaire), none of them can be construed as a unitary entity that might be localized to one brain system. And so, from this perspective, because conscientiousness isn't any one thing, it doesn't seem reasonable to expect it to localize to one part of the brain.

The exercise of thinking through the phrenological treatment of conscientiousness highlights two concepts that are highly relevant to contemporary cognitive neuroscience. The first is fundamentally "cognitive": a model of the neural instantiation of a cognitive function depends critically on the validity of the function that it seeks to explain. The question of construct validity will be important for every domain of behavior that we consider in this book. For many, the formal models of the construct under study will come from one of the "different scholarly disciplines" invoked in the *Introduction to Section I*.

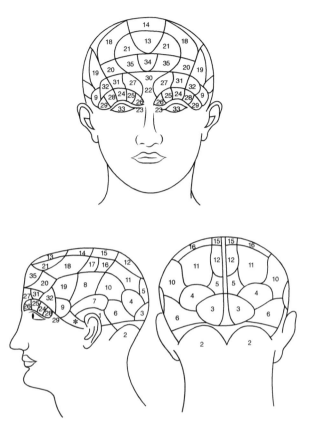

FIGURE 1.1 A phrenological map of the anatomical organization of mental faculties. They are organized into two primary categories (bold), with subcategories (underlined) under each. The first primary category is **Affective Faculties:** Propensities: * – Alimentiveness; 1 – Destructiveness; 2 – Amativeness; 3 – Philoprogenitiveness; 4 – Adhesiveness; 5 – Inhabitiveness; 6 – Combativeness; 7 – Secretiveness; 8 – Acquisitiveness; 9 – Constructiveness; Sentiments: 10 – Cautiousness; 11 – Approbativeness; 12 – Self-esteem; 13 – Benevolence; 14 – Reverence; 15 – Firmness; 16 – Conscientiousness; 17 – Hope; 18 – Marvelousness; 19 – Ideality; 20 – Mirthfulness; 21 – Imitation. The second primary category is **Intellectual Faculties:** Perceptive: 22 – Individuality; 23 – Configuration; 24 – Size; 25 – Weight and resistance; 26 – Coloring; 27 – Locality; 28 – Order; 29 – Calculation; 30 – Eventuality; 31 – Time; 32 – Tune; 33 – Language; Reflective: 34 – Comparison; 35 – Causality. Source: Spurzheim, Johann C. 1834. Phrenology or the Doctrine of the Mental Phenomenon, 3rd ed. Boston: Marsh, Capen, and Lyon. Public Domain.

A second concept invoked by phrenology is fundamentally "neuroscience," and its legacy from the Age of Reason is more complicated. On the one hand, we now know that subtle, idiosyncratic variations in gross shape from one brain to another have little, if anything, to do with the "kind of person" that one is. We also recognize that the assignments of function that Gall gave to various parts of the brain were not based on rigorous science and turned out to be altogether wrong. A third point is that the very selection and definition of functions that phrenologists mapped onto the brain lacked systematicity and rigor. There was, however, at the core of the phrenological enterprise, a powerful idea that has continued to animate many debates about brain function up to the present time – the idea of localization of function.

Localization of function vs. mass action

The principle of localization of function refers to the idea that different aspects of brain function, such as visual perception vs. the control of our emotions vs. our talents as musicians, are governed by, and therefore localizable to, different "centers" in the brain. An analogy might be that different functions of the body – extracting oxygen from blood vs. pumping blood vs. filtering blood – are each accomplished by different organs (i.e., the lungs, the heart, and the kidneys) that are located in different parts of the body. This notion can be contrasted with an alternative idea, **mass action**, according to which a particular function can't necessarily be localized to a specific area of the brain, and, conversely, any given area of the brain can't be thought of as a "center" that is specialized for any one function. To stick with our analogy to familiar parts of the body below the neck, we can illustrate the principle of mass action by zeroing in on the kidney. The overall function of the kidney – filtering blood – is carried out in the same way by the top portion and the middle portion and the bottom portion. To understand how the kidney does its job, one could study in detail the inner workings of only the top, or only the middle, or only the bottom of this organ, and one would learn the same thing from each. In effect, then, different zones of the kidney are "interchangeable" with respect to understanding their functions. Now project yourself back in time a few centuries to a time when what I've just written hasn't yet been discovered. It is an era when biomedical research techniques are limited, and the best tool that you have for studying the function of an organ is to damage a portion of it and then observe the consequent impact of this damage on its function. Your kidney research would indicate that damaging comparable-sized regions of the upper vs. middle vs. lower kidney has the same effect in all cases: an overall decline in the efficacy of blood filtration. Thus, you would have discovered that a principle of mass action applies to the kidney: a larger **lesion** results in a larger diminishment of the rate of blood filtration; smaller lesions result in a smaller diminishment of the rate of blood filtration; and, critically, the effects of damage to different parts of the kidney are the same.

Now, let's return to the functions of the brain. In the decades following the introduction of phrenology, and in some instances in reaction to it, scientists and physicians began pursuing the idea of localization of function in the brain with methods reflecting the maturation of the scientific method that was occurring in many branches of science, from biology to chemistry to physics. At this general level, this entailed the a priori articulation of falsifiable hypotheses (i.e., devising experiments that, if successful, can rule out previously plausible ideas and stating how various possible outcomes would be interpreted prior to performing the experiment) and the design of controlled laboratory experiments that could be replicated in other laboratories. An important advance for studies of the brain,

in particular, was the careful analysis of the behavioral consequences resulting from damage to a particular brain structure. This method has come to be known as **neuropsychology**. Armed with this approach, nineteenth-century scientists began to disprove many specific localizationist claims from phrenology. Perhaps most influential were the studies of French scientist Pierre Flourens (1794–1867). A tireless critic of phrenology, Flourens did much of his experimental work with pigeons and dogs. This research was influential on two levels. First, particular experiments of Flourens disproved specific phrenological claims, such as his demonstration that damage to the **cerebellum** disrupted locomotor coordination, but had no effect on amativeness (i.e., predilection for sexual arousal), as Gall's model would have predicted. Secondly, and at a broader level, Flourens' studies of the brain largely failed to find evidence for localization of many functions. Thus, although damage to the brain invariably produced marked disruption of behaviors associated with judging, remembering, and perceiving, these impairments seemed to occur regardless of what part of the brain had been damaged. By inference, such results seemed to indicate that all regions of the brain contributed equally to these behaviors. (Note that the same was not true of Flourens' studies of the brainstem, to which, for example, the cerebellum belongs.) (*Figure 1.2*).

Another important concept to come out of the work of Flourens was derived from the fact that, over time, animals with experimental damage to a part of cortex often recovered to presurgical levels of functioning. Because this occurred without evident repair of the damaged tissue itself, it was assumed that intact areas of the brain had taken over this function. This gave rise to the concept of **equipotentiality**, the idea that any given piece of cortical tissue had the potential to support any brain function.

Roughly 50 years prior to the writing of this textbook, and 130 years after the heyday of phrenology, neuroscientists Charles Gross and Lawrence Weiskrantz wrote that "The 'heroic age of our field' was opened by Gall (1835) . . . [who] stimulated the search for centers and gave the mass action-localization pendulum its first major swing" (Gross and Weiskrantz, 1964). Implied in this quote was that understanding the localization–mass action dichotomy would provide insight into understanding key contemporary problems in neuroscience. Indeed, this theme will prove to be useful for understanding many of the concepts and controversies that are prominent in contemporary cognitive neuroscience.

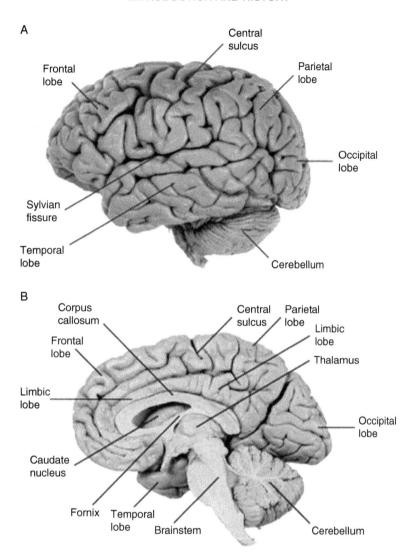

FIGURE 1.2 The human brain (see *Chapter 2* for labels and terminology). **A.** Lateral view. **B.** Medial view. Source: Dr. Colin Chumbley/Science Source.

The first scientifically rigorous demonstrations of localization of function

Although the concepts advocated by Gall, on the one hand, and Flourens, on the other hand, still resonate today, the same cannot be said for most of the "facts" that arose from their work. Rather, it was in the mid-to-late 1800s, during what can be seen as the first return of the pendulum back toward localization, that we see the emergence of principles of brain function that, at least to a first order

of approximation, have held up through to the present day. These involved the functions of motor control, vision, and language.

The localization of motor functions

Beginning in the 1860s, British neurologist John Hughlings Jackson (1835–1911) described the systematic trajectory of certain focal **seizures** that appear to start in the fingers and spread along the arm toward the trunk, sometimes ending with a loss of consciousness. From this distinctive pattern, which has since come to be known as the

Jacksonian march (it's as though the seizure is "marching" along the body), Jackson proposed that the abnormal brain activity presumed to be the cause of this progression of abnormal muscle contractions begins in a part of the brain that controls the fingers, then moves continuously along the surface of the brain, progressively affecting areas that control the palm of the hand, the wrist, the forearm, and so forth. There were two important implications of Jackson's theory. The first was quite simply the proposal that the capacity for movement of the body (i.e., motor control) is a function that is localized within the brain. The second was what has come to be understood as a fundamental principle underlying the organization of function within many portions of the brain, which is that their organization can mirror the organization of the body (or, as we shall see, of a particular part of the body). Specifically, in this case, the proposal was that the area of the brain that controls the muscles of the fingers is adjacent to the area of the brain that controls the muscles of the palm, which is adjacent to the area of the brain that controls the muscles of the wrist, and so forth. Thus, the functions of what came to be known as the *motor cortex* are physically laid out on the surface of the brain in a kind of map of the body (that is, in a **somatotopy**). In this way, the idea of a lawful, **topographic organization** of function was introduced in the void left by the (by-now-largely-discredited) arbitrary, willy-nilly scheme of the phrenologists. (The principle and characteristics of somatotopy will be considered in detail in *Chapter 5* and *Chapter 8*.)

Although the ideas that Jackson proposed were based on careful observation of patients, an idea such as the somatotopic organization of motor cortex couldn't be definitively evaluated without either direct observation or, preferably, manipulation of the brain itself. The ability to undertake such definitive empirical investigation became possible because Age-of-Enlightenment advances in thinking about how science should be conducted, such as the importance of the scientific method, were being paralleled by technical advances that afforded improved experimental methods. Of particular importance was the development of methods for performing aseptic (i.e., sterile) surgery. These enabled experimenters to keep animals alive for weeks or longer after performing the **craniotomy** that was necessary to create a lesion or to manipulate the functioning of the brain. Prior to this, infection would often limit postsurgical survival to a matter of hours or days. This technical advance had several important consequences, one of which was that it opened the way for direct

stimulation of, and subsequently recording from, the brain of an intact animal (techniques that fall under the category of **neurophysiology**). Thus, it was that German physicians Gustav Fritsch (1838–1927) and Eduard Hitzig (1838–1907) reported that electrical stimulation of **anterior** portions of the cerebral cortex of a dog, in the **frontal lobe**, produced movements on the opposite side of the body (Fritsch and Hitzig, 1870). Noteworthy were two facts. First, comparable stimulation of a more **posterior** brain region, the parietal lobe, did not produce body movements (*Figure 1.3*). Based on this demonstration of the **anatomical specificity** of their effect, Fritsch and Hitzig explicitly challenged the idea of equipotentiality that had been advocated by Flourens. Second, the part of the body affected by electrical stimulation (e.g., neck, forelimb, hind limb) varied systematically with the positioning of the stimulating electrode. This observation very directly supported Jackson's idea of a somatotopic organization of motor functions in the brain.

The localization of visual perception

A second function that was the focus of intense scientific investigation during this period was visual perception. Here, too, the work of Flourens had been influential. Although it had brought to light the principle of crossed lateralization of function – in that lesions of the left hemisphere produced visual impairments in the right visual field, and vice versa – it had found no evidence that this general pattern varied as a function of *where* in the hemisphere the

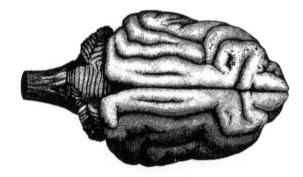

FIGURE 1.3 Illustration of a top-down view of a dog's brain, with symbols indicating, from rostral to caudal, the areas whose stimulation produced muscle contractions on the right side of the body of the neck, forelimb, and hind limb, respectively. Source: Fritsch, Gustav, and Eduard Hitzig. 1870. "Über die elektrische Erregbarkeit des Grosshirns." Archiv für Anatomie, Physiologie und wissenschaftliche Medicin 37: 300–332. Public Domain.

lesion was made. That is, Flourens had failed to find evidence for localization of visual function within a **cerebral hemisphere**. As with other of his null findings, however, this one was overturned thanks to the advent of newer, better, empirical methods. In the case of vision research, the refinement of methods of experimental lesioning was the critical development that led to an important discovery. As noted earlier, the development of aseptic surgical techniques resulted in longer and healthier postsurgical survival of experimental animals than had previously been possible. This, in turn, afforded considerably more sophisticated and conclusive assessments of behavior. Indeed, for a time, it was not uncommon for a researcher to bring a lesioned animal to a scientific conference so as to demonstrate the behavioral alterations produced by a particular lesion. Perhaps for related reasons, this period also saw an increase in research with monkeys, whose brain and behavior are more comparable to those of humans than are those of, say, birds and carnivores. It is much more expensive to acquire and house monkeys relative to other species, such as birds, rodents, and carnivores. Thus, research with monkeys would not have been considered practical prior to nineteenth-century refinements of surgical techniques.

The discovery of a region specialized for motor control was quickly followed by intense research on the neural bases of visual perception. Using electrical stimulation techniques and surgical lesioning, initial research in dogs and monkeys indicated a privileged role for posterior regions in supporting visual perception. That is, whereas it had been shown conclusively that motor control did *not* depend on posterior regions, the opposite was true for vision. The final decades of the nineteenth century witnessed vociferous debate about whether the "visual centers," a characterization of British physiologist David Ferrier (1843–1928), were localized to a region of the parietal lobe or to a region of the occipital lobe (*Figure 1.4*). This research, carried out primarily in monkeys and most prominently by Ferrier, by Berlin-based physiologist Hermann Munk (1839–1912), and by London-based physiologist Edward Schäfer (1850–1935), gradually converged on the conclusion that the region whose destruction produced frank and lasting blindness – as opposed to more nuanced and transient visual deficits – was the occipital cortex. This conclusion, reinforced by observations of human patients with head injuries, led to the universally accepted localization of **primary visual cortex** to the occipital lobe.

The localization of speech

The last function of the brain that we will consider in this introductory chapter is language, more specifically, the ability to speak. The faculty of language had been localized by phrenologists to a region of the frontal lobe directly behind the eyes. (This idea is said to have derived from Gall's observation of a classmate who had a prodigious verbal memory and protruding eyes, the latter presumed by

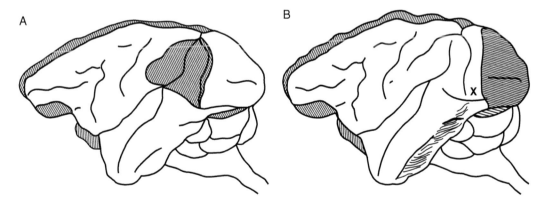

FIGURE 1.4 Two experimental lesions produced by Ferrier in experiments designed to localize the cortical locus of visual processing. Panel **A** illustrates a lesion of the parietal lobe that Ferrier interpreted in 1876 as producing lasting blindness, and, a decade later, as producing only temporary blindness. Panel **B** illustrates a lesion of the occipital lobe that he interpreted as producing no visual impairments. Neuroscientist Charles Gross (1998) has marked with an X the region that we now know to represent central (i.e., foveal) vision, a topic to be explored in depth in *Chapter 4*. These lesions, and their interpretation, illustrate how difficult it was for nineteenth-century neuroscientists to localize cognitive functions in the brain.

Gall to be the result of a bulging frontal lobe that pushed the eyes forward.) As was the case with motor control, the post-phrenology study of language began with clinical observations. In the decades prior to the 1860s, isolated reports suggested, in some cases, that damage to the left side of the brain was associated with impairments of speech, in others, that damage to anterior portions of the brain had this effect. As these cases accumulated, the more specific idea emerged that anterior portions of the left hemisphere were important for speech. During the 1860s, French surgeon Paul Broca (1824–1880) published a series of case studies that confirmed this idea. Most celebrated was the case of the **stroke** patient "Tan," so nicknamed because this was the only sound that he could make with his mouth. (Phonetically, "tan" is pronounced in French as /tôn/.) Importantly, Tan's impairment was specific to *speech production* (i.e., talking), because he could understand and follow verbal instructions that he was given. Additionally, there wasn't any obvious paralysis of the speech apparatus, in that, despite some right-side-of-the-body motor impairment, he could eat and drink and make his famous verbal utterance. Upon Tan's death, Broca examined the patient's brain and concluded from the prominent damage that he observed (*Figure 1.5*) that the ability to speak was localized to the "posterior third of the third

convolution" (a portion of the left inferior frontal gyrus that today is known as Broca's area). It's worth taking a moment to consider why Broca, rather than some of his predecessors and contemporaries who had made similar observations of patients, gets the credit for this discovery. One important factor is that Broca's reports were seen as confirming an idea that had been predicted ahead of time – a sequence of events that fit with the era's increasing emphasis on hypothesis testing as an important element of the scientific method.

WHAT IS A BRAIN AND WHAT DOES IT DO?

As we conclude this whirlwind review of the birth of modern brain science, it is instructive to consider why *motor control*, *vision*, and *language* were among the first three functions of the brain to yield to newly developed ways of thinking about and doing science. With particular relevance to the first two, let's pose the general questions of *what is a brain?* and *what does it do?* For example, what properties do animals with brains share that living organisms without brains (trees, for example) do not? One answer is that brains confer the ability to detect changes in the environment – let's say a falling rock – and to take an appropriate action: to get out of the way. The hapless tree, on the other hand, has no means of acquiring the information that something potentially dangerous is happening, nor the ability to do anything about it. So it gets crushed. Thus, the ability to see (i.e., vision) and the ability to move (i.e., motor control) are functions that are relatively easy to observe and measure. The same is not true for more nuanced examples of brain-mediated events that occur in the world. Let's take the example of writing the textbook that you're reading right now. Certainly, vision and motor control were involved. (For example, when the author received an emailed inquiry from the publisher about writing this textbook, he used vision to read the pattern of black and white characters on the computer screen that conveyed the inquiry from the publisher.) There were also, however, many additional steps. These included judgment and decision making (*Is it worth all the extra work to take this on? Are the terms of the contract with the publisher acceptable?*), retrieving long-term memories (*There's that great quote from Gross's chapter in the Warren & Akert book that I should use here*), making (and retrieving) new long-term memories (*Where did I put my copy of the* MIT Encyclopedia of the Cognitive Sciences *the last time I was working on this chapter?*),

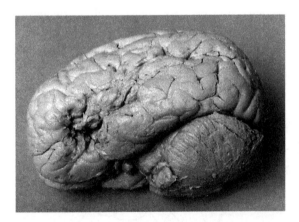

FIGURE 1.5 The left external surface of the brain of Broca's patient Tan. Note the extensive damage in the posterior inferior frontal gyrus, a region that has come to be known as Broca's area. Source: N. F. Dronkers, O. Plaisant, M. T. Iba-Zizen, E. A. Cabanis, Paul Broca's historic cases: high resolution MR imaging of the brains of Leborgne and Lelong, Brain, Volume 130, Issue 5, May 2007, Pages 1432–1441, https://doi.org/10.1093/brain/awm042; © 2019 Oxford University Press

and much, much, more. These latter operations are "internal" or "mental," in the sense that they happened in the author's mind without any obvious way of seeing or measuring them. Thus, although many aspects of so-called high-level cognition – including decision-making and different aspects of memory – are the focus of intense research in contemporary cognitive neuroscience, we shall see that the ability to define what constitutes a distinct function (analogous to vision or motor control), and the ability to localize the brain region(s) on which these functions depend, become much more complicated propositions. The principles that explain how such functions work will be less accessible to direct observation and measurement than are brain functions that are more closely tied to either the "input" of information to the brain (for which vision provides one important channel) or the "output" of the brain that is expressed via movements of the body (i.e., the actions of the organism). (However, as we shall see, there are influential perspectives on brain function as a whole that construe it as a hierarchically organized series of sensorimotor circuits, with the more abstract ones [e.g., all the processes that went into writing a textbook] having been superimposed, over the course of evolution, onto more basic ones [e.g., reflexive avoidance of a threat].)

Okay, so the neural bases of vision and motor control may have been among the first to be studied "scientifically" because they are easy to observe. But the same cannot be said about language. Although its production clearly has motoric components, we've already reviewed ways in which the language impairment experienced by Tan and other patients was fundamentally different than paralysis. Indeed, in ways, language can be construed as being at the opposite end of the concrete-to-abstract continuum of human faculties that spans from such "concrete" (that is, easily observable and relatively easily measurable) functions as vision and motor control to the most abstract aspects of conscious thought. Although language is readily observable, there is also a sense in which it epitomizes abstract, high-level cognition. One reason is that language entails the use of arbitrary, abstract codes (i.e., a natural language) to represent meaning. Intuitively, because many of us have the sense that we "think in words," language can be viewed as "the stuff of thought," and thus epitomizes an "internal," indeed, a *cognitive*, function. And so I'm reticent to conclude that language was among the first human faculties to be studied with modern neuroscientific techniques because it is easy to observe. Instead, what I'll offer is that it is language's intuitive "specialness," and its seeming

uniqueness to humans, that has made it a focus of interest for millennia. The Age of Enlightenment, then, may simply be the epoch of human history when developments in the scientific method caught up with an age-old focus of human curiosity.

LOOKING AHEAD TO THE DEVELOPMENT OF COGNITIVE NEUROSCIENCE

To conclude this introductory chapter, we see that by the dawn of the twentieth century, brain scientists were beginning to investigate the brain from both ends of the continuum that captures its functions in relation to the outside world. In the process, we can think of modern brain science as developing along two paths. The first, pursued by physiologists and anatomists who studied brain structure and function in nonhuman animals, focused on neural *systems* that support various functions (e.g., visual perception and motor control), and has come to be known as systems neuroscience. The second took as its starting point a human behavior, such as language, and proceeded with the logic that careful study of the way that this behavior is altered by insults to the brain can be informative, with respect both to how the behavior is organized and to how the workings of the brain give rise to this organization. This latter approach has matured into the allied disciplines of behavioral neurology (when carried out by physicians) and neuropsychology (when carried out by nonphysician scientists). Now, of course, the notion of two categories of brain science developing in parallel at the dawn of modern brain science is an overly facile dichotomization imposed by this author, and many instances can be found that do not fit neatly into such a taxonomy . Nonetheless, we will find this distinction to be useful as we proceed. It is these two scientific traditions summarized in this chapter, together with a third, experimental psychology (which also got its start in the second half of the nineteenth century, but which won't be covered in depth here [see *Further Reading* section]), that provide the foundation from which modern cognitive neuroscience has emerged. Before we plunge full-on into cognitive neuroscience itself, however, we need to review some facts about the brain – how it is put together (gross anatomy) and how it works (cellular physiology and network dynamics). These will be the focus of *Chapter 2*. Then *Chapter 3* will offer an overview of the methods used for studying brain structure and function.

END-OF-CHAPTER QUESTIONS

1. How does cognitive neuroscience differ from the related disciplines of cognitive psychology and systems neuroscience?

2. Although most of the specific claims arising from phrenology turned out to be factually incorrect, in what ways did they represent an important development in our thinking about the brain?

3. One of the major flaws of phrenology concerns the functions that it sought to relate to the brain: in some cases, they were not valid constructs; in others, they weren't amenable to neuroscientific explanation at the level that Gall was seeking to capture. From the phrenological bust in *Figure 1.1*, select at least one putative function (other than conscientiousness) to which each of these critiques applies, and explain your reasoning.

4. In the nineteenth century, what kind(s) of scientific evidence was (were) marshaled to support mass-action models of the brain? To support localizationist models of the brain?

5. Might it be possible that some tenets of mass-action and localization-of-function models could both be true? If yes, how might this be possible?

6. How does the concept of anatomical specificity relate to testing hypotheses of mass action and localization of function?

7. The neural bases of what domains of behavior were the first to be systematically explored in the second half of the nineteenth century? For each, what is a likely explanation?

8. What is the principle of topographic representation in the brain? Although this chapter emphasized the topographic organization of the motor system, according to what principles/dimensions might the topography of various sensory modalities (e.g., vision, somatosensation, audition) be organized?

9. Apart from one study that employed electrical stimulation, this chapter described experiments that relied on inferring the neural bases of a function from observation/measurement of the consequences of damage to different parts of the brain. Nonetheless, there are two fundamentally different ways in which one can pursue such research. Describe them, and name the disciplines of science and/or medicine that are associated with each. Which of the two requires the use of nonhuman experimental animals?

REFERENCES

Dronkers, Nina F., O. Plaisant, M. T. Iba-Zizen, and E.A. Cabanis. 2007. "Paul Broca's Historic Cases: High Resolution MR Imaging of the Brains of Leborgne and Lelong." *Brain* 130 (5): 1432–1441.

Fritsch, Gustav, and Eduard Hitzig. 1870. "Über die elektrische Erregbarkeit des Grosshirns." *Archiv für Anatomie, Physiologie und wissenschaftliche Medicin* 37: 300–332. [Available in English translation as: Fritsch, Gustav, and Eduard Hitzig. 2009. "Electric Excitability of the Cerebrum (Über die elektrische Erregbarkeit des Grosshirns)." *Epilepsy and Behavior* 15 (2): 123–130. doi: 10.1016/j.yebeh.2009.03.001.]

Gross, Charles G. 1998. *Brain, Vision, Memory*. Cambridge, MA: MIT Press.

Gross, Charles G., and Lawrence Weiskrantz. 1964. "Some Changes in Behavior Produced by Lateral Frontal Lesions in the Macaque." In *The Frontal Granular Cortex and Behavior*, edited by John M. Warren and Konrad Akert, 74–98. New York: McGraw-Hill.

Spurzheim, Johann C. 1834. *Phrenology or the Doctrine of the Mental Phenomenon*, 3rd ed. Boston: Marsh, Capen, and Lyon.

OTHER SOURCES USED

Calvin, William H., and George A. Ojemann. 1994. *Conversations with Neil's Brain*. Reading, MA: Addison Wesley.

Finger, Stanley. 1994. *Origins of Neuroscience*. Oxford: Oxford University Press.

Gross, Charles G. 2007. "The Discovery of Motor Cortex and Its Background." *Journal of the History of the Neurosciences: Basic and Clinical Perspectives* 16 (3): 320–331. doi: 10.1080/09647040600630160.

Holyoak, Keith J. 1999. "Psychology." In *The MIT Encyclopedia of the Cognitive Sciences*, edited by Robert A. Wilson and Frank C. Keil, xxxix–xlix. Cambridge, MA: MIT Press.

Kolb, Bryan, and Ian Q. Whishaw. 2003. *Fundamentals of Human Neuropsychology*, 5th ed. New York: Worth Publishers.

Whitaker, Harry A. 1999. "Broca, Paul." In *The MIT Encyclopedia of the Cognitive Sciences*, edited by Robert A. Wilson and Frank C. Keil, 97–98. Cambridge, MA: MIT Press.

FURTHER READING

Boring, Edwin G. 1929. "The Psychology of Controversy." *Psychological Review* 36 (2): 97–121.

A classic overview of key developments in nineteenth-century psychology and psychophysics, adapted from Boring's Presidential Address to the American Psychological Association, delivered on December 28, 1928. Boring was a professor at Harvard University.

Finger, Stanley. 2000. *Minds Behind the Brain*. Oxford: Oxford University Press.

A history, stretching from ancient Egypt to twentieth-century North America and Europe, of scientists who have made seminal contributions to the understanding of the structure and functions of the brain. It is longer and covers a broader scope of topics than Gross (1998). Finger has authored many authoritative books on the history of neuroscience and is Editor-in-Chief of the Journal of the History of the Neurosciences.

Gross, Charles G. 1998. *Brain, Vision, Memory*. Cambridge, MA: MIT Press.

A highly engaging collection of "Tales in the history of neuroscience." It spans the same range of human history as does Finger (2000), but with fewer chapters and a narrower focus of the functions named in the title. The author is, himself, a highly respected visual neuroscientist, now Emeritus Professor at Princeton University.

Ringach, Dario L. 2011. "The Use of Nonhuman Animals in Biomedical Research." *American Journal of Medical Science* 342 (4): 305–313.

Ringach is a UCLA-based neuroscientist whose person, family, and house have been the targets of harassment and intimidation by animal rights activists. This fact, however, does not color this even-toned consideration of this debate. Among other things, he urges "civil discourse . . . free of threats and intimidation" and concludes that "The public deserves an open and honest debate on this important topic." The article contains extensive references to original writings by individuals offering many different perspectives on this debate.

CHAPTER 2
THE BRAIN

KEY THEMES

- The central nervous system (often abbreviated as CNS) is organized into four parts: the spinal cord, the brainstem, the cerebellum, and the cerebrum.

- The computational elements of the brain are its neurons, whose cell bodies are primarily located in cortex, on the outside surfaces of the cerebrum and cerebellum, as well as in subcortical nuclei.

- The neocortex of the cerebrum is organized in six layers, each of which has a distinct functional role.

- The neuron has three principal components: the cell body; the branching dendrites, which receive signals from other neurons; and the projecting axon, which sends signals to other neurons.

- Neurons maintain electrical and chemical disequilibria in relation to the surrounding extracellular space, and many neuronal properties and processes can be understood in terms of the maintenance of these disequilibria.

- Neurons transmit information to other neurons via electrical action potentials that are propagated along axons.

- These electrical action potentials are converted to chemical signals – the release of neurotransmitter – at the synapse, the interface between the "sending" and the "receiving" neuron.

- Different types of neurons release different neurotransmitters, the principal "driving" neurotransmitters have either excitatory (depolarizing) effects or inhibitory (hyperpolarizing) effects on the postsynaptic neuron, whereas the "modulatory" neurotransmitters influence the state of the neuron, and can promote longer timescale processes such as learning, or influencing the overall state of arousal of the organism.

- The net effect on a neuron of receiving different types of neurotransmitter that influence it at different moments in time is that its membrane potential oscillates.

- The coordination and control of oscillations in different populations of neurons is an important way that the flow of information in the brain is controlled.

CONTENTS

PEP TALK

Thinking, consciousness, decision making, language – these and topics like them – are what you're expecting to learn about in this book. And you will. But for our explorations to be substantive, it's necessary that we establish a foundation of knowledge about how the brain is organized, structurally, and how it works, as well as how measurements of brain structure and function are acquired and interpreted. This necessitates the present chapter (which will cover neuroanatomy and cellular and circulatory neurophysiology) and *Chapter 3 (Methods of cognitive neuroscience)*, which contain little to no consideration of behavior. Thus, the following pages may be tough sledding for some readers who, for example, may have some background in psychology, but not in biology or biomedical engineering. But it will be good for you! And we just can't get around it. For example, understanding the physiology of the neuron is critical for interpreting neuroimaging data, because while some methods measure electrical fluctuations, others measure changes in blood oxygenation, and others measure changes in the reflectance of light. Thus, valid interpretation of data from a neuroimaging study of, say, face recognition, or emotion regulation, will depend critically on understanding what aspect of brain activity it is that is generating the signal that we are interpreting.

Later in this book, we'll see that one of the traits supported by the prefrontal cortex (PFC) is the ability to choose to engage in behavior for which the reward may not be immediate, but for which it will be worth the wait. And so, in that spirit, let's keep our eyes on the prize, and dive in!

GROSS ANATOMY

Like much of the rest of the body, most parts of the brain come in twos. There are two hemispheres, and within each hemisphere there are five lobes of the cerebrum – four of them named after the four bones of the skull under which they lie (frontal, parietal, occipital, and temporal), plus the deep-lying limbic lobe, which is not adjacent to the skull (*Figure 1.2*). The brain belongs to the CNS, which also comprises the spinal cord and the motor neurons that leave the cord to innervate the muscles of the body. All of these elements derive, during development of the embryo, from the neural plate, a 2-D sheet of tissue that folds until its edges touch and merge to form the neural tube. In primates, which are bipedal, the CNS makes a right-angle bend between the midbrain and the forebrain such that, when standing erect, the spinal cord, brainstem, and midbrain are oriented vertically, and the diencephalon (thalamus) and cerebrum are oriented horizontally. Thus, for a standing human, while the directional terms *dorsal* and *ventral* refer to the "back" and "front" of the spinal cord, they refer to the "top" and "bottom" of the brain (*Figure 2.1*).

When you look at a brain that's been dissected out of a skull the first thing that you notice is how wrinkly it is (*Figure 1.2*, *Figure 1.4*, and *Figure 2.2*). The reason for this makes sense once you understand two facts. The first is that the computational power of the mammalian brain is in the cerebral cortex – the thin (roughly 2.5-mm-thick) layer of tissue that covers the outside surface of the five lobes of the cerebrum – and in the cortex of two other structures, the hippocampus and the cerebellum. (For ease of exposition, this introductory section will focus on the cerebral cortex.) The second is that the cerebral cortex is, in effect, a 2-D sheet, with a surface area of roughly 2500 cm². Indeed, if one could "unfold" and flatten out the cortex, it would take roughly the same surface area as a tabloid-style newspaper. (In Madison, WI, where I live and work, I use the example of the weekly news and culture paper *The Isthmus*, but you can pick your favorite rag.) Now we can return to the wrinkliness, noting that "crumpled" might be a better descriptor: over the course of evolution, as the computational power of the brains of vertebrate species has increased, the ever-larger cerebral cortex has needed to fold ever-tighter onto itself in order to fit inside the roughly spherical skull. The alternative, evolving a thin, 50 cm × 50 cm square (or a comparably sized disk) for a skull might look okay on a creature from a sci-fi anime cartoon, but would clearly be highly inefficient, particularly from the standpoint of the amount of "wiring" that would be needed to connect different parts of the brain together. In contrast, it has been calculated that the "crumpled-up" solution with which we have evolved yields a cerebrum that is almost optimally shaped to minimize the connection distances between its roughly 80 billion neurons, which make trillions of synapses with each other.

When the brain is sliced, you see that the cortex looks darker (gray matter, although to me it looks more taupe than gray) than the white matter that takes up most of the volume of the brain (*Figure 2.2*). The gray matter is the aggregation of hundreds of thousands of brain cells – neurons and "supporting" cells known as glial cells. All neurons have a branch via which they communicate with

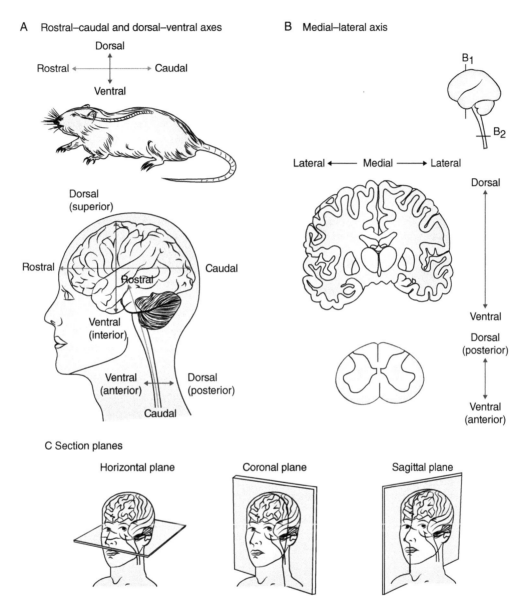

A Rostral–caudal and dorsal–ventral axes

B Medial–lateral axis

C Section planes

Horizontal plane   Coronal plane   Sagittal plane

FIGURE 2.1 Terminology for anatomical orientation. **A.** The quadruped. **B**. In the bipedal human, the terminology is complicated by the right-angle bend that the neuroaxis makes between the midbrain and forebrain. **C.** The cardinal planes of section. Source: Kandel, Eric R., James H. Schwartz, and Thomas M. Jessell, eds. 2000. Principles of Neural Science. New York: McGraw-Hill. Copyright 2000. Reproduced with permissions of McGraw-Hill Education.

other neurons. These branches, called **axons**, conduct electrical signals that are fundamental to how neurons communicate. Just like the electrical power lines that carry electricity from power plants to the places where the electricity is needed, axons are insulated, so as to conduct electrical current efficiently. In the brain, this insulation is a cholesterol-laden substance called **myelin**, and the aggregation of millions of myelinated axons is what makes the white matter look white. As illustrated in *Figure 2.1*, there are three cardinal planes along which brains are sliced: in vertical cross section ("coronally"); in horizontal cross section ("axially"); or vertically along the long axis ("sagittally").

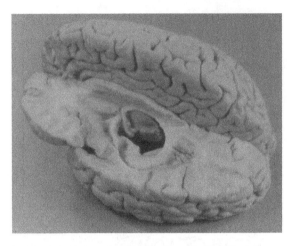

FIGURE 2.2 Human brain with dorsal half of the left hemisphere dissected away with a cut in the axial plane to reveal several subcortical anatomical features. Along the medial surface of the temporal lobe is the hippocampus, and the fibers projecting from its medial edge make up a tract called the fornix, which forms an arc as it courses rostrally from the caudal end of the hippocampus to terminate at the mammillary bodies of the hypothalamus. Immediately rostral to the rostral-most extent of the fornix, and to the rostral end of the hippocampus, is the anterior commissure, running in a mediolateral-oriented tract carrying transhemispheric axons between the anterior temporal lobes. Finally, the medial aspect of thalamus of the right hemisphere appears as though it is immediately ventral to the arc of the fornix. The surface of the axial cut also reveals the depth of many sulci (lined with gray matter), including the intraparietal sulcus at the caudal end of the brain. Source: Warwick and Williams, 1973. Reproduced with permission of Elsevier.

From the coronal section, in particular, as well as from *Figure 2.2*, one can appreciate that many of the brain's "wrinkles" are quite deep. The technical term for the wrinkles is sulci. Indeed, if you were to hold a human brain in your hand and rotate it so that you could inspect it from every angle, you'd only be able to see about one third of the cortical surface. Two thirds are "hidden" along the walls of sulci, or lining the "middle" walls of each hemisphere. In inspecting a brain like this, you'd be viewing its *lateral* (i.e., "toward the outside edge") surface. Regions nearer to the center of the volume are said to be *medial*. The "bumps" and "ridges" that are visible on the surface of the intact brain are referred to as gyri. The dissection featured in *Figure 2.2* provides a "window" revealing several regions of gray

matter that are medial to the cortex (i.e., deep "inside" the brain). Most of these subcortical structures are nuclei – collections of cell bodies that all perform a related function. Also visible in *Figure 2.2* are large bundles of myelinated axons. The largest, the corpus callosum, carries millions of axons projecting from one hemisphere to the other.

Let's return to the visual inspection of the intact brain that we are holding in our hands. The most obvious landmark is the midline, the gap separating the two hemispheres. From the canonical "side view" (as in *Figure 1.2*), we see the two prominent sulci that more-or-less divide the rostral from the caudal half, and we see the distinctive temporal lobe that is located ventrally and oriented parallel to the rostral-caudal axis. The sulcus that forms the dorsal boundary of the temporal lobe is the Sylvian fissure (a fissure being a really big sulcus); the sulcus that's parallel to the caudal-most section of the Sylvian fissure, and roughly two gyri rostral to it, is the central sulcus. As its name implies, the central sulcus divides the "front half" from the "back half," the frontal lobe from the parietal lobe. We shall see that, in functional terms, this anatomical landmark also demarcates, to a first order of approximation, the fundamental distinction of "action" (frontal) vs. "perception" (parietal, temporal, and occipital). The occipital lobe, the caudal-most in the brain, sitting atop the cerebellum, has the least obvious boundaries with the two lobes with which it makes contact on its lateral surface, the temporal lobe, ventrally, and the parietal lobe, dorsally.

The tissue connecting the two hemispheres is the thick band of fibers (i.e., axons) that connect neurons of the left hemisphere to those of the right, and vice versa. This is the corpus callosum. If one severs the corpus callosum, so as to separate the two hemispheres, one can then inspect the medial surface of either hemisphere (a so-called midsagittal view; *Figure 1.2.B*). This reveals the fifth lobe of the brain, the limbic lobe, which consists primarily of the anterior cingulate gyrus and the posterior cingulate gyrus, the gyri that surround the corpus callosum in its rostral and caudal aspects, respectively (*Figure 1.2.B* and *Figure 2.2*).

Very prominent in the middle of the brain, whether sliced axially, coronally, or sagittally, is a large gray-matter structure called the thalamus. Fittingly, the thalamus acts as a hub for the brain: all incoming sensory information (with the exception of olfaction) is relayed through the thalamus on its way to the cortex. Additionally, the thalamus acts as a node in several circuits, sometimes called "loops," in which information cycles from the cortex to the thalamus and back to the cortex again (in some cases with several other subcortical connections interposed). More details about the anatomy and functions of these "corticothalamic" loops will be considered in subsequent chapters of the book. Because it is so intricate, and is organized anatomically according to the different brain systems with which it connects (e.g., the visual system, the auditory system, the PFC), the thalamus is itself considered to be made up of several nuclei. Throughout the book, as we consider different systems and/or functions, we'll also look more closely at the relevant thalamic **nucleus** (e.g., the lateral geniculate nucleus [LGN] for vision, the medial geniculate nucleus [MGN] for audition, and the mediodorsal [MD] nucleus for PFC).

Finally, from any of the dissections that reveal the thalamus one can also see the hollow chambers that take up a considerable portion of the intracranial volume. These are the **ventricles**, the largest being the lateral ventricle. In living organisms, the ventricles are filled with the **cerebrospinal fluid (CSF)** that bathes the brain on its inside (i.e., ventricular) and outside surfaces. On the outside, it circulates between a thin layer of tissue that lies on the surface of the cortex – the pia mater – and the arachnoid membrane, so named because it suggests the appearance of a spider web. Enclosing the arachnoid membrane is the rubbery dura mater, and, finally, the bone of the skull. Together, the dura, arachnoid, and pia are referred to as the meninges. (Meningitis is the infection and consequent swelling of this tissue.) The CSF is under pressure, such that it gently inflates the brain from within the ventricles, and it is in continuous circulation through the ventricles and down the entire length of the spinal cord. For this reason, the CSF extracted via a procedure called a lumbar puncture (a.k.a. a "spinal tap") can be assayed to assess certain measures of brain health. For example, elevated levels of certain proteins in the CSF may signal excessive cell death, as happens in Alzheimer's disease.

A full appreciation of the section that you have just read would likely benefit from spending some time visiting a website with an interactive 3-D model of a brain. (In addition, I strongly encourage you seek out an opportunity to have the hands-on experience of holding a brain and, if possible, dissecting it. Often, especially in a formal class, the dissection will be performed with the brain of a sheep. At the level of the gross anatomy, you'll find all the structures that have just been described, and the experience will generalize to any vertebrate species [including humans] that you may subsequently study in more depth.)

The cerebral cortex

The cerebral cortex is also referred to as neocortex, in reference to the fact that it is a relatively new feature of the CNS in evolutionary terms. It is made up of six layers, which are visible when a very thin slice of cortex is viewed under a microscope. The reason that these layers look different is that each is made up of different types of cells and/or tissues that serve different functions (*Figure 2.3*).

The central layer, layer IV, is the layer that receives input from the thalamus. Thus, in the case of primary visual cortex ("area V1"), for example, the neurons of layer IV are the first cortical neurons to begin processing signals relating to, say, the photons reflecting off this page of the book (or, perhaps, emanating from the screen you are looking at). Layers II and III have subtle differences, but for the purposes of this textbook they can be thought of as being functionally equivalent, and will often be referred to as "layer II/III." The property of layer II/III that will be emphasized here is that neurons from this layer tend to be those that send information to higher levels of cortex. As illustrated in *Figure 2.4*, it is neurons of layer II/III of the second visual processing area ("V2") that project to a higher-level visual processing area ("V4"). (More details on what is meant by "project to" are considered in the next section, *The Neuron*; more details on the organization of the visual system are in *Chapter 4*.) Because the flow of visual information can be construed as traveling from the retina to the LGN of the thalamus, from thalamus to V1, from V1 to V2, and from V2 to V4, the layer II/III projections from V2 to V4 are considered feedforward connections. The neurons of layers V and VI tend to be those that send feedback projections to earlier levels of cortex. (Depending on the region of the brain, corticocortical feedback projections may arise predominantly from layer V or from layer VI.) Finally, neurons from layer VI (and sometimes also layer V) tend to be those that send their projections to the thalamus, or to other subcortical structures (*Figure 2.4*). (Note that this and the subsequent paragraph are referring to **pyramidal cells**, one of the two general classes of cortical neurons that will be described in more detail in the ensuing section, *The Neuron*.)

As we shall see in the next paragraph, it's often convenient to refer to one layer of cortex as being "above" or "below" another (as in *layer IV is located below layer II/III*). The reason that this can be problematic, however, is evident from inspecting the cartoon of a coronal slice illustrated in *Figure 2.1.B*. If we consider the cortex located on

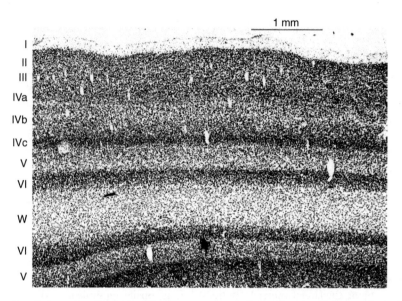

FIGURE 2.3 Cross-sectional view of cortical area V1 of the macaque, illustrating the six layers, including several subdivisions of layer IV. The tissue is from a narrow gyrus such that, at the bottom of the image, below the region of white matter (labeled "W"), one sees the deep layers of a nearby area of cortex. Source: Hubel and Wiesel, 1977. Reproduced with permission of the Royal Society.

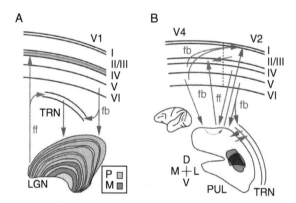

FIGURE 2.4 Schematic illustration of thalamocortical and corticocortical feedforward (ff) and feedback (fb) projections. Panel **A** illustrates the projections from the LGN (diagrammed in cross section) of the thalamus to layer IV of V1, and from layer VI of V1 back to the LGN. Note that there are six primary layers of the LGN, four containing parvocellular (P) neurons and two containing magnocellular (M) neurons (to be detailed in *Chapter 4*), and that these two types of LGN neuron project to different sublayers of cortical layer IV (to be detailed in *Chapter 6*). Also to be detailed in *Chapter 4* will be the role of the thalamic reticular nucleus (TRN), which receives collaterals from the axons carrying feedforward and feedback signals. Panel **B** illustrates the organization of feedforward and feedback pathways between two higher levels of visual cortex, areas V2 and V4. "Direct" corticocortical feedforward projections go from layer II/III of V2 to layer IV of V4. "Indirect" thalamically mediated feedforward projections go from layer V of V2 to the pulvinar nucleus of the thalamus (PUL) and then from pulvinar to layer IV of V4. Direct corticocortical feedback projections are illustrated as going from layers V and VI of V4 to layer I of V2 (the laminar specificity for this and other interareal projections varies across brain systems and across species). Indirect feedback projections go from layer VI of V4 to the pulvinar nucleus, and then from pulvinar to layer I of V2. The small diagram of a lateral view of the macaque brain relates to how these data were generated and will be more easily understood after reading *Chapter 3*'s section on *tract tracing*. For now, we'll just state that it shows where cell bodies in area V4 (pink) and area V2 (blue) each send axons to an overlapping projection zone of the pulvinar nucleus of the thalamus (diagrammed in coronal section). Source: Saalmann, Yuri B. and Sabine Kastner. 2011. "Cognitive and Perceptual Functions of the Visual Thalamus." Neuron 71 (2): 209–223. doi: 10.1016/j.neuron.2011.06.027. Copyright 2011. Reproduced with permissions of Elsevier.

the ventral surface of the Sylvian fissure, and thus defining the dorsal boundary of the temporal lobe, layer II/III is, indeed, above layer IV (e.g., it is closer to the top of the page). For the cortex just dorsal to it, however, on the dorsal surface of the Sylvian fissure, the opposite is true, in that layer IV in this bit of cortex is closer to the top of the page than is layer II/III. This is a simple consequence of the fact that the "crumpled" nature of cortex means that portions of it are oriented in virtually every direction possible. A constant property of the layers of cortex, however, no matter in what part of the brain, or whether the owner of that brain is standing upright, lying on their back, and so on, is that layer II/III is always closer to the surface of the cortex and layer IV is always closer to the white matter. Thus, their positions relative to each other are most accurately referred to as being superficial (closer to the surface) or deep (further from the surface).

And what about layer I? Layer I is distinctive from all other layers of cortex in that it contains very few cell bodies. Instead, it is made up primarily of the "antennae" (dendrites) extending from neurons in deeper layers so as to receive signals being sent by neurons in other regions. Indeed, this property of neurons with cell bodies located in one layer having dendrites that extend to another layer is common in all layers of cortex. Another important fact about cortical neurons is that, in addition to the properties of receiving thalamic input, sending feedforward projections, or sending feedback projections, many also send projections to other neurons located in the layer(s) above or below. As a result, groups of neurons that are "stacked" on top of each other (from layer II/III to layer VI) can often be characterized as forming cortical columns. As this designation implies, neurons are more strongly interconnected with other neurons within the same column than they are with neurons that are immediately adjacent but do not belong to the same column. This principle of "columnar organization" will be important for understanding the principles of visual information processing (*Chapter 4* and *Chapter 6*) and auditory and somatosensory information processing (*Chapter 5*) and may also be important in brain areas that process higher-level information (e.g., the PFC, which features prominently in the final section of this book). A final fact about the layers of cortex: depending on the region of the brain, corticocortical connections can terminate in layers I, II/III, and/or V. At this point, however, we've gone about as far as we can go without addressing the following questions: *What is a neuron? What are its functional units? How does it work?*

THE NEURON

The elements that give the brain its unique properties are the cells that make it up. Recall from earlier in this chapter that there are many, many different types of cells in the brain, and that many of these are not neurons. For the time being, however, we'll focus on just two classes of cortical neurons: pyramidal cells and interneurons. The pyramidal cell is the canonical neuron. Its large cell body has a 3-D shape that resembles a pyramid, with antenna-like dendrites sprouting from the cell body, particularly its pointy "top," and the axon leaving from the base of the cell (*Figure 2.5*). Typically, the dendrites, like the branches of a tree, spread themselves out over an area much larger than that taken up by the cell body itself (the cell body being the trunk in this analogy). The function of dendrites is to receive messages from other neurons (i.e., it is along dendrites that connections are formed with axons from other neurons). The axon, as we've already discussed, conveys signals from the neuron in question to other neurons. So let us now consider the nature of these electrical signals.

Electrical and chemical properties of the neuron

It may be difficult to find a single definition of "life" that any two people would find fully acceptable. Nonetheless, one (perhaps narrowly reductive) stab at a definition would be a system that is able to actively maintain itself in a high-energy state. This is a definition that draws on thermodynamics, in that it invokes the general tendency for objects in our universe to dissipate (or acquire) energy so that their own level of energy matches that of the surrounding environment. As one example, food that is removed from a hot oven and left on the countertop will cool until its temperature matches that of the kitchen. Similarly, food taken out of the freezer and left on the countertop will warm up until it achieves room temperature. In terms of their temperature, these nonliving objects passively give off or take on energy until they reach an equilibrium, which is the temperature of the kitchen. Neurons maintain life by maintaining a *dis*equilibrium along two dimensions, electrical and chemical. With regard to the former, a neuron in its baseline state will maintain a negative voltage of approximately −70 mV relative to the extracellular space that surrounds it. It achieves this by maintaining a lower concentration of positively charged ions relative to

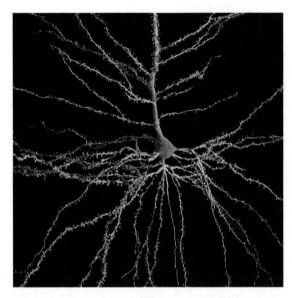

FIGURE 2.5 A pyramidal neuron from layer III of the PFC of a monkey. Branching out from what would be the corners at the base of the pyramid-shaped soma are the basal dendrites, and projecting up toward the top of the page is the trunk of the apical dendrite, which will branch extensively when it reaches layer I (not shown). Each dendritic branch is studded with spines, submicron-sized protrusions at the end of which a synapse is formed with an axon. Extending from the base of the soma and projecting straight down and off the bottom of the image is this neuron's axon, distinguishable as the only process that is not studded with spines. To generate this image, the neuron was first microinjected with a dye, and then images were acquired with a laser confocal microscope. For details about the method, see Dumitriu, Rodriguez, and Morrison (2011); Morrison and Baxter (2012, *Further Reading*) demonstrates the use of this method to study the effects of aging in the PFC. Source: Image courtesy of John H. Morrison, PhD. Reconstruction done by The Visual MD.

the extracellular space. In addition to this electrical disequilibrium, the neuron at rest maintains a chemical disequilibrium by "hoarding" a high concentration of potassium (K^+) ions relative to the extracellular space, and maintaining a relatively low concentration of sodium ions (Na^+) and calcium (Ca^{++}) ions relative to the extracellular space. (Recall that the membranes of all cells in the body are made up of a lipid bilayer that, because it repels water, separates the watery cytoplasm [and all dissolved or floating in it] inside the cell from the watery extracellular fluid [and all dissolved or floating in it].)

The consequence of the electrical and chemical imbalances in the neuron is that if a hole were to be poked in the membrane of a neuron, two events would happen: (1) the electrical gradient would be reduced by an influx of positively charged ions into the cell (including Na^+ and Ca^{++} ions); (2) simultaneously, the chemical gradient would be reduced by an efflux of K^+ from the high K^+ concentration inside the cell to the low K^+ concentration outside the cell. (And note that the influx of Na^+ ions from (1) would also be reducing the chemical gradient.) As you might imagine, however, the workings of the neuron are much more orderly and regulated than the poking of holes. Indeed, the membrane of the neuron is punctuated with channels that can open and close, thereby allowing the passage of certain types of ions in a controlled manner. Most of these channels have differential permeability to different types of ions, and are only activated under certain conditions. One example is the voltage-gated Na^+ channels that are concentrated at (or near, depending on the type of neuron) the **axon hillock**, the region where the bulky cell body funnels down into the long tube that is the axon. These channels are closed when the neuron is at rest (i.e., with a membrane potential of -70 mV). However, when the neuron has been sufficiently **depolarized** (e.g., the voltage raised to roughly -20 mV, varying depending on cell type), these channels open and allow for the very rapid influx of Na^+ ions. Recall that this influx is propelled by both the electrical gradient and the chemical gradient that exist across the cell membrane. This influx of Na^+ ions gives rise to the action potential (or "spike" in neurophysiology vernacular), which is the basis for neuronal communication.

Ion channels

The membrane-spanning channels of neurons are made up of complexes of proteins that span the membrane from the intra- to extracellular space. The channel's walls are formed by several subunits, as is illustrated in *Figure 2.6*. A channel's specificity for one type of ion is determined by the physical size of its pore (e.g., Na^+ channels are simply too small to allow large ions or molecules to pass through), as well as by the properties of binding sites located in the pore of some channels that result in preferential permeability to some ions over others. When closed (i.e., "at rest"), the pore is too narrow for anything to pass through. When activated, the channel twists (due to a change in conformational state of one or more of the subunits), thereby enlarging the pore and allowing ions to pass through.

A

B

C

D

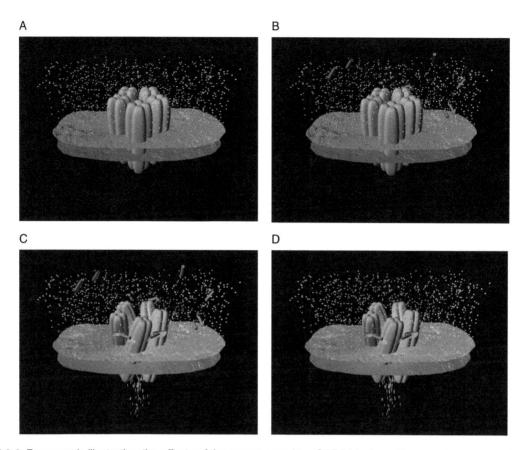

FIGURE 2.6 Four panels illustrating the effects of the neurotransmitter GABA binding with a membrane-spanning $GABA_A$ receptor (a subtype of GABA receptor). **A**. The receptor is composed of five separate proteins, each one consisting of four membrane-spanning domains (green "cigars"). As a functional unit, the receptor spans the neuronal membrane (gray planes) and provides a pathway for the inflow of ions (yellow dots represent a cloud of chloride (Cl–) ions in the extracellular space). The receptor is shown in its resting, closed state, with no neurotransmitter bound, and no available path for ions to flow across the membrane. **B**. Shorter, darker green cylinders represent GABA molecules having been released into the synaptic cleft. (That is, an action potential in a GABAergic neuron has triggered the release of GABA from the presynaptic terminal [not shown] into the cleft.) **C**. Binding of GABA molecules at two binding sites results in conformational change of receptor proteins (illustrated here as a mechanical twisting of subunits), which opens the pore of the receptor, thereby allowing Cl⁻ ions to enter the neuron. **D**. For some period of time after unbound GABA has been cleared from the synapse, the receptor remains active. Source: Images courtesy of Mathew V. Jones, University of Wisconsin-Madison. Reproduced with permission.

Depending on the type of channel, activation is accomplished by the binding of a molecule of neurtransmitter to the binding site, by a change of membrane voltage, or by activation of the channel via chemical signaling from within the cell. The former is characteristic of the so-called "driving" neurotransmitters **glutamate** and **GABA**, which will be considered in detail in the upcoming section on Different neurotransmitters and different receptors can produce different effects on the postsynaptic cell. The latter is typical of many channels that are influenced by a

neurotransmitter other than glutamate or GABA. In the case of the neurotransmitter dopamine, for example, the dopamine receptor is not part of a ligand-gated channel. Instead, the binding of dopamine to a receptor triggers a cascade of biochemical events inside the postsynaptic terminal, and a downstream consequence of this can be the opening of a membrane-spanning channel. Note that such a multistep cascade is likely to take longer to influence the voltage of a neuron than is a ligand-gated channel. This is one of the reasons that neurotransmitters such as dopamine are considered to be neuromodulators – their effect is to modulate the responsivity of the neuron to the faster-acting, "driving" effects of glutamate and GABA.

How neurons communicate: the action potential

A neuron initiates an action potential by taking advantage of the local depolarization produced by the influx of Na^+ ions that occurs at the axon hillock. It has to "act fast," because although this event produces a short-lived "cloud" of depolarization in the vicinity of the axon hillock, this "cloud" will quickly dissipate due to the rapid diffusion of Na^+ ions throughout the intracellular space (recall that there's a low concentration of Na^+ ions inside the cell). The neuron exploits this temporary depolarization by having another cluster of voltage-gated Na^+ channels that is "strategically" located just a little way along the axon. The channels of this cluster of receptors experience a brief pulse of depolarization from the hillock and they, in turn, open. Next, guess what? There's *another* cluster of voltage-gated Na^+ channels even further along the axon. This pattern repeats, and in this way an action potential is propagated

along the length of the axon until it reaches the dendrites of the next neuron. (You may be wondering, "But how did the initial depolarization that triggered the activity at the axon hillock come to pass?" Stick with me, and we'll come to that.)

When the action potential arrives in the vicinity of a dendrite from another neuron, it encounters a change in the morphology (i.e., shape) of the axon. This is a swelling of the axon into what is referred to as a synaptic bouton, a synaptic varicosity, or a presynaptic terminal (*Figure 2.7*). For a neural signal to pass from one cell to another, there needs to be a synapse linking them. At the presynaptic terminal, one finds the machinery necessary for converting the electrical signal of the action potential into the chemical signal that will be transmitted to the postsynaptic terminal. (Note from *Figure 2.7* that the pre- and postsynaptic cells do not touch, there is a gap [the **synaptic cleft**] between them.) That is, the electrical signal does *not* jump to the next neuron. Instead, the rapid depolarization triggered by the action potential opens voltage-gated calcium-ion (Ca^{++}) channels located in the membrane of the presynaptic terminal. Not unlike Na^+, Ca^{++} ions are in

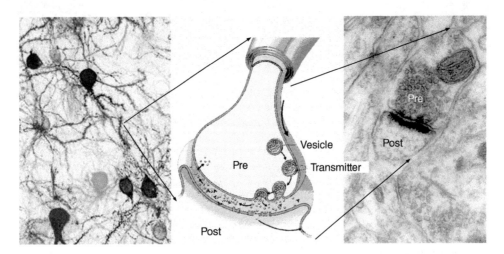

FIGURE 2.7 Left-side panel: light microscope image of neural tissue with cell bodies and processes visible. Many of the dendrites are studded with dendritic spines, including the one from which a tiny region is blown up to illustrate a synapse in the center panel. Center panel: diagram of a synapse, illustrating (with arrows) the events triggered by the arrival of an action potential: vesicles filled with neurotransmitter in the presynaptic terminal ("pre") are transported to the membrane, where they fuse with it to release neurotransmitter into the synaptic cleft; subsequently, molecules of neurotransmitter (or, depending on the type of synapse, a catabolite) are reuptaken back into the presynaptic terminal. Right-side panel: an electron microscope image of a synapse, with vesicles visible in the presynaptic terminal, and the "postsynaptic density," partly composed of membrane-spanning receptors, visible as a dark band at the postsynaptic terminal.
Source: Buszáki, György. 2006. Rhythms of the Brain. Oxford: Oxford University Press. Copyright 2006. Reproduced with permission of Oxford University Press.

much higher concentration outside than inside the neuron, and thus the opening of voltage-gated Ca^{++} channels results in an influx of Ca^{++} ions into the presynaptic terminal. Ca^{++} ions have a special status in the nervous system, in that they can initiate many biochemical processes within the cell. At the presynaptic terminal, the influx of Ca^{++} ions triggers a cascade of events that result in the release of neurotransmitter into the synaptic cleft (*Figure 2.7*). (Note that it is almost never the case that a neuron makes contact with just one other neuron. Rather, axons typically have multiple branches [in some cases, thousands], each of which ends in a synaptic terminal and many of which also have synaptic varicosities along their length. The action potential propagates equally along each branch.)

How neurons communicate: the postsynaptic consequences of neurotransmitter release

On the postsynaptic side of the synapse, the membrane is studded with *neurotransmitter receptors*, each of which can bind with a molecule of the just-released neurotransmitter. This, in turn, will, depending on the type of receptor, produce one of any number of possible effects in the postsynaptic neuron. The most straightforward of these to describe, and one that's satisfactory for our present purposes, is to open a **ligand**-gated Na^+ channel – that is, a molecule of neurotransmitter acts like a key by fitting into a special slot in the receptor and "unlocking" it. With its channel now open, the receptor allows the influx of Na^+ ions into the postsynaptic terminal of the dendrite. Thus, in our example, the binding of a molecule of neurotransmitter to a receptor has the effect of generating a small "cloud" of depolarization in the vicinity of this receptor within the postsynaptic terminal. Now if only one such event happens, the cloud of depolarization quickly dissipates. If, however, many molecules of neurotransmitter have bound with many receptors, particularly if this has happened more-or-less simultaneously at many synapses (e.g., if this neuron is on the receiving end of a volley of many action potentials arriving simultaneously from many different presynaptic neurons), then all these individual pulses of depolarization may combine, via their summation, to the point at which the cell's membrane voltage is raised to the neighborhood of −20 mV, in which case an action potential may be triggered in this postsynaptic neuron. (A mechanism by which the action potentials from many neurons can be coordinated to all arrive at a postsynaptic neuron at the same time is considered near the end of this chapter, in the section *Synchronous oscillation*.)

Different neurotransmitters and different receptors can produce different effects on the postsynaptic cell

The postsynaptic effects of neurotransmission that were described in the previous paragraph apply to the **excitatory neurotransmitter** glutamate binding to a particular kind of glutamate receptor called the **AMPA receptor**. (Neurotransmitter receptors are often named after chemicals that can mimic the effects of their naturally occurring ligands. Thus, the *AMPA* receptor is named after the chemical that binds this type of glutamate receptor, but not other types of glutamate receptor: α-amino-3-hydroxy-5-methyl-4-isoxazole propionic acid.) There are other glutamate receptors in the brain that work differently, and we will consider them in some detail when we discuss memory functions. **Glutamate** is released by cortical pyramidal cells, and because these are the most numerous neurons found in the cortex, glutamate is the predominant excitatory neurotransmitter in the CNS. Equally important for the functioning of the brain is **inhibitory neurotransmission**, and this is most commonly carried out by the neurotransmitter GABA (gamma-aminobutyric acid). GABA is the primary neurotransmitter released by **inhibitory interneurons**, neurons whose primary function is to **hyperpolarize** neurons (i.e., to make their voltage more negative). (Because the resting potential of the neuronal membrane is negative, the convention is to say that a current whose effect is to move this potential closer to 0 mV is a depolarizing current, and a current whose effect is to move the potential further from 0 [i.e., making it more negative] is a hyperpolarizing current.) The most common GABA receptor produces its hyperpolarizing effects by opening channels that pass chloride ions (Cl^-; *Figure 2.6*), and because the extracellular concentration of Cl^- ions is considerably higher than the intracellular concentration, opening these channels results in an influx of negatively charged Cl^- ions, and, thus, a negative shift in the membrane potential. The reason that this effect on the postsynaptic neuron is inhibitory is because it pulls the membrane voltage away from the trigger potential of −20 mV, thereby decreasing the neuron's probability of firing an action potential.

Synaptic plasticity

All examples of learning, as well as of maturation in the developing organism, are outward products of synaptic plasticity, the changing of strength of synapses in response to changes in what might be thought of as their input/output regime. "Input" in this context refers to the frequency with which action potentials prompt a presynaptic terminal to release neurotransmitter and/or to the chemicals that bind with receptors at or near the presynaptic terminal in conjunction with these action potentials. (The presynaptic terminals of many synapses also have receptors, sometimes "autoreceptors" that bind with the chemicals that they themselves release, and sometimes receptors that bind with molecules released by other nearby cells.) "Output" in this context refers to the effect that neurotransmitter release has on the postsynaptic neuron, more specifically, on the probability that activity in the presynaptic neuron is associated with an action potential in the postsynaptic neuron. We'll consider different examples of synaptic plasticity going forward, but for now it's sufficient for us to know that it is a fundamental property of most synapses in the nervous system.

Housekeeping in the synaptic cleft

Although the pre- and postsynaptic terminals of a synapse are very close together, there is no guarantee that any one molecule of neurotransmitter will bind with a receptor. Rather, binding is a probabilistic event, with the probability of binding influenced by many factors, including the amount of neurotransmitter released by an action potential, and the density of receptors in the postsynaptic membrane. It is also important to know that those molecules of neurotransmitter that do bind with a postsynaptic receptor do not enter the postsynaptic cell. Rather, after some period of time, they disengage from the receptor and float back into the synaptic cleft. The nervous system thus has to solve the problem of clearing out the synapse after each action potential, lest the lingering molecules of neurotransmitter in the synapse continue to influence the postsynaptic neuron for longer than intended. The nervous system has evolved many different solutions to this problem that are implemented at different synapses. Common to many is the process of **reuptake**, whereby molecules of neurotransmitter are taken back into the presynaptic terminal via specialized complexes of membrane-spanning proteins. Reuptake is an active process that requires energy, because it entails moving molecules of neurotransmitter from a low-concentration environment (the extracellular space) into a high-concentration environment (the presynaptic terminal). That is, it entails working against a chemical gradient. At glutamate synapses, the reuptake occurs both at the presynaptic terminal and at astrocytes

that are located adjacent to the synapse. (In astrocytes, glutamate is converted into a precursor metabolite, which is transferred back to the presynaptic pyramidal cell for conversion back into glutamate.) At GABA synapses, the neurotransmitter is reuptaken into the presynaptic terminal by the GABA transporter. At synapses for many of the neuromodulatory transmitters, there are enzymes that either break down the molecules of neurotransmitter into smaller, constituent molecules or transform them into a different molecule, the result in either case being a product that can no longer bind with the postsynaptic receptors. These now-inert molecules are then taken back into the presynaptic neuron, where they are reconverted into molecules of neurotransmitter. There are entire textbooks devoted to the details of how neurotransmitters are synthesized and metabolized (often falling under the rubric of "neuropharmacology"), and others to the details of the trafficking, release, and reuptake of neurotransmitters at the presynaptic bouton. Indeed, we'll need to revisit this in more detail in *Chapter 3*, when we consider the metabolic bases of some neuroimaging methods, but for now we can move on.

OSCILLATORY FLUCTUATIONS IN THE MEMBRANE POTENTIAL

Neurons are never truly "at rest"

As previously noted, the small cloud of depolarization produced by the opening of any single ligand-gated Na^+ channel typically won't, by itself, be enough to trigger an action potential, particularly when the channel is located on a distal dendritic branch, far from the cell body and its axon hillock. If this one channel were to periodically open then close, open then close, what one would observe at the cell membrane in the immediate vicinity of the receptor would be a slight depolarization – say, from −70 mV to −69 mV – followed by a return from −69 mV back to −70 mV, followed by another rise back to −69 mV, and so on. That is, the local membrane voltage would oscillate. Now, of course, we already know that the factors influencing the membrane voltage are much more complicated than the behavior of a single ligand-gated ion channel. For starters, a single synaptic event (i.e., an action potential in the presynaptic neuron) is likely to activate hundreds, if not thousands, of individual channels. These individual depolarizing events will add together (they will "summate"). So, too, will the depolarizing effects produced by activity at other synapses. As a result, the overall effect that is experienced by the synaptic membrane will depend on the synchrony of incoming action potentials from glutamate-releasing axons. That is, if many (glutamatergic) action potentials arrive at about the same time, the aggregated depolarizing effect will be much larger than if the same number of incoming action potentials are spread out in time, because the depolarizing effect of the earliest-arriving action potentials will have started to dissipate (also referred to as "decay") before the depolarizing effect of the later-arriving ones starts to kick in. A second factor that influences fluctuations in overall membrane potential is the hyperpolarizing effects of action potentials at GABAergic synapses. What emerges from this cacophony of thousands of events at individual channels, some depolarizing and some hyperpolarizing, is an ongoing pattern of fluctuations in the dendritic membrane potential that can be measured with an electrode in the extracellular space. Because the dendritic branches from many, many neurons are located in close proximity to each other, it is typically not possible to isolate the fluctuations attributable to a single neuron. Rather, what is recorded is the local field potential (LFP), a reference to the fact that it's the local region of cortex where these dendrites are located (i.e., the "dendritic field") from which the signal is being measured (the method for measuring the LFP, together with an example, is illustrated in *Figure 3.7*).

Despite the fact that there can be literally hundreds of thousands of individual channels distributed across the tens to hundreds of neurons contributing to the LFP recorded by any one electrode, the LFP is nonetheless often observed to fluctuate in an orderly oscillation. Indeed, different oscillatory frequencies, or "states," are so typical of the cortex that they've been given names. One common state is when the LFP oscillates with roughly 40 up–down cycles per second (i.e., 40 Hz); the range of oscillatory frequencies within which this falls, from roughly 30 Hz to roughly 80 Hz, is referred to as the gamma frequency band (or, "the gamma band," or, sometimes, in the shorthand slang of the laboratory, just "gamma"). (This and other frequency bands are illustrated in *Figure 3.10*.) How can the brain orchestrate what we've just referred to as a cacophony of individual single-channel events into a steady, predictable up-and-down rhythm? One important factor is that neurons are wired up in a way that facilitates the control of their oscillations. For many pyramidal cells, for example, inhibitory inputs from interneurons tend to synapse near the base of the dendrite, or even on the cell body itself. This positioning

allows these inhibitory synapses to act as a kind of "gate" that either does or does not allow the depolarizing currents from more distally located glutamatergic synapses to influence the membrane voltage at the cell body. As we'll see in *Chapter 5*, the biophysical properties of one common type of inhibitory interneuron make it tend to oscillate within a narrow band of frequencies in the gamma range. A second factor that leads to the orderly oscillation of membrane potentials is that many interneurons, in addition to sending axons to local pyramidal cells, are also "wired together" via axonal connections that they make with each other (i.e., they are reciprocally connected). As a result, the activity of these interneurons tends to be synchronized. The consequence for pyramidal neurons is that they tend to experience a regular pattern of the delivery, then dispersion, of hyperpolarizing currents.

Oscillatory synchrony

A consequence of receiving a synchronized delivery of hyperpolarizing GABA 40 times per second is that the membrane potential of a pyramidal cell won't actually remain at the steady −70 mV that was implied at the beginning of this chapter. Rather, it will oscillate between, say, −90 mV and −50 mV. And because a neuron is much more likely to reach its firing threshold when its membrane potential is in the neighborhood of −50 mV than that of −90 mV, this results in a situation in which the pyramidal neuron has, in effect, 40 "windows" per second during which it might be able to fire an action potential. Given this state of affairs, if a pyramidal neuron in area X wants to make a pyramidal neuron in area Y fire an action potential, there are two things that it needs to do. First, it needs to time the delivery of its own action potential so that this action potential arrives during a depolarized phase of the membrane oscillation of neuron Y. Second, it needs many of its neighbors in area X to also send their action potentials at the same time, such that the area Y neuron receives many depolarizing impulses all at the same time. From this, it is apparent that the synchronization of membrane oscillations in different populations of neurons is a critical factor in the smooth operation of the brain. Within a local group of cells, there's a degree of synchrony between interneurons and pyramidal neurons. Across regions, oscillatory synchronization of ensembles of neurons is necessary if they are to effectively communicate. Indeed, in many contemporary models of neuronal function, synchrony plays a critical role.

To illustrate how synchrony might be important for brain function, let's walk through a hypothetical example. Let's pretend that, while you're walking through the woods, three things happen simultaneously: (a) the sun breaks through the clouds; (b) a bird in an overhead tree takes flight; and (c) a bear comes ambling along the path in your direction. Now, because individual elements in your brain don't "know" that one of these events has more immediate relevance to you, the organism, than the other two, information about all three is propagated from the visual system to "downstream" brain systems that might implement one of three behaviors: (a) stopping to enjoy the warm caress of the sun; (b) reaching for the binoculars to watch the bird in flight; or (c) RUNNING AWAY(!). According to synchrony models, one way that

the nervous system has evolved to let choice (c) be the "winner" is that action potentials carrying information about the bear are timed to arrive during a depolarized phase in the oscillation of their downstream target system, whereas action potentials carrying information about the sun and about the bird arrive during a hyperpolarized phase of their downstream targets. Thus, the information about the bear is more potent, in a neural sense, and it thereby "wins" the "competition" to control your behavior. Such coordination and control of the LFP in different regions is referred to as **phase synchronization** (*Figure 2.8*). We shall see in subsequent chapters that some psychiatric disorders might be attributable to abnormal control of synchronization between neural systems.

FIGURE 2.8 Illustration of how phase synchrony may govern the effectiveness of neural communication.

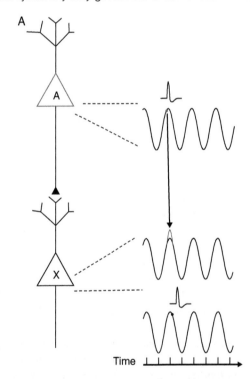

FIGURE 2.8.A Synchrony between an upstream and a downstream neuron. Neuron A and neuron X are oscillating with a phase lag of near 0 (i.e., membrane voltage of neuron A reaches the maximum depolarization point of its cycle just a fraction of a second more quickly than does neuron X). A consequence of this is that an action potential fired by neuron A (red dot superimposed on neuron A's membrane oscillation) arrives at neuron X (illustrated by arrow) just as neuron X is at its maximally depolarized state, and thus the action potential from neuron A boosts the level of depolarization of neuron X at a moment in time when neuron X is already close to threshold, thereby triggering an action potential (black dot superimposed on neuron X's membrane oscillation).

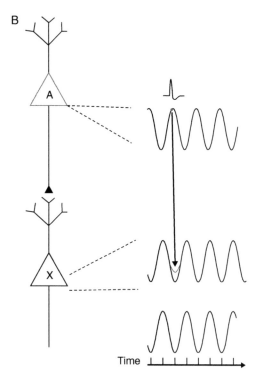

FIGURE 2.8.B A scenario in which neuron A and neuron X are oscillating out of phase with each other. Here, the action potential fired by neuron A arrives at neuron X at its maximally hyperpolarized state. Thus, although the action potential from neuron A does boost the level of depolarization of neuron X, it does so at a moment in time that it's unlikely to push neuron X above its threshold. Thus, the signal propagated by A "dies" at X and has no further influence on perception or behavior.

It should be noted that although no neuroscientist questions the existence of oscillations in membrane potentials and in the LFP, there is vigorous debate about the extent to which these oscillations actually play a functional role in the workings of the brain. From one perspective, these oscillations could be nothing more than byproducts of the complex pattern of synaptic events experienced constantly by every neuron in the brain. This perspective might be paraphrased as *They happen, yes, but they may not actually serve any useful function.* On the other hand, with each passing month, it seems that there are new discoveries of specific neural systems (sometimes in the visual system [*Chapter 4*], sometimes in the systems that govern motor control [*Chapter 8*], sometimes in the hippocampus [*Chapter 11*]) whose functioning seems to depend on the precise synchronization of oscillations in different ensembles of neurons. Keeping this controversy in mind, the phenomenon of neuronal oscillations is one that we will return to at several points in this book.

COMPLICATED, AND COMPLEX

The human brain is truly an amazing thing. It is made up of billions of neurons and trillions of synapses. What we have considered in this chapter only scratches the surface of all there is to know about the brain's molecular and cellular makeup. Entire careers have been, and will continue to be, devoted to understanding these elemental components of the brain. With the facts that we have considered, and the questions we have posed, however, we have enough foundational knowledge that we can now move on to our final "introductory" task, which is to learn about the methods that are used to study brain structure and function. But first, while all these facts about how the brain is put together and how it works are still fresh in the mind, let's take a moment to consider a distinction made very eloquently by György Buzsáki in his 2006 book *Rhythms of the Brain*. The nuts-and-bolts facts of how cells in the brain work are, indeed, *complicated*. How this all works together to give rise to cognition, however, is not only complicated, it's *complex*. "The term 'complex' does not simply mean complicated but implies a nonlinear relationship between constituent components," and refers to properties often termed "spontaneous, endogenous, . . .

self-organized, self-generated, self-assembled, and emergent" (p. 11). Stated another way, what we shall see again and again in our exploration of cognitive neuroscience is that the whole is different than the sum of its parts. This chapter has introduced many of the critical "parts" – parts of the brain, neurons, and neurotransmitters – as well as some of their properties – ion channels, oscillating membrane potentials, action potentials, and oscillating LFPs. Now we can consider how they are measured and, in the process, start to consider how they combine and interact.

END-OF-CHAPTER QUESTIONS

1. In what ways does the anatomical organization of the adult brain follow from its developmental origin as the neural tube?

2. What factors explain the "crumpled-up" shape of the brain? What evolutionary pressures may have produced this shape?

3. What is a distinctive property of each layer of neocortex? Why might there be some advantages to segregating these properties/functions by layer?

4. What are the principal disequilibria of the neuron that were emphasized in this chapter, and what role do they play in the generation of the action potential?

5. It can be said that neurons communicate via signals that, at one level of analysis, can be considered digital, yet

from another level of analysis can be considered analog. What phenomena that were discussed in this chapter are an example of each?

6. What explains the oscillation of voltage across the cell membrane? What explains the oscillation of voltage in the LFP?

7. Although some scientists have argued that neuronal oscillations are primarily byproducts of the workings of neurons and networks of neurons (i.e., they may be "epiphenomenal"), there is a growing consensus that they may play an important role in governing the effectiveness of communication between one brain area and another. Describe the concept of phase synchronization and the role it may play in neuronal communication.

REFERENCES

Buszáki, György. 2006. *Rhythms of the Brain*. Oxford: Oxford University Press.

Dumitriu, Dani, Alfredo Rodriguez, and John H. Morrison. 2011. "High-Throughput, Detailed, Cell-Specific Neuroanatomy of Dendritic Spines Using Microinjection and Confocal Microscopy." *Nature Protocols* 6 (9): 1391–1411. doi: 10.1038/nprot.2011.389.

Hubel, David H., and Torsten Wiesel. 1977. "Functional Architecture of Macaque Monkey Visual Cortex." *Proceedings of the Royal Society of London. Series B, Biological Sciences* 198 (1130): 1–59.

Kandel, Eric R., James H. Schwartz, and Thomas M. Jessell, eds. 2000. *Principles of Neural Science*. New York: McGraw-Hill.

Saalmann, Yuri B., and Sabine Kastner. 2009. "Gain Control in the Visual Thalamus during Perception and Cognition." *Current Opinion in Neurobiology* 19 (4): 408–414. doi: 10.1016/j.conb.2009.05.007.

Saalmann, Yuri B. and Sabine Kastner. 2011. "Cognitive and Perceptual Functions of the Visual Thalamus." *Neuron* 71 (2): 209–223. doi: 10.1016/j.neuron.2011.06.027.

Warwick, Roger, and Peter L. Williams. 1973. *Gray's Anatomy*. 35th British ed. Philadelphia, PA: W. B. Saunders.

OTHER SOURCES USED

Cooper, Jack R., Floyd E. Bloom, and Robert H. Roth. 1991. *The Biochemical Basis of Neuropharmacology*, 6th ed. New York: Oxford University Press.

Diamond, Marian C., Arnold B. Scheibel, and Lawrence M. Elson. 1985. *The Human Brain Coloring Book*. New York: Harper Perennial.

Hubel, David H. 1988. *Eye, Brain, and Vision*. Scientific American Library, no. 22. New York: W. H. Freeman.

FURTHER READING

Buszáki, György. 2006. *Rhythms of the Brain*. Oxford: Oxford University Press.

If there's only one book that this textbook inspires you to read, this might be it. Provides a rigorous neurophysiological and anatomical basis for much of the dynamical systems-influenced content of this book, including, for example, the final section of this chapter.

Diamond, Marian C., Arnold B. Scheibel, and Lawrence M. Elson. 1985. *The Human Brain Coloring Book*. New York: Harper Perennial.

In science, it's often said that "the best way to ensure that you understand a topic is to teach a class about it." The same might be said about neuroanatomy and drawing.

Morrison, John H., and Mark G. Baxter. 2012. "The Ageing Cortical Synapse: Hallmarks and Implications of Cognitive Decline." *Nature Reviews Neuroscience* 13 (4): 240–250. doi: 10.1038/nrn3200.

The PFC is one of the first regions of the brain to show age-related decline in healthy, middle-aged individuals. This paper shows that, in the monkey, this decline is largely due to a dramatic loss of dendritic spines in layer III, the cortical layer whose integrity is most closely tied to age-related cognitive impairment.

CHAPTER 3
METHODS FOR COGNITIVE NEUROSCIENCE

KEY THEMES

- Most cognitive neuroscience research employs the rationale (as well as the methods) from cognitive psychology that behavior in controlled experimental settings provides an effective, albeit indirect, way to measure mental activity.

- Neuropsychology, the study of the effects of brain lesions on behavior, can support strong inference about the *necessity* of a region's contributions to a function of interest, provided that the specificity of the effect(s) of the damage can be established.

- Many of the principles of experimental design and interpretation that apply to neuropsychology are equally important for noninvasive transcranial neurostimulation techniques, such as transcranial alternating current stimulation (tACS), transcranial direct current stimulation (tDCS), and transcranial magnetic stimulation (TMS).

- Many methods for studying the structure of the brain (i.e., neuroanatomy) exploit properties of physiological and cell biological processes.

- Depending on the exact method used to record electrical activity in the brain, one can either measure action potentials, a high-frequency proxy for action potentials referred to as multiunit activity (MUA), or oscillating field potentials at different spatial scales.

- There are two broad classes of data analysis that can be applied to signals recorded at the scalp with electroencephalography (EEG) or magnetoencephalography (MEG): event related (event-related potential [ERP] with EEG) and

time–frequency. Each makes assumptions about how the signals were generated by the brain, thereby limiting the kinds of phenomena that can be measured.

- Different applications of magnetic resonance imaging (MRI) can be used to measure gross anatomy, to estimate the trajectory (and integrity) of white-matter tracts, to measure relative changes in metabolic activity (over timescales ranging from 10s of minutes to individual cognitive events), and to measure the chemical makeup of different brain regions, including level of neurotransmitters and their metabolites.

- Because functional MRI (fMRI) is inherently constrained by the brain's hemodynamics, its temporal resolution is relatively poor, and its low-pass filtering qualities must be taken into account when making choices about experimental design and analysis.

- The brain is a massively interconnected nonlinear dynamical system, and neural network modeling offers one way to simulate some of these properties in ways that are simply not possible with empirical measurements of brain activity.

- Parallel distributed processing (PDP) can be an effective way to evaluate the plausibility of theoretical models about the computations underlying cognitive functions.

- Deep neural networks (DNNs) apply powerful techniques from machine learning and artificial intelligence (AI) to attain remarkably high levels of categorization of visual images, speech sounds, and other classes of stimuli. Close links between research on DNNs and on the neural bases of sensory systems (e.g., visual perception) are generating important insights leading to advancements in both fields.

- Because the brain is a massively interconnected system, analytic tools from network science, including graph theory, provide powerful ways to learn about this fundamental aspect of brain function.

CONTENTS

BEHAVIOR, STRUCTURE, FUNCTION, AND MODELS

Studying the brain, and how its functioning gives rise to thought and behavior, requires the ability to measure behavior, anatomy, and physiology, each at several different scales, and to articulate and test models of neural computations and of mental activity that explain behavior and physiology. In order to learn about cognitive neuroscience in a substantive way, it's necessary to have an understanding of the methods that are used to acquire these measurements. The goal of this chapter, therefore, is to equip the reader with enough of an understanding of the methods used by cognitive neuroscientists that, for the remainder of the book, we won't need to go into a lengthy aside every time a new method is introduced. (On the other hand, if you would prefer not to power through this chapter right now, and would prefer to "get to the good stuff" right away, references throughout the remainder of the book have been inserted to point the reader to this chapter's description of whatever method it is that needs to be understood in order to fully understand the experiment that's being described.)

We can divide the topic of methods of cognitive neuroscience into two general categories: how a method works (i.e., the physics, physiology, and engineering on which it depends); and how the data that it generates are analyzed and interpreted. Sometimes it's not really possible to talk about one without the other, but to the extent possible, in this chapter we'll limit ourselves to "how it works." This is because the "how the data are analyzed and interpreted" is so central to understanding cognitive neuroscience that to include it here would be, in effect, to cram much the content from our remaining 17 chapters into this one.

BEHAVIOR

Many neuroscientists consider the ability to move – so as to acquire food and/or to avoid becoming someone else's food – to be the factor whose adaptive value has been the primary driving force behind the evolutionary development of nervous systems. Put another way, the need to move, and to constantly improve the efficacy with which movements are carried out, may be the driving force behind why we have such complex brains. From this, some would argue that the only reason that we can see, hear, and

taste, the only reason that we can entertain abstract thoughts about situations that we may never directly encounter in our lives, may be because these capacities help us carry out actions more effectively. At a more concrete and pragmatic level, because no one has figured out a way to directly measure *thinking* per se, behavior offers indirect access to mental activity, because it can be an overt and measurable consequence of our mental activity. Furthermore, as was noted in *Introduction to Section I*, not all neural information processing that influences cognition is accessible to conscious awareness. It is nonetheless often the case that such "unconscious" brain processes will influence behavior in systematic ways that we can measure and interpret.

Neuropsychology, neurophysiology, and the limits of inference

Neuropsychology relies on the logic that one way to understand the workings of a system is to systematically damage (or inactivate) parts of it and observe how damaging each part affects the system's functioning. A neuropsychological study (sometimes also called a "lesion study") can address the question *Does* region A *make a necessary contribution to* behavior X? Note that this will be a necessary, but often not sufficient, step in addressing the broader question of *How does the brain generate* behavior X? To give a concrete example, *Chapter 1* describes how neuropsychological experimentation was used to determine definitively that visual perception is localized to the occipital lobe. Not until *Chapter 4*, however (and, chronologically, not until roughly 80 years later), will we see that electrophysiological experiments revealed how it is that neurons in the occipital lobe process visual information. Without the knowledge of where in the brain to record neuronal activity, however, the experiments addressing the "how" questions could never have been carried out. It's also the case that a carefully conducted neuropsychological experiment can support stronger inference about the *necessity* of a region's integrity for a particular behavior than can experiments that measure neurophysiological variables, because the latter are correlational. (For example, one can imagine a situation in which activity in a brain area varies systematically with changes in a particular behavior, but that the lesioning of that brain area doesn't alter the behavior. Such a scenario could come about if the brain area in question makes a redundant contribution to the behavior, or if it is a downstream recipient of output from the region whose activity *is* necessary for that behavior.)

Many stimulation-based techniques share inferential principles with neuropsychology

We'll get to the technical details of different methods for neurostimulation further along in this chapter. These will include tDCS and tACS, TMS, and optogenetics. But it is important to note that much of the reasoning that we are applying to neuropsychology will also apply to any method that involves direct manipulation of some aspects of brain function. Indeed, *Chapter 1*'s summary of the electrical microstimulation experiments of Fritsch and Hitzig (1870; in section on *The localization of motor functions*) offers a nice example with which to consider specificity.

The importance of specificity

In both neuropsychology and stimulation experiments, the strength of one's conclusions depends importantly on being able to demonstrate anatomical specificity. What made the Fritsch and Hitzig (1870) experiment so powerful, for example, was the specificity with which it demonstrated that stimulation of a particular portion of the frontal lobe in the left hemisphere produced movement of the right fore-limb, and that it did not produce movement of the right hind limb, nor of the left forelimb. Indeed, it was stimulation of a different, although nearby, area that selectively produced movement of the right hind limb, but no longer of the right forelimb. Had the authors limited their report to the fact that the stimulation of this left frontal region produced movement of the right forelimb, two important questions would have been left unanswered. First, *might stimulation of other regions also produce movement of the right forelimb?* Hinging on the answer to this would be an understanding of how localized is the control of the right fore-limb. Second, *can stimulation of this region of the brain also produce movement of other parts of the body?* The answer to this would give an indication of how specific is the function of brain area in question. (Note that this illustration of speci-ficity is somewhat contrived, because simple observation would be expected to indicate whether or not other parts of the body also moved as a result of microstimulating this one region. But imagine that one was interested in a less overtly observable aspect of behavior than that of skeleto-motor control. For example, *Does stimulation of this region of the brain only disrupt the ability to produce speech, or does it also disrupt verbal working memory?* Establishing specificity in a situation like this would require administering two (or more) tests while delivering the same stimulation protocol, and observing whether it does/does not disrupt performance on the verbal working memory task.)

These considerations can be particularly important in neuropsychological studies of humans, such as those described in *Chapter 1*'s section on *The localization of speech*, because these studies often rely on "accidents of nature," such as stroke, neurodegenerative disease, or head injury. In such studies, scientists have no control (other than patient selection) over the extent of damage in the patients that they study, neither in terms of how many structures may be damaged, nor of the overall volume of tissue that is affected. When lesions are large, there can be an inherent difficulty in determining which of the damaged structures is responsible for the observed behavioral deficit. One log-ical way to tackle this problem, the "double dissociation of function," will be introduced in *Chapter 6*.

Different kinds of neuropsychology address different kinds of questions

Cognitive neuropsychology

A one-sentence summary of cognitive psychology might be that it seeks to understand the organization and princi-ples underlying different aspects of cognition. A cognitive psychologist will typically carry out an experiment by varying some independent variables (perhaps the speed at which stimuli are presented to the experimental subject, or the number of stimuli concurrently presented to them) and then comparing performance at different levels of that independent variable. Cognitive neuropsychology can be thought of as a specific type of cognitive psychology in which the independent variable is integrity of the brain. That is, the cognitive neuropsychologist will administer the same task to healthy control subjects and to neurologi-cal patients, and then compare performance between the two groups. One set of examples that we will consider in detail in *Chapter 10*, on *Visual object recognition and knowl-edge*, are studies of object recognition deficits observed in patients with brain damage due to a wide range of causes (e.g., blunt trauma, encephalitis, stroke). One patient who suffered carbon monoxide poisoning retained the ability to discriminate colors, brightness, and size, but was "unable to distinguish between two objects of the same luminance, wavelength, and area when the only difference between them was shape" (Benson and Greenberg 1969). Other patients have been described as someone who can copy drawings of objects with a high level of fidelity (e.g., a guitar or an owl), but who cannot generate the name of these objects. These patterns of spared vs. impaired perfor-mance were used to develop models positing a series of discrete stages of processing that support visual object recognition. Importantly, these models didn't address with any detail *where* in the brain these processing stages may be implemented, just that these stages may be indicative of how object recognition is accomplished. (This is sometimes referred to as the "cognitive architecture" of object recognition.)

Experimental neuropsychology

Experimental neuropsychology is a method used to study what brain areas support a particular function, or, conversely, what functions are supported by a particular brain area. A nice example comes from Baldo and Dronkers (2006), who wanted to test the hypothesis that two brain areas – left inferior parietal lobule (IPL) and left inferior frontal gyrus (IFG) – support different functions related to verbal working memory: storage and rehearsal, respectively. (An example of verbal working memory is when someone tells you a telephone number, and you need to remember it while you first look for your phone, then look for a quiet place to make the call – we'll consider it in detail in *Chapter 14*.) In the introduction to their study, the authors highlight the fact that their approach for recruiting patients was to identify patients with damage to one of these regions or the other, without any knowledge of the behavioral impairments that may have been caused by their stroke. The alternative approach would be to start by seeking patients with a behavioral deficit in either the ability to store or to rehearse information in verbal working memory. The logic of their approach was that if one recruited patients based on their impairment, and then observed that, for example, most of the recruited patients with a rehearsal deficit had damage in left IFG, one couldn't conclude that the left IFG is necessary for rehearsal in verbal working memory. This is because one couldn't know whether, among the population of stroke cases from which the patient selection was carried out, there may have been additional patients who had comparable damage to left IFG, but who did not have a verbal working memory impairment. Thus, the emphasis here was on assessing the necessity of specific brain areas for a specific kind of behavior.

Clinical neuropsychology

For both cognitive neuropsychology and experimental neuropsychology, it's not uncommon that the researcher will develop novel behavioral tasks in order to test the question of interest. This is almost never the case for clinical neuropsychology, which is concerned with patient care. Clinical neuropsychology is used for the diagnosis of a neurological or psychiatric disorder, the assessment of the severity of a disorder, and/or for the assessment of the efficacy of a treatment and/or the rate of recovery of a patient. Therefore, a clinical neuropsychologist will almost invariably use standardized tests that have been normed on healthy subjects, typically across a range of ages, and for which an individual patient's level of performance can be used for diagnosis, staging, or assessment. The methods of clinical psychology involve having extensive knowledge of what kinds of "neuropsych tests" (as they're often known) are most appropriate for different kinds of patients, and how to interpret the results of these tests. (Because the emphasis of this book is on basic science, clinical neuropsychology won't come up very often.)

How does behavior relate to mental functions?

It is important, at this juncture, to remind ourselves that many aspects of cognition don't have to generate overt behavior. For example, you can read two words and decide whether or not they rhyme without having to say them out loud. Stated another way, we can generate phonology, and reason about it, without having to generate, nor hear, speech. (Indeed, assuming that you are not reading aloud, you may be experiencing what is informally known as "inner speech" – "covert speech" to a scientist – right now.) When behavior is covert, as it is with covert speech, mental imagery, or imagined activity, one must be careful about interpreting the results of cognitive neuroscience experiments. In particular, a form of reasoning known as "reverse inference" can often pose challenges.

Reverse inference has a formal definition in the scholarly discipline of logic, but for our purposes we can use an example. Let's say that you are looking at the results from an fMRI study in which you acquired brain scans while your subjects were shown black and white drawings and asked to indicate whether each drawing depicted a living or a nonliving thing, by pressing one of two buttons. (We'll get to fMRI further along in this chapter.) The results indicate that one of the regions that showed elevated activity during task performance is a region of posterior IFG in the left hemisphere – Broca's area from *Chapter 1*. It might be tempting to conclude from this that, because Broca's area was engaged by the task, subjects must have been covertly naming the pictures. (And you would be doing so, in part, because of the "forward inference" from previous studies, in which overt speech was seen to be reliably accompanied by elevated activity in Broca's area; that, and the fact that lesion studies indicated that Broca's area is necessary for speech production.) The potential problem with this conclusion, however, is that it depends on the proposition that "Broca's area is specialized for speech production," and the consequent assumption that activity in Broca's area can mean only one thing. Another way of saying this would be to say that activity in Broca's area is specific for speech production. But what if this isn't true? Indeed, this very question was addressed in an influential review paper by Poldrack (2006), who sorted through a database that contained the results from 749 neuroimaging studies that contained patterns of brain activity from 3222 experimental comparisons; 869 of these comparisons were language related, and of these, only 166 showed language-related change in activity in Broca's area. (That is, activity in Broca's area would seem to not be a very sensitive indicator of using language.) Furthermore, of the 2353 comparisons that were not thought to emphasize language, 199 nonetheless showed a task-related change in Broca's area. (That is, activity in Broca's area is also not specific to language.) We'll see, as we go forward, examples of analysis methods that can raise the sensitivity and the specificity of a comparison, such that reasoning from reverse inference is more tenable. But for now, we'll leave this as a cautionary tale that impels us to think critically when drawing conclusions about brain–behavior relations.

Methods for lesioning targeted areas of the brain

We have already seen, in *Chapter 1*, that there are many circumstances that a scientist can either create or capitalize on to study the effects of altered brain function on behavior. Here, we'll limit ourselves to considering instances of damage incurred by an adult brain that had developed typically prior to the incident that produced the damage.

Stroke

For studies of human behavior, the classic source of patients with brain damage is the neurology clinic. Stroke, in which either the blockage or the rupture of an artery deprives an area of the brain of oxygenated blood for an extended period of time, results in often stereotyped patterns of cognitive and behavioral impairment that can be very similar across patients due to the fact that the vasculature of the brain, like every other part of our anatomy, is very similar across individuals. (For example, infarction of the superior branch of the middle cerebral artery [MCA] in the left hemisphere will typically produce weakness or paralysis on the right side of the body, and often an aphasia similar to that described by Broca [*Chapter 1, The localization of speech; Figure 1.5*]). This can mean that, regardless of whether a poststroke scan may be needed for diagnostic and/or treatment purposes, the neuropsychologist may not need detailed neuroimaging to know what brain areas were likely impacted by the stroke.

Nonlocalized trauma

Brain damage can result from factors that produced damage that produces behavioral deficits that are less predictable than those resulting from stroke. *Viral encephalitis*, for example, occurs when a viral infection causes widespread swelling in the brain. (Variants of the herpes virus are often the culprit.) When the acute stage has passed and the swelling has subsided, any of a number of brain areas may turn out to have suffered irreversible damage. One notorious outcome of viral encephalitis can be damage to a subfield of the hippocampus, which, if bilateral (i.e., affecting the hippocampus in both hemispheres), can lead to severe memory impairment. (We'll get to this starting in *Chapter 11*.) In other patients, however, the acute phase of the encephalitis can have spared the hippocampus, but created lasting damage in circuits that are important for the sense of smell, or for vision, or other functions.

Environmental hypoxia/anoxia results from extended periods of time in a low-oxygen environment. (Hypoxia – access to low levels of oxygen – or anoxia – total deprivation of oxygen – can be the proximal causes of tissue damage due to stroke or other events "internal" to the body. Here, the reference is to the lack of oxygen in the breathable air.) As with encephalitis, because the impact is systemic, lasting damage can be highly variable. For example, environmental hypoxia is the cause of damage to patients who we'll study further along in the book who have very specific impairments in the ability to name (and, perhaps, to think about) living objects in the world (but not nonliving objects), and to others who have difficulty recognizing individual faces, but not other complex and only-subtly-differing kinds of objects, like automobiles, or flowers.

Focal lesions

In the neurology clinic, focal lesions can result from penetrating head wounds, such as those caused by a bullet or by shrapnel in warfare, or from any of a variety of gruesome accidents that I won't review here. (Although we will take up the case of Phineas Gage in *Chapter 17*!) Depending on how circumscribed is the damage, careful assessment of patients with focal lesions can yield valuable insight about structure–function relations. One early example is that, because the Russo-Japanese war (1904–1905) has the ignoble distinction of being one of the first conflicts to feature the widespread use of high-velocity bullets, Japanese ophthalmologist Tatsuji Inouye was able to make pioneering discoveries about the organization of visual cortex by studying wounded veterans with visual impairments. (Among other attributes, these high-velocity bullets produced clear entry and exit wounds in the skull, which facilitated inferring the locus of brain damage despite the nonexistence of medical scanning technology.) Note that, although a stroke can produce focal lesions, there's always a concern when interpreting the effects of a stroke that it may have caused other damage in addition to the focal lesion, perhaps more diffuse damage too subtle to pick up with a brain scan. For this reason, many studies of stroke patients with damage to area X will include a control group of patients with damage that spares area X, in an effort to control for possible nonspecific effects contaminating the behavior of the experimental group.

In laboratories where experiments are carried out on nonhuman animals, many methods are used to disrupt brain function. Reversible methods are often used when the behavior of interest might involve learning, or periodic changing of behavior. In these studies, it can be particularly important to be able to show that normal behavior is restored when the intervention is not being applied. Methods for reversible disruption include electrical stimulation, the application of chemical agents whose effects dissipate after a finite period of time (sometimes called "behavioral pharmacology"), and cryogenic disruption (informally known as "cortical cooling"). Reversible pharmacological intervention often uses compounds that mimic the effects of neurotransmitters or other endogenous substances. Cooling takes advantage of the fact that all living tissues in warm-blooded organisms functions optimally within only a very narrow range of ambient temperature. There are specific protocols whereby brain tissue can safely be cooled until its ongoing activity ceases, and with functions returning to normal once the source of

the cooling is removed. Note, however, that cortical cooling entails a craniotomy and inserting the cooling device (sometimes a probe, and sometimes a coil that is positioned on a gyral surface or inside a sulcus), and so while the impairment from the cooling per se is reversible, it is not a noninvasive procedure.

Methods for producing permanent damage, for which there is often a scientific rationale, range from producing electrical damage (increasing the voltage in an electrode) to chemical/pharmacological damage (sometimes targeting, for example, only neurons with a certain subtype of neurotransmitter receptor), to surgical intervention.

Noninvasive methods to produce "virtual lesions"

One of the exciting, and powerful, developments that has accompanied the explosion of cognitive neuroscience over the past quarter century has been the introduction and refinement of methods to alter brain function that don't require surgical intervention. Many of these methods involve either magnetic or electrical energy, and each will get its own section(s) in ensuing parts of this chapter.

TRANSCRANIAL NEURO-STIMULATION

The methods summarized in this section are classified as noninvasive, because they don't involve a surgical procedure. Indeed, although many of them are in varying stages of consideration as potential treatments for a variety of neurological and psychiatric maladies, all are widely used to carry out studies of brain function in healthy young adults. All of the methods that we will consider here work on the basis of manipulating the electrical properties of the nervous system.

The methods that we'll summarize in this section are emblematic of a trade-off that often confronts scientists who want to study the neural bases of human cognition: although we are studying the cognition of the species that is ultimately of primary interest to most of us – ourselves – we are often limited to methods that don't offer the same resolution as do invasive methods that are often used for research carried out on other species. In the case of neurostimulation, the precision of extracranial methods is orders of magnitude more coarse than are invasive microstimulation of the kind already summarized and deep brain stimulation that is used as a surgically implanted treatment for conditions such as Parkinson's disease (to be discussed in detail in *Chapter 8, Skeletomotor control*). Nonetheless, these are powerful tools that have provided important insights about the neural bases of human cognition.

The importance of specificity (again)

Transcranial neurostimulation techniques are similar to neuropsychology in that they have poor anatomical specificity. Therefore, careful design of experiments is necessary if they are to support strong inference. Most often this means inclusion of a so-called "active control" condition, in which the same stimulation parameters that were delivered to the brain area of interest are also delivered to a control brain area that is not believed to contribute to the behavior of interest. A pattern of results in which stimulation of the brain area of interest has the predicted effect on behavior, but stimulation of the control area does not have this same effect, rules out the possibility that the former may have just been a nonspecific effect of stimulation of the brain. Depending on the experimental hypothesis, it can sometimes also be useful to include a control task.

Transcranial electrical stimulation

This method entails applying two largish electrodes (typically on the order of 5 cm × 5 cm) to opposite sides of the cranium, and passing current from one – the anode – to the other – the cathode.

Transcranial direct current stimulation

Direct current is a steady flow of electrical current in one direction. Most batteries that power portable devices, from flashlights (a.k.a. torches) to electric lawn mowers, deliver DC power. Like a battery, tDCS has one contact (one electrode) that is the anode – from which electrons flow – and a second that is the cathode – to which electrons are drawn. Functionally, tDCS works on the basis that brain tissue located close to the anode will exhibit an increase in excitability during the application of tDCS, and for an extended period of time after tDCS has been applied. As a result, the typical expectation is that if the brain area under the anode is important for the performance of a task, performance will improve with anodal tDCS. Because the converse is true for tissue located close to the cathode, the electrode montage for a tDCS study will often place the critical electrode over the brain area of interest, and the other electrode over a non-brain part of the head, such as the cheek (see *Figure 3.1*). This avoids a situation in which

it's unclear whether a tDCS-related change in behavior resulted from excitation of the region under the anode or inhibition of the region under the cathode.

tDCS is typically delivered at an intensity of 1–2 mA, and this has been estimated to produce a relatively weak electrical field that may change the membrane potential of neurons most affected by the tDCS by less than 1 mV. Therefore, the effect of tDCS on neural tissue is almost certainly a modulatory one. (That is, it doesn't directly trigger action potentials in the neurons that it affects, but it changes the conditions under which these neurons will generate action potentials.) It's noteworthy that the behavioral effects of tDCS can extend for considerable period of time after the stimulation is no longer being delivered. The duration of these residual effects can vary from tens of minutes to more than hours, and depends on the strength and duration that tDCS is applied, as well as the cortical area being targeted. The reason for this isn't fully understood, but the assumption is that some factors that influence excitability, like the density of postsynaptic receptors at certain types of synapses, must be retaining the influence of the tDCS during this time. As a result, some studies will have subjects performing one or more behavioral tasks while tDCS is being delivered, and others will begin the behavioral testing immediately after the administration of tDCS. With either procedure, repeated measures designs that call for subjects to perform the same task(s) with different areas targeted will typically impose a separation of a day or more between tDCS sessions targeting different areas.

There are many more variants of how tDCS has been administered, and how it can influence neural activity and behavior, than we can summarize here. The details of just one seemingly simple study, however, give a sense of some of the nuances of the method. In their study, Thirugnanasambandam and colleagues (2011) delivered anodal or cathodal tDCS over motor cortex for 20 minutes, and then assessed its effect by delivering single pulses of TMS to the motor cortex and measuring the resultant electrical response in the muscles of the hand (the "motor-evoked potential," MEP). (This procedure, using TMS to evoke MEPs, is conceptually similar to the microstimulation study of Fritsch and Hitzig [1870] – directly stimulates a region of the motor cortex and observes the resultant effect on the muscles of the body part controlled by the stimulated tissue. We'll get to the details of how TMS stimulates cortical tissue further along in this section.) An important detail about the experimental procedure is

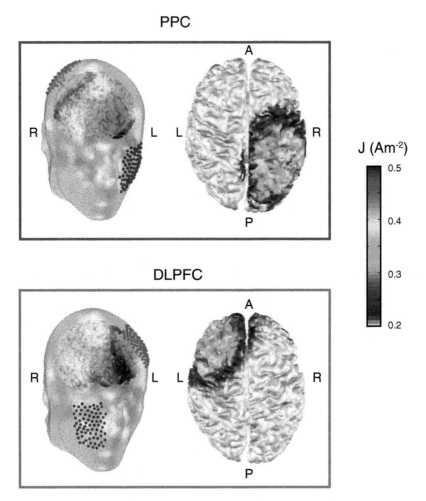

FIGURE 3.1 Illustration of electrode montages, and estimated current density distributions, from a tDCS study of working memory. Positioning of the anodes on the scalp, over the posterior parietal cortex (PPC) of the right hemisphere and over the dorsolateral prefrontal cortex (DLPFC) of the left hemisphere, is illustrated with clusters of red dots. Positioning of the cathodes, on the contralateral cheek, is illustrated with clusters of blue dots. For both montages, the current density distributions are illustrated in two formats: relative to the two electrodes on the left and superimposed on top-down views of the brain on the right. In this experiment, tDCS of PPC was hypothesized to influence task performance, and tDCS of DLPFC served as an active control condition. A = anterior; P = posterior; R = right; L = left. Source: From Wang, S., Itthipuripat, S., & Ku, X. (2019). Electrical stimulation over human posterior parietal cortex selectively enhances the capacity of visual short-term memory. The Journal of Neuroscience, 39, 528–536. Copyright 2019. Reproduced with permission of Society for Neuroscience.

that the researchers included a gap of 120 seconds after the offset of tDCS before delivering single pulses of TMS. When this 120-second gap was unfilled, the effect of anodal tDCS was to increase the amplitude of the TMS-evoked MEP, and the effect of cathodal tDCS was to decrease the amplitude of the TMS-evoked MEP. An interesting twist, however, was that the authors included a

second condition during which subjects were asked to squeeze their hands at 20% maximum possible force during the 120-second gap. Interestingly, in this "voluntary muscle contraction" (VMC) condition, the anodal tDCS-related increase in the TMS-evoked MEP was markedly reduced, as was the cathodal tDCS-related decrease in the TMS-evoked MEP.

The results from Thirugnanasambandam et al. (2011), and others like these, highlight two important points. First, the no-VMC results show directly that tDCS does, indeed, have effects on cortical excitability, when it is measured directly. (That is, if stimulation of motor cortex with an identical amount of energy produces a larger effect in muscles of the hand when preceded by 20 minutes of anodal tDCS, it must be the case that the excitatory effect of that pulse of TMS was larger [and, therefore, that the targeted tissue was more excitable] in the tDCS condition.) This is often just assumed, and inferred from effects of tDCS on performance of cognitive tasks (as illustrated in *Figure 3.1*), but it's impotant to have direct evidence for this assumption. The second point is really two points. Most generally, the effects of tDCS are highly dependent on the state that the brain was in at the time of the stimulation. This is illustrated by the fact that VMC had the effect of blocking the effects that tDCS would have otherwise had. At a finer level of granularity, this property of state dependency is not limited to just the time that the tDCS is being delivered, as established by the fact that the VMC intervention occurred during the gap between tDCS offset and measurement of the MEPs.

The final point to make here is a caveat about precision of targeting. Most tDCS systems come with software that estimates the current density experienced by different brain areas as a function of stimulator output and the configuration of the electrodes, as illustrated in *Figure 3.1*. Interpretation of these estimates, however, needs to be tempered by a few factors. One is that the bone of the skull has high resistance, and so half or more of the current leaving the anode does not make it through the skull and into the brain. The second is that typically the software estimating the path of current flow between the two electrodes does not have information specific to the brains of the individuals participating in that experiment. Rather, it uses assumptions about the typical human brain that were baked into the software long before the researchers acquired the equipment. And because different tissues and fluids (importantly, CSF) have different conductive properties, any differences between the heads of one's research subjects and the assumed "average head" will necessarily produce error in the estimate.

Transcranial alternating current stimulation

tACS works on the same physical principles as does tDCS, and, indeed, many transcranial electrical stimulation devices are capable of operating in either mode. The difference is that, with tACS, the status of the two electrodes as the anode and the cathode switches back and forth several times per second. Therefore, the frequency of stimulation is an important parameter in tACS. The most common research questions that are addressed with tACS relate to endogenous oscillations in the brain. Recall that near the end of *Chapter 2* we considered the phenomenon of oscillatory synchrony. Further along in this chapter, in the sections on *Electroencephalography (EEG)* and on *Analysis of time-varying signals*, we'll begin to see how patterns of oscillation at different frequencies are associated with different physiological states, and sometimes with different cognitive functions. In this context, tACS is often used to try to influence internally generated (i.e., endogenous) oscillations – perhaps their frequency, or their amplitude, or their phase – as a way to test hypotheses about the functional role of these properties of oscillatory dynamics. When we get to the section on EEG, we'll learn that the neurogenic (i.e., "brain-derived") electrical signals measured at the scalp are tiny, much smaller than those generated by tACS. As a consequence, the real-time effects of tACS on endogenous oscillations can only be inferred, but not measured directly. Thus, although one might hope that an effect of tACS delivered at, say, 10 Hz, might be to entrain an endogenous oscillation that typically cycles at 9.5 Hz (i.e., to "speed it up"), that's only one possibility. Another is that the endogenous oscillation and the exogenously delivered one will simply coexist. Or the tACS may weaken the endogenous oscillation, which may or may not, in turn, have an influence on the function of a region not targeted by the tACS, but that is influenced by the 9.5 Hz oscillation. Careful experimental design is needed to anticipate and rule out as many alternative scenarios as one can reasonably think of.

Transcranial magnetic stimulation

Transcranial magnetic stimulation works on the principle of electromagnetic induction – the same principle at work in the turbines at electric power generation plants, or in the pickup of an electric guitar. In secondary school physics, many of us learned this as the "right-hand rule" – if you curl the fingers of your right hand and extend your thumb, current flowing in the direction that your thumb is pointing will generate a magnetic field oriented in the direction of your fingers. Conversely, electrical current will be induced in a length of conductive wire (pointing in the same orientation as your thumb) experiencing a changing magnetic field (in the same orientation as your curled fingers). With TMS, the TMS coil produces a magnetic field that, when held near the scalp, is experienced by fibers in the brain (*Figure 3.2*). Because these fibers have the properties of conductive wire, the magnetic pulse induces a "pulse" of current in these fibers. The biophysics of cortical tissue is such that current induction is maximal at places where there are geometric changes in fiber pathways (such as where efferent axons bend at right angles to plunge into white matter, or where afferent axons bend at right angles upon emerging from white matter to run parallel to the cortical surface) and where there are discontinuities in cell membranes, such as at the

axon hillock and at synaptic terminations. Thus, the effects of TMS on a region of cortex are decidedly nonspecific, in that it induces current in incoming fibers, in local circuitry, and in fibers exiting the region.

A TMS system consists of the "stimulator," essentially a large capacitor that can hold a large amount of electrical energy and then discharge it all at once, a cable through which the current from the stimulator travels to the coil, and the coil through which the conduction of this current generates the magnetic field. Although coils come in different configurations intended for different purposes, the

FIGURE 3.2 Transcranial magnetic stimulation.

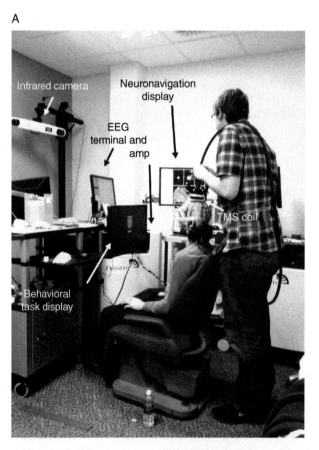

A

FIGURE 3.2.A An experimental setup in the author's laboratory. The subject is seated and viewing stimuli on the behavioral task display screen (responding via keyboard in his lab), and is wearing an EEG electrode cap. On his face, and on the TMS coil, are infrared-reflective beacons that are detected by the infrared camera, and used to coregister the location of the TMS coil and the subject's brain, so as to compute a real-time estimate of where TMS current induction will be maximal in his brain. The experimenter uses this real-time display for "neuronavigation" – positioning the coil so as to precisely target the intended region of the brain. (The TMS stimulator, located to the right of the subject and experimenter and therefore obscured from view in this photo, is roughly the same size as the housing enclosing the EEG amplifier.) Source: Bradley R. Postle.

B

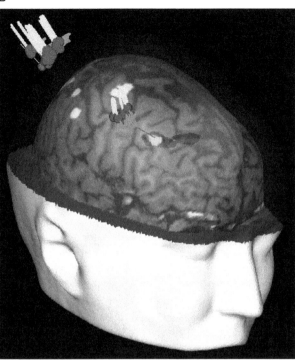

FIGURE 3.2.B An image taken from a neuronavigation display at the end of an experimental session in which hotspots of activity in DLPFC and PPC were targeted with rTMS. The 3-D head-and-brain image was derived from an fMRI experiment during which the subject performed a cognitive task, and then a thresholded statistical map from the fMRI data was merged with the T1-weighted anatomical images. In this image, the scalp and some of the cortex have been "peeled" away to expose the hotspots (white blobs), two of which were targeted with rTMS. Each red ball and yellow "spike" shows where the TMS coil was positioned on the scalp when a train of pulses was delivered, and the purple shadow covering the hotspot in right middle frontal gyrus is the estimated focus of electromagnetic induction. The estimated extent of the weaker "penumbra" of the magnetic field is shown in green.

most common is the 70 mm "figure-of-8," with the crossing-over point of the "8," the point of maximal induction, and therefore the part of the coil used for targeting. (Because the cable and coil invariably come as one piece, the two together tend to be referred to as "the coil.")

The effects of TMS on the brain depend on the strength of the pulse, with moderately strong pulses capable of generating action potentials not only in neurons that directly experience the magnetic flux of the TMS, but also in neurons that are one or more synapses downstream from the stimulated tissue. For example, as we saw in the preceding section on tDCS, single pulses of TMS delivered to the motor cortex of humans can produce a twitch in a finger or thumb which, as we'll see in *Chapter 8*, requires signaling across a minimum of two synapses. That is, because it is certain that the magnetic field generated by a pulse of

TMS delivered at the skull does not physically reach the muscles of the hand, the only explanation for the systematic production of muscle twitches by TMS of the motor cortex is that it generates action potentials in cortical neurons that travel from the brain to the spinal cord, and are sufficiently robust to trigger action potentials in the spinal motor neurons that, in turn, are the proximal cause of contraction of muscles in the hand. *Figure 20.3* illustrates the effects of single pulses of TMS on brain activity, as measured with EEG.

Although the technique often elicits questions about safety (e.g., *Can it elicit seizures?*), the method has a lengthy and well-documented history of safe application in cognitive neuroscience research (see, for example, the report by Rossi et al. 2009 that is listed at the end of this chapter).

Single-pulse TMS

Although some protocols for delivering multiple TMS pulses in structured series can have the effect of disrupting brain function, and therefore behavior, this is NOT the typical application for TMS in single-pulse mode. Indeed, one common use effectively assumes that TMS has no effect on ongoing brain activity, and so can be used as a kind of physiological probe. A second takes advantage of the fact that, under the right conditions, single pulses of TMS delivered to different parts of the visual system can produce the percept (the "illusion" in a sense) of a flash of light.

First, the physiological probe, for which we can find an analogy from computer networking. When troubleshooting a computer network, it's common to send out a "ping" signal to determine what other computers are actively connected with the one sending out the ping. Other computers in the network will send a return ping that also indicates (in milliseconds), the amount of time that it took the ping to travel out and back. In the study by Thirugnanasambandam et al. (2011), we saw that single pulses of TMS were used to measure the excitability of cortical tissue that had been targeted with tDCS. Note that, because the effects of tDCS are modulatory, one wouldn't necessarily be able to detect any "before/after" change in the targeted tissue without probing it. Although this kind of probing could also, in principle, have been carried out by asking subjects to move their hands in a specific way and measuring whether some aspect of the movement was different after tDCS, using this kind of "behavioral challenge" can, in practice, produce noisier data, and data that are more difficult to interpret. The reason has to do with control over the stimulation of motor cortex. Recall that we noted previously that the energy delivered by the TMS coil is identical on every trial. If, in contrast, Thirugnanasambandam et al. (2011) had used a procedure in which subjects were asked to, say, press a button every time they heard a beep, how could they know whether the processing of the auditory signal was the same on each trial, and that it resulted in the delivery of an identical neural command to the motor cortex on every trial (e.g., carried by an identical number of action potentials, and with the identical latency relative to the beep)? How could they know whether the subject's state of vigilence was identical on each trial?

While the Thirugnanasambandam et al. (2011) study probed peripheral (i.e., "not the brain") neurophysiology, TMS can also be used to probe physiology of the central

nervous system. This is done by measuring the TMS-evoked response via concurrent measurement of brain activity, typically with EEG or with fMRI. One study, for example, tested a hypothesis that a nucleus of the thalamus is dysfunctional in schizophrenia. Although a behavioral challenge could also have been used for this study, added to the complicating factors considered in the previous paragraph was the uncertainty about whether the overall level of compliance (or motivation, or distractability) might be different for the patients vs. the nonschizophrenic control groups.

Single-pulse TMS is also useful if one wants to study neurophysiology in different states of arousal. In *Chapter 20*, for example, we'll consider a study in which single pulses of TMS were delivered to subjects, late at night, as they transitioned from quiet wakefulness into sleep. The motivation was to test the prediction that a state of connectivity between neural circuits, hypothesized to be necessary for conscious awareness, would transform into a state of disconnectivity as people fell asleep (and, therefore, out of consciousness). Note that not only would a behavioral challenge not work for this kind of experiment, but a "sensory challenge" would also be problematic. People's eyes close when they fall asleep, and so comparing the brain's responses to flashes of light when awake vs. asleep wouldn't work. Our ears "stay open," however, so what about comparing the brain's response to tones? The complication here is that any differences in the brain response to tones during sleep vs. wake might be due to a sleep-related change in how the auditory signal is processed and transmitted from the ear to the brain, rather than to a change in the functioning of the brain per se. The physics of electromagnetic transduction, in contrast, "don't care" if a nearby brain is awake or asleep when a pulse of current travels through a TMS coil.

A second common application of single pulses of TMS is to produce functional effects. We've already seen that single pulses of TMS delivered to the motor cortex can produce muscle twitches in the periphery. Within the domain of vision, single pulses of TMS delivered to occipital cortex can produce a perceptual phenomenon known as a phosphene — a cloudlike flash of light that a subject perceives even when the eyes are closed and covered with a blindfold (*Figure 4.7*). (Recall from Chapter 1's section on *The localization of visual perception*, that visual perception was localized to the occipital lobe in the late 1800s.) Phosphenes are presumed to be generated by the propagation of action potentials through the visual cortex. The generation of phosphenes points out an interesting paradox about TMS, which is that although its anatomical specificity is poor, what we might consider as its "functional specificity" can be quite, well, persnickety. Simply put, it's difficult to generate visual phosphenes with TMS. The process requires a great deal of trial and error of stimulating at multiple locations on the scalp until the subject eventually reports seeing a phosphene. (On unlucky days, a subject may never see a phosphene, and eventually the experiment is aborted.) Importantly (and this also applies to generating MEPs via TMS of the motor cortex), the orientation of the coil at a particular location on the scalp also matters. Thus, even if you knew exactly where on the scalp to position the crossing-over point of the coil, you'd still need to rotate it through several orientations (relative to the scalp) to find the specific orientation for which, at that location on the scalp, a pulse of TMS will produce a phosphene. (Operationally, the orientation of the coil is described with reference to the handle (i.e., where the cable feeds into the coil) relative to a gross anatomical landmark, such as the midline.)

In addition to the positioning of the TMS coil on the scalp, the other important factor that determines its effects is the amount of energy used to generate the pulse. Although the output of a stimulator can be expressed in physical units, it almost never is. Instead, the "common currency" for reporting TMS stimulator intensity is a value normalized to the functional effects of TMS. That is, for any two healthy individuals, the physical signal intensity needed to generate a muscle twitch (when applied to motor cortex) or a phosphene (when delivered to visual cortex) will vary considerably. What's most important, therefore, is to ensure that the same "relative" stimulation intensity is delivered to each subject participating in the study. This will be achieved by, for example, determining the minimal stimulator intensity that will evoke a muscle twitch on 5 of 10 pulses, and then increasing that intensity by, say, 15%, for the experiment. In this way, although the stimulator intensity may be different for each subject, each is nonetheless experiencing TMS at "115% of motor threshold."

"Virtual-lesion" TMS protocols

While single-pulse TMS protocols are based on measuring the effects of each pulse, the "virtual lesion" protocols are more akin to cathodal tDCS: TMS is delivered to a brain area for an extended period of time, and the subject then performs a behavioral task, with the assumption that functioning of the targeted area has been altered. More specifically, two TMS protocols are intended to produce local hypometabolism (and, consequently, a decrease in excitability) that lasts for several tens of minutes after TMS: 1 Hz repetitive (r)TMS and continuous theta burst (tb)TMS; 1 Hz rTMS produces hypometabolism in the targeted tissue that persists beyond the period of stimulation for roughly as long as was the duration of the stimulation (e.g., 20 minutes of 1 Hz rTMS will produce a virtual lesion that lasts for 20 minutes beyond the termination of the rTMS). Continuous tbTMS is administered as triplets of TMS pulses spaced apart by 20 ms, and delivered once every 200 ms. A 40-second train of continuous tbTMS produces local hypometabolism for roughly 30 minutes.

For any variant of "virtual-lesion" TMS, the rules of inference from neuropsychology apply.

High-frequency repetitive (r)TMS

The physiological effects of high-frequency rTMS – here understood as trains of pulses delivered at a frequency of 5 Hz or higher – are the most variable and context dependent of any TMS protocol. Although some have summarized the effects as "excitatory," it's not that straightforward. In some studies by this author's group, for example, trains of 10 Hz rTMS delivered during the delay period of a working memory task (the subject is presented some information to remember, like a short list of letters, or an array of shapes, and then, after a several-second delay, is tested for their memory for that information [see *Chapter 14*]) impair performance. In another, however, it improved performance. In yet another, it impaired the performance of some subjects but improved the performance of others, and did so in a way that correlated with how it affected the concurrently measured EEG. It's a powerful tool, but proceed with care.

ANATOMY AND CELLULAR PHYSIOLOGY

Many of the details of gross anatomy that the reader of this book will need to understand were already covered in the first two chapters, and were discovered via dissection of postmortem tissue. For the contemporary cognitive neuroscientist, gross anatomy isn't as much a "discovery science" as it is a body of knowledge over which one must have considerable mastery in order to be able to interpret, for example, the effect that subtle differences in the location and/or the spatial extent of a lesion are likely to have on some domain of behavior, or whether two nearby foci of activity are part of the same or different structures.

Thematically, many methods that might be best classified as *anatomical* vs. *physiological* can be closely related, either because of physical and/or biological principles that they exploit, or because of the kind of question they are used to address. For that reason, this section will jump back and forth between anatomy and physiology.

Techniques that exploit the cell biology of the neuron

No matter how long, or multiply branched, an axon is part of one cell. Thus, there needs to be a mechanism, for example, for proteins that are synthesized from DNA in the nucleus of the neuron to be transported to synapses at far reaches of each axonal branch. (For me, the "gee whiz" factor is particularly high when I think of the giraffe, or of massive whales . . . .) And so, intracellular mechanisms for **axonal transport** are critical for the healthy functioning of the neuron.

Tract tracing

Axons are microscopically small. Although bundled tracts of thousands of axons can be visible to the naked eye (e.g., the optic nerve, tract, and radiations; *Figure 3.14*), this sheds no light on the destination of individual axons. One strategy for determining precisely where axons project to is to inject a substance into a region of the brain, let the natural processes of intracellular transport "do their thing" (i.e., transport the substance from the region of uptake to other parts of the neuron), and later look for where the substance has been transported. Over the past 100 years or so, sometimes through trial and error and sometimes as a result of sophisticated understanding of cell biology, anatomists have devised many ways to address different questions about neuroanatomy. For our purposes we can think of three categories of cell biology-based (as opposed to neuroimaging-based) neuroanatomy: anterograde transport, retrograde transport, and whole-cell staining.

Anterograde transport would be used, for example, to determine whether, and how, neurons from somatosensory area S1, the first cortical stage of processing of somatosen-sory information, projects to somatosensory area S2, the second stage in the somatosensory processing hierarchy. (**Somatosensation** refers to the sense of touch, with *soma* derived from the Greek for *body*.) To do the experiment, one would inject tracer into S1, wait for a set period of time, then sacrifice the animal, slice the brain, process the tissue, and, finally, inspect the tissue under a microscope to observe where the tracer ended up. Such an experiment depends on several factors. First, the tracer needs to be taken up by the dendrites and/or cell bodies of the neurons of interest. Second, one needs to know the amount of time it takes for the tracer to be transported to the ends of the neuron. Third, there needs to be a reliable way to detect the tracer under a microscope. With regard to the passage of time, this becomes particularly important if the tracer can "jump" the synaptic cleft, and thereby be passed from one neuron to another. Exquisitely detailed studies have been performed with such tracers such that, when different animals are sacrificed at different times relative to the injection, the anatomist can determine the monosynaptic connectivity of a region, its disynaptic connectivity, and so on. With regard to detection, some tracers are visible in and of themselves, and others require the tissue to be stained or otherwise prepared so as to make the tracer visible.

Retrograde transport is the opposite of from S1 to antero-grade transport. It relies on a tracer that is taken up at the presynaptic terminal, and then carried "backward" along the axon to the cell body and dendrites. *Figure 3.3* is an illustration of how this approach can be used to determine the laminar organization of projections from S1 to S2 vs. the nuclei of the "somatosensory thalamus," a set of nuclei that we will introduce in *Chapter 5*.

Whole-cell staining is just what it sounds like. A stunningly beautiful example of one application can be admired in *Figure 2.5*. The trick with this method is to use a stain that is selective for a particular kind of cell. For example, we'll see in *Chapter 5* that to study the local architecture of pyramidal vs. interneurons in primary somatosensory cortex, the researchers take advantage of the fact that one class of interneurons contains the protein parvalbumin, but other interneurons and pyramidal cells do not.

Functional tissue imaging

Of the two general categories of tissue imaging – structural and functional – we will consider three examples of the latter and, within the category of functional imaging,

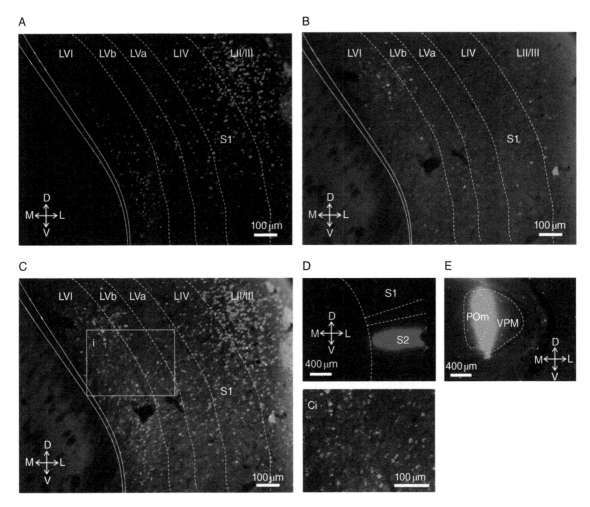

FIGURE 3.3 A neuroanatomical tracer study of the laminar organization of projections from primary somatosensory cortex (S1). To follow the logic of the experiment, look first at panel **D** – this shows the region of secondary somatosensory cortex (S2) where the (purple-colored) tracer was injected. It is a fluorescent, retrograde tracer called "Retrobeads." Panel **A** shows the distribution of labeling resulting from the injection shown in panel **D**. That is, the purple Retrobeads were taken up at presynaptic terminals in S2 (panel **D**) and retrogradely transported back to the cell bodies (panel **A**) of the neurons whose axons formed those presynaptic terminals (that is, of the neurons whose axons projected to S2). Note that the labeling is heaviest in layer II/III of S1, meaning that it is the neurons in this layer that send the heaviest feedforward projection from S1 to S2. (Note, also, that this laminar organization is not absolute – biological systems are rarely as tidy and lawful as one might want them to be). Next, let's go to panel **E**, which illustrates the injection site of green Retrobeads into the posterior medial (POm) nucleus and ventral posterior medial (VPM) nucleus of the thalamus. The green spots concentrated in the upper left-hand corner of panel **B** are the retrograde labeling resulting from this thalamic injection. This indicates that the S1 neurons projecting to thalamus are located primarily in layers VI and Vb. Panel **C** illustrates an overlay of A and B, to emphasize the distinct projection profiles of the superficial vs. the deep layers of S1. Panel **Ci**, a blowup of the box in panel **C**, shows the absence of "double labeling," the presence of which would have meant that individual neurons project to both S2 and the thalamus. (Addressing this question was, in fact, the primary reason that this experiment was performed.) The key in the lower left corner of each image indicates the dorsoventral and mediolateral axes. Source: Petrof, Viaene, and Sherman, 2012. Reproduced with permission of Wiley.

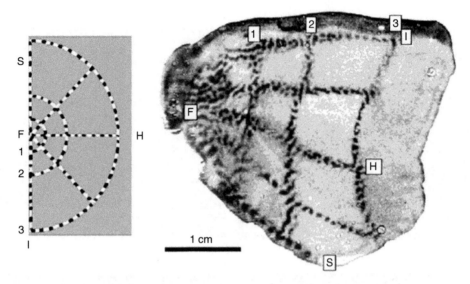

FIGURE 3.4 Retinotopic organization of V1, visualized with 14C 2DG labeling. The stimulus (left side of the figure) was presented to one eye of a monkey, with the "F" label indicating the portion projected on to the fovea, "S" and "I" indicating superior and inferior aspects of the vertical meridian, respectively, and "H" indicating the horizontal meridian. Black and white coloring on the lines and concentric circles of the stimulus flickered back and forth at a rate of three times per second, to vigorously stimulate activity in neurons of the primary visual cortex of the occipital lobe (V1) into whose receptive fields the stimulus fell. (*Chapter 4* will define these anatomical and functional properties of the visual system.) The right side of the figure shows an autoradiograph of a section of cortex dissected from primary visual cortex, with the letters (F, S, I, H) indicating the cortical representation of these demarcations of the visual field and the digits (1, 2, 3) indicating the cortical representation of the three concentric rings of the stimulus display. Source: Tootell, Silverman, Switkes, and De Valois, 1982. Reproduced with permission of AAAS.

there exists another distinction between those using invasive vs. noninvasive methods. We will use the term "functional tissue imaging" to refer to the former, because each entails performing a craniotomy and taking measurements directly from the tissue of interest. (With noninvasive brain scanning [a.k.a. "functional neuroimaging"], in contrast, one "takes pictures" of tissue and/or activity inside the cranial cavity without having to surgically access the brain.)

The oldest of the three, 2-deoxyglucose (2DG) imaging, entails the systemic administration of glucose molecules that have been tagged with a radioactive ion (such as ^{14}C) and further modified so that they can't undergo glycolysis. It exploits the fact that glycolysis is the primary source of energy for all cells in the body. Once administered, 2DG blends with naturally circulating glucose in the bloodstream, and, just like endogenous glucose, is taken up by neurons in an amount proportional to their activity. Once taken into a cell, 2DG is trapped. Upon

completion of the experiment, the animal is sacrificed, and the tissue of interest dissected out of the brain and placed against a radiographic plate that detects the radiation given off by the 2DG that has been trapped in previously active neurons (a process called autoradiography). The results from a 2DG experiment are illustrated in *Figure 3.4*. (More details about the physiological bases for 2DG imaging can be found in subsection further along in this chapter on *The metabolic demands of the action potential and of synaptic activity*.)

A second functional imaging method uses voltage-sensitive dye. It differs from 2DG in three important ways. First, it is sensitive to fluctuations in membrane voltage. Second, it gives a real-time measure of activity, whereas 2DG requires extensive after-the-fact processing in order to obtain the results from the experiment. Third, voltage-sensitive dye can be used to measure cortical activity in several different conditions (e.g., responses to viewing bars that are angled at several different orientations, spanning

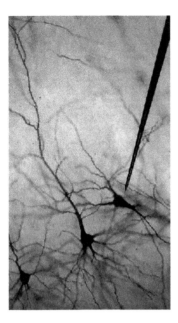

FIGURE 3.5 Extracellular electrophysiology. Photograph of a light microscope-magnified image of an extracellular microelectrode in the vicinity of a pyramidal neuron in the visual cortex of a monkey. The entire height of the image corresponds to roughly 1 mm (Hubel, 1988). Source: Max Planck Institute for Biological Cybernetics.

the 180° of possible rotation). 2DG, on the other hand, requires the sacrifice of the animal after just one experimental condition. Thus, to determine the pattern of brain responses to the neural processing of six different orientations, each spaced by 30°, one would need to acquire data from (a minimum of) six animals, and then try to assemble a composite map across these.

Finally, optical imaging of intrinsic signals entails shining light of a particular wavelength on the tissue of interest, and measuring how the makeup of the reflected light changes as a function of experimental condition. The somewhat coy reference to "intrinsic signals" reflects the fact that there remains some uncertainty about all of the factors that produce changes in tissue reflectance – although the majority of the variance in the signal is due to different reflectance properties of oxygenated vs. deoxygenated blood, there may also be contributions from membrane voltage (despite the absence of externally applied dye), and even from the subtle stretching and compressing that neuronal membranes exhibit under different physiological states.

ELECTROPHYSIOLOGY

Invasive recording with microelectrodes: action potentials and local field potentials

Although many methods have been devised to study the electrical properties of neurons, most have made use of some variant of the electrode. A microelectrode is a fine-tipped (of the order of 1 µ across) conductor of electricity, typically a glass micropipette filled with a conductive fluid or a platinum or tungsten wire, that can be inserted into brain tissue (thus, a "penetrating electrode"), and is connected to an amplifier. Because it is conductive, it detects the changes in voltage that were detailed in *Chapter 2*. Some electrodes have multiple contacts along their shafts, the signal from each of which will feed into a different channel in the amplifier.

Many of the properties of the neuron that were described in *Chapter 2* were discovered via **intracellular recordings** obtained with in vitro methods, in which individual neurons, or more often slices of CNS tissue, are surgically removed from an animal and kept alive "in a dish" by bathing it in artificial CSF. Only by recording intracellularly can one measure, for example, the activity of a single postsynaptic channel, or the kinetics of voltage-gated ion channels. However, because cognitive neuroscience, by definition, deals with the neural processes that underlie behaviors (that we associate with cognition), the emphasis in this book will be on **extracellular recordings** obtained in vivo – that is, recordings made outside of the cell in the brain of a living animal (*Figure 3.5*).

Extracellular electrophysiology can measure two types of signals: action potentials and *local field potentials* (LFPs). Because action potentials are all-or-none events – an individual neuron cannot alternate between generating "big" vs. "little" action potentials – they are treated as binary events, and a standard way to quantify a neuron's activity is by its firing rate. A neuron typically has a baseline firing rate, maybe 5 "spikes per second" (i.e., 5 Hz) for a generic pyramidal cell. The neuron's involvement in a particular function, let's say visual perception, can then be assessed by determining whether this firing rate changes systematically with a systematic change in the environment, say, flashing a light in an area that the animal is looking at. Depending on the neuron, and on the state of the brain, a

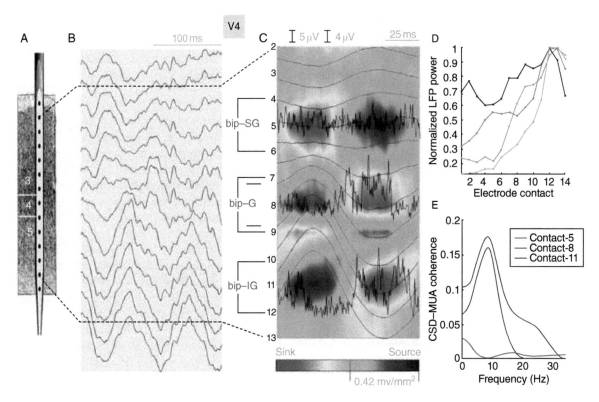

FIGURE 3.6 The local field potential (LFP). **A.** Schematic diagram showing an electrode with 14 equally spaced (by 200 μm) contacts ("3," "4," and "5" refer to layers of cortex). **B.** 200 ms of LFP recorded at each of these 14 contacts. **C.** A current source density map, computed from the pattern of LFPs in panel **B**. What this shows is the pattern, over time, of where in this cortical column current is flowing from intracellular space to extracellular space (a "source"), or from extracellular space to intracellular space (a "sink"). To compute the current source density map, bipolar recordings are needed, and "bip-SG" indicates the two contacts used to obtain these recordings in the supragranular layer, "bip-G" in the granular layer, and "bip-IG" in the infragranular layer. The jagged black traces show the MUA. The smooth blue lines are transformations of the raw LFP that we will disregard. **D.** Peak LFP power at 10 Hz (vertical axis) as a function of cortical depth (horizontal axis) from four separate recording sessions, each plotted in a different color. **E.** CSD-MUA coherence, an index of the rhythmicity of spiking at different frequencies of the LFP, of the recording session illustrated in **C**. This plot indicates that the spiking in granular and infragranular layers is most strongly "clocked" in the alpha frequency range, whereas spiking in supragranular layers is relatively unrelated to the LFP. These recordings were made in an area called "V4" (as referenced in *Figure 2.4*). Source: From Bollimunta, Anil, Yonghong Chen, Charles E. Schroeder, and Mingzhou Ding. 2008. "Neuronal Mechanisms of Cortical Alpha Oscillations in Awake-Behaving Macaques." Journal of Neuroscience 28 (40): 9976–9988. doi: 10.1523/JNEUROSCI.2699-08.2008. Copyright 2008. Reproduced with permission of Society for Neuroscience.

neuron involved in the perception of this flash of light might exhibit either an increase or a decrease from its baseline firing rate. Depending on the cell type, an activated neuron's firing rate can be anywhere from 5 to 10 Hz to upward of 200 Hz.

To record the LFP, the signal(s) from an electrode are filtered to remove contamination from individual action potentials, leaving the LFP – the aggregated fluctuation of dendritic membrane potentials from cells in the vicinity of the electrode. Importantly, action potentials from

neurons that are very close to the electrode will produce large-amplitude deflections in the signal that are relatively easily detected and filtered out of the LFP recordings. Action potentials from neurons more distal to the electrode, however, produce tiny blips that are not differentiable from subtle fluctuations in the aggregated membrane potentials of nearby cells. Although these tiny blips would be expected to cancel out if they were uncorrelated with respect to one another, when a large population of neurons is firing in synchrony, this will show up as small-amplitude fluctuations across a broad band of high frequencies in the LFP recording. (The idea is that, because the synchronization of the precise moment of firing between individual neurons within a large population of neurons won't be perfect, the deflections in the LPF signal that are attributable to these individual action potentials would be offset by fractions of hundredths of milliseconds, and so would show up in the LFP signal as fluctuations at higher frequencies than is generally thought to be attributable to an oscillating membrane potential. And because these tiny fluctuations are "noisy" imprecision of precise synchronization, they will show up across a broad range of frequencies.) As a general rule, oscillations in the LFP signal that are higher frequency than ~120 Hz are assumed to reflect the so-called multiunit activity (MUA), and are taken as a proxy for firing rate in the local population sampled by that electrode (*Figure 3.6*).

Electrocorticography

An alternative to recording electrical activity with penetrating electrodes is to use a grid of electrodes that is positioned on the surface of the cortex, a technique called **electrocorticography** ("ECoG" [pronounced "ee-cog"] for short). Because ECoG electrodes lie on the surface of the cortex (typically beneath the thick [and electrically resistive] dura matter, i.e., "subdurally"), they lack the spatial resolution that can be obtained with penetrating electrodes. Additionally, they are too far from any cell bodies to measure action potentials, and so can only measure field potentials. Although ECoG signals can, in principle, be measured in any species, they are particularly relevant for cognitive neuroscience when used by neurosurgeons and neurologists for presurgical planning in human patients being evaluated for neurosurgery. In cases of epilepsy that cannot be controlled with antiepileptic medication, for example, a treatment of last resort is to first isolate, then surgically remove, the tissue that is responsible for triggering systemic seizures. That is, it can sometimes be the case that only a small region of brain is responsible for initiating the disordered activity that, when it spreads to other parts of the brain, results in **grand mal** seizures. The reasoning in such cases is that if the pathological tissue can be surgically removed without causing too much "collateral damage," the patient may experience many fewer seizures. The "without causing too much collateral damage" part is key, of course, because recall from *Chapter 1* that damage to the wrong brain areas can produce, for example, irreversible paralysis, blindness, or disruption of spoken language. And so the neurosurgical removal of brain tissue is often preceded by presurgical planning, in which the surgeon places electrodes in the vicinity of the suspected focus of the seizure activity. (Depending on the surgeon, the electrodes will either be strips and/or grids of ECoG electrodes placed on the cortical surface after a partial craniotomy, or penetrating electrodes inserted through small holes drilled through the skull, a technique called stereotaxic (s)EEG.) The physicians then wait for the patient to have a seizure, and analyze the location and timing of seizure activity at different electrodes to localize the area(s) responsible for initiating seizures. Because the first postimplantation seizure may not occur for several days, if the patient is willing to participate in a research study, scientists can ask the patient to perform experiments. In this way it is possible to collect ECoG data recorded from the human brain while the patient/subject is engaged in a task.

We will return to ECoG later in this chapter when we discuss analyses of phase–amplitude coupling (PAC) that can be measured in ECoG data. We'll consider detailed examples of ECoG studies in *Chapter 10* (see *Figure 10.7*) and *Chapter 19*.

Electroencephalography

The EEG is recorded from the scalp (*Figures 3.7* and *3.8*). There's a sense in which the EEG signal measured at the scalp is a kind of "macro LFP," in that it reflects the summation of many, many local cortical LFPs. This is necessarily the case, due to the conductive properties of all the tissues lying between the cortex and the scalp. Note that while the changes in the membrane potential of a single neuron vary within a range spanning roughly 100 mV (e.g., from −90 to +20 mV), the LFP, which is recorded from the extracellular space adjacent to groups of neurons, typically varies within a range of roughly 10 µV (i.e., 2–3 orders of magnitude smaller). As measured at the scalp, a cortically derived electrical signal needs additionally to pass through blood vessels, CSF, meninges, skull, muscles, and skin, all of which attenuate and blur it. This has at least three implications for the EEG. First, to be detectable, the EEG necessarily reflects the summation of signals from tens of thousands (or more) neurons. (Hence, the characterization as a "macro LFP" − it pools the signals from many, many, many more neurons than does the invasively recorded LFP.) Second, the fluctuations in the membrane potential of these neurons need to be synchronized − otherwise, negative and positive potentials would cancel out each other. Finally, the EEG signal is disproportionately influenced by signal from superficial layers of cortex, with smaller contributions from deeper layers and, in most cases, only negligible contributions from subcortical structures.

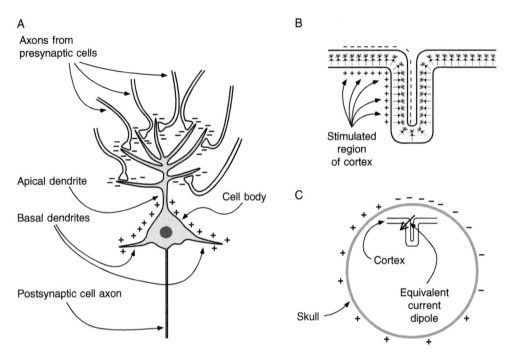

FIGURE 3.7 Principles of EEG generation. Source: From Luck, Steven J. 2005. An Introduction to the Event-related Potential Technique. Cambridge, MA: MIT Press. Copyright 2005. Reproduced with permission of The MIT Press.

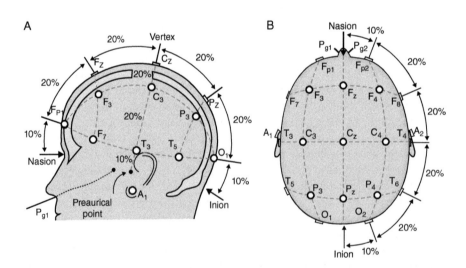

FIGURE 3.8 Placement and names of the 21 electrodes in the internationally standardized 10–20 system, as viewed from the left (A) and looking down on the top of the head (B). Reference points are the nasion, which is the indentation at the top of the nose, level with the eyes; and the inion, which is the bony lump at the base of the skull on the midline at the back of the head. From these points, the skull perimeters are measured in the transverse plane (i.e., the trajectory through O1-T5-T3-F7 in the side view) and the median plane (i.e., the trajectory through PZ-CZ-FZ in the top view). Electrode locations are determined by dividing these perimeters into 10% and 20% intervals. Since the development of this convention, the electrodes in many EEG systems now come embedded within caps that fit snugly over the subject's head (rather like a swimming cap), often with an array of 64 or 128 electrodes. High-density systems come as "nets" of 256 electrodes. EEG systems whose electrode placement doesn't correspond to the 10–20 systems will typically provide formulae in their software that will convert the system's native electrode placement scheme into 10–20 equivalents. Source: From Malmivuo, Jaakko, and Robert Plonsey. 1995. Bioelectromagnetism: Principles and Applications of Bioelectric and Biomagnetic Fields. New York: Oxford University Press. Copyright 1995. Reproduced with permission of Oxford University Press.

Scientists studying human cognition began recording the EEG decades before the term "cognitive neuroscience" came into common usage. A big part of its appeal is that it is noninvasive, and so can be measured from the same healthy young adult population that is sampled disproportionately for studies of cognitive psychology (i.e., the "convenience-sampled" population of university students). A second appeal is that, despite the attenuation of the amplitude of the extracranial EEG, its temporal properties are virtually the same as if they had been measured from within the brain, and this high temporal resolution provides exquisitely precise information about the timing of neural events. A major limitation of EEG, however, is that its spatial resolution is quite poor. This is due to the averaging that the signal undergoes between the brain and the scalp, and also to a fundamental mathematical truth: for any given pattern of signals that is measured from the surface of a sphere (e.g., electrical signals on the scalp), there is an infinite number of configurations of sources located within the sphere that could have produced that pattern. This is known as the **inverse problem**, and its nature is such that it can never be solved with certainty. Nonetheless, there are algorithmic approaches that estimate "solutions" to the inverse problem by making a variety of assumptions about which of the infinite number of possible "source solutions" is the most likely for any particular dataset.

Magnetoencephalography

Let's be clear: according to the literal definition, MEG is not a type of EEG. However, the tight coupling of magnetic and electrical energy, as manifested by the principle of electromagnetic induction, allows us to get away with summarizing MEG in this section. MEG takes advantage of the fact that electrical activity in neural tissue, including where axons arriving from other regions arrive at dendritic fields, produces weak magnetic fields. MEG measures fluctuations in these magnetic fields with sensors that are built into a kind of helmet that is lowered over the scalp. As suggested in *Figure 3.7*, due to the orientation of neuronal dipoles in cortex in gyri vs. in cortex lining the walls of sulci, it's generally thought that the EEG signal emphasizes activity in gyri, whereas the MEG signal emphasizes activity in sulci. (This is because the magnetic field induced by current moving radially with respect to the skull would be oriented along the surface of the skull, and therefore difficult to detect. The magnetic field induced by current moving in parallel with the surface of the skull, in contrast, would oriented radially, and therefore relatively easy to detect.)

Because magnetic energy has the same temporal properties as electrical energy, MEG also boasts high temporal resolution. However, because, unlike electricity, magnetic fields do not interact with biological tissue, the MEG signal is believed to have superior spatial resolution in comparison to EEG. (The inverse problem applies to MEG just as much as it does to EEG, but the absence of any smearing by the tissue between the brain and the scalp means that best estimates of sources have less spatial uncertainty than do source solutions with EEG data.)

INVASIVE NEUROSTIMULATION

Electrical microstimulation

We were introduced to electrical microstimulation via *Chapter 1*'s consideration of Frisch and Hitzig's (1870) studies of motor control. These studies demonstrated the use of microstimulation that is most common in neuroscience: the controlled excitation of a circuit to study the effects of activity in this circuit. We'll reencounter this method several times as we make our way through this book, including in the study of eye movement control and the control of attention, and in the study of perceptual decision-making. Microstimulation can also be used to alter ongoing activity by, for example, imposing an oscillatory rhythm. Finally, microstimulation can be used to disrupt ongoing activity by either disrupting an endogenous rhythm or just overriding activity in the affected area with a strong flow of current.

Interventions using the same logic as what we've described here can also be carried out with ECoG electrodes, by, for example, passing current between adjacent electrodes in a grid or a strip that is lying on the pial surface. Technically, this should be referred to as "macrostimulation," because of the spatial scale over which it's carried out, but conceptually it fits here.

Optogenetics

The logic of optogenetics in neuroscience can be described in four steps: (1) Clone a gene that codes for a protein that one wants to have expressed in a cell. (2) Insert this strand of DNA into a target population of cells (a process called transfection). (3) Wait for a suitable period of time for the protein to be synthesized and trafficked to the part of the cell where it will have its effects. Up until this point, the words written here could have also been used to describe gene therapy. The "opto-" part, however, is what makes this such a powerful experimental tool.

In an experiment that we'll describe in considerable depth in *Chapter 5*, for example, the transfected DNA codes for channelrhodopsin-2 (ChR2), a cation channel activated by light of a very narrow wavelength centered at 470 nm. (The protein occurs naturally in the single-celled algae *Chlamydomonas reinhardtii*, in which it serves the function of phototaxis – light-triggered movement.) And now we can return to the last of our four steps: (4) Shine light of the critical wavelength on the transfected cells (e.g., 470 nm light for ChR2), thereby experimentally controlling their activity. (For neurons transfected with ChR2, shining 470 nm light has the effect of depolarizing them, thereby causing them to spike.)

To date, the technique of optogenetics has been particularly impactful in research with rodents, especially mice (see section *Optogenetic exploration of gamma-band oscillations in somatosenory cortex, Chapter 5*). The cell biology of mice is such that their neurons are considerably more amenable to transfection and the subsequent synthesis and trafficking of the light-sensitive protein product than is that of NHPs. For example, this author has been told about several experiments with NHPs that have been abandoned, or at least put on hold, because the expression of light-sensitive channels only occurred in a subset of the targeted population of neurons, rendering the effects of shining light weak and difficult to interpret.

ANALYSIS OF TIME-VARYING SIGNALS

Once one has collected data with one of the methods reviewed in the previous section, one has to analyze them. The choice of analysis will often depend on the question that one wants to ask, but it can also be influenced by assumptions being made about how the brain works.

Event-related analyses

Fundamental to understanding how the workings of the brain give rise to the mind is the identification of signals that correspond to the aspect of cognition that one is studying. We've already reviewed how minutely small are the signals that can be detected by EEG and MEG. It also turns out that another property of the brain that makes its study difficult is that it always carries out several functions simultaneously. This means that the signal that one is hoping to measure is often embedded in noise. Let's take the example from this chapter of the visual presentation of oriented bars. At the same time that your research subject is viewing these visual stimuli while you record, let's say, the EEG, another part of her brain, is monitoring her blood sugar level, another part is keeping her fingers poised over the response keys, another is thinking "try not to blink . . . the researcher told me to not to blink . . . ," and so on. (Or another way to think about it is that probably only a small fraction of your subject's 80 billion neurons are directly involved in the visual perception of your stimuli, but all 80 billion are nonetheless alive and are doing *something*!)

Advantages of the ERP technique

Given the considerations laid out in the previous paragraph, one powerful way to analyze neural data is to average them in a way that is temporally aligned (a.k.a. "time locked") with the event that you're interested in. That is, if you show your subject the same stimulus 50 times, or 100 times, and average the data from each of these trials, it's unlikely that she'll be thinking exactly the same thing with exactly the same timing relative to your stimulus on any two presentation events, let alone on 50 or 100 of them. And so the logic is that, after averaging the signals from many trials, any signals that are not closely time locked to the onset of your visual stimulus will "average out," and all that will be left will be signals that occurred with the same timing on all (or, at least, most of) your trials. (*Figure 5.4* shows a remarkable example of this, in which time locking the EEG signal to the onset of a soft click sound produces nearly identical auditory-evoked responses, even though in one of the two conditions the subject was also reading a book!)

When this time-locked averaging method is applied to EEG data, it's called the event-related potential (ERP) technique, and the averaged signals that isolate processes of interest are called ERPs. With MEG, there is not a single universally used abbreviation. (Some groups use "MEP," but this can be confusing, because for decades this abbreviation has also been used for the motor-evoked potential.)

Caveats about the ERP technique

When using the ERP technique (or evaluating ERP data), it's important to keep in mind that the time-locking step used to generate ERPs will obscure any change in task-related activity that differs from trial to trial. For example, we've already seen that the LFP captures ongoing oscillations in neural tissue. The process of generating an ERP typically removes any evidence of these oscillations. This would be fine if these oscillations only corresponded to noise that is unrelated to cognitive functioning, but what if this weren't true? Let's go back to the example of studying how the brain processes a soft click. What if it were the case that oscillations in the LFP of tissue that's involved in auditory processing are always present whether the momentary state of the environment is filled with loud sound or if it is silent? Furthermore, what if it were the case that the amplitude of this oscillation always gets bigger the moment that auditory information gets to the brain? In this scenario, an ERP analysis would not detect this effect, even though it is clearly an effect that is related to the cognitive process that we're trying to study. This is because, in a typical experiment, the delivery of a stimulus like the soft click in our current example is jittered in time, to prevent the subject from being able to predict its precise moment of onset. A consequence of this is that the oscillation of the LFP will be random in phase on a trial-by-trial basis, and averaging these phase-random signals will remove any

evidence of these oscillations. This could be thought of as a kind of "Type II error" in which the analysis failed to detect an effect of interest (*Figure 3.9*).

Perhaps a more pernicious outcome would be one in which the event-related change in the oscillation is one of phase resetting, rather than a change of amplitude. That is, what if it were the case that although the trial-by-trial phase of the oscillation in the LFP is random prior to the onset of the click, the oscillation always resets to the same the moment that auditory information gets to the brain? In this scenario, an ERP analysis would remove any evidence of the oscillation during the pre-click period, but time-locked realignment of these oscillations would create the artifactual impression of an ERP that was time locked to the onset of the click. That is, the analysis would "show" that the click started a process that hadn't been active prior to its onset, when in truth it didn't start any new process but, rather, shifted the timing of processing that was already underway (*Figure 3.9*).

FIGURE 3.9 Cartooned diagrams of two hypothetical situations in which an ERP analysis would misrepresent an event-related change in oscillatory dynamics.

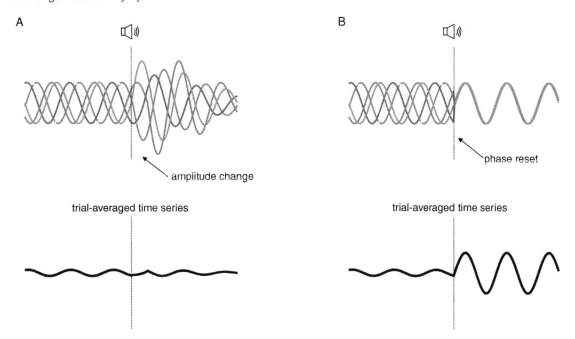

FIGURE 3.9.A Top panel illustrates hypothetical EEG signals from five trials (color-coded and superimposed) in a scenario in which an auditory click produces a change in the magnitude of the ongoing oscillation. The trial-averaged time series (below) does not show any change associated with the click.

FIGURE 3.9.B Top panel illustrates five trials in which an auditory click produces a reset of the phase of the ongoing oscillation. Trial averaging (below) creates an artifactual "ERP" that does not exist in any of the individual trials.

Considerations

Having listed some caveats about how the ERP technique doesn't deal well with oscillations in the EEG signal, the next methods for us to introduce are ones that measure oscillatory dynamics. Before we move on to that though, let's take a moment to consider some of the "philosophical" issues that this discussion engages ("Philosophical" in the sense of *My philosophy about how to study the brain is . . .*). An oversimplistic caricature of the ERP technique is that it assumes that a primary design principle of the brain is that it is composed of systems that are specialized for carrying out specific tasks – let's say perceiving auditory clicks, or visually presented bars – and that these systems are quiet when not actively engaged. This is sometimes summarized as treating the brain as a passive system that processes information when information is fed to it. An alternative is to construe the brain as an active system that is constantly churning, always processing information so that the organism can achieve and maintain a state that is an optimal trade-off between its goals and the current conditions in the world. (One specific theory that advances this idea is called predictive coding, and will be considered at the end of this chapter.) Because living brain cells will often change the patterns of their oscillations, sometimes drastically, but they will rarely (if ever) stop oscillating, analyses that emphasize oscillations over ERPs can sometimes be construed as working from the assumption of the brain as an active system.

Time–frequency analysis

In the early 1800s, the French mathematician Jean Baptiste Joseph Fourier (1768–1830) proposed that any fluctuating waveform, no matter how complicated, can be represented as a finite collection of sines and cosines. The mathematical operation of transforming a fluctuating time series into a collection of sinusoids has come to be known as performing a **Fourier transform**. The resultant collection of sinusoids can be represented in a **power spectrum**, as illustrated in panel D of *Figure 3.10*, and it is thus also known, in jargon, as a spectral transform. (The term "spectral transform" is also a more general term that can apply to other mathematical operations that accomplish the kind of transform, such as the Hilbert transform, or any of various wavelet transforms). It would be difficult to overstate the impact that Fourier's theorem has had on the natural and physical sciences, and on many branches of engineering. From a practical perspective, sinusoids and power spectra

can be easier to work with analytically than are complicated signals that fluctuate over time. From a conceptual standpoint, the idea that many systems can be thought of as having cyclical (or oscillatory) properties has provided many new insights for how these systems might work.

As we proceed through this book, it will be helpful to be familiar with some concepts, conventions, and terminology related to the spectral analysis of time-varying signals. (For starters, a more efficient way to say "the spectral analysis of time-varying signals" is "signal processing," a term that originated in electrical engineering. In practice, a Fourier transform is often accomplished with an analytic technique called a fast Fourier transform [FFT].) *Figure 3.10* illustrates many of these. It illustrates the re-representation of an EEG signal into three summary sinusoids corresponding to three narrowband ranges of frequencies: theta (4–7 Hz); alpha (8–12 Hz); and beta (15–25 Hz).

Phase–amplitude coupling

Now that we've covered some important basics of signal processing, we can return to the concept of **phase–amplitude coupling (PAC)**, which refers to the fact that sustained spiking activity in populations of neurons is often not literally an unbroken, high-frequency train of activity, but, rather, a pulsatile pattern of activity that waxes and wanes at the frequency of a prominent low-frequency train in the LFP. At the level of a single neuron, it can be seen that this comes about because a neuron tends to fire bursts of action potentials at the trough of the LFP and to fire relatively few action potentials at the peak. This makes sense if we reflect on the fact that the LFP measures the extracellular voltage, and so the trough of the LFP (i.e., the most negative part of its cycle) corresponds to the peak of the (intracellular) fluctuation of voltage of the neuron. Thus, a neuron firing at, say, 50 Hz, in a region with a prominent peak of 6 Hz in the theta band of the LFP may fire in clusters of 6–9 action potentials spaced by just a few milliseconds each, and then pause for roughly 10 ms, its spiking effectively being synchronized by the LFP.

In ECoG data, PAC manifests as "stripes" in high frequencies of the TFR that are spaced by gaps in time corresponding to the frequency to which the high-frequency "stripes" are synchronized (e.g., Figure 3.12). The "stripes" correspond to bursts of power in the "high gamma" range (typically defined as >90 Hz), a range in which ECoG data are believed to index MUA rather than true oscillations. Thus, the *phase* in phase–amplitude coupling refers to the

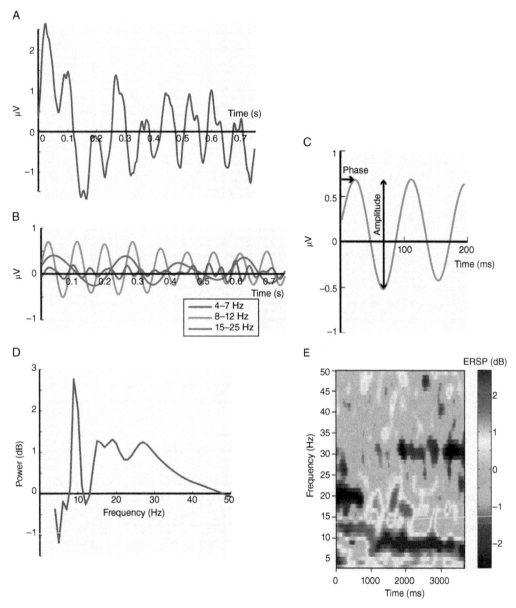

FIGURE 3.10 Spectral transform of an EEG signal. **A.** A 700-ms snippet of EEG data from electrode P1 during the delay period of a working memory task (something we'll take up in detail in *Chapter 14*). **B.** Illustration of the decomposition of waveform in panel **A** into three sinusoids corresponding to three canonical narrowband ranges of frequencies. (Note that a much finer grained set of sinusoids would be needed to represent the waveform in panel **A** with high fidelity.) Note that, in addition to frequency, the sinusoids illustrated here differ from one another in two other ways. One is that each differs in terms of its magnitude: the height of its peaks and the depth of its troughs. In the parlance of signal processing, the magnitude of a sinusoid is referred to as its **power**. The second is that each of the sinusoidal waves is at a different phase of its cycle at time 0. (That is, they are shifted in phase relative to one another.) **C.** Graphical representation of phase and magnitude. Note that, because one full cycle of a sinusoid is equivalent to one complete rotation of a needle around a dial (e.g., what the second hand of an analog clock does once every 60 seconds), phase can be expressed in degrees, ranging from 0° to 359°. **D.** The full complement of sinusoids required to completely represent the waveform in panel **A** is

(Continued on page 64)

phase of the low-frequency field potential that clocks the waxing and waning of the *amplitude* of the high-gamma signal (*Figure 3.11*).

Figure 3.12 illustrates the principal finding from Voytek et al. (2010), which is that the low-frequency band that organizes high gamma (and therefore, by inference, MUA) in posterior cortical regions varies depending on whether subjects are carrying out a visual or a nonvisual task.

Finally, an implication of this section is that although ECoG offers nothing close to single-unit resolution, it can, at least under some conditions, index a proxy for spiking activity.

Other measures of phase dynamics

Although PAC gets its own name and its own acronym, it is one example of a broader range of ways in which one can assess the synchrony between two signals. Interareal phase synchrony refers to the extent to which oscillations in two regions are phase locked. This does not need to mean that the two signals are cycling with the same phase angle (i.e., that they are each at 0° at the same time, at 15° at the same time, etc.), but would also include the (much more frequently observed) scenario when, for example, signal A is at 15° at each instant that signal B is at 0°. In this case, the two signals would be highly synchronous despite being phase lagged relative to each other. Measures of synchrony are also not limited to oscillations of the same frequency. If signal A is oscillating at 20 Hz, for example, and signal B is oscillating at 10 Hz, one can still assess whether signal A is always at 15° when signal B is at 0°; it will just be the case that for every other cycle of A, B will be at 180°. When signals at distal electrodes show a high degree of synchrony, it is assumed that one of two things is happening: either the areas giving rise to these signals are communicating with each other, or a third area is exerting similar control over these two areas.

Caveats and considerations for spectral analyses

An important caveat about spectral analyses is that it's often the case that neuronal oscillations, whether measured intracranially or extracranially, are not sinusoidal. To some extent this is evident from looking back-and-forth between panels A and B of *Figure 3.10*. Note how the peaks and the troughs of the raw EEG data (panel A) are alternately either flattened off or very sharp, relative to the "smoothed off" curves of the band-pass-filtered wave forms of panel B. This is what a Fourier transform does: you give it a time series and, regardless of its "true" shape, it gives you back the set of sinusoids that can be used to represent that time series. And while this transformation will be mathematically correct, one cannot assume that any given component of this transformation, in isolation, corresponds to a discrete component that exists in the original time series. *Figure 3.13* illustrates how the nature of a neuronal oscillation's deviation from purely sinusoidal can be quantified (in *Chapter 8, Skeletomotor control*, we'll see an example where this caveat has real-world implications for how we understand the neurophysiological mechanisms underlying the symptom-reducing effects of dopamine replacement medication on Parkinson's disease.)

A second caveat relates to the hypothetical example illustrated in *Figure 3.9.B*, a scenario in which an auditory click produces a phase reset, but no change in power. Although a spectral analysis would not produce an artifactual component (as would an ERP analysis), a conventional TFR would nevertheless be insensitive to the event-related phase shift. This is simply because an analytic procedure that involves computing power on each trial and then averaging these power values across trials cannot also carry information about phase. To detect the event-related phase shift, a different kind of analysis is required, one that ignores power but that assesses the extent to which the phase at each frequency in the spectrogram is the same or different across trials. For the scenario in

---

FIGURE 3.10 (*Continued*) represented in this power spectrum. (Note that this power spectrum reflects the task-related change in oscillatory power vs. baseline; thus, the fact that it dips below 0 at low frequencies means that there is more low-frequency power in the EEG during baseline than during task.) **E.** Another way to represent the spectral transform of the data from A is to, in effect, compute the Fourier transform at many narrower slices in time, and thereby illustrate the evolution of the spectral representation of the signal over time, in a time–frequency **spectrogram** (sometimes also referred to as a time–frequency representation [TFR]). Note that although frequency is represented by position relative to the vertical axis, and power by color, the TFR does not represent phase. Therefore, one can't know, by inspecting panel **E**, whether the prominent oscillation in the theta band is phase locked with the more transient burst of power in the lower beta band that is visible at 2000 seconds. The reason for this insensitivity to phase is that the TRF is generated by computing the spectral transform on a trial-by-trial basis, and then averaging (across trials) the power at each frequency for each time point.
Source: Figure courtesy of B. Kundu, derived from data from Johnson, Kundu, Casali, and Postle, 2012.

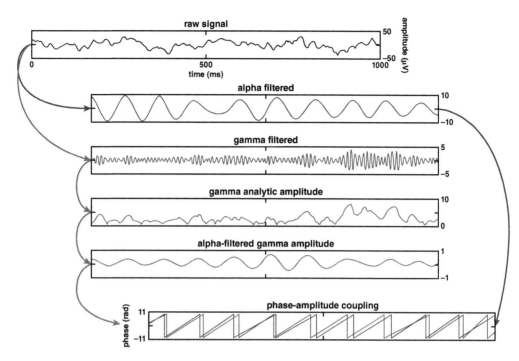

FIGURE 3.11 Data processing schematic for calculating PAC, from Voytek et al. (2010). "Raw signal" panel: self-explanatory; "alpha-filtered" panel: the raw signal band-pass filtered to isolate its components with frequencies from 8 to 12 Hz; "gamma-filtered" panel: the raw signal band-pass filtered to isolate its components with frequencies from 80 to 150 Hz; "gamma analytic amplitude" panel: the envelope of fluctuating power of the "gamma-filtered" time series; "alpha-filtered gamma amplitude": the gamma analytic amplitude band-pass filtered to isolate its components with frequencies from 8 to 12 Hz. This corresponded to the extent that high-gamma power fluctuated in the alpha frequency band. "Phase–amplitude coupling" panel: plotted together are the phase of the "alpha-filtered" component from the raw signal and the phase of the "alpha-filtered gamma amplitude." Conceptually, phase locking is the extent to which these two phase time series are "in sync." Source: From Bradley Voytek, Ryan T. Canolty, Avgusta Shestyuk, Nathan E. Crone, Josef Parvizi, Robert T. Knight Front Hum Neurosci. 2010; 4: 191. Published online 2010 Oct 19. doi: 10.3389/fnhum.2010.00191. Copyright 2010. Reproduced with permission of Frontiers Media S.A.

Figure 3.9.B, intertrial phase synchrony would be low prior to the click, and then would increase markedly for the period after the reset when the phase in all trials is temporarily aligned. (And a caveat to this caveat is that if one's data contain a true ERP, that ERP would produce an artifactual increase of intertrial phase synchrony.) This measure is often referred to as intertrial coherence (ITC).

MAGNETIC RESONANCE IMAGING

Physics and engineering bases

Nuclear magnetic resonance (NMR) refers to the fact that the *nucleus* of an atom (a proton, in the case of hydrogen) *resonates* at a preferred frequency in the presence of a strong

magnetic field. For our purposes, we can understand this as owing to the fact that an "unpaired" nucleus (i.e., no neutron paired up with the H atom's lonely proton) rotates. In the jargon of nuclear physics, this rotation is a "spin." NMR imaging, now commonly referred to as MRI, is based on the fact that the application of a radiofrequency (RF) pulse at just the right frequency – the resonant frequency of the H atom – can perturb its axis of rotation.

Physicists and biomedical engineers figured out how to take advantage of these properties of the H atom by designing a tube big enough for a human body to slide into, and running a massive amount of current through the tube so as to create a very strong magnetic field that's aligned to the long axis of the tube; this is called the "B_0" field (pronounced "bee naught"). The strength of the B_0 field is measured in units of tesla, resulting in scanners

FIGURE 3.12 High gamma amplitude couples preferentially with different low-frequency bands as a function of brain region and behavioral task, from Voytek et al. (2010). Source: From Voytek, B., Canolty, R. T., Shestyuk, A., Crone, N., Parvizi, J. and Knight, R. T. (2010). Shifts in gamma phase-amplitude coupling frequency from theta to alpha over posterior cortex during visual tasks. Frontiers in Human Neurosciences 4: 1–9. Copyright 2010. Reproduced with permission of Frontiers Media S.A.

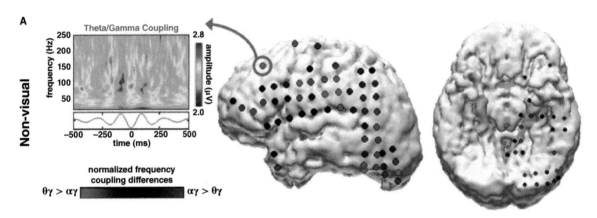

FIGURE 3.12.A "Theta PAC" at a frontal electrode during performance of a nonvisual task. The spectrogram labeled "Theta/Gamma Coupling" is a TFR aligned to a trough in the theta component of the ECoG signal, which is the red waveform below the TFR. The theta component displayed here is equivalent to the "alpha-filtered" panel in *Figure 3.11*, except that here the band-pass filtering was for a different range of frequencies (4–8 Hz), and it's averaged over all the nonvisual trials from the experiment. The "striping" of high-frequency power reflects the fact that the power at these frequencies is bursting periodically, rather than being sustained at a steady level. (Note that, because these data are trial averaged, and because the signal that is band-pass filtered from 4 to 8 Hz "speeds up" and "slows down" over time, within this range [as the relative influence of higher vs. lower frequencies within this range fluctuates], aligning the signal at any point in time during a trial will necessarily produce the "flattening out" of the waveform at more distant time points.) Although the electrodes are plotted on a single brain volume, they reflect the aggregate coverage from the two patients whose data were analyzed for this experiment.

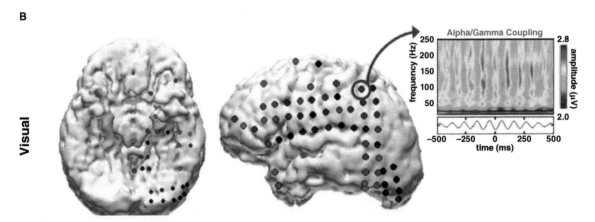

FIGURE 3.12.B "Alpha PAC" at a parietal electrode during performance of a visual task. Graphical conventions are the same as *Figure 3.12.A*; the only difference is that the low-frequency oscillation plotted under the TFR is in the alpha band (i.e., band-pass filtered from 8 to 12 Hz).

FIGURE 3.13 Quantifying the shape of a neuronal oscillation. Source: From Jackson, N., Cole, S. R., Voytek, B. and Swann, N. C. (2019). Characteristics of waveform shape in Parkinson's disease detected with scalp electroencephalography. eNeuro 6. doi.org/10.1523/ENEURO.0151-19.2019. Copyright 2019. Reproduced with permission of Society for Neuroscience.

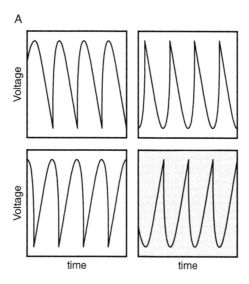

FIGURE 3.13.A Four possible ways that an oscillation can deviate from purely sinusoidal, with sharper peaks or sharper troughs accompanied by steeper rises or steeper falls. The quadrant shaded blue is the shape characteristically seen in EEG signal from electrodes C3 and C4, which emphasize signal from the underlying sensorimotor cortex (*Chapter 5, Audition and somatosensation*; *Chapter 8, Skeletomotor control*).

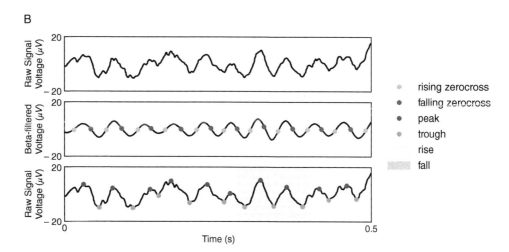

FIGURE 3.13.B Top panel: a segment of raw EEG signal. Middle panel: the signal from the top panel after being band-pass filtered to isolate its β-band component (13–30 Hz). Conceptually, the implementation of this filtering can be understood as first performing an FFT, and then extracting the range of frequencies that are of interest. Note how band-pass filtering generates a perfectly sinusoidal signal. The zero crossings are marked with lighter and darker green dots, to identify half-cycle boundaries – only one peak or trough can occur between any two of these boundaries. Bottom panel: Peaks and troughs in the original signal that lie between each zero crossing.

C

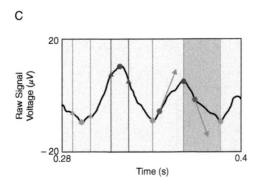

FIGURE 3.13.C Peak sharpness is calculated as the mean of the difference between the peak (purple circle) and the voltage points three samples before and after the peak (purple triangles). The trough sharpness is calculated in a similar fashion, indicated with blue circles and triangles that cycle's peak and "ending trough". The rise and fall of one cycle are highlighted in yellow and green, with the maximum slope of each illustrated with arrows.

being labeled, for example, "1.5 tesla" (T), or "3T" scanners. Importantly, when a body is slid into this tube, the axes of spin of the H atoms in the body align with the "B_0" field, and this becomes their low-energy state. The reason for aligning the spin of an H atom along a known axis of rotation is that when the angle of this axis of rotation is changed by a brief "push" delivered by a pulse of RF energy, its natural tendency is to return to its initial state. (That is, after the push, the B_0 magnetic field pulls the spin back into alignment.) This process is referred as **relaxation**, and because relaxation amounts to the H atom transitioning from a temporarily higher-energy state back to the low-energy state of alignment with the B_0 field, it involves a release of energy. It is this energy that creates the MR image.

MRI methods for in vivo anatomical imaging

Pulses of RF energy can be delivered to the body in different configurations (e.g., number and duration of pulses; frequency of pulses), each referred to as a **pulse sequence**. Different pulse sequences influence nuclear spins in different ways, and, because of this, they can be designed to emphasize different types of tissue. Additionally, different ways of measuring the relaxation signal can yield different kinds of information about the tissue being imaged.

T1-weighted imaging
One component of the relaxation signal, the T1 component, is particularly well suited for distinguishing white matter from gray. This is illustrated in *Figure 3.14*, which features

D

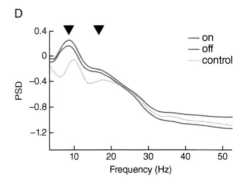

FIGURE 3.13.D The group-average power spectra (here, the "power spectral density" [PSD]) from EEG electrodes C3 and C4 from a sample of Parkinson's disease patients when they are optimally medicated with dopamine-replacement medication ("on"; blue), when they have no medication in their systems ("off"; red), and from a group of neurologically healthy control subjects (yellow). The black arrows point to peaks in the PSDs that correspond to elevated power in the alpha band and in the beta band (in the latter, the elevated power manifests as a "shoulder" in the overall trend in the PSD of decreasing power with increasing frequency). An important point made by this panel is that although the dopamine-replacement medication has a marked effect of "smoothing out" the shape of oscillations in the beta band (from the sawtoothed, asymmetric shape illustrated in the highlighted quadrant in panel A to a more sinusoidal shape), this change in shape is not captured in measures that only index power.

a "T1-weighted" image of a virtual slice through a brain. T1-weighted images ("T1 images," or, sometimes, "T1s" for short) are the most common structural MR images that one encounters in the cognitive neuroscience literature.

Voxel-based morphometry
Voxel-based morphometry (VBM) is carried out on T1-weighted anatomical MR images to estimate the density of gray matter or the volume of a structure. Importantly, VBM calculations are made after differences in the shape of the structure in question are removed, by calculating a deformation field (Ashburner and Friston, 2000). Thus, what VBM compares is T1-weighted signal intensity values in each voxel of gray matter – effectively, the average "darkness" of each voxel in group A vs. group B. Differences in VBM are often summarized as differences in "volume." (An example of a VBM analysis, comparing the hippocampi of London taxi drivers vs. bus drivers, can be seen in *Figure 12.4*.)

It should be noted that there is some controversy about exactly what it is that VBM measures, and the extent to which it might be sensitive to "microstructural" differences in the anatomy of different subjects. The website of

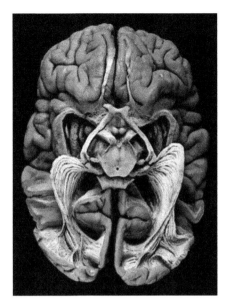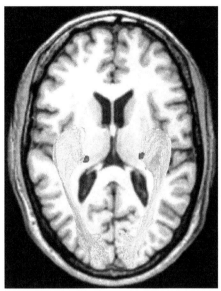

FIGURE 3.14 Image on left shows a human brain dissected to reveal the pathways from retina, via the lateral geniculate nucleus (LGN) of the thalamus, to cortical area V1 (all to be detailed in *Chapter 4*). Particularly striking are the elegant optic radiations, the myelinated fibers running between the thalamus and the primary visual cortex. To expose these pathways and structures, the anatomist dissected away tissue from the ventral surface of the brain, up to and including the ventral bank of the calcarine fissure. Thus, only the dorsal bank of the calcarine and the fibers of the optic radiations that project to it are seen here. The image on the right shows a slice from a T1-weighted anatomical MR scan (gray scale) oriented to include both the LGN (pseudocolored red) and the calcarine sulcus. Superimposed, in yellow, is an image of the path of the optic radiations as estimated via MR-derived diffusion-tensor imaging (DTI) tractography. Source: Wandell and Winawer, 2011. Reproduced with permission of Elsevier.

a suite of neuroimaging software tools called FSL includes this caveat on its VBM page:

> It is sometimes not possible to determine if the results you may find are the consequence of an effective reduced thickness or atrophy in the gray matter, or rather an indirect reflection of a different gyrification pattern. Indeed, it might be possible that a misalignment of the gyri/sulci or even different folding patterns may lead to the difference of gray matter distribution that you have found.

Diffusion tensor imaging – in vivo "tract tracing"

Within tracts of myelinated axons there are teeny gaps between each axon. As a consequence, water molecules travel much more quickly along such a bundle than across it. This motion, or *diffusion*, of water molecules can be represented as a 3-D volume (called a *tensor*) that indexes how symmetrically, or otherwise, water molecules are diffusing. Diffusion tensor imaging (DTI) uses a pulse sequence that measures the shape of diffusion tensors in different parts of the brain. A perfectly spherical tensor, said to be "isotropic," would indicate that water is diffusing freely in all directions. Fiber tracts, in contrast, produce "anisotropic" tensors shaped more like rugby footballs. By estimating the most probable route that a series of such tiny end-to-end "footballs" makes through a section of brain, one can infer the existence of a fiber tract, as illustrated in *Figure 3.14*.

Functional magnetic resonance imaging

While T1-weighted and diffusion tensor images measure aspects of anatomy that are assumed to be static during the time it takes to acquire the images, the "f" in fMRI indicates that its images index the physiology of the brain. That is, it measures levels of activity that change over time. To be able to interpret fMRI data, it is necessary to understand it at several levels: the cellular processes that give rise to the activity that fMRI is measuring; the hemodynamic response that these processes trigger; and the physics of its signal.

The metabolic demands of the action potential and of synaptic activity

The generation of an action potential entails reversing the electrical and chemical imbalances that are characteristic of the neuron at rest. How the neuron restores these imbalances turns out to be the basis for many neuroimaging techniques, including fMRI. One key mechanism is the so-called Na⁺/K⁺ pump, a membrane-spanning complex of proteins that act as a "reverse channel." At the cost of one molecule of adenosine triphosphate (ATP) per cycle, the Na⁺/K⁺ pump expels three Na⁺ ions and draws in two K⁺ ions. Also highly energy demanding is the process of releasing molecules of neurotransmitter into the synaptic cleft (exocytosis). ATP, the "fuel" that is needed for these operations, is derived from the processing of glucose in mitochondria inside the synaptic bouton of the axon. Glucose, in turn, is delivered to cells via the bloodstream. Herein lies the linkage to the signal that is measured by fMRI: elevated activity in neurons creates a need for arterial blood, and the resultant increase of oxygenated blood to active parts of the brain is what these methods detect. (The principle is the same for all tissues of the body. Think about whether more blood flows to the muscles of your arm when you are lifting something heavy, or when your arm is just hanging by your side. Note that this section is also relevant for the 2DG method of functional tissue imaging that we have already considered in this chapter, and for some applications of positron emission tomography [PET], which we will take up further along in this chapter.)

Interestingly, neurotransmitter reuptake, although a metabolically expensive process, isn't directly responsible for the consumption of ATP. Instead, it leverages electrochemical ion gradients across the membrane via a process called cotransport. A useful analogy is to consider a revolving door. To get it "inside," a molecule of neurotransmitter is yoked to a Na⁺ ion. Now we have a situation in which the chemical gradients of the yoked components are in opposition – the low intracellular concentration of Na⁺ provides an inward "pull," and the opposite is true for the neurotransmitter. It's as though two people have entered a door that's rotating counterclockwise (anticlockwise), and one (the "Na⁺ ion") is pushing the door in the counterclockwise direction so as to enter, while the other (the "molecule of neurotransmitter") is pushing in the opposite way. (The role of the neurotransmitter can be played by a 2-year-old in our analogy: *No, Mommy, I don't want to go into that store!*) The stalemate is broken by adding a third "person" who is inside and who wants to leave: cotransport also entails the simultaneous transport of a

molecule of K⁺ along its concentration gradient and its electrical gradient, from inside the cell to outside. "Two against one," the molecule of neurotransmitter "loses," and it ends up back inside the neuron.

If cotransport "passively" takes advantage of chemical gradients, why is it metabolically expensive? The answer is that although the reuptake process itself doesn't consume ATP, its operation decreases the resting-state imbalance of higher intracellular concentration of K⁺ and higher extracellular concentration of Na⁺. Thus, each cycle of the reuptake processes prompts a compensatory, energy-expending cycle of the Na⁺/K⁺ pump in order to restore the imbalance.

T2*-weighted fMRI

In 1990, biophysics researcher Seiji Ogawa and colleagues, working at AT&T Bell Laboratories in Murray Hill, NJ, made a breakthrough discovery: Leveraging the fact that oxygenated hemoglobin (Hb) and deoxygenated hemoglobin (dHb) have different magnetic profiles, they were able to coax from MRI a "blood oxygen level-dependent" (BOLD) signal. That is, with the right pulse sequence they demonstrated that a different MRI measure of relaxation, called T2*, varied with the balance between Hb and dHb. The reason that this was important comes from the principle of brain physiology that we've already seen: elevated activity in neurons creates a need for arterial blood. This is accomplished by the dilation of arterioles, and the consequent increase of arterial blood carrying Hb in the vicinity of the active cells. A distinctive feature of mammalian physiology is that this hemodynamic response is overcompensatory, in that the increased supply of Hb-rich blood exceeds the demand and, as a result, the ratio of Hb:dHb in venules carrying blood away from the capillary bed in the region of elevated activity temporarily increases (*Figure 3.15*). Neurosurgeon Wilder Penfield, who we'll read about when we get to somatosensation (*Chapter 5*), observed this with his naked eye over 70 years ago, as venous blood becoming redder during localized seizures (seizures being a pathological cause of increased activity).

fMRI is an extremely popular tool for cognitive neuroscience research, primarily because it offers the highest spatial resolution of any noninvasive method for measuring activity in the brain. The typical fMRI experiment includes the collection of a high-resolution T1-weighted volume, with a resolution of .75 mm³ or better. The T2*-weighted data (from here on out, the "fMRI data") acquired during the same experiment are coregistered to and merged with the T1-weighted anatomical image

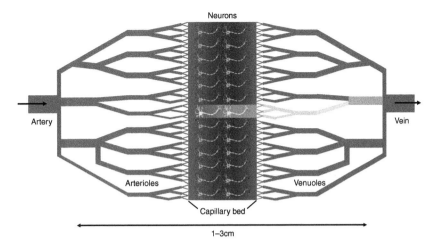

FIGURE 3.15 The hemodynamic response to neuronal activity. In this cartoon, the elevated activity of a small cluster of neurons (depicted in yellow) results in an increase in the delivery of Hb-rich blood to the local capillary bed (depicted by redder tint of capillaries), and an increase in Hb:dHb in the venules draining this blood away from the capillary bed (also depicted by a redder tint).

(the "anatomicals"), thereby allowing one to see what structure(s) in the brain are generating the activity measured with fMRI. The physics of the T2*-weighted signal is such that T2*-weighted images are of lower resolution than are T1-weighted images and typically the fMRI data acquired during an experiment will have one-quarter the resolution of the anatomical data. For example, a scanning session with an anatomical resolution of .75 mm³ will have a functional resolution of 3.75 mm³. Although MRI data can be acquired as individual 2-D slices, they are more typically acquired as 3-D "wholebrain" images. Nonetheless, to show "the insides," not just the cortical surface, the data are often displayed as mosaics of 2-D slides that, when stacked, would reconstitute the 3-D volume of space that the physical brain occupies (see *Figure 3.16*). When one views a slice taken from an image with .75 mm³ resolution, the "picture elements" (i.e., pixels) that make up the image are .75 mm². Importantly, however, with MRI even a single slice has thickness to it, and so it is better construed as a slab. Scientists acknowledge this by referring to the "3-D pixels" that make up MRI images as "volumetric pixels," or voxels.

Although acquisition of a T1-weighted anatomical scan requires roughly 5 minutes of scanning time, the resultant image does not contain any temporal information. fMRI data, in contrast, do typically contain temporal information, because a scan typically entails the acquisition of several 3-D volumes. Let's think back to EEG for a moment. If one is acquiring EEG data with a 60-electrode cap, a 5-minute session of recorded task performance will yield

data from 60 channels, each one a time series with (let's say) 1024 recorded values per second, which could be described a 2-D "slice" made up of 60 "pixels," each one of which contains a time series with (1024 time points/second * 60 seconds/minutes *5 minutes =) 307,200 values. (EEG data are typically down sampled to 500 Hz or less, but that would still be a lot of values!) fMRI, rather than using 60 electrodes, is collected in (brain-shaped) volumes of roughly 30,000 voxels (depending on the size of the individual's head). The typical fMRI scan lasts anywhere from 4 to 10 or more minutes, with a new volume acquired roughly once every 2 seconds. (The time required to acquire a single volume of fMRI data, called the "time to repeat" [TR], will vary depending on the pulse sequence, the size of the volume being imaged, and other factors.) And so, although fMRI's sampling rate of .5 Hz is orders of magnitude lower than that of EEG (and, consequently, its temporal resolution is dramatically inferior to that of EEG), a 5-min-long scan with a TR of 2 seconds nonetheless acquires (.5 time points/second * 60 seconds/minutes *5 minutes=) 150 values per voxel.

Task-related functional imaging

To illustrate how MRI and fMRI data are acquired to study the activity of the brain associated with cognition, *Figure 3.16* walks us through a study of one of the simplest behaviors I can think of, pressing a button (. . . over and over and over, once every 20 seconds). The simplicity of the task will allow us to emphasize a couple of fundamental

FIGURE 3.16 fMRI data from a 6-minute-40-second-long scan during which the subject pushed a button with each thumb, once every 20 seconds.

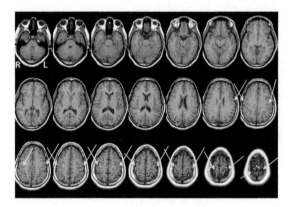

FIGURE 3.16.A Voxels in the pre- and postcentral gyrus for which the increase in fMRI signal in response to the button presses exceeded a statistical threshold. This "thresholded statistical map" is superimposed on a T1-weighted anatomical image acquired during the same scanning session. This mosaic displays axial slices arrayed from ventralmost (upper left) to dorsalmost (lower right). Arrows point to the central sulcus in each slice in which it is visible. (Primary motor cortex is located in the precentral gyrus [*Chapter 8*], and primary somatosensory cortex is located in the postcentral gyrus [*Chapter 5*].) Note that although these data were acquired during a whole brain scan, they have been masked so that activity in only precentral and postcentral gyrus can be observed.

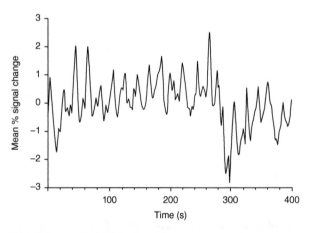

FIGURE 3.16.B BOLD signal fluctuations pooled across the voxels identified in *Figure 3.16A*. The data have been mean centered and scaled to units of percentage change from baseline, but are otherwise "raw." A button press occurred at each of the tick marks on the time axis (except 400).

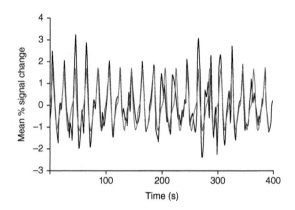

FIGURE 3.16.C Same data as *Figure 3.16.B*, but with drift removed, and an overlay (red waveform) of the fits of the statistical model to the data. With regard to "drift," the overall drop in signal at time 280 in *Figure 3.16.B* reflected some factor other than the once-every-20-seconds button press, and so was removed statistically. The independent variable in the statistical model was a series of delta functions (a.k.a. "stick functions") placed at time 0 seconds, 20 seconds, and so on up to 380 seconds – i.e., at each time point when a button press was to be executed – and convolved with a basis set of sine waves and cosine waves, so as to capture the predicted shape of the HR. While the red waveform shows the least-squares best fit of the model to the data, the data themselves illustrate the trial-to-trial variability of the fMRI response to each of the 20 presumably identical events.

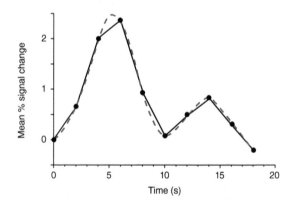

FIGURE 3.16.D In black, the trial-averaged response to a bimanual thumb press made at time 0 (i.e., the average of time segments 0–20, 20–40, . . . 380–400 from *Figure 3.16.C*). This trial-averaged time series can be thought of as an empirically derived estimate of the hemodynamic response function (HRF) for this subject. In red, the HRF has been interpolated, to represent the fact that it is a continuous function, despite only being sampled at a 0.5-Hz rate in this study.

properties of the BOLD signal that we'll need to keep in mind for every fMRI study that will be considered during the remainder of this book. First, note from *Figure 3.16.A* that this presentation of the anatomical localization of button press-related activity does not display the BOLD data themselves, but, rather, the results of a statistical analysis of the data. This will be true of almost all presentations of fMRI data that one is likely to encounter. Second, note from *Figure 3.16.D* how sluggish and smeared out in time the BOLD response to the button press is. That is, although the neural activity producing the button press was probably completed by time 0.5 seconds or so, this subject's fMRI response didn't peak until time 6 seconds. And, importantly, if this fMRI response to the button press were only delayed, but not otherwise transformed, it would be expected to produce a fairly steep ramp-up at roughly time 6 seconds, and a similarly steep drop-off roughly 500 ms later. What we see, however, is a relatively shallow rise to its peak, and an equally shallow return to baseline – although we're assuming that the neural activity of interest lasted only half a second, the full-width and half maximum of this response

(a common way to estimate the width of this kind of response) looks to be roughly 5 seconds long. Importantly, these "sluggish and smeared-out" qualities of the fMRI response are not due to technical limitation – the BOLD response would look like this regardless of whether it was sampled at 1000 Hz or .1 Hz. Rather, the slow rise and slow fall of the fMRI signal are consequences of the hemodynamic response to neural activity that it measures. The dilation and subsequent constriction of arterioles simply lag behind the neural signals that trigger them. Indeed, these properties of the BOLD response are so fundamental to the signal that the waveform derived in *Figure 3.16* is an empirically derived example of what is called the hemodynamic response function (HRF).

Because cognitive neuroscientists are typically interested in the activity of neurons, not in hemodynamics per se, the smearing properties of the HRF can be thought of as a filter that transforms the signal of primary interest in our experiments. More specifically, it has the properties of a low-pass filter, which render it poorly suited to tracking and resolving rapidly fluctuating signals. *Figure 3.17* gives a sense for why this is.

FIGURE 3.17 The low-pass filtering properties of fMRI. Source: Adapted from Buckner, R. L., & Dale, A. M. (1997). Selective averaging of rapidly presented individual trials using fMRI. Human Brain Mapping, 5, 329–340. doi:10.1002/(SICI)1097-0193(1997)5:5<329::AID-HBM1>3.0.CO;2-5

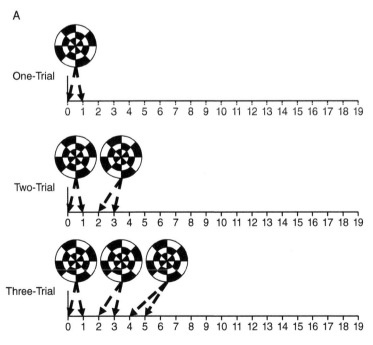

FIGURE 3.17.A Diagram of three types of trials, in which a 1-second-long visual presentation of a flickering checkerboard stimulus – chosen to evoke a strong response from visual cortex – was presented once (One-Trial), twice (Two-Trial), or three times (Three-Trial), with a 1-second intertrial interval (ITI) dividing presentations on the two- and three-event trials.

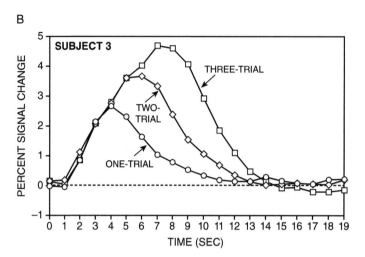

FIGURE 3.17.B The BOLD response, from the visual cortex of a single subject, from each of the three trial types, all plotted on the same graph, and all time locked to the onset of the first checkerboard stimulus. Note that, for the Two-Trial and Three-Trial responses, although they are of higher magnitude, it's not evident that they were generated by more than one discrete event.

Functional connectivity

Although it is a fairly general term, "functional connectivity" is most commonly used to refer to analyses in which, rather than measuring how performance of a task produces changes in activity in different regions in the brain (e.g., *Figure 3.16.A*), one measures how performance of a task produces changes in the correlation of activity between different regions. In the following section we'll look in some detail at a method for assessing such patterns of interregional correlation, but doing so while the subject is explicitly *not* performing a task: resting state functional correlations (RSFCs). RSFCs measure the extent to which the fluctuations in activity between regions covary over an extended period of time. For functional connectivity analyses – i.e., task-related patterns of correlation – the RSFC data effectively provide the baseline, because experimenters are interested in knowing whether, and if so, how, correlations between regions may strengthen (or weaken) above and beyond their baseline level of correlation. Conceptually, functional connectivity is interpreted in a manner similar to interareal phase synchrony (see earlier section on *Time–Frequency Analysis, Other measures of phase dynamics*), in that if two or more regions show a task-related increase in functional connectivity, it is assumed that their activity is coordinated, and/or that they are sharing information. Conversely, if the functional connectivity

between two areas decreases, for example, engaging in a task produces increases in activity in area A and decreases in B, and furthermore, A and B are negatively correlated; this might be interpreted as suggestive that A has an inhibitory effect on B. Importantly, as is also the case with interareal phase synchrony, changes in functional connectivity between A and B might also result from a third area, C, driving both of them (or C and D, or . . .).

Resting state functional correlations

Important insights into brain function, how it supports behavior, and how it varies with age and with disease state, have been gained by studying brain activity when the subject is "at rest" – awake, but not performing any experimentally controlled task. This was first reported by Marcus Raichle and colleagues at Washington University, when they observed in a meta-analysis of PET studies of a wide variety of cognition tasks that a common set of areas reliably decreased their activity in response to engagement in almost any cognitive task. (PET will be introduced in section *Tomography*.) These regions, including most prominently the posterior cingulate cortex and precuneus (on the medial surface of the parietal lobe), medial frontal cortex, and inferior parietal lobule, have come to be known as the default mode network (DMN), because they are active "by default," when no experimenter-designed task is being carried out (*Figure 3.18.C*, Association System labeled "Default"). Shortly thereafter, Bharat Biswal and colleagues, then at the Medical College of Wisconsin, demonstrated with fMRI that anatomically disparate areas that increase their activity in common during performance of goal-directed tasks often also exhibit correlations in the low-frequency fluctuations (<.1 Hz) that are typical of the brain "at rest." These two developments were initially very influential because they provided potent examples of intrinsic activity in the brain – ongoing activity that is not tied in any direct way to events in the outside world. Subsequent advances have established that resting state functional correlation (RSFC) can be used to "parcellate" the entire brain into networks that are named after the domains of cognition that engage them in "task-positive" contexts. In summary, an RSFC map can be construed as reflecting "a statistical history of coactivation that [has been] sculpted over the lifespan of the individual" (Wig, 2017, p. 983).

To generate an RSFC map, MRI scanning entails acquiring a T1-weighted anatomical image of the subject's brain, and then an fMRI scan while the subject is instructed to lie still and stare at a fixation point during the entirety of the scan (here we're illustrating the procedure with the 6-minute-long scan used by Han et al., 2018). Then, the data processing is carried out as follows:

1. Perform a virtual "inflation" of the (3-D) T1-weighted anatomical volume (*Figure 3.18.A*) to facilitate creation of a 2-D surface map of the resting state (RS) BOLD data, and the visualization of the results.
2. Transform the 3-D volume of BOLD data acquisition space into a 2-D virtual mesh that covers the inflated anatomical images. This 2-D mesh is made up of 32,000 vertices per hemisphere, each of which has a time series made up of the 180 images acquired during the fMRI scan (360 seconds of scanning with a TR of 2 seconds).
3. Compute the pairwise correlation between the time series of each vertex and that of every other surface vertex, thereby generating a $32,492 \times 64,984$ (32k $\times$ 64K) correlation matrix for each hemisphere.
4. Transform the correlation matrix from each subject into a common atlas space so that results can be combined across subjects.
5. Convert individual-subject correlation matrices into a group-level similarity matrix by correlating each vertex's RSFC map with that of every other vertex (e.g., two adjacent vertices that are going to end up belonging to the same region will have very similar patterns of correlation with the rest of the brain (i.e., similar correlation matrices) and so will be highly correlated in the similarity matrix).

FIGURE 3.18 Generating an RSFC map. Source: Adapted from Han, L., Savalia, N. K., Chan, M. Y., Agres, P. F., Nair, A. S. and Wig, G. S. (2018). Functional parcellation of the cerebral cortex across the human adult lifespan. Cerebral Cortex 28: 4403-4423.

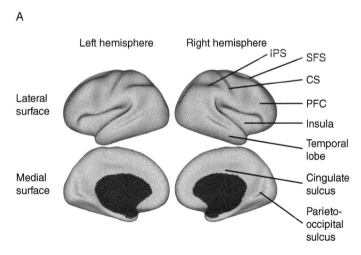

FIGURE 3.18.A An inflated anatomical volume that has been warped into atlas space, illustrating step 1. Note that in addition to exposing the insula, the inflation process has separated the dorsal bank from the ventral bank of the Sylvian fissure. Some landmarks are identified in the right-hemisphere images: CS = central sulcus; IPS = intraparietal sulcus; PFC = prefrontal cortex; SFS = superior frontal sulcus.

B

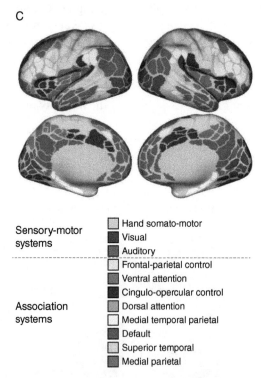

Stable RSFC pattern ▮▮▮▮▮ RSFC pattern transition

Edge density

FIGURE 3.18.B RSFC-based boundary maps, with warmer colors indicating higher probability of a boundary between areas, and darker colors indicating collections of vertices with high RSFC similarity, and therefore belonging to the same area.

C

Sensory-motor systems
- ▮ Hand somato-motor
- ▮ Visual
- ▮ Auditory

Association systems
- ▮ Frontal-parietal control
- ▮ Ventral attention
- ▮ Cingulo-opercular control
- ▮ Dorsal attention
- ▮ Medial temporal parietal
- ▮ Default
- ▮ Superior temporal
- ▮ Medial parietal

FIGURE 3.18.C RSFC parcellation into functional networks of data from 88 healthy young adults, aged 20–34 years.

6. Generate a gradient map by identifying abrupt transitions in the similarity matrix.
7. Apply a "watershed" edge-detection algorithm to the gradient map to convert the "abrupt transitions" into boundaries between RSFC-defined regions. Note that in the terminology of graph theory, "edge" refers to the connections between nodes (and that for RSFC, vertices act as nodes) (*Figure 3.18.B*).

8. Label the resulting RSFC-based parcellation according to the functional systems that have been identified and validated several times since their introduction in 2011.

The section of this chapter on *Network Science and Graph Theory* will consider some of the ways that RSFC is used in cognitive neuroscience.

MAGNETIC RESONANCE SPECTROSCOPY

Just as T2*-weighted pulse sequences take advantage of the fact that hemoglobin's magnetic resonance properties vary with its state of oxygenation, many other compounds in the world have distinct magnetic resonance profiles that can be detected and measured with MRI, including neurotransmitters and their metabolites. Magnetic resonance spectroscopy (MRS) can be used, for example, to assess the level of GABA in someone's visual system, or in the vicinity of their frontal eye fields. The signals from neurotransmitters and their metabolites are very weak, and so lengthy scans are required (on the order of minutes). This means that nothing approaching real-time measurement of fluctuations in the levels of, say, the synthesis or the breakdown of a certain neurotransmitter, perhaps in conjunction with task performance, is possible at this time.

TOMOGRAPHY

The word tomography refers to the acquisition of an image of a 2-D section through a volume, via some kind of penetrating wave. Although X-ray technology was used for medical imaging throughout the twentieth century, it was limited to the acquisition of single 2-D sections (or "slices") until the advent of computer-implemented algorithms that allowed for the effectively simultaneous collection of all the slices needed to reconstruct a volume. (Note that, as is the case with MRI, although these sections are referred to as being "2-D slices," each actually has some thickness to it, and thus they might be better construed as thin "slabs.")

X-ray computed tomography

Computed tomography (CT) scanning is only capable of generating anatomical images, and has markedly poorer spatial resolution relative to MRI, and so is of limited utility as a research tool for cognitive neuroscience. It is a workhorse in Emergency Medicine and Radiology departments in hospitals, however, for reasons of safety: sliding an unconscious accident victim into an MRI scanner could be catastrophic if the patient has any metal in their body (including brain), whether due to a previous injury or in the form of a medical device. Consequently, the brain images used by neurologists and neuropsychologists to estimate the damage incurred by brain-injured patients are frequently CT images acquired in the first hours after the patient's admission to the hospital.

Positron emission tomography

PET entails the injection of a short-lived (of the order of minutes to hours) radioactive isotope into the bloodstream, and subsequently measuring where in the body this substance accumulates. It works on the principle that the decay of the isotope releases subatomic particles called positrons, and that when a positron encounters an electron the resultant annihilation releases two photons (gamma rays) that travel in opposite directions. The photons are detected by detectors that are arrayed in a ring around the body, and sophisticated math is performed to estimate where in space the annihilation event must have occurred in order to produce the pattern of photons detected by the scanner.

In principle, any biologically active molecule that can be rendered radioactive can be used as a PET "tracer." Thus, PET can be used to assess patterns of energy metabolism in the brain (via radiolabeled glucose, a method that can be thought of as "in vivo 2-DG imaging;" e.g., *Figure 3.5*), or patterns of activity of certain neurotransmitter systems (as with, e.g., radiolabeled dopamine agonists). For now, we'll limit our discussion to $H_2^{15}O$ – a.k.a. "radiolabeled water," or just "^{15}O" (pronounced "oh fifteen"). (Regardless of what kind of tracer we use, PET invariably has a low signal-to-noise ratio (SNR), which will impose constraints on experimental design, as we'll see in the next section of this chapter.) In a "cognitive activation study" (as illustrated in *Figure 3.19*), the subject lies in the scanner, with an IV in her arm. Upon the delivery of $H_2^{15}O$ into the bloodstream, she begins performing a cognitive task. At this point, it is useful to think back to the earlier section in this chapter on *The metabolic demands of the action potential and of synaptic activity*. In a "hearing words" task, for example, increases in the activity of the brain tissue engaged in processing this auditory information will trigger an increase in the delivery of arterial blood to this region. With $H_2^{15}O$ PET, such changes in regional cerebral blood flow (rCBF) are quantified as changes in "PET counts" detected by the scanner (i.e., a high number of photons emanating from a location in the brain indicates that there is a high concentration of $H_2^{15}O$ molecules at that location).

Because the entire brain is always consuming large amounts of arterial blood (20% of the energy consumption of the body), task-based imaging requires image subtraction in order for the results from the scanning session to be interpretable. To identify brain areas involved in "hearing words," for example, Petersen and colleagues

(1988) acquired two scans per subject: one was an "experimental" scan acquired while the subject is fixating (visually) a point on the screen and listening to the auditory presentation of words, and the second was a "control" scan acquired while the subject was only fixating a point on the screen. Then they subtracted the resultant control image from the experimental image (*Figure 3.19*). Note that although some form of subtraction is required by most brain-measurement techniques (including EEG, MEG, fMRI, and some applications of PET) in order to effect a baseline normalization, the term "cognitive subtraction" refers to a specific experimental strategy that will come up periodically in the book, and that we will consider in more detail in the next section.

FIGURE 3.19 PET scans of word processing, illustrating the logic of image subtraction.

A

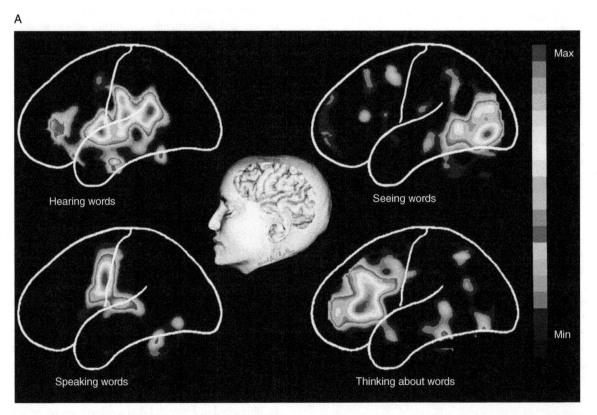

FIGURE 3.19.A PET subtraction images of rCBF when subjects perform the labeled cognitive operation vs. a matched control condition. For *Hearing words* and *Seeing words*, the contrast is vs. fixation of a point on the screen. For *Speaking words*, the contrast is vs. *Seeing words*. For *Thinking about words*, the task was to think about a use for the presented word (e.g., for the word "cake," the subject might think "eat"), and the contrast is vs. *Speaking words*. Source: From Petersen, Steven E., Peter T. Fox, Michael I. Posner, Mark Mintun, and Marcus E. Raichle. 1988. "Positron Emission Tomographic Studies of the Cortical Anatomy of Single-Word Processing." Nature 331 (6157): 585–589. doi: 10.1038/331585a0. Reproduced with permission of Nature Publishing Group.

B

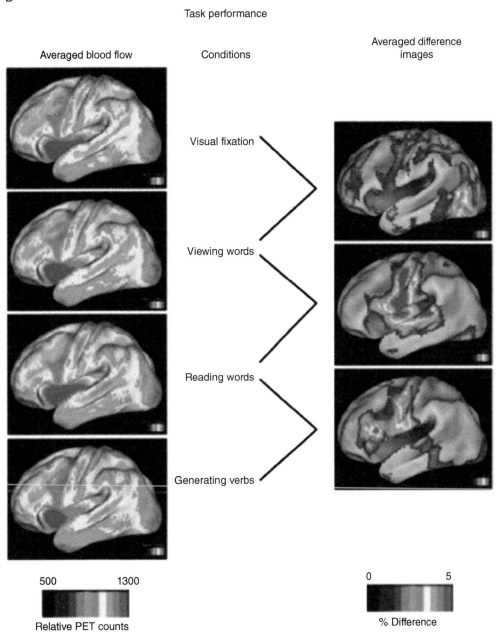

FIGURE 3.19.B A different way to represent the Petersen et al. (1988) data. The right-hand column presents three of the rCBF contrast images from *Figure 3.19.A* (now looking more "modern" because the activity maps are warped onto inflated left hemispheres). The top image is *Seeing Words*, the middle image is *Speaking Words*, and the bottom *Thinking About Words*. Note that no task-related changes in rCBF exceed 5% of the baseline level of rCBF. The left-hand column shows the unsubtracted rCBF images associated with each of the four conditions used to generate the subtraction images in the right-hand column. Try covering up the labels, as Dr. Raichle has been wont to do during presentations, and see if you think you'd be able to identify which unsubtracted rCBF image corresponds to which behavioral condition. Source: From Raichle, Marcus E. 2010. "Two Views of Brain Function." Trends in Cognitive Neuroscience 14 (4): 180–190. doi: 10.1016/j.tics.2010.01.008. Reproduced with permission of Elsevier.

NEAR-INFRARED SPECTROSCOPY

Near-infrared light passes through the scalp, and different wavelengths are then absorbed by oxygenated vs. deoxygenated hemoglobin, thereby producing a BOLD signal. Although the optics are such that only superficial regions of cortex can be imaged, portability and ease of use relative to fMRI make it a preferred tool for pediatric neuroimaging.

SOME CONSIDERATIONS FOR EXPERIMENTAL DESIGN

The generic cognitive neuroscience experiment can be summarized (and simplified) as *recording brain activity from a subject while they perform a cognitive psychology task*. For the purposes of this consideration of experimental design, let's return to a variant of a fairly straightforward task that we've already considered: one-at-a-time visual presentation of drawings that each require a button press to indicate if they depict a living or nonliving object. This task lends itself to addressing many different kinds of questions about cognition. For example, one might use it, or a variant, to study visual perception, perceptual decision-making, categorization, semantics (i.e., the representation of meaning), or motor control. Independent of the cognitive domain of interest, however, the method that one uses to measure neural activity will constrain how the task can be designed and the data analyzed.

What if we were interested in the visual perception of these drawings? With EEG or with fMRI we could carry out an event-related design in which we separated and averaged signals from trials presenting living vs. nonliving things, and time lock them to stimulus onset. The EEG data could be analyzed to generate ERPs or spectrally transformed to generate TFRs. With fMRI, because of the sluggishness of the HRF, we would need to either separate each trial by 16-or-more seconds, to allow the signal to return to baseline after each trial, or randomize the order in which living and nonliving trials are presented. The benefit of randomization is that the ITI can be shortened, and so more trials can be presented during a fixed period of time; the drawback is that the effect size will be much smaller with a shorter ITI (*Figure 3.20*). With ¹⁵O PET, however, the SNR is sufficiently low that only one number (per voxel) can be acquired per several-minute-long scan (e.g., "the visual cortex had an activity level of 1153 PET counts during the 6 minute-long block of performing 'living things' trials"). This necessitates the use of a block design, in which many trials of the same kind are performed in quick succession (i.e., short ITIs), and the signal is aggregated to yield one value.

Designing an experiment with a block-design procedure raises an additional set of considerations. The most obvious issue for our experiment is that a block of trials that each involves *view*, *decide*, and *press button* stages will necessarily generate brain activity that reflects more than just visual perception – it will be "contaminated" (as we say) by contributions from the decision-making and the button-pressing portions of each trial. Here's where cognitive subtraction comes in. One could design a control task in which, for example, a nonsense pattern of lines is presented at the beginning of each trial, and the subject judges whether the pattern is more dense on the right or left half of the image and pushes a button accordingly. The logic of this approach is that resultant subtractions of [Living − Control] and [Nonliving − Control] would reflect the differences in visual processing of viewing a real object vs. viewing a meaningless pattern of lines, but no contamination from decision-making, nor from button pressing, because both the experimental and the control tasks included these, and so the signal associated with those processes would be subtracted out. One caveat from applying this logic of cognitive subtraction that should be obvious right away is that if the elements being subtracted out in an analysis aren't perfectly matched, components of them might be left over. Let's say, for example, that the left/right judgment required in the control task is simply a lot easier than the living/nonliving judgment required in the experimental task, such that the latter takes an average of 500 ms longer for the subject to complete. Assuming that the stimulus onset asynchrony (SOA) of the two tasks is identical (as it should be for any self-respecting control condition), the amount of "task" contributing to the aggregated-across-the-block signal from the experimental task will exceed that from the control task. How can one be sure, then, whether any differences that may be revealed by the [Experimental−Control] comparison are due to differences in the perception of a real object vs. a pattern, vs. some other factor, such as increased anxiety on the part of the subject of getting responses to the more difficult living/nonliving pictures incorrect, or even something as seemingly trivial as simply having spent more time on task in the experimental condition? The point of both of these

FIGURE 3.20 Simulations of fMRI signal in an experiment with multiple identical events (1-second long visual checker-board presentations, as in *Figure 3.17*), and with different-length interstimulus intervals (ISI). Source: From Burock, M., Buckner, R. L., Woldorff, M. G., Rosen, B. R., & Dale, A. M. (1998). Randomized event-related experimental designs allow for extremely rapic presentation rates using functional MRI. NeuroReport, 9, 3735–3739. Copyright 1998. Reproduced with permission of Lippincott Williams & Wilkins/Wolters Kluwer.

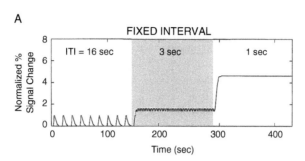

FIGURE 3.20.A fMRI signal intensity across each of three concatenated blocks of 144 seconds each, illustrating the responses to nine events separated by 15-second ISIs, to 36 events separated by 3-second ISIs, and to 72 events separated by 1-second ISIs. Note the two countervailing trends as the number of events increases and the SOA decreases: the amplitude of the average response to individual events decreases (indeed, the signal saturates and responses to individual events are no longer detectable with an ISI of 1 second), and the mean value of the signal, aggregated across each 144-second block, increases (from roughly 1% signal change, to nearly 2%, to slightly above 4%).

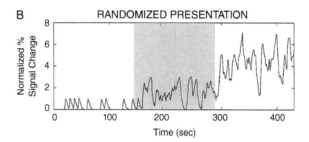

FIGURE 3.20.B Responses to the same number of events as in *Figure 3.20.A*, with the same average ISI, but with the actual ISI randomized from event to event.

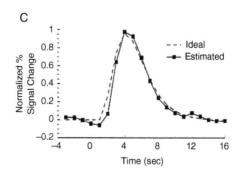

FIGURE 3.20.C An estimate of the HRF derived from the time course of the 1-second ISI condition from *Figure 3.20.B*. The analysis that generated this estimate is a variant of multiple regression that has come to be known as "linear deconvolution." This estimated HRF is superimposed on the "ideal" HRF that was used to generate the responses in *Figures 3.20A* and *3.20B* (Burock et al., 1998).

possible concerns is to highlight a fundamental caveat about cognitive subtraction: the results of a cognitive-subtraction analysis are only amenable to straightforward interpretation if the two conditions *only* differ by the factor that is explicitly changed for the control task.

Another issue that we haven't addressed with our hypothetical block design, but that may be problematic, is that we've been assuming that our experimental task would consist of blocks of only living objects and blocks of only nonliving objects, in addition to blocks of the control task. After all, if we combined living and nonliving within the same block, how could we ever hope to learn about how the processing of one differs from the processing of the other? A possible problem with this, however, is the behavioral strategy that might arise from seeing only a series of living things (e.g., a cat, a tree, an inchworm, a dolphin, . . .) and pressing the "living" button each time. I'll leave it to the reader to think about why this might be problematic, and how one might try to address this concern.

Figure 3.20 also addresses a fact that is specific to the low-pass-filtering properties of the fMRI signal: the SNR of fMRI block designs is inherently superior to the SNR of fMRI event-related designs. This is evident from the fact that the magnitude of the response associated with single events in the simulation (a maximum of 1% signal change relative to baseline) is markedly smaller in magnitude than the change in signal intensity in the 144-second blocks featuring the shorter ITIs. Therefore, absent any other considerations, a block design will always generate fMRI data with greater statistical sensitivity than will an event-related design. This can be a particularly compelling factor if one is carrying out a study in which the stimuli don't lend themselves to randomization. Consider if, in our living/nonliving scenario, for example, our primary interest was in the neural bases of the decision and resulting motor response. There'd be no way to isolate the button press from the remainder of the trial, because one can't meaningfully randomize the order of picture presentations and button presses.

COMPUTATIONAL MODELS AND ANALYTIC APPROACHES

I've done my best in this chapter to avoid, as much as possible, straying into specific content domains (e.g., visual perception or speech production). This will be more difficult to pull off, however, in this final section, because it covers aspects of our science that are difficult to discuss without reference to the domains of cognition with which they engage. Beyond the specific examples used here to introduce them, these are approaches and ideas that have become increasingly prominent and influential during the past several years, and that are playing important roles in shaping the direction of contemporary cognitive neuroscience.

Neural network modeling

Neural network models are computer programs composed of multiple simulated processing units (sometimes referred to as "nodes") that are interconnected so that they can pass signals from one to another. As the name implies, these networks take inspiration from what is known about operating principles of networks of neurons in the brain. In particular, individual units are built with response properties that mimic those of neurons (i.e., the functions that determine what will be the output of a unit given different levels of input), and the connections between units are designed to be able to change their weights in a manner that mimics rules of synaptic plasticity. The units in neural network models are organized into layers, the simplest being two-layer feedforward networks in which patterns of activation are presented to an input layer and units in the input layer then each send their level of activity to the units in the output layer with which they are connected. The "feedforward" designation means that activity can only flow in one direction; a unit in the output layer cannot send signals to the input layer. The network will typically carry out some kind of transformation of the original pattern, such that the pattern of activity that emerges in the output layer is different from the pattern of activity that had been fed to the input layer. Importantly, this transformation is construed as a model for "information processing" as it is carried out by the brain system(s) being modeled.

In addition to input and output layers, most neural network models have one or more "hidden layers," so-called because they neither receive input from the (simulated) outside world, nor do they send output to it. Rather, they are internal to the workings of the network, receiving input from an upstream layer and sending output to a downstream layer. Before this gets too abstract, let's illustrate some of these points with a specific example.

A PDP model of picture naming

The model illustrated in *Figure 3.21* was developed to assess a theory of how the brain processes knowledge about the visual world, as operationalized by seeing an object and generating the appropriate verbal label, or hearing a name and knowing what object is being referred to. It is an example of a class of neural network models called parallel distributed processing (PDP) models. Its units have nonlinear response properties, like neurons, in that their activation level is determined by a nonlinear function, such as a sigmoid. (That is, if the activation state of a unit is really low or really high, a small change in input won't make much difference; its midrange, in contrast, is highly sensitive to small changes in the input.) The architecture of this particular model (but not of all PDP models) differs from the *input→hidden→output* organization of the generic neural network model in that activity can propagate in either direction, so as to capture the fact that humans can process information about objects "in both directions." At a gist level, to simulate naming, a pattern of activity is input

to the visual layer, and that activity propagates to the hidden layer (called the "semantic layer" in this specific model); the resultant input triggers recurrent activity within the hidden layer, the resultant pattern of which then propagates to the verbal layer, and the resultant pattern of activity in the Names division of the verbal layer is interpreted as the network's attempt to name the object (as though it were speaking aloud). To simulate understanding a spoken word (i.e., "going in the other direction"), a pattern of activity is input to the verbal layer, activity then flows to the hidden layer, the pattern of activity resulting from recurrent activation in the hidden layer then propagates to the Perceptual, Functional, and Encyclopedic divisions of the verbal layer and to the visual layer, and the resultant patterns of activity in these divisions of the verbal layer and in the visual layer are interpreted as the network's attempt to "think about" the visual characteristics of the object whose name has just been spoken. A higher level of detail about the PDP model from Rogers et al. (2004) is provided in *Figure 3.21*.

FIGURE 3.21 Architecture of the model from Rogers et al. (2004). Source: Adapted from Rogers, Timothy T., Matthew A. Lambon Ralph, Peter Garrard, Sasha Bozeat, James L. McClelland, John R. Hodges, and Karalyn Patterson. 2004. "Structure and Deterioration of Semantic Memory: A Neuropsychological and Computational Investigation." Psychological Review 111 (1): 205–235. doi: 10.1037/0033-295X.111.1.205.

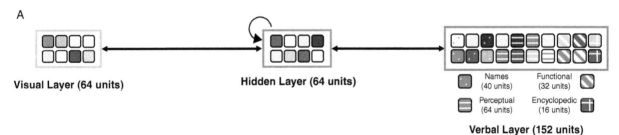

FIGURE 3.21.A 40 objects were selected as stimuli, each belonging to one of six semantic categories, three of these corresponding to living things (birds, mammals, and fruits) and three to nonliving things (vehicles, tools, household objects). The semantic properties of these stimulus items had been derived empirically by asking a group of human subjects to list attributes for each of them. Based on these "norms," the verbal layer was divided into four divisions, each corresponding to a category of verbal statements that can be made about an object: names, perceptual properties, functional properties, and encyclopedic properties. Within each division of the verbal layer, individual units corresponded to a feature that applied to one or more of the 48 stimulus items. Units in the Names division could refer to different levels of specificity (e.g., *animal*, *bird*, *swan*); example units from the Visual Properties division included *has eyes* and *has wheels*; from the Functional Properties division included *can fly* and *can roll*; and from the Encyclopedic Properties division included *lives in Africa* and *found in kitchen*. Within the visual layer's 64 units, each unit corresponded to a feature that had been identified in drawings that human subjects were asked to make of each of the stimulus items (e.g., *has stripes* or *is round*). In this way, the visual layer represented similarity structure: the 64-D visual representation of, for example, the stimulus items "sparrow" and "finch" would be more similar to each other in terms of shared units than to "eagle," but the visual representations of any pair from these three would be much more similar than would any one of them with "airplane."

B

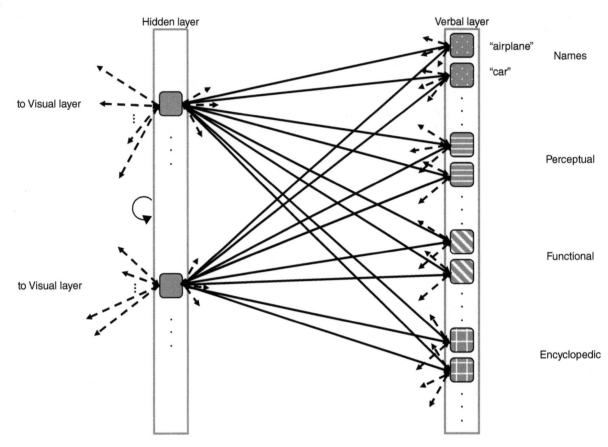

FIGURE 3.21.B Zoom-in on two units in each division of the verbal layer and two units in the hidden layer, illustrating the reciprocal connection of these units, the absence of connections between units in the verbal layer, and the interconnectivity of all units within the hidden layer (indicated by arrow "circling back on itself").

The motivation for the Rogers et al. (2004) model (*Figure 3.21*) was to demonstrate the plausibility of the idea that an information processing system like the brain can carry out simple tasks that we think of as requiring knowledge about the world (e.g., answering the question *Can an ostrich fly?*) without there being explicit representations of this knowledge stored in the system. (We'll get to the context for this model, and some alternative accounts, in *Chapter 13: Semantic Long-Term Memory*.) Importantly, it wouldn't be very compelling, nor very informative, if one were to simply build a network in which the Name unit for *ostrich* was deliberately wired up such that it would activate the Functional Properties units for *lives in Africa* and the Visual unit for *has feathers*. Such a model would be no different than a talking doll that one can buy at a toy store. What *would* be impressive, however, and potentially informative

for generating ideas about how the brain might solve the task of interest, would be if one could demonstrate that a network with the architecture described in *Figure 3.21*, starting out with no "knowledge" about what might differentiate an ostrich from an airplane, or whether or not most birds can fly, could *learn* these facts, such that, with enough training, it could perform at near-human levels of accuracy. And it would be even more impressive if, once it had been so trained, it could perform at a similarly high level with new items that it had never seen during training.

To initiate a training trial that would simulate the verbal presentation of the name "ostrich," one would set the *ostrich* unit in the Names division of the verbal layer to a value of 1, all the other units in the Names division of the verbal layer to a value of 0, and then observe the resultant patterns of activation taken on by all the remaining units in the network. (Each unit can vary in value between 0 and 1, and the resting level of activity of any unit in this model was approximately .19). The next step would be to compare the final activation level in every unit in the verbal and visual layers vs. the ideal activation level that should have been produced in order for a complete representation of "ostrich" to have been activated, and to register the error (i.e., the difference between the ideal activation level and the actual activation level on that trial). (The ideal activation level for "ostrich," for example, would have values of 1 for the nodes *animal*, *bird*, *has feathers*, *can walk*, etc., and values of 0 for such nodes as *has wheels*, *can fly*, *can roll*, and *found in kitchen*.) For each unit, the error value would be used to calculate how the weight of each connection to that unit would need to be changed in order to decrease the magnitude of the error at that unit the next time "ostrich" was presented. The final step in this trial, then, would be to deliver the "teaching signal" that actually implements these adjustments of the weights of connections within the model.

To illustrate this process at a very concrete level, let's walk through this trial again, but now starting from the perspective of the *lives in Africa* unit in the Encyclopedic properties division of the verbal layer. Because the model starts out with randomized connection weights, the activation level of this unit will have only changed slightly, and in an arbitrary way, in response to the presentation of "ostrich." But let's say that on this trial its activation level increased slightly. Important for understanding the process of training the model is to realize that, although the *ostrich* unit and the *lives in Africa* unit are not connected directly, they do connect indirectly through 64 paths, because each connects directly to every unit in the hidden layer. Because the level of activity of each unit in the hidden layer will also have changed in response to the presentation of "ostrich," the teaching signal would tend to strengthen the hidden-to-*lives in Africa* connections for the hidden units whose activation also increased. The consequence of these adjustments would be that the next time "ostrich" was spoken, the resultant activation in these critical hidden units would have a stronger influence on the activation of the *lives in Africa* unit. Taking such a restricted view of the dynamics of the model, however, misses key principles of how it works. What if one were to also consider, for example, the *ostrich*-to-hidden connections? Although at first one might assume that the connections to all hidden units that increased in activation would also be strengthened, this fails to take into account that each of those hidden units is also connected to EVERY other node in the

Encyclopedic, the Perceptual, the Functional, and the Visual layers. Thus, whether any individual *ostrich*-to-hidden connection gets strengthened or weakened will also depend on how that hidden unit's response to "ostrich" relates to the concurrent change in the activation of each of the other 215 units in those layers with which it is connected (e.g., of the *found in kitchen* unit, of the *has wheels* unit, of the *can walk* unit, of the *meows* unit . . .). The software that's implementing the training of the network applies a gradient descent algorithm to modify each of the $[64^2 + (216 * 64) =] 17,920$ connections involved on that trial such that, the next time "ostrich" is presented to the network, the overall error from the response of these 216 units will be smaller. (In relation to an actual human brain, which has trillions of synaptic connections, the total of 17,920 connections in this model is quite modest. Nonetheless one can readily see how, even in such a small network, we have to grapple with the complexity of the kind invoked by Buszsáki (2006) in *Chapter 2*.)

Zooming back out to the overall process of carrying out this experiment, the training of the model would proceed with the serial presentation to the model of each of the remaining patterns corresponding to a feature represented in the verbal and the visual layers, adjusting the network's weights after each presentation. Once the full set of features had been presented, the process would start all over again, this time with a different order of presentation. Typically, a PDP model like this will be trained with hundreds or thousands of full cycles of training. Sometimes the entire process of conducting thousands of training cycles will, itself, be carried out several times, each time starting with a new random set of initial connection weights.

How is it determined if a PDP modeling experiment has been successful? The answer will differ depending on the context, but let's consider this model from Rogers et al. (2004). As indicated earlier, the researchers' motivation for carrying out this work was to test the plausibility of a broader theory that posits, in part, that knowledge about objects may not be stored in the brain as discrete, fully encapsulated representations of, say, "ostrich-ness" or "airplane-ness," but, rather, as "nothing more" than sets of connections that link together all of an object's attributes. Thus, the fact that they could successfully train their model to simulate successful performance on a large number of cognitive tasks that depend on object knowledge (e.g., picture naming, drawing, word-to-picture matching) was taken as evidence for the plausibility of their theory. At a more rigorous level of assessment, one important characteristic of a successful model can be its ability to generalize to new stimuli. That is, one might argue that this model's performance wouldn't have strong implications for theories of semantics if, after all those thousands of training iterations, it was only able to demonstrate a high level of performance with those same 40 items with which they had been trained. More impressive would be if one could feed the fully trained network a stimulus that it had never seen before, and it would be able to correctly answer questions about it. For example, if one were to present a set of encyclopedic, functional, and perceptual features that the network had never encountered on a single trial (corresponding to an object that it had never encountered before, like a turkey, or a helicopter), would it generate a reasonable visual representation? A second criterion that is often applied to neural network models, including PDP models, is if their behavior matches human behavior in principled ways. In *Chapter 13*, for example, we'll see how selectively "lesioning" elements in models similar to that from Rogers et al. (2004) can produce deficits in performance that resemble the deficits observed in human patients with particular patterns of brain damage.

Deep neural networks

The "deep" in deep neural networks (DNNs) refers to the fact that these networks employ more than one hidden layer. Unlike the PDP model that we considered in the previous section, DNNs invariably have an explicit directionality, or hierarchy, with an input layer feeding into early processing layers, followed by intermediate layers, and so on. Although DNNs have applications outside of computational neuroscience, the majority of research with DNNs that will be relevant for our purposes has been on the analysis of sensory signals, including visual image classification and auditory speech perception. Providing more specific information about how DNNs work would be difficult to do without first introducing many concepts about neural sensory processing, so we'll hold off on getting into details until we get to *Chapter 6, Using deep neural networks (DNNs) to understand computations underlying object recognition; Figure 6.11*. For now, we'll just consider one example of the kind of problem that DNNs are good at solving.

One highly influential example of "deep learning" with a DNN was published in 2012 by Krizhevsky, Sutskever, and Hinton, then all at the University of Toronto, who trained of a deep convolutional neural network (DCNN) on 1.2 million photographic images taken from the internet (the ImageNet database), each of which had been labeled as belonging to one of 1000 different categories by

human labelers. Next, they presented to the trained network 150,000 images that it had never previously processed. Performance on these 150,000 was on the order of 85% correct, which was record setting at the time. (The performance of DNNs has gotten much better since then.) Important points to emphasize here are that the data fed to the input layer of the DCNN were not "hand coded" with predetermined labels like "has wings" or "can roll," as had been the case in the PDP model that we considered in the previous section. Rather, the input layer of the Krizhevsky et al. (2012) DCNN was a 256 × 256 array of sensors, each of which was fed the color and brightness value of the corresponding pixel from the image. In *Chapter 6*, we will go into detail about how DNNs have been used to study the computations underlying primate visual perception. Thus, one generalization that can often (thought not always) be true is that while PDP models test ideas at a more abstract or conceptual level, DNNs can be used as models of neural computations at a circuit level of analysis.

Network science and graph theory

Largely spurred by the discovery of the default mode network (DMN) and resting state functional correlations (RSFCs), for the past decade neuroscientists have been analyzing brain networks with tools originally developed in other branches of science, including mathematics, sociology, and statistics. To get a quick sense for the relevance of these other branches of science, consider the following overview from cognitive neuroscientist Gagan Wig (2017):

> Research on complex networks has revealed that many social, biological, and man-made networks exhibit organizational properties that support efficient design and resilience to perturbation. . . For example, networks often contain distinct and segregated subnetworks, and effective network function is supported by maintaining subnetwork segregation while simultaneously allowing integration across the subnetworks. Excess in either direction can be harmful for a network. Too much interconnectedness between subnetworks can lead to rapid spreading of disease, and has rendered financial networks vulnerable to sudden crisis. Conversely, too much segregation between subnetworks can result in their diminished interaction or isolation, as has been argued in reference to the increasing polarization of the network of US political interactions. (p. 981)

Many of the measures and phenomena that are commonly encountered in network neuroscience are illustrated in *Figure 3.22*.

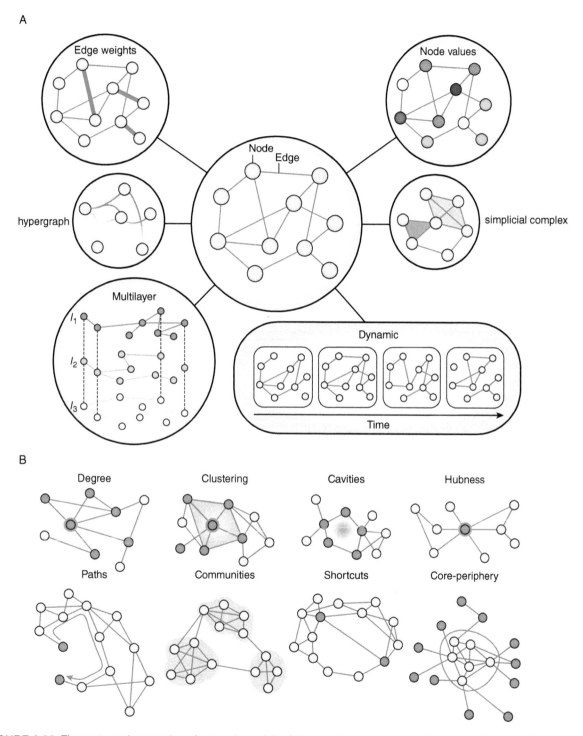

FIGURE 3.22 Elements and properties of network models. **A** Brain networks are typically represented as patterns of connections (*edges*) between neural units (*nodes*). Depending on the scale of the model, a node may represent a single neuron, a nucleus, or an entire brain region. Edges can have different weights, and nodes can have different functional

(Continued on page 88)

Tools from graph theory can be applied to RSFC data, as well as to functional connectivity data. With RSFC it has been established that healthy young adult brain networks exhibit fairly segregated communities characterized by dense interactions within each community and sparser interactions between communities. With normal aging (i.e., absence of any age-related neurological disease), community segregation weakens, with relative decreases of interactions within and increases of interactions between communities. (Before the introduction of graph theory to the analysis of neuroimaging data, a similar phenomenon had already been described as a process of "dedifferentiation.") Furthermore, independent of age, individual differences in community segregation among the networks classified as "Association systems" in *Figure 3.18.C* predict cognitive ability, with greater segregation associated with superior episodic long-term memory (to be covered in *Chapters 11* and *12*) and fluid intelligence (*Chapter 15*). Higher community segregation is also associated with better outcomes in rehabilitation after brain injury.

The examples given in the preceding paragraph treat network configuration as a trait — a set of stable properties that are characteristic of a person's brain for an extended period of time. One can also employ methods from "dynamic network neuroscience" to measure the changes in network structure that occur in real time during cognitive task performance. For example, one analysis of the data from 344 healthy individuals performing a working-memory task found that the degree of reconfiguration in networks involving the frontal cortex (frontal flexibility) was positively related to performance on the task. Measures included the probability that individual nodes changed allegiance between networks such as those illustrated in *Figure 3.18.C* during the task.

END-OF-CHAPTER QUESTIONS

1. Why is the question of specificity so crucial for interpreting the results of a lesion study? Try to think of an example of an experimental finding for which a lack of evidence for specificity might weaken its interpretation. How might this weakness be addressed, perhaps by administering additional behavioral tasks, or by acquiring data from one or more subjects with damage to a different area of the brain?

2. For what kind of research question might tDCS be a better experimental tool than TMS? For what kind of research question might the opposite be true?

3. How can one distinguish MUA from LFP in an extracellular recording? What phenomenon, at the level of the activity of one or a group of neurons, is each of these believed to be measuring?

4. Describe the measure of PAC. Why is it of interest to cognitive neuroscientists?

5. Name a brain process that the ERP analysis technique is better able to detect and measure than a spectral analysis technique. Name a brain process for which the opposite is true.

6. What is an assumption made by the Fourier transform that is often not true of the data to which it is applied?

7. What does fMRI measure? What are the processes, at the cellular level, that can produce fluctuations in the fMRI signal?

8. What constraints do the low-pass filtering qualities of the fMRI signal impose on experimental design and analysis?

9. How do PDP models differ from DNN models? What kinds of questions are each well suited to address?

FIGURE 3.22 (*Continued*) properties. *Multilayer networks* can be used to represent interconnected sets of networks, and studying the *dynamic* properties of networks helps to understand the reconfiguration of network systems over time. A *hypergraph* is a graph in which a single edge can connect more than two nodes. A *simplicial complex* is a graph in which higher-order interaction terms become nodes. **B** Common measures of interest include *degree*, the number of edges emanating from a node; *clustering*, which is related to the prevalence of triangles; *cavities*, portions of a graph where there are no edges; *hubness*, a node's influence on other nodes; *paths*, which determine the potential for information transmission; *communities*, local groups of densely interconnected nodes; *shortcuts*, a possible marker of global efficiency of information transmission; and *core-periphery structure*, which facilitates local integration of information gathered from or sent to more sparsely connected areas.

REFERENCES

Baldo, Juliana V. and Dronkers, Nina F. 2006. "The role of inferior parietal and inferior frontal cortex in working memory." *Neuropsychology* 20: 529–538.

Bassett, D. S., P. Zurn, and J. I Gold. 2018. "On the Nature and Use of Models in Network Neuroscience." *Nature Reviews Neuroscience* 19: 566–578.

Benson, D. Frank, and John P. Greenberg. 1969. "Visual Form Agnosia: A Specific Defect in Visual Discrimination." *Archives of Neurology* 20 (1): 82–89. doi: 10.1001/archneur.1969.00480070092010.

Biswal, B., F. Z. Yetkin, V. M. Haughton, and J. S. Hyde .1995. "Functional Connectivity in the Motor Cortex of Resting Human Brain Using Echo-planar MRI." *Magnetic Resonance in Medicine* 34: 537–541.

Bollimunta, Anil, Yonghong Chen, Charles E. Schroeder, and Mingzhou Ding. 2008. "Neuronal Mechanisms of Cortical Alpha Oscillations in Awake-Behaving Macaques." *Journal of Neuroscience* 28 (40): 9976–9988. doi: 10.1523/JNEUROSCI.2699-08.2008.

Buckner, R. L., and Dale, A.M. 1997. "Selective Averaging of Rapidly Presented Individual Trials Using fMRI." *Human Brain Mapping* 5: 329–340. doi:10.1002/(SICI)1097-0193(1997)5:5<329::AID-HBM1>3.0.CO;2-5

Burock, M., R. L. Buckner, M. G. Woldorff, B. R. Rosen, and A M. Dale. 1998. "Randomized Event-Related Experimental Designs Allow for Extremely Rapic Presentation Rates Using Functional MRI." *NeuroReport* 9: 3735–3739.

Han, L., N. K. Savalia, M. Y. Chan, P. F. Agres, A. S. Nair, and G. S. Wig. 2018. "Functional Parcellation of the Cerebral Cortex across the Human Adult Lifespan." *Cerebral Cortex* 28: 4403–4423.

Hubel, David H. 1988. *Eye, Brain, and Vision*. New York: Scientific American Library: Distributed by W. H. Freeman.

Jackson, N., S. R. Cole, B. Voytek, and N. C. Swann. 2019. "Characteristics of Waveform Shape in Parkinson's Disease Detected with Scalp Electroencephalography. eNeuro 6. doi.org/10.1523/ENEURO.0151-19.2019.

Krizhevsky, A., I. Sutskever, and G. E. Hinton. 2012. "ImageNet classification with Deep Convolutional Neural Nets." *Advances in Neural Information Processing Systems*, 1: 1097–1105.

Luck, Steven J. 2005. *An Introduction to the Event-related Potential Technique*. Cambridge, MA: MIT Press.

Malmivuo, Jaakko, and Robert Plonsey. 1995. *Bioelectromagnetism: Principles and Applications of Bioelectric and Biomagnetic Fields*. New York: Oxford University Press.

Menon, Ravi S., and Seong-Gi Kim. 1999. "Spatial and Temporal Limits in Cognitive Neuroimaging with fMRI." *Trends in Cognitive Sciences* 3 (6): 207–215. doi: 10.1016/S1364-6613(99)01329-7.

Petersen, Steven E., Peter T. Fox, Michael I. Posner, Mark Mintun, and Marcus E. Raichle. 1988. "Positron Emission Tomographic Studies of the Cortical Anatomy of Single-Word Processing." *Nature* 331 (6157): 585–589. doi: 10.1038/331585a0.

Petrof, Iraklis, Angela N. Viaene, and S. Murray Sherman. 2012. "Two Populations of Corticothalamic and Interareal Cortico-cortical Cells in the Subgranular Layers of the Mouse Primary Sensory Corticies." *Journal of Comparative Neurology: Research in Systems Neuroscience* 520(8): 1678–1686. doi: 10.1002/cne.23006.

Poldrack, R. A. 2006. "Can Cognitive Processes Be Inferred from Neuroimaging Data?" *Trends in Cognitive Sciences* 10: 59–63.

Raichle, Marcus E. 2010. "Two Views of Brain Function." *Trends in Cognitive Neuroscience* 14 (4): 180–190. doi: 10.1016/j.tics.2010.01.008.

Raichle, M. E., A. M. MacLeod, A. Z. Snyder, W. J. Powers, D. A. Gusnard, and G. L. Shulman. 2001. "A Default Mode of Brain Function." *Proceedings of the National Academy of Science (USA)* 98: 676–682.

Rogers, Timothy T., John R. Hodges, Matthew A. Lambon Ralph, and Karalyn Patterson. 2003. "Object Recognition under Semantic Impairment: The Effects of Conceptual Regularities on Perceptual Decisions." *Language and Cognitive Processes* 18 (5–6): 625–662. doi: 10.1080/01690960344000053.

Rogers, Timothy T., Matthew A. Lambon Ralph, Peter Garrard, Sasha Bozeat, James L. McClelland, John R. Hodges, and Karalyn Patterson. 2004. "Structure and Deterioration of Semantic Memory: A Neuropsychological and Computational Investigation." *Psychological Review* 111 (1): 205–235. doi: 10.1037/0033-295X.111.1.205.

Rossi, Simone, Mark Hallett, Paolo M. Rossini, Alvaro Pascual-Leone, and The Safety of TMS Consensus Group. 2009. "Safety, Ethical Considerations, and Application Guidelines for the Use of Transcranial Magnetic Stimulation in Clinical Practice and Research." *Clinical Neurophysiology* 120 (12): 2008–2039.

Shulman, G. L., J. A. Fiez, M. Corbetta, R. L. Buckner, F. M. Miezin, M. E. Raichle, and S. E. Petersen. 1997. "Common Blood Flow Changes across Visual Tasks: II. Decreases in Cerebral Cortex." *Journal of Cognitive Neuroscience* 9: 648–663.

Thirugnanasambandam, N., R. Sparing, M. Dafotakis, I. G. Meister, W. Paulus, M. A. Nitsche, and G. R. Fink. 2011. "Isometric Contraction Interferes with Transcranial Direct Current Stimulation (tDCS) Induced Plasticity – Evidence of State-Dependent Neuromodulation in Human Motor Cortex." *Restorative Neurology and Neuroscience* 29: 311–320.

Tootell, Roger B. H., Martin S. Silverman, Eugene Switkes, and Russell L. De Valois. 1982. "Deoxyglucose Analysis of Retinotopic Organization in Primate Striate Cortex." *Science* 218 (4575): 902–904. doi: 10.1126/science.7134981.

Voytek, B., R. T. Canolty, A. Shestyuk, N. E. Crone, J. Parvizi, and R. T. Knight. 2010. "Shifts in Gamma Phase-Amplitude

Coupling Frequency from Theta to Alpha over Posterior Cortex during Visual Tasks. *Frontiers in Human Neurosciences* 4:1–9.

Walsh, Vincent, and Alan Cowey. 2000. "Transcranial Magnetic Stimulation and Cognitive Neuroscience." *Nature Reviews Neuroscience* 1 (1): 73–80.

Wandell, Brian A., and Jonathan Winawer. 2011. "Imaging Retinotopic Maps in the Human Brain." *Vision Research* 51 (7): 718–737. doi: 10.1016/j.visres.2010.08.004.

Wang, S., S. Itthipuripat, and X. Ku .2019. "Electrical Stimulation over Human Posterior Parietal Cortex Selectively Enhances the Capacity of Visual Short-Term Memory." *The Journal of Neuroscience* 39: 528–536.

Wig, G. 2017. "Segregated Systems of Human Brain Networks." *Trends in Cognitive Sciences* 21: 981–996.

OTHER SOURCES USED

Aguirre, Geoffrey K., Eric Zarahn, and Mark D'Esposito. 1998. "The Variability of Human, BOLD Hemodynamic Responses." *NeuroImage* 8 (4): 360–369. doi: 10.1006/nimg.1998.0369.

Bassett, D. S. and Sporns, O. 2017. "Network Neuroscience." *Nature Neuroscience* 20: 353–364.

Braun, U., A. Schäfer, H. Walter, S. Erk, N. Romanczuk-Seiferth, L. Haddad, J. I. Schweiger, and D. S. Bassett. 2015. "Dynamic Reconfiguration of Frontal Brain Networks during Executive Cognition in Humans." *Proceedings of the National Academy of Sciences, USA* 112: 11678–11683.

Guller, Y., F. Ferrarelli, A. J. Shackman, S. Sarasso, M. J. Peterson, F. J. Langheim, . . . B. R. Postle. 2012. "Probing Thalamic Integrity in Schizophrenia Using Concurrent Transcranial Magnetic Stimulation and Functional Magnetic Resonance Imaging." *Archives of General Psychiatry* 69: 662–671.

Haxby, James V., Cheryl L. Grady, Barry Horwitz, Leslie G. Ungerleider, Mortimer Mishkin, Richard E. Carson, . . . Stanley I. Rapoport. 1991. "Dissociation of Object and Spatial Visual Processing Pathways in Human Extrastriate Cortex." *Proceedings of the National Academy of Sciences USA* 88 (5): 1621–1625. doi: 10.1073/pnas.88.5.1621.

Oztas, Emin. 2003. "Neuronal Tracing." *Neuroanatomy* 2: 2–5. http://www.neuroanatomy.org/2003/002_005.pdf.

Polanía, R., M. A. Nitsche, and C. C. Ruff. 2018. "Studying and Modifying Brain Function with Non-invasive Brain Stimulation." *Nature Neuroscience* 21: 174–187.

Reinhart, R. M., J. D. Cosman, K. Fukuda, and G. F. Woodman. 2017. "Using Transcranial Direct-Current Stimulation (tDCS) to Understand Cognitiver Processing." *Attention, Perception, and Psychophysics* 79: 3–23.

Snyder, A. Z., and M. E. Raichle. 2012. "A Brief History of the Resting State: The Washington University Perspective." *Neuroimage* 62: 902–910.

Villringer, Arno. 2000. "Physiological Changes during Brain Activation." In *Functional MRI*, edited by C. T. W. Moonen and Peter A. Bandettini. Heidelberg: Springer-Verlag.

Woods, A. J., A. Antal, M. Bikson, P. S Boggio, A. R. Brunoni, P. Celnik, . . . M. A. Nitsche. 2016. "A Technical Guide to tDCS, and Related Non-invasive Brain Stimulation Tools." *Clinical Neurophysiology* 127: 1031–1048.

FURTHER READING

Aguirre, G. K., and M. D'Esposito. 1999. "Experimental Design for Brain fMRI." In *Functional MRI*, edited by C. T. W. Moonen and P. A. Bandettini, 369–380. Berlin: Springer-Verlag.
A comprehensive and accessible treatment of how and why the BOLD fMRI signal imposes constraints on experimental design, and of various design strategies that take these constraints into account.

Oztas, Emin. 2003. "Neuronal Tracing." *Neuroanatomy* 2: 2–5. http://www.neuroanatomy.org/2003/002_005.pdf.
A very concise overview of some of the physiology, and biochemistry, underlying anatomical tracing techniques.

Wandell, Brian A., and Jonathan Winawer. 2011. "Imaging Retinotopic Maps in the Human Brain." *Vision Research* 51(7): 718–737. doi: 10.1016/j.visres.2010.08.004.
Nice treatment of the use of MRI for tractography. Additionally, some of the content of this paper, particularly in the section on "Functional specialization and maps," will be relevant in subsequent chapters, particularly Chapter 9.

SECTION II
SENSATION, PERCEPTION, ATTENTION, AND ACTION

INTRODUCTION TO SECTION II
SENSATION, PERCEPTION, ATTENTION, AND ACTION

Near the end of *Chapter 1* we discussed the fact that brains confer the ability to detect changes in the environment and to perform actions based on this information. The chapters in this section will delve into the wondrous amount of intricate detail and organization that underlies this simple premise. As a preamble, let's briefly consider the four words in the title of this section, the concepts that they encompass, and how these classes of function relate to one another.

If we think of the functions of the brain as lying along a detection-to-action continuum, sensation is the term given to the neural processes that correspond most closely to the concept of detection. Under this rubric falls the translation of information from the environment into a neural signal that includes a representation of the physical properties of the environmental information (e.g., brightness for vision, loudness and pitch for audition, force and texture for touch, and concentration for olfaction). Perception refers to the identification of features of what is being sensed, and, often, to its recognition (e.g., *Am I looking at a house cat or a raccoon? Is the voice on the telephone that of my daughter or of my wife? Does the blade of this ice skate feel sufficiently sharp as I run my finger along it? Does this milk smell as though it has gone sour, or can I drink it?*). Attention can refer to many things, but in the present context it's how we mentally prioritize the perception (and related actions) of just one or a few of the multitudes of stimuli that typically clutter our environment. In *Chapter 2*, oscillatory synchrony was introduced as a mechanism for focusing one's attention on an approaching bear rather than on sunlight or on a bird. Action can refer to isolated acts of motor control (e.g., reaching for the computer mouse while surfing the Web, or turning one's head to see who is walking into the room), but also to goals or plans that can be abstracted from isolated movements, such as *making dinner* or *finishing my homework*.

The definition of sensation has raised a concept that will be of central importance to all of the chapters of this section, and that's the idea of translating information into signals that the nervous system can understand and manipulate. This is fundamental, because when you see a cat, for example, it's not the case that the photons reflecting off this animal are traveling into your brain to be projected against a "neural screen." Nor is it the case that when you smell bread baking in the oven, molecules of heated dough are entering your brain through your nose. Rather, the nervous system represents information about the seen cat and about the smelled bread with what we call a neural code.

A code is a system for representing information in a format that's different from the physical properties of what's being represented. The English language is a code by which, for example, this author can convey to you, the reader, the idea of a cat without having to plop a furry, domesticated feline into your lap. Similarly, the nervous system uses a code made up of action potentials and neurotransmitters to convey information about patterns of light or the strength and quality of a smell. (It also uses a code to generate the mental image of a cat, like the one you may have just experienced a few moments ago, but we won't get to that until *Section III*.) It uses a "neural code" to trigger the click of your finger on the computer mouse, and the speed and manner (quickly and with alarm, or slowly and with insouciance) with which you turn your head toward the door. There is much about neural coding that we do not yet understand, and discovering the rules that govern them is an important goal of cognitive neuroscience. (By analogy, without knowledge of the rules governing the English language, one wouldn't be able to understand the difference between the sentences "Dog bites man" and "Man bites dog.") At the front end of "translating information into a neural code" is the translating part, that is, the transduction of physical energy into the neural code. For vision, we'll consider how the retina transduces light of different wavelengths; for audition, how the cochlea transduces fluctuations in air pressure; for somatosensation, how receptors in the skin transduce different mechanical forces, chemical agents, and temperatures that the body encounters.

It's a matter of organizational convenience that this section presents the topics of *sensation*, *perception*, *attention*, and *action* in the order that it does. However, an important fact to keep in mind as one reads through these chapters is that it would be overly simplistic, to the point of being inaccurate, to think of neural information processing as proceeding in a strictly feedforward series of stages from sensation to perception to attentional selection, to action. We saw in *Chapter 2* that the cortex is wired up for the feedback as well as for the feedforward conveyance of signals, and in *Chapter 4* we'll begin to see some of the functional consequences of this. A related point is that it would be equally inaccurate to think of the brain as "sitting passively" until a stimulus comes along to "activate it," to kickstart it into action. Instead, we'll see that a better model might be to think of a brain that is constantly "churning" along, actively representing the world, and that the sensation of a newly presented stimulus corresponds to the "nudging" of this already-active system into a different state of activity. That is, the brain is a dynamic system, and the waking brain is never truly "at rest." I raise these points now in part because, even though I believe them myself, I often find myself slipping into a more straightforward, intuitively appealing way of thinking about brain function. By this intuitive account, a newly occurring stimulus evokes activity in a previously quiescent brain, and this event triggers the sensation-through-action cascade that has been previewed in this preamble. I also raise these points to be explicit about the fact that this "feedforward activation" perspective, by virtue of its intuitive appeal, offers a more straightforward way of introducing many aspects of brain function. This is evident in much of the terminology that we use to describe the brain and its functions, in figures that appear in many textbooks, and, indeed, in the order of chapters in textbooks such as this one. What we shall see in this section, however, is that a feedforward-activation view of brain function can limit one's ability to appreciate all its nuances. To give just one example, what one perceives can often be influenced, if not determined, by one's action plan.

But that's enough "pre"; let's dive in!

CHAPTER 4
SENSATION AND PERCEPTION OF VISUAL SIGNALS

KEY THEMES

- The brain's representation of the visual world is dictated by the optics of the eye and, in particular, where light from the visual scene falls on the retina.

- The topography of the retina is reflected in the retinotopic organization of primary visual cortex.

- The disproportionate retinal representation of central (i.e., foveal) vision results in its cortical magnification in terms of the surface area of V1.

- A central principle for understanding the response properties of retinal photoreceptors, retinal ganglion cells, neurons of the lateral geniculate nucleus (LGN) of the thalamus, and neurons of primary visual cortex (area V1) is that of the receptive field.

- Individual neurons of V1 assemble input from multiple LGN neurons to construct specific features to which they respond preferentially, and thereby act as feature detectors.

- At least five classes of features emerge at the level of V1: edge orientation; direction of movement; binocular disparity; end-stopping; and the representation of color; as well as an increase in acuity.

- Functional tissue imaging, which permits visualization of a large surface of cortex, has revealed that columns representing smooth transitions through the full 180° of orientation are organized in pinwheel-like configurations.

- Visual information processing does not consist of simply a serial feeding forward of signals from retina to LGN to V1, because V1 sends heavy feedback to the LGN, as do higher level visual areas to V1.

- Functions of feedback include biasing the LGN to emphasize the orientation vs. the direction-of-movement of visual stimuli, and determining how effective signals from LGN will be at driving cortical responses.

- At a broader spatial scale, the processing of incoming visual information is influenced by the magnitude of ongoing oscillations in the alpha frequency band (~8–13 Hz) of the EEG.

- A principle underlying the overall functioning of the visual system may be one of predictive coding – the construction and continual updating of an internal model of the current state of the external world and of what is likely to happen next.

CONTENTS

THE DOMINANT SENSE IN PRIMATES

Of all our senses, one could make a strong case that vision is singularly important for successfully navigating through and interacting with our environment. (As a thought experiment, imagine being dropped off in an unfamiliar city where an unfamiliar language is spoken and written, with no one to guide you, while [a] blindfolded, [b] "earplugged," [c] "noseplugged," etc. Which would make it most difficult for you to find your way to a friend's address?) It is also probably the case that vision is the sense that has been most studied from each of the three traditions that have informed contemporary cognitive neuroscience most heavily: systems neuroscience; neuropsychology; and experimental psychology. As discussed in *Chapter 1*, visual perception is a function characterized by close linkage between the environment and the brain. Indeed, a useful first step in considering the cognitive neuroscience of vision is to learn the conventions of how scientists describe this interface.

ORGANIZATION OF THE VISUAL SYSTEM

The visual field

The **visual field** refers to the spatial extent of everything that you can see right now. While you're reading this page, it's likely that your visual field is largely filled by this page, with blurry images of what's behind the book (or screen) surrounding the more in-focus percept of the page itself. Note that when you pay attention to the sharpness-of-focus of your vision for words that are one or two lines above or below where you're currently reading, it's clear that your acuity falls off precipitously in all directions. The center of gaze, where your acuity is sharpest, is referred to as foveal vision, because it corresponds to the part of your retina called the **fovea**. Now move your center of gaze over to the crease that's formed between the left and right page of the book (or to the edge of the screen that you're reading from – from here, for simplicity, we'll just assume a book). Assuming that you are sitting or standing upright, and the crease is vertical, then the crease is passing through your fovea and defining the boundary between the left side and the right side of your visual field, dividing it into a left

visual hemifield and a right visual hemifield. Note that this remains true regardless of the position of your eyes in your head. To demonstrate, start by holding your book in front of you and fixating the central crease (to fixate is to hold one's center of gaze on a selected location, often on an object that is stationary at that location). Now, while holding the book still and holding your head still, move your eyes as far to the left as you can, and hold them there, your center of gaze falling on an object to the left of the book. The book is still in your field of view, but it now occupies the right visual hemifield. Conversely, if you keep your center-of-gaze fixated on one point while slowly moving your head back and forth, or up and down, your visual field remains stationary. Thus, your visual field is centered on the location of gaze your eyes, not on the orientation of your head.

The retina and the LGN of the thalamus

Phototransduction, the process of converting light into neural signals, is accomplished by a class of cells called photoreceptors. In a sentence, these cells contain a pigment called rhodopsin that, when impacted by light, undergoes a conformational change that triggers an intracellular molecular cascade that results in the closing of membrane-spanning ion channels. Thus, a type of physical energy (electromagnetic radiation with wavelengths between 400 and 700 nm [i.e., "visible light"]) is converted into an electrical signal; in this instance, a change in the membrane potential of the photoreceptor cell. This change in membrane potential produces currents within the complicated cellular circuitry of the retina, the end result of which is that action potentials are generated by retinal ganglion cells, the output neurons of the retina whose axons project to the LGN (Figure 4.1).

There are two classes of photoreceptors, cones and rods. Cones are sensitive to color, with the rhodopsin in "blue," "green," and "red" cones being preferentially sensitive to light at short, medium, and long wavelengths, respectively. Rods, on the other hand, are sensitive to this entire range of wavelengths (i.e., they are "colorblind") and contribute primarily to vision at extremely low levels of illumination, such as at night.

Although containing an array of detector elements, not unlike the grid of photodiodes in a digital camera, the packing density of photoreceptors in the retina is not uniform. Rather, at the fovea, one finds the highest density of cones, which drops off very quickly as one moves in any direction toward the periphery. For this reason, the fovea provides the highest acuity, the highest spatial resolution, of any part of the retina. This explains why, for example, we move our eyes across the page as we read: only the highest acuity portion of the retina can deliver the detail that is needed to resolve small type. The same is true for recognizing faces, for appreciating a painting, for reading road signs, and so on. As one moves away from the fovea toward the periphery, three properties begin to change. One is the progressive decrease in the density of cones. The second is that the surface area that each cone occupies on the retina increases. The third is that an increasing number of rods surround each cone. It is a consequence of the fact that there are fewer cones devoted to each square unit of surface in the periphery of the visual field that our peripheral vision is of much lower resolution than is foveal vision.

A reasonable supposition as to why the retina is organized this way comes from considering the incredible amount of processing power required to analyze information from the fovea (as will be detailed further along, beginning with the consideration of cortical magnification). If the packing density of the entire surface of the retinae were the same as that at the fovea, it's been estimated that one's optic nerves would need to be as thick as one's arms (to accommodate so many more axons), and one's primary visual cortex would need to be the size of a large room (to accommodate all the neurons needed to process so much high-resolution visual information).

Now move your eyes back to the vertical crease dividing the left vs. right pages of the book. The imaginary line that lies on top of the crease and extends to the uppermost and lowermost reaches of the visual field is called the **vertical meridian** of the visual field. In the same way that one can vertically divide the visual field into left vs. right hemifields, one can also divide it into upper vs. lower hemifields, a divide defined by the **horizontal meridian**. Together, the two meridians divide the visual field into quadrants: upper right, lower right, lower left, and upper left.

Before we start to consider how these components of the visual field are mapped onto the brain, let's try one more simple exercise. Close your left eye. Note that, with this eye closed, although your visual field is somewhat narrower than it is with both eyes open, you can still see much of the left visual hemifield. That is, *the two hemifields do not correspond to each of the two eyes*. Rather, each eye represents most of the visual field, with considerable overlap in the center, and only the most eccentric (i.e., farthest-from-the-center) portions of each hemifield are represented by

only one eye. The reason that one eye can represent most of the visual field follows from its structure. If one thinks of the eye as a transparent globe, the retina, which is the portion of the eye that transduces light energy into neural signals, is attached to the inside surface of the half of the globe that's inside the eye socket, as though it was painted on. That is, it is concave. Thus, as illustrated in *Figure 4.1*, light that is emanating from the left side of the visual field strikes the right side of the retina, and light that is emanating from the right side of the visual field strikes the left side of the retina. For there to be a "one eye = one visual hemifield" arrangement, one of two things would have to be different about our eyes. Either (1) the retinae would have to be flat, instead of concave; or (2) the eyes would have to be farther apart, halfway between their present location and the temples. Animals that are "wall eyed," like most fish and birds, have two nonoverlapping visual fields.

The retinotopic organization of primary visual cortex

We concluded *Chapter 1* with three branches of science – systems neuroscience; neuropsychology; and experimental psychology – poised to make important discoveries about brain structure and function at the dawn of the twentieth century. With regard to vision, once occipital cortex was defined as its necessary substrate, the next task was to work out its organization. The fact that there is systematicity to the organization of visual cortex was soon established by neuropsychologists and neurologists whose studies of soldiers with penetrating head injuries indicated that damage to specific areas of the occipital lobe invariably produced partial blindness in the same regions of the visual field. Thus, for example, damage to dorsal regions of the right occipital lobe produced blindness in the lower-left quadrant of the visual field. A large lesion would produce a quadrantanopia – blindness in the entirety of the quadrant – or a hemianopia – blindness in one half of the visual field – whereas a smaller lesion might produce a scotoma within that quadrant. An implication of this finding was the realization that everything in the primary visual cortex is "flipped" with respect to the visual field. Things that appear to the left of the vertical meridian are initially processed in the right hemisphere of the brain, and things that appear above the horizontal meridian are initially processed in ventral portions of the occipital lobe. To understand why, we must turn to the anatomy of the visual system.

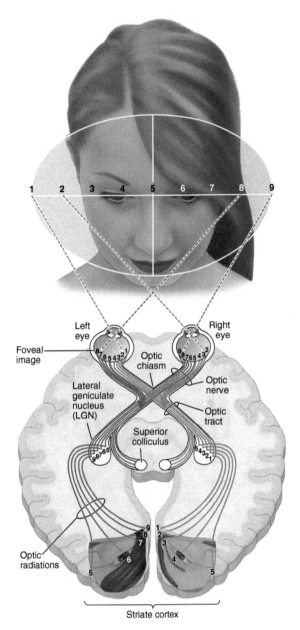

FIGURE 4.1 Illustration of the visual field, emphasizing the right (shaded red) and left (shaded blue) hemifields, and how information from these hemifields is conveyed to primary visual cortex (labeled as "striate cortex," for reasons explained in the text). Source: From Wolfe, Jeremy M., Keith R. Kluender, Dennis M. Levi, Linda M. Bartoshuk, Rachel S. Herz, Roberta Klatzky, . . . Daniel M. Merfeld. 2012. Sensation and Perception, 3rd ed. Sunderland, MA: Sinauer Associates. Copyright 2012. Reproduced with permission of Oxford University Press.

The left-right "flip" and the up-down "flip" can each be understood as resulting from the optics of the eye. For now, let's keep the left eye closed, and consider just the right. The concave shape of the retina on the back of the eyeball dictates that light reflecting off stimuli appearing below the point of fixation will be projected onto the upper retina, and that stimuli appearing to the right of fixation will be projected onto the left side of the retina. Because we're only considering the right eye for the moment, this left half of the retina is referred to as the nasal retina, because it is closest to the nose. (The half of the retina that is closest to the temple is the temporal retina.) And so, already in the eye, what's "up" in the world is "down" in the retina, and what's "left" in the world is "right" in the retina. This flipped state of affairs is preserved as axons from the retina project posteriorly to the thalamus. Along the way, the fibers that correspond to the nasal retina cross over to the left hemisphere, so that half of the axons from the right eye – those from the temporal retina – project to the right LGN of the thalamus, and the half from the nasal retina project to the left LGN (*Figure 4.1*). Another important fact about these axons is that their trajectory preserves the layout of the retina, such that axons from adjacent retinal ganglion cells synapse onto adjacent neurons of the LGN. Similarly, adjacent neurons of the LGN project to adjacent neurons in primary visual cortex (a.k.a. *V1*). In this way, the spatial organization of V1 reflects the spatial organization of the retina (*Figure 4.1*). If you're looking at the trunk of a tree, green light from the leaves is reflecting onto the lower half of the retina, and the dirt where the trunk meets the ground is reflecting onto the upper half of the retina. When V1 receives this information via the LGN, ventral portions of this region begin processing information about the color, texture, etc. of the leaves, whereas dorsal portions of this region begin processing information about the dirt. Thus, the functional topography of V1 mirrors the functional topography of the retina. This is referred to as **retinotopic organization**, or *retinotopy*.

Inspection of *Figure 4.1* (and of *Figure 3.14*) identifies the fiber pathways that convey information from the retina to the cortex. Although all axons project uninterruptedly from retina to LGN, their bundles from eye to optic chiasm are referred to as the *optic nerve*, whereas downstream of this decussation (crossing over) it's the *optic tract*. Thus, the optic nerves carry eye-specific information, whereas the optic tracts carry visual field-specific information.

The LGN itself gets its name from the fact that an early anatomist thought it looked like a bent knee, and geniculate derives from Latin for "knee-like." The fibers of the optic tract terminate in one of several layers of the LGN, according to a principle that we'll return to further along this chapter. In the human, the projection from LGN to V1 runs along the dramatically curving and fanned-out optic radiations, a striking consequence of the mechanical twisting and stretching of this fiber tract brought about by the rapid growth of the brain during in utero development (*Figure 3.14*). Note that these radiations terminate along the medial surface of the occipital lobe. More specifically, the majority of them terminate inside a sulcus that runs parallel to the rostrocaudal axis of the brain, starting near the occipital pole and extending for roughly the length of the occipital lobe. This is the calcarine fissure, named after the shape that its indentation makes in the adjacent posterior horn of the lateral ventricle when viewed from a dissection similar to *Figure 2.2*. This indentation resembles the spur of a rooster's leg; in Latin, *calcar avis*.

The axial view depicted in the MR image in *Figure 3.14* is not dissimilar to the one that Italian medical student Francesco Gennari inspected, and later described in a 1782 book, noting a thick white stripe in cross-sectional views of the cortex lining the calcarine fissure. This "stripe of Gennari," as it has come to be known, was later shown to correspond to myelinated fibers coursing through layer IV of this cortex. Its primary significance for this book is that the term *striate cortex* (i.e., "striped cortex") has come to be one of the most common terms according to which this region is known. Relatedly, the visual areas "outside" of V1, but to which it projects, are known collectively as *extrastriate cortex*.

Another consequence of the fact that the organization of V1 is dictated by the organization of the retina is that roughly 80% of V1 is devoted to processing information from the central 10% of the retina, a simple consequence of the inhomogeneity of cone density on the retinal surface. What this means for V1 is that there is a much larger portion of cortical area devoted to processing information from the fovea than from the peripheral retina, thereby leading to the distortion of the retinal image that is "projected" onto the cortex. That is, columns of neurons are spaced the same distance apart at the rostralmost extent of the calcarine fissure as they are at the occipital pole. Thus, if there is to be a one-to-one correspondence between photoreceptors and V1 neurons (which strictly speaking, there isn't, but the approximation is fine for our purposes),

it's clear why the cortical representation of the fovea, with its many more columns of neurons, covers a large area relative to the cortical representation of the periphery, with its relatively fewer columns of neurons. Such disproportionate representation of high-acuity portions of a sensory system is referred to as **cortical magnification**. This property will be seen again when we consider other systems, prominently so in the somatosensory (*Chapter 5*) and motor (*Chapter 8*) systems. (Indeed, the depiction of cortical magnification is what gives the statue that graces the cover of this book its distorted features!)

An additional implication of retinotopy for V1 neurons is that they will be sensitive to stimulation within a larger or smaller portion of the visual field, depending on what portion of the retina a neuron represents. This fundamental property of a sensory neuron is referred to as its **receptive field**. In V1, receptive fields are smallest at regions receiving input from the fovea, and largest at regions receiving input from the periphery, properties that follow from the spatial resolution of foveal vs. the peripheral regions of the retina. Because a V1 neuron's receptive field properties are determined by the region of the retina that projects to it, the most direct way to quantify the dimensions of its receptive field would be to do so in terms of the surface area of retina from which it is receiving input. Thus, as an approximation, one might estimate that a V1 neuron representing the fovea has a receptive field corresponding to a 5×10^{-5} mm² surface area of the retina, whereas a V1 neuron representing the periphery might have a receptive field corresponding to a 5×10^{-4} mm² surface area of the retina. Although these values reflect the causal factors determining V1 receptive field properties, they are neither intuitively graspable nor very meaningful with regard to the function of primary visual cortex. Instead, it is much more useful to think of receptive field size in terms of **degrees of visual angle**.

The receptive field

Two adjacent photoreceptors in the retina are likely to receive information from adjacent portions of a visual scene. For example, when you're reading the word "adjacent," there is a viewing distance at which photons reflecting off the dot over the *j* are impacting one photoreceptor and the photons reflecting off the space between the dot and the letter are impacting a photoreceptor just above it. These two photoreceptors have different receptive fields, in that they *receive* information from different portions of the visual *field*. Now, let's make the dot on the letter *j* blink on and off. If your eye were perfectly still, the photoreceptor whose receptive field is restricted to the tiny space under the dot wouldn't detect anything about the blinking dot. That's because this dot is not in this photoreceptor's receptive field. The principle of the receptive field holds for retinal ganglion cells, and for neurons of the LGN and V1 (indeed, for many other areas of the brain).

One feature of V1 receptive fields is that they are larger than those of foveal photoreceptors. For example, a V1 neuron with a receptive field centered on the tiny space below the dot of the *j* might also encompass the area corresponding to the dot. The reason for this is that each V1 neuron receives input from several retinal photoreceptors (*Figure 4.3*). One consequence of such "pooling" across the inputs from several retinal photoreceptors is that a V1 neuron necessarily has a lower spatial resolution than do any of the individual photoreceptors that feed into it. A simplified way to think about this is as follows: if five photoreceptors feed into a single V1 neuron, and that V1 neuron fires each time any of these five photoreceptors signals that something has changed in its receptive field, anyone "listening" to that neuron can know when something has happened within the area covered by those five photoreceptors, but can't know precisely where within that area of five photoreceptor receptive fields the change has occurred.

There is, however, an important trade-off for the loss of spatial resolution that occurs between the retina and V1, and that is a gain in information about *what* has changed in the visual world. In effect, by sampling across the input of multiple photoreceptors, neurons in V1 can "infer" information that was not conveyed by any one photoreceptor alone. This process is already evident at the level of the retinal ganglion cell. Although it may fire robustly when a tiny spot of light is projected into the center of its receptive field, it will respond at an even higher rate (of action potentials) if that central spot is surrounded by a dark ring. This property of an "ON center" and an "OFF surround" means that the neuron signals when there is something bright in the center of its receptive field, but does so even more robustly when that bright "thing" has a dark "thing" right next to it. *Figure 4.3* illustrates how this principle, by pooling across a series of neurons with ON center/OFF surround receptive fields, can yield an "edge detector."

To get a feel for the concept of visual angle, set down your book for a moment and fixate a point on the wall across the room. Next, hold your arms out straight to each side, as though you were about to start flapping them like wings. Now slowly move your arms toward each other, as though you were going to clap your hands with locked

elbows (in the way that people do to mimic a trained seal) and stop moving them when each hand is just visible in the right and left periphery of your field of view. A person with normal vision in both eyes will find that their arms don't get far before they have to stop. As you hold this position with arms slightly bent at each shoulder, you can think of your arms as defining a very wide angle; let's estimate it to be 170°. Next, swing your right arm into your field of view until it is pointing at the same spot that you are fixating. The imaginary line passing through your right hand and stretching from floor to ceiling defines the vertical meridian, and now your right and left arms are defining an angle that is one half of 170°, or 85°. And so, by these rough calculations, the visual field subtends 170° of visual angle along the horizontal meridian. Finally, let's make one more observation before you relax your right arm. With your elbow straight, hold up your thumb (as if indicating "1" on Continental Europe, or signaling "I am cool" in the United States). As you stare at it, the width of the joint in the middle of the thumb subtends approximately 2° of visual angle. Having established this more functional metric of size in the visual system, we can return to receptive fields in V1. In regions of V1 representing the fovea, receptive fields can be as small as 0.5° × 0.5° of visual angle, or smaller, whereas as in regions of V1 representing the periphery, they can be an order of magnitude or more larger.

At this point, we've considered some principles of the early stages of visual sensation: how light projects onto the retina and how it is transduced into a neural code; how visual signals are carried to the cortex; and how these facts give us embodied constraints for describing the visual world. With this knowledge, we can now start to address the question *How does the brain accomplish visual perception?* Or, in the parlance of previous sections of this chapter, *What do we mean by visual information processing in cortex?* We'll start by traveling to Johns Hopkins University in the late 1950s, where a revolutionary set of experiments set the course for the next half-century of visual neuroscience.

INFORMATION PROCESSING IN PRIMARY VISUAL CORTEX – BOTTOM-UP FEATURE DETECTION

The V1 neuron as feature detector

The anatomy of projections from retina, via LGN, to the cortex lining the calcarine fissure, combined with the systematicity of visual impairments produced by damage to the calcarine cortex, makes it clear that this region plays an important role in visual perception. By 1958, David Hubel (1926–2013) and Torsten Wiesel also knew, from extracellular electrophysiological experiments by themselves and others, that neurons in the LGN responded with a burst of action potentials when a small spot of light was projected in specific regions of the visual field (different regions of the LGN for neurons receiving input from retinal ganglion cells projecting from different parts of the retina). That is, LGN neurons had retinotopic receptive fields. They began studying the functions of primary visual cortex, therefore, by projecting small spots of light into the eye of a cat with an ophthalmoscope (not unlike the apparatus whose eye pieces you look through when having your vision examined by an eye doctor) while recording from neurons with the extracellular electrophysiological method described in *Chapter 3*'s section on *Invasive recording with microelectrodes: action potentials and local field potentials*. Initially, experimental sessions during which the scientists stimulated the retina with spots of light failed to generate any responses from V1 neurons. The breakthrough discovery came when, in the process of changing slides in the ophthalmoscope, they found that a neuron responded with bursts of action potentials each time the shadow of the edge of the glass slide passed across the retina at a particular angle (*Figure 4.2*).

The importance of this discovery is that it led to a profound conceptual insight about the mechanisms whereby the primary visual cortex begins the computations that result in visual perception. More specifically, as illustrated in *Figure 4.3*, it led to the insight that neurons in primary visual cortex construct elemental features in the visual scene via the integration of information from the simpler center-surround receptive fields of LGN neurons.

Having made this initial observation, Hubel and Wiesel abandoned tiny spots of light in favor of bars of light (as can be made by cutting a long, narrow slit in a piece of cardboard, and then shining a bright light through the slit and onto a projection screen). With this new stimulus, they quickly confirmed that many neurons in V1 responded robustly to a straight-line boundary defined by light on one side and darkness on the other. Importantly, for any given cell, this only occurred when that boundary was oriented in a particular direction. That is, one neuron might respond preferentially to a boundary oriented vertically, another to a boundary tilted 10° to the right of vertical, another to a boundary oriented horizontally, and so on, such that, with enough recording from enough neurons,

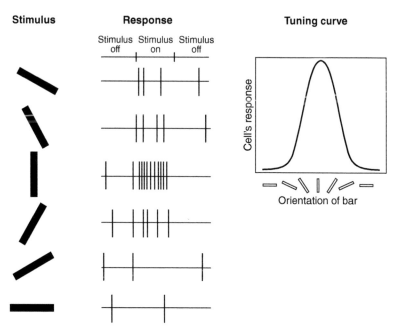

FIGURE 4.2 Orientation selectivity of a simple cell in V1, illustrated by action potentials ("Response" column) generated in response to a bar presented at six orientations ("Stimulus" column), and summarized in a tuning curve.

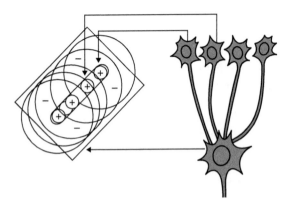

FIGURE 4.3 Constructing orientation selectivity in V1 from LGN neurons with ON center/OFF surround receptive fields. Hubel and Wiesel's idea was that a V1 neuron receives input from four LGN neurons whose receptive fields are overlapping, and whose centers are aligned along an imaginary line tilted 45° to the right of vertical. By holding a flat edge at different orientations across this array of receptive fields, you can prove to yourself that a bar of light oriented at 45° would (1) maximally excite the ON centers of each LGN neuron while (2) minimally stimulating their OFF surrounds. The consequence for the V1 neuron is that this stimulus produces the strongest net input from the LGN, thereby making it (the stimulus) optimal for eliciting a robust response from this V1 neuron. Source: From Hubel, David H. 1988. Eye, Brain, and Vision. New York: Scientific American Library: Distributed by W. H. Freeman. Copyright 1988. Reproduced with permission of Scientific American Library.

they collected evidence that the full 180° of possible orientations would be covered. The property of firing action potentials most robustly in response to a narrow range of orientations has come to be known as "orientation selectivity," or "orientation tuning," and is a hallmark property of many V1 neurons. (The term "tuning" is an analogy to the concept, from signal processing, of a tuned filter, a circuit used to separate signals in relatively narrow bands of frequencies from a broadband source.)

Orientation tuning is a specific example of a broader phenomenon that arises in V1, which is the construction of information (the "drawing of inferences," if one were to anthropomorphize) from more elemental information (e.g., the presence or absence of spots of light). Indeed, including orientation, there are at least five other such "transforms" of elemental information into visual features: direction selectivity; higher acuity; disparity between the images projected into the two eyes (i.e., binocular disparity); the convergence of ON center/OFF surround and

OFF center/ON surround (there are both types of LGN neurons), which gives rise to more complex response properties than the 1-D orientation-tuned "simple cell" illustrated in *Figure 4.2*); and the construction of a single color axis from three cone types. A detailed consideration of how each of these transforms is computed is beyond the scope of this book. But let's briefly consider just one more example: direction of motion. Imagine receiving information that the receptive field of a neuron in the LGN is responding to visual stimulation at time *t*, and that an LGN neuron with an immediately adjacent receptive field (let's say, 90° from vertical to the left) subsequently signals visual stimulation at time *t* + 1. Voilà: a neuron receiving such time-lagged inputs from these two cells could "deduce" that an object in this part of the visual field is moving to the left. (To prove to yourself that the brain is "set up" to infer motion from temporally sequential and spatially offset events, consider the lighted marquee of a theater: a row of light bulbs lines the edges of the marquee, and flashing them rapidly, in succession, creates the illusion of a spot of light "running in circles" around the marquee.)

What might be the utility of some of the properties that emerge from processing in V1? Orientation tuning allows for the detection of edges, a fundamental property of most objects that we encounter in our daily lives. Direction selectivity is a necessary element of motion perception. Binocular disparity is used to see in depth (i.e., 3-D). Combining inputs from the two classes of LGN receptive fields can produce complex tuning properties such as responding to the end of a line (a property referred to as being "end-stopped"). Finally, with regard to color, it has been estimated that humans can distinguish among millions of hues of color, a feat that surely requires considerable processing of the initial inputs to cortex from just three types of wavelength-sensitive cones. These properties (e.g., edges, motion, binocular disparity, end-stopping, color) can be thought of as low-level visual features that, when combined in different ways, are the constituent elements of the complex objects that we recognize and interact with, "effortlessly," on a daily basis. Thus, an influential way of understanding information processing in V1 is to view these neurons as **feature detectors** – units in the system whose function is to signal the presence of their particular feature each time it appears in their receptive field, and to pass this information forward to downstream networks in the visual system that will aggregate information from many such feature detectors to assemble a representation of what object is currently in the visual field. *Chapter 6* will give more consideration to what we know about object perception, but for now we'll maintain our focus on the organization of V1.

Columns, hypercolumns, and pinwheels

A common practice during single-electrode extracellular recording sessions is, having characterized the response properties of the first neuron encountered, to advance the electrode further into (or, depending on the angle of penetration, along) the region under investigation. (In later chapters we'll introduce the technique of simultaneously recording from multiple electrodes.) The results of several perpendicular penetrations (i.e., from layer I to layer VI within a single column) led Hubel and Wiesel to conclude that neurons within a cortical column share the same orientation selectivity. (Hence the term "orientation column.") Penetrations that were parallel to the cortical surface, in contrast, revealed that adjacent columns tended to be tuned to orientations that were slightly rotated from those of the neighboring columns. This led to the proposal that primary visual cortex was organized into "**hypercolumns**," a hypothesized grid of orientation columns that, together, represented every possible orientation that could fall within a receptive field. Of course, to be equipped to analyze every possible feature that could occupy a receptive field, such hypercolumns would additionally need to represent all possible features (or, minimally, the constituent elements needed to construct all possible features) that could fall into this receptive field. Consistent with the idea of omni-featured hypercolumns was the fact that parallel penetrations would periodically encounter columns of neurons that did not show orientation selectivity, but that did seem to process information about color. Thus, Hubel and Wiesel conjectured that each hypercolumn also contained nonselective, chromatically tuned neurons that are important for the perception of color.

A few decades later, advances in functional tissue imaging techniques (section *Functional tissue imaging, Chapter 3*) led to further specifications, and then modifications of the hypercolumn model. It is important to note that, in their studies from the 1950s and 1960s, Hubel and Wiesel never directly observed a hypercolumn in toto. Rather, they inferred the existence of this organizing structure from after-the-fact interpretation of the results of tens, if not hundreds, of experimental sessions. An advantage of functional tissue imaging over single-electrode recording is that it can assess the activity of a relatively large region of cortex (on the order of cm^2) all at once. A critical early discovery, triggered by the research of Margaret Wong-Riley (1989), was that superficial layers of V1, when assayed to detect activity of the enzyme cytochrome oxidase, revealed "blobs" of tissue that stained more darkly than the surrounding tissue. Cytochrome oxidase is a membrane-spanning protein found in mitochondria, and is critical in the generation of adenosine triphosphate (ATP) via oxidative metabolism. Therefore, darker staining, which indicates elevated levels of cytochrome oxidase activity, is a marker for elevated activity in the cells containing it. Subsequent extracellular recordings from neurons located within blobs, by Margaret Livingstone and David Hubel, discovered these to primarily contain the orientation-nonselective, chromatically tuned neurons described in the previous paragraph.

More recently, optical imaging methods have revealed that, instead of being organized into cube-like modules of tissue organized in tidy rows of columns, the functional organization of V1 is more reminiscent of a series of **pinwheels**, with an orientation-insensitive blob at the center and several "petals" made up of orientation-selective columns of neurons emanating from it.

Inside the pinwheel

Before leaving this section, a few more words about how blobs come to be blobs, and orientation-selective regions come to be orientation selective. There are two major classes of retinal ganglion cells (RGCs), which take their names from the *magnocellular* and *parvocellular* neurons of the LGN to which they project. (As is illustrated in *Figure 2.4* and *Figure 6.1*, layers 1 and 2 of the LGN contain magnocellular ["M"] neurons, and layers 3, 4, 5, and 6 contain parvocellular ["P"] neurons.) M RGCs have dendritic arbors that span a larger surface area, and thus sample from a larger number of photoreceptors than do P RGCs. Two consequences of such broad sampling are that (1) M RGCs have relatively larger receptive fields, and (2) they do not carry color-specific information (whereas P RGCs do). M RGCs also have faster axonal conduction velocities than do P RGCs. It is magnocellular neurons (of the LGN) that project to orientation-selective regions of V1, and parvocellular neurons that project to blobs. This information will be particularly germane in *Chapter 6*, when we consider feedforward projections from V1 to different regions of extrastriate cortex, and how these regions come to have different functional specializations.

INFORMATION PROCESSING IN PRIMARY VISUAL CORTEX – INTERACTIVITY

At this point, the discussion from *Chapter 2* about the projection profiles of different layers of neocortex (*Figure 2.4*) becomes particularly germane. We'll begin with one of the more astounding facts about the anatomy of the visual system; one that, until this point, has been glossed over.

Feedforward and feedback projections of V1

Take a moment to revisit *Figure 3.14*, noting the elegant sweep of the optic radiations that connect the LGN with V1. A new piece of information that will prompt renewed consideration of this tract is that *an estimated 75–90% of the fibers comprising the optic radiations are traveling from V1 back to the thalamus.* That is, for each axon conveying signals from the LGN to V1 (i.e., in the feedforward direction), there are anywhere from three to nine axons conveying feedback signals from V1 to LGN. One hint at why our visual systems evolved in this way has come from a series of elegant (and technically very challenging) experiments performed by London (UK)-based neuroscientist Adam Sillito and his colleagues. Their studies, summarized in *Figure 4.4*, provide evidence that ongoing activity in extrastriate and striate cortex can, in effect, shape the signals that the LGN, will deliver to them. Cudeiro and Sillito (2006) summarize the more general principle as follows: "For sensory input to be rapidly assessed and to guide the behaviour of an organism, it needs to be continually placed in the context of hypotheses formulated by higher brain centres" (p. 304).

Why does this circuitry require so disproportionate a balance of feedback vs. feedforward projections? One reason is that implementing functions like the anticipatory "wave" of activity in the LGN that is illustrated in *Figure 4.4.C* requires a lot of circuitry. Related to this is the fact that layer VI projections to subcortical structures are much more complicated than what is illustrated in *Figure 4.4.B*. For example, the region labeled "PGN" in this figure, the perigeniculate nucleus, is part of a larger structure known as the thalamic reticular nucleus (TRN; *Figure 2.4*). Although the TRN is present in many of the photographic images in this book that include the thalamus (e.g., *Figure 1.2.B* and *Figure 2.2*), it can't actually be

seen in these figures due to its microscopic thinness. Unlike most subcortical nuclei, which are clumps of cell bodies whose shape often resembles a familiar object (e.g., an almond, a knee, a tail), the TRN is a thin sheet that wraps around much of the body of the thalamus. (The familiar object that it evokes for me is plastic kitchen wrap.) The name "reticular" derives from the fact that it is reticulated – punctuated with holes necessitated by the fact that hundreds of thousands of axons pass through it while traveling from thalamus to cortex or from cortex to thalamus. The TRN is made up exclusively of GABAergic neurons whose output, in addition to collateral projections within the TRN itself, is exclusively focused on the thalamus. As a result, the TRN is ideally suited to implement an **inhibitory surround**.

Corticothalamic loops, center-surround inhibition, and signal-to-noise ratio

To explain the concept of an inhibitory surround, let's return for a moment to the anecdote from *Chapter 2* of the encounter with the bear. This anecdote was first used to illustrate how oscillatory synchrony might contribute to prioritizing the processing of the bear over the bird or the dappling sunlight. Another goal for the brain in this situation, however, would be to generate as sharp a representation of the bear as is possible. One way to accomplish this goal would be to minimize competition from the neural representation of other largish, hirsute quadrupeds. (Now would not be the time, for example, to contemplate the similarities and differences of bears to, say, deer, raccoons, and dogs; as in *Hmmm. Is it really true that bears are evolutionarily more closely related to dogs than to cats?*) When a cortical neuron involved in the neural representation of the bear (let's call it neuron *A*) sends a volley of action potentials to another neuron in the hypothetical "bear network" (we'll call it *B*), a layer V or VI neuron in the same column as *A* may send a parallel volley of action potentials to a thalamic neuron (*T*) that, itself, is connected with neuron *B*. Thus, *B* receives excitatory (i.e., glutamatergic) inputs from both *A* and *T*. (This is effectively what's illustrated in the feedforward projections from V2 to V4 that are illustrated in *Figure 2.4*.) The real value added of this corticothalamic circuit, however, is that where the *A*-to-*T* axon passes through the TRN, it sends collateral branches to TRN neurons that, in turn, project to thalamic neurons that surround *T* (but don't synapse with *T* itself). The result is that action potentials from layer VI of cortical column *A* have the effect of inhibiting activity in thalamic neurons

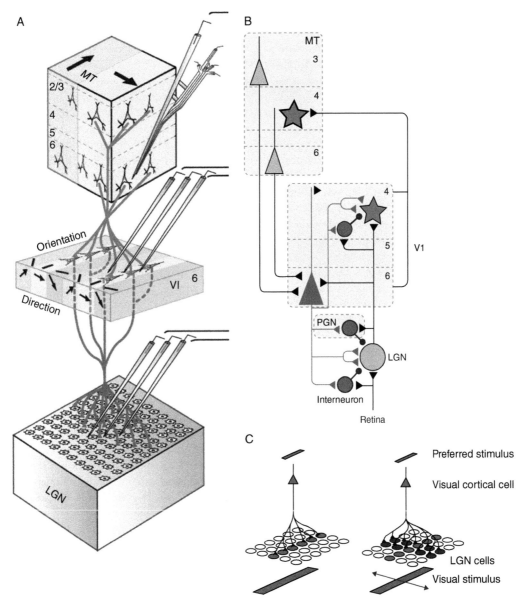

FIGURE 4.4 The role of feedback in shaping what we see. **A.** An experimental setup in which neurons with overlapping receptive fields are simultaneously recorded from in the LGN, layer VI of primary visual cortex (V1), and layer VI of the motion-sensitive region of extrastriate visual cortex known as MT. **B.** Schematic diagram of some of the circuitry that is important for motion perception. The feedforward pathway begins at the bottom of the panel, with an axon from a retinal ganglion cell projecting to and synapsing on a neuron in the LGN of the thalamus, and a branch from this axon synapsing on an adjacent inhibitory interneuron. The LGN sends its "primary" axon to a neuron in layer IV of cortical area V1 (green star), with collaterals to other cell types. The final feedforward step that is illustrated here is from V1 to MT, cartooned as axons projecting from layers VI and IV of V1 to a neuron in layer IV of MT (green star). (Note that this deviates from the "general rule" described in *Chapter 2*, that corticocortical feedforward projections tend to originate from layer II/III – the brain is complicated; one can find an exception to almost all "general rules.") The principle feedback pathway illustrated here starts with projections from MT (layers 3 and 6) to a layer-VI neuron in V1, which, in turn, projects to the same LGN

(Continued on page 102)

surrounding T (because excitatory collaterals from the A-to-T pathway activate TRN neurons that, in turn, release GABA onto these neurons). Back in the cortex, the consequence for B (and, by extension, for the neural representation of the bear) is lowered activity in competing neurons (representations), due to a reduction of their excitatory thalamic "drive." A more formal way to summarize this is that engagement of this corticothalamic loop increases the signal-to-noise ratio (SNR) in the information carried by A and B. The same network dynamics described here (minus the part about the bear) have been demonstrated in studies of attention, which we will take up in *Chapter 7*. (The more general mechanism of center-surround inhibition will also be central to understanding functional contributions of the basal ganglia to motor control [*Chapter 8* and *Chapter 9*].)

And so, to return to the question of why the disproportionate ratio of corticothalamic to thalamocortical fibers in the optic radiations, another part of the answer is the implementation of cortical control of center-surround dynamics of the LGN. Note that, although the example in the previous paragraph used just one layer V or VI neuron in cortical column A, the inhibitory surround is, in fact, implemented by descending fibers from many neurons. A second function implemented by corticothalamic fibers, and therefore also implicated in the makeup of the optic radiations, is controlling the "thalamic wake-up call."

Setting the state of thalamic neurons: tonic mode vs. burst mode

Another function of corticogeniculate fibers is to modulate the effectiveness with which LGN neurons can convey signals to the cortex. They can do this because of a property of thalamic neurons that distinguishes them from most cortical pyramidal neurons: thalamic neurons can display two distinct modes of firing action potentials, tonic and burst modes. Burst-mode spiking has been referred to

as "a wake-up call" from thalamus to cortex, for reasons that are evident in *Figure 4.5*: burst-mode action potentials are markedly more effective at transmitting information than are tonic-mode action potentials. Thalamic neurons move into tonic firing mode when their membrane potential is maintained at a depolarized level for 100 ms or longer, and into a burst mode when maintained at a hyperpolarized level for 100 ms or longer. Research from Sillito's group, using a procedure not unlike that illustrated in *Figure 4.4.A*, has established that the layer VI neurons in V1, via projections to GABAergic neurons in the TRN and within the LGN itself, play a role in controlling whether visual information from the LGN is conveyed to V1 via tonic or bursting spike trains. Thus, in effect, the functions of some of the fibers descending from cortex to thalamus via the optic radiations enable the cortex to decide when to give itself a wake-up call.

Circularity? It can depend on your perspective

On the surface, some of the phrases used in the previous two sections, such as that a "function of [layer VI of cortex] is to modulate the effectiveness with which LGN neurons can convey signals to the cortex" can sound circular. Indeed, it was physically difficult for me to type that sentence – some stage of the process of motor preparation "knew" that it didn't sound right. When given more consideration, however, one can see that what may make one-sentence summaries of findings like these seem paradoxical is the absence of context. To interpret these findings on their own illustrates that it can be a thorny proposition to attempt to isolate discrete "processes" (or "functions") within the spectacularly complex, nonlinear system that is the brain. For example, in defining a process, where is one to stop? Take the case of the proximal control that V1 exerts over whether LGN activity emphasizes the static orientation of a stimulus or its motion (*Figure 4.4.C*). While this could be characterized as an instance of

---

FIGURE 4.4 (*Continued*) neuron with which we started (with collaterals to inhibitory interneurons that also synapse on our LGN neuron). (Note that these MT-to-V1 projections also deviate from the "general rule.") **C.** A schematic "zoom-in" on the LGN, illustrating a functional consequence of this bidirectional network architecture with the example of the processing of information about a bar moving through the visual field (labeled "preferred stimulus"). Depending on the balance of top-down signals from MT to V1, the output of V1-to-LGN feedback projections (schematized as just a single neuron, but in reality coming from a local population of neurons), either emphasize the LGN's processing of the orientation of the visual stimulus (left side of panel), or its processing of the motion in which the bar is moving (right side of panel). Indeed, in the latter case, corticothalamic feedback signals can produce a "wave" of activity in LGN that precedes the arrival of the stimulus into the receptive field, in effect "preparing" the LGN with a predicted trajectory of the stimulus.

FIGURE 4.5 A schematic interpretation of the effects of thalamic action potentials on layer IV cortical cells in V1.

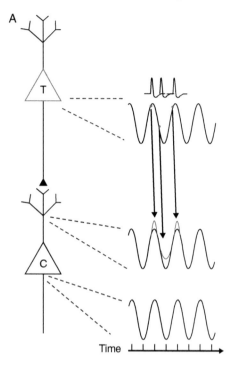

FIGURE 4.5.**A** A thalamic neuron (T) firing in tonic mode. Individual action potentials are sufficiently spaced out in time that, in the dendtrites of the postsynaptic cortical neuron (C), the ESPSs from earlier spikes dissipate by the time the next spike arrives.

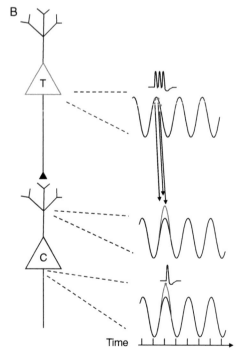

FIGURE 4.5.**B** A thalamic neuron firing in burst mode. Here, the EPSPs from individual action potentials summate in the postsynaptic cortical neuron, and this has the effect of depolarizing this neuron above threshold.

feedback control of V1 over the LGN, we saw that this network is embedded in a larger network that includes MT, with inputs from the latter influencing layer VI of V1. And, although we won't get there until *Chapter 6*, the reader won't be surprised to learn that there are analogous networks operating at even larger scales, and using more abstract information (e.g., the behavioral context) to influence whether MT should influence V1 to influence the LGN as to whether orientation or direction information should be prioritized. Thus, the "local" examples of feedforward/feedback interactions that we've considered here are a part of the active "churning" of the brain that was invoked in the *Introduction to Section II*. Cudeiro and Sillito (2006) make this point explicitly by invoking the "[continual placing] in the context of hypotheses." We will return to these ideas at the conclusion of this chapter, but first, an example of physiological "churning" that operates over the scale of the entire visual system.

The relation between visual processing and the brain's physiological state

In the 1920s, a German physician and physiologist named Hans Berger (1873–1941) began studying the electrical properties of the brain via recordings made with an electrode affixed to the scalp (i.e., as far as we know, Berger invented EEG; see section *Electroencephalography, Chapter 3*). In a 1929 publication that was widely dismissed in his native country, he reported that the EEG was strongest when recorded over the occipital cortex, and that it had two key properties. First, it was characterized by "continuous oscillations in which . . . one can distinguish larger first order waves with an average duration [i.e., cycle] of 90 milliseconds and smaller second order waves of an average duration of 35 milliseconds." Thus, he was describing a complex wave form that comprised the summation of oscillations at two different frequencies. (See *Figure 3.7* for

analogous properties of the local field potential [LFP] and *Figure 3.10* for the decomposition and quantification of different frequency components of the EEG signal.) He named the more prominent "first-order" rhythm *alpha* and the "second-order" rhythm *beta*. The second key property that Berger described was that the alpha rhythm of the EEG was considerably larger when the subject's eyes were closed than when they were open (*Figure 4.6*). This fact can be interpreted in at least two ways. One interpretation is that brain systems (in this case, the visual system) may slip into default "idling" states when they are not actively processing information. From this perspective, the opening of the eyes, and consequent delivery of visual information to the cortex, gives the visual system "something to do," and thereby pushes it out of its idling mode. A second interpretation is that the brain may configure itself into different states to optimize performance on the currently prioritized task. Visual perception is often (although not always) prioritized when the eyes are open. We shall also see, in *Chapter 7*, that the dynamics of alpha-band oscillations vary systematically with the control of visual attention.

A contemporary experiment that can be construed as elaborating on Berger's original observations was carried out by Romei et al. (2008), who took advantage of the fact that single pulses of TMS delivered to occipital cortex can produce a perceptual phenomenon known as a phosphene – a cloud-like flash of light that a subject perceives even when the eyes are closed and covered with a blindfold (*Figure 4.7*). Phosphenes are presumed to be generated by the propagation of action potentials through the visual cortex. The generation of phosphenes with TMS is probabilistic, with the probability that any given pulse of TMS will generate a phosphene increasing with the magnitude of the TMS pulse. The procedure of Romei et al. (2008) was quite simple, in that they determined for each subject the intensity of TMS that would produce a phosphene on

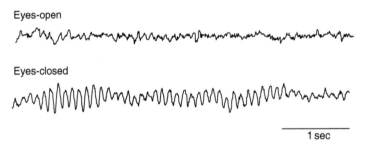

Eyes-open

Eyes-closed

1 sec

FIGURE 4.6 EEG traces illustrating the "Berger rhythm," recorded from channel O1 while the subject sits quietly with eyes open (top row) and closed (bottom row). Note how, in the eyes-closed trace, the magnitude of the alpha-band oscillation fluctuates over time.

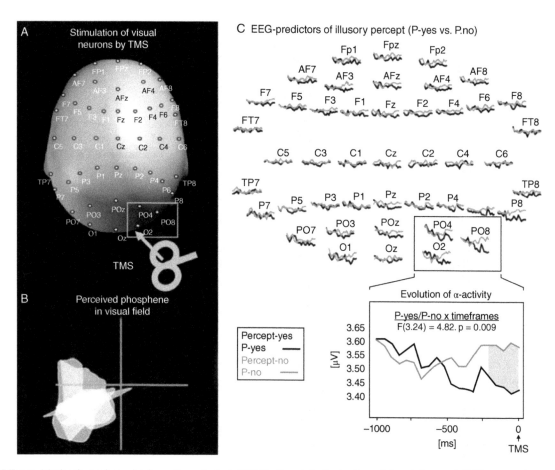

FIGURE 4.7 Methods and results from Romei et al. (2008). Panel **A.** Illustration of the delivery of TMS to a region of the scalp medial to electrode O2, thus targeting a dorsal portion of visual cortex. (Small circles and labels correspond to EEG electrodes.) Panel **B.** An "artist's rendering" of the shape and location of TMS-evoked phosphenes reported by subjects in this study. Note that the location in the visual field corresponds to the known functional organization of the early visual system. Panel **C.** Illustration of the temporal evolution in the magnitude of alpha-band oscillations across the 1000 ms prior to TMS delivery on trials when subjects did (Percept-yes) vs. did not (Percept-no) report perception of a phosphene. The upper portion of this panel shows these measures at each of the electrodes illustrated in panel **A**; the lower portion of this panel shows the average of these measures from the three electrodes at which the alpha-band magnitude for the roughly 250 ms prior to TMS delivery predicted phosphene perception. Source: From Romei, Vincenzo, Verena Brodbeck, Christoph Michel, Amir Amedi, Alvaro Pascual-Leone, and Gregor Thut. 2008. "Spontaneous Fluctuations in Posterior [alpha]-Band EEG Activity Reflect Variability in Excitability of Human Visual Areas." Cerebral Cortex 18 (9): 2010–2018. doi: 10.1093/cercor/bhm229. Copyright 2008. Reproduced with permission of Oxford University Press.

50% of trials, and then delivered many pulses at that intensity. After each pulse they asked the subject to report whether or not they experienced a phosphene, and sorted trials accordingly. A final detail is that the EEG was being recorded for each subject right up until the time that TMS was delivered. Thus, upon completion of an experiment, the researchers could determine whether the EEG immediately prior to the delivery of TMS differed

systematically on trials when TMS did vs. did not produce a phosphene.

As illustrated in the top and bottom panels of *Figure 4.7.C*, there was, indeed, a systematic difference in the EEG that predicted whether or not the subject would experience a phosphene on each trial: on average, the magnitude of alpha-band oscillations was lower on trials when subjects did detect a phosphene than on trials when they did not. From this, the authors concluded that "momentary levels of posterior alpha-band activity [reflect . . .] distinct states of visual cortex excitability, and [. . . this] spontaneous fluctuation constitutes a visual operation mode that is activated automatically even without retinal input" (p. 2010). This study thus reinforces the idea, built up through the second half of this chapter, that visual perception depends, in part, on the internally determined state of the visual system; a state that exists prior to the initial delivery of information from the outside world.

WHERE DOES *SENSATION* END? WHERE DOES *PERCEPTION* BEGIN?

We started this chapter by describing the visual system from what I've described as the intuitive, feedforward perspective: photons impact the retina and are transduced into a neural signal; retinal ganglion cells carry this neural signal to the LGN; and the LGN relays this neural signal to V1, where features are detected (or, if you prefer, constructed). At one level of analysis, this is necessarily accurate: it's well established that damage to the retina and/or to various locations along the feedforward pathway from retina to cortex produces frank blindness. Additionally, in *Chapter 6* we will see how regions at higher stations of the visual system (including V4 and MT, plus many regions further downstream) construct image and object representations from the features first detected in V1. That is, extrastriate regions accomplish perception with the features first detected in V1. From this perspective, then, one may want to draw the sensation–perception boundary for vision at V1. Complicating this clear-cut, tidy demarcation, however, is the influence of feedback projections from higher regions on the functions of V1 and the LGN. That is, we have seen that many neural operations that, from one perspective, are very clearly associated

with sensation, are nonetheless heavily influenced by concurrent *perceptual* processing at higher levels of the system. This complicates the labeling of any given stage of processing as "belonging" to sensation or to perception. (And we shall see, in *Chapter 6* and *Chapter 10*, that top-down influences from long-term memory influence perception in a manner analogous to how we've just documented top-down influences of perception on sensation.) These are hallmarks of a complex system, the idea introduced at the end of *Chapter 2*.

The section of this chapter on *interactivity* described two experiments illustrating how the cortex can exert control over how the thalamus delivers sensory information to the cortex. These two examples can be seen as concrete implementations of the principle of **predictive coding** – the idea that a core function of the brain is to construct, and constantly update, a model of the external world and of what's likely to happen in the immediate future. (Such a scheme is what Cudeiro and Sillito [2006] are referring to as "hypotheses formulated by higher brain centres".) We have already discussed how this principle can be seen to play out across multiple spatial scales (i.e., LGN–V1 bidirectional interactions; LGN–V1–extrastriate bidirectional interactions; and fluctuations in the alpha-band of the EEG recorded over occipital cortex). This principle can also play out across multiple time scales. In the example of the anticipatory wave of activity across the LGN described in *Figure 4.4*, the prediction is about where a moving stimulus will be in the next few milliseconds. In an example of, say, how the author will choose to invest whatever proceeds he may earn from sales of this book, the prediction may be about what he thinks his life will be like after retirement. In *Chapter 10*, after we've gained familiarity with other sensory systems, and then gotten deeper into vision and how visual perception is integrated with systems that generate action, we'll take a closer look at formalization of the idea of predictive coding in the visual system.

END-OF-CHAPTER QUESTIONS

1. What is the principle of retinotopic organization? What is cortical magnification and why does it occur?

2. What is a receptive field? How does the size of receptive fields vary as one moves from the fovea to the periphery? As one moves from retina to LGN to V1?

3. In what sense do neurons in V1 act as feature detectors? What kinds of features are represented in V1? These features are not explicitly represented at the level of the

retina or the LGN, so how do they come to be represented in V1?

4. Any of the features invoked in Question 3 can be present at any location in the visual field. How is this reflected in the "macro-level" organization of V1?

5. Provide evidence of the statement that *the cortex can exert control over how the thalamus delivers sensory information to the cortex.*

6. What functions are supported by feedback projections from V1 to the LGN? How, at the level of network mechanisms, are they implemented?

7. What factors influence the electroencephalogram (EEG) signal as it is measured at the scalp? What neurophysiological processes can be inferred from the EEG?

8. What is the phenomenon of alpha-band oscillations in the EEG? How do dynamics in the magnitude of these oscillations relate to visual processing?

9. What is the principle of predictive coding? What evidence is there that this principle may underlie many aspects of the functioning of the visual system that were considered in this chapter?

REFERENCES

Cudeiro, Javier, and Adam M. Sillito. 2006. "Looking Back: Corticothalamic Feedback and Early Visual Processing." *Trends in Neurosciences* 29 (6): 298–306. doi: 10.1016/j.tins. 2006.05.002.

Luck, Steven J. 2005. *An Introduction to the Event-related Potential Technique*. Cambridge, MA: MIT Press.

Romei, Vincenzo, Verena Brodbeck, Christoph Michel, Amir Amedi, Alvaro Pascual-Leone, and Gregor Thut. 2008. "Spontaneous Fluctuations in Posterior [alpha]-Band EEG Activity Reflect Variability in Excitability of Human Visual Areas." *Cerebral Cortex* 18 (9): 2010–2018. doi: 10.1093/ cercor/bhm229.

Sekuler, Robert, and Randolph Blake. 1985. *Perception*. New York: Alfred A. Knopf.

Sherman, S. Murray. 2001. "A Wake-Up Call from the Thalamus." *Nature Neuroscience* 4 (4): 344–346. doi: 10.1038/85973.

Wolfe, Jeremy M., Keith R. Kluender, Dennis M. Levi, Linda M. Bartoshuk, Rachel S. Herz, Roberta Klatzky, . . . Daniel M. Merfeld. 2012. *Sensation and Perception*, 3rd ed. Sunderland, MA: Sinauer Associates.

Wong-Riley, Margaret T. T. 1989. "Cytochrome Oxidase: An Endogenous Metabolic Marker for Neuronal Activity." *Trends in Neurosciences* 12 (3): 94–101. doi: 10.1016/0166-2236(89) 90165-3.

OTHER SOURCES USED

Adrian, Edgar D. 1934. "Electrical Activity of the Nervous System." *Archives of Neurology and Psychiatry* 32 (6): 1125–1136. doi: 10.1001/archneurpsyc.1934.02250120002001.

Blasdel, Gary G., and Guy Salama. 1986. "Voltage-Sensitive Dyes Reveal a Modular Organization in Monkey Striate Cortex." *Nature* 321 (6070): 579–585. doi: 10.1038/321579a0.

Bonhoeffer, Tobias, and Amiram Grinvald. 1991. "Iso-Orientation Domains in Cat Visual Cortex Are Arranged in Pinwheel-Like Patterns." *Nature* 353 (6343): 429–431. doi: 10.1038/353429a0.

Fritsch, Gustav, and Eduard Hitzig. 1870. "Über die elektrische Erregbarkeit des Grosshirns." *Archiv für Anatomie, Physiologie und wissenschaftliche Medicin* 37: 300–332. [Available in English translation as: Fritsch, Gustav, and Eduard Hitzig. 2009. "Electric excitability of the cerebrum (Über die elektrische Erregbarkeit des Grosshirns)." *Epilepsy and Behavior* 15 (2): 123–130. doi: 10.1016/j.yebeh.2009.03.001.]

Grinvald, Amiram, Edmund Lieke, Ron D. Frostig, Charles D. Gilbert, and Torsten N. Wiesel. 1986. "Functional Architecture of Cortex Revealed by Optical Imaging of Intrinsic Signals." *Nature* 324 (6095): 361–364. doi: 10.1038/324361a0.

Gross, Charles G. 1998. *Brain, Vision, Memory*. Cambridge, MA: MIT Press.

Hirsch, Sven, Johannes Reichold, Matthias Schneider, Gábor Székely, and Bruno Weber. 2012. "Topology and Hemo-dynamics of the Cortical Cerebrovascular System." *Journal of Cerebral Blood Flow and Metabolism* 32 (6): 952–967. doi: 10.1038/jcbfm.2012.39.

Hubel, David H. 1996. "David H. Hubel." In volume 1 of *The History of Neuroscience in Autobiography*, edited by Larry R. Squire, 294–317. Washington, DC: Society for Neuroscience. http://www.sfn.org/about/history-of-neuroscience/~/ media/SfN/Documents/Autobiographies/c9.ashx

Malmivuo, Jaakko, and Robert Plonsey. 1995. *Bioelectromagnetism: Principles and Applications of Bioelectric and Biomagnetic Fields*. New York: Oxford University Press.

Sillito, Adam M., Javier Cudeiro, and Helen E. Jones. 2006. "Always Returning: Feedback and Sensory Processing in Visual Cortex and Thalamus." *Trends in Neurosciences* 29 (6): 307–316. doi: 10.1016/j.tins.2006.05.001.

Tootell, Roger B. H., Martin S. Silverman, Eugene Switkes, and Russell L. De Valois. 1982. "Deoxyglucose Analysis of Retinotopic Organization in Primate Striate Cortex." *Science* 218 (4575): 902–904. doi: 10.1126/science.7134981.

Walsh, Vincent, and Alan Cowey. 2000. "Transcranial Magnetic Stimulation and Cognitive Neuroscience." *Nature Reviews Neuroscience* 1 (1): 73–80.

Wang, Wei, Helen E. Jones, Ian M. Andolina, Thomas E. Salt, and Adam M. Sillito. 2006. "Functional Alignment of Feedback Effects from Visual Cortex to Thalamus." *Nature Neuroscience* 9 (10): 1330–1336. doi: 10.1038/nn1768.

FURTHER READING

Friston, Karl. 2010. "The Free-Energy Principle: A Unified Brain Theory." *Nature Reviews Neuroscience* 11 (2): 127–138. doi: 10.1038/nrn2787.
A readable summary of many models from theoretical neuroscience addressing high-level questions of what the brain does and why it does it that way.

Glickstein, Mitchell, and David Whitteridge. 1987. "Tatsuji Inouye and the Mapping of the Visual Fields on the Human Cerebral Cortex." *Trends in Neurosciences* 10 (9): 350–353. doi: 10.1016/0166-2236(87)90066-X.
A brief history of the identification and mapping of human visual cortex, including "the brilliant contributions of a young Japanese ophthalmologist, Tatsuji Inouye, which have gone largely unrecognized."

Hubel, David H. 1988. *Eye, Brain, and Vision*. New York: Scientific American Library: Distributed by W. H. Freeman.
A summary of principles of visual neuroscience from one of the field's pioneering researchers.

Luck, Steven J. 2005. *An Introduction to the Event-related Potential Technique*. Cambridge, MA: MIT Press.
Although this book emphasizes event-related potentials (ERP), which are generated by the averaging of EEG signals that are time-locked to events of interest within trials, rather than EEG per se, it is nonetheless an excellent introduction to how EEG data are collected and processed. (Additionally, ERPs will be directly relevant for this textbook, beginning with Chapter 5.*)*

Rossi, Simone, Mark Hallett, Paolo M. Rossini, Alvaro Pascual-Leone, and The Safety of TMS Consensus Group. 2009. "Safety, Ethical Considerations, and Application Guidelines for the Use of Transcranial Magnetic Stimulation in Clinical Practice and Research." *Clinical Neurophysiology* 120 (12): 2008–2039.
At the time of this writing, the definitive, authoritative compilation of safety and ethical information relating to the use of TMS in cognitive neuroscience and clinical research. It replaces an analogous report that had been published in 1998, and so one can imagine that the next update will appear a year or so preceding 2020.

Walsh, Vincent, and Alvaro Pascual-Leone. 2003. *Transcranial Magnetic Stimulation: A Neurochronometrics of Mind*. Cambridge, MA: MIT Press.
A thorough, readable overview of the physics and physiology underlying TMS, and the cognitive neuroscience applications of this technique.

CHAPTER 5
AUDITION AND SOMATOSENSATION

KEY THEMES

- Auditory perception depends on exquisitely delicate and precise mechanical transmission in the inner ear, including the cochlea.

- The physical properties of the basilar membrane within the cochlea effect a spectral decomposition of acoustic energy, which is read out as a place code by the auditory nerve.

- The multiple synaptic connections of the auditory brainstem produce the stereotyped early- and middle-latency components of the auditory-evoked potential (AEP).

- Deriving event-related potentials (ERPs) from the EEG, such as the AEP, is a powerful way to measure neural events that are difficult to isolate in the "noisy" EEG.

- Primary auditory cortex (A1), in the superior temporal gyrus, contains a 1-D tonotopic map.

- Many different types of receptors are responsible for the detection of mechanical vs. thermal vs. painful sensations on the skin.

- S1 contains somatotopic maps, distributed across four Brodmann areas in postcentral gyrus of the primate.

- Optogenetic manipulation of S1, combined with electrophysiology, has provided insight into the functions of, and mechanisms underlying, gamma-band oscillations in the LFP.

- Discoveries about somatosensory plasticity in the rodent and nonhuman primate have provided insight into the neural bases of learning, as well as the neurological disorders of phantom limb and phantom pain.

CONTENTS

APOLOGIA

If sensory systems were people whose feelings could be hurt, I would feel apologetic, and guilty, toward audition and somatosensation, and even more so toward olfaction and gustation, for the short shrift that they receive in this book. In this author's defense, however, it can fairly be said that such "visuocentrism" is characteristic of how most cognitive neuroscience texts are written and courses taught, with the exception of those that focus on a particular sensory modality or a domain of cognition, such as language, that is heavily dependent on a sense other than vision. (And, of course, within language, reading is a subdomain that is highly reliant on vision …) The hope for this chapter, as, indeed, is true for all chapters in this text, is that the necessarily selective overview that it offers will prompt some to want to study these problems in more detail. Where possible, the intent will be to emphasize principles and themes that generalize to other domains of cognitive neuroscience.

AUDITION

Audition has been studied from many perspectives, including speech perception, sound localization, and plasticity. This chapter will focus on the mechanical and signal-processing elements of sound transduction, an overview of auditory system anatomy, and the cortical representation of sound.

Auditory sensation

What is it that we hear?

Just as vision results from the transduction of energy from a relatively narrow band of the electromagnetic spectrum (what we call "visible light"), audition is the transduction of energy in the form of changes in air pressure. Many are familiar with the question *If a tree falls in a forest, but there are no animals (including humans) within one hundred miles, does it make a sound?* Auditory science tells us unequivocally that the answer is "No." Consider this: if a human were on the surface of an asteroid floating in the vacuum of space, and it were possible to wear a spacesuit that left the ears uncovered, the most violent of collisions (falling tree, exploding supernova, whatever) would not make a sound. This is because the auditory system works on the principle of detecting changes in energy in the medium making contact with the ear. In everyday life, this corresponds to the movement of molecules of the gases that make up the air that's all around us; or, when underwater, to collisions between and movement of molecules of water (plus whatever else may be dissolved in it). We will refer to this as **acoustic energy**. (Publicity for the sci-fi movie *Alien* had it right: "In space, no one can hear you scream.")

Mechanical transmission and neural transduction

From a mechanical standpoint, the steps leading to transduction of this energy are much more complicated than are those involved in the transduction of light energy. These are illustrated in *Figure 5.1*, but, in summary, a membrane in the outer ear (the tympanic membrane) vibrates in response to vibrations in the air, and its vibration is transferred through a series of tiny bones (the ossicles) to another membrane, the oval window, whose vibration produces changes of pressure in the fluid-filled cochlea. From the outside, the cochlea resembles a tube that has been rolled into a spiral. (Indeed, *cochlea* derives from the Greek for snail.) There are, in fact, three long and narrow chambers rolled up within the cochlea, and we'll focus on the (long and narrow) membrane that separates two of these chambers – the **basilar membrane**. The basilar membrane vibrates in response to the pressure changes caused by the "pushing" and "pulling" of the ossicles on the oval window. Note that, up to this point, what we've summarized is a series of mechanical *transfers* of energy, from vibrations in the atmosphere, through several intermediate steps, to vibrations of the basilar membrane. *Transduction* hasn't yet occurred.

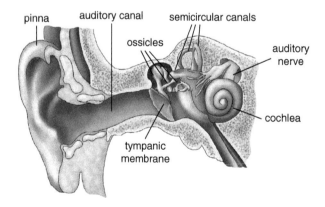

FIGURE 5.1 Major structures of the outer, middle, and inner ear. Source: miha de/Shutterstock.com.

Transduction of the mechanical, vibrational energy of the basilar membrane into a neural signal occurs in **inner hair cells** that are attached all along the long-axis of the basilar membrane. At a simplified level of description, the base of each inner hair cell is secured to the basilar membrane. At the other end of the cells are tufts of hair-like cilia extending into the fluid of the cochlea. Movement of the basilar membrane causes the cilia of inner hair cells to bend, "like tiny clumps of seaweed swept back and forth by invisible underwater currents" (Sekuler and Blake, 1985, p. 311). This back-and-forth movement produces mechanical shearing forces in the membrane of the hair cell, which, in turn, opens or closes mechanically gated ion channels. The resting potential of inner hair cells is approximately −60 mV, and the opening of these channels results in the inflow of K^+ and Ca^{2+} ions, the consequent depolarization of the cells, and the consequent release of neurotransmitter (probably glutamate) at synapses linking the base of the inner hair cell to ganglion cells of the cochlear nucleus. These ganglion cells are the first neurons of the auditory system, analogous to the retinal ganglion cells of the visual system, and they carry electrical signals about auditory information into the central nervous system. Together, the fibers of these cochlear ganglion cells are known as the **auditory nerve**. (We'll return to **outer hair cells** further along in this chapter.)

Two important properties of sound are its volume and pitch. Volume is encoded in this system in a fairly straightforward way: increasing the volume of a sound results in greater displacement of the oval window (i.e., "bigger" vibrations) and, consequently, in greater deflection of the basilar membrane, greater shearing forces on the cilia, greater depolarization of the inner hair cells, and, finally, the release of more glutamate onto cochlear ganglion cells. The end result is that the volume of a sound is encoded by the rate at which neurons of the cochlear nucleus fire (i.e., a **rate code**). To understand pitch, we have to understand another fundamental mechanical property of the basilar membrane, which is the exquisite specificity with which different regions vibrate as a function of frequency.

Spectral decomposition performed by the basilar membrane

Having drawn the analogy between cochlear ganglion cells and retinal ganglion cells, it is important to note differences in how they encode information, differences that arise from fundamental differences in the physical properties of photons vs. acoustic energy. For our purposes,

photons travel in a straight line, a property that our visual system exploits to detect objects regardless of where they appear in the visual field. Thus, for example, a visible rustling of leaves in a tree will produce a pattern of changing light energy on any small patch of the retina, regardless of whether one's center of gaze is directed at the tree, to the right of it, above it, or below it. Thus, retinal ganglion cells carry signals that, when pooled, create a 2-D map of space. Acoustic energy, in contrast, expands out from its source in all directions. (A commonly used metaphor is the pattern of expanding, concentric ripples that result from dropping a pebble into a previously placid pond – just keep in mind that acoustic "ripples" expand in 3-D, in concentric spheres.) Therefore, acoustic energy would not project in an orderly way against a 2-D receptor sheet in the way that visible light energy projects onto the retina. Consequently, the inner ear has evolved to discriminate differences in auditory stimuli along just one dimension (other than volume), and that's vibrational frequency. The physical property of vibrational frequency corresponds to the auditory percept of pitch, such that low-frequency vibrations (as produced, e.g., by a tuba) are perceived as having low pitch, and high-frequency vibrations (as produced, e.g., by a flute) are perceived as having high pitch.

Although the cochlea is a spiral structure, it's useful to think of it as a ("unrolled") straight tube. The end of the tube where the oval window is located is referred to as the base. The tube is widest at this end, and it progressively narrows until it terminates at the apex. Interestingly, however, the basilar membrane within this tube is thick and narrow at the base, and becomes progressively thinner and wider as one moves toward the apex. The consequence of this configuration is that different regions of the basilar membrane will vibrate preferentially to acoustic vibrations of different frequencies. In effect, the basilar membrane is tuned like a piano, with the regions near the base vibrating preferentially to the highest-frequency sounds that we can hear, and regions near the apex vibrating preferentially to the lowest-frequency sounds that we can hear (*Figure 5.2*). This results in a principle for representing sounds of different frequencies with what is referred to as a **place code**, reflecting the fact that auditory nerve fibers arising from different *places* along the basilar membrane preferentially carry information about one relatively narrow range of frequencies.

If you understand the ideas that have been described in the two previous paragraphs and accompanying figures, then you understand the concepts underlying the "spectral decomposition" referred to in this section's title. What's left for us to do is to match those concepts onto this jargony phrase. The frequency at which a signal oscillates, such as the physical vibrations that produce acoustic energy, can be characterized as falling along a continuum, or a *spectrum*, that ranges from low frequencies to high frequencies. A sound source that vibrates at just one frequency is referred to as a pure tone. (This would be represented mathematically by a sine wave.) As we've established, a pure tone with a relatively low pitch would produce a traveling wave whose largest peak occurs at a single location near the apex end of the basilar membrane, whereas a pure tone with a relatively high pitch would produce a peak at a location near the base end. A new piece of information to add to this discussion is that if we were to play both of these tones

simultaneously, the resultant "combined" traveling wave would produce two peaks, one at each of these two locations along the basilar membrane. Indeed, most sounds in nature are made up of a combination of many, many vibrational sources, each vibrating at a different frequency, as well as with a different amplitude. Consider for instance the human voice (a sound source that we won't consider in detail until *Chapter 19*). Speech is produced by forcing air across many flaps of tissue (the vocal cords), each of which vibrates at a different frequency; vibrations of the uvula, tongue, and lips also contribute. This mélange of vibrations at multiple frequencies is conveyed by a single complicated pattern of pushes and pulls by the oval window. This is a point that is important to emphasize: because there is only one oval window, there is no sense in which many *distinct* vibrational signals are being delivered to the

FIGURE 5.2 Mechanical spectral decomposition of acoustic energy in the cochlea. Source: From Wolfe, J.M., Kluender, K.R., Levi, D.M. et al. (2012) Sensation and Perception (3rd Ed), Sinauer Associates. Copyright 2012. Reproduced with permission of Oxford University Press

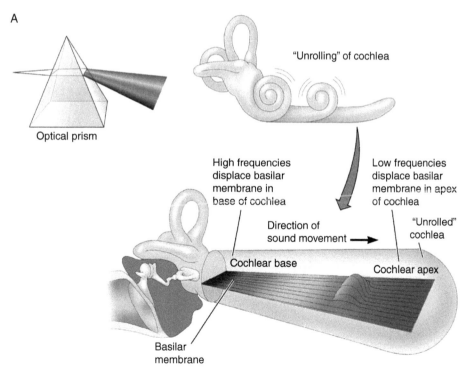

FIGURE 5.2.A On the top left is an illustration of an analogy from optics: just as a beam white light can be separated into its constituent lights of different wavelengths by shining it onto an optical prism; the basilar membrane separates the acoustic signal into vibrations at different frequencies. On the top right, it's easiest to understand the functional layout of the basilar membrane if one first imagines unrolling it. Botton: Diagram of the basilar membrane in the "unrolled" cochlea, color coded to refer back to the analogy to the optical prism.

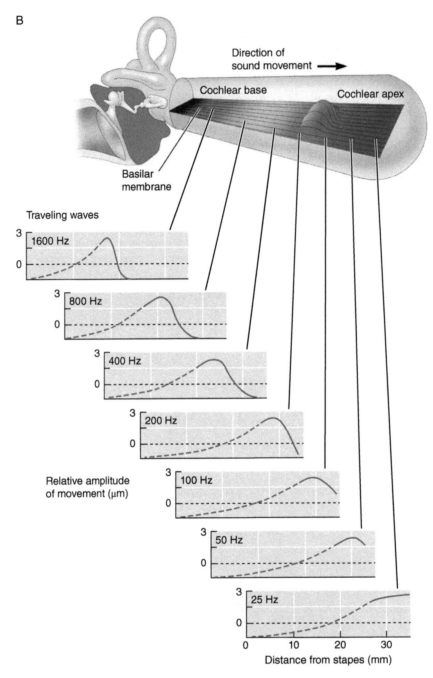

FIGURE 5.2.B Pure tones of different frequencies would produce differently shaped traveling waves on the basilar membrane, with the wave peaking progressively further from the base as frequency decreases. Source: From Wolfe, J.M., Kluender, K.R., Levi, D.M. et al. (2012) Sensation and Perception (3rd Ed), Sinauer Associates. Copyright 2012. Reproduced with permission of Oxford University Press

cochlea from different sources. Rather, what the cochlea receives is a *single* signal in which many different frequencies are embedded. Although this amounts to a much more complicated version of our scenario of playing two pure tones simultaneously, the basilar membrane responds to it in the same way – the multiply-embedded traveling wave contains multiple peaks along the length of the basilar membrane, one corresponding to each vibrational frequency in the sound. It is in this sense that the basilar membrane can be thought of as decomposing the sound signal: the signal comes in as a single "complex waveform" in which many different vibrational frequencies are embedded, and it "leaves" the cochlea (via the auditory nerve) having been separated ("decomposed," in the parlance of signal processing) into its constituent frequencies. (Because each frequency can be represented as occurring at some point along a spectrum, the decomposition of a complex waveform into constituent frequencies can be called a "spectral decomposition.") In effect, the cochlea has performed an operation on the sound signal that is equivalent to the Fourier transform that we introduced in section *Time–Frequency Analysis, Chapter 3*.

Auditory perception

The auditory-evoked response

As *Figure 5.3* illustrates, the auditory signal is processed in many subcortical structures, collectively referred to as the **brainstem auditory nuclei**, before arriving at primary auditory cortex, A1. Synaptic activity associated with these many nodes produces the many deflections (the "peaks") in the early-latency components of the auditory-evoked potential (AEP; *Figure 5.4*). It is in these brainstem nuclei that many of the computations are performed that allow us, among other things, to localize where a sound in the environment is coming from. Importantly, at each of these nodes, the topography of the "place code" for frequency that was established by the properties of the basilar membrane is preserved. The remarkable speed at which all of this processing takes place can be appreciated from inspection of the AEP (*Figure 5.4*), each peak of which is associated with a synaptic event (and, therefore, a stage of processing) along the auditory pathway.

To foreshadow some questions that we'll consider in future chapters, let's note that sensory-evoked responses (the AEP and its visual and somatosensory analogues) are stereotyped, stable responses that are effectively identical in all neurologically healthy individuals. This makes them very effective "tools" for studying the neural bases of many

different behavioral and physiological states. To study attention, for example, one can examine how the AEP may differ on trials when one is attending to a concurrent stream of visual stimuli vs. trials when one is attending to the auditory stimuli. (For example, does inspection of *Figure 5.5* suggest that some steps in the neural processing of sound are vs. are not influenced by attention?) To study different stages of sleep, one can examine how the AEP may differ when tones are played when the research subject is awake (and quietly resting), vs. in stage 1 (i.e., "light") sleep, vs. stage 2 (i.e., "heavy") sleep, vs. rapid-eye-movement (REM) sleep.

The organization of auditory cortex

In the AEP (*Figure 5.4*), the N1 and subsequent components are the first that unambiguously correspond to cortical processing. Thus, despite all the circuitry of the auditory brainstem, auditory information gets from the ear to the cortex in considerably less than 100 ms! The primary auditory cortex, A1, is located in posterior superior temporal gyrus, bilaterally, on the ventral wall of the Sylvian fissure (*Figure 5.6*). In a manner that echoes the functional organization of primary visual cortex, the place code that originates in the cochlea is preserved all the way to A1. But although the analogy to retinotopy might make one expect that A1's functional organization would be called "basilotopy" or "cochleotopy," the standard term is **tonotopy**, a reference to the fact that it is organized in terms of the frequency of the tone(s) being processed. That is, as one moves along the long axis of the surface of A1, one finds that neurons respond to progressively higher (or lower) frequencies. (Note that A1 can be considered to have a "1-D" representation of sound, in that preferred frequency does not vary appreciably along the mediolateral axis of A1, but only along its rostrocaudal axis.) Also analogous to the visual system is the fact that neurons in A1 project to anatomically adjacent areas – referred to as "belt" and "parabelt" cortex – for subsequent processing. Unlike the visual system, however, auditory cortex has systematic differences between the two hemispheres, both in terms of its anatomy and the kind of information that it processes. To provide a hint about one of the important factors behind these differences, the question of lateralization will be addressed in some detail when we consider the neural bases of speech perception, in *Chapter 19*.

Top-down control of auditory transduction

Before leaving the auditory system, we revisit the principle of feedback in sensory processing. Not unlike the optic radiations, which carry more axons traveling from cortex to LGN than from LGN to cortex, the auditory nerve carries

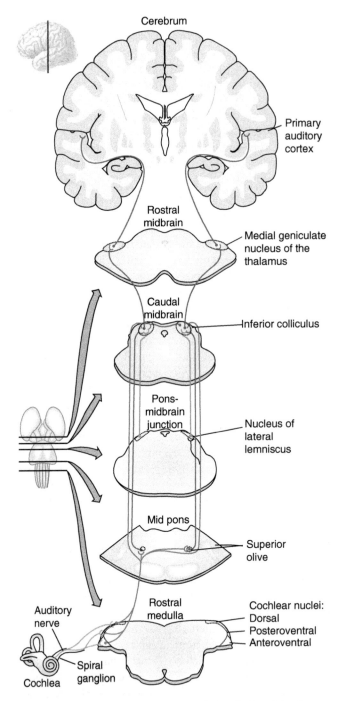

FIGURE 5.3 The auditory pathways from cochlea to cortex. On the left-hand side of the figure are three 3-D renderings of anatomical structures: the bottom is the cochlea; in the middle is a view from behind of the brainstem, midbrain, and thalamus (with cerebellum removed), and horizontal lines showing the locations of the cross-sectional slices illustrated on the right; and at the top is a view of the cerebrum, with a line indicating the level of the coronal slice. Note that there are multiple pathways, with synapses at each of the five levels of brainstem and midbrain that are illustrated (although not every pathway synapses at each level). Pathways originating from only the left ear are shown. Source: From Wolfe, Jeremy M., Keith R. Kluender, Dennis M. Levi, Linda M. Bartoshuk, Rachel S. Herz, Roberta Klatzky, … Daniel M. Merfeld. 2012. Sensation and Perception, 3rd ed. Sunderland, MA: Sinauer Associates. Copyright 2012. Reproduced with permission of Oxford University Press.

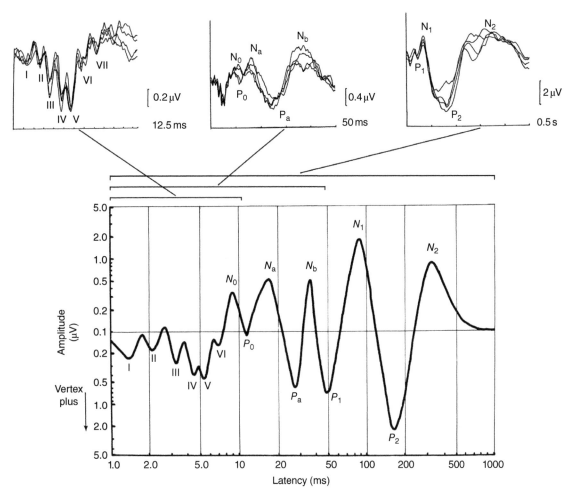

FIGURE 5.4 The auditory-evoked potential (AEP) to a 60-dB$_{SPL}$ click of 50-µs duration. Each of the panels in the top row shows four traces, each an ERP computed by averaging the EEG from 1024 trials, and each acquired from the same subject, but in four different recording sessions. (This was done to demonstrate stability of this response within an individual). Although each of these waveforms is timelocked to the click, each differs with regard to the length of time following the click over which the EEG traces were averaged. Thus, the panel on the left illustrates the first 12.5 ms post-click; the middle panel includes these initial 12.5 ms, but also the ensuing 37.5 ms (i.e., a time span of 50 ms); the panel on the right a time span of 500 ms. Note that the scale on the vertical axis (i.e., the amplitude of the ERP) changes by an order of magnitude from the 12.5-ms time window to the 500-ms time window. These three panels correspond to what are known as the *early*, *middle*, and *long* latency components of the AEP, with the early-latency components known to correspond to processing in the auditory brainstem structures illustrated in *Figure* 5.3. The bottom panel illustrates the AEP (windowed at 1000 ms) averaged across eight individuals, and accommodates the fact that these three groups of components play out on different scales of time and magnitude by plotting the "log/log" transform of the AEP (that is, both the temporal and amplitude values are plotted on logarithmic scales). Source: From Picton, Terrence W., Steven A. Hillyard, Howard I. Krausz, and Robert Galambos. 1974. "Human Auditory Evoked Potentials. I: Evaluation of Components." Electroencephalography and Clinical Neurophysiology 36: 179–190. doi: 10.1016/0013-4694(74)90155-2. Copyright 1974. Reproduced with permission of Elsevier.

more fibers projecting to the cochlea than projecting from it. Of course, an important difference between these two systems is that the cochlea is the site of transduction of acoustic energy into a neural code, more analogous to the retina than to the LGN. (The structural analogue of the LGN in the auditory system is the medial geniculate nucleus of the thalamus [MGN; *Figure 5.3*].) Indeed, while the role of top–down projections to the LGN is to implement neural filtering of the afferent visual signal, in the auditory nerve, the fibers projecting into the cochlea are directly influencing

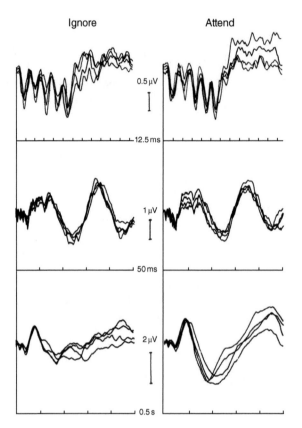

FIGURE 5.5 How attention to sound does, and doesn't, influence the AEP. Clicks (60 dB$_{SPL}$) were delivered once per second, and once every 5–30 seconds the click was slightly softer. In the IGNORE condition the subject was instructed to read a book and pay no attention to the clicks. In the ATTEND condition, they were instructed to detect each softer click and report the number detected at the end of the session. This figure illustrates the AEP to the soft click using the same three time windows as *Figure* 5.4, but here they are arranged vertically to facilitate comparison between the IGNORE and ATTEND conditions. An effect of attention was only seen reliably on the N1 and P2 components of the AEP (bottom row). The figure illustrates two important points, one relating to the effect of attention in the ATTEND condition, and the second to the absence of detectable effects of reading in the IGNORE condition. In the ATTEND condition, because there was not an external event occurring roughly 100 ms after the click, the effects on the N1 and P2 ERP components are assumed to result from a *neural event* that was not directly evoked by an external stimulus. That is, they reflect an endogenous influence on the ERP. (We'll consider the nature of this neural event, and other attention-related ERP components, in *Chapter 7*.) The ERP from the IGNORE condition, in contrast, illustrates how trial averaging can remove components from the EEG that derive from events that one is not interested in. Note how the waveforms from the two conditions look nearly identical (apart from N1 and P2), even though only one includes reading. Reading involves regular eye movements, which produce enormous artifacts in the EEG, as well as neural processing associated with visual perception (*Chapter 4* and *Chapter 6*) and analyzing the grammar and the meaning of sentences (*Chapter 19*). However, because none of these reading-related events was timelocked to the clicks, their influence on the EEG was averaged out of these ERPs.
Source: From Picton, Terrence W., Steven A. Hillyard, Howard I. Krausz, and Robert Galambos. 1974. "Human Auditory Evoked Potentials. I: Evaluation of Components." Electroencephalography and Clinical Neurophysiology 36: 179–190. doi: 10.1016/0013-4694(74)90155-2. Copyright 1974. Reproduced with permission of Elsevier.

the very mechanism of transduction. They do so by synapsing on the outer hair cells, thereby controlling the behavior of these cells. Outer hair cells, although also attached to the basilar membrane at one end and sprouting cilia at the other, differ from inner hair cells in that their cilia attach to another part of the cochlea. A result of the coordinated activity of many outer hair cells, therefore, can be to sharpen the movement of the basilar membrane by dampening its vibrations

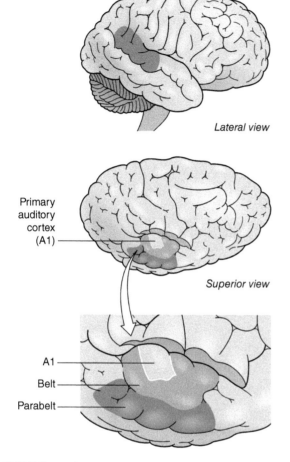

Lateral view

Primary
auditory
cortex
(A1)

Superior view

A1

Belt

Parabelt

FIGURE 5.6 A1, belt, and parabelt cortex of the human auditory system (shown for the right hemisphere). Because these lie on the ventral bank of the Sylvian fissure, the overlying parietal cortex has been "cut away" to expose them in the superior view.

on both sides of the peak of the traveling wave. In effect, the outer hair cells implement a mechanical center-surround inhibition of the movement of the basilar membrane. An important focus of ongoing research in auditory perception is determining the source and the nature of the control signals to the outer hair cells.

Adieu to audition

This concludes our whirlwind introduction to auditory perception. Some of the principles that we've covered here will be relevant to this chapter's next section, on somatosensation, and others to many topics that will be covered

further along in this book. Additionally, of course, the auditory system will be front and center in *Chapter 19*'s focus on the cognitive neuroscience of language. There, among other topics, we will consider how speech is processed relative to non-speech-related sounds. But now, onto the sense of touch.

SOMATOSENSATION

Transduction of mechanical and thermal energy, and of pain

The sensations that we "feel with our skin" result from the activation of one or several of the many types of somatosensory receptor cells embedded in the skin. These fall into three broad categories: **mechanoreceptors, thermoreceptors**, and pain receptors. Different mechanoreceptors are specialized for different mechanical forces that can act upon the body: pressure, texture, flutter, vibration, and stretch. Distinct thermoreceptors respond to hot and to cold. Distinct pain receptors generate the sensation of pain due to extreme temperature, due to mechanical stress (e.g., pinching or cutting the skin), or due to contact with chemicals that damage tissue. In addition to these functional properties, the different types of receptors summarized here also differ in terms of various physiological properties, such as size of receptive field, and speed of transduction and signal propagation. Each receptor is located at the end of a long process (technically, an axon) that projects past its cell body and into the spinal cord. The cell bodies of sensory receptor cells are clustered in ganglia that are spaced out along the long axis of the spinal cord. Because these ganglia are outside the spinal cord proper, these cells are classified as being a part of the peripheral nervous system (PNS; as opposed to the CNS). Indeed, unlike the prototypical CNS neuron, sensory receptor neurons do not generate action potentials at an axon hillock of the cell body. Rather, the action potential is generated in the membrane of the axon that is immediately adjacent to the receptor, and this action potential bypasses the cell body, propagating directly from the peripheral receptor into the spinal cord, and then rostrally along the cord to a synapse on one of two nuclei in the brainstem. Next, the ascending sensory signal is transmitted to the ventral posterior lateral (VPL) nucleus of the thalamus, and from here to the primary somatosensory cortex (S1; *Figure 5.7*). In the primate, S1 is located in parietal cortex immediately posterior to the central sulcus. Unlike either V1 or A1, S1 is divided across four distinct Brodmann areas, some of which support different functions (*Figure 5.8*).

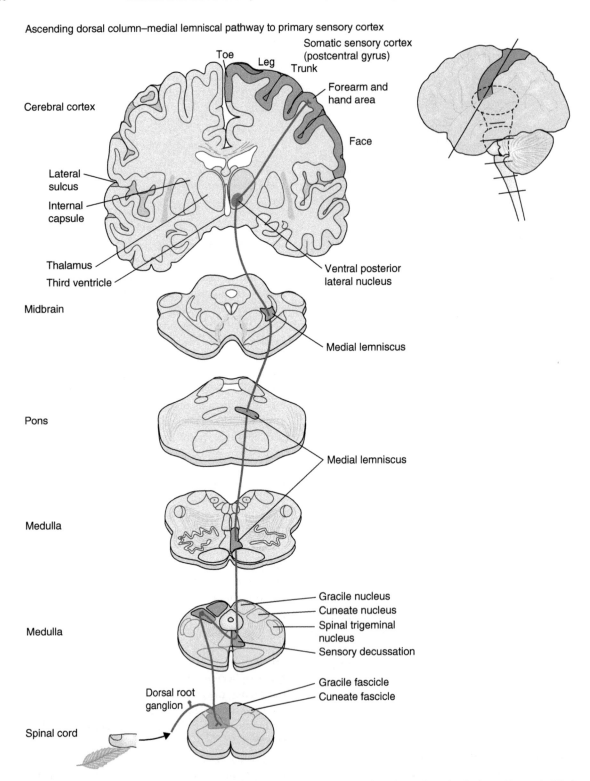

FIGURE 5.7 Somatosensory pathways from periphery to dorsal roots, ascending along dorsal spinal cord to ventral thalamus, and projecting into cortex. Source: From Kandel, Eric R., James H. Schwartz, and Thomas M. Jessell, eds. 2000. *Principles of Neural Science.* New York: McGraw-Hill. Copyright 2000. Reproduced with permission of McGraw-Hill Education.

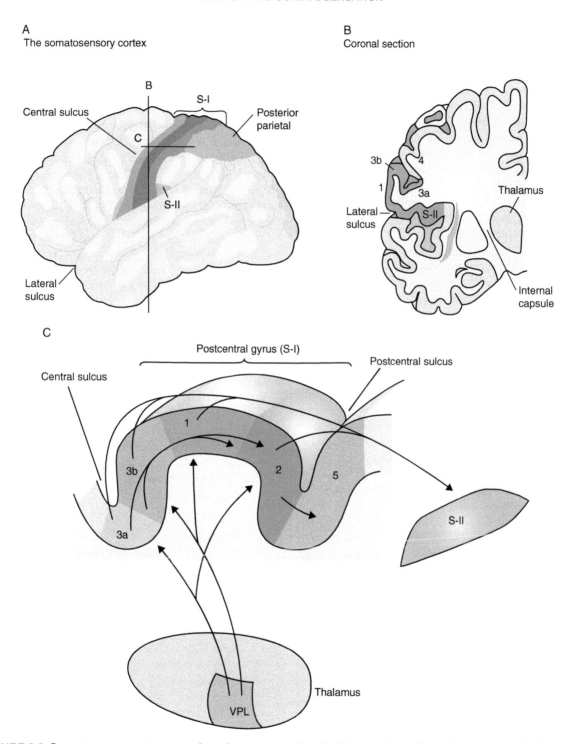

A
The somatosensory cortex

B

Central sulcus

C

S-I

Posterior parietal

S-II

Lateral sulcus

B
Coronal section

3b   4

1   3a

Lateral sulcus

S-II

Thalamus

Internal capsule

C

Postcentral gyrus (S-I)

Central sulcus

Postcentral sulcus

1

3b

2   5

3a

S-II

Thalamus

VPL

FIGURE 5.8 Somatosensory cortex, seen from three perspectives. **A.** External view of the left hemisphere, highlighting the location of S1 and S2 (here labeled "S-I" and "S-II"), and adjacent posterior parietal cortex. (Note that "lateral sulcus" is an alternate name for Sylvian fissure.) **B.** Location of S1 and S2 as viewed in coronal section. **C.** Parasagittal cross section through central sulcus and postcentral gyrus, illustrating the four Brodmann areas that make up S1. One consequence of this organization, and the circuitry schematized here, is that there are more than one somatotopic map in S1.
Source: From Kandel, Eric R., James H. Schwartz, and Thomas M. Jessell, eds. 2000. Principles of Neural Science. New York: McGraw-Hill. Copyright 2000. Reproduced with permission of McGraw-Hill Education.

Somatotopy

The principle of topographic organization of sensory cortex is perhaps most intuitively demonstrated in S1: adjacent parts of the body ("soma," from the Greek) are, apart from a few necessary discontinuities, represented in adjacent regions of cortex. Although this organizational principle had been deduced from assessment of patients with focal brain lesions, it is most enduringly associated with the pioneering electrical stimulation studies of the Montreal-based neurosurgeon Wilder Penfield (1891–1976) and his colleagues. Because the brain itself is not **innervated** with somatosensory receptors, neurosurgery is often performed while the patient is awake, and can report what they are experiencing, and/or can be asked to perform actions that give the surgeon important moment-to-moment information about the status of the surgical intervention. Penfield was an epilepsy surgeon. His approach was guided by the dictum, still by-and-large in force today, that *no brain is better than bad brain*. That is, if a region of cortex is so dysfunctional that its periodic episodes of uncontrolled activity lead to systemic (in this context, "brain-wide") patterns of abnormal activity (i.e., seizures), the surgical removal of this **epileptogenic** tissue might bring an end to debilitating seizures, thereby allowing the remaining, healthy brain tissue to carry out its functions without interruption. An important corollary to this dictum, however, is that the surgeon does not want to damage cortex that supports functions that are critical for the patient's quality of life. Procedures used to abide by this corollary have led to important discoveries about brain function.

As an example, one case that Penfield and Rasmussen (1950) described was an 18-year-old male whose **epilepsy** was determined, upon inspection of the exposed cortex, to stem from a tumor located in the right precentral gyrus. Thus, they wrote, "The surgical problem was to remove the tumor without damage to the precentral gyrus and thus to avoid paralysis" (p. 6). (*Chapter 8* will detail the organization and function of primary motor cortex of the precentral gyrus.) To accomplish this, they stimulated the surface of the brain at many locations that were near the region where they expected to find the "epileptogenic focus." In the process, different stimulation sites produced different responses in the patient's body, or different verbal reports of what sensation he felt. Examples of observed effects included "Closure of the hand and pronation of the forearm" and "Dorsiflexion of the left foot." Examples of patient reports included "Contraction of my left hand and arm" (even though the doctors didn't observe any movement and "A little of that feeling in the leg as though it were going to contract." After performing many such surgeries, Penfield and Rasmussen (1950) were able to generate the "sensory **homunculus**" illustrated in *Figure 5.9*. The precision of their measurements allowed them to document quantitatively the cortical magnification of various parts of the body, a feature analogous to the disproportionally larger cortical representation of the central vs. peripheral visual field in V1. Indeed, recall from *Chapter 4* that the fovea has a higher "packing density" of photoreceptors than do peripheral parts of the retina, and that foveal receptive fields are consequently much smaller than are peripheral receptive fields. Similarly, from *Figure 5.9*, we can deduce that the skin of the lower lip is much more densely innervated with sensory receptors than is the skin from the elbow, and that, consequently, cortical receptive fields representing the lower lip (and other parts of the face, and many parts of the hand, etc.) are much smaller than are cortical receptive fields representing the elbow (and the small of the back, and the knee, etc.). The enterprising reader can appreciate this, after a quick search on the internet, from the full-monty view of the sculpture of the somatosensory homunculus that graces the cover of this book. (As an aside, the word homunculus is Latin for "little man." In psychology and neuroscience, it is used in two contexts. Here, it refers to the representation of the human body as illustrated in *Figure 5.9*. It can also be used (often facetiously) in relation to the idea that a "little man inside our heads" is responsible for the conscious perception of sensory information, or for the coordination and control of brain activity.)

Somatotopy in the rodent: barrel cortex

This chapter will consider data from several species, and now we turn to the mouse, from which scientists have learned a great deal about somatosensation and, more broadly, many principles of mammalian neural function. Many species of rodent, including the mouse and the rat, are nocturnal and/or spend much of their time underground. For such animals, somatosensation via the whiskers is as important for navigation and exploration as is vision for humans. Indeed, an alert mouse actively explores its environment by "whisking," a rhythmic back-and-forth sweeping of the whiskers that can vary in rate from 3 to 25 or more whisks per second. There is also considerable evidence that animals coordinate whisking rate and phase (i.e., where the whiskers are in the back-and-forth cycle) with movements of the head and paws. Whisking for a rodent can be thought of as analogous to the near-constant eye movements that are characteristic of primates (human

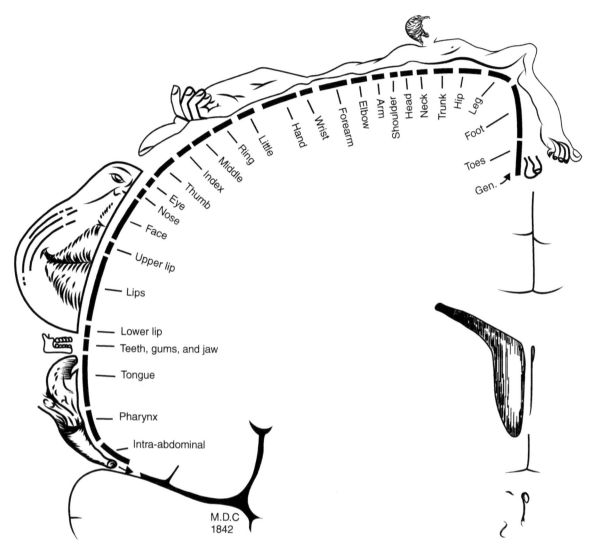

FIGURE 5.9 Depiction of the somatotopy of S1, as distributed along a paracoronal plane of section. The relative size of each body part reflects the surface area of cortex that represents it, as does the length of the line to which each label points. Three additional points are worthy of note. First, this scheme is a statistical composite derived from many individual cases. Second, this somatotopy can undergo rapid reorganization as a function of behavioral context or deafferentation (as will be described further along in this chapter, and is also demonstrated by Kaas (2005, *Further Reading*)). Third, the discontinuities in the body representation of the somatosensory homunculus (e.g., the abrupt transition from face to fingers) underlie some fascinating neurological phenomena, also to be detailed further along in this chapter. Source: Penfield and Rasmussen, 1950.

and nonhuman), something that we will consider in detail in *Chapter 9*.

The importance of whiskers to the rodent is reflected in the organization of its somatosensory cortex, in which one finds disproportionate representation of the snout, and within the "snout area" of the cortex, the distinctive **barrel cortex** that processes information from the whiskers

(*Figure 5.10*). The elegant simplicity of the barrel system, including its "one-barrel-to-one-whisker" organization, makes it an ideal system for studying many principles of sensory processing. For example, there is considerable evidence that the phase of LFP oscillations within the barrel cortex is synchronized with the phase of whisking. The importance of such brain–body coordination is illustrated

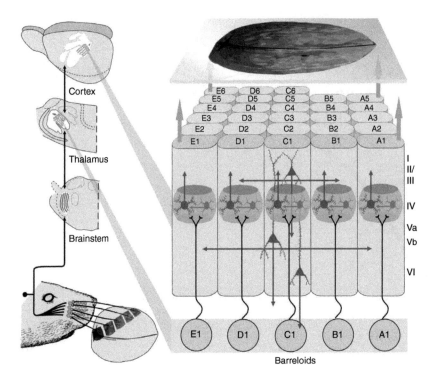

FIGURE 5.10 A diagram of the barrel system in a rodent. The column on the left illustrates a simplified rendition of the primary pathway from periphery to cortex. In the lower left, a cartoon rat is exploring an object, with the brown squares corresponding to the region of the object touched by each whisker. Each vibrissa follicle is innervated by mechanoreceptors that send their signals via the trigeminal nerve (of the PNS), in parallel, to one of many trigeminal nuclei in the brainstem (the first synapse of the system; only one nucleus illustrated here). Afferent fibers from the principal brainstem nucleus then project into the multiple "barreloids" of the VPM nucleus of the thalamus (second synapse), the arrangement of which mirrors the topography of the animal's face – that is, each barreloid receives information relating to just one whisker. Finally, thalamic afferents project into cortical area S1 (third synapse), which contains a map of the entire body surface of the animal (top left). Note that, as we have already seen with the human somatosensory homunculus, this map is distorted in a manner that corresponds to the innervation density of various parts of the animal's body. The right side of the figure illustrates more detail of the thalamocortical projection and the cortex. Thalamic afferents project into layer IV, where both excitatory (red) and inhibitory (blue) neurons make up each barrel. Also illustrated for cortical column C1 (which corresponds to whisker C1 on the animal's face) are pyramidal cells in superficial and deep layers of cortex. Floating fancifully "above" the barrel cortex is the artist's rendition of the animal's neural representation of the object being whisked (a walnut?), which is emergent from cortical processing in S1 and downstream areas. Source: Image courtesy of Dr. Jochen Staiger

by a study from Cardin, Carlén et al. (2009), into which we will now take a deep dive.

Optogenetic exploration of gamma-band oscillations in somatosensory cortex

In *Chapter 2*, the phenomenon of oscillatory synchrony was introduced as a possible mechanism for organizing neural activity, such as the selection of to-be-emphasized information. The study that we will consider here is noteworthy because it provides *causal,* mechanistic insight into how one type of neuronal oscillation – oscillation in the gamma frequency band – is generated. Further, it demonstrates the importance of these oscillations for the cortical processing of afferent sensory signals. It accomplishes this by using optogenetic methods to exogenously control gamma-band oscillations in one class of neurons located in barrel cortex – fast-spiking (FS) inhibitory interneurons. In doing so, it illustrates how experimental methods that exert external control over an aspect of the brain or brain function can support stronger inferences than would have an experiment that "merely observed" the kind of activity that is described here.

Cardin, Carlén et al. (2009) used optogenetic methods to test the hypothesis that inhibitory interneurons of barrel cortex (the "blue" neurons from *Figure 5.10*) govern the gamma-band oscillations that had previously been observed in LFPs recorded from barrel cortex. The experiment proceeded in three stages. First, as illustrated in *Figure 5.11*, the authors established that their preparation gave them (a) direct optical control over these FS neurons and, by extension, (b) indirect inhibitory control over "regular-spiking" (RS) excitatory

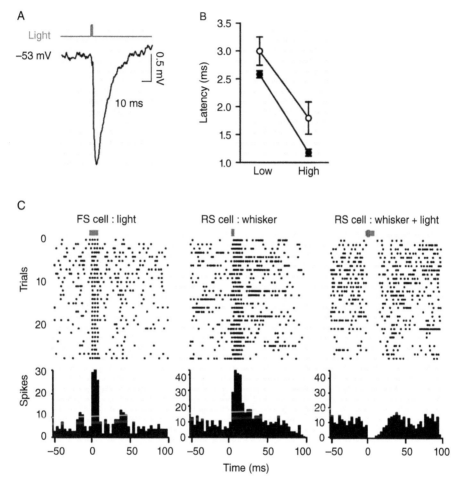

FIGURE 5.11 **A** Intracellular recording from an RS neuron shows a near-instantaneous inhibitory postsynaptic potential (IPSP) in response to a pulse of 470-nm light onto this tissue via an optical fiber. Note that because RS neurons were not transfected, the IPSP could not have been caused directly by the light. Rather, it must have been the result of the release of GABA from transfected FS neurons onto this pyramidal cell (i.e., the pulse of light produced an action potential in one or more FS neurons). **B.** The latency between pulses of light delivered at low or high intensity, and two measures: action potentials in FS neurons (filled symbols) and IPSPs in RS neurons (unfilled symbols). These data indicate that the propagation of the action potential of an interneuron, the resultant release of GABA, and the consequent postsynaptic influx of Cl⁻ ions in an adjacent RS neuron all transpire in less than 1 ms. **C.** A functional consequence of sustained optical stimulation of FS inhibitory interneurons is to block sensory-evoked responses in RS neurons to a whisker deflection. Left-hand panel shows sustained firing of an FS cell in layer 2/3 in response to a 10-ms light pulse (blue line). Center panel shows the response of an adjacent RS cell to a whisker deflection (red bar). Right-hand panel shows that when light is delivered coincident with whisker deflection, the result is to suppress activity in the RS neuron to below its baseline level. Source: From Cardin, Jessica A., Marie Carlén, Konstantinos Meletis, Ulf Knoblich, Feng Zhang, Karl Deisseroth, ... Christopher I. Moore. 2009. "Driving Fast-Spiking Cells Induces Gamma Rhythm and Controls Sensory Responses." Nature 459 (7247): 663–667. doi: 10.1038/nature08002. Copyright 2009. Reproduced with permission of Springer Nature.

neurons (the "red" neurons from *Figure 5.10*). Second, they established that FS cells have a preferred oscillatory frequency in the 40–50-Hz range (*Figure 5.12*). Third, they implemented a test of the idea that "gamma oscillations . . . have a functional impact on cortical information processing by synchronizing the output of excitatory neurons" (p. 666). To do so, they imposed an oscillation at this frequency by stimulating FS cells with light pulses delivered at 40 Hz, and recorded the responses of RS cells to deflections of a whisker performed at five different phases relative to the gamma cycle (*Figure 5.13.A*). The logic was that, if gamma-band oscillations serve a gating function, one should see optimal responses to afferent sensory information by layer IV excitatory neurons when that information arrives at the optimal time relative to the ongoing gamma-band oscillation in this neuron's membrane potential (*Figure 5.13.B*). The results, illustrated in *Figure 5.13.C–E*, demonstrated the predicted gating function of gamma-band oscillations, in that the response to whisker stimulation of a layer IV RS neuron was sensitive to the phase of the optically entrained LFP. That is, unless the whisker stimulation was delivered to coincide with the peak of the LFP, it lacked the magnitude of response (in spikes per trial) of the baseline response.

Why weren't optimally timed whisker stimulations stronger than the baseline? One possible answer comes from looking back at *Figure 5.10*, and noting that thalamo-cortical axons carrying sensory signals from barreloids to barrels also send collaterals that synapse directly on layer IV interneurons. This suggests that, because the circuitry is in place to coordinate interneuron activity with the processing of the afferent sensory signal, the best that an experiment like this might be able to achieve is to push the system away from its optimal-performing regime, which is in place on "baseline" trials when no optical stimulation is being delivered.

Somatosensory plasticity

Plasticity refers to the ability of the nervous system to change its function, and sometimes its structure, in response to a change in the internal or external environment. The somatosensory system, including rodent barrel cortex, is a system in which a great deal of research has been conducted on the principles and mechanisms underlying plastic change in the brain. For example, many studies have been performed in which an individual whisker is clipped close to the skin, so that it no longer comes in contact with objects during whisking. This results in a reorganization of the representation of this, and adjacent whiskers, in barrel cortex. Such research has provided important insights into developmental, genetic, molecular, and pharmacological factors underlying

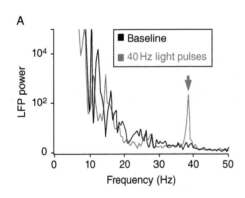
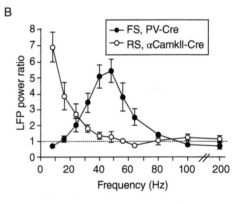

FIGURE 5.12 FS cells have a preferred oscillatory frequency in the 40–50-Hz range. **A.** A power spectrum of the LFP in barrel cortex as recorded in the absence of optical stimulation ("Baseline"; black) and during the delivery of 40-Hz light pulses (blue). **B.** Across a wide range of frequencies of optical stimulation, the LFP power ratio ($LFP_{optical stimulation}/LFP_{baseline}$) for transfected FS neurons (filled symbols) is highest in the 40–50-Hz range, suggesting that this is the "naturally preferred" frequency (or **resonant frequency**) of these cells. For comparison, in a different group of animals, RS cells (unfilled symbols) were transfected with the ChR2 gene. The highest LFP power ratio in these animals, and thus the presumed resonant frequency of RS cells, was at 8 Hz (i.e., in theta frequency band). Source: From Cardin, Jessica A., Marie Carlén, Konstantinos Meletis, Ulf Knoblich, Feng Zhang, Karl Deisseroth, … Christopher I. Moore. 2009. "Driving Fast-Spiking Cells Induces Gamma Rhythm and Controls Sensory Responses." Nature 459 (7247): 663–667. doi: 10.1038/nature08002. Copyright 2009. Reproduced with permission of Springer Nature.

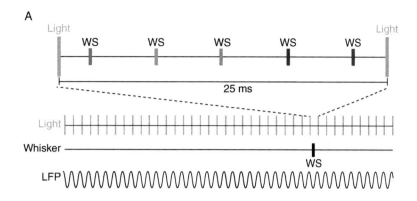

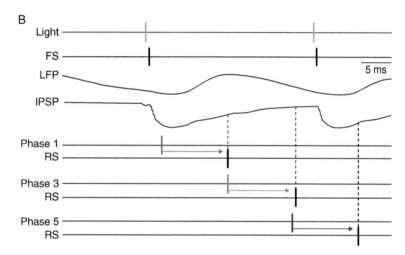

FIGURE 5.13 Optogenetically controlled gamma-band oscillations influence cortical processing of afferent sensory signals. **A.** Gamma-band oscillation is imposed by delivering light pulses at 40 Hz to ChR2-transfected FS cells, and whisker deflections (WS) are performed at five different phases relative to the gamma cycle. **B.** Schematic model of the expected effects of this procedure: the top trace shows the once-every-25 ms (i.e., 40 Hz) pulses of light; the next, the consequent action potential in FS neurons; the next, the resultant gamma-band oscillation of the LFP; and the next, the concurrent fluctuations in membrane potential experienced by a RS pyramidal cell (IPSP). An afferent signal from a WS delivered at phase 1 (red WS in panel **A**) would arrive at the RS cell at the point of maximal inhibition in the gamma-phase, and would produce a spike roughly 10 ms later; a WS delivered at phase 3 (green) would arrive at a point of rising membrane potential, and produce a spike near the peak membrane potential; and a WS delivered at phase 5 (purple) would arrive near the peak of the membrane potential, the resultant spike occurring near the trough of the membrane potential. **C.** Responses of a layer IV RS neuron to WSs in the absence of optical stimulation (baseline). **D.** Responses of the same neuron to WSs delivered at each of the five phases portrayed in *A*. **E.** Average spikes/trial for baseline (B) and each of the five phases. Source: From Cardin, Jessica A., Marie Carlén, Konstantinos Meletis, Ulf Knoblich, Feng Zhang, Karl Deisseroth, ... Christopher I. Moore. 2009. "Driving Fast-Spiking Cells Induces Gamma Rhythm and Controls Sensory Responses." Nature 459 (7247): 663–667. doi: 10.1038/nature08002. Copyright 2009. Reproduced with permission of Springer Nature.

experience-dependent plasticity. Here, however, we turn to systems-level research in the nonhuman primate to examine the phenomenon of somatosensory plasticity and some of the principles underlying it. As we shall see, this research sheds light on fascinating (human) neurological syndromes that can result from damage to the somatosensory periphery.

Use-dependent plasticity

Some of the pioneering work in this area has been carried out by Michael Merzenich and colleagues at the University of California, San Francisco. One of the motivations for these studies was an observation made while mapping the cortical representation of fingers within the hand area: the

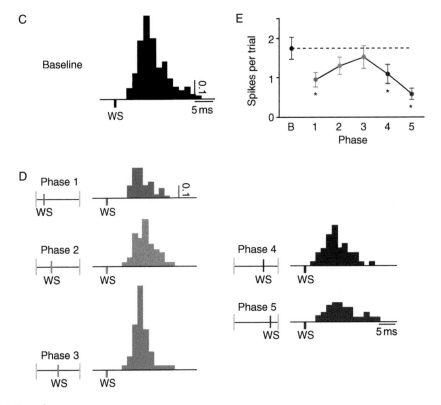

FIGURE 5.13 (Continued)

surface area of cortex representing a given finger could differ dramatically from one animal to the next. This gave rise to the idea that cortical representation of the body might not be rigidly fixed by a genetically determined anatomical structure, but may, instead, be a dynamic property that, even in the adult, can change to reflect changes in experience (i.e., two monkeys might have differently organized hand areas because they use their hands differently.) To test this idea, Merzenich and colleagues devised experiments in which the somatosensory experience of the animal was altered in systematic ways, and the cortical representation of the implicated body part was examined for evidence of experience-dependent change. One type of study involved training the animal to make discriminations between vibrations of different frequencies delivered to one specific location on one finger (*Figure 5.14*). Results indicated that the cortical representation of the targeted area grew dramatically after extensive training. Importantly, this effect was specific to the piece of skin that directly experienced this stimulation, because "before-and-after" mapping of the cortical representation of the contralateral fingertip, which hadn't been trained, showed no change. Additionally, this plasticity was shown to be sensitive to

cognitive factors, because little change would occur in the cortical representation of a finger that received the same amount of vibrational stimulation as the one illustrated in *Figure 5.14*, but did so while the monkey was paying attention to a difficult auditory discrimination task. The fact that attention to tactile stimuli seems to be necessary for somatosensory plasticity to occur is yet another demonstration of a critical role for top-down input onto primary sensory cortex.

The brain mimics the body

A second type of study involved surgically fusing the skin of two fingers in order to transform them into effectively a single, "fat" finger in the middle of the hand. Here, the result was that the discrete, discontinuous boundary between the two fingers, which had existed prior to the surgery, was replaced by a "blurred" cortical representation that did not discriminate between the two previously separate fingers. This is a concrete example of a principle that we will consider in some depth in *Chapter 11*: neural plasticity can be construed as the process of the brain changing its structure so as to more closely model the structure of the world that it is representing. In the case summarized here, the monkey's

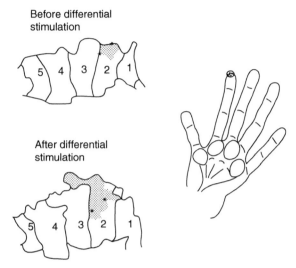

Before differential stimulation

After differential stimulation

1mm

FIGURE 5.14 Somatosensory reorganization after training. The upper-left figure shows the representation of the five fingers of the hand (thumb = "1"; index finger = "2"; etc.) in primary somatosensory cortex of area 3B of an adult owl monkey, with the shaded portion indicating the location of neurons with receptive fields corresponding to the tip of the index finger (drawing on right). The bottom-left figure shows the dramatic increase in the number of neurons with this same receptive field, the asterisks indicating the same anatomical location in both images. Note that this reorganization occurs primarily within the preexisting representation of digit 2, with the expansion at the top of the image corresponding to a rostral "invasion" of area 3A. The boundaries between finger zones, however, are largely unchanged. Source: From Merzenich, Michael M., Gregg H. Recanzone, William M. Jenkins, Terry T. Allard, and Randolph J. Nudo. 1988. "Cortical Representational Plasticity." In Neurobiology of Neocortex: Report of the Dahlem Workshop on Neurobiology of Neocortex Berlin, 1987 May 17–22, edited by Pasko Rakic and W. Singer, 41–67. Chichester: John Wiley and Sons. Copyright 1988. Reproduced with permission of John Wiley & Sons, Inc.

physical reality was that a hand that previously had five independent fingers now had four. The consequent plastic reorganization in somatosensory cortex can be seen as its brain reshaping itself to better represent this new state of affairs. In *Chapter 8*, we shall see that analogous plasticity in the motor system is characteristic of learning new motor skills.

Mechanisms of somatosensory plasticity

In the demonstrations of cortical plasticity that we have considered in the previous two subsections, there are at least two possible explanations for how the cortical representation of a part of the body increases in size (*Figure 5.14*) or merges with another. One is that afferent axons that drove one discrete region of cortex have "sprouted" into the new cortical territory (e.g., by sending newly sprouted axon collaterals into the "conquered" region). A second is that previously present, but for some reason quiescent, axon collaterals representing the expanding part of the body were "unmasked" as a result of the changes inducing the plasticity. Evidence for the latter comes from amputation/transection studies in which either a part of the body (e.g., a finger) or a sensory nerve innervating that part of the body is experimentally severed. In these situations, the expansion of representation of adjacent, undamaged skin surfaces (e.g., the adjacent finger) is evident almost immediately, a timescale much too fast for sprouting (which would require new protein synthesis, the growth of the axon collateral, etc.). Thus, unmasking of preexisting axonal terminals is the most plausible explanation for at least the earliest stages of cortical plasticity. But by what mechanism was the now-influential axon previously "masked"? One explanation is by lateral inhibition driven by the previously dominant excitatory input, although alternative models have also been proposed.

One take-away from this section is that, although illustrated depictions of somatotopic representation in S1 (e.g., *Figure 5.9*) suggest a static homunculus, the cortical representation of the body's surface may, instead, be quite dynamic. The anatomical substrate that affords rapid reorganization in somatosensory cortex may be somewhat different from the one-to-one projection zones suggested by the "one-whisker-one-barrel" scheme for rodent barrel cortex. Instead, a manifold of overlapping afferent inputs, the majority of which are typically "masked" due to local inhibitory dynamics, may be the norm, at least in the primate brain.

Phantom limbs and phantom pain

Although the types of experiments described in preceding paragraphs may seem particularly sadistic from the standpoint of the experimental animal, such studies have given important insight into human neurological disorders, such as the "phantom limb" syndrome that can be experienced by patients who have either lost a limb or suffered damage to the spinal cord such that afferent sensory information from that limb no longer reaches the brain. These patients often report experiencing tactile sensations, and sometimes pain, that feel as though they are coming from the limb that they have lost. Although it was long assumed that these sensations were due to scrambled firing in nerve fibers at the stump of the lost limb, this explanation has been ruled out by the fact that injecting local anesthetic into the stump, which silences these fibers, does not quiet the phantom sensations. A synthesis of much of the research presented in this chapter, however, suggests an alternative: it may be that plastic reorganization of the cortex, resulting from the loss of innervation from the amputated limb, results in cortical signals that get "misinterpreted" as coming from the no-longer-existing limb.

A hypothesized mechanism

Evidence for a cortical reorganization of phantom sensations comes from the patient shown in *Figure 5.15*, whose left arm had been amputated 4 weeks previously. In a deceptively simple but enormously influential case study, the neurologist and cognitive neuroscientist V. S. Ramachandran brushed several parts of this patient's body with a cotton swab. Only when two parts of the body were brushed – the side of the face ipsilateral to the amputation and the upper arm above the amputation site – did the patient report two simultaneous tactile sensations: one on the skin that was actually touched, and one on a part of the amputated hand. Ramachandran argued that the location of these "reference zones" was not arbitrary. Recall from the "Penfield homunculus" (*Figure 5.9*) that the cortical representation of the hand is located immediately dorsal to that of the face, and immediately ventral to that of the arm.

Thus, the region of cortex that previously represented this patient's left hand may have been "invaded" by afferents carrying information about the face, as well as about the upper arm. More specifically, from *Figure 5.15*, one can infer that afferent signals from the left cheek may now be driving cortical tissue that had previously represented the left thumb, that afferent signals from the left upper lip are now driving cortical tissue that had previously represented the left index finger, and so on. The phantom sensations experienced by the patient, then, may indicate that the reference zones illustrated in *Figure 5.15* may, in fact, correspond to newly formed receptive fields of neurons occupying what had been the hand area.

Helpful or harmful?

Ramachandran's experiments, and follow-up work by his group and others, have sparked enthusiastic discussion and debate, among philosophers as well as cognitive neuroscientists. One unresolved problem posed by the cortical-reorganization model is that of reconciling the beneficial, indeed necessary, functions of plasticity with the obviously undesirable, and sometimes debilitating, clinical phenomena of phantom limbs and phantom pain. How is it that a capacity that serves us so well when it comes to learning to read Braille or learning a musical instrument can also

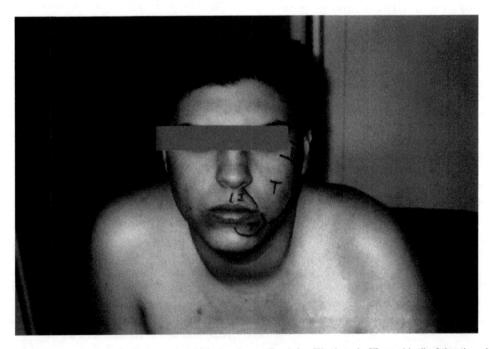

FIGURE 5.15 The reference zones of the "phantom" index finger (I), pinky (P), thumb (T), and ball of the thumb (B) of the left hand, on the face of a left-arm amputee. Source: Ramachandran and Hirstein, 1998. Reproduced with permission of Oxford University Press.

produce distracting, and sometimes excruciatingly painful, sensations that seem to be coming from a limb that no longer exists? A second relates to the conscious perception of touch. The cortical-reorganization model of the phantom-limb phenomenon must assume that the read-out of signals from the cortex that had represented the hand is still being "interpreted" as corresponding to the hand. Where and how does this interpretation take place? And why can't this erroneous interpretation be overridden by the patient's simultaneous awareness that the hand from which s/he seems to be experiencing sensations no longer exists?

Proprioception

Before concluding this section, we need to introduce the somatosensory modality of **proprioception**, which refers to the sense of where different parts of the body are relative to one another. Let's think back for a moment about the earlier section of this chapter on audition and the problem of localizing the source of a sound. It was noted in passing that it is in auditory brainstem nuclei that the computations are performed that allow us to localize where a sound in the environment is coming from. In a sentence, this is accomplished by the fact that we have two detectors (i.e., two ears), and that a sound emanating from a single source will be sensed as having slightly different amplitudes (the **interaural intensity difference**, IID), and will be sensed at slightly different times (the **interaural timing difference**, ITD), by the two ears. One need not understand the nuts and bolts of how sound localization works to appreciate the following: a sound originating 10 feet (3 meters) behind you will produce markedly different IIDs and ITDs depending on whether your head is pointed straight ahead or whether it is turned to the side (e.g., if you are looking off to your left); yet the proverbial tiger that is stalking you is in the same location relative to your body, and so your most appropriate response is the same regardless of how your head happens to be oriented at the moment. The same is true for vision: if you are looking straight ahead, the tiger that is directly in front of you would occupy your foveal vision, whereas if your head is turned to the left, the tiger would occupy the extreme right side of your visual field. Thus, to accurately process information coming in through the ears, or through the eyes, the brain needs to know how the head is positioned relative to the torso. (And for vision, an additional factor that will receive more attention in *Chapter 9* is that the brain also needs to know the position of the eyes relative to the head.) Information about the position of the head relative to the torso is provided by proprioceptive receptors located in the neck. For our purposes, these are stretch receptors that gauge the state of the neck muscles.

Adieu to sensation

The final section of *Chapter 4* already disabused us of the idea that sensation and perception can be construed as discrete, serial processes, the latter "taking over" *where* and *when* the former terminates. This was reinforced in the present chapter, via evidence for top-down control of the very process of transduction of the acoustic signal in the auditory system, and for the synchronization of whisking with the oscillatory state of somatosensory cortex. Beginning with *Chapter 6*, however, we will be fully "in the brain," and squarely in the domain of perception. Or will we? Those who prefer their science to come in tidy, clearly demarcated superordinate categories may be dismayed to learn that we're transitioning out of the frying pan of the interactive continuum along which sensation and perception lie, and into the fire of an interactive continuum linking perception and memory. It'll be messy at times. We're going to get dirty. But it will be so worth it!

END-OF-CHAPTER QUESTIONS

1. How might the mechanical properties of the cochlea result in the fact that the tuning of auditory nerve fibers broadens with increasing sound intensity?

2. What is meant by the term "spectral decomposition"? How is the operation carried out in the auditory system?

3. What components of the AEP are sensitive to attentional state? What can one infer from this about which stages of auditory processing are, vs. are not, influenced by attention?

4. Which part of your body is more densely innervated with mechanoreceptors, your tongue or your forearm? What are the corresponding differences in how these two parts of your body are represented in S1?

5. What are the two properties of optogenetics that make it such a powerful tool for studying the brain?

6. How specific or nonspecific is use-dependent plasticity? How dependent on attention? Why are the answers to these questions important?

7. What kind of data has been used to assess the contributions of unmasking vs. sprouting to somatosensory reorganization after deafferentation? What might be a broader implication to this question?

8. Does the cortical-reorganization hypothesis of phantom-limb syndrome assume that a neuron can simultaneously have two receptive fields? Explain.

REFERENCES

Cardin, Jessica A., Marie Carlén, Konstantinos Meletis, Ulf Knoblich, Feng Zhang, Karl Deisseroth, . . . Christopher I. Moore. 2009. "Driving Fast-Spiking Cells Induces Gamma Rhythm and Controls Sensory Responses." *Nature* 459 (7247): 663–667. doi: 10.1038/nature08002.

Kandel, Eric R., James H. Schwartz, and Thomas M. Jessell, eds. 2000. *Principles of Neural Science.* New York: McGraw-Hill.

Merzenich, Michael M., Gregg H. Recanzone, William M. Jenkins, Terry T. Allard, and Randolph J. Nudo. 1988. "Cortical Representational Plasticity." In *Neurobiology of Neocortex: Report of the Dahlem Workshop on Neurobiology of Neocortex Berlin, 1987 May 17–22,* edited by Pasko Rakic and W. Singer, 41–67. Chichester: John Wiley and Sons.

Penfield, Wilder, and Theodore Rasmussen. 1950. *The Cerebral Cortex of Man.* A Clinical Study of Localization of Function. New York: Macmillan.

Picton, Terrence W., Steven A. Hillyard, Howard I. Krausz, and Robert Galambos. 1974. "Human Auditory Evoked Potentials. I: Evaluation of Components." *Electroencephalography and Clinical Neurophysiology* 36: 179–190. doi: 10.1016/0013-4694 (74)90155-2.

Ramachandran, V. S., and William Hirstein. 1998. "The Perception of Phantom Limbs: The D. O. Hebb Lecture." *Brain* 121 (9): 1603–1630. doi: 10.1093/brain/121.9.1603.

Sekuler, Robert, and Randolph Blake. 1985. *Perception.* New York: Alfred A. Knopf.

Wolfe, Jeremy M., Keith R. Kluender, Dennis M. Levi, Linda M. Bartoshuk, Rachel S. Herz, Roberta Klatzky, . . . Daniel M. Merfeld. 2012. *Sensation and Perception,* 3rd ed. Sunderland, MA: Sinauer Associates.

Yantis, Steven. 2014. *Sensation and Perception.* New York: Worth.

OTHER SOURCES USED

Ganguly, Karunesh, and David Kleinfeld. 2004. "Goal-Directed Whisking Increases Phase-Locking between Vibrissa Movement and Electrical Activity in Primary Sensory Cortex in Rat." *Proceedings of the National Academy of Sciences of the United States of America* 101 (33): 12348–12353. doi: 10.1073/pnas. 0308470101.

Morel, Anne, Preston E. Garraghty, and Jon H. Kaas. 1993. "Tonotopic Organization, Architectonic Fields, and Connections of Auditory Cortex in Macaque Monkeys." *Journal of Comparative Neurology* 335 (3): 437–459. doi: 10.1002/cne. 903350313.

Picton, Terrence, and Steven A. Hillyard. 1974. "Human Auditory Evoked Potentials. II: Effects of Attention." *Electroencephalography and Clinical Neurophysiology* 36: 191–199. doi: 10.1016/ 0013-4694(74)90156-4.

Recanzone, Gregg H., Michael M. Merzenich, William M. Jenkins, Kamil A. Grajski, and Hubert R. Dinse. 1992. "Topographic Reorganization of the Hand Representation in Cortical Area 3b of Owl Monkeys Trained in a Frequency-Discrimination Task." *Journal of Neurophysiology* 67 (5): 1031–1056.

Schubert, Dirk, Rolf Kötter, and Jochen F. Staiger. 2007. "Mapping Functional Connectivity in Barrel-Related Columns Reveals Layer- and Cell Type-Specific Microcircuits." *Brain Structure and Function* 212 (2): 107–119.

Woolsey, Thomas A., and Hendrik Van der Loos. 1970. "The Structural Organization of Layer IV in the Somatosensory Region (SI) of Mouse Cerebral Cortex: The Description of a Cortical Field Composed of Discrete Cytoarchitectonic Units." *Brain Research* 17 (2): 205–242. doi: 10.1016/0006-8993(70)90079-X.

FURTHER READING

Belluck, Pam. 2013. Interview with Katherine Bouton. "Could Hearing Loss and Dementia Be Connected?" *New York Times,* February 11. http://wwwf.nytimes.com/2013/02/12/science/could-hearing-loss-and-dementia-be-connected.html?ref=disabilities&_r=0.
An interview with author Katherine Bouton, whose column in The New York Times *is also referenced here.*

Bouton, Katherine. 2013. "Straining to Hear and Fend Off Dementia." *New York Times,* February 11. http://well.blogs.nytimes.com/2013/02/11/straining-to-hear-and-fend-off-dementia/.
Newspaper column from a 65-year-old woman that eloquently captures the difficulties in daily life, as well as broader concerns, of a person experiencing age-related hearing loss.

Heckert, Justin. 2012. "The Hazards of Growing Up Painlessly." *New York Times Magazine*, November 18, 27–31, 54–55, 4.

An in-depth look at the life-altering consequences of being physiologically unable to experience a somatosensory modality that most of us take for granted, and often try to suppress.

Kaas, Jon H., ed. 2005. *The Mutable Brain: Dynamic and Plastic Features of the Developing and Mature Brain*. Amsterdam: Taylor and Francis.

A definitive collection of chapters on the plasticity of cortex, including such topics as time course of reorganization, and context dependent expansion/contraction of sensory and motor representations.

Lin, Frank R., E. Jeffrey Metter, Richard J. O'Brien, Susan M. Resnick, Alan B. Zonderman, and Luigi Ferrucci. 2011. "Hearing Loss and Incident Dementia." *Archives of Neurology*, 68 (2): 214–220. doi: 10.1001/archneurol.2010.362.

Scientific study investigating hypothesized link between hearing loss and dementia.

Lin, Frank R., Kristine Yaffe, Jin Xia, Qian-Li Xue, Tamara B. Harris, Elizabeth Purchase-Helzner, . . . Eleanor M. Simonsick, for the Health ABC Study Group. 2013. "Hearing Loss and Cognitive Decline in Older Adults." *JAMA Internal Medicine* 173 (4): 293–299. doi: 10.1001/jamainternmed.2013.1868.

Scientific study investigating hypothesized link between hearing loss and dementia.

Luck, Steven J. 2005. *An Introduction to the Event-related Potential Technique*. Cambridge, MA: MIT Press.

An accessible, authoritative introduction to ERPs, the reading of which will make one an informed and critical consumer of ERP research, regardless of whether or not one uses the technique.

Padden, Carol, and Tom Humphries. 2005. *Inside Deaf Culture*. Cambridge, MA: Harvard University Press.

A very compelling "insider's account" of what it's like to be deaf in contemporary society, written by two professors of communication at the University of California, San Diego.

Ramachandran, V. S., and William Hirstein. 1998. "The Perception of Phantom Limbs: The D. O. Hebb Lecture." *Brain* 121 (9): 1603–1630. doi: 10.1093/brain/121.9.1603.

A review from this celebrated neurologist and cognitive neuroscientist on the history, phenomenology, and neurobiology of this fascinating disorder.

Van Rullen, Rufin, and Julien Dubois. 2011. "The Psychophysics of Brain Rhythms." *Frontiers in Psychology* 2 (203): 1–10. doi: 10.3389/fpsyg.2011.00203.

A review of recent research on the idea that many perceptual and attentional processes may be regulated by neuronal oscillations.

Wolfe, Jeremy M., Keith R. Kluender, Dennis M. Levi, Linda M. Bartoshuk, Rachel S. Herz, Roberta Klatzky, . . . Daniel M. Merfeld. 2012. *Sensation and Perception*, 3rd ed. Sunderland, MA: Sinauer Associates.

Chapter 14 and Chapter 15 of this textbook provide excellent introductions to the neurobiology, psychology, and cultural anthropology of olfaction and taste, respectively.

CHAPTER 6
THE VISUAL SYSTEM

KEY THEMES

- The anatomical connectivity of the visual system creates a substrate for two broad classes of visual information processing: determining *what* an object is and determining *where* it is.

- Carefully designed lesion studies have localized these two categories of function to the temporal and parietal cortex, respectively.

- Within this organizing principle, there is considerable cross talk and integration of information.

- Visual object recognition is accomplished via a progressive hierarchy of processing steps, beginning in V1 and extending into inferotemporal (IT) cortex.

- Although one logical terminus of such a hierarchy might be an apex level of extreme localization, with one neuron representing each object that we've encountered in our lives, the data suggest a more plausible distributed combination-coding scheme.

- Since the 1960s, the processing of faces in the temporal cortex has been a focus of interest, and a source of puzzlement, because faces seem to follow different "rules" than other types of complex objects.

- Some populations of IT neurons seem to encode "the idea of a face," rather than the identity of any one specific face, thereby demonstrating an explicit interface between perception and memory.

- Training deep neural networks (DNNs) to categorize visual images can be an effective approach for studying principles of object recognition.

- An important role for feedback is to disambiguate conflicting bottom-up signals from early stages of the visual system.

CONTENTS

FAMILIAR PRINCIPLES AND PROCESSES, APPLIED TO HIGHER-LEVEL REPRESENTATIONS

This chapter will address the question of how we recognize objects that we encounter in the world. In one sense, it will represent our first foray into the unequivocally cognitive, in that recognizing what an object is necessarily entails *thinking* about where you've seen it before and what its significance is for you (e.g., *I see my [spouse/friend/professor/etc.]*), or, if it's an exemplar that you've never seen before, *thinking* about how it relates to what you already know about the world (e.g., *It's the face of a man/woman/young person/old person*) and what an appropriate action when encountering such an object might be. And so, as this preamble implies (as, indeed, does the word *recognition*), this is a domain that necessarily engages memory – stored representations of knowledge about the world; the very stuff of thought.

The emphasis of this chapter will be on the sensory modality of vision, with consideration given to how information from multiple sensory modalities can interact to influence how we perceive the world. At a detailed, mechanistic level, we shall see that a principle that we examined in *Chapter 4* also underlies object recognition. In that earlier chapter, we saw that features such as orientation, depth, and direction of motion are constructed by neurons in V1 via selective sampling across inputs from LGN neurons that have much simpler response properties. At a simplified level of description, the same principle, playing out in parallel in hundreds of thousands of neurons, is believed to underlie object recognition – just the right set of features, located in just the right part of the visual field, combine to represent a complex object. Thus, there is necessarily a hierarchical element to object recognition, in that many simpler elements are combined at lower levels to construct progressively more complex representations at higher levels. This idea of many feeding into one will lead us to revisit the tension introduced in *Chapter 1*, between localization vs. mass action (reframed here as localist vs. distributed representation). We shall also see that the principle of feedback from higher to lower levels of representation, first considered in detail in *Chapter 4*, is also characteristic of the hierarchies involved in recognizing objects. This will also invoke the principle of predictive coding, the idea that a neural representation of the world is already present in the brain, and that the brain uses incoming sensory inputs to update this representation.

As we have done previously, we'll start by considering the anatomical structure of the systems that support high-level vision.

TWO PARALLEL PATHWAYS

A diversity of projections from V1

You won't be surprised to learn that *Chapter 4*'s treatment of area V1, and of that region's interactions with MT, glossed over some of the intricacies both of V1's functional organization and of its feedforward projections. As is illustrated in *Figure 6.1*, there is both a direct route from V1 to MT, which carries signals primarily from the magnocellular pathway, and an indirect route, with neurons from V1 first projecting to V2 and V2 neurons projecting to MT. The indirect pathway carries information from both the magno- and parvocellular pathways, a result of the convergence in layer IV of V1 of inputs from magno- and parvocellular pathways on the neurons projecting to V2. V2, like V1, also shows heterogeneous cytochrome oxidase staining, with regions appearing as thick stripes, thin stripes, or interstripes (*Figure 6.1*). The indirect pathway from V1 to MT synapses in the thick stripe region of V2. MT, in turn, sends feedforward projections to area MST (located in *medial superior temporal* cortex in the macaque brain), which, in turn, sends feedforward projections to several areas in the posterior parietal cortex (PPC).

Separate from the direct and indirect projections to MT, different populations of V1 neurons project to the thin- and interstripe regions of V2. These regions of V2, in turn, send feedforward projections to area V4 of the temporal cortex, with V4's feedforward projections extending into the inferotemporal (IT) cortex. More detail about the anatomical connectivity of the visual system appears further along in this section, but at this point, we'll focus on the fact that there are two pathways originating in V1: one that projects through the thick stripes of V2 to MT, MST, and PPC, and the second that projects through the thin- and interstripe regions of V2 to V4, and then IT. As suggested by this anatomy, there are important, and importantly dissociable, visual functions supported by the posterior parietal and temporal lobes.

Insights into the visual functions of the PPC and IT have come from both of the scientific traditions highlighted at the end of *Chapter 1*: behavioral neurology/neuropsychology and systems neuroscience. Those clearly associated with the former include case studies detailing visual impairments in patients whose strokes have

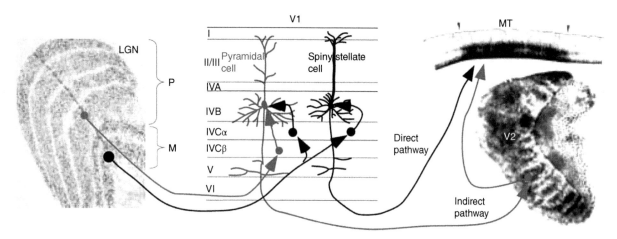

FIGURE 6.1 Dual routes from LGN to MT. The left-hand panel illustrates a cross section through the LGN, highlighting the two layers containing magnocellular (M) neurons, and the four layers containing parvocellular (P) neurons. Superimposed in black ink is a cartooned magnocellular neuron projecting to two neurons in layer IVCα of cortical area V1. Superimposed in red ink is a cartooned parvocellular neuron projecting to a neuron in layer IVCβ of cortical area V1. Of the two IVCα neurons receiving magnocellular input, one projects to a "spiny stellate" neuron in layer IVB, which, in turn, projects directly to area MT, and the other to a pyramidal neuron in layer IVB. The layer IVCβ neuron also projects to this layer IVB pyramidal neuron, which can thus be characterized as "blending" input from both the M and P channels. The IVB pyramidal neuron projects to a neuron in the thick stripe region of cortical area V2 which, in turn, also projects to MT. This figure shows a portion of the surface area of V2 that has been stained for cytochrome oxidase. Note that immediately to the right of the thick stripe highlighted in this image is an interstripe and, to the right of the interstripe, a thin stripe. Also note that, in this image, the portion of V2 that is illustrated is shaped like a boomerang, with a clear discontinuity between it and the cortical tissue bordering it to the right. The region to the right is V1, identifiable from the small spots (i.e., *blobs*) of cytochrome oxidase staining. Finally, MT is illustrated here in cross section, with deep layers at the bottom of the image, and has been stained to emphasize one of its distinguishing characteristics, which is heavy myelination of fibers even after they have entered the cortex. (Note that this dark staining ends abruptly near the middle of the cortex; why should that be?). Source: From Born, Richard T. 2001. "Visual Processing: Parallel-er and Parallel-er." Current Biology 11 (14): R566–R568. doi: 10.1016/S0960-9822(01)00345-1. Copyright 2001. Reproduced with permission of Cell Press.

damaged circumscribed regions of cortex. Those clearly associated with the latter include extracellular electrophysiological studies in monkeys performing visual discrimination tasks, and detailed studies of neuroanatomy. Many examples of this research will be considered further along in this chapter, but we will start examining these functions via a study performed in the late 1960s at a laboratory firmly grounded in both scientific traditions: the Laboratory of Neuropsychology at the National Institutes of Health.

A functional dissociation of visual perception of what an object is vs. where it is located

If one were forced to choose a label for the Section of Neuropsychology (as it was then called) in the 1960s and

1970s, one would classify it as a systems neuroscience laboratory – the primary experimental approach was one of testing hypotheses about the neural bases of cognitive functions by surgically lesioning specific brain regions and assessing the effects on carefully designed behavioral tasks. Many of those in the lab, however, also had formal training in human neuropsychology, studying behavioral abnormalities in neurological patients. In this environment, a graduate student named Walter Pohl conducted a series of experiments that, in retrospect, seem to have been "made to order" for this textbook writer. I say this for three reasons: first, they illustrate an important principle of the functional organization of the visual system; second, they illustrate a key principle of inference in neuropsychological research, the double dissociation of function; and third, they illustrate the value of performing thorough, carefully controlled experiments, in that an experiment that was intended at the time as a "mere" control condition turned out to provide critical, substantive information.

The primary question of interest for Pohl was different from the one that we are currently addressing. It was motivated by the idea that "The spatial position of an object may be defined in either of two ways, [. . .] either by reference to the position of the observer himself who serves as the spatial anchor, or by reference to the position of another object which functions as an external landmark" (Pohl, 1973, p. 227). In particular, a previous study of brain-injured war veterans conducted by neuroscientist Josephine Semmes and colleagues had shown that while some had difficulty viewing a mark on a picture of a body and pointing to the corresponding location on their own body, others had difficulty following maps. The former is referred to as **egocentric localization** (i.e., using a self-centered frame of reference), and the latter as **allocentric localization** (i.e., using an other-centered frame of reference). Thus, Pohl's experiments weren't intended to be about the visual functions of posterior parietal and IT cortex, per se, but rather about two ways of perceiving space, and of acting on objects in the environment. (The topic of spatial frames of reference will be revisited in greater detail in *Chapters 7–9*, on spatial cognition and skeleto- and oculomotor control, respectively.) It is the inclusion of a critical control experiment that makes them so relevant for our present purposes.

As illustrated in *Figure 6.2*, Pohl (1973) collected data from monkeys performing a "landmark discrimination" task, intended to assess the ability to judge the location of a "landmark" stimulus, and an "object discrimination" task, intended to assess visual object recognition. Depending on the experiment (four are reported in Pohl [1973], but only two are emphasized here), the monkeys had surgical lesions of PPC, IT, and prefrontal cortex, or were unoperated control subjects. The behavioral tasks, administered in a Wisconsin General Testing Apparatus (WGTA) (*Figure 6.2.A*), were variants of the reversal learning procedure, which accommodates the fact that, unlike human research subjects, one cannot give animals verbal instructions. Instead, the rate at which an animal can learn the rules for the task by trial-and-error guessing is used as one of the primary dependent measures. In the landmark reversal task, for example, *Figure 6.2.B* illustrates what the monkey might see on the very first trial of the experiment. The food reward was always hidden in the well adjacent to the landmark stimulus. If the monkey reached correctly, it retrieved the reward, the screen was lowered, and the tray prepared for the subsequent trial. If the monkey incorrectly reached for the well that was distant from the landmark, it would find this well to be

empty. On such error trials, feedback was given on a subsequent correction trial by re-presenting to the animal the same stimulus array, but with the incorrect food well now uncovered, so that the animal could see that it contained no reward. Regardless of whether the prior trial had been performed correctly or incorrectly, on the subsequent trial there was an equal likelihood that the landmark would appear on the right or on the left.

Each day the monkey performed 30 trials in this fashion (not counting correction trials) until the experimenters deemed it to have learned the rule, defined as getting at least 28 of 30 trials correct on one day's testing (i.e., ~93%), followed by a subsequent day of getting at least 24 of 30 trials correct (i.e., ~80%). The primary measure of performance was the number of errors that monkeys made before reaching this level of performance. Note that ending the experiment here, however, would leave interpretation of the results ambiguous, for two reasons. One is that this criterion for learning, although seemingly reasonable, is also arbitrary. Second, what if, for example, a given monkey was just "slow to catch on" to the rule, independent of the cognitive function that the task is intended to capture? (Note from *Figure 6.2.H*, for example, that, on the landmark-reversal task, the IT-lesioned group performed worse than the PPC-lesioned group on the first set of trials, as, indeed, did the unoperated control animals!) As *Figure 6.2.H* and *Figure 6.2.I* illustrate, however, including the reversal element to the experimental design largely alleviates such concerns about subjectivity and experimental error: on the day after the monkey had achieved criterion-level performance, the rule was reversed. That is, from this day forward, the reward would be placed in the food well that was *not* adjacent to the landmark. As on day one, the monkey had to learn the new rule by process of trial and error. In this way, the experiment proceeded, through a total of seven reversals.

The object reversal task (*Figure 6.2.E*) followed the same procedure, with the exception that the animals had to learn, on each set of trials, under which of two distinctive objects (a red striped cylinder vs. a silver cube) the reward was located.

The results from these experiments showed persuasively that PPC is selectively important for performance on the landmark task, but not for performance on the object discrimination task. Conversely, IT cortex was shown to be selectively important for the object discrimination task, but not the landmark task. The particular pattern of these results – that damage to area A affected performance on task X, but not on task Y, whereas damage to area B affected performance on task Y, but not on task X – constitutes a

FIGURE 6.2 The experimental procedures used by Pohl (1973).

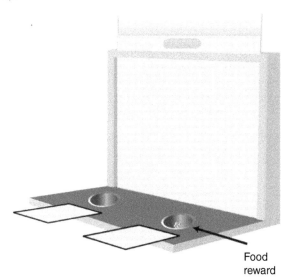

Food
reward

FIGURE 6.2.A Photograph of a macaque monkey performing a task in a WGTA. Note that the animal is in a cage, and can interact with items on the wooden testing tray by reaching through the bars in the cage. The tray is on wheels and can be pulled away or toward the animal via ropes on pulleys. A screen can be lowered such that the animal can neither see, nor reach, through the bars of the cage while a trial is being prepared. On the right side of the image is a one-way mirror, though which an experimenter can observe the animal's behavior, and from where he or she can control screen and tray. Source: Harlow Primate Laboratory, University of Wisconsin, Madison.

FIGURE 6.2.B Illustration of a two-food-well configuration of the testing tray, with a food reward being placed in one of the two wells. (Pohl [1973] used a half peanut on each trial.) Source: Adapted from Curtis and D'Esposito, 2004. "The Effects of Prefrontal Lesions on Working Memory Performance and Theory." Cognitive, Affective, and Behavioral Neuroscience 4(4): 528–539.

Correct
choice

FIGURE 6.2.C Illustration of a monkey reaching for one of the two covered food wells. Because the animal cannot see the contents of either food well, it must make its choice based on some second-order factor not related to the wells themselves (e.g., a guess, a memory of the previous trial [these would be considered "internal" factors, because they originate with the animal itself], or, as illustrated in panels **D** and **E**, the configuration of stimuli in relation to the two wells.) Source: From Curtis, C. E., and D'Esposito, M. 2004. "The Effects of Prefrontal Lesions on Working Memory Performance and Theory." Cognitive, Affective, and Behavioral Neuroscience 4(4): 528–539. Copyright 2004. Reproduced with permission of Springer Nature.

FIGURE 6.2.D The monkey's view of the testing tray during a trial of the landmark reversal (i.e., spatial discrimination) task. In the actual experiment, the landmark was a red cylinder with vertical white stripes. Source: From Pohl, Walter. 1973. "Dissociation of Spatial Discrimination Deficits Following Frontal and Parietal Lesions in Monkeys." Journal of Comparative and Physiological Psychology 82 (2): 227–239. doi: 10.1037/h0033922. Reproduced with permission of APA.

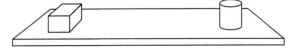

FIGURE 6.2.E The monkey's view of the testing tray during a trial of the object reversal (i.e., object discrimination) task. Each object is placed directly on top of a food well. Source: From Pohl, Walter. 1973. "Dissociation of Spatial Discrimination Deficits Following Frontal and Parietal Lesions in Monkeys." Journal of Comparative and Physiological Psychology 82 (2): 227–239. doi: 10.1037/h0033922. Reproduced with permission of APA.

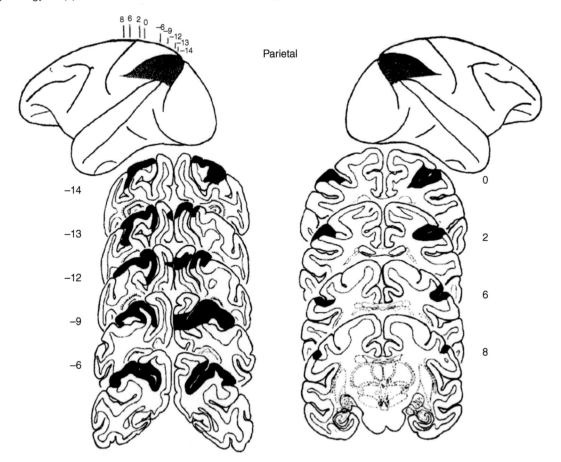

FIGURE 6.2.F Illustration of a PPC lesion. The numbers in the lateral view of the right hemisphere correspond to each of the coronal sections shown below, and also refer to the distance, in millimeters, from an origin point in an atlas of the monkey brain. Note that although the figure is labeled "parietal," great care was taken that the lesions did not invade more anterior areas known to be important for somatosensation (as detailed in *Chapter 5*). Source: From Pohl, Walter. 1973. "Dissociation of Spatial Discrimination Deficits Following Frontal and Parietal Lesions in Monkeys." Journal of Comparative and Physiological Psychology 82 (2): 227–239. doi: 10.1037/h0033922. Reproduced with permission of APA.

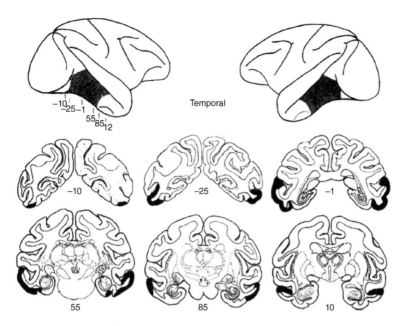

FIGURE 6.2.G Illustration of an IT lesion, same conventions as *Figure 6.2.F.* Source: From Pohl, Walter. 1973. "Dissociation of Spatial Discrimination Deficits Following Frontal and Parietal Lesions in Monkeys." Journal of Comparative and Physiological Psychology 82 (2): 227–239. doi: 10.1037/h0033922. Reproduced with permission of APA.

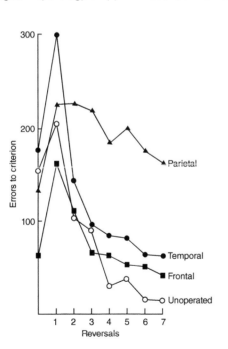

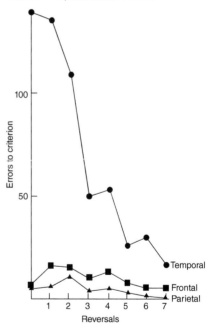

FIGURE 6.2.H Behavioral performance of monkeys from four experimental groups on the landmark reversal (i.e., spatial discrimination) task, illustrating markedly inferior performance by the parietal-lesioned animals. Source: From Pohl, Walter. 1973. "Dissociation of Spatial Discrimination Deficits Following Frontal and Parietal Lesions in Monkeys." Journal of Comparative and Physiological Psychology 82 (2): 227–239. doi: 10.1037/h0033922. Reproduced with permission of APA.

FIGURE 6.2.I Behavioral performance of monkeys from three experimental groups on the object reversal (i.e., object discrimination) task, illustrating markedly inferior performance by the inferior temporal-lesioned animals. (An unoperated control group was not included in the object reversal testing.) Source: From Pohl, Walter. 1973. "Dissociation of Spatial Discrimination Deficits Following Frontal and Parietal Lesions in Monkeys." Journal of Comparative and Physiological Psychology 82 (2): 227–239. doi: 10.1037 /h0033922. Reproduced with permission of APA.

double dissociation of function. (In the context of an analysis of variance [ANOVA], it would be identified as a crossover interaction.) This was important, because it supported conclusions about the specificity of the functions supported by these two brain areas. To illustrate why this is important, imagine for a moment that Pohl had only tested these three groups of monkeys on the landmark task. Although the larger deficit in the PPC-lesioned animals would be consistent with a role for that brain region in spatial perception, it would also be consistent with any number of equally plausible alternative interpretations. What if, for example, the true effect of the PPC lesions was to produce an incessant, loud ringing in the ears, such that the animals would have difficulty holding their concentration on a difficult task for a long period of time? If this were true, it could be that the PPC-lesioned animals would also have performed worse on the landmark task. However, the fact that a different lesion group, the IT-lesioned group, was impaired on the object task, rules out the possibility that the impairment of the PPC group on the landmark task was only due to the difficulty of the task.

In interpreting his findings, Pohl (1973) also drew on many previous theories and findings to advance the interpretation that regions of PPC may be "specialized for extrafoveal visual functions, as in the perception of spatial relations among objects in the visual field [. . . whereas IT cortex may be] specialized for perceptual functions related to foveal vision, as in object and pattern discrimination" (p. 238). (*Figure 6.3* illustrates what is meant by extrafoveal vs. foveal vision.)

The profound importance of the Pohl (1973) study was manifest a few years later when Pohl's mentor at the Laboratory of Neuropsychology, Mortimer Mishkin, together with colleague Leslie Ungerleider, published a thorough review of what was known about the anatomical connections of occipital-to-temporal and occipital-to-parietal cortex, about the effects of lesions to IT vs. PPC, and about the electrophysiological properties of neurons in IT vs. PPC (*Figure 6.3*), and argued that the former constituted a system for analyzing *what* an object is, and the latter a system for analyzing *where* an object is (*Figure 6.4*). This model has been highly influential in cognitive neuroscience, to the point that it has acquired two sets of nicknames, a functional one – *what* vs. *where* – and an anatomical one – *ventral stream* vs. *dorsal stream*.

Interconnectedness within and between the two pathways

As is often the case with conceptual models of brain function, the details of the functional organization of the visual system turn out to be much more complicated than what is

FIGURE 6.3 Electrophysiological properties of neurons characteristic of the *what* and *where* systems. For both panels, the vertical and horizontal axes represent the meridia of the visual field, the intersection of the two axes thus corresponding to the fovea, and each tick mark indicating 10° of visual angle.

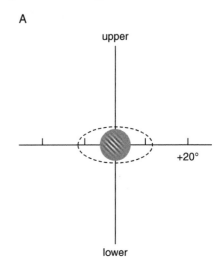

FIGURE 6.3.A For IT neurons, 60% have receptive fields (dashed oval) that span the vertical midline (i.e., they are bilateral), and the receptive fields of 100% of them include the fovea. These neurons are typically selective for particular visual features (here, an oriented grating).

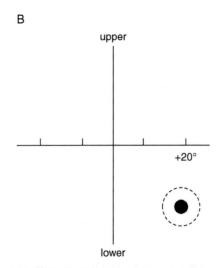

FIGURE 6.3.B For PPC neurons, the receptive fields of >60% of them are restricted to the contralateral visual field, and the receptive fields of >60% of them do NOT include the fovea. At the time it was believed that these neurons did not show selectivity for stimulus features, meaning that a PPC neuron would respond to any stimulus falling within its receptive field. (This assumption has since been disproven for at least some regions of the PPC.)

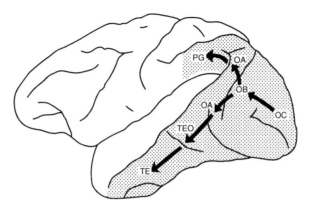

FIGURE 6.4 Schematization of the *what* and *where* pathways of the dorsal and ventral visual processing streams in the macaque monkey. The labels correspond to names given to these regions in a neuroanatomy atlas: OC is striate cortex; OB and OA are extrastriate occipital areas; PG is posterior and inferior parietal cortex; TEO and TE are IT cortex. Note that the sulcus forming the dorsal boundary of area PG is the intraparietal sulcus, which will feature prominently in chapters on attention, skeleto- and oculomotor control, and decision-making. Source: From Mishkin, Mortimer, Leslie G. Ungerleider, and Kathleen A. Macko. 1983. "Object Vision and Spatial Vision: Two Cortical Pathways." Trends in Neurosciences 6: 414–417. doi: 10.1016/0166-2236(83)90190-X. Copyright 1983. Reproduced with permission of Elsevier.

suggested, for example, by *Figure 6.4*. One example relates to information carried by the M vs. the P channels from the retina through V1, and onto higher levels of the visual system. The two-parallel-pathways model seems to fit quite well with the differential projection patterns, considered earlier in this section, of neurons in thick- vs. thin- and interstripe regions of area V2. Indeed, this led some to suggest that the dorsal visual processing stream may carry primarily magnocellular signals, whereas the ventral stream may carry primarily parvocellular signals. This logic led to a research strategy that sought to exploit this presumed correspondence of the M and P channels with the dorsal and ventral pathways, respectively, by presenting stimuli under **isoluminant** conditions. That is, the stimulus display was controlled such that all objects, including the background, were presented with the same brightness (the same *luminance*), and so only differences in chrominance (i.e., color) could be used to discriminate objects from each other, and from the background. The logic was that an isoluminant display would require exclusive reliance on the ventral stream, because the dorsal stream was presumed to be dominated by the M channel. Isoluminance, therefore, might be a way to study the functions of the ventral stream while experimentally keeping the dorsal stream "offline."

Follow-up studies, however, have called the logic of testing at isoluminance into question. One persuasive line of work, by Peter Schiller and his colleagues, examined the effects of selectively lesioning one channel or the other (via small lesions in selective layers of the LGN), and then testing performance on tasks that require visual processing characteristic of the dorsal or the ventral stream. These studies did, indeed, find preferential impairment on "ventral-stream tasks" after lesions of the P system, and on "dorsal-stream tasks" after lesions of the M system, but the impairments were relatively subtle, not the devastating effects that one would predict from frank lesions of the temporal or parietal cortex. Thus, for example, parvocellular lesions impaired discrimination of textures with high spatial frequencies, but not those with low spatial frequencies. Magnocellular lesions, in contrast, impaired the ability to discriminate a flickering light from a steady light at high, but not low, rates of flicker. (Note that, as the rate of flicker increases, at some point, it becomes so fast that it's no longer possible to discriminate from a steady signal.) Similarly, they impaired the ability to detect a difference in the speed of motion of a small group of dots within a larger field of moving dots, but only when the difference in speed was small, and thus difficult to detect. In both cases, deficits were exacerbated when the to-be-detected stimulus was presented at low contrast (e.g., the impairment was worse for a dim light flickering against a gray background than for a bright light flickering against a black background). Schiller and colleagues concluded that the important contribution of the M and P channels is to extend the range of performance of the visual system in two dimensions – spatial frequency and temporal – under challenging conditions. Thus, the P channel extends the visual system's resolution for fine-grained discriminations of texture and color, and the M channel extends its temporal resolution.

One implication of the findings of Schiller and colleagues was that the M and P pathways each contribute to functions of both the dorsal and the ventral streams. Consistent with this is what we already saw in *Figure 6.1*, that the anatomy indicates that there is already considerable blending of inputs from magno- and parvocellular neurons at the level of V1. Additionally, as portrayed in *Figure 6.5*, we now know that there are many instances of direct, reciprocal interconnections between regions of the two streams considered to be at comparable levels of the visual processing hierarchy, such as between V4 of the ventral stream and MT of the dorsal stream.

Two final comments about the *what* vs. *where* model before we delve into the functions of the ventral stream. First, it's noteworthy that *Figure 6.4* omits the feedback projections that we already know from earlier chapters make fundamental contributions to how the brain works. This will be taken up near the end of this chapter. Second, an additional level of complexity that we won't consider in detail until *Chapter 8* (*Skeletomotor control*) hearkens back to the question raised in *Chapter 1* of *what is a function*: some argue that the dorsal stream is better construed as carrying out the function of *how* the body carries out actions in space, rather than of determining *where* objects are located in space.

Having now established some gross anatomical boundaries for the visual system, and a framework for thinking about how the functions of this system are organized (i.e., *what* vs. *where*), we can proceed to consider the question *How does visual object recognition work?* In the process, some of the ideas left underspecified in the preceding paragraph, such as that of the "visual processing hierarchy," will be fleshed out. For the remainder of this chapter, periodically returning to *Figure 6.5* will be helpful, and, in the process, its "subway system wiring diagram" will become progressively more familiar and less daunting!

THE ORGANIZATION AND FUNCTIONS OF THE VENTRAL VISUAL PROCESSING STREAM

Hand cells, face cells, and grandmother cells

Electrophysiological recordings in temporal cortex

By the 1960s, decades of clinical observations in humans and lesion studies in experimental animals had established that the temporal lobes play a critical role in object recognition. What remained unknown, however, was *how* this region carried out that function. *What were the mechanisms that gave rise to object recognition?* It was the pioneering studies of Hubel, Wiesel, and their colleagues, of the functions of V1 (*Chapter 4*), that provided the methodological and conceptual framework that could be applied to this question. Among the first to do so were Charles Gross and his colleagues, who began recording from neurons in IT cortex, and subsequently the superior temporal sulcus, in the monkey in the 1960s and 1970s. Because the previous literature indicated that lesions of IT cortex produced profound object recognition deficits, they began by measuring responses to a wide array of complex objects.

In a narrative reminiscent of that from Hubel and Wiesel's early experiments, Gross and colleagues initially had difficulty finding any type of object that would reliably elicit a response. Their first successful recording came when, after failing to activate a neuron with a large set of stimuli, an experimenter waved his hand in front of the display screen, and the neuron responded to it vigorously. *Figure 6.6* illustrates a series of stimuli that the researchers subsequently presented to this neuron. There is a subjective sense from this stimulus set that the closer the stimulus was to resembling a monkey's hand, the more effective it was at activating the neuron. Thus, it didn't seem to be the case that this neuron's preferred stimulus was a more "primitive," or generic shape, and that the hand just happened to have many of the same characteristics of this neuron's preferred stimulus characteristics. Rather, it seemed that this neuron's preferred stimulus was truly the hand, and other stimuli could "drive" it with more or less effectiveness as a function of how closely they resembled the hand. Another noteworthy property of this neuron was that it responded to the hand stimulus with a comparably vigorous response when it was positioned in any of many locations in the visual field. That is, it had a very large receptive field. Although this experiment did not find other neurons with the same selectivity for the hand stimulus shown in *Figure 6.6*, it did find many visually responsive neurons in IT that responded better to "complex" stimuli than to simple geometric shapes. Typically, each of these neurons had its own unique preferred stimulus. The property that IT neurons had in common, however, was large receptive fields – often larger than $10° \times 10°$ of visual angle – within which the neuron would respond equally well to its preferred stimulus. (This property was highlighted by Ungerleider and Mishkin [1982], and is illustrated in *Figure 6.3*.) One way of describing this property of an IT neuron is that its sensitivity to its preferred stimulus is relatively **invariant** to the location in which it appears.

Perhaps even more provocative than the finding of an IT neuron with responses selective to a hand was a

FIGURE 6.5 Anatomical connectivity of the visual system. 1991. Source: Adapted from Felleman, Daniel J., and David C. Van Essen. 1991. "Distributed Hierarchical Processing in Primate Visual Cortex." Cerebral Cortex 1 (1): 1–47. doi: 10.1093/cercor/1.1.1.

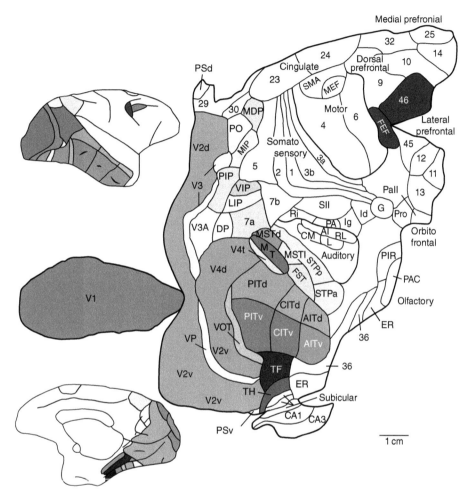

FIGURE 6.5.A Color-coded views of the right hemisphere of the macaque cerebral cortex (plus hippocampus) as seen from lateral (top-left inset) and medial (bottom-left inset) views, and as a virtual "flat map" that illustrates the entire cortical surface. To create a flat map that preserves the relative size of each region (e.g., that doesn't "stretch" or otherwise distort), it is necessary to make several virtual "cuts." Most obvious in this view is that V1 has been "cut out" along the entirety of its border with V2, with the exception of its anterior-most point on the lateral surface. The next major cut is on the medial surface, "carving" around the roughly circular boundary of cortex and non-cortex (revisiting *Figure 1.2.B* and *Figure 2.2* may be helpful here). At this point, one can imagine approaching the brain from the lateral view and reaching one hand, palm down, over the parietal cortex and down along the midline, and the other hand, palm up, under the ventral surface of the temporal lobe with fingers curling up and in along its medial surface, and then pulling up with the top hand and down with the bottom so as to "unroll" this 3-D shape and flatten it. If, in the process, one were also to gently stretch by pulling the two hands further apart, the interior of the sulci would also be exposed. It is thus, for example, that the red and yellow bands of the tissue that is not visible in the lateral view, by virtue of lying within the superior temporal sulcus, are exposed in the flat-map view. Finally, two additional cuts are needed to complete the task, each separating regions that are adjacent in the intact brain. One is a cut separating prefrontal areas 46 from 45 (running approximately along the fundus of the principal sulcus), and the second is a slit starting at the temporal pole and proceeding caudally along the medial surface, thereby artificially dividing the entorhinal cortex (ER) into two. This type of flat map is also often used to display results from fMRI studies of the human brain, as we shall see further along in this book.

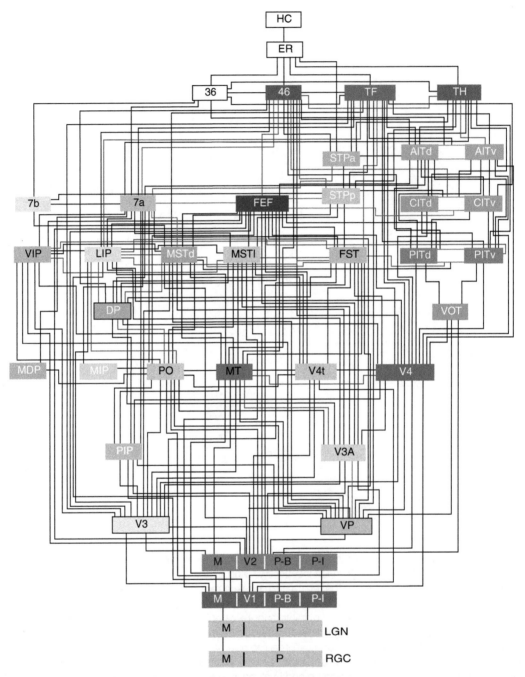

FIGURE 6.5.B Connectivity diagram of the visual system as a hierarchically organized set of interconnected regions, with each region's position in the hierarchy determined by its anatomical connectivity with all the others, as determined by laminar patterns of its inputs and outputs (see *Figure 2.4*, and associated text from *Chapter 2*). Color coding corresponds to *Figure 6.5.A*. Each line is understood to correspond to bidirectional connections between two connected regions, with feedforward connections going from lower to higher on the diagram (e.g., from MT to one of the MST subregions) and feedback connections the opposite. Thus, each horizontal level of the diagram corresponds to a putative "hierarchical level of cortical processing." The bottom level is the retina, with retinal ganglion cells (RGCs) divided into M and P channels, which project to the magnocellular (M) or parvocellular (P) layers of the LGN (second level). The first two levels of cortex, V1 (purple) and V2 (pink), are each divided into sectors that carry unequivocally M or P-blob (P-B) or P-interblob (P-I) signals; connections from regions that blend these signals are labeled "V1" and "V2."

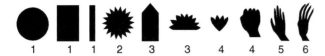

FIGURE 6.6 Examples of shapes that Gross, Rocha-Miranda, and Bender (1972) presented to a neuron in IT cortex, rank ordered by their ability to elicit activity in it, from none (1) or little (2 and 3) to maximum (6). Source: From Gross, Charles G., C. E. Rocha-Miranda, and D. B. Bender. 1972. "Visual Properties of Neurons in Inferotemporal Cortex of the Macaque." Journal of Neurophysiology 35 (1): 96–111. Reproduced with permission of The American Physiological Society (APS).

subsequent report by this group of the results of recording from the superior bank of the superior temporal sulcus. Referring back to *Figure 6.5*, this corresponds to the regions labeled STPp (superior temporal polysensory – posterior), FST (**fundus** [or "floor"] of the superior temporal gyrus), and STPa (superior temporal polysensory – anterior). Gross and colleagues named this area the **superior temporal polysensory area (STP)** because many neurons in this area respond to stimuli from other sensory modalities in addition to visual stimuli. In one report, they found that nearly all neurons responded to visual stimuli, and of these, 21% responded to visual and auditory stimuli, 17% to visual and somesthetic stimuli, 17% to stimuli from all three modalities, and 41% exclusively to visual stimuli. The most provocative finding from this study was that a small number of neurons (7) was found that responded selectively to images of monkey and human faces. More specifically, these neurons seemed to respond to the category of *face*. For example, *Figure 6.7* illustrates a neuron that responded equally robustly to images of different monkeys' faces, as well as to an image of a human face. Psychologists make the distinction between **type** (or category) of stimulus, such as the category of faces, and a **token**, a specific exemplar drawn from within this category, such as the face of a particular person. Thus, the "face cells" discovered by Gross and colleagues seemed to be selective for stimuli of this particular type, but not for specific tokens within this category.

The two properties of the neurons of IT that we have emphasized here – a high degree of selectivity for complex stimuli and large receptive fields – were in marked contrast to the known properties of neurons from V1. Neurons in V1 respond with high specificity to well-defined, low-level features, such as a boundary or line within a narrow range of orientation or the end of a line. Additionally,

within a module in V1 (i.e., within a pinwheel), there is an organization such that adjacent columns of neurons represent a smooth transition in the feature being represented. In temporal cortex, in contrast, the hand and face cells responded to the complete shape, but, for the most part, not to its component parts. Furthermore, there was no evident topography to suggest that similar kinds of stimuli might be represented in adjacent areas of IT or STP. With regard to receptive fields, in V1, foveal receptive fields are quite small, on the order of $0.5° \times 0.5°$. The majority of neurons in temporal cortex, in contrast, had very large receptive fields. In STP, many of the receptive fields approached the size of the entire visual field. An additional feature that distinguished these temporal cortex neurons was that they could also be invariant to the size of their preferred image.

The invariance to location and to size of visually responsive temporal-lobe neurons had two implications. The first was that temporal cortex may contain the machinery to allow for the recognition of objects across a wide range of viewing conditions. This is a necessary property for vision in everyday life, because we rarely see an object under exactly the same viewing conditions on more than one occasion. For example, as I sit here typing this text, I recognize my mobile telephone lying on the desk to the left of my computer monitor. If I pick up the phone and bring it near my face, I will recognize it as the same unchanged, familiar object, this despite the fact that entirely different populations of cells in the retina, LGN, and V1 are processing low-level featural information about the object when it is in these two locations. Similarly, when my son walks into the room and turns on the light (I'm currently writing in my study at home, on an overcast day), I still recognize the telephone despite the fact that the colors that my retina detects differ considerably depending on the level of illumination in the room. The second implication was, in effect, a "problem" for the visual system that is created by these invariances. Note that when I look at my phone, pick it up, and set it down on the other side of my monitor, I not only recognize *what* it is but I also know *where* it is (*my phone is to the left of the monitor; my phone is near my face*; etc.). A mobile phone-tuned neuron with a large receptive field that spans the width of my computer monitor, however, won't carry information about where the phone is located. This follows from the fact that the response of this neuron is equivalent whether the phone is to the left of the monitor, to the right, or near my face. This sets up what has been referred to as the **binding problem**: *if ventral-stream neurons encode stimulus identity*

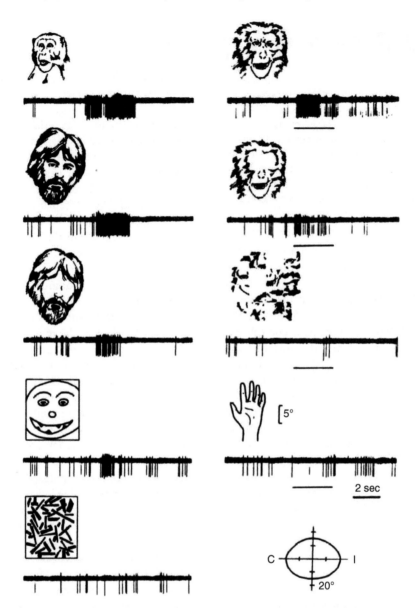

FIGURE 6.7 Responses of a face cell from area STP. Under each stimulus image is a trace from the electrode recording the neuron's activity, with each vertical bar corresponding to an action potential. Note that the neuron's response is comparably vigorous for the faces of two different monkeys and for a human face. The response is markedly weakened, though not abolished, when eyes are removed from the faces, as well as for a hand-drawn cartoon of a face. Changing the physical arrangement of the elements of an image, however, such as by scrambling the parts so that it no longer resembles a face, abolished the response. The lack of response to the hand is emblematic of the fact that this neuron did not respond to many other stimuli, ranging from bars, spots, and edges, to complex objects such as hands and brushes. The lower right illustrates the neuron's receptive field (C=contralateral; I=ipsilateral). Source: From Bruce, Charles, Robert Desimone, and Charles G. Gross. 1981. "Visual Properties of Neurons in a Polysensory Area in Superior Temporal Sulcus of the Macaque." Journal of Neurophysiology 46 (2): 369–384. Reproduced with permission of The American Physiological Society (APS).

but not location, and if dorsal-stream neurons encode stimulus location, but not identity, how is this information brought together to create the unitary, "bound" percept that we experience of an object occupying a location in space? The binding problem is of central interest in many branches of cognitive neuroscience. Although we will not arrive at definitive solution to it in this chapter (nor, indeed, in this book), we will return to it periodically.

Broader implications of visual properties of temporal cortex neurons

In one sense, the studies by Gross and colleagues fit well with the findings from prior decades, that lesions of the temporal lobe produced profound deficits in the ability to recognize objects. They also brought to light many thorny questions about precisely how visual object recognition is accomplished by the brain. Recall, for example, that we noted that face cells seemed to respond to any token drawn from this category, rather than to a specific face. This raises two interesting puzzles. The more concrete of the two is: what supports our ability to recognize and discriminate among the tens of thousands of faces that we've encountered across our lives? This presents a puzzle because the neurons reported by Gross and colleagues didn't seem to have the specificity to support this ability. We will return to this in the next paragraph. The second question follows from the fact that these face cells seemed to signal the presence of an item from the category "face," rather than the presence of any specific face. Thus, in a sense, these neurons seemed to be conveying information about the idea of what it means to be a face – a face is the part of the body of many animals that includes the eyes, nose, and mouth; a face is associated with the individual identity of an animal more so than other parts of the body; a face conveys the mood or state of mind of the individual; and so on – rather than about the particular object being viewed. If the IT and STP are, indeed, parts of the visual object recognition system, how does one get from the low-level operations carried out in the retina, LGN, and V1 to cells in temporal cortex processing "the idea of a face"? Processing information about "the idea of a face," if that's what these neurons were doing, involves cognitive processes that are outside the domain of vision, strictly defined. Engaging knowledge about what a face is, what it can do, how a face is typically configured, and so on falls under the rubric of semantic memory, a topic that we will address in detail in *Chapter 13*. That is, it entails accessing knowledge about the world that was already present in the brain prior to the particular instance of seeing a face that occurred during that experimental session. These are deep questions that make explicit how the study of perception is central to building an understanding of cognition. For the time being, we'll leave it that the discovery of face cells made explicit the fact that a full understanding of visual object recognition will need to take into account the interface between the processing of newly encountered information in the world with our long-term store of knowledge about the world.

Let's return now to the fact that the face cells described by Gross and colleagues did not seem to show a specificity for any individual face. The failure to find any such neurons nonetheless required scientists to confront the implications of what it would mean if they *had* found such neurons. The expectation/assumption that there will be neurons that are selective for individual faces implies a visual system in which the processing of information at each stage becomes progressively more specific until, at the highest level, there is one neuron that represents each of the millions of objects in the world that we can recognize. Such a scheme for organizing the visual system, and the existence of highly specialized "gnostic" units at the highest level, had been contemplated by the prominent Polish neuroscientist Jerzy Konorski (1903–1973). (The term "gnostic" has its roots in the Greek word for knowledge.) It had also been caricatured and disparaged by American neurophysiologist Jerome Lettvin (1920–2011) as the "grandmother cell theory." The idea was that if there is a neuron specialized for recognizing every object with which one is familiar, among these must be a neuron that is specialized for recognizing one's grandmother.

At first blush, the grandmother cell theory might have an intuitive appeal as the logical terminus of a hierarchical system of progressively increasing specialization. To be taken seriously, however, it would need to account for (at least) two major concerns. The first is vulnerability: if the system relies on highly specialized neurons at the apex of a processing hierarchy to represent each specific token that the organism can recognize, what would happen if one or more of these neurons were to be damaged? The second is the pragmatics of **phylogeny** and **ontogeny**: how, from the perspective of evolutionary or organismal development, could a system "know" a priori how many top-level gnostic units would need to be created and "set aside" such that the organism would have enough to represent each of the distinct objects that it would encounter throughout its life?

As we have seen, the research of Gross and colleagues in IT and STP provided some clues, and stimulated many ideas, about how visual object recognition might work. To make further progress, however, it would be necessary to "close the gap" between the properties of neurons at these most advanced stages of the ventral visual pathway and those of neurons at the earliest stages of cortical processing. Thus, it is to intermediate levels of the ventral stream that we turn next.

A hierarchy of stimulus representation

In the years between the early 1970s and today, many, many laboratories have performed experiments to "close the gap" between V1 and IT/STP. The picture that has emerged is generally consistent with the principle first introduced in *Chapter 4*: progressively higher levels of stimulus representation are constructed at progressively higher levels of the system by selective integration of more elemental information from earlier (or, in the terminology of a hierarchy, "lower") stages of processing. Ungerleider and Bell (2011) have likened this to the process of assembling words from syllables (*Figure 6.8*), which are, themselves, assembled from an even more elemental unit of written language, individual letters. In this analogy, words correspond to high-level, complete object representations, syllables to intermediate-level conjunctions of features, and letters to low-level features (such as those detected by V1). As detailed in *Chapter 4*, simple cells in V1 respond preferentially to specific luminance-defined orientations, and complex cells can signal the termination of lines. Neurons in area V2, located one synapse downstream, can respond to corners (which can be constructed from two differently oriented line segments). Neurons in V4, which receives

projections from both V1 and V2, have a wide array of response profiles: some V4 neurons respond preferentially to curved contours (as illustrated in *Figure 6.9*); some to a specific combination of a degree of curvature at a particular orientation; some to outlines of basic geometric shapes; and some to conjunctions of features, such as a line that is oriented at a particular angle *and* is of a particular color. At each of these stages, one sees that the stimulus information processed at that stage is more complicated than the information that was processed at the preceding stage(s) from which it receives feedforward inputs. Thus, one can say that neurons at higher stages in the ventral visual pathway become progressively more specialized. (For example, while a simple cell in V1 might respond to the "leading edge" of a bar tilted 45° to the right, regardless of color so long as the bar is darker than the background, a neuron in V4 might respond only to red bars of this orientation.) In most cases, one can see how the preferences of neurons at a particular stage of processing can be assembled, or, in some cases, inferred, from the convergence of inputs from an earlier stage. *Figure 6.9* offers an illustration of how this scheme of progressively more complex and specific stimulus representation might be engaged in the perception of a twisting tree branch. To return to the assembling-a-word analogy of visual perception (*Figure 6.8*), individual letters might be represented in V1 and V2, syllables in V4, and the "final" percept, in this case an automobile, in IT.

Figure 6.9 also illustrates the fact that, in parallel to the trend of progressively narrowing selectivity of the visual features to which a neuron will respond, neurons at progressively downstream stages of processing in the ventral stream have progressively larger receptive fields. Thus, although a neuron in V4 will respond to many fewer stimuli than will a neuron in V1, the V4 neuron will respond to

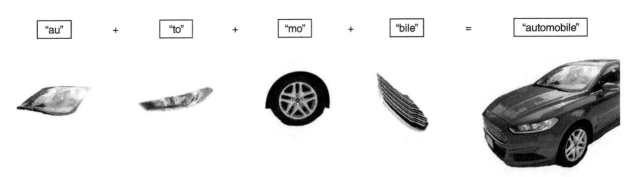

FIGURE 6.8 One analogy to object recognition is that just as words can be assembled from syllables, object representations may be assemblable from more primitive visual features.

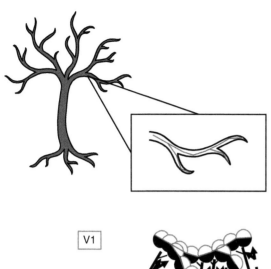

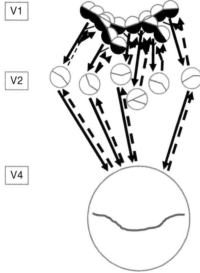

FIGURE 6.9 Hierarchical processing in the visual system. The perception of the branch of a tree (outlined with red line in call-out box) proceeds from the fine-grained representation of the dense orientation structure by orientation-selective neurons in V1, to the representation of corners and straight sections of varying length in V2, to a representation of the overall contour of the branch in V4. Note that receptive fields increase in size at each level of processing. Solid arrows represent the feedforward direction of information processing, and dashed arrows the feedback.

its preferred stimuli when they appear in a much larger range of the visual field. As we saw in *Chapter 4* and *Chapter 5*, this is a necessary by-product of integrating over several smaller receptive fields from a lower level. It is also important to note one of the efficiencies of a feature-based hierarchical object recognition system, which is that the lower levels of the system can contribute to the perception

of many, many objects. Returning again to the assemble-a-word analogy (*Figure 6.8*), if the word being presented were *apple*, instead of *automobile*, the "a" and "e" feature detectors (from our fanciful V1) could also contribute to the perception of this very different word. For *apple*, the *a* would feed forward to a neuron in V4 representing the syllable *app*, and the *e* detector to a neuron in V4 representing the syllable *le*. Neurons at the level of our fanciful area V4 could also contribute to the perception of multiple words, the *mo* detector, for example, also contributing to the perception of *motion*, or *dynamo*.

Paradoxically, however, what makes for an efficient system at lower levels might lead to a hopelessly inefficient, and perhaps impossible, situation at the highest level. That is, the logical end of a hierarchical chain of progressively more specialized feature detectors is a highest level of representation at which individual neurons are each dedicated to one of the seemingly infinite number of objects that we can recognize – one for an *automobile*, one for an *apple*, one for *motion*, one for a *dynamo*, and so on, ad infinitum.

One solution to the problems voiced here is to devise a scheme for object representation that does not rely on a "grandmother level," and that's what we consider next.

Combination coding, an alternative to the "grandmother cell"

A breakthrough in thinking about how object recognition might be accomplished at the highest levels came from Japanese neurophysiologist Keiji Tanaka and his colleagues at the Institute of Physical and Chemical Research (RIKEN). The key to their breakthrough came from their "reduction method" of studying stimulus selectivity in IT neurons. They made their recordings in anterior IT cortex, the regions labeled CITv and AITv (central IT, ventral portion; anterior IT, ventral portion) in *Figure 6.5*. Upon isolating a single neuron with a microelectrode, they presented the monkey with dozens of 3-D objects, drawn from many categories, so as to find the complex stimulus that was most effective at driving the cell. Having done so, they next created a 2-D facsimile of the 3-D object via the superposition of many simple features. *Figure 6.10* illustrates this process for an IT neuron that was discovered to be most responsive to a top-down view of a tiger's head. (Presumably, in the first step of the process, the investigators presented a toy tiger and the neuron responded vigorously. Upon holding the tiger at different distances, and rotating it to present different views, they determined that the strongest responses were elicited by displaying the head from this top-down view.) The key to the 2-D facsimile

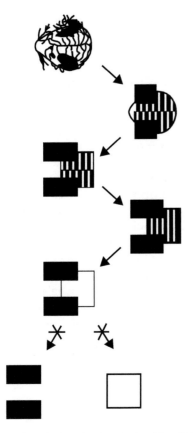

FIGURE 6.10 Tanaka's reduction method for determining the tuning properties of an IT neuron. After the neuron was determined to be maximally responsive to the top-down view of a tiger's head (top image), the investigators confirmed that it responded equally vigorously to a 2-D facsimile image comprising an oval, a pattern of black and white bars, and two black rectangles. The facsimile image was then "reduced" in a step-by-step manner until the investigators reached the simplest configuration of elements beyond which they could not further decompose the image without eliminating the neuron's response. Source: From Tanaka, Keiji. 1993. "Neuronal Mechanisms of Object Recognition." Science 262 (5134): 685–688. doi: 10.1126/science.8235589. Copyright 1993. Reproduced with permission of American Association for the Advancement of Science - AAAS.

image is that it was assembled to be "decomposable" into progressively simpler configurations. For the tiger head, for example, the 2-D facsimile image consisted of an oval (a stand-in for the shape of the head), a superimposed array of a texture of black and white bars ("stripes"), and two black rectangles (stand-ins for the ears). Critically, the

investigators confirmed that the neuron's response to this 2-D facsimile was comparable to its response to the tiger head itself. The "reduction" process, then, proceeded via the progressive removal, or in some cases simplification, of individual elements of the facsimile image, until reaching the point at which the neuron's response was no longer as robust as it had been to the original 3-D object.

For the IT neuron featured in *Figure 6.10*, this procedure revealed that this neuron's true selectivity was not for a top-down view of a tiger's head (which would be a highly specific configuration of features indeed!) but, rather, for a simpler configuration of a white square with two superimposed black rectangles. As the investigators heroically repeated this process with hundreds of real-world 3-D objects, they confirmed, again and again, that IT neurons that were initially found to respond vigorously to one of these objects were, in fact, tuned to a simpler, reduced set of generic features. Thus, the visual system may not accomplish the recognition of a top-down view of a tiger's head via a hierarchy of progressively more specialized neurons that terminates with a "top-down-view-of-a-tiger-head neuron" in a putative "grandmother cell" level at the apex of this hierarchy. Instead, the tiger's head may be recognized by the simultaneous (i.e., the "combined") activation of several tens (if not hundreds or more) of neurons that each represent a different set of features that uniquely make up this stimulus: one group of cells representing the configuration of two black rectangles superimposed on one white square, as illustrated in *Figure 6.10*, another group representing the black and white stripes, another group representing the moist, glistening skin of the nose, another group representing the whiskers, and so on. Note that, in contrast to the extreme localizationism implied by a hierarchy ending in a "grandmother level," the principle of combination coding proposed by Tanaka and colleagues has two hallmarks of a distributed system. First, there is no one place where a tiger head is represented. Instead, its representation entails the simultaneous activation of many elements that are distributed throughout IT cortex. Second, any one of these elements can participate in the representation of many different objects. For example, although the recognition of a top-down view of the head of an otter or a warthog is sure to activate many elements not activated by the tiger, it may nonetheless activate the same IT neurons that can represent whiskers.

It is noteworthy that *faces* was the one class of stimuli that Tanaka and colleagues found that were generally not amenable to decomposition by the reduction method. That is, when an IT neuron was found that responded vigorously to a face stimulus, it was generally not possible to decompose the image into a simpler collection of features that would still drive the cell equally well. (The one exception was a single neuron described in one study for which a face could be simplified to two eyes and a nose.) This finding was generally consistent with what Gross and colleagues had reported roughly 20 years earlier.

Object-based (viewpoint-independent) vs. image-based (viewpoint-dependent) representation in IT

Implied in the *assemble-a-word* analogy to object recognition (*Figure 6.8*) is the assumption that object representations are viewpoint independent. That is, it need not matter whether one is viewing, say, an automobile, from the front, side, or back; so long as one can identify enough constituent parts (e.g., one or more tires, a bumper, the red, curved metal of a part of the body), one can identify it as an automobile. The visual system would seem to be well constructed to perform such viewpoint-independent – or object-based – recognition. For example, whether viewing the car from the side, and therefore the curved chrome shape of the bumper appears in the left visual field, or from the front, with bumper occupying central vision just below the fovea – both regions of the visual field are covered by V1 feature detectors that will detect and pass along information about the contour and color of the bumper to higher levels of the system that will, progressively, complete the process of recognition. Although we've already discussed how the work of Tanaka and colleagues helped to dispel the idea that there may exist a "grandmother level" of representation, such a model nevertheless implies that there exists an abstract representation of "automobile" that is, well, independent of any specific angle from which it can be viewed.

There is, however, evidence consistent with alternative, *viewpoint-dependent* encoding of objects in IT. Behaviorally, it has been shown that people are faster, and more accurate, at recognizing objects from previously seen than from previously unseen orientations. This has been done, for example, with "bent paperclip" shapes that subjects would never have seen prior to participating in the experiment. Using these shapes, neurophysiologist Nikos Logothetis and colleagues have recorded from neurons in IT cortex while monkeys viewed the same type of stimuli. An intriguing finding from these studies is that some IT neurons showed their strongest response to a particular preferred viewing angle of a novel object, and these responses dropped off monotonically as the object rotated further from this preferred view.

It is obvious from the Logothetis studies (reviewed in Logothetis and Sheinberg, 1996), because they were performed with stimuli that their animals had never seen prior to the beginning of the experiment, that the neurons' viewpoint preferences with these stimuli were acquired through experience. Thus, viewpoint dependence for certain stimuli may be a property that can be acquired through experience. Indeed, this may be the basis for some forms of expertise. Having made these points, however, one must also consider the fact that, early in development, the establishment of viewpoint-dependent representations *must* precede that of viewpoint-independent representations, for the simple reason that our first encounter with any object is necessarily from a particular point of view. Indeed, it would seem that a viewpoint-independent representation can't be constructed until a minimum number of distinct views has been acquired. The extent to which viewpoint-dependent and -independent representations coexist, and interact, remains an important question for the cognitive neuroscience of object recognition.

A critical role for feedback in the ventral visual processing stream

Using deep neural networks to understand computations underlying object recognition

In recent years, as **deep neural networks (DNNs)** have begun to approach (and, in some cases, exceed) human levels of performance on categorization tasks, scientists have begun to apply deep learning to the study of visual and auditory perception, with the rationale being that, unlike with a real nervous system, every computation carried out by a DNN can be observed and quantified. Thus, learning how a DNN achieves levels of performance approaching that of a human or a monkey might provide novel insights about how this performance is achieved in biological systems. One study, from Yamins and colleagues (2014) in the laboratory of James DiCarlo at MIT, used a database of images, not unlike the one described in section *Deep neural networks, Chapter 3*, and presented these images to monkeys and to several different DNNs. The database contained 5760 black-and-white photographic images of objects drawn from eight categories: animals, boats, cars, chairs, faces, fruits, airplanes, and tables. The researchers used arrays of electrodes to record simultaneously from 168 neurons in IT while their monkeys viewed these images. Rather than "ask the monkey" to report what it had seen by requiring a behavioral response after each image, the researchers effectively "asked IT" what it was seeing by assembling a 168-D vector from each image (the activity of each neuron being a dimension), and then determining whether all the images from the *animals* category tended to cluster in the same region of 168-D space, if all the *boats* clustered, etc. Successful "decoding" of neural activity is achieved if a pattern classifier that has been trained on a subset of the 5760 images (let's say on 5000 of them) can successfully classify a picture that it has never seen before (one of the 760 images that was "held back" from training). For example, does it successfully classify an image of a banana as belonging with the fruit, instead of one of the other categories? (This approach to pattern classification was developed in a discipline of engineering called machine learning. For this particular study, the authors used a computer program called a "support vector machine" [SVM].)

How do DNNs differ from PDP models, such as we detailed in section *A PDP model of picture naming, Chapter 3*? To start with, recall that the DNN that was described in *Chapter 3* had an input layer analogous to a retina – a 256 × 256 array of sensors. Indeed, in panel A of *Figure 6.11*, the input layer of the "feedforward DCNN" is equated to the retina of the primate visual system. Also different from a PDP model, each layer of a DNN applies a series of operations to its input. One of these is **convolution**, which consists of applying a series of filters to the input and passing along to the next layer the output from these filters. It's conceptually very similar to the operations performed by a hypercolumn in V1: if an aspect of the stimulus matches a feature detector, the level of activity in that feature detector will increase, thereby passing along to the next layer that its preferred feature is present in the image. Another operation applied at each layer of a DNN is called pooling, which is analogous to the increase in receptive field size when multiple neurons in V1 feed into the same neuron in V2, and when multiple neurons in V2 feed into the same neuron in V4 (*Figure 6.9*). It's a step that trades spatial resolution for integration, and the creation of more complex filters at the next layer. And this brings us back to convolution again: one of the really powerful properties of DNNs is that the filters applied during the convolution operation at each layer don't have

to be prespecified, they can be learned during the training process.

Training of the DNNs in the Yamins et al. (2014) study used a procedure comparable to the one described in section *A PDP model of picture naming, Chapter 3*, in that after activity triggered by an image propagated through the network, the resultant pattern of activation in the output layer indicated the extent to which the DNN "wanted" to classify that image as belonging to each of the eight categories. The error between this pattern in the output layer and what the perfect answer should have been (e.g., a "1" for fruit and a "0" for every other category) was then back-propagated through the network. Importantly, as alluded to above, this error signal was used to modify the filters applied in the convolution operation at each layer.

To determine empirically what normative performance on this image database should be, Yamins et al. (2014) first used "web-based crowdsourcing methods" (undoubtedly Amazon Mechanical Turk) to determine that humans can successfully classify over 90% of the images in the database. Decoding signals from monkey IT cortex yielded roughly 80% accuracy, whereas decoding signals from V4 (from which they also recorded) yielded accuracy below 70%. The best-performing DNN among those that they tried achieved classification performance that was numerically higher than that of humans, almost at 100% correct. A general finding from this study – and, importantly, it didn't have to turn out this way – was that the better the performance of a particular DNN, the closer the responses of units in its final layer resembled those of neurons in monkey IT when compared on an image-by-image basis. Furthermore, units in this DNN showed progressively higher levels of invariance to changes in low-level image features at progressively "deeper" layers in the network, comparable to the progression observed from V1 to V2 to V4 to IT in the monkey.

In a follow-up study, the DiCarlo group addressed a more specific question, which is whether feedback projections in the ventral visual stream play an important role in object recognition. Although there is indisputable evidence for strong reciprocal connectivity between each region in the ventral visual stream (*Figure 6.5.B*), there has been a debate about the necessity of feedback for real-time perception, based in part on the fact that the average fixation during natural viewing only lasts between 200 and 300 ms. This may not leave enough time for feedback to contribute meaningfully to the highly accurate object recognition that we all know that we experience on a moment-to-moment basis. From a different perspective, this experiment also addressed a broader point about this kind of research that may have occurred to you while reading this section: *just because a DNN* can *perform image classification at levels achieved by primates doesn't mean that it accomplishes this in the same way.* Indeed, the standard DNN architecture is exclusively feedforward, including that from Yamins et al. (2014). (And note that although the best-performing DNN identified by Yamins et al. (2014) showed unit-level responses that most closely resembled those observed in IT, the highest level of predictivity

FIGURE 6.11 Methods and stimuli for the study "Evidence that recurrent circuits are critical to the ventral stream's execution of core object recognition behavior." Source: From Kar, K., Kubilius, J., Schmidt, K. et al. Evidence that recurrent circuits are critical to the ventral stream's execution of core object recognition behavior. Nat Neurosci 22, 974–983 (2019). https://doi.org/10.1038/s41593-019-0392-5. Copyright 2019. Reproduced with permission of Springer Nature.

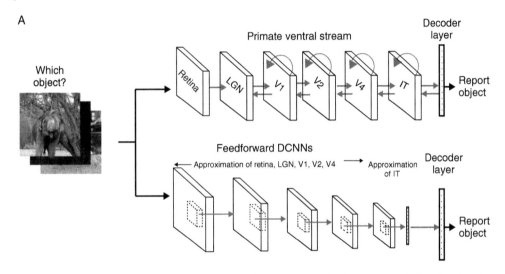

FIGURE 6.11.A The gist of the experiment was to compare image processing by neurons in the visual system of the monkey with the processing steps carried out in each layer by a DNN. On the diagrammed hierarchy of the ventral stream in the monkey, known feedforward connections are illustrated with blue arrows, known feedback connections with straight red arrows, and known lateral connections between neurons in the same regions are illustrated with curved red arrows. In the matched diagram of a "feedforward DCNN" (deep convolutional neural network), the smaller "slab" rendered with dotted lines indicates one area over which pooling will occur. That is, every unit inside that slab will converge on a single unit in the next layer. Note that the resultant decrease in the number of units in each successive layer is indicated by drawing each layer progressively smaller.

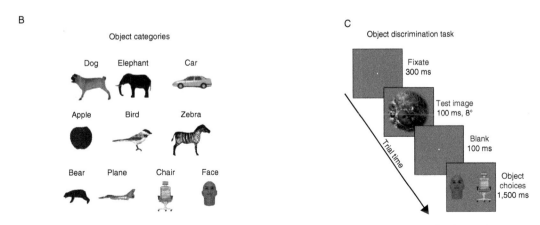

FIGURE 6.11.B The 10 categories.

FIGURE 6.11.C The categorization task.

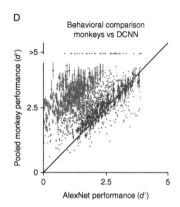

FIGURE 6.11.D Performance on each image, for the two monkeys (averaged) and for the DCNN, was computed as a *d'* value, and images for which monkeys performed at a superior level (red) were used as "challenge images," and images for which the two performed at the same level (blue) were used as "control images." (Images in gray were not used for the analyses.)

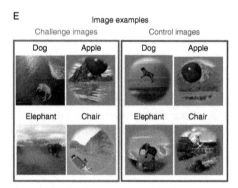

FIGURE 6.11.E Examples of images from the two groups. Note that each object was superimposed on a randomly selected "cluttered" background, to ensure that background was not correlated with object identity, and therefore couldn't be used to help performance.

achieved was that knowing all the responses in the final layer of this DNN to a particular image predicted 48.5% of the variance of the IT response to this same image. That is, they were far from identical.)

To carry out this study, Kar et al. (2019) presented two monkeys and one DCNN (deep convolutional neural network) with a large set of images, and for each sample image followed it with two test images, requiring the observer (monkey or machine) to indicate which of the two test images belonged to the same category as the sample (*Figure 6.11.C*). After testing, they sorted the images into a group that was classified more accurately by the monkeys than by the DCNN ("challenge images") and one for which monkey and DCNN performance was comparable ("control images"; *Figure 6.11.D*). They had repeated the procedure from their previous study of recording from IT neurons while the monkeys viewed the images, and decoding from these recordings indicated that the time required for image classification from IT neurons was longer for challenge images (median 145 ms) than for control images (median 115 ms). This finding replicated what was known from studies of neural recordings and from behavior, and was consistent with the idea that additional processing was needed for the challenge images. The likelihood that this additional processing included "**recurrent processing**" (i.e., processing involving feedback) was suggested by the fact that the high-dimensional pattern of activity in IT evolved considerably during these additional 30 ms. That is, it didn't take longer for activity triggered by the presentation of an image to get to IT for the complex vs. the control images; rather, once the signals arrived, it took longer for activity in IT to achieve a stable pattern. Finally, the comparison of activity patterns in the DCNN vs. IT indicated that DCNN predictivity of IT activity was highest, close to 50% predicted variance, for the "leading edge" of the IT response – what's often referred to as the "initial feedforward sweep" of stimulus-evoked activity – but then dropped to below 20% predicted variance by the time IT achieved its stable pattern of optimal image decoding. Thus, one way in which primate object recognition differs from image classification carried out by DNNs may be that the former depends importantly on recurrent processing (*Figure 6.11*).

One takeaway from this approach to studying visual perception is that it uses a conceptualization of what it means "to recognize something" that's different from how we have described it up until now. At the beginning of this chapter, we noted that the process of recognition necessarily entails engaging with long-term memory, with preexisting knowledge about the world. Here, although memory is involved in the sense that the DNN had to have learned from past experience (indeed, it had to be taught), successful performance doesn't entail matching perceived information to a long-term memory representation of, say, an Air France 747. Rather, it entails being able to classify this image as belonging with other airplanes, rather than with cars, or fruit, or any other category. Indeed, throughout this literature you'll read about "classification" and "categorization," but you'll rarely encounter the word "recognition." The same is true for how activity in neurons in the temporal cortex of the monkey was processed and interpreted. As we proceed through the book, particularly when we get to *Chapter 10, Visual object recognition and knowledge*, ask yourself whether "recognizing" an image (e.g., of an Air France 747) entails anything more than concluding that it looks more similar to airplanes that you have previously seen in your life relative to the seemingly infinite number of "not-airplane" categories with you are also familiar.

Top-down feedback can disambiguate the visual scene

Although the multiplicities of small receptive fields in the retina, LGN, and V1 that "tile" the visual field offer high acuity, they also create challenges for the visual system, one of which is known as the **aperture problem**. At a generic level, what this refers to is the fact that each neuron with a small receptive field is, in effect, viewing the visual scene through a very small aperture. (Imagine looking through the peephole in a door if the fish-eye lens has been removed from it.) *Figure 6.12* illustrates this problem for motion perception: when an irregularly shaped object is moving in one direction, the direction of motion of small portions of the edge of the object, when seen through a small aperture, can be ambiguous. To illustrate, the figure shows one receptive field on the neck of the deer at which at least three direction-sensitive feature detectors (each tuned for one of the directions indicated by the red and green arrows) might signal that its preferred direction of motion is occurring within this receptive field. Although this would result in a noisy, ambiguous message being sent "up the chain" to MT, we know from *Chapter 4* that there is circuitry in place whereby MT can control activity in V1 and the LGN, thereby reducing this feedforward cacophony and emphasizing the motion signals that it would "prefer" to receive (see *Figure 4.4*). Indeed, one of the points made in that earlier discussion was that, in the same way that MT can control V1 and the LGN, there must be a higher level of representation feeding back onto MT. We are now in a position to consider what that higher level of representation might be.

Recall from *Figure 6.5.B* that there are lateral interconnections between MT and V4. In our example of the

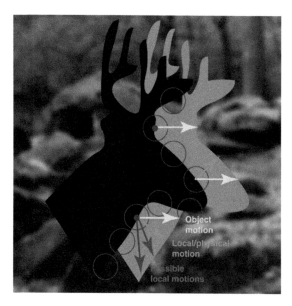

FIGURE 6.12 An example of how high-level visual object recognition is needed to influence earlier stages of visual processing. The black silhouette of the deer head represents its current location, and the gray silhouette a location it will occupy in the very near future. The purple circles represent V1 receptive fields, with yellow arrows indicating the true direction of motion of the object (i.e., of the deer), green arrows the physical motion within two receptive field "apertures," and the red arrows two other possible interpretations of the direction of motion within one of these receptive fields. Source: From Dakin, Steven C. 2009. "Vision: Thinking Globally, Acting Locally." Current Biology 19 (18): R851–R854. doi: 10.1016/j.cub.2009.08.021. Copyright 2009. Reproduced with permission of Cell Press.

running deer, then, networks in IT that have assembled the representation of the deer can feed this information back onto V4 (as well as, perhaps, laterally to area MST), V4 can then share it with MT, and MT is now equipped to "know" how to control the bottom-up flow of motion signals coming from V1. One implication of this is that, to fully perceive an object in the environment, it's probably not the case that the information flows from retina to LGN to V1, etc., all the way up to IT, and that a stable percept of the object can then persist while the activity in the regions earlier in the hierarchy returns to baseline. Rather, activity at every stage of processing is likely needed to experience a coherent percept. Keep this in mind when, in a few chapters, we talk about how a predictive coding model would account for this.

TAKING STOCK

Although there is so, so much more to know about visual object recognition and the functioning of the ventral visual stream, I'm sure that my editor is already getting uneasy about the length of this chapter. Therefore, let's just take a moment to review what we've covered and, in particular, how it relates to what's yet to come. The anatomical connectivity of the visual system, beginning at the retina if one considers the M and P classes of RGCs, creates a substrate for two broad classes of visual information processing: determining *what* an object is and determining *where* it is. (And we've already noted that a question for *Chapter 8* is whether *where* is perhaps better construed as *how*.) Lesion studies have confirmed, in the form of a double dissociation of function, that the temporal lobe, particularly IT cortex, contributes importantly to object recognition, and the PPC contributes to spatial processing. This chapter's focus on the ventral stream has emphasized the hierarchical nature of object recognition. After flirting with a grandmother layer of object representation, a distributed, combination-coding scheme was seen to offer a more plausible account of stimulus processing in IT cortex. Particularly intriguing is the neural processing of faces. For one thing, faces don't lend themselves to image decomposition as readily as do other complex objects. Additionally, IT (and STP) responses to faces suggest two interesting properties of this system. First, IT may support both object- and image-based representations. Second, some neurons in the temporal lobe may represent "the idea of a face," rather than any one exemplar face. Finally, we've seen that careful assessment of how image processing is accomplished by DNNs can be a powerful way to learn about how it is, and isn't, accomplished by biological systems, with this approach highlighting an important role for recurrent processing in the primate ventral visual processing system.

After the next three chapters conclude this section's emphasis on *perception*, *attention*, and *action*, *Chapter 10* will address the question (with apologies to *Chapter 4*) *Where does perception end, and where does memory begin?* As with the earlier chapter's question (which was posed about sensation vs. perception), we'll see that the boundary is a blurry one, one having more basis in arbitrary labeling than in neurobiological or psychological principle. For example, this chapter will consider agnosias, syndromes of "perception stripped of its meaning" that result from damage to the ventral temporal lobe. It will consider the phenomenon of "category specificity" in ventral temporal cortex (e.g., the *fusiform face area*, the *parahippocampal place area*, the *extrastriate body area*, and the *visual word form area*): how did these areas come into being (nature vs. nurture); what does their existence tell us (or not) about the organization of knowledge in the brain; and, more concretely, what are the limits of what we can infer from activity in one of these areas?

END-OF-CHAPTER QUESTIONS

1. In what sense does function follow from structure in the visual system?

2. Design an experiment that, if your hypothesis is correct, might produce evidence for a double dissociation of function. In what way would your hoped-for results support stronger inference than a single dissociation?

3. Define the functions of the *what* and *where* pathways. Summarize at least three pieces of evidence for this organizing principle of the visual system.

4. Summarize the type of image processing that occurs at each level of the ventral-stream hierarchy.

5. Define the concept of the "grandmother cell." For what reasons is it logically appealing? For what reasons, practical or theoretical, is it problematic?

6. What is a plausible alternative to a "grandmother level" of object recognition? Summarize the evidence for it.

7. Are faces processed differently than other objects by the visual system?

8. Define object-based vs. image-based models of object recognition. Name an advantage, and a limitation, of each.

9. What are the similarities and differences between DNNs developed for image categorization and the primate ventral visual processing stream?

10. Although we know of the anatomical existence of feedback connections between areas in the ventral visual processing stream, what's the evidence that they play an important role in visual-scene processing?

REFERENCES

Born, Richard T. 2001. "Visual Processing: Parallel-er and Parallel-er." *Current Biology* 11 (14): R566–R568. doi: 10.1016/S0960-9822(01)00345-1.

Bruce, Charles, Robert Desimone, and Charles G. Gross. 1981. "Visual Properties of Neurons in a Polysensory Area in Superior Temporal Sulcus of the Macaque." *Journal of Neurophysiology* 46 (2): 369–384.

Curtis, C. E., and D'Esposito, M. 2004. "The Effects of Prefrontal Lesions on Working Memory Performance and Theory." *Cognitive, Affective, and Behavioral Neuroscience* 4 (4): 528–539.

Dakin, Steven C. 2009. "Vision: Thinking Globally, Acting Locally." *Current Biology* 19 (18): R851–R854. doi: 10.1016/j.cub.2009.08.021.

Felleman, Daniel J., and David C. Van Essen. 1991. "Distributed Hierarchical Processing in Primate Visual Cortex." *Cerebral Cortex* 1 (1): 1–47. doi: 10.1093/cercor/1.1.1.

Gross, Charles G., C. E. Rocha-Miranda, and D. B. Bender. 1972. "Visual Properties of Neurons in Inferotemporal Cortex of the Macaque." *Journal of Neurophysiology* 35 (1): 96–111.

Hubel, David H. 1988. Eye, Brain, and Vision. New York: Scientific American Library: W. H. Freeman.

Hubel, David H. 1996. "David H. Hubel." In *The History of Neuroscience in Autobiography*, edited by Larry R. Squire, Vol 1 294–317. Washington, DC: Society for Neuroscience.

http://www.sfn.org/about/history-of-neuroscience/~/media/SfN/Documents/Autobiographies/c9.ashx.

Kar, K., J. Kubilius, K. Schmidt, E. B. Issa, and J. J. DiCarlo. 2019. "Evidence That Recurrent Circuits are Critical to the Ventral Stream's Execution of Core Object Recognition Behavior." *Nature Neuroscience* 22: 974–983.

Logothetis, Nikos K., and David L. Sheinberg. 1996. "Visual Object Recognition." *Annual Review of Neuroscience* 19: 577–621. doi: 10.1146/annurev.ne.19.030196.003045.

Mishkin, Mortimer, Leslie G. Ungerleider, and Kathleen A. Macko. 1983. "Object Vision and Spatial Vision: Two Cortical Pathways." *Trends in Neurosciences* 6: 414–417. doi: 10.1016/0166-2236(83)90190-X.

Pohl, Walter. 1973. "Dissociation of Spatial Discrimination Deficits Following Frontal and Parietal Lesions in Monkeys." *Journal of Comparative and Physiological Psychology* 82 (2): 227–239. doi: 10.1037/h0033922.

Tanaka, Keiji. 1993. "Neuronal Mechanisms of Object Recognition." *Science* 262 (5134): 685–688. doi: 10.1126/science.8235589.

Yamins, D. L. K., H. Hong, C. F. Cadieu, E. A. Solomon, D. Seibert, and J. J. DiCarlo. 2014. "Performance-Optimized Hierarchical Models Predict Neural Responses in Higher Visual Cortex." *Proceedings of the National Academy of Sciences USA*, 111: 8619–8624.

OTHER SOURCES USED

Blake, Randolph, and Robert Sekuler. 1985. *Perception*. New York: Alfred A. Knopf.

Gross, Charles G., D. B. Bender, and C. E. Rocha-Miranda. 1969. "Visual Receptive Fields of Neurons in Inferotemporal Cortex of the Monkey." *Science* 166 (3910): 1303–1306. doi: 10.1126/science.166.3910.1303.

Kandel, Eric R., James H. Schwartz, and Thomas M. Jessell, eds. 2000. *Principles of Neural Science*. New York: McGraw-Hill.

Kell, A. J. E., and J. J. McDermott. 2019. "Deep Neural Network Models of Sensory Systems: Windows onto the Role of Task Constraints." *Current Opinion in Neurobiology* 55: 121–132.

Schiller, Peter, H. 1993. "The Effects of V4 and Middle Temporal (MT) Area Lesions on Visual Performance in the Rhesus Monkey." *Visual Neuroscience* 10 (4): 717–746. doi: 10.1017/S0952523800005423.

Schiller, Peter H., and Nikos K. Logothetis. 1990. "The Color-Opponent and Broad-Band Channels of the Primate Visual System." *Trends in Neurosciences* 13 (10): 392–398. doi: 10.1016/0166-2236(90)90117-S.

Semmes, Josephine, Sidney Weinstein, L. Ghent, John S. Meyer, and Hans-Lukas Teuber. 1963. "Correlates of Impaired Orientation in Personal and Extrapersonal Space." *Brain* 86 (4): 747–772.

Tanaka, Keiji. 1996. "Inferotemporal Cortex and Object Vision." *Annual Review of Neuroscience* 19: 109–139. doi: 10.1146/annurev.ne.19.030196.000545.

Tanaka, Keiji. 1997. "Mechanisms of Visual Object Recognition: Monkey and Human Studies." *Current Opinion in Neurobiology* 7 (4): 523–529. doi: 10.1016/S0959-4388(97)80032-3.

Ungerleider, Leslie G., and Mortimer Mishkin. 1982. "Two Cortical Visual Systems." In *Analysis of Visual Behavior*, edited by David J. Ingle, Melvin A. Goodale, and Richard J. W. Mansfield, 549–586. Cambridge, MA: MIT Press.

Ungerleider, Leslie G., and Anderw H. Bell. 2011. "Uncovering the Visual 'Alphabet': Advances in Our Understanding of Object Perception." *Vision Research* 51 (7): 782–799. doi: 10.1016/j.visres.2010.10.002.

Wolfe, Jeremy M., Keith R. Kleunder, Dennis M. Levi, Linda M. Bartoshuk, Rachel S. Herz, Roberta Klatzky, . . . Daniel M. Merfeld. 2012. *Sensation and Perception*, 3rd ed. Sunderland, MA: Sinauer Associates.

FURTHER READING

Logothetis, Nikos K., and David L. Sheinberg. 1996. "Visual Object Recognition." *Annual Review of Neuroscience* 19: 577–621. doi: 10.1146/annurev.ne.19.030196.003045.

A very extensive review emphasizing behavioral, neuropsychological, and electrophysiological evidence for viewpoint-dependent mechanisms of object recognition, as well as for different populations of IT neurons specialized to perform different functions, such as categorization, recognizing living things, and supporting the action of object grasping.

Mishkin, Mortimer, Leslie G. Ungerleider, and Kathleen A. Macko. 1983. "Object Vision and Spatial Vision: Two Cortical Pathways." *Trends in Neurosciences* 6: 414–417. doi: 10.1016/0166-2236(83)90190-X.

A concise summary of the two-parallel-pathways model of the visual system.

Schiller, Peter H., and Nikos K. Logothetis. 1990. "The Color-Opponent and Broad-Band Channels of the Primate Visual System." *Trends in Neurosciences* 13 (10): 392–398. doi: 10.1016/0166-2236(90)90117-S.

A thorough summary, and synthesis, of the relative contributions of the M and P channels to visual perception.

Tanaka, Keiji. 1996. "Inferotemporal Cortex and Object Vision." *Annual Review of Neuroscience* 19: 109–139. doi: 10.1146/annurev.ne.19.030196.000545.

An overview of this investigator's pioneering work on the neural principles underlying object recognition in IT cortex.

Ungerleider, Leslie G., and Anderw H. Bell. 2011. "Uncovering the Visual 'Alphabet': Advances in Our Understanding of Object Perception." *Vision Research* 51 (7): 782–799. doi: 10.1016/j.visres.2010.10.002.

A remarkably comprehensive overview of the cognitive neuroscience of object recognition, ranging from nineteenth-century studies of brain and behavior (some of which we considered in Chapter 1*) to an insightful critique of contemporary methods for multivariate pattern analysis of fMRI data (which we will introduce in* Chapter 9*).*

CHAPTER 7
SPATIAL COGNITION AND ATTENTION

KEY THEMES

- Neuropsychology can be an important tool for hypothesis generation, as evidenced by the models of attention derived from the study of unilateral neglect.

- The neglect syndrome provides evidence that attention (a) can operate within an egocentric frame of reference; (b) can be location-based or object-based; (c) may operate differently in peripersonal vs. extrapersonal space; and (d) may act to resolve competition among neural representations.

- The cortical regions whose damage produces unilateral neglect have been studied in neurologically healthy individuals with positron emission tomography (PET) and functional magnetic resonance imaging (fMRI), and show elevated activity during tasks requiring spatial- vs. object-based visual discriminations, as well as retinotopic organization.

- Neurons of posterior parietal cortex (PPC) of the dorsal visual stream integrate retinotopically encoded visual information with proprioceptive information about the position of gaze relative to the head, the head relative to the trunk, and so on, to effect coordinate transformations that are necessary for making appropriate visually guided actions with various parts of the body.

- The transformation from an egocentric to an allocentric (i.e., environment-centered) reference frame is observed in the activity of place cells in the hippocampus.

- The metaphor of attention as a spotlight highlights many of its properties: increasing the salience of one (or a few) among many stimuli; the distinction between attending to a location vs. attending to an object; the distinction between the *site* of action of attention vs. its *source*.

- ERP studies have isolated the timing, and electrophysiological signature, of the attentional selection of a visual feature.

- Extracellular electrophysiological and fMRI studies have established that spatial attention works, in part, by elevating baseline activity in the neural pathways that process information about stimuli occupying particular locations.

- An important function of attention may be to bias the neural competition for representation that results from the simultaneous perception of multiple stimuli in the environment.

CONTENTS

UNILATERAL NEGLECT: A FERTILE SOURCE OF MODELS OF SPATIAL COGNITION AND ATTENTION

Previous chapters have emphasized the importance of lesions, or other factors that perturb normal brain structure or function (e.g., electrical stimulation, optogenetic manipulation), as a tool for drawing inferences about the necessity of a brain region (or physiological phenomenon) for a particular aspect of behavior or function. In effect, the emphasis has been on the importance of perturbation methods for *hypothesis testing*. We begin this chapter by highlighting a different role that neuropsychology can play in cognitive neuroscience: brain lesions can be a fertile source for *hypothesis generation*. Stated another way, careful observation of the deficits experienced by patients with

brain damage can suggest organizational principles that were not evident to introspection or observation of normal behavior. With particular relevance to our present focus, the neurological syndrome of unilateral neglect has raised many intriguing questions about the fundamental workings of spatial cognition and attention.

Unilateral neglect: a clinicoanatomical primer

Unilateral neglect (sometimes referred to as "unilateral spatial neglect," or "hemispatial neglect") is a neurological syndrome in which damage in one hemisphere, typically due to a stroke that affects the right hemisphere, results in a pronounced unawareness of information from the side opposite the lesion (typically, the left side; *Figure 7.1*). The syndrome is highly heterogeneous, in that it can vary

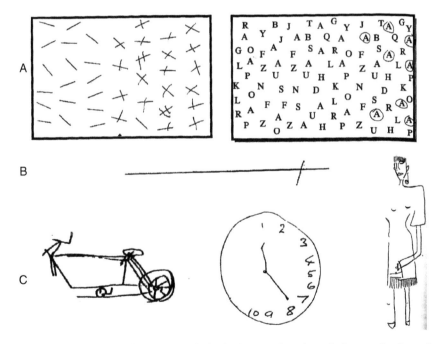

FIGURE 7.1 Patient performance on classic neuropsychological tests of neglect. **A.** One patient's performance is superior on a simple cancelation task that only requires marking each stimulus vs. on a more difficult task that requires identifying and circling targets ("A"s) among distractors (other letters). **B.** Line bisection, in which the subject was asked to mark the center point of the horizontal line. Note that for "bedside paper-and-pencil" tasks such as these, direction of gaze, orientation of body, and so on are typically not controlled. **C.** Drawing from memory. Note that neglect is evident at multiple scales, in that not only are the left side of the overall pictures of the bicycle and the woman neglected, but so too are the leftward-facing handlebar and leftward-facing pedal of the bicycle, and the left side of the handbag strap and left-sided earring of the woman. (Note that "left" is defined here in retinotopic terms, in that, e.g., the neglected handlebar and earring are both on the right in object-centered terms.) Source: From Driver, Jon, Patrik Vuilleumier, and Masud Husain. 2004. "Spatial Neglect and Extinction." In The Cognitive Neurosciences III, edited by Michael S. Gazzaniga, 589–606. Cambridge, MA: MIT Press. Reproduced with permission of The MIT Press.

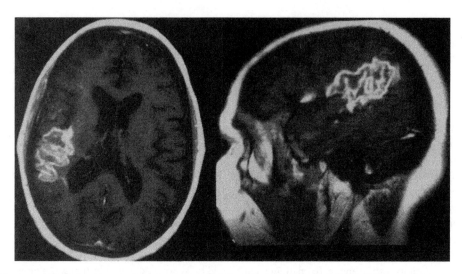

FIGURE 7.2 MR scan of unilateral neglect patient described by Chatterjee (2003), with bright regions corresponding to the damaged tissue in the right inferior parietal lobule and superior temporal gyrus. (Note that radiological convention is to display the right side of the brain on the left side of the image.) Source: Chatterjee, 2003. Reproduced with permission of The MIT Press.

considerably from patient to patient. Nonetheless, we have to start somewhere, and we'll do so with a patient whose case is representative of many described in the literature. The patient, described by neurologist and cognitive neuroscientist Anjan Chatterjee (2003), suffered a stroke when occlusion of the right middle cerebral artery deprived portions of the posterior inferior parietal lobule (Brodmann areas [BAs] 39 and 40) and the posterior superior temporal gyrus (BA 22) of the right hemisphere of arterial blood (*Figure 7.2*). This patient's consequent neglect of the left side of her world was dramatic: when her left hand was held in front of her face, she opined that she was looking at the hand of the examining doctor; she ate only the food that was located on the right side of her plate; unswallowed food sometimes collected in the left side of her mouth; despite the disproportionate weakness of the left side of her body, when asked about why she was in the hospital she would refer to weakness in her right arm. Thus, her neglect of "the left" applied to both her own body and to stimuli in her immediate environment (e.g., the plate immediately in front of her).

Unilateral neglect can result from damage to many brain areas, including the temporoparietal junction, the frontal eye fields, and the caudate nucleus, as well as to white matter tracts that innervate these regions. Although there remains some controversy about precisely which areas must be damaged to produce neglect, *Figure 7.3* illustrates the results of one large group study.

Hypotheses arising from clinical observations of neglect

Countless clinical observations such as those summarized in *Figure 7.1* have given rise to many ideas about spatial cognition and attention. Perhaps the most obvious is that humans maintain an *egocentric* reference frame that is defined with respect to the midline of the body, with the right hemisphere responsible for representation of the left side of our body. (Furthermore, the fact that unilateral neglect of the left side of space occurs much more often after right hemisphere damage than does unilateral neglect of the right side of space after left hemisphere damage has led to the supposition, although without much additional rationale, that right hemisphere structures must allocate attention across both sides of the midline, whereas left hemisphere structures do so only for contralateral body and space.) Underlying this undisputable observation, however, is the more vexing question of *what mechanism(s) underlie this phenomenon*. Perhaps the most intuitive account is to appeal to attention, and the possibility that attention to our bodies and immediately adjacent areas is handled by posterior contralateral cortex. Although attention has many definitions, here it refers to our ability to prioritize one stream of afferent sensory information over others. (In *Figure 5.5*, for example, we considered the effects on the auditory evoked response [AEP] of attending to visual information from a book while clicks were simultaneously

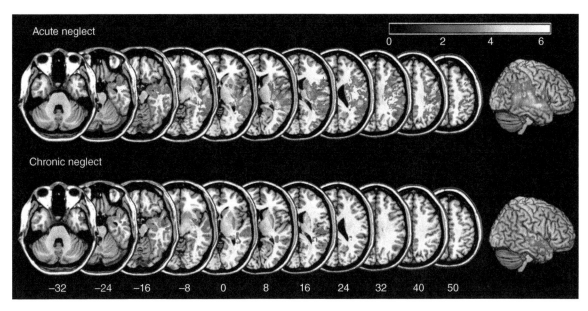

FIGURE 7.3 The neuroanatomy of unilateral neglect. This figure illustrates statistical maps of the regions most likely to be damaged in patients suffering from acute neglect (as assessed an average of 12.4 days after stroke onset) and chronic neglect (as assessed an average of 490.8 days after stroke onset). The prominence of subcortical regions in these analyses suggests important roles in spatial attention for not only gray-matter structures like the basal ganglia but also white-matter connections between cortical regions, in particular, the inferior occipitofrontal fasciculus. The results displayed here were generated by correlating lesion location (as measured on MR scans, such as those shown in *Figure 7.2*) with neglect symptoms in a sample of 54 patients with right hemisphere stroke, an approach known as voxel-based lesion-symptom mapping (VLSM). In this sample of 54 patients, 24 had neglect symptoms during the acute stage, and 8 of the 24 still exhibited neglect symptoms when reassessed in the chronic stage. Color coding indicates the *t*-value of the lesion-to-deficit correlation; numbers under the slices in the "chronic neglect" row indicate the distance (in mm) of each slice from the conventionally defined origin, along the dorsoventral (a.k.a. "z") axis. Source: Karnath, Rennig, Johannsen, and Rorden, 2011. Reproduced with permission of Oxford University Press.

presented auditorily, vs. attending to the clicks.) An alternative to attention, however, is suggested by noting that all of the examples referred to thus far – neglect of one's own hand, of food before one on a plate (or in one's mouth!), of the left side of lines on a piece of paper – can be portrayed as involving the motor system. Such observations have led to the postulation that the egocentric reference frame may arise from the intention to perform an action either with or on a part of our body. Thus, a second theme that will recur in this chapter is the debate over whether the principle of *intention* or *attention* better captures the functions of the parietal lobes. (Relatedly, in subsequent chapters, we shall consider the theoretical proposition that attention is a consequence of the intention to act – the "premotor theory" of attention.)

The *intention vs. attention* interpretation of unilateral neglect was the explicit focus of a highly influential study

by Italian neuropsychologists Edoardo Bisiach and Claudio Luzzatti, who asked two stroke patients, lifelong residents of Milan, to describe their city's most famous outdoor plaza, the Piazza del Duomo. Note that these sessions took place in the hospital, and, therefore, the patients' descriptions were generated not from visual perception, but from memory. A simple, but breathtakingly insightful, manipulation that they included was to ask their patients to describe the plaza from two perspectives: first, while "facing" the cathedral from the opposite side of the plaza; and second, while "standing on the steps" of the cathedral and "looking" across the plaza at the location where they had previously been "standing." Both patients, while imagining facing the cathedral from across the plaza, named several landmarks located on its south side, which would be on one's right when standing at that location and facing east. Only one of the two patients also named three landmarks on the north side (he had named nine on the south). When imagining standing on the steps of the cathedral, and therefore imagining facing in the opposite direction (i.e., facing west), in contrast, the patients each spontaneously named several landmarks on the north side of the square (i.e., on what was now their "right"), and neglected to name any of those on the south side, despite the fact that they had named several of these just moments before (*Figure 7.4*)! Thus, it wasn't that their mental representations (i.e., their memories) had been somehow cleaved in two by their strokes. Rather, it was their ability to direct their attention to (i.e., to think about) the left side of the representation as they were mentally interrogating it at that moment. From these findings, Bisiach and Luzzatti (1978) concluded that "unilateral neglect cannot be reduced to a disorder confined to the input-output machinery of the organism interacting with its physical environment" (p. 132). That is, they argued that neglect is a disorder of attention, not intention.

In their writings, Bisiach and Luzzatti (1978) emphasized that neglect revealed a deficit in mental representation. Certainly, when the performance of their patients is compared with that of the patient illustrated in *Figure 7.1*, one sees that patients can neglect the left side of individual objects, even when their drawings suggest that they are aware that other objects and/or parts of the environment exist to the left of the object in question. This provides at least a hint for the existence of an *object-based attention* that operates at the level of individual objects, rather than (spatially defined) large swaths of the visual field.

Yet a different concept – that we may represent "near" vs. "far" space with different mechanisms – is illustrated by the case of a patient who very well may not have produced the same results as those reported by Bisiach and Luzzatti in 1978. This patient was a British mechanic, tested in a stroke rehabilitation center in Oxford, UK. The research team first tested the patient on line bisection using the standard method: a horizontal line drawn on a piece of paper was held 45 cm from his eyes, centered on his midline, and he was asked to mark the midline with a pen. The results (from several trials, with lines of varying length) were comparable to that shown in *Figure 7.1*, demonstrating marked neglect of the left side of the line. Next, the patient was tested with lines placed on a whiteboard that was located 2.44 m distant from the patient (a distance chosen because it corresponded to the standard distance between dart board and dart thrower that one

finds in British pubs). The thickness and length of the lines were scaled up, such that they subtended roughly the same amount of visual angle as had the lines on paper from the "45 cm" test. Instead of drawing with a pen, the patient was asked to indicate the midpoint of the across-the-room lines by pointing with a light pointer (equivalent to a laser pointer, but this was 1990) or, in a cheeky follow-up test, by throwing darts at it. Whether with light pointer or darts, the patient showed no systematic bias away from the midpoint: his errors clustered near the true midpoint, and half of them fell to the left of it. Thus, this case demonstrated concretely a dissociation between the processing of **peripersonal space** – the space surrounding one's body that is within one's grasp – vs. **extrapersonal space** – those portions of perceived space that are farther than we can reach.

A final concept that we will consider here is that studies of neglect also provide compelling evidence for *competition* within the brain. **Extinction**, in the neurology clinic, refers to the fact that a patient can sometimes be aware of a single object presented on their left, but will lose awareness of it if a second item is then presented on their right. It's as though the item on the right extinguishes awareness of the item on the left.

Each of these ideas about how the brain may represent space and allocate attention will be revisited in this chapter as we consider the anatomy and physiology of spatial processing and attention in neurologically intact organisms (be they humans, monkeys, or rats).

THE FUNCTIONAL ANATOMY OF THE DORSAL STREAM

There have been many, many important experiments using fMRI to study the neural bases of human vision. At this juncture, we'll consider just one that succinctly illustrates the two aspects of cognition that are the focus of this chapter: the representation of space and attention. In it, Ayse Saygin and Marty Sereno (2008) employed a variant of a method that is commonly used to perform retinotopic mapping – the "rotating wedge." The logic is that, while the subject fixates a central location, a wedge shaped like a slice of pie is slowly rotated around the fixation point. The wedge looks different from the background – classically it might contain an alternating black-and-white checkerboard pattern presented against a gray background – and the logic is

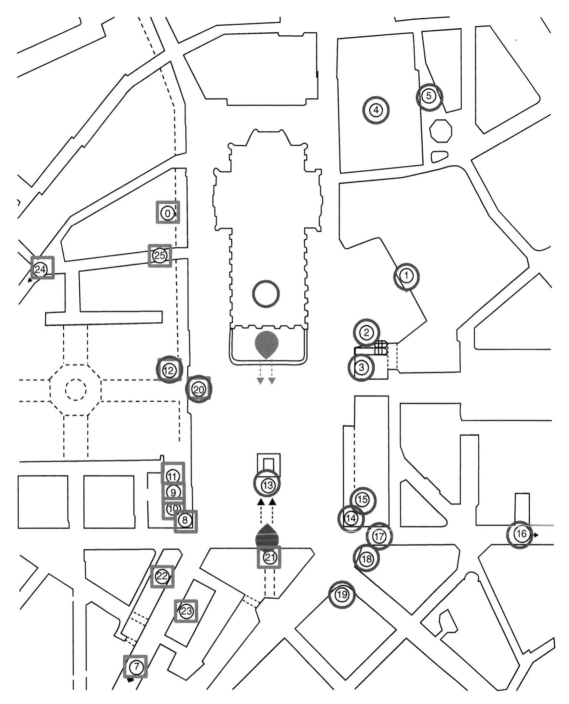

FIGURE 7.4 Landmarks named by two unilateral neglect patients of Bisiach and Luzzatti (1978). When imagining themselves looking at the cathedral from across the plaza (i.e., facing east, as cartooned by blue head and two arrows), they named the landmarks highlighted with blue circles; when imagining themselves standing on the cathedral steps and looking in the opposite direction (i.e., facing west, red head and two arrows) they named the landmarks highlighted with red squares. Source: From Bisiach, Edoardo, and Claudio Luzzatti. 1978. "Unilateral Neglect of Representational Space." Cortex 14 (1): 129–133. Copyright 1978. Reproduced with permission of Elsevier.

simply that areas of the brain that represent a particular region of retinotopic space will only show increases of activity when the region that they represent is occupied by the wedge. Saygin and Sereno employed a clever riff on this basic logic, by presenting clusters of colored dots on the entire screen, rotating the entire display around the fixation cross, and having the clusters of dots defining the wedge take on recognizable shapes (*Figure 7.5.A*). Specifically, the recognizable shapes were "point-light walker" images, created by having an actor wear a black bodysuit fitted with small lights on the arms, legs, head, and torso, and filming them as they performed a variety of actions (walking,

jumping rope, etc.). Their rationale was that many areas of the brain outside low-level visual cortex are more likely to process information about "biologically relevant" stimuli than something as mundane as a checkerboard pattern.

An important factor to point out about the Saygin and Sereno (2008) results is that, while it's obvious how their experiment systematically stimulated different locations in the visual field, thereby allowing them to identify retinotopically selective parts of the visual system, there was nothing in the experiment that explicitly isolated each of the regions identified in *Figure 7.5*. That is, their display didn't somehow only activate V1,

FIGURE 7.5 Retinotopy and attention in human occipital, temporal, parietal, and frontal cortex. Source: Saygin and Sereno, 2008. Reproduced with permission of Oxford University Press.

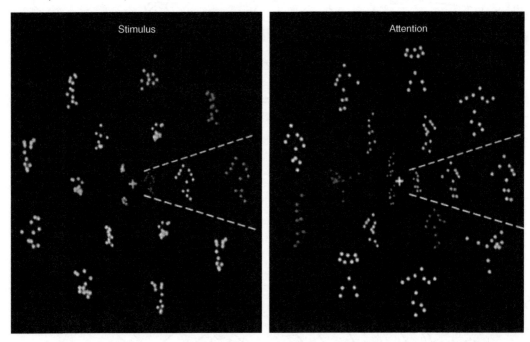

FIGURE 7.5.A A static screenshot of the stimulus displays from the two conditions. The entire display rotated about the fixation point (like a speeded-up depiction of the night sky over the course of 12 hours), with clusters continually fading in and out. In the *Stimulus* condition, the wedge was defined by clusters taking on the configuration of a point-light walker (the wedge is indicated here, for clarity of exposition, with white dotted lines that were not present during the experiment). In the *Attention* condition, all clusters took on the shape of point-light walkers, the to-be-attended wedge differentiated by being the only sector in which three of the same color were aligned along a single radius. In the *Stimulus* condition, the color of the fixation cross changed every second, and subjects were to report each time a change constituted a "2-back" repetition (e.g., *red-yellow-red* would be a hit, but *red-yellow-blue-red* and *red-red-yellow* would not). Note that, although the *Stimulus* condition placed heavy demands on attention, an analysis identifying regions responding to the periodic sweep of the wedge would not detect activity corresponding to steady attention to the central point on the display. The *Attention* condition, in contrast, can be construed as the attention-to-the-rotating-wedge condition, because subjects, although fixating centrally, were to respond each time the three same-color point-light walkers in the wedge were *not* all moving identically (e.g., if two were jumping rope and one was doing karate kicks).

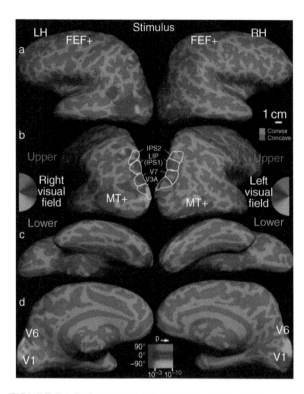

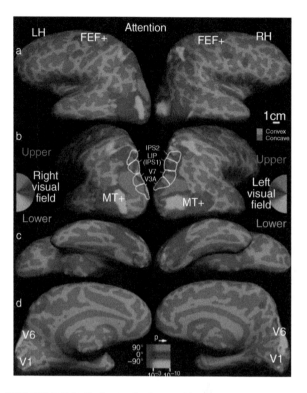

FIGURE 7.5.B Retinotopically sensitive cortex in the *Stimulus* condition, displayed on three views of an inflated hemisphere (×2, because both hemispheres are displayed); lighter gray corresponds to gyri. Superimposed on the anatomical images are patches of red-blue-green color identifying cortical regions whose activity fluctuated (either increased or decreased) each time the wedge passed through a particular region of the display, the specific color indicating the part of the visual field to which that patch of cortex was responsive (keys are located to left and right of view b for each hemisphere). Statistical significance of a region's retinotopic responses is represented by the translucency of the superimposed color. The identity of each region is labeled with white letters and, along the intraparietal sulcus (IPS), borders.

FIGURE 7.5.C Retinotopically sensitive cortex in the *Attention* condition.

V3A's retinotopy transitions from the horizontal to upper visual field, V7's from upper to horizontal, IPS1's from horizontal to upper, and IPS2's from upper to horizontal to lower. Known anatomical landmarks were also used to determine precisely where to draw boundaries.

Coordinate transformations to guide action with perception

Although *Figure 7.5* indicates that many areas of occipitoparietal cortex are organized retinotopically, they, too, follow the principle of hierarchical processing that we saw with the ventral visual processing stream in *Chapter 6*. Different from much the ventral stream, however, is the fact that many dorsal stream regions are characterized by multisensory integration, in particular, the integration of proprioceptive with visual information. Consider the following scenarios: as I sit writing this section, I occasionally lift my right hand from the keyboard to reach for the glass that is located to the right of the keyboard. (1) Sometimes I turn my head to look at the glass before reaching for it, (2) sometimes my face remains directed at the screen and

then V3, then IPS2, and so on. So how were they able to discriminate these regions and determine their boundaries? The answer is that they used their knowledge of the physiology and anatomy of the primate visual system, including the fact that retinotopic maps "reverse" at each interareal boundary. This is particularly evident along the strip of tissue extending from V3A to V7 to IPS1 to IPS2 in the right hemisphere *Attention* image. Moving in a caudoventral-to-rostrodorsal direction,

I move my eyes to fixate the glass, and (3) sometimes neither head nor eyes are moved, because while both are directed toward the screen, the glass occupies the outer edge of my right visual field. In each of these three scenarios, the glass is in the same location in allocentric coordinates (i.e., it's always in the same location on the arm of the couch in my living room), yet information that my nervous system processes about its location is different in all three. In the first two scenarios, it occupies the same retinotopic coordinates, but different head-centered coordinates, in that it is in a different location relative to my head (straight ahead in scenario 1, at "2 o'clock" in scenario 2). In the second and third scenarios, it is at the same head-centered coordinates, but in different retinotopic coordinates. And in each of these three, what ultimately matters is that the glass is located in precisely the same location relative to my hand. What this example illustrates is that virtually all of the actions that we carry out over the course of our lives, certainly all of our sensory-guided actions, entail an ongoing process of translation between sensory-based and motor effector-based coordinate frames. To accomplish this, the nervous systems needs to know the position of each part of the body relative to the others (e.g., *Are the eyes pointing right or straight ahead?*). And recall from the final section of *Chapter 5* that this proprioceptive information is supplied by muscle stretch detectors.

As we shall review here, an important function carried out by the PPC is to compute the **coordinate transformations** that are necessary to know where an object is located relative to different parts of the body. The study illustrated in *Figure 7.5* provides no insight about how these coordinate transformations are computed, because it didn't involve movement of the body. Many studies that have provided important information about these processes have come from electrophysiological recordings of neurons in the awake, behaving monkey.

Extracellular electrophysiological studies of monkeys

Area 7a of the PPC and the immediately superior IPS (see *Figure 6.5.A*) have been the focus of intensive investigation by neurophysiologists since the 1980s. In subsequent chapters, we will consider the role of these areas in such functions as the control of movement, the control of attention, and decision-making. Presently, however, we'll focus on the role of these regions in the integration of visual and proprioceptive information. Neurophysiologist Richard Andersen and colleagues have studied this by first finding the retinotopically defined receptive field of a neuron in

PPC, then systematically varying the location at which the monkey has to fixate, all the while keeping a visual stimulus in that neuron's receptive field (*Figure 7.6*). A property of these PPC neurons that this procedure reveals is that the gain of their response is modulated by eye position. *Figure 7.6*, for example, illustrates a neuron with a retinotopic receptive field centered 20° to the right and 20° below fixation, and with the additional property that the magnitude of its response is sensitive to the position of the eyes within the head, with maximal gain when the animal is looking up and to the left. What this means is that, in effect, this neuron is processing information not just about the location of a stimulus relative to the retina but also about the location of a stimulus relative to the head. Note that, in panel B of this figure, although the stimulus is in the same retinotopic location during "Fix center" and "Fix left" displays, it is in a different location relative to the animal's head, which is in the same location for both conditions.

As alluded to at the end of *Chapter 5*, coordinate transformations such as those illustrated in *Figure 7.6* are critically important, because an organism needs to know where a stimulus (say, a predator, or prey) is located relative to its body, regardless of the position of its eyes in their sockets, or of its head on its neck. It may also be the case that transformation of visual percepts into body-centered coordinates is important for our sense that our visual world, and the objects in it, are stable, despite the fact that we move our eyes an average of three times per second. This is illustrated, for example, when you watch a video shot from a camera mounted on the forehead of a downhill skier, or a mountain biker, or a surfer – if you've ever engaged in any of these activities, you know that your perceptual experience while actually doing them isn't as jerky as is the perceptual experience of watching a video captured by a head-mounted camera. The difference is that when you are actually moving through the world, your vestibular system registers all the bumps and jolts that cause your head and body to jerk around, and stretch receptors register all the rapid changes of eye position (and neck position, etc.), and the brain uses this information to "smooth out" your perceptual experience.

Data such as those illustrated in *Figure 7.6* indicate that neurons in area 7a of PPC must receive information about eye position from somatosensory cortex, in addition to retinotopic visual information. (Eye position can be computed from signals from stretch receptors that are sensitive to the state of contractedness vs. relaxedness of the muscles that control the movement of the eyes.) Evidence for such integration of information from different sensory domains,

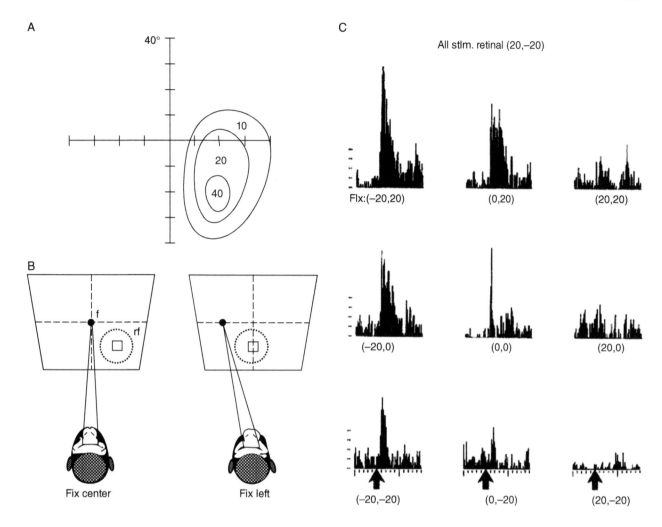

FIGURE 7.6 Eye position modulates spatial signaling of PPC neurons. **A** illustrates a contour plot of the receptive field of a neuron in area 7a, with the numbers indicating firing rate when the monkey is fixating centrally (corresponding to the point of intersection of the two axes) and a stimulus appears within each ring. **B** shows a monkey fixating two locations, with a stimulus presented within the receptive field for both fixations. **C** shows the responses of this neuron to the onset of the stimulus (arrow) in its receptive field when the animal is fixating at each of nine locations. Source: From Andersen, Richard A., Greg K. Essick, and Ralph M. Siegel. 1985. "Encoding of Spatial Location by Posterior Parietal Neurons." Science 230 (4724): 456–458. Reproduced with permission of American Association for the Advancement of Science - AAAS.

and the coordinate transformation that this allows, suggest that the PPC is well positioned to mediate the transition between sensory attention and motor intention. Indeed, more explicit evidence for PPC involvement in motor control comes from delayed-response tasks. In these tasks, a cue stimulus is briefly presented to identify for the subject (in the present example, a monkey, but the same procedure is also used with humans) the region on a screen to which it will need to make either an eye movement (a **saccade**) or

a reaching movement. The subject has been trained, however, to withhold this movement until the offset of the fixation point, which lags behind the flash of the cue by several hundred milliseconds, or longer. *Figure 7.7* illustrates that extracellular recordings collected during a task that varied whether the animal was to make an eye movement or a reach identified two regions of the PPC that showed preferentially elevated delay-period activity as a function of which body part was to be used on that trial. (Note that,

FIGURE 7.7 The response of PPC neurons to delayed response with the eyes vs. the hand. Source: From Snyder, Larry H., Aaron P. Batista, and Richard A. Andersen. 1997. "Coding of Intention in the Posterior Parietal Cortex." Nature 386 (6621): 167–170. doi: 10.1038/386167a0. Reproduced with permission of Nature Publishing Group.

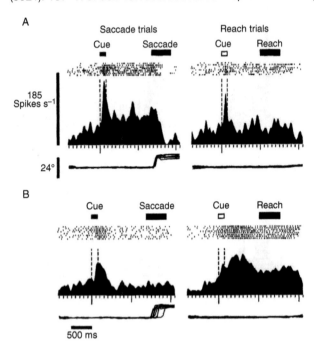

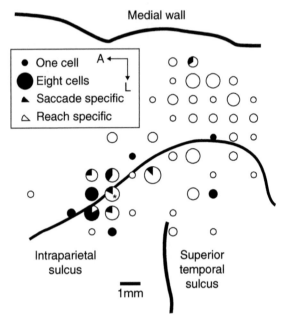

FIGURE 7.7.A Row **A** illustrates the responses of a neuron to a red-colored cue (filled bar in this diagram), signaling that a saccade will be required, and to a green-colored cue (open bar), signaling that a reach will be required. The eight rows of rasters (tick marks) each illustrate the timing of action potentials recorded during a single trial; the histogram below the spike rate computed from those trails; and the traces below that the position of the eye along the vertical meridian as a function of time. Note the near-instantaneous "jumps" in eye position characteristic of a saccade, as opposed to the smooth movement of tracking a moving object. This neuron responds more during the pre-saccade than the pre-reach delay period. Row **B** illustrates responses of a "reach-specific" neuron on the same trials.

FIGURE 7.7.B The anatomical segregation of saccade-specific neurons (filled circles, found primarily along the walls of the IPS) and reach-specific neurons (unfilled circles, found more posteriorly and medially, most prominently on the gyrus separating the IPS from the midline ("Medial wall")). A = anterior; L = lateral; medial wall = midline.

FROM PARIETAL *SPACE* TO MEDIAL-TEMPORAL *PLACE*

Let's return for a moment to the patients of Bisiach and Luzzatti (1978; *Figure 7.4*). The fact that they were able to correctly conjure up several landmarks surrounding all sides of the Piazza, despite the fact that both required two perspectives from which to do it, means that their long-term memories contained a complete representation of this locale, one in which the position of each element relative to the others was stable – an allocentric representation. Their deficit evidently came from the fact that, in order to think about this representation, they had to do so from an egocentric perspective. First, let's consider a neural substrate for this allocentric representation of space. In the final section of this chapter, we'll take up how one might begin to address the notion of "thinking about" it.

because the cued location was the same on both types of trial, these neurons were presumably computing information relevant to their preferred motor effector, rather than to a sensory representation of where in space the cue had been presented.) In *Chapter 8*, we will return to some implications of findings such as these for consideration of whether the dorsal stream is better construed as processing *where* a stimulus is located or as processing *how* the organism is going to interact with it.

Place cells in the hippocampus

Important insights about the formation of allocentric representations of space have come from research in the rat, much of it propelled by the groundbreaking research of (and, more recently, "Nobel Laureate") neurophysiologist John O'Keefe and colleagues. The initial discovery was made while acquiring extracellular recordings from the dorsal hippocampus of a rat while it freely moved about, exploring an elevated maze (essentially a small tabletop with "runway" arms, and no side walls; *Figure 7.8*). What they found was that whereas some pyramidal neurons of the hippocampus responded to certain behaviors, such as grooming or sniffing and whisking, others showed a qualitatively different profile, seeming to encode where the animal was in the environment, independent of the orientation of its body. The difference was so striking that they categorized hippocampal pyramidal neurons into two groups: "the *displace* units are those whose firing pattern relates to the behavior of the animal, irrespective of where it occurs in an environment, while *place* units are those whose firing pattern is dependent on the location of the animal in an environment" (O'Keefe

and Nadel, 1978, pp. 196–197; italics added). Their choice to label "displace" units in reference to the property that they did not display indicates which of the two they found more interesting.

As shown in *Figure 7.8*, "place cells," as they have come to be known, are qualitatively different from cells with body-centered receptive fields, because the fundamental piece of information that they represent is where the animal is located in space, not what it is currently seeing or doing. Thus, the neuron illustrated in *Figure 7.8* has a *place field* (by analogy to receptive field), in that it fires selectively whenever the animal passes through or stops in this location, regardless of the direction in which it is facing. For example, its discharge during "Run A→B" is similar to its discharge during "Run C→B)," even though the former entails turning to the right and the latter turning to the left. Another interesting property is that the magnitude of this neuron's discharge, in a pattern reminiscent of the gain fields of the lateral intraparietal area (LIP), is influenced by factors other than place in and of itself. For example, the response is visibly stronger when the animal is entering the B arm than when leaving and seems to be stronger when passing through the place field than when sitting in it.

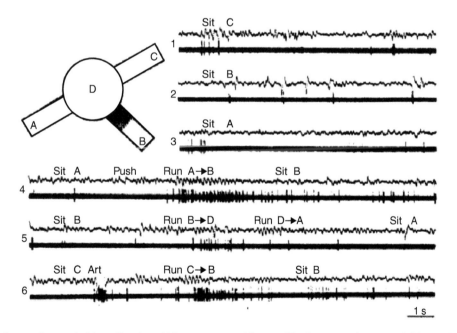

FIGURE 7.8 A place cell recorded from the dorsal hippocampus of the rat. The image on the upper left is a schematic top-down view of a three-arm maze, with the shaded area in B the location of the cell's place field. A pair of electrode traces is shown from each of six epochs of behavior, the upper the LFP (bandpass filtered to only display the theta band), the lower the spiking. Note the neuron produced place-related bursts of activity regardless of whether the animal had turned right to enter the B arm (epoch 4), turned left to enter the B arm (epoch 6), sat in the B arm (epochs 4 and 6), or was leaving the B arm (epoch 5). (The label "Art" in epoch 6 identifies an artifact in the recording.) Source: From O'Keefe, John, and Lynn Nadel. 1978. The Hippocampus As a Cognitive Map. Oxford: Oxford University Press. Reproduced with permission of Oxford University Press.

Note that, in isolation, a single place cell's output wouldn't be very informative to the organism, in that it can only signal, for example, that *I am currently in the corner of my living room near the fireplace*, or *I am* not *currently in the corner of my living room near the fireplace*. Thus, when walking around in my kitchen, say, or from my bedroom to the study, this neuron, through its silence, would be continually, implicitly, conveying that *I am* not *currently in the corner of my living room near the fireplace*. Thanks a lot. But, of course, individual neurons don't operate in isolation. Indeed, as *Figure 7.9* illustrates, a population of place cells, by each representing a different location, is ideally suited to

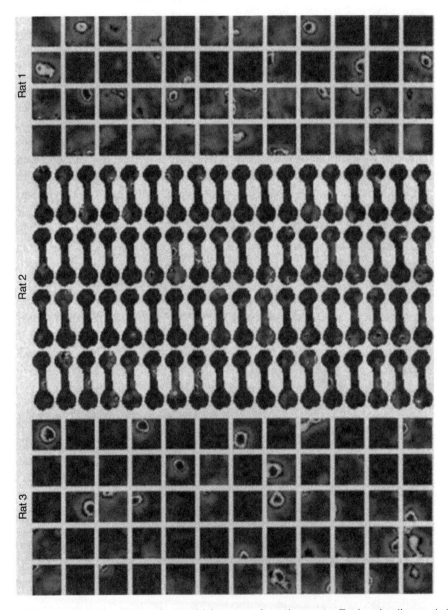

FIGURE 7.9 Place fields of multiple hippocampal pyramidal neurons from three rats. Each animal's panel shows multiple top-down views of the environment that the animal was exploring (a square-shaped enclosure for rats 1 and 3, a dumbbell-shaped enclosure for rat 2). Each individual square or dumbbell image represents the place tuning of a single neuron, the heat map indicating the neuron's firing rate at each location of the enclosure (range: blue = 0 Hz; red = ≥10 Hz). These illustrate how, by sampling across a population of place cells, the hippocampus can be thought of as having a "map" of a locale. Source: Wilson and McNaughton, 1994. Reproduced with permission of AAAS.

represent an environment. Indeed, one can see how, just as is the case with receptive fields in the visual or somatosensory system, the smaller the hippocampal place field, the finer-grained the representation of that region of space.

How does place come to be represented in the hippocampus?

The detailed allocentric representation of a locale with which one is highly familiar (say, the Piazza del Duomo for natives of Milan) comes into existence over time, stitched together from multiple egocentric "snapshots." Only by having stood on the steps of the cathedral and looked out across the plaza, having strolled along the shops, having had a coffee at one of the outdoor cafes, and so on, could one construct a representation of how each of the discrete elements in this locale relate to one another. Because these egocentrically defined "snapshots" each depend on processing in the dorsal stream, it is reasonable to ask how it comes to pass that allocentric space is represented in a structure that lies along the rostromedial wall of the *temporal* lobe (*Figure 2.2*). From *Figure 6.5.B*, one might conjecture that the hippocampus is at the terminus of the ventral stream, located adjacent, as it is, to vAIT. Indeed, this reasoning wouldn't be incorrect; just incomplete.

Figure 6.5.B indicates that, in fact, BA 7 in the PPC and AIT are equidistant from the hippocampus in terms of connectivity, each only two synapses removed. To understand how this organization came to pass, it's useful to return to the rodent. In both the rodent and the primate, the hippocampus is slightly curved, like a banana. Whereas in the primate, its long axis parallels that of the temporal lobe, whereas in the rodent, it is oriented dorsoventrally, effectively "cupping" the caudal surface of the thalamus. Thus, in the rodent, the dorsal hippocampus (where O'Keefe and colleagues initially discovered place cells) lies directly below the cortical surface of the parietal lobe (*Figure 7.10*). Scanning across species in *Figure 7.10* makes it seem plausible that, over the course of evolution, the primate hippocampus has "migrated" ventrorostrally from a previous dorsoventral orientation, and that it has "taken its parietal connections along with it."

THE NEUROPHYSIOLOGY OF SENSORY ATTENTION

The deficit (of the patients of Bisiach and Luzzatti [1978]) evidently came from the fact that, in order to think about this representation, they had to do so from an egocentric perspective. It follows from this sentence from the previous section that cognition often requires us to select just a subset of the totality of information that's available to us, in order to submit this subset of information to the processing that will enable

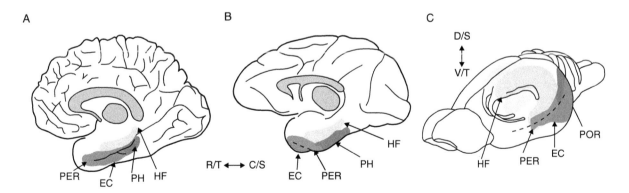

FIGURE 7.10 Location of the hippocampal formation (HF) and adjacent cortex in the human (**A**), the rhesus macaque (**B**), and the rat (**C**). The cortical areas are the perirhinal cortex (PER), the entorhinal cortex (EC), the parahippocampal cortex (PH) and, in the rat, the postrhinal cortex (POR), a homologue to primate PH. The convention for directional labels is general anatomical label/hippocampus-specific label, thus: R/T = rostral/temporal; C/S = caudal/septal; D/S = dorsal/septal; V/T = ventral/temporal. Note that area 36, which *Figure 6.5.B* shows to be the temporal-lobe target of PPC projections, corresponds to the blue-colored POR/PH in this figure. An intriguing disorder of ego- and allocentric processing that results from damage to this region is described by Aguirre (2003, *Further Reading*). Source: From Burwell, Rebecca D., and Kara L. Agster. 2008. "Anatomy of the Hippocampus and the Declarative Memory System." In Memory Systems, edited by Howard Eichenbaum. Vol. 3 of Learning and Memory: A Comprehensive Reference, edited by John H. Byrne. 4 vols., 47–66. Oxford: Elsevier. Reproduced with permission of Elsevier.

us to achieve a behavioral goal. The operations of selection and the prioritization for subsequent processing are operations ascribed to the function of *attention*. Among the influential late-nineteenth-century psychologists whose research and writing laid important foundations for contemporary cognitive neuroscience (see the concluding section of *Chapter 1*) was William James (1842–1910), who famously wrote that "Everyone knows what attention is. It is the taking possession by the mind, in clear and vivid form, of one out of what seem several simultaneously possible objects or trains of thought" (James, 1950, pp. 403–404). This nicely captures our intuition. The difficulty of understanding the cognitive and neural bases of attention, however, was cheekily captured by cognitive scientist Harold Pashler, who wrote roughly a century later that "no one knows what attention is, and . . . there may even not be an 'it' to be known about (although of course there might be)" (1999, p. 1). Here, we'll scratch the surface of this intriguing, notoriously slippery phenomenon that will be relevant for almost every aspect of cognition and behavior that we consider from here on out.

A day at the circus

The metaphor of attention as a spotlight has been a useful one for researchers, in that it captures the phenomenon of directing a resource (light) so as to prioritize one out of many possible targets. At the circus, for example, the lighting director may want the audience to attend to the diversionary pratfalls of a juggling clown, rather than to the trapeze artist who is climbing a ladder up to a high platform, but who isn't yet performing, or to the animal handler wrangling the recalcitrant tiger back into its cage, after it has jumped through the Hoop of Fire. To accomplish this, the spotlight is shined on the clown, thereby making him brighter and easier to see than the climbing acrobat or the retreating tiger. What if Clown 1 is joined by two or three additional jugglers? The area being illuminated by the spotlight can be enlarged or reduced, depending on the number of clowns performing. When it's time for the trapeze performance, the spotlight can jump from the floor, where the clowns were juggling, to a platform high overhead – the platform is currently empty, but the director wants the audience to be attending to the location onto which the trapeze artist will soon make her dramatic, graceful entrance.

This scenario captures two properties of attention that we will briefly consider here: the function of *increasing the salience* of one or more selected stimuli and the distinction between *attending to an object* vs. *attending to a location*. A third, the ability to change the aperture of attention, will be considered in *Chapter 14* (under the rubric of *capacity limitations* of working memory). A final, critical element in our circus scenario is the person who controls the spotlight, turning it on and off, and pointing it first at the clowns, then at the platform. The invocation of the spotlight operator highlights the distinction between the *site* of attention's effects (in our scenario, the clowns, then the platform), vs. the *source* of attention (the spotlight itself, which generates the bright light, and the person controlling its operation). Our present focus will be on the site of attention. We will consider the source(s) of attention in *Chapter 9*.

Attending to locations vs. attending to objects

One place where we've already seen the effects of attention was in *Chapter 5*, when considering the difference in the AEP when the subject attended to a series of beeps, vs. when they read a book and ignored the beeps. In that example, because the effect of attention was to increase the magnitude of the N1 and P2 components of the AEP (*Figure 5.5*), we can say that the site of attention's effects included primary auditory cortex. And what might have caused this "amplification" of these ERP components? There are at least three plausible explanations. One is that the neurons responding to the afferent auditory information may have done so more vigorously in the count-the-beeps condition than in the book-reading condition. A second is that the level of synchrony among the neurons carrying afferent auditory information may have increased (a scenario illustrated in *Figure 3.9.B*). And a third is that more neurons may have been recruited when the subject attended to the auditory channel. (Recall, as you think about these different scenarios, that in *Chapter 5* we learned that loudness is encoded in the auditory pathway by firing rate – the louder a sound, the higher the rate of firing of neurons representing that sound's "places" along the basilar membrane.)

Thus, by any of these accounts, a reasonable supposition is that paying attention to a particular sensory channel may be equivalent, in neural terms, to increasing the energy of the stimulus itself (such as by making it louder, or brighter). The phenomenon of an attention-related boost in the magnitude of sensory processing signals is also seen with the visual-evoked potential (VEP) and somatosensory-evoked potential (SEP). *Figure 7.11* illustrates a test of visual

FIGURE 7.11 EEG of visual spatial attention, as studied by Di Russo, Martinez, and Hillyard, 2003.

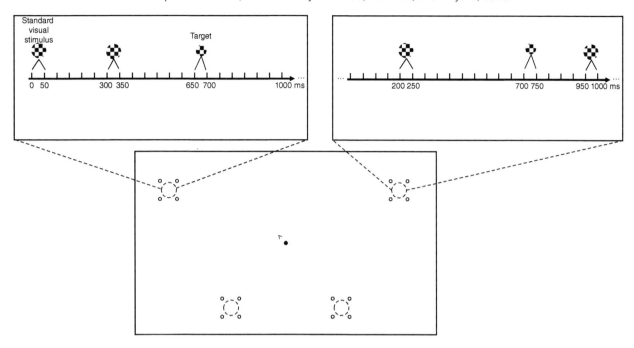

FIGURE 7.11.A The display: subjects were instructed to fixate the central dot while checkerboard stimuli flashed rapidly and randomly at two locations (illustrated here are upper left and upper right) and to count the number of infrequently appearing targets appearing in the cued location. (This is a variant of the famous Posner [1980] spatial cuing paradigm.)

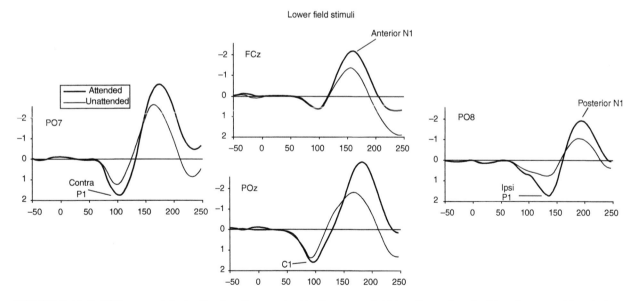

FIGURE 7.11.B ERPs associated with stimuli presented to the lower-right location, when that location was *Attended* and when it was *Unattended* (i.e., when visual presentation was identical, but subjects were attending to the left visual field location). The four plots show the signal recorded at electrodes FCz, PO8, POz, and PO7 (in relation to *Figure 3.8*, FCz is located midway between Fz and Cz, and the PO electrodes midway between the P and O electrodes). Source: From Di Russo, Fracesco, Antigona Martínez, and Steven A. Hillyard. 2003. "Source Analysis of Event-Related Cortical Activity During Visuo-Spatial Attention." Cerebral Cortex 13 (5): 486–499. doi: 10.1093/cercor/13.5.486. Reproduced with permission of Oxford University Press.

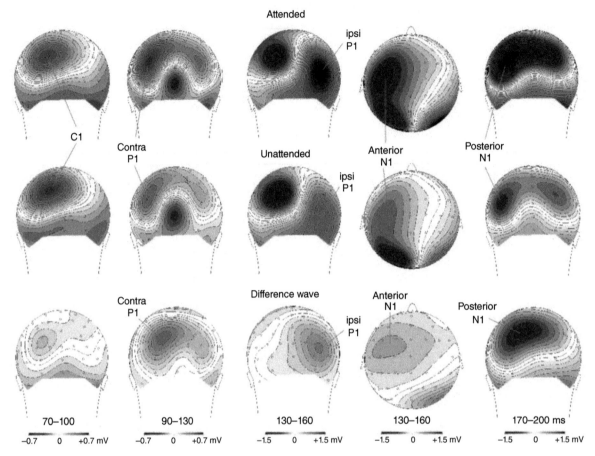

FIGURE 7.11.C Four "snapshots in time" of the ERP from *Figure 7.11.B* as is it distributed across the scalp (as viewed from behind, except fourth column from left is as viewed from above, with the nose at the top of each image.). The top and middle rows show the ERP vs. baseline, the bottom row shows the subtraction images of [Attended – Unattended]. Although recorded from an array of 64 discrete electrodes, interpolation was applied to generate this continuous 2-D voltage map. Source: From Di Russo, Fracesco, Antigona Martínez, and Steven A. Hillyard. 2003. "Source Analysis of Event-Related Cortical Activity During Visuo-Spatial Attention." Cerebral Cortex 13 (5): 486–499. doi: 10.1093/cercor/13.5.486. Reproduced with permission of Oxford University Press.

spatial attention, in that subjects fixate centrally but focus their attention on a location and wait for events to occur there. (This is referred to as covert attention. Overt attention would be to look directly at the location of interest.) From it, one can see that the effect of spatial attention on early components of the VEP is to magnify them. Further, it seems to do so in a particular way. Looking at the traces from the PO electrodes, it's as though the voltage of the *Attended* ERP at each time point was computed from the corresponding time point of the *Unattended* waveform by multiplying the latter by some constant value (perhaps 1.7 at PO7, and a value closer to 2.0 at PO8). Based on this observation, in isolation, one might hazard that attention

works by applying a multiplicative gain factor to sensory information processing. We'll return to this idea in a moment, after first considering two additional pieces of evidence that complicate things a bit.

Although it's clear that we can attend to a location in space and "wait for something to happen" there, our intuitions, confirmed by decades of experimental psychology, tell us that we can also attend to an object, regardless of where it is located in space. (The author does this, for example, when proudly admiring his daughter at a gymnastics competition, despite the fact that she's surrounded by a milling gaggle of many, many other adolescent girls, all wearing similar leotards and hairstyles.) A clever ERP experiment designed by the same group featured in the previous paragraph, led by electrophysiologist Steven Hillyard, directly compared the effects of spatial attention vs. those of nonspatial feature-based attention. (For our purposes, it's simpler to start by considering single features, rather than complex objects.) In it, subjects attended to a nonspatial feature of the stimulus, its color or its direction of motion, which, from trial-to-trial, appeared in one of the two visual hemifields (*Figure 7.12.A*). Comparison of the VEP for the attended vs. the unattended hemifield revealed the amplification of the P1, N1, and subsequent visual-evoked components that we have

FIGURE 7.12 EEG of feature-based visual attention, as studied by Anllo-Vento and Hillyard, 1996.

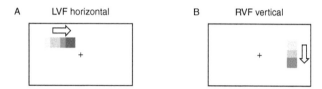

FIGURE 7.12.A There were three relevant stimulus dimensions, color (red, blue) direction of motion (horizontal, vertical), and visual field (right, left). Any of the six possible stimulus types (e.g., red horizontal motion in the LVF) could occur on each trial. On any block of trials, subjects were to indicate when the *speed* of the moving stimulus was slower than typical, but only when this slowing occurred for the instructed stimulus type (i.e., red, blue, horizontal, or vertical) in the instructed visual field.

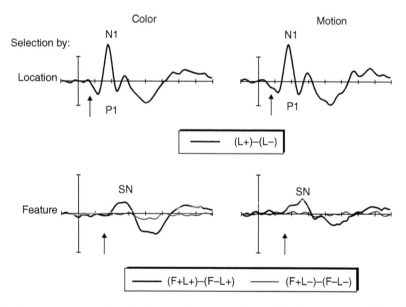

FIGURE 7.12.B ERP difference waves showing the effects of spatial attention (top row) and feature-based attention (bottom row). The waveforms in the top row are conceptually what would result if, in *Figure 7.11.B*, one were to subtract the "Unattended" (here, labeled "L–") from the "Attended" (L+) waveforms at any electrode – they illustrate the extent to which the VEP is magnified by spatial attention while the subject was attending to a color (left-hand column) or a direction of motion. In the bottom row, the waveforms drawn in heavy lines illustrate the subtraction of trials when the instructed color (or direction) was present in the attended visual field ("F+L+") vs. when the uninstructed feature was present in the attended visual field ("F–L+"), thereby emphasizing the SN component. The faint lines are the analogous subtractions from the unattended visual field. Arrows indicate the points at which attentional effects first achieved statistical significance. Source: Adapted from Anllo-Vento, Lourdes, and Steven A. Hillyard. 1996. "Selective Attention to the Color and Direction of Moving Stimuli: Electrophysiological Correlates of Hierarchical Feature Selection." Perception & Psychophysics 58 (2): 191–206.doi: 10.3758/BF03211875. Reproduced with permission of Springer.

already seen (compare *Figure 7.12.B* to *Figure 7.11.B*). The comparison of the attended nonspatial feature vs. the unattended nonspatial feature, in contrast, revealed an ERP component that was not "merely" an amplification of the VEP (*Figure 7.12.B*). Instead, this "selection negativity" (SN) must correspond to the enhanced processing of the critical stimulus feature (e.g., of its redness, or of its horizontalness), rather than to the indiscriminant amplification of all activity in the visual system that seems to result from spatial attention. That is, it seems to reflect a neural process corresponding to the *selection* of one feature from the two-feature stimulus. It is important to note that although the shape and timing of the SN for attention to a color or to a direction of motion appear to be very similar, the two were extracted from electrodes on different parts of the scalp: the color SN was greatest at contralateral occipitotemporal sites, whereas the motion SN was greatest at bilateral temporal and parietal sites.

Mechanisms of spatial attention

Contradictory findings from fMRI vs. EEG?

Having established the distinction between location- vs. feature-based attention, let's now consider how they might work. Recall that *Figure 7.11.B* suggested a mechanism applying "multiplicative gain." *Huh?* Consider, for a moment, your mobile phone (or whatever device you kids use to listen to your "music" these days): you set the volume to a larger number and the music is louder; you set it to a smaller number and it's softer. But what about during a silent gap in the music? Whether the volume is set at *2* or at *11*, the silence will sound the same. This is because the products of [*2*0*] and of [*11*0*] are both *0*. Your volume controller only influences the gain of the signal being produced by your music player, by *multiplying* it by some factor. It doesn't *add* anything to the signal. (Now look back at *Figure 7.11.B* and note how the Attended and Unattended waveforms are identical until afferent visual signals get to

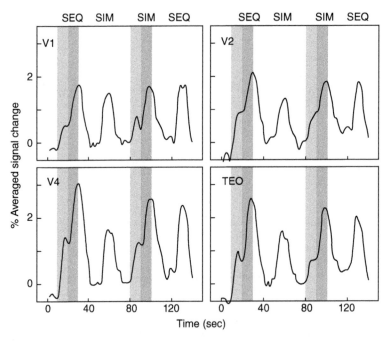

FIGURE 7.13 Visual attention studied with fMRI. Source: Kastner, Pinsk, De Weerd, Desimone, and Ungerleider, 1999. Reproduced with permission of Elsevier. Time courses of fMRI activity in four regions of the visual system (V1, V2, V4, TEO). The labels "SEQ" and "SIM" indicate when stimuli were presented (SEQ and SIM being simultaneous and sequential presentation, respectively, which we can disregard for our purposes). The gray band highlights the *expectation* period, the blue band periods of attention to the stimuli, and the "unbanded" regions of the plot portions of the scan when subjects attended central fixation, effectively ignoring stimulus presentation. Source: From Source: Kastner, Sabine, Mark A. Pinsk, Peter De Weerd, Robert Desimone, and Leslie G. Ungerleider. 1999. "Increased Activity in Human Visual Cortex During Directed Attention in the Absence of Visual Stimulation." Neuron 22 (4): 751–761. doi: 10.1016/S0896-6273(00)80734-5. Reproduced with permission of Elsevier.

V1, at which point the attended vs. unattended VEPs diverge in a multiplicative manner.) An *additive* process, on the other hand, might look like what is illustrated in *Figure 7.13*. This illustrates an fMRI study of both location- and feature-based attention, the former being our interest at the moment. Note how, in *Figure 7.13*, there is an increase in fMRI signal, relative to baseline, during the *expectation* period of the task. That is, although the baseline and *expectation* periods of the task are identical with respect to visual presentation (empty gray screen) and overt behavior (fixation), the signal in brain areas representing the area of the visual field being attended to nonetheless increases when spatial attention is directed there. This phenomenon is sometimes referred to as a "baseline shift," in that "nothing has happened yet" in the *expectation* portion of the trial, and so an analysis that ignored the cognitive state of the subject ("not attending" vs. "attending") would conclude that the *expectation* period was a baseline period with a level of activity that was shifted upward.

And so the EEG studies featured in *Figure 7.11* and *Figure 7.12* suggest that spatial attention operates via a multiplicative mechanism, whereas the fMRI study of *Figure 7.13* suggests that spatial attention operates via an additive mechanism. Is one wrong? Of course, it's not that straightforward. We need to take into account that the two sets of experiments were intended to address different questions ("early vs. late selection" being an important question motivating the ERP studies, "biased competition" [which we'll take up presently] motivating the fMRI study), and that these two techniques are sensitive to different neurophysiological signals. Additionally, how the data are analyzed can determine what one is capable of detecting in one's data. Thus, although the ERP studies summarized here don't provide evidence for a baseline shift associated with covert spatial attention, many studies analyzing spectral elements of the EEG do find such evidence. Most prominently, covertly directing spatial attention to one visual field is known to produce a decrease in alpha-band power in the contralateral hemisphere and an increase in alpha-band power in the ipsilateral hemisphere. An ERP analysis would not detect such an effect, because alpha-band dynamics represent a major component of the "noise" that ERP averaging is intended to minimize (see *section Event-related analyses, Chapter 3,* and *Figure 3.9*). (For a study that did a head-to-head comparison of EEG/ERP vs. fMRI measures of the same subjects performing the same tasks, see Itthipuripat et al. [2019], *Further Reading*.) These considerations also highlight the fact that, for some questions, to "really know what's going on," one needs to get inside the cranium.

Effects of attention on neuronal activity

Spatial attention

Figure 7.14 illustrates a study performed in the laboratory of Robert Desimone (the neurophysiologist whose work we first encountered in *Chapter 6*, when, as a trainee of Charles Gross, he studied the visual functions of IT cortex). After Desimone joined the fabled Laboratory for Neuropsychology at the National Institutes of Health, he and his trainees carried out many pioneering studies on the neurophysiological bases of visual attention. As with studies that we've already considered, the subject (in this case, a monkey) covertly attended a specific location and waited for a predesignated target to appear there. Recordings from neurons in areas V2 and V4 clearly showed that covertly attending to a location within their receptive fields produced a sustained increase in their firing rate – a baseline shift. From these findings, the authors concluded that "visual-spatial attention operates, at least in part, by creating a preset sensory bias that modulates the initial volley of sensory information as it passes through area V4" (Luck, Chelazzi, Hillyard, and Desimone, 1997, p. 41).

Object-based attention

As we near the end of this chapter, we turn to a final theme invoked by the study of unilateral neglect – the *competition* for stimulus representation inferred from the phenomenon of extinction. Evidence for such competition at the neuronal level has come from research from Desimone's group (with collaboration from the British experimental psychologist John Duncan), illustrated in *Figure 7.15*. The experiment that they carried out was modeled on visual search tasks that have been used in (human) experimental psychology for decades – you first present the target, and then an array of stimuli within which subjects must either find the target or indicate that it's not present in the search array. For the monkeys that were tested by Chelazzi, Duncan, Miller, and Desimone (1998), the task was simplified by presenting arrays of only two or one stimuli. When considering the data, the first thing to note is that these recordings were made in vAIT, the anterior-most region of the ventral visual processing stream where, as we saw in *Chapter 6,* neurons show a high degree of preference for the conjunctions of features that define specific objects. To carry out these experiments, the researchers first isolated a neuron with an extracellular electrode and then presented many images to find one to which it responded preferentially (they refer to it as the "good" stimulus for that neuron; in *Figure 7.15*, it's the butterfly). Then they had the animal perform several

FIGURE 7.14 The effects of spatial attention on neural activity in the macaque visual system, as studied by Luck, Chelazzi, Hillyard, and Desimone, 1997.

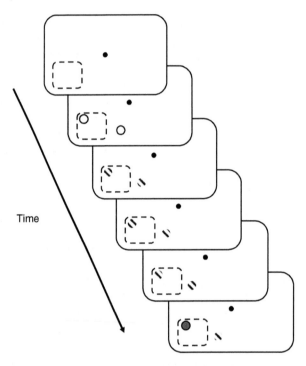

Time

FIGURE 7.14.A The behavioral task, in which stimuli were presented sequentially or simultaneously in two "marked" locations (indicated here by unfilled circles). In different conditions, these locations were either both placed within the receptive field of the neuron in V2 or V4 that was being recorded from or one was inside and one outside the receptive field (dashed line). At the beginning of each block of trials, the animal was cued as to which of the two marked locations was to be attended and was rewarded for indicating when the stimulus in the cued location changed. (Cartooned here is a trial in which the stimulus located inside the receptive field changes.) Of interest here is activity from neurons in conditions when one of the marked locations was located outside the receptive field.

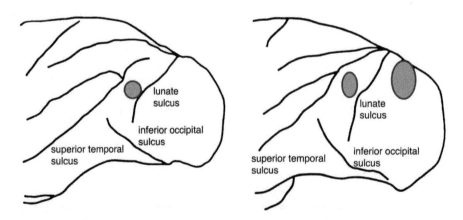

FIGURE 7.14.B Locations of electrode penetration and recording sites in two monkeys. V4 recordings were made within shaded oval regions anterior to lu in both animals. In the animal illustrated on the right, recordings were also made in V1 (shaded oval posterior to the lunate sulcus) and in V2, reached through penetrations extending rostrally from the V1 recording site to the caudal bank of the lunate sulcus where V2 is located. Source: From Luck, Steven J., Leonardo Chelazzi, Steven A. Hillyard, and Robert Desimone. 1997. "Neural Mechanisms of Spatial Selective Attention in Areas V1, V2, and V4 of Macaque Visual Cortex." Journal of Neurophysiology 77 (1): 24–42. Reproduced with permission of The American Physiological Society (APS).

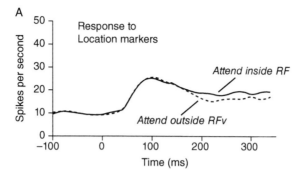

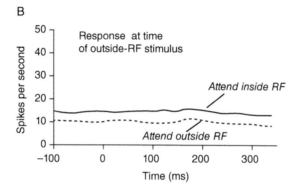

FIGURE 7.14.C Two examples of a "baseline shift" effect of spatial attention in the average firing rate of 40 V4 neurons. **A** illustrates the prestimulus period: onset of location markers occurred at time 0, and approximately 175 ms later the neurons' firing rates diverged depending on whether attention was focused inside or outside the receptive field (RF). **B** illustrates the middle of the trial: a nontarget stimulus is presented outside the RF at time 0, when the animal is attending to a marked location inside vs. outside the RF. Source: From Luck, Steven J., Leonardo Chelazzi, Steven A. Hillyard, and Robert Desimone. 1997. "Neural Mechanisms of Spatial Selective Attention in Areas V1, V2, and V4 of Macaque Visual Cortex." Journal of Neurophysiology 77 (1): 24–42. Reproduced with permission of The American Physiological Society (APS).

kinds of trials, the results from four of which are illustrated in *Figure 7.15.B*: (1) the search array only contained one item, which matched the target, and was the "good" stimulus; (2) the search array contained two items, with the match being the "good" stimulus; (3) the search array contained two items, including the "good" stimulus, but the other stimulus (a "poor" stimulus for that neuron) matched the target; (4) the search array only contained one item, which matched the target, and was the "poor" stimulus for that neuron. Additionally, there were catch trials in which the target was not present in the search array (not shown). In *Figure 7.15.B*, the first 150 ms after array onset demonstrates evidence for competition between the neural representations of objects: the response to a "good" target stimulus is suppressed when it is paired with a second stimulus ("Target = good stim. in 2-stim. array") relative to when it is presented alone ("Target = good stimulus alone"). This is presumably due to lateral inhibition that

occurs when collateral branches of neuron A's axon synapse on inhibitory interneurons located in cortical columns that are tuned for different stimulus configurations, and these interneurons, in turn, release GABA to suppress activity within that column. That is, the "initial [feedforward] volley of sensory information" associated with each of the objects in the two-item array results in the neural representations of each being activated, and these, in turn, suppress the neural representation of the other. The influence of attention is seen at the inflection, at ~175 ms, of the response to the target when it is the "good stim. in 2-stim. array" (as labeled in *Figure 7.15.B*). From that point forward, the neuron's response is identical in the "alone" and "2-stim." conditions – the suppression from the competing stimulus representation has been overcome. In the parlance of Desimone and colleagues, the *competition* has been *biased* by the top-down influence of selective attention.

FIGURE 7.15 Neural evidence for biased competition as a mechanism of object-based attention. Source: (B) From Chelazzi, Leonardo, John Duncan, Earl K. Miller, and Robert Desimone. 1998. "Responses of Neurons in Inferior Temporal Cortex During Memory-Guided Visual Search." Journal of Neurophysiology 80 (6): 2918–2940. Reproduced with permission of The American Physiological Society (APS).

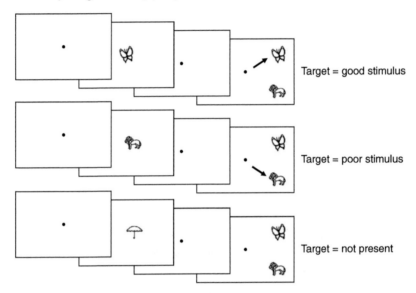

FIGURE 7.15.A The search task. The target object is presented at the beginning of the trial, and after the delay, an array of two stimuli is presented and the monkey makes a saccade to the target (no saccade if the target is not in the array). Not shown are trials with one-item arrays, which could also be "target-present" or "target absent."

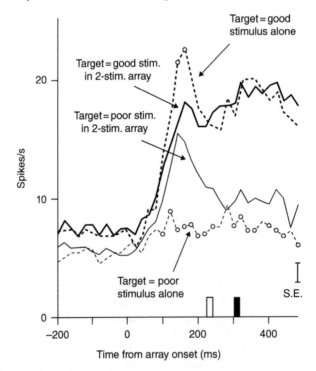

FIGURE 7.15.B A summary of the results from 58 neurons, contrasting good vs. bad targets in 2-stim.-array and 1-stim.-array ("alone") trials. Neural activity is time-locked to the onset of the search array (i.e., array appeared at time 0). Circles indicate time points for which the two good traces or the two poor traces differ statistically from each other. Unfilled and filled bars on the horizontal axis indicate average response times to one-item and two-item arrays, respectively.

The competition is also seen in responses when the "poor" stimulus is the target. When presented alone, the poor stimulus, by definition, barely influences the activity of the neuron. When the neuron's good stimulus is presented along with the poor target, however, the neuron initially responds to its preferred stimulus. After ~125 ms, however, the neuron's response to its preferred stimulus begins to be tamped down, presumably because the neurons representing the target have begun winning the competition, and are suppressing the neural representation of the nontarget stimulus.

These results from Chelazzi, Duncan, Miller, and Desimone (1998) suggest a neural mechanism that implements the psychological process invoked by William James 100 years before: "[Attention] implies withdrawal from some things in order to deal effectively with others" (James, 1950, p. 404). A topic to be taken up in *Chapter 9* will be: "What is the source of this biasing influence?"

TURNING OUR ATTENTION TO THE FUTURE

In this chapter, we started by considering how aspects of cognition that many of us take for granted (e.g., being aware that there's food on both the right *and* the left side of our dinner plate) can be severely disrupted as a result of a stroke affecting temporoparietal cortical regions and ended with a consideration of neural phenomena that may explain some of the mechanisms underlying visual awareness; mechanisms that may break down in some neurological conditions. In the next two chapters, we'll build on what we've learned about the properties of the parietal cortex and of spatial cognition to introduce several new ideas about the neural organization of behavior and cognition. One, already hinted at on more than one occasion, is that an alternative way to construe the functions of the dorsal stream is that it governs *how* the body can interact with objects that necessarily occupy different locations in the environment. A second will be how parietal and frontal cortical systems coordinate with subcortical systems to produce the finely controlled actions that we execute countless times each day. A third will be to consider the source(s) of attentional control. These include the ideas that PPC and other regions support a dynamic "priority map" that helps guide attentional selection, and that the volitional vs. the reflexive control of attention are governed by different neural systems.

At a literal, and also profound, level, we are about to make our first foray across the central sulcus, into the frontal lobe.

END-OF-CHAPTER QUESTIONS

1. What symptom(s) of unilateral neglect inform ideas about the lateralization of attention; about how we process "near" vs. "far" space; about the competition for representation within the brain; and about location-based vs. object-based attention?

2. In what ways have neuroimaging studies supported, and extended, neuropsychology-based ideas about the functions of the dorsal visual processing stream?

3. Summarize the idea of a coordinate transformation. Why is this a critical principle underlying the guidance of action with perception? What evidence is there for the computation of coordinate transformations in the PPC?

4. We saw how the integration of retinotopic sensory information with eye-position information results in a coordinate transformation from retina-centered to head-centered coordinates. Using the same logic, how could the brain effect a coordinate transformation from head-centered to trunk-centered coordinates?

5. What is the difference between an egocentric vs. an allocentric representation of space? Summarize neurophysiological evidence that both are maintained in the brain.

6. In what ways is the analogy to a spotlight appropriate for understanding visual attention? Describe (at least) two aspects of attention that do not conform to this analogy.

7. How do location-based and object- (or feature-) based attention differ neurophysiologically?

8. Summarize the evidence that biased competition may be an important principle for visual attention.

REFERENCES

Andersen, Richard A., Greg K. Essick, and Ralph M. Siegel. 1985. "Encoding of Spatial Location by Posterior Parietal Neurons." *Science* 230 (4724): 456–458.

Anllo-Vento, Lourdes, and Steven A. Hillyard. 1996. "Selective Attention to the Color and Direction of Moving Stimuli: Electrophysiological Correlates of Hierarchical Feature Selection." *Perception & Psychophysics* 58 (2): 191–206. doi: 10.3758/BF03211875.

Bisiach, Edoardo, and Claudio Luzzatti. 1978. "Unilateral Neglect of Representational Space." *Cortex* 14 (1): 129–133.

Burwell, Rebecca D., and Kara L. Agster. 2008. "Anatomy of the Hippocampus and the Declarative Memory System." In *Memory Systems*, edited by Howard Eichenbaum. Vol. 3 of *Learning and Memory: A Comprehensive Reference*, edited by John H. Byrne. 4 vols., 47–66. Oxford: Elsevier.

Chatterjee, Anjan. 2003. "Neglect: A Disorder of Spatial Attention." In *Neurological Foundations of Cognitive Neuroscience*, edited by Mark D'Esposito, 1–26. Cambridge, MA: MIT Press.

Chelazzi, Leonardo, John Duncan, Earl K. Miller, and Robert Desimone. 1998. "Responses of Neurons in Inferior Temporal Cortex during Memory-Guided Visual Search." *Journal of Neurophysiology* 80 (6): 2918–2940.

Di Russo, Fracesco, Antigona Martínez, and Steven A. Hillyard. 2003. "Source Analysis of Event-Related Cortical Activity during Visuo-Spatial Attention." *Cerebral Cortex* 13 (5): 486–499. doi: 10.1093/cercor/13.5.486.

Driver, Jon, Patrik Vuilleumier, and Masud Husain. 2004. "Spatial Neglect and Extinction." In *The Cognitive Neurosciences III*, edited by Michael S. Gazzaniga, 589–606. Cambridge, MA: MIT Press.

James, William. 1950. *The Principles of Psychology*, vol. 1. New York: Dover. First published 1890.

Karnath, Hans-Otto, Johannes Rennig, Leif Johannsen, and Chris Rorden. 2011. "The Anatomy Underlying Acute versus Chronic Spatial Neglect: A Longitudinal Study." *Brain* 134 (3): 903–912. doi: 10.1093/brain/awq355.

Kastner, Sabine, Mark A. Pinsk, Peter De Weerd, Robert Desimone, and Leslie G. Ungerleider. 1999. "Increased Activity in Human Visual Cortex during Directed Attention in the Absence of Visual Stimulation." *Neuron* 22 (4): 751–761. doi: 10.1016/S0896-6273(00)80734-5.

Luck, Steven J., Leonardo Chelazzi, Steven A. Hillyard, and Robert Desimone. 1997. "Neural Mechanisms of Spatial Selective Attention in Areas V1, V2, and V4 of Macaque Visual Cortex." *Journal of Neurophysiology* 77 (1): 24–42.

O'Keefe, John, and Lynn Nadel. 1978. *The Hippocampus as a Cognitive Map*. Oxford: Oxford University Press.

Pashler, Harold E. 1999. *The Psychology of Attention*. Cambridge, MA: MIT Press.

Posner, Michael I. 1980. "Orienting of Attention." *Quarterly Journal of Experimental Psychology* 32 (1): 3–25. doi: 10.1080/00335558008248231.

Saygin, Ayse Pinar, and Martin I. Sereno. 2008. "Retinotopy and Attention in Human Occipital, Temporal, Parietal, and Frontal Cortex." *Cerebral Cortex* 18 (9): 2158–2168. doi: 10.1093/cercor/bhm242.

Snyder, Larry H., Aaron P. Batista, and Richard A. Andersen. 1997. "Coding of Intention in the Posterior Parietal Cortex." *Nature* 386 (6621): 167–170. doi: 10.1038/386167a0.

OTHER SOURCES USED

Burgess, Neil. 2006. "Spatial Memory: How Egocentric and Allocentric Combine." *Trends in Cognitive Sciences* 10 (12): 551–557. doi: 10.1016/j.tics.2006.10.005.

Halligan, Peter W., and John C. Marshall. 1991. "Left Neglect for Near but not Far Space in Man." *Nature* 350 (6318): 498–500. doi: 10.1038/350498a0.

Thier, Peter, and Richard A. Andersen. 1998. "Electrical Micro-stimulation Distinguishes Distinct Saccade-Related Areas in the Posterior Parietal Cortex." *Journal of Neurophysiology* 80 (4): 1713–1735.

Weintraub, Sandra, and M.-Marsel Mesulam. 1989. "Neglect: Hemispheric Specialization, Behavioral Components and Anatomical Correlates." In *Handbook of Neuropsychology*, vol. 2, edited by Francois Boller and Jordan Grafman, 357–374. Amsterdam: Elsevier.

Wilson, Matthew A., and Bruce L. McNaughton. 1994. "Reactivation of Hippocampal Ensemble Memories during Sleep." *Science* 265 (5172): 676–679. doi: 10.1126/science.8036517.

FURTHER READING

Aguirre, Geoffrey K. 2003. "Topographical Disorientation: A Disorder of Way-Finding Ability." In *Neurological Foundations of Cognitive Neuroscience*, edited by Mark D'Esposito, 90–108. Cambridge, MA: MIT Press.
Thorough overview of the clinical aspects of this intriguing disorder, and its implications for our understanding of spatial cognition.

Burgess, Neil. 2006. "Spatial Memory: How Egocentric and Allocentric Combine." *Trends in Cognitive Science* 10 (12): 551–557. doi: 10.1016/j.tics.2006.10.005.
Synthesis of human behavioral studies of navigation and spatial representation with the neurophysiological representation of space in the rat, monkey, and human to argue that parallel, interacting ego and allocentric reference frames guide spatial cognition.

Desimone, Robert, and John Duncan. 1995. "Neural Mechanisms of Selective Visual Attention." *Annual Review of Neuroscience* 18: 193–222. doi: 10.1146/annurev.ne.18.030195.001205.
The highly influential articulation of the principle of biased competition that draws equally from theory, experimental psychology, and neurophysiology.

Itthipuripat, S., T. C. Sprague, and J. T. Serences. 2019. "Functional MRI and EEG Index Complementary Attentional Modulations." *Journal of Neuroscience* 39: 6162–6179. doi: https://doi.org/10.1523/JNEUROSCI.2519-18.2019.
A head-to-head comparison of EEG/ERP vs. fMRI signals measured from the same individuals performing the same visual attention tasks.

Lezak, Muriel D. 1995. *Neuropsychological Assessment*, 3rd ed. New York: Oxford University Press.
This is the authoritative compendium of seemingly every neuropsychological test ever developed, how it is administered, and what behavioral and/or cognitive functions that test measures. Among these are many tests of spatial cognition and of attention.

CHAPTER 8
SKELETOMOTOR CONTROL

KEY THEMES

- Motor control is governed by three principal circuits: the corticospinal system that carries efferent motor commands from motor cortex to the spinal circuits that carry out motor commands; and the cortico-cerebellar and cortico-basal ganglia systems, each of which processes copies of the efferent signal and feeds back to motor cortex so as to refine and/or facilitate future actions.

- The motor cortex plans and generates commands in abstract x,y,z coordinates, and the spinal cord translates these commands (often nonlinearly) into a set of muscle tensions needed to carry out the action.

- Despite its generally homuncular functional topography, primary motor cortex differs from sensory analogues in that its representation of the body surface is not highly precise, and it also represents actions that are not tied to any one part of the body.

- M1 neurons have relatively broad tuning, and precise control of movement is achieved via population coding – the summation of input from thousands of units firing at varying intensities.

- Electrophysiological and microstimulation evidence that PPC neurons contribute to planning and execution of eye movements and reaches, combined with neuropsychological evidence, have been taken

as evidence that a core function of PPC may be to determine *how* visual information can be used to guide behavior.

- The unique wiring of the cerebellum makes it a key neural substrate for motor learning, fine-tuning ongoing movements, and, perhaps, encoding and refining cognitive models of the world.

- The cortico-basal ganglia-thalamic circuits integrate information from many cortical territories and use the output to gate frontal cortex activity.

- Common principles of network and neurochemical dynamics (including a key role for dopamine [DA]) across "affective," "cognitive," and "motoric" basal ganglia systems suggest that many principles of motor control generalize to these other domains of behavior.

- The acquisition and modification of stereotyped sequences of movements, including what we think of as habits, rely on principles of reinforcement learning.

- The discovery of mirror neurons in premotor cortex (PMC) has generated intriguing theories ranging from the understanding of actions and intentions of others, to atypical neurological development (e.g., autism spectrum disorder [ASD]), to the evolution of human culture.

CONTENTS

Our introduction to the frontal lobes came in *Chapter 1*, where we considered a nineteenth-century breakthrough in thinking about how the brain controls movement of the body, and one relating to the production of speech. The former is most relevant here. The observations of the neurologist Jackson gave rise to the idea of a somatotopic organization of the brain system controlling movement in the human. The electrical stimulation experiments of Fritsch and Hitzig indicated that motor cortex, in the dog, was localized to frontal cortex, and that distinct **motor effectors** contralateral to stimulation sites were represented by discrete regions within this lobe. In addition to fleshing out more detail about the structure and function of the motor system, this chapter will introduce many concepts related to the control of action that will be highly germane to many of the remaining topics to be covered in this book, including *cognitive control*, *decision making*, *language*, *social behavior*, and *consciousness*. Indeed, some have argued that much of our thought and behavior, no matter how abstract and "high level," can be construed as more-or-less abstract implementations of the same principles of motor control that we will consider in detail in this chapter.

THE ORGANIZATION OF THE MOTOR SYSTEM

The anatomy of the motor system

Primary motor cortex (M1) occupies the anterior bank of the central sulcus and the caudal portion of the precentral gyrus, a region classified as BA 4 that stretches from the medial wall of the hemisphere to the dorsal lip of the Sylvian fissure. In loose analogy to cortical sensory systems, there are two adjacent motor-related fields with "second-level functions," both sharing BA 6: the premotor cortex (PMC) of the rostral half of the precentral gyrus and the supplementary motor area (SMA) located superior to PMC and rostral to the superior-most and medial portions of M1. Unlike with sensory systems, however, the functions of M1, PMC, and SMA aren't as hierarchically organized as are, say, V1, V2, and MT. For example, each of these three motor areas sends projections from cortex to the areas of the spinal cord that directly trigger movement. (Thus, for simplicity, when describing general properties that apply to each, the term "motor cortex" will be used to refer nonspecifically to M1, PMC, and SMA.)

The motor cortices are a central hub in three circuits that carry action-related signals. The fibers carrying the signals that descend along the ventral spinal cord to trigger movements of the body are called **motor efferents**. The other two carry copies of these motor signals, but are considered **recurrent circuits** (often called "loops") in that they eventually feed back onto the frontal cortex. They implement the principle of **efference copy**, whereby a copy of the efferent signal can be used to, for example, compare the intended outcome of the outgoing motor command against information about the actual outcome (conveyed by ascending sensory signals) and, if the two don't match up, send a corrective signal back to the motor cortex so that the next movement will be more successful. Let's consider each circuit in turn.

The corticospinal tract

The giant pyramidal cells of layer V of motor cortex send axons that descend through the brain, funneling together as the internal capsule that passes between the (laterally located) lentiform nucleus of the basal ganglia and the (medially located) caudate nucleus and thalamus, decussating in the caudal brain stem, and descending in the ventral spinal cord as the corticospinal tract (*Figure 8.1*). Each corticospinal fiber activates motor neurons in the cord either via direct synaptic connections or, most typically, indirectly via local circuitry in the cord. Motor neurons send axons that innervate muscles, but because their cell bodies are located within the spinal cord, they are considered part of the CNS. Their functional organization can be considered from two perspectives: from that of the muscle, each muscle is innervated by a grouping of neurons called a motor neuron pool; from that of the individual motor neuron, it and each of the motor fibers that it innervates make up a motor unit. Motor neurons release the neurotransmitter acetyl choline (ACh) at their synapses, which are often referred to as the neuromuscular junction. More on muscles in the upcoming section on *The biomechanics of motor control*.

The cortico-cerebellar circuit

The cerebellum is an intriguing structure, in that it contains 50 billion of the roughly 80 billion neurons of the brain, yet, as the neurobiologist and theorist Giulio Tononi has noted, its complete removal has effectively no impact on the subjective sense of consciousness. It is made up of hundreds of thousands of parallel, modular (i.e., not interconnected) circuits that are constantly comparing motor output to sensory feedback so as to fine-tune the targeting and smooth execution of our movements, and to help us

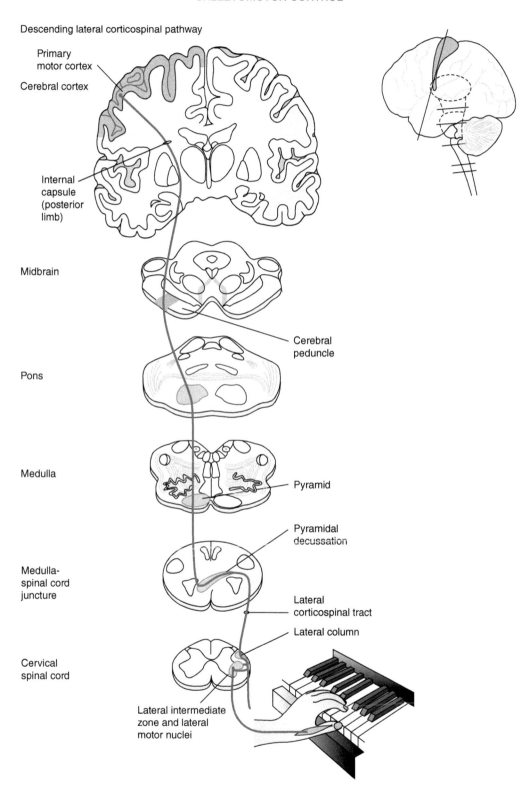

FIGURE 8.1 The descending corticospinal pathway. Source: From Kandel, Eric R., James H. Schwartz, and Thomas M. Jessell, eds. 2000. Principles of Neural Science. New York: McGraw-Hill. Reproduced with permission of McGraw-Hill.

maintain postural stability. The circuit is one-way, linking frontal cortex to pons to cerebellar cortex to cerebellar deep nuclei to thalamus and back to frontal cortex. The densely foliated cerebellar cortex is shaped like one half of a sawed-in-two tennis ball, but with thicker walls due to deep folds between the lobes of the cerebellar cortex, and the fibers projecting from this cortex to the deep nuclei lying near the "deepest" point of this concavity. Also projecting into the cerebellum are ascending projections carrying information about the periphery (touch on the skin, body configuration from muscle stretch receptors and joint-position receptors) and overall body position and balance from the vestibular system. The workings and

functions of the cerebellum will be considered in the upcoming section on *Cerebellum: motor learning, balance, … and mental representation?*

The cortico-basal ganglia-thalamic circuits

This circuit and its nuclei play important roles in motor control, learning, and motivated behavior, and a variety of dysfunctions within it can result in disorders ranging from Parkinson's disease (PD) (impairing movement and high-level cognition) to unilateral neglect to pathological gambling. As illustrated in *Figure 8.2*, its unidirectional circuits

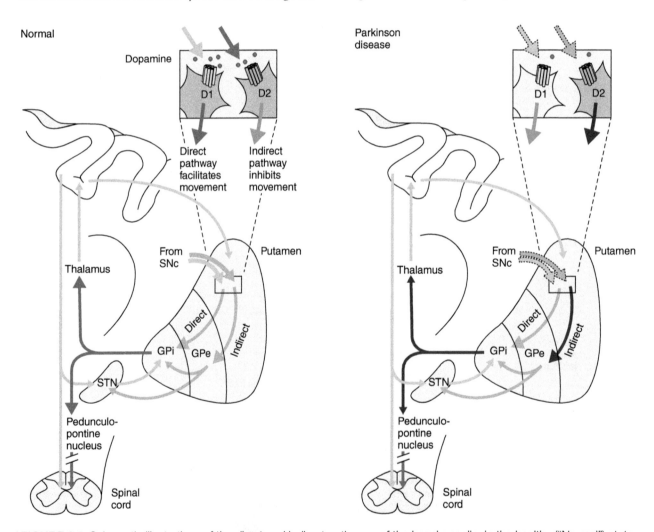

FIGURE 8.2 Schematic illustrations of the direct and indirect pathways of the basal ganglia, in the healthy ("Normal") state on the left, and in PD on the right. Red arrows connote excitatory pathways, gray inhibitory, and the change of saturation in the PD figure whether that pathway strengthens or weakens as a result of disease processes. SNc = substantia nigra pars compacta; GPe = external segment of the globus pallidus; GPi = internal segment of the globus pallidus; STN = subthalamic nucleus. Source: From Kandel, Eric R., James H. Schwartz, and Thomas M. Jessell, eds. 2000. Principles of Neural Science. New York: McGraw-Hill. Reproduced with permission of McGraw-Hill.

comprise a "direct pathway" (cortex to neostriatum to internal segment of the globus pallidus [GPi]/substantia nigra pars reticulata [SNpr] to thalamus back to cortex), an "indirect pathway" (cortex to neostriatum to external segment of the globus pallidus (GPe) to subthalamic nucleus (STN) to GPi/SNpr to thalamus back to cortex), and a "hyperdirect pathway" (cortex to STN; not shown in Figure 8.2). At the gross anatomical level, there are five relatively segregated *cortico-basal ganglia-thalamic circuits*, each associated with dissociable classes of behavior: *motor, oculomotor, dorsolateral prefrontal, lateral orbitofrontal,* and *limbic* (*Figure 8.3.A*). *Figure 8.3.B* illustrates the connectivity of the cortico-basal ganglia-thalamic motor circuit. Although the focus during this chapter will be the motor circuit, a noteworthy property of each of these circuits is the way each integrates inputs from multiple cortical regions and funnels the integrated result back onto the originating region of frontal cortex. Another common feature of these circuits is that the neostriatal node of each is densely innervated by dopaminergic fibers from a mid-brain nucleus called the substantia nigra pars compacta (SNpc). (The distinctive black pigmentation of the sub-stantia nigra ["black substance" in Latin], visible to the naked eye, derives from the high concentration in its neurons of the enzyme tyrosine hydroxylase, which is critical for the synthesis of DA.)

FUNCTIONAL PRINCIPLES OF MOTOR CONTROL

The biomechanics of motor control

Although moving the body in a coordinated fashion is something that we do effortlessly, often "thoughtlessly," this subjective facility masks a very complicated engineering problem. To summarize just one example, the movements of our arms from almost any points A to B are remarkably straight, despite the fact that executing such movements entails the precisely coordinated, simultaneous changing of joint angles – at the shoulder and the elbow (*Figure 8.4*). How does the nervous system accomplish this?

Muscles behave like springs

The tissue that puts the "motor" in "motor control," of course, is muscle. To move one's hand from point A to point B, it is necessary to contract the appropriate sets of muscles such that the joints that connect the bones of the arm move

FIGURE 8.3 The five cortico-basal ganglia-thalamic circuits. Source: From Kandel, Eric R., James H. Schwartz, and Thomas M. Jessell, eds. 2000. Principles of Neural Science. New York: McGraw-Hill. Reproduced with permission of McGraw-Hill.

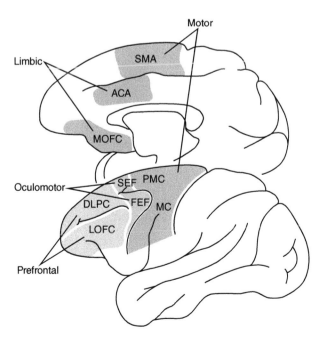

FIGURE 8.3.A Frontal targets the five loops. ACA = anterior cingulate area; MOFC = medial orbital frontal cortex; LOFC = lateral orbitofrontal cortex.

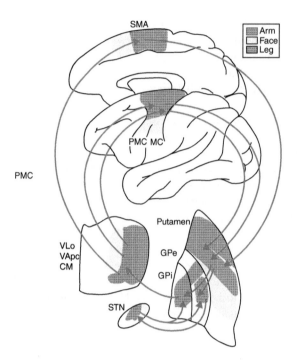

FIGURE 8.3.B Diagram of segregation of somatotopic information at each stage of the motor circuit. GPe = external segment of the globus pallidus; GPi = internal segment of the globus pallidus; STN = subthalamic nucleus; VLo = pars oralis segment of the ventrolateral (VL) nucleus of the thalamus; VApc = parvocellular segment of the ventral anterior (VA) nucleus of the thalamus; CM = centromedian nucleus of the thalamus.

in the ways illustrated in the lower columns of *Figure 8.4*. Note that at this most proximal level of motor control, any one muscle can only influence a single joint angle. For example, the antagonistically paired biceps and triceps muscles can change the angle of the elbow – nothing more. Thus, to execute even a simple "hand movement" as in *Figure 8.4*, the appropriate neural commands have to be delivered to at least four sets of muscles (biceps and triceps for the elbow, and the analogous muscles for the shoulder) in a precise, exquisitely choreographed sequence that will produce a smooth, straight motion. Note from this figure that movements to the same location, but starting from different locations, require nonlinearly different transformation of joint angles and joint angular velocities.

Fortunately, the mind-bogglingly complicated sets of nonlinear computations required each time we move our bodies are simplified by a felicitous mechanical property of muscles: they behave like springs. That is, one has only to set the tensions on a set of muscles and they will "automatically" adjust to a new equilibrium point, pulling the bones to which they are attached into a new position. Therefore, in principle, each unique location in x,y,z, coordinates within the peripersonal space of an individual has a unique set of muscle tensions associated with it. For example, from where you are currently sitting (or standing; or reclining; or whatever), there is a unique set of tensions required of the muscles controlling your shoulder, elbow, wrist, and index finger in order for you to touch your finger to this period →.← (Go ahead, do it. No one is watching.) And so, in principle, your cortex only needs to issue the command *Touch that period* and the requisite muscle-tension settings will automatically deliver your index finger to that point. The nonlinear transformations of joint angles that enable your finger to travel from wherever it was to the period that you are now pointing to "just happen" as a consequence of moving from A to B; they don't have to be explicitly calculated and commanded.

Motor computations in the spinal cord

If the motor cortex needs only to specify a new endpoint for, say, the index finger, what translates between the presumed command to *Go to X,Y,Z* and the resetting of muscle tensions? (That is, to execute a movement, each individual muscle only experiences a change in the amount of ACh being released onto it, it can't "know" anything about point *X,Y,Z*.) The answer is that this translation happens in the networks of interneurons in the spinal cord to which most corticospinal fibers project. Evidence for this was first produced in the frog,

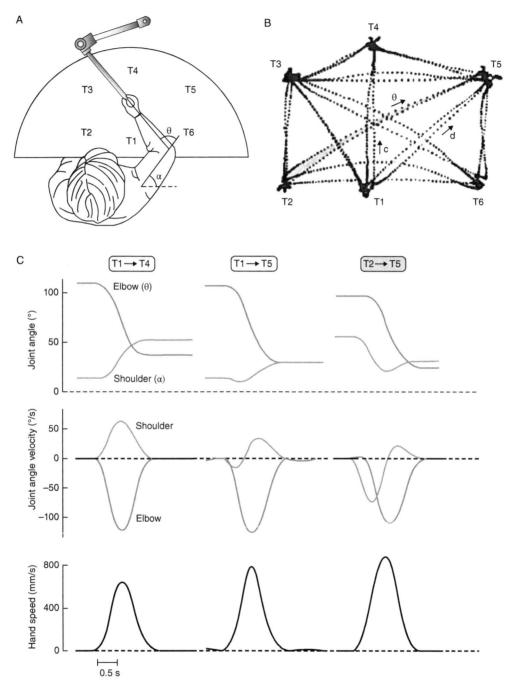

FIGURE 8.4 **A.** Top-down view of a subject grasping the handle of a manipulandum, which can be moved in a 2-D plane to each of the six targets indicated on the table. **B.** Movement paths (two between each target) for a subject. **C.** Joint angle, joint angular velocity (i.e., speed at which joint angle is changing), and hand speed plotted against time to execute each of three movements from **B.** Source: From Kandel, Eric R., James H. Schwartz, and Thomas M. Jessell, eds. 2000. Principles of Neural Science. New York: McGraw-Hill. Reproduced with permission of McGraw-Hill.

subsequently in the rat, in an experimental preparation in which all descending skeletomotor and ascending somatosensory connections between spinal cord and brain are severed. Microstimulation is then applied to the portion of the spinal cord whose motor neurons innervate, say, the left forelimb. The intriguing finding is that stimulation in this single location always brings the forepaw to the same location in space, regardless of what its starting location was. The area being stimulated, therefore, is converting that burst of electrical stimulation into the new tension settings of the muscles controlling the shoulder and elbow joints that bring the limb into that position in space. Stimulate a second spot in the cord, the limb moves to a different location. Further, a property whose importance will become evident further along in this chapter is that the simultaneous stimulation of both spots moves the limb to a third location. Therefore, in effect, the motor cortex issues the general command, and the spinal cord "takes care of the details."

In addition to being able to translate cortical commands into precise settings of muscle tensions, the spinal cord also contains what are referred to as "pattern generators." Thus, for example, early-to-mid twentieth-century experiments established that if the connections between an animal's brain and spinal cord are severed, and the animal is suspended in a harness over a treadmill, it will walk with an appropriate gait when the treadmill starts moving. Not only that, its gait adjusts appropriately to treadmill speed, from a trot to a gallop to a walk. The only explanation for such demonstrations is that remarkably sophisticated sensorimotor control circuitry exists within the spinal cord itself.

Motor cortex

Okay, so we've established that an animal can effectively walk, gallop, and trot, as appropriate, without any input from the brain. So what does the motor cortex do? (Why do we need it?) In a sentence, it controls *volitional* movement of the body. Let's take a closer look.

The motor homunculus

Unless you have just picked up this book for the first time, and opened to this chapter, you won't be surprised to learn that M1 has a rough somatotopic organization (as do at least four other regions in motor cortex). *Figure 8.5* is also taken from surgical mapping studies of Penfield and Rasmussen (1950). One important difference between the primary somatosensory and motor cortices, however, is that the latter's map lacks the same high-fidelity correspondence to different parts of the body. This is in part because the smallest unit that the motor cortex can represent corresponds to bones, rather than the surface of the skin. For

example, it wouldn't make sense for M1 to represent different parts of the forearm in the way that S1 does, with a set of receptive fields that cover its entire surface. Rather, M1 represents the forearm as a unit that can be moved. Another, more fundamental, reason is that M1 represents *actions*, not just parts of the body. For example, it is often pointed out that a person's handwriting looks similar (albeit progressively sloppier) when the pen is held in the right hand vs. the left hand vs. in the mouth vs. between the toes. Thus the representation of the shape of one's handwriting must exist in a format that is accessible by any set of effectors that needs to implement it. Nonetheless, the fact that there is a mapping between different parts of M1 and different parts of the body means that motor cortex doesn't only employ purely abstract representations like "point X,Y,Z," or the shape of one's handwriting. Rather, it clearly chooses which effector(s) will be used to carry out any particular action. Further, it also controls the **kinematics** and **dynamics** of movement, such as the speed with which the arm will move, or the tightness with which the fingers will grip.

In addition to these considerations, neuroscientist Jon Kaas (2012) has proposed that if one "zooms in" from the overall somatotopic organization of M1 to look at smaller regions in greater detail, the coherent body map is "fractured" into clusters of cortical columns that are organized according to movement sequences, such as reaching, grasping, or defending the head against a blow. Thus, for example, within the large expanse of cortex that we label the "hand area," there will be columns of neurons corresponding to the arm, to the wrist, and to individual fingers, that, when activated in the appropriate sequence, produce the "reaching out and picking up food on my plate" movement sequence that that we each execute several times each day.

How do PMC and SMA differ from M1? For now, we'll content ourselves with the short-hand summary that they contribute to high-level aspects of motor control, such as planning to make a movement, imagining making a movement, and coordinating motor sequences made up of many individual actions. So that's an overview of what motor cortex does. Now, how does it work?

The neurophysiology of movement

EEG and LFP correlates

In the decade following Berger's discovery of the alpha rhythm (section *The relation between visual processing and the brain's physiological state, Chapter 4*), one of several rhythms to be discovered in the scalp EEG was the mu. Although the mu rhythm occupied the same frequency band as alpha, it differed in morphology – as a series of thin, spiky peaks

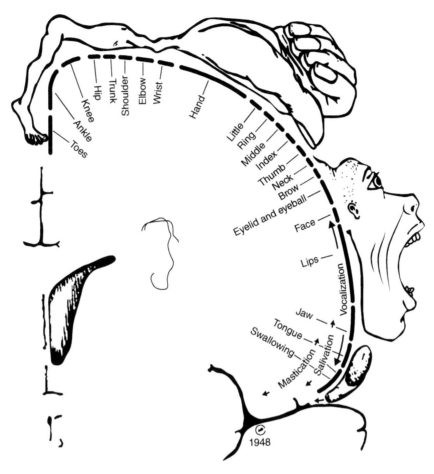

FIGURE 8.5 Penfield and Rassmussen (1950) motor homunculus. Same graphical conventions as *Figure 5.9*. Source: From Penfield, Wilder, and Theodore Rasmussen. 1950. The Cerebral Cortex of Man. A Clinical Study of Localization of Function. New York: Macmillan.

surrounded by rounded troughs (suggesting the shape of the greek letter "μ") – in topography – being maximal over central electrodes – and in function – being insensitive to perceptual factors, but being related to movement. Specifically, the amplitude of the mu rhythm is maximal when the body is at rest and begins to decline roughly 2 seconds prior to a movement. Such "mu desynchroniza-tions" can be local to the cortical representation of the effec-tor being moved, if one moves, for example, just the finger, toe, or tongue. Also prominent in the functioning of the motor system are EEG oscillations in the beta band. Although also inversely related to motor activity, the timing of changes in the beta band tend to be much more tightly coupled to the initiation in movement. Thus, there's a sense in which both the central mu and the central beta rhythms are analogous, for skeletomotor control, to posterior alpha for visual perception, in that they reflect a state of suppression

and/or "idling" in the thalamocortical system producing them. (*Figure 8.6* illustrates mu and beta, recorded intracra-nially from motor cortex of a monkey performing visually guided movements of a handle from a central location to one of six radial locations [not unlike the task illustrated in *Figure 8.4*]. It also nicely illustrates how an oscillating signal can be trial-averaged to generate an ERP [as discussed in section *Analysis of time-varying signals, Chapter 3*].)

(In the 1950s, the French neurologist/neurophysiologist Henri Gastaut [1915–1995] and his colleagues made the discovery that mu desynchronization also occurs when the subject is immobile, but watching the movements of others. As we shall see further along in this chapter, this phenom-enon has acquired considerable attention of late in relation to hypotheses linking autism with "mirror neurons.")

And so, we've established from scalp and LFP record-ings that motor cortex is maintained in a functionally

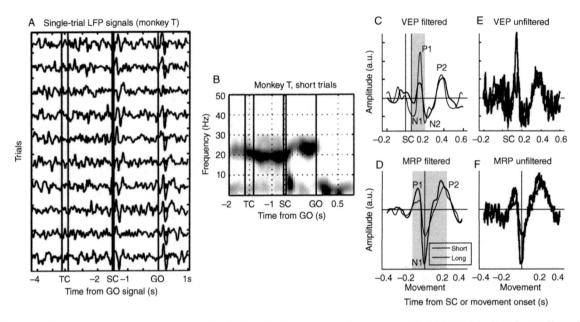

FIGURE 8.6 Movement-related dynamics in the LFP. **A** illustrates traces from several individual trials, bandpass filtered from 1 to 15 Hz (to emphasize the sensory- and motor-related evoked potentials), time-locked to the GO signal, and ordered by increasing RT (RT for each trial is indicated by a filled circle appearing in each trace; they are difficult to see, but occur at ~250 ms following the GO signal). The axis along the bottom of the figure shows the timeline of the task: TC is the "temporal cue" that informed the animal whether delay would be short (700 or 1000 ms, depending on the monkey) or long (1500 or 2000 ms); SC = the spatial cue, a visual cue that indicated the target to which the handle must be moved; GO = the signal to initiate movement; RT = "response time," defined as the lag between the GO signal and the instant when the handle began to move (i.e., movement-initiation time); MT = movement time, the time between the RT and the acquisition of the target. **B.** Time–frequency representation of trials from **A**, but without lowpass filter. Darker color corresponds to higher power. Pre-movement power is dominated by activity in the beta band, the frequency shifting from ~19 Hz to ~23 Hz after the SC, then stopping abruptly with the onset of the GO signal. Brief increases in low-frequency power correspond to evoked responses. **C** and **D.** Visual-evoked potential (VEP) and movement-related potential (MRP), respectively, time-locked to SC and RT, respectively, filtered from 1 to 15 Hz. **E** and **F.** Same data as **C** and **D**, but unfiltered. Source: From Kilavik, Bjorg Elisabeth, Joachim Confais, Adrián Ponce-Alvarez, Markus Diesmann, and Alexa Riehle. 2010. "Evoked Potentials in Motor Cortical Local Field Potentials Reflect Task Timing and Behavioral Performance." Journal of Neurophysiology 104 (5): 2338–2351. doi: 10.1152/jn.00250.2010. Reproduced with permission of The American Physiological Society (APS).

inactive state that is lifted just prior to the execution of a movement, that sensory signals triggering a movement are registered in this region, and that a large ERP immediately precedes the movement. Next, let's "drill down" to the level of the individual neuron.

Single-unit activity reveals population coding

Among the pyramidal cells of layer V of M1, one finds massive Betz cells, discovered by the Ukrainian anatomist Vladimir Alekseyevich Betz (1834–1894) only a few years after the pioneering microstimulation studies of Fritsch and Hitzig (1870; section *The localization of motor functions, Chapter 1*). These are the largest neurons in the nervous system, with large-diameter axons seemingly specialized to send

neural impulses very rapidly and over long distances. (Subsequent research has established that many other neurons contribute to the corticospinal tract, with axons from Betz cells contributing only between 3% and 10% of the total.) The breakthrough insight about how these neurons actually generate coordinated movement had to wait another 100 years, for the research of Greek-born, US-based neurophysiologist Apostolos Georgopoulos and his colleagues.

Georgopoulos, Kalaska, Caminiti, and Massey (1982) recorded from the arm area of M1 from five hemispheres of four different monkeys while they performed a "center-out" task virtually identical to the one used in the experiment from *Figure 8.6*. The results, illustrated in *Figure 8.7*, revealed another important difference between M1 and

FIGURE 8.7 The electrophysiology of primary motor cortex.

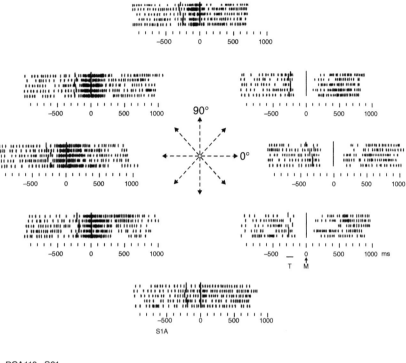

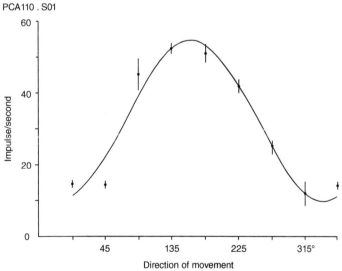

FIGURE 8.7.A Top panel displays responses of a single neuron from the arm region of M1 for each of five reaches made in each of eight directions, aligned to the time of initiation of movement (arbitrarily labeled "0"). Labels under the raster plot corresponding to movement in the 315° direction indicate the time of the GO signal (labeled "T") and initiation-of-movement time (M). Bottom panel is the tuning curve derived from these data. Source: From Georgopoulos, Apostolos P., John F. Kalaska, Roberto Caminiti, and Joe T. Massey. 1982. "On the Relations between the Direction of Two-Dimensional Arm Movements and Cell Discharge in Primate Motor Cortex." Journal of Neuroscience 2 (11): 1527–1537. Reproduced with permission of the Society of Neuroscience.

FIGURE 8.7.B Vector representation of the activity of 241 neurons recorded from the arm region of M1, for each of eight directions of movement (center of figure). Dashed lines with arrowheads are vector sums derived from each of the eight directions of movement. Source: From Georgopoulos, Apostolos P., Roberto Caminiti, John F. Kalaska, and Joe T. Massey. 1983. "Interruption of motor cortical discharge subserving aimed arm movements"." Experimental Brain Research Supplementum 7: 327–336. DOI:10.1007/BF00238775 Reproduced with permission of the Springer Nature.

the primary sensory areas that we have studied to date: instead of narrowly tuned response properties mirroring the precision of receptive fields of, for example, V1 and S1, M1 neurons had remarkably broad tuning. *Figure 8.7.A,* for example, illustrates a neuron whose responses are nearly indistinguishable for movements of 135°, 180°, and 225° (as, indeed, they are for movements of 45°, 0°, and 315°). Different neurons had different overall preferred directions: for example, if *Figure 8.7.A* illustrates a "to-the-left" neuron, others were "away-and-to-the-right" (270° being toward the animal's body), and so forth.

Upon first consideration, it's difficult to see how the activity in neurons with such broad tuning could be responsible for arm movements that we know can be much, much more fine-grained than the separated-by-45° movements required of the monkeys in this experiment. The principle must be different from what we've seen with sensory systems, for which it's customary to think of individual neurons as acting as feature detectors. Although no one really believes that a single neuron (or column) in V1

can be responsible for a single visual percept, one can none-theless imagine, in "shorthand," that viewing a vertical line preferentially activates neurons tuned for vertical orientations, and that rotating that line ever so slightly to the right will preferentially activate different sets of neurons tuned for a slight tilt to the right. With M1 tuning properties, however, it'd be as though the V1 cell were equally responsive to orientations from 135° to 225°. How could one construct a system with elements whose tuning is so sloppy? The answer came to the research team after they considered the entire sample of 241 neurons from which they recorded in the arm area in M1: *population coding.*

Instead of thinking of the motor system as being made up of discrete "channels," each one specialized for its own narrowly defined features, *Figure 8.7.B* illustrates a fundamentally different principle of neural coding. It presents a graphical representation of the response of each of the 241 neurons to each of the eight directions of motion. Each line in each of the eight, radially arrayed "explosion" plots is a vector whose length corresponds to the firing rate of a neuron during motion in that direction. (Responses of the neuron illustrated in *Figure 8.7.A,* therefore, correspond to one of the longish vectors at the 135°, 180°, and 225° positions, to a somewhat shorter vector at 90°, to one shorter still at 270°, and to vectors of length 0 at 45°, 0°, and 315°.) The "vectorial hypothesis" proposed by Georgopoulos, Caminiti, Kalaska, and Massey (1983) makes the assumption that each neuron's activity on any given trial corresponds to the magnitude of its influence on the direction of movement on that trial. Now recall from mathematics that a vector is a way of representing a force that has a magnitude and a direction, and that if two or more forces are simultaneously present, their vectors can be combined and the resultant "vector sum" (or "vector average," depending on which operation one performs) will indicate what their combined effects will yield. This is what the dashed lines with arrowheads amid each vectorial plot represent: the empirically derived summation of all 241 vectors. Given that each vector was estimated from only five or so movement trials, and that only 241 neurons were sampled out of the millions that make up motor cortex, it's astoundingly impressive that an analysis of these neural data could produce an estimated direction of motion (the eight vector sums) that so closely corresponds to the actual direction in which the monkey moved its hand on each trial (the eight arrows in the center of the plot). What Georgopoulos and colleagues accomplished was to decipher a component of the neural code for skel-etomotor control! And the key insight, one to which we will return often for the remainder of this book, is that the principle underlying this function of the nervous system can only be understood by simultaneously considering the activity of a large population of neurons. Serially inspecting one after the next after the next would never reveal it.

I can imagine that to some readers this may seem too abstract. *Vector summation? What, is there some little homuncular mathematician perched up on the precentral gyrus, legs dangling into the central sulcus, calculating vector sums prior to each movement?* Nope. No need in this scheme to resort to the homunculus. Assuming that these are all layer V pyramidal cells, all located in the arm area, each neuron's action potentials travel to the same general area in the spinal cord, many of them destined for the same motor neuron pool. These action potentials all activate the very same networks of interneurons that were described in the earlier section on *Motor computations in the spinal cord*. In effect, it's as though

241 teensy weensy microelectrodes have been simultaneously lowered into the spinal cord. Recall that we saw previously that while stimulation of point A in the cord moved the forepaw to location X,Y, and stimulation of point B in the cord moved the forepaw to location X',Y', simultaneous stimulation of A and B moved the forepaw to X'',Y''. Thus, the vector summation that a scientist can compute through the arduous analyses and computations summarized here turns out to be a more formal description of what we had glibly summarized earlier: *the motor cortex issues the general command* (in the form of a population code), *and the spinal cord "takes care of the details."*

One of the most exciting consequences of "cracking the code," as Georgopoulos and colleagues have done with motor cortex, is that one can then *decode* new signals as they are generated. *Figure 8.8* illustrates how neural decoding is used for brain–computer interface (BCI).

FIGURE 8.8 A closed-loop BCI system for direct neural control of a robotic arm. Source: From Carmena JM, Lebedev MA, Crist RE, O'Doherty JE, Santucci DM, Dimitrov DF, et al. (2003) Learning to Control a Brain–Machine Interface for Reaching and Grasping by Primates. PLoS Biol 1(2): e42. https://doi.org/10.1371/journal.pbio.0000042.

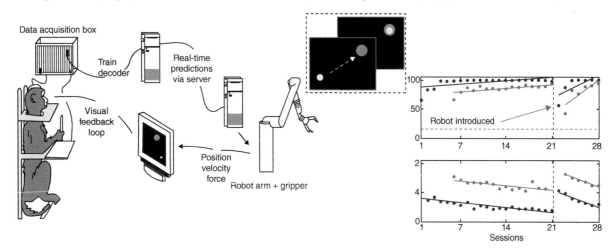

FIGURE 8.8.A The experiment started with the monkey moving a joystick that controlled the movement of a cursor on a screen (inside dotted box), and while the monkey learned to perform this task, a computer analyzed task-related neural activity from its frontal and parietal cortices, in order to learn to decode it. The experimental setup is illustrated on the left side of the image, with "Real-time predictions via server" referring to the statistical assessment of the quality of decoding on a trial-by-trial basis (i.e., *How well would we have done on this trial had we guided the cursor with neural signals rather with the joystick?*). Once decoding performance was high enough, the researchers disconnected the joystick from the system, such that the movement of the cursor was now being controlled by the monkey's neural output – "brain control mode." Performance is plotted across sessions 0–21, in joystick control mode (blue dots and regression line) vs. with brain control mode (red dots and regression line). "Robot introduced," at session 22, indicates the switch to robot control mode, when the joystick was reconnected to the display, but now it was controlled by a robotic arm, and the monkey's neural signals controlled the robotic arm. Performance on the task dropped precipitously on the first session of "robot control mode," but already by the subsequent session performance improved markedly.

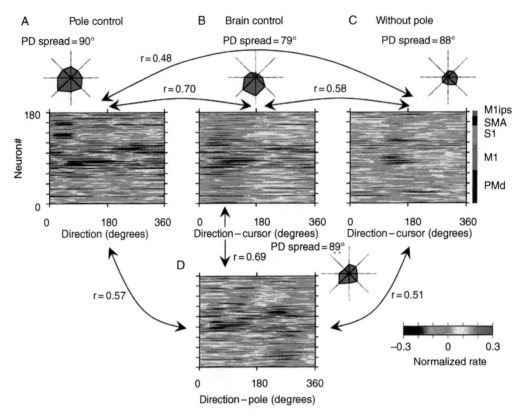

FIGURE 8.8.B Tuning curves for 180 neurons recorded from dorsal premotor cortex (PMd), M1, S1, and SMA (all contralateral to the joystick-controlling hand), and ipsilateral M1 (M1ips). (To understand the graphical convention in these plots, consider the tuning curve in *Figure 8.7.A*. If you were to look down on it from above, you'd want to color code its height at each point along the direction axis, so as to not lose this information. Otherwise it would just look like a straight black line. [The same logic applies for 2-D geographical maps that indicate the elevation of mountains vs. valleys with different colors.] Here, red indicates the highest firing rate ["altitude" in the geographical map analogy], violet the lowest. The reason for adopting this "viewed from above" display convention is that it allows the simultaneous display of an entire "stack" of tuning curves in one panel.) The plot on the left shows the tuning curves for this set of 180 neurons when the animal is performing the task in joystick (a.k.a. "pole") control mode, tuning calculated from cursor movements; the plot in the middle is from brain control mode but with the joystick still in the hand, tuning calculated from cursor movements; the bottom plot is also from brain control mode with the joystick still in the hand, but with tuning calculated from joystick movements; and the plot on the right from brain control mode with no joystick (tuning necessarily computed from cursor movements). Visual inspection suggests that the tuning of many of the neurons in this sample changes as a function of task condition. The *r* values are coefficients produced by correlating the overall similarity of direction tuning of these 180 neurons in each of four conditions. Finally, the purple polar plots illustrate the tuning of the entire sample in each condition, a "PD spread" of 90° indicating a uniform distribution, and the magnitude of the values at each axis (relative to the origin) the strength of tuning preference in that direction.

MOTOR CONTROL OUTSIDE OF MOTOR CORTEX

Parietal cortex: guiding how we move

One of the facts made clear by studies such as that summarized in *Figure 8.8* is that the neural code involved in carrying out complex actions, such as a reach followed by

a grasp followed by a pull, entails much more than the vector summation of 300 or so M1 neurons. (For example, with just that information, I could be made to swing my arm so as to knock the cup of coffee off the arm of my couch, but not to grasp it and bring it to my lips.) Many regions other than M1 can contribute valuable signals for decoding movement-related neural activity. Although the study of Carmena et al. (2003; *Figure 8.8*) didn't venture posterior of S1, others have done so, emphasizing that a greater range of cognitive factors – including intention, expectation, and reward value – might be best sampled outside of traditional motor areas. "For example, PRR [parietal reach region; see *Figure 7.7*] neural activity codes the intention to reach to an object at a particular location in space, whereas motor cortex codes the direction to move the hand" (Andersen, Musallam, and Pesaran, 2004, p. 271). The critical role for PPC for the planning and execution of visually guided action, through its contributions to sensorimotor integration and coordinate transformation, also figures prominently in the alternate framing of the ventral vs. dorsal pathways as being one of *what* vs. *how*.

A neurological dissociation between perceiving objects and acting on them

A reassessment of the "what vs. where" framework for understanding the functional specialization of the visual system (section *A functional dissociation of visual perception of*

what an object is vs. where it is located, Chapter 6) was proposed by Canada-based researchers Melvin Goodale and David Milner after their detailed studies of a patient who developed a profound impairment in the ability to recognize objects (a form of visual agnosia, a syndrome that we will define and consider in detail in *Chapter 10*) after carbon monoxide inhalation produced extensive damage in lateral occipital BAs 18 and 19, but largely sparing V1. Despite this severe impairment, this patient (known by her initials "D.F.") seemed remarkably unimpaired when "directing accurate reaching movements toward objects in different orientations" (Goodale, Milner, Jakobson, and Carey, 1991, p. 155). To formalize and quantify this dissociation, which was reminiscent of that demonstrated by monkeys with IT lesions in the Pohl (1973) study (section *A functional dissociation of visual perception of what an object is vs. where it is located, Chapter 6*), Goodale and colleagues (1991) carried out the experiment illustrated in *Figure 8.9*.

In addition to the experiment illustrated in Figure 8.9, Goodale et al. (1991) also carried out an experiment that

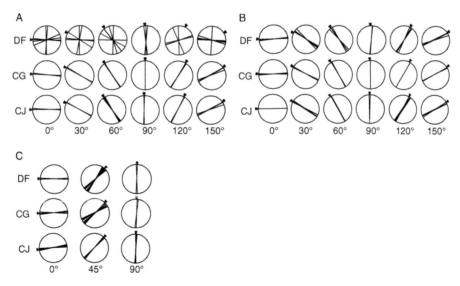

FIGURE 8.9 Is the dorsal stream important for processing "where" an object is, or "how" to interact with it? **A.** Responses on five trials at each of six different orientations in a "perception" task, in which D.F. and two control subjects were asked to match the orientation of a slot in a disk that was 45 cm distant. The true orientation of each of the six is illustrated by the small arrowheads. At all angles except 90° (i.e., vertical) D.F. was profoundly impaired. (And because each trial was started with the card held at 90°, her relatively intact performance at this orientation may be artifactual.) **B.** When asked to "post" the card by reaching to insert it into the slot (as though putting a postcard into a mail slot), D.F.'s performance was nearly as precise as that of control subjects. **C.** Performance on a mental imagery version of the matching task, in which subjects were asked to close their eyes and imagine the slot at each of three orientations. This indicates that poor performance illustrated in **A** was not due to a misunderstanding of instructions, or some other uninteresting factor. Source: From Goodale, Melvyn A., A. David Milner, Lorna S. Jakobson, and David P. Carey. 1991. "A Neurological Dissociation between Perceiving Objects and Grasping Them." Nature 349 (6305): 154–156. doi: 10.1038/349154a0. Reproduced with permission of Nature Publishing Group.

looked at a second kind of visually guided action, grasping. First subjects were asked to indicate with finger and thumb the front-to-back size of each of five plaques, ranging in shape from a 5 × 5 cm square to a 2.5 × 10 cm rectangle. These two digits were fitted with infrared light–emitting diodes so that precise measurements could be made of finger-to-thumb aperture. The results showed that, unlike control subjects, D.F.'s responses did not differ with shape. When asked to pick each of the shapes up, in contrast, she was just as accurate as control subjects.

In interpreting their findings, the authors conjectured that they did "not correspond to the two streams of output from primary visual cortex generally identified in the primate brain – a ventral one for object identification and a dorsal one for spatial localization [. . . instead, they] indicate separate processing systems not for different subsets of visual information, but for the different uses to which vision can be put." Therefore, to return to a theme from the previous chapter, this view has more in common with *intention* rather than *attention* interpretations of the neglect syndrome and of the functions of the parietal lobe.

The *Further Reading* list at the end of this chapter provides references to more recent installments of the lively debate about this idea that has continued more than two decades after its initial articulation. Furthermore, one possible resolution to the "where" vs. "how" debate is that both may be true. Kravitz, Saleem, Baker, and Mishkin (2011) have proposed that three major pathways originating in parietal cortex provide body-centered spatial information that supports three broadly defined functions: a parieto-prefrontal pathway that supports "conscious visuospatial processing," including spatial working memory (and corresponding to the classically defined "where" pathway); a parieto-premotor pathway that supports "non-conscious visuospatial processing," including visually guided action (and corresponding to Goodale and Milner's "how" pathway); and a parieto-medial temporal lobe pathway, including intermediate connections in posterior cingulate and retrosplenial regions, that supports navigation. (The last was highlighted in section *From parietal space to medial-temporal place, Chapter 7.*)

Cerebellum: motor learning, balance, . . . and mental representation?

From the standpoint of aesthetics, the cerebellum may be the most beautiful part of the brain: the tight foliation of its cortex (housing 10% of the volume but > 50% of the neurons of the brain); the remarkably profuse, yet "flat,"

dendritic arbors of its Purkinje cells; the elegant simplicity of its circuitry. As summarized previously, the cerebellum is a hub in one of the two major recurrent circuits of the motor system, serving an online error-monitoring function and feeding real-time performance information back onto frontal cortex.

Cerebellar circuitry

The cerebellum is highly modular, consisting essentially of a bank of hundreds of thousands of parallel circuits that all perform the same computational operations, but each on a different, segregated channel of information. The central processing element around which cerebellar circuits are built is the Purkinje cell. These are among the largest cells in the nervous system, with massive dendritic arbors that afford an incredibly large number of synaptic connections (estimated to be as high as 200,000 per neuron!). Despite their size and extensive and intricate branching, however, these arbors only extend through a 2-D plane. A row of Purkinje cells aligned in a "strip" that is parallel to the plane of their dendritic arbors makes up a circuit (called a "microzone"), the output of which, via Purkinje cell axons, is focused on neurons of the cerebellar deep nuclei that were discussed in the introductory section of this chapter. At a simplified level, the copy of the efferent motor command is carried into the cerebellum by climbing fibers, and information about the sensory inputs – the "context" in which an action is taken and the sensory consequences of that action – by mossy fibers. Climbing fibers are axons originating in the inferior olive in the brainstem, their name deriving from the fact that a single climbing fiber wraps around and around the basal dendrites of a single Purkinje cell, like a vine climbing a tree trunk. Climbing fibers make between 300 and 500 synaptic connections with a single Purkinje cell. Mossy fibers arise from the pons (in the brainstem), their name deriving from the distinctive "rosettes" in which they form synaptic connections with granule cells. Granule cells then send "parallel fibers" that run perpendicular to, and therefore through, the dendritic arbors of many Purkinje cells (i.e., they cut across many microzones). Individual parallel-fiber synapses on Purkinje cells are relatively weak, whereas a single action potential in a climbing fiber will reliably trigger an action potential in a Purkinje cell.

Cerebellar functions

An influential model of the computations carried out by the cerebellum, articulated by the British neuroscience polymath David Marr (1945–1980), and subsequently refined by the American electrical engineer and roboticist James Albus (1935–2011), is summarized as follows. The premise, rooted in control theory from engineering, is that the Purkinje cells in the cerebellum contain a representation of a prediction of what the effects of a movement should be (i.e., an "internal model"). The sensory report of what the consequences of the action actually were is delivered, via parallel fibers, to the cerebellum. To the extent that the internal model and the actual consequences differ, a modification is made to the cerebellar circuitry, and the results of this assessment get forwarded to the motor cortex so that it can adjust how it commands this action the next time. In effect, the idea is that the cerebellum houses one instantiation of the predictive coding models introduced back in *Chapter 4* (section *Where does* sensation *end? Where does* perception *begin?*). In his influential 1971 paper, Albus summarized the role for the modification of cerebellar circuits in motor learning as follows:

> It is an obvious fact that continued training in motor skills improves performance. Extended practice improves dexterity and the ability to make fine discriminations and subtle movements. This fact strongly indicates that learning has no appreciable tendency to saturate with overlearning. Rather, learning appears to asymptotically approach some ideal value. This asymptotic property of learning implies that the amount of change that takes place in the nervous system is proportional to the difference between actual performance and desired performance. A difference function in turn implies error correction, which requires a decrease in excitation upon conditions of incorrect firings. (1971, p. 52)

By the Marr–Ablus model, "desired performance" is represented by the climbing fibers, which carry the copy of the efferent motor signal. When discharge of a climbing fiber triggers an action potential in a Purkinje cell, there is a refractory period that follows, during which the Purkinje cell cannot fire another action potential. If action potentials from parallel fibers arrive during this refractory period, the consequence of this is a weakening of synapses between them and the Purkinje cell. (The reason is introduced in the next section, on *Synaptic plasticity*.) This modification of synaptic strength in the cerebellum is a neural instantiation of motor learning, which we will address in the upcoming subsection on *Synaptic plasticity*. One way to think of this process is that subtle changes in cerebellar circuitry might produce subtle changes to the length of a few of the M1 neuronal activity vectors illustrated in *Figure 8.7.B*, which, in turn, would produce an ever-so-subtle

change in the end point of that movement the next time it is executed.

The circuitry of the cerebellum continues to fine-tune our movements throughout our lives. Evidence for this comes from the fact that damage to the cerebellum produces abnormalities in posture, gait, reaching, and other skeletomotor functions regardless of when in life the damage is incurred.

Interestingly, some channels of the cortico-cerebellar circuit link PFC with the cerebellum, and so this computational principle presumably also carries out fine-grained, online corrective processing of high-level cognitive functions, such as those mentioned in the introductory paragraph to this chapter. The Japanese neurophysiologist Masao Ito has proposed an intriguing model that we can't detail here, but whereby, in addition to its internal models of motor behaviors, the cerebellum may also encode internal models "that reproduce the essential properties of mental representations in the cerebral cortex . . . [thereby providing a] mechanism by which intuition and implicit thought might function" (Ito, 2008, p. 304).

Synaptic plasticity

All learning and memory (with the possible exception of working memory [*Chapter 14*]) results from changes in the strength of synaptic connections between neurons.

The example of Pavlovian conditioning

What is a memory? Fundamentally, it is the association of two previously unlinked pieces of information. Let's make this more concrete by walking through a cartoon example that will be familiar to most. In the early twentieth century, the Russian physiologist Ivan Pavlov (1849–1936), whose research on the physiology of digestion garnered the Nobel Prize in 1904, noticed that dogs in his laboratory would often begin salivating in advance of the arrival of their food. Formal experimentation established that the animals could be taught to associate an arbitrarily selected stimulus (the ticking of a metronome – the *conditioned stimulus* [CS]) with the act of getting fed – the *unconditioned stimulus*. The essence of the protocol was to play the metronome sound at the time of food delivery. After several such trials, it would be sufficient to play the metronome sound, alone, and the dog would begin to salivate. In doing so, the dog was expressing a memory, because prior to the experiment the playing of a metronome sound did not produce this behavior. Therefore, it was the influence of its recent experience that was being expressed in its behavior. If one wants to project human-like cognition onto the dog, one can speculate that hearing the metronome makes the dog think of being fed (i.e., makes it *remember* being fed the last time it heard this metronome).

Staying at this schematic, cartoon level, what must have happened within the brain of our subject? Prior to the experiment, the sound of a metronome would activate its auditory system, and would perhaps trigger an exogenous shift of attention, but certainly no activity in the digestive system. During training, this auditory circuit was active at the same time as were the olfactory circuits that processed the smell of the food and then activated what we'll call the "salivary neurons" that trigger the salivary reflex. Something about the coactivation of these two circuits led to a strengthening of the link between the auditory representation of the metronome and the salivary neurons. In effect, the dog's brain perceived an association in the world that hadn't existed previously, and modified the connectivity in its brain to represent this association. Thus, at this neural level, the memory *is* this new connection. Now, let's drill down to consider how this might come to pass.

Hebbian plasticity

One of the most influential and enduring principles in neuroscience was articulated by the Canadian Psychologist Donald Hebb (1904–1985), in his book *The Organization of Behavior* (1949): "When an Axon of cell A is near enough to excite a cell B and repeatedly or persistently takes part in firing it, some growth process or metabolic change takes place in one or both cells such that A's efficiency, as one of the cells firing B, is increased" (p. 62). This has come to be known as "Hebb's postulate," and it has been fundamental for understanding synaptic plasticity in biological systems, as well as being the basis for learning in virtually all computational neural network models, including those introduced in section *Computational models and analytic approaches, Chapter 3*.

We won't do a deep dive into the physiological correlates of Hebb's postulate until *Chapter 11*, but for now it will be useful to apply it to our Pavlovian conditioning example. What it is saying is that there was something special about the fact that metronome-related activity in the auditory system coincided with the activity in the salivary neurons. Thus, even though the auditory activity initially had no causal role in driving the salivary neurons, something about the coincidence of activity in auditory neurons and in salivary neurons led to a strengthening of the

ability of the auditory processing of the sound of the metronome to drive salivary activity, and this has to have been through the strengthening of connections. (As you might have inferred from *Chapter 5*'s treatment of *Somatosensory plasticity* and of *Phantom limbs and phantom pain*, it could not happen that, in our cartoon example, auditory neurons would send new long-range connections to salivary centers in the brain to create this memory; rather, the association must come about through the modification of the strengths of connections that already existed before the learning episode occurred.)

A final, important fact that we need to consider here is that Hebbian plasticity works in two directions. That is, what we have described is a situation in which activity in neuron A that coincides with activity in neuron B leads to a strengthening of the A-to-B synapse. As we'll see in *Chapter 11*, this strengthening is referred to as long-term potentiation (LTP). It is also the case, however, that when activity in neuron A is anticorrelated with activity in neuron B, A-to-B synapses will be weakened. Thus, when action potentials in A occur shortly *after* B has been depolarized, long-term depression (LTD) of this synapse occurs. It is LTD that implements the "error correction" in cerebellar circuits invoked by Albus (1971) in the previous section on *Cerebellar functions*.

Basal ganglia

The basal ganglia define the second major recurrent circuit of the motor system. This circuit, too, is not limited to motor control. Rather, as *Figure 8.3.A* highlights, the skeletomotor circuit is just one of five (frontal) cortical-basal ganglia-thalamic circuits. One major difference between the skeletomotor circuit and the other four is that its cortical efferents synapse in the putamen of the neostriatum, whereas three of the others target different regions of the caudate nucleus of the neostriatum, and the ACC and MOFC of the limbic circuit target the nucleus accumbens of the neostriatum (also referred to as the ventral striatum). Subsequently, each of the five circuits, although remaining segregated, bifurcates into a direct and an indirect pathway, the direct pathway linking striatum to the GPi, the GPi to thalamus, and thalamus back to cortex.

A subcortical gate on cortical activity

Although learning the wiring diagram of this circuit has been known to induce dizziness and headache in some students, the basic operating principles are actually quite straightforward. The key to understanding its function is remembering a distinctive property of the GPi. The GPi is a tonically *active* structure. That is, its baseline state, when the circuit is not being engaged, is to steadily fire at a high frequency (~85 Hz in one study in the monkey; this will vary with species and with an animal's physiological state). Because GPi neurons are GABAergic, the default state of affairs is that the GPi maintains a steady inhibitory "clamp" on the thalamus. Now, moving one step upstream, the cells of the neostriatum that project to the GPi are also GABAergic. Thus, engaging the direct pathway in a basal ganglia circuit has the effect of phasically inhibiting the GPi's activity, thereby phasically lifting the GPi's inhibitory clamp on a small region of thalamus, and thereby allowing it to send an excitatory signal to its frontal cortical target. In this way, the basal ganglia can be construed as "gating" cortical activity.

The role of DA

Figure 8.2 illustrates another important component to the operation of the basal ganglia: the delivery of the neurotransmitter DA. DA, along with norepinephrine, serotonin, and others, is classified as a neuromodulatory neurotransmitter. That is, it's typically not thought to convey information in the way that glutamatergic neurons do, but it sets the "tone" within networks, thereby biasing the way in which glutamatergic and GABAergic signals are processed. In particular, the level of DA at a synapse influences the rate of strengthening/weakening that glutamatergic and GABAergic synapses will undergo as a result of activity. DA is synthesized in three adjacent nuclei in the midbrain. Depending on who is describing them, they are either identified as *A8, A9,* and *A10* or as the SNpc and the ventral tegmental area (VTA). The SNpc (or A9) delivers DA to other nuclei of the basal ganglia, most densely to the neostriatum via the nigrostriatal tract, and the other nuclei project primarily to neocortex.

The influence of DA on the basal ganglia can be considered at many levels, and here we'll consider three: setting the balance between the direct and indirect pathways; controlling the synchrony between subnuclei; and reinforcement learning.

Balancing between the direct and indirect pathways. We'll begin with what's illustrated in *Figure 8.2*. Although all fibers of the nigrostrial tract deliver "the same" molecule throughout the neostriatum (DA), its effects can be either excitatory or inhibitory,

depending on the zone receiving this input. How can this be? The answer is that, unlike glutamate and GABA, DA receptors are not linked to membrane-spanning channels that directly gate the passage of ions. Rather, they are so-called "second-messenger-linked" receptors. It means that the binding of a molecule of DA to a DA receptor initiates a cascade of chemical events within the postsynaptic neuron, the ultimate consequences of which might be excitation or inhibition, depending on which second-messenger pathway is triggered. There are two classes of DA receptors: D1-type and D2-type. (D1-type DA receptors include D1 and D5, and D2-type include D2, D3, and D4.) In general, D1-type receptors produce a depolarizing effect and D2-type receptors a hyperpolarizing effect.

The neostriatum is made up of a checkerboard of compartmentalized zones that are dominated by D1-type or by D2-type receptors. The two types of zones are the basis for the direct vs. indirect pathways, because their outputs project to either the GPi or the GPe. One of the ways that we understand the influence of DA on this system is from decades of research on Parkinson's disease (PD), which is characterized by progressive degeneration of the SNpc, and, therefore, a progressive lowering of DA levels in the neostriatum. Motorically, PD is associated with three cardinal clinical symptoms: resting tremor in one or more limbs, muscular rigidity, and slowness to initiate movement ("bradykinesia"). One way to understand these symptoms (and, by extension, to understand the role of DA in the healthy system) is illustrated in *Figure 8.2*: the reduction of DA in the neostriatum throws off the balance of its output, such that the descending signal of the direct pathway is weakened, and that of the indirect pathway is strengthened. The direct pathway is consequently less able to impose phasic inhibition on the GPi, which, in turn, reduces excitatory drive from thalamus to cortex. This can account for the rigidity and bradykinesia seen in PD. By applying the same analysis to the Gpe one can see that a similar end point is reached.

Although it seems as though we've only considered the indirect pathway in relation to its dysfunctional activity in disease states, it is, of course, the case that it has an important role in normal brain function. One is to help control movement by sending "no-go" signals to the cortex when a possible action would be a bad choice. Another important function may be that of learning about the negative value of some behaviors. To understand why, we need to wait until we get to the section on *Habits and reinforcement learning.*

Synchrony between nuclei. For decades, neuropathologists have known that PD is associated with degeneration of the SNpc, because postmortem brains of PD patients have visibly less black pigmentation in the midbrain than do brains from non-neurological autopsies. Although DA replacement is an obvious idea for treatment, development of an effective drug was held up for years by the fact that DA cannot cross the blood–brain barrier, the layer of endothelial cells that prevents diffusion from capillaries of many substances carried in the bloodstream. A breakthrough occurred when the Swedish physician and neuropharmacologist, Arvid Carlsson, discovered that a precursor to DA, L-3,4-dihydroxyphenylalanine (commonly referred to as levodopa, or L-dopa), does cross the blood–brain barrier. Once in the brain, enzymes present in dopaminergic neurons convert L-dopa to DA. Shortly after the initial discovery, it was shown that L-dopa can be effective at alleviating motor symptoms of PD. To this day, L-dopa remains the primary pharmacological treatment for PD. (For this discovery, Carlsson was awarded the Nobel Prize in 2000.)

However, L-dopa has limitations. One is a result of the fact that the loss of DA neurons does not occur evenly across dopaminergic neurons of the midbrain. Rather, neurons of the SNpc precede those of the VTA, meaning that optimal treatment of motor symptoms with L-dopa (*Figure 8.10*) can result in an excess of DA in other brain regions, such as the PFC, which is innervated by the VTA. Dopamine's effect on the brain, like that of many neurochemicals, follows an "inverted U-shaped" function, meaning that neural function is suboptimal when the concentration is too low, optimal in a mid-range, but impaired again as the concentration gets too high. A complication with titrating the dose of L-dopa for PD patients, therefore, is finding the "Goldilocks dose" that provides optimal relief from motor symptoms, yet that won't produce cognitive impairment. This becomes increasingly difficult to do as the disease progresses, and there are fewer and fewer viable neurons remaining in the SNpc that can convert L-dopa into DA and deliver it to its striatal targets. At advanced stages of the disease, surgery is often considered the most viable option for treatment.

The first surgical treatments developed were "-otomies"; for example, a *pallidotomy* entails lesioning the GPi so as to reduce its pathologically strong inhibitory clamp on the thalamus. A *thalamotomy* most often targets the ventral intermediate nucleus of the thalamus, which has the specific effect of reducing tremor. At present, however, the most common surgical treatment for PD is so-called deep

brain stimulation (DBS), in which one or more stimulating electrodes are implanted in a basal ganglia nucleus (often the STN, but, as *Figure 8.10* illustrates, other nuclei can also be targeted). The electrode is driven by a battery that is surgically implanted in the chest and can be controlled via a handheld magnet. As *Figure 8.10* makes clear, brain function is exquisitely sensitive to the precise oscillatory dynamics within and between basal ganglia nuclei. Steady, high-frequency electrical stimulation delivered to one or more of these can be an effective counter to disease-related loss of DA. Recordings from the OFF condition, when patients did not have L-dopa in their systems (i.e., they were "off their meds"), showed that the power in the LFP recorded from the STN and GPi was shifted to lower frequencies relative to the ON condition (i.e., when they were taking the clinically optimal dose of L-dopa). Even more strikingly, oscillatory coherence of LFPs between the two structures was highest at 6 Hz and 20 Hz in the OFF condition, but shifted to 70 Hz in the ON condition. It may be, therefore, that the resting tremor in PD results from a disruption of optimal oscillatory synchrony in this system. Note, also, that these indices of the oscillatory

FIGURE 8.10 The effects of dopamine-replacement therapy on oscillatory dynamics within and between basal ganglia nuclei. Source: (B): From Brown, Peter, Antonio Oliviero, Paolo Mazzone, Angelo Insola, Pietro Tonali, and Vincenzo Di Lazzaro. 2001. "Dopamine Dependency of Oscillations between Subthalamic Nucleus and Pallidum in Parkinson's Disease." Journal of Neuroscience 21 (3): 1033–1038. Reproduced with permission of the Society of Neuroscience.

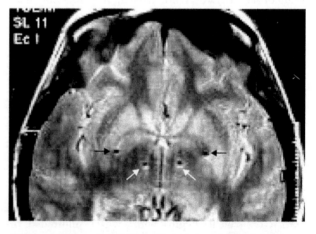

FIGURE 8.10.A Axial MR image of a patient after bilateral implantation of electrodes in the GPi (black arrows) and the STN (white arrows).

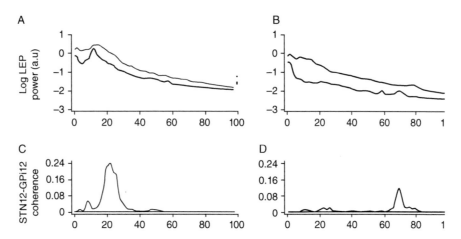

FIGURE 8.10.B Panels **A** and **C** illustrate measurements averaged over four patients when they were OFF L-dopa and **B** and **D** when they were ON. **A** and **B** show power spectra in the GPi (higher power) and the STN; **C** and **D** coherence between the two structures (i.e., the degree to which oscillations in the two structures are synchronized, at each frequency from 0 to 100 Hz, as indicated on the horizontal axis of each plot).

dynamics within and between structures give a concrete example of what is meant by references to a region's "dopaminergic tone."

Another line of research has examined oscillations in field potentials measured on the surface of M1 with ECoG electrodes, and with scalp EEG, in patients with PD. The ECoG recordings were made in patients with DBS electrodes implanted and compared cortical activity when DBS was on vs. off. The initially published finding was that ECoG recordings showed an atypically high level of phase–amplitude coupling (PAC) between the phase of the beta-band oscillation and high gamma power when DBS was off, and that the amplitude of this PAC decreased markedly when DBS was on. (For a refresher on PAC, see section Caveats and considerations for spectral analyses, Chapter 3; Figure 3.13.) This was interpreted as evidence that PD motor symptoms may be caused by cortical pyramidal cells being pathologically "locked-in" to a highly synchronous pattern of firing that prevented the quick and flexible modification of large-scale patterns of activity that underlie healthy motor control. Superficially, it suggested an electrophysiological correspondence to the bradykinetic aspects of PD. This conceptualization of the cortical pathophysiology of PD soon had to be revised, however, when it was discovered that the atypically high levels in PAC in the DBS-off recordings weren't due to periodic bursting of MUA, the standard interpretation of the high-gamma

component of PAC, but were instead an artifact of the shape of the wave forms in the cortical oscillation. Specifically, the peaks of the beta oscillation were abnormally sharp in the DBS-off condition (resembling the upper right-hand quadrant in *Figure 3.13.A*), and this resulted in the spectral transform (for our purposes, the FFT) needing to add power at high frequencies to be able to represent the steep slope of the rising edge of each cycle. The effect of DBS, as it turned out, was to smooth out the peaks in the beta-band oscillation (i.e., to render the slope of the rising edge of the cycle less steep). (See Cole et al., 2017 from *Further Reading*.)

A follow-up study with scalp EEG recorded from a different group of patients showed similar results for DA-replacement medication: the beta-band oscillation recorded at central electrodes had an atypically sawtoothed shape when patients were OFF their medication, and a smoother, more sinusoidal shape when they were ON their medication. From a systems level of analysis, this suggests that one consequence of the elevated coherence at 20 Hz between GPi-STN when patients are OFF medication (*Figure 8.10.B*) may be abnormally high synchrony in the thalamic signalling to motor cortex that is gated by the cortico-basal ganglia-thalamic circuit. From a clinical perspective, this finding may have important implications for diagnosis and tracking of disease progression, because the medication-related change in the shape of the beta-band oscillation was not accompanied by a change in beta power. That is, conventional EEG analyses that ignore waveform shape are insensitive to this effect.

Habits and reinforcement learning. For decades, the basal ganglia have been associated with "habit learning," a term that refers to the stereotyped behaviors that we engage in while carrying out everyday tasks, often multiple times each day, often with little or no awareness that we are carrying them out. One example from this author's personal experience is throwing away garbage while preparing food in the kitchen (how's that for mundane?). For years, the garbage can in our kitchen was at the end of the counter, and when working at the sink, throwing something away required a 110° (or so) pivot to the right (clockwise) and then three steps to the garbage can. Recently, however, our kitchen was renovated, and the garbage can is now (much more conveniently) located in a pull-out drawer that's immediately to the right of the sink (now you don't have to take even a step!). For the first few weeks in the newly renovated kitchen, however, when working at the sink and needing to throw something away, I would find myself having pivoted 110° to the right and having taken a step or two toward where the garbage can used to be located, before realizing that I was "following an old habit" that was no longer appropriate for that context. In terms of my subjective experience of this, I was probably thinking about the goal of whatever it was that I was doing (like what I was going to do next with this food, once I discarded the packaging that I had just removed it from) but I was obviously not volitionally "pre-planning" my route to the garbage can. Rather, knowing that I was holding something that needed to be discarded prompted the automatic, unconscious triggering of this suite of behaviors.

What processes in my brain led to the formation of this habit? It's almost certainly the case that, upon moving into the house and beginning to cook in its unfamiliar kitchen, for the first several times that I threw something away while working at the sink, I did think explicitly about what actions I needed to take. Over time, however, there developed a "fracture" in my motor homunculus (perhaps in the hand area?) that represented the behavioral sequence of throwing-something-away-when-working-at-the-sink. Before the kitchen renovation, then, anytime circumstances called for it, a signal from a higher-level motor area, likely in SMA, would engage the circuits in my putamen that would temporarily lift the gate on the cortical representation of throwing-something-away-when-working-at-the-sink, and off I would go. The formation of this habit was produced by a particular kind of learning called reinforcement learning.

Reinforcement learning is a computational approach to learning that is enormously influential in automation, robotics, and artificial intelligence (AI). More recently, it has also become highly influential as a model for understanding certain types of learning in the nervous system, and it will make appearances in many chapters after this one, particularly in the final section of the book on *High-Level Cognition*. Here we will introduce some of its key principles, and how these inform how we understand the contributions of the basal ganglia to motor control.

Interestingly, the discipline of reinforcement learning was first inspired by findings from studies of Pavlovian conditioning, particularly the phenomenon of "blocking," and the Rescorla–Wagner model (Rescorla and Wagner, 1972) that was developed to explain it. We can illustrate blocking by returning to the metronome-and-food procedure that we considered in the previous section. Now, after the animal has learned the association of *metronome predicts food*, we introduce a second arbitrary "CS," let's say a flashing light. If we pretend, in this example, that 10 trials of presenting metronome + food had been sufficient to produce active salivation when the metronome was played alone, in our new experiment, for trials 11 through 20, we'll play the metronome and flash the light when food is presented. The critical question is, if, on the 21st trial, only the light is flashed, does it produce salivation? (That is, has the dog now learned to also associate the light with the presentation of food?) The answer is no. In the parlance of the animal learning literature, the previous learning of the association of the metronome with food has "blocked" the association of the light with food. The reason that this effect is important is that it demonstrates that the mere temporal co-occurence of two events is not sufficient for associative learning to occur. What Rescorla and Wagner proposed was that learning only happens when an event violates the animal's expectations, i.e., when its prediction of what should happen in a particular situation is wrong. In the parlance reinforcement learning, this prediction error is called a "reward," and the goal of reinforcement learning algorithms is to maximize the reward.

What does this all have to do with DA and the basal ganglia? To answer this, we need to jump from New Haven, CT, in the 1970s (Rescorla and Wagner, 1972) to Fribourg, Switzerland, in the 1980s, where understanding the linkage between motor control and DA was the goal of the Mexican neurophysiologist Ranulfo Romo and his then-postdoctoral mentor the Swiss neurophysiologist Wolfram Schultz. Despite the well-characterized motor deficits of patients with PD, and the equally well-known palliative effects of L-dopa treatment, no direct linkage had been made between midbrain DA neurons and motor behaviors. To their surprise, Romo and Schultz (1990) failed to find evidence that these cells responded to movement; instead, they seemed to respond to motivational and reward factors. Romo later noted: "It took us about 5 years to publish the paper because nobody was going to buy this idea that those neurons were not associated with movement, but reward" (Downey, 2006, p. 14264). Once they did finally get it published, however, it was quickly realized that these findings had profound implications for such topics as learning, motivated behavior, decision making, and addiction. And, importantly for our current focus on reinforcement learning, midbrain DA neurons seemed to be broadcasting information about prediction errors to the neostriatum and to cortex (*Figure 8.11*).

Before leaving this section on reinforcement learning, there's one more important point to be made. Although one will often read about DA as being related to reward, it's not the case that the release of DA at a synapse is some-

how akin to, say, a grade-school student being given a $10 bill for every A she brings home on her report card. Rather, the cell biological effects of DA are such that, when mediated by D1-type receptors, they enhance membrane excitability. Thus, a positive RPE (i.e., an elevated level of DA) increases the likelihood that concurrent activity at corticostriatal synapses will have potentiating effects, whereas a negative RPE (i.e., a pause in firing that lowers the level of DA) increases the likelihood that this same activity will have depressing effects. The fact that the opposite is true for neurons with a higher density of D2-type receptors allows us to finish the explanation of how the indirect pathway may have an important role in learning the negative value of some actions. The idea is that because the indirect pathway is governed by the inhibitory effects of DA on D2 receptors, it is during the pauses in the tonic release of DA that constitute negative RPEs (*Figure 8.11*, group B) that neostriatal neurons in the indirect pathway experience transient increases in excitability and, consequently, a boost in the plasticity of their synaptic connections with incoming cortical fibers. Thus, it is in conjunction with "disappointments," or situations in which outcomes are worse than expected, that reinforcement learning is strongest in the indirect pathway. It follows from this fact that experiments have shown that interfering with D1 functioning impairs reward learning and interfering with D2 functioning impairs avoidance learning.

COGNITIVE FUNCTIONS OF THE MOTOR SYSTEM

As suggested at the beginning of this chapter, many have argued that high-level cognition may have evolved over millennia as increasingly abstracted versions of motor control. Ito's (2008) proposal of mental models represented in the cerebellum would fall into this category. Also consistent with this idea is the fact that the cortico-basal ganglia-thalamic circuits of the PFC and the ACC/LOFC operate on the same principles as does the motor circuit that we have focused on in this chapter. Two additional examples of this idea have come from the Italian neurophysiologist Giacomo Rizzolatti and his colleagues. One is the "premotor theory of attention," the idea that attention may "have evolved from"/"be a byproduct of" motor preparation. This theory will receive considerable attention in the next chapter, when we take up oculomotor control and the control of attention. In the meantime, this chapter concludes with another of Rizzolatti's discoveries: mirror neurons.

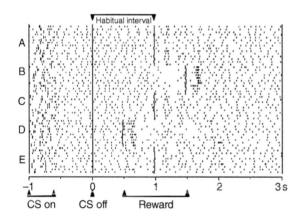

FIGURE 8.11 The reward prediction error (RPE) signal. Each row shows activity in a midbrain DA neuron during a well-learned Pavlovian conditioning task in which the offset of the CS is followed by a reward 1 second later (the habitual interval). When the expected outcome occurs (groups A, C, and E), there is no change in firing rate. When the reward is unexpectedly delayed (B), there is a brief pause in firing – a negative RPE signal – then an elevated burst upon the unexpected, later delivery – a positive RPE signal. When the reward is unexpectedly delivered early (D), there is a positive RPE – an elevated burst signaling its delivery – and an absence of the spike that is perfectly aligned with reward delivery on habitual trials. Source: From Wolfram Schultz, "Behavioral Theories and the Neurophysiology of Reward", Annual Review of Psychology. Vol. 57:87–115. Copyright 2005. Reproduced with permissions of Annual Reviews.

Mirror neurons

Sector F5 is a subregion of PMC, located immediately caudal to the ventral branch of the arcuate sulcus, that has been shown in recording and microstimulation experiments to be involved in the high-level representation of actions involving the hand and the mouth. Different neurons, for example, respond selectively to grasping, tearing, or holding movements. In a study carried out by Gallese, Fadiga, Fogassi, and Rizzolatti (1996), upon isolating a neuron in

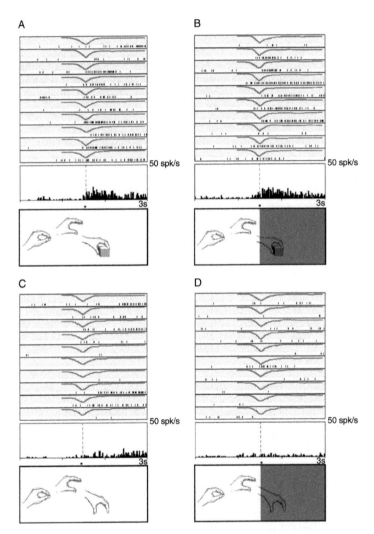

FIGURE 8.12 Mirror neurons in PMC of the monkey. Responses of an F5 neuron in a study carried out by Dr. Maria Alessandra Umiltà, in which the monkey watched while an experimenter reached to either grasp an object (panels **A** and **B**) or to mimic the same action but in the absence of the object. In between the monkey and the table on which the object was displayed was a metal frame that could either be empty (panels **A** and **C**) or hold an opaque screen, thereby blocking the monkey's view of this location (panels **B** and **D**). For all four trial types, neural activity is aligned to the moment in time when the reaching hand passed the boundary of the frame (and, therefore, became no longer visible in **B** and **D**). (The waveforms above each row of rasters illustrate the location of the hand relative to the boundary of the frame as a function of time, aligned the same as neural data.) Note how the neuron only responds to the act of grasping a physical object, not to the identical movement when the object was absent, and that it does so regardless of the presence or absence of the screen. Source: From M.A. Umiltà, E. Kohler, V. Gallese, L. Fogassi, L. Fadiga, C. Keysers, G. Rizzolatti, "I Know What You Are Doing A Neurophysiological Study", Neuron. Copyright 2001. Reproduced with permission of Elsevier.

F5, the investigators would present the monkey with an item of food on a tray, either within or outside of the animal's reach, as well as with objects of various sizes. The size manipulation enabled distinguishing between three types of grip: "precision grip" (thumb and index finger); "prehension grip" (thumb opposite fingers to extract an object from a deep, narrow container); and "whole-hand prehension." This procedure can be construed as analogous to the procedure for determining the visual response properties of a neuron in IT cortex by presenting it with a series of objects.

The novel innovation of this study was that the testing battery also included an experimenter performing several actions while the monkey watched: grasping a piece of food from a tray, bringing the food to their mouth, giving the food to a second experimenter, etc. Of 532 neurons recorded from F5, 92 discharged to specific actions by the monkey *and* to the observed actions of an experimenter. These were dubbed "mirror neurons." The actions that mirror neurons responded to were categorized as shown in *Table 8.1*.

The responses summarized in *Table 8.1* were insensitive to manipulation of factors related to the animal's reference frame, such as side of the body the actions were performed on, or distance of the observed actor. Importantly, however, these neurons did not respond when their preferred action was performed as mimicry (e.g., experimenter pretending to pick up a piece of food when there was no object on the tray; *Figure 8.12*), or when observing the same action performed with tools (forceps or pliers) instead of the hand.

In this study, 31.5% of mirror neurons were classified as being "strictly congruent," meaning that they responded preferentially to the same action, regardless of whether the animal was observing an experimenter performing the action or executing the action itself; and 61% were classified as "broadly congruent" – preferred observed and executed actions were similar, but not identical. For example, one such neuron responded vigorously during *observation* of a precision grip and observation of whole-hand prehension, but only responded to *execution* of a precision grip. The remainder were "noncongruent" with regard to preferred observed vs. executed action(s).

The authors suggested that mirror neurons are sensitive to the meaning of an action, rather than to its precise sensory or motoric properties. They further speculated that mirror neurons may underlie learning through imitation (prevalent in humans as well as in nonhuman primates), as well as, possibly, speech perception in humans.

TABLE 8.1 Actions that mirror neurons responded to in Gallese et al. (1996) study

Observed hand actions	Number of neurons
Neurons with one preferred action	
Grasping	30
Placing	7
Manipulating	7
Hands interaction	5
Holding	2
Neurons with two preferred actions	
Grasping/placing	20
Grasping/manipulating	3
Grasping/hands interaction	3
Grasping/holding	5
Grasping/grasping with the mouth	3
Placing/holding	1
Hands interaction/holding	1
Neurons with three preferred actions	
Grasping/placing/manipulating	1
Grasping/placing/holding	4

Holding a mirror up to nature?

Mirror neurons have been among the "hottest" topics of research in contemporary cognitive neuroscience. One factor driving this interest is that many believe that they offer a neural basis for many popular schools of thought in cognitive psychology, including embodied cognition, common coding theory, and the motor theory of speech perception. The last one was referenced by Gallese and colleagues (1996). It is a theory, first proposed in the 1960s, that an important component of speech perception is the "listener" watching the mouth of the talker, and simulating in her mind the larynx, tongue, and lip movements being made by the talker. Gallese and colleagues (1996) speculated that the F5 region in the monkey may bear homologies to Broca's area in the human – the region of left inferior PFC that we first encountered in *Chapter 1*, and will do so again in *Chapter 19*.

With regard to ASD, some have proposed that dysfunction of the mirror system underlies the disorder. One high-profile study (Dapretto et al., 2005) compared fMRI activity in typically developing adolescents vs. "high-functioning" children with ASD on tasks requiring passive observation vs. mimicking of visually presented facial expressions. The findings indicated that, although behavioral performance was comparable between the two groups, "the ASD group showed no activity in the mirror area in the pars opercularis" (Dapretto et al., 2005, p. 29). Note that, strictly speaking, the assertion of "no activity" cannot, literally, be true. If it were, pars opercularis in these subjects would have been necrotic. Indeed, despite this statement, the authors further reported that the level of activity in this region that purportedly showed "no activity" correlated negatively with ASD subjects' scores on two scales, indicating that "the greater the activity in this critical component of the [mirror neuron system] during imitation, the higher a child's level of functioning in the social domain" (pp. 29–30).

In response to this embrace of mirror neurons as a factor accounting for important human behaviors, and their pathological dysfunction, cautionary voices have been raised. For these, the reader is directed to two opinion pieces, one entitled "A Mirror up to Nature" (Dinstein, Thomas, Behrmann, and Heeger, 2008), and a second entitled "Eight Problems for the Mirror Neuron Theory of Action Understanding in Monkeys and Humans" (Hickok, 2009), both listed under *Further Reading*.

IT'S ALL ABOUT ACTION

The principles of motor control lie at the core of understanding human behavior. On one level, this must literally be true, because we cannot *behave* without the release of ACh at the neuromuscular junction causing the contraction of muscle, nor all the steps that precede this event. At a deeper level, we have also seen that circuits identical to those responsible for the triggering of a motor command and for the fine-tuning of its execution apply those same computations to signals from prefrontal areas associated with our most abstract thoughts and our most powerful feelings. It will serve us well, therefore, as we delve further from the sensory and motor peripheries, in the exploration of "higher levels" of cognition, to be at the ready with such principles and concepts as population coding, efference copy, neuromodulatory tone, and RPE signals.

END-OF-CHAPTER QUESTIONS

1. What is the principle of efference copy? Name two major systems that work on this principle.

2. What does it mean to say that muscles behave like springs? How does the motor system take advantage of this property?

3. In what way(s) is the motor homunculus similar to the somatosensory homunculus? In what way(s) are they different?

4. Explain how population coding across hundreds of broadly tuned units can produce exquisitely precise movements.

5. What principle do you think better captures the function of the parietal cortex, processing *where* objects are located in the visual world, or computing *how* to act on the world based on visual input? Or, alternatively, can both be valid? Support your answer with empirical evidence.

6. How is a copy of the efferent motor signal processed in the cerebellum? What is the result?

7. What are two consequences of basal ganglia functioning of reducing levels of DA in the neostriatum, as occurs in PD?

8. What are the fundamental principles of reinforcement learning? In what way do these seem to be carried out by the basal ganglia nuclei?

9. What properties distinguish mirror neurons of PMC from "conventional" neurons in this region? What functions might such neurons enable?

REFERENCES

Albus, James S. 1971. "A Theory of Cerebellar Function." *Mathematical Biosciences* 10 (1–2): 25–61. doi: 10.1016/0025-5564(71)90051-4.

Andersen, Richard A., Sam Musallam, and Bijan Pesaran. 2004. "Selecting the Signals for a Brain–Machine Interface." *Current Opinion in Neurobiology* 14 (6): 720–726. doi: 10.1016/j.conb.2004.10.005.

Brown, Peter, Antonio Oliviero, Paolo Mazzone, Angelo Insola, Pietro Tonali, and Vincenzo Di Lazzaro. 2001. "Dopamine Dependency of Oscillations between Subthalamic Nucleus and Pallidum in Parkinson's Disease." *Journal of Neuroscience* 21 (3): 1033–1038.

Carmena, Jose M., Mikhail A. Lebedev, Roy E. Crist, Joseph E. O'Doherty, David M. Santucci, Dragan F. Dimitrov, . . . Miguel A. L. Nicolelis. 2003. "Learning to Control a Brain–Machine Interface for Reaching and Grasping by Primates." *PLoS Biology* 1 (2): e42. doi: 10.1371/journal.pbio.0000042.

Cole, S. R., R. van der Meij, E. J. Peterson, C. de Hemptinne, P. A. Starr, and B. Voytek. 2017. "Nonsinusoidal Beta Oscillations Reflect Cortical Pathophysiology in Parkinson's Disease." *Journal of Neuroscience* 37: 4830–4840.

Dapretto, Mirella, Mari S. Davies, Jennifer H. Pfeifer, Ashley A. Scott, Marian Sigman, Susan Y. Bookheimer, and Marco Iacoboni. 2005. "Understanding Emotions in Others: Mirror Neuron Dysfunction in Children with Autism Spectrum Disorders." *Nature Neuroscience* 9 (1): 28–30. doi: 10.1038/nn1611.

Gallese, Vittorio, Luciano Fadiga, Leonardo Fogassi, and Giacomo Rizzolatti. 1996. "Action Recognition in the Premotor Cortex." *Brain* 119 (2): 593–609. doi: 10.1093/brain/119.2.593.

Gastaut, Henri J., and J. Bert. 1954. "EEG Changes during Cinematographic Presentation." *Electroencephalography and Clinical Neurophysiology* 6 (3): 433–444. doi: 10.1016/0013-4694(54)90058-9.

Georgopoulos, Apostolos P., John F. Kalaska, Roberto Caminiti, and Joe T. Massey. 1982. "On the Relations between the Direction of Two-Dimensional Arm Movements and Cell Discharge in Primate Motor Cortex." *Journal of Neuroscience* 2 (11): 1527–1537.

Georgopoulos, Apostolos P., Roberto Caminiti, John F. Kalaska, and Joe T. Massey. 1983. "Spatial Coding of Movement: A Hypothesis Concerning the Coding of Movement Direction by Motor Cortical Populations." *Experimental Brain Research Supplementum* 7: 327–336. doi: 10.1007/978-3-642-68915-4_34.

Goodale, Melvyn A., A. David Milner, Lorna S. Jakobson, and David P. Carey. 1991. "A Neurological Dissociation between Perceiving Objects and Grasping Them." *Nature* 349 (6305): 154–156. doi: 10.1038/349154a0.

Ito, Masao. 2008. "Control of Mental Activities by Internal Models in the Cerebellum." *Nature Reviews Neuroscience* 9 (4): 304–313. doi: 10.1038/nrn2332.

Kaas, Jon H. 2012. "Evolution of Columns, Modules, and Domains in the Neocortex of Primates." *Proceedings of the National Academy of Sciences of the United States of America* 109 (Supplement 1): 10655–10660. doi: 10.1073/pnas.1201892109. Kandel, Eric R., James H. Schwartz, and Thomas M. Jessell, eds. 2000. *Principles of Neural Science.* New York: McGraw-Hill.

Kilavik, Bjørg Elisabeth, Joachim Confais, Adrián Ponce-Alvarez, Markus Diesmann, and Alexa Riehle. 2010. "Evoked Potentials in Motor Cortical Local Field Potentials Reflect Task Timing and Behavioral Performance." *Journal of Neurophysiology* 104 (5): 2338–2351. doi: 10.1152/jn.00250.2010.

Kravitz, D. J., K. S. Saleem, C. I. Baker, and M. Mishkin. 2011. "A New Neural Framework for Visuospatial Processing." *Nature Reviews Neuroscience* 12: 217–230.

Penfield, Wilder, and Theodore Rasmussen. 1950. *The Cerebral Cortex of Man.* A Clinical Study of Localization of Function. New York: Macmillan.

Pohl, Walter. 1973. "Dissociation of Spatial Discrimination Deficits Following Frontal and Parietal Lesions in Monkeys." *Journal of Comparative and Physiological Psychology* 82 (2): 227–239. doi: 10.1037/h0033922.

Rescorla, R. A., and A. R. Wagner. 1972. "A Theory of Pavlovian Conditioning: Variations in the Effectiveness of Reinforcement and Nonreinforcement." In *Classical Conditioning II* edited by A. H. Black and W. F. Prokasy, 64–99. New York: Appleton-Century-Crofts.

Umiltà, Maria Alessandra. 2000. "Premotor Area F5 and Action Recognition." PhD diss., University of Parma.

OTHER SOURCES USED

Albus, James S. 1989. "The Marr and Albus Theories of the Cerebellum: Two Early Models of Associative Memory." Paper presented at COMPCON Spring '89. Thirty-fourth IEEE Computer Society International Conference: Intellectual Leverage, Digest of Papers. San Francisco, California, February 27–March 1: 577–582. doi: 10.1109/CMPCON.1989.301996.

Blakeslee, Sandra. 2008. "Monkey's Thoughts Propel Robot, a Step That May Help Humans." *The New York Times*, January 15. http://www.nytimes.com/2008/01/15/science/15robo.html?pagewanted=all&_r=0.

Cohen, M. X., and M. J. Frank. 2009. "Neurocomputational Models of Basal Ganglia Function in Learning, Memory and Choice." *Behavioural Brain Research* 199: 141–156.

de Hemptinne, C., N. Swann, J. L. Ostrem, E. S. Ryapolova-Webb, N. B. Galifianakis, and P. A. Starr. 2015. "Therapeutic Deep Brain Stimulation Reduces Cortical Phase-Amplitude Coupling in Parkinson's Disease." *Nature Neuroscience* 18: 779–786.

Dinstein, Ilan, Cibu Thomas, Marlene Behrmann, and David J. Heeger. 2008. "A Mirror up to Nature." *Current Biology* 18 (1): R13–R18. doi: 10.1016/j.cub.2007.11.004.

Eccles, John C., János Szentágothai, and Masao Itō. 1967. *The Cerebellum as a Neuronal Machine*. Berlin: Springer-Verlag.

Frank, M. J. 2005. "Dynamic Dopamine Modulation in the Basal Ganglia: A Neurocomputational Account of Cognitive Deficits in Medicated and Nonmedicated Parkinsonism." *Journal of Cognitive Neuroscience* 17: 51–72.

Frank, M. J. 2011. "Computational Models of Motivated Action Selection in Corticostriatal Circuits." *Current Opinion in Neurobiology* 21: 381–386.

Georgopoulos, Apostolos P., Andrew B. Schwartz, and Ronald E. Kettner. 1986. "Neuronal Population Coding of Movement Direction." *Science* 233 (4771): 1416–1419.

Goodale, Melvyn A., and A. David Milner. 1992. "Separate Visual Pathways for Perception and Action." *Trends in Neurosciences* 15 (1): 20–25. doi: 10.1016/0166-2236(92)90344-8.

Hikida, T., Kimura, K., Wada, N., Funabiki, K. and Nakanishi, S. (2010). Distinct roles of synaptic transmission in direct and indirect striatal pathways to reward and aversive behavior. *Neuron* 66: 896–907.

Jackson, N., Cole, S. R., Voytek, B. and Swann, N. C. (2019). Characteristics of waveform shape in Parkinson's disease detected with scalp electroencephalography. eNeuro 6. doi. org/10.1523/ENEURO.0151-19.2019

Jakobson, Lorna S., Y. M. Archibald, David P. Carey, and Melvyn A. Goodale. 1991. "A Kinematic Analysis of Reaching and Grasping Movements in a Patient Recovering from Optic Ataxia." *Neuropsychologia* 29 (8): 803–805. doi: 10.1016/0028-3932(91)90073-H.

Leveratto, Jean-Marc. 2009. "La revue internationale de filmologie et la genèse de la sociologie du cinéma en France." *Cinémas: revue d'études cinématographiques* 19 (2–3): 183–215. doi: 10.7202/037553ar.

Mink, J. W., and W. T. Thach. 1991. "Basal Ganglia Motor Control. I. Nonexclusive Relation of Pallidal Discharge to Five Movement Modes." *Journal of Neurophysiology* 65 (2): 273–300.

Romo, Ranulfo, and Wolfram Schultz. 1990. "Dopamine Neurons of the Monkey Midbrain: Contingencies of Responses to Active Touch during Self-Initiated Arm Movements." *Journal of Neurophysiology* 63 (3): 592–606.

Schultz, Wolfram, and Ranulfo Romo. 1990. "Dopamine Neurons of the Monkey Midbrain: Contingencies of Responses to Stimuli Eliciting Immediate Behavioral Responses." *Journal of Neurophysiology* 63 (3): 607–624.

Tononi, Giulio. 2008. "Consciousness as Integrated Information: A Provisional Manifesto." *Biological Bulletin* 215 (3): 216–242.

Umiltà, Maria Alessandra, E. Kohler, V. Gallese, L. Fogassi, L. Fadiga, C. Keysers, and G. Rizzolatti. 2001. "I Know What You Are Doing. A Neurophysiological Study." *Neuron* 31(1):155–165.

FURTHER READING

Cole, S. R., R. van der Meij, E. J. Peterson, C. de Hemptinne, P. A. Starr, and B. Voytek. 2017. "Nonsinusoidal Beta Oscillations Reflect Cortical Pathophysiology in Parkinson's Disease." *Journal of Neuroscience* 37: 4830–4840.

Paper that reassessed ECoG data of DBS in PD and demonstrated that elevated measures of PAC were artifactually produced by the sharpness of beta-band oscillations and that the real effect of DBS is to smooth out the shape of this waveform.

Dinstein, Ilan, Cibu Thomas, Marlene Behrmann, and David J. Heeger. 2008. "A Mirror up to Nature." *Current Biology* 18 (1): R13–R18. doi: 10.1016/j.cub.2007.11.004.

A critique of the mirror neuron literature, from neurophysiological and functional perspectives, from two leading cognitive neuroscientists and their trainees.

Hickok, Gregory. 2009. "Eight Problems for the Mirror Neuron Theory of Action Understanding in Monkeys and Humans." *Journal of Cognitive Neuroscience* 21 (7): 1229–1243. doi: 10.1162/jocn.2009.21189.

The title says it all.

Hickok, Gregory. 2014. *The Myth of Mirror Neurons: The Real Neuroscience of Communication and Cognition*. New York: W. W. Norton.

An update and elaboration on the 2009 opinion piece that not only critiques the "action understanding" theory but also reviews alternative proposals about mirror neuron contributions to motor control. Some of this book's content is also relevant to the work of Hickok, and others, that is detailed in Chapter 19: Language.

Himmelbach, Marc, Rebecca Boehme, and Hans-Otto Karnath. 2012. "20 Years Later: A Second Look on D. F.'s Motor Behaviour." *Neuropsychologia* 560 (1): 139–144. doi: 10.1016/j.neuropsychologia.2011.11.011.
A comparison of more recently acquired control data with the original results from D.F. that concludes that "the widespread and popular presentation of strong dissociations between distinct visual systems seems to be exaggerated."

Jenkinson, Ned, and Peter Brown. 2011. "New Insights into the Relationship between Dopamine, Beta Oscillations, and Motor Function." *Trends in Neurosciences* 34 (12): 611–618. doi: 10.1016/j.tins.2011.09.003.
Review that nicely demonstrates how clinical research (in this case, with PD) can inform basic science-level knowledge, and vice versa.

Kravitz, Dwight J., Kadharbatcha S. Saleem, Chris I. Baker, and Mortimer Mishkin. 2011. "A New Neural Framework for Visuospatial Processing." *Nature Reviews Neuroscience* 12 (4): 217–230. doi: 10.1038/nrn3008.
A proposal that discrete anatomical pathways support the processing of spatial information in the service of where*-based behavior (PPC-FEF and PFC),* how*-based behavior (PPC-premotor cortex), and navigation behavior (PPC-medial PPC-MTL).*

Rizzolatti, G., and L. Craighero. 2004. "The Mirror-Neuron System." *Annual Review of Neuroscience* 27: 169–192. doi: 10.1146/annurev.neuro.27.070203.144230.
This review has, at the time of this writing, accrued an astounding 3588 citations in the peer-reviewed literature.

Schenk, Thomas, and Robert D. McIntosh. 2010. "Do We Have Independent Visual Streams for Perception and Action?" *Cognitive Neuroscience* 1 (1): 52–62. doi: 10.1080/17588920903388950.
A "20-years later" critical review of the what *vs.* how *model, followed by commentary/rebuttals from many leading researchers, including Goodale and Milner.*

CHAPTER 9
OCULOMOTOR CONTROL AND THE CONTROL OF ATTENTION

KEY THEMES

- Four regions that contribute importantly to oculomotor control are the superior colliculus, regions of occipital and parietal cortex (particularly V1 and LIP), regions of frontal cortex (the frontal eye field [FEF] and the supplementary eye field [SEF]), and the brainstem oculomotor nuclei that send commands to the extraocular muscles.

- Although superficial layers of the superior colliculus receive direct projections from the retina, the deep layers contain multimodal sensory neurons and movement neurons to which posterior cortical neurons project, and which, in turn, project to the brainstem oculomotor nuclei.

- FEF neurons, which project directly to the brainstem oculomotor nuclei, encode vector saccades, whereas SEF neurons encode goal-oriented saccades.

- Saccade planning and the endogenous control of attention both engage a dorsal frontoparietal network that includes FEF and IPS/LIP.

- The reflexive, exogenous capture of attention is associated with a right hemisphere–biased ventral cortical system, including the temporal–parietal junction and the inferior PFC, regions also associated with hemispatial neglect.

- Microstimulation of monkey FEF produces retinotopically specific behavior that is equivalent to the effects of cuing the endogenous control of attention, and also produces attention-like effects in V4 neurons whose receptive fields overlap with the movement field of the stimulation site.

- Multivariate pattern analysis (MVPA) can dramatically increase the sensitivity of neuroimaging data analyses and provide novel insights into information processing in the brain.

- MVPA has been used to demonstrate that the same systems that control eye movements are also used to control attention.

CONTENTS

ATTENTION AND ACTION

We have reached the final chapter of Section II, themed *Sensation, perception, attention, and action*. While we have seen in previous chapters how strongly perception and attention are linked (e.g., the attentional amplification of one sensory input channel over others, the dependence on attention of experience-dependent plasticity, unilateral neglect), this chapter will emphasize even more strongly than did the previous two how closely intertwined are the constructs of attention and action. At another level, this is the last chapter for which it will be necessary to begin with an overview of the anatomy of the system of interest. We haven't, of course, covered "everything," but upon completion of this chapter, the reader will have had an introduction to all of the major systems and principles of functional neuroanatomy, and will have the foundation on which to build a more detailed understanding of the intricacies of any particular component. Similarly, if the author has done his job, we'll see in every chapter following this one – i.e., in every chapter detailing some aspect of "high-level cognition" – that the principles underlying memory storage and retrieval, cognitive control, decision making, language, emotion, and so on can be understood as variations on the principles introduced in this section on the more "elemental" functions of the brain.

WHYS AND HOWS OF EYE MOVEMENTS

Let's take a moment to recall one of the reasons why the ability to move the eyes is so important: it enables the visual system to disproportionately devote processing resources and, therefore, acuity, to just a small area of the visual field. If you need detailed information about an object located 15° to the left of your current point of fixation, you can obtain it within~300 ms by moving your center of gaze to this location. The tradeoff for the "cost" of those 300 ms is not having to drag around a visual cortex that's the size of the room that you're currently sitting in, or, alternatively, not having to survive with a visual system with uniformly horribly low resolution. Because there are some instances when a few tens of milliseconds can matter, however, we'll see that one feature of the oculomotor system is multiple pathways that control the eyes in different contexts.

Three categories of eye movements

Eye movements fall into three categories: smooth pursuit, vergence shifts, and saccades. We won't spend much time considering the first two, but will define them briefly here. If you hold out your thumb and fixate the nail, then slowly move your head from side to side, all the while maintaining fixation on your thumbnail, you are performing smooth-pursuit eye movements. (If your eyes hadn't been moving to compensate for the back-and-forth motion of your head, your thumbnail wouldn't have remained in your center of gaze.) The equivalent eye movements are generated if you now hold your head still and move your thumb back and forth. Thus, smooth-pursuit eye movements enable you to maintain fixation on an object while it, you, or both, are moving. During smooth pursuit, the two eyes move "together" (i.e., the movements of one eye are identical to those of the other). These are referred to as **conjugate eye movements**.

Vergence shifts occur when an object of interest is moving toward or away from you, and in order to fixate it with both eyes, the two must move in opposite directions: toward each other if the object is looming; away from each other if it is receding. Signals from stretch receptors on the extraocular muscles convey information about the positions of the eyes relative to each other (i.e., how parallel vs. converging). This information contributes to the perception of depth (i.e., seeing in 3-D) and of motion in depth (i.e., motion toward and away from you, as opposed to motion along a 2-D surface, such as when looking at a movie screen or a computer monitor).

The third category, saccadic eye movements, is the focus of this chapter. The word *saccade* is French for a jerky movement and was first associated with eye movements by the French ophthalmologist Emile Javal (1839–1907), who held a mirror at an angle while reading and observed that his eyes made a series of brief fixations, punctuated by abrupt conjugate movements, rather than a smooth, continuous drift across the page as most had previously assumed. Saccades are of primary interest here because of their central importance to many domains of cognition (reading being just one of them), and because they have been used extensively in neuroscience and psychology research as an indirect measure of cognitive operations. Because monkeys cannot be verbally instructed how to respond on experimental tasks, nor can they tell you what they are thinking, saccades are a way for the animal to indicate its choice in an experiment. (For example, *Which is the most recent item that you saw?* or *Which of these stimuli is different from the others?*) And although it would be disingenuous to suggest that training a monkey to perform a cognitive task is ever "easy," it is the case that saccadic eye movements are natural behaviors for monkeys (and humans), and their use as a dependent measure can at least give the researcher a "head start" in the training process. For humans, the saccade itself can be of fundamental interest, if it is a fundamental part of the behavior of interest (e.g., reading, or visual search). Additionally, scientists have devised clever ways to use eye movements to assay the dynamics of cognitive operations as they are being carried out, as opposed to, or in addition to, retrospective reports that come at the end of each trial. (For example, whether the eyes start to drift toward word *X* or word *Y*, both displayed on a screen, can provide insight into the mental operations engaged while a person is listening to a semantically ambiguous sentence.)

THE ORGANIZATION OF THE OCULOMOTOR SYSTEM

An overview of the circuitry

There are many factors that can prompt an eye movement. I can decide, while taking a pause as I type this sentence, to look up at a painting hanging on the wall of my living room. My eyes can move reflexively toward the source of a sound, or to a touch (the cat walks into the room, then jumps up onto the couch next to me). I can scan the shelves of a grocery store, looking for that particular brand of soy sauce. Much of the discussion here, however, will be framed in terms of visually guided saccades. That is, what are the steps that transpire between the detection on the retina of a visual stimulus and the saccade that brings that stimulus into foveal vision. Along the way, we'll note instances where the principles may be similar or different for these other factors that can also trigger saccades.

Although the principal pathway for visual processing in most mammalian species is the retino-geniculo-striate pathway, it "wasn't always that way." In fish and amphibians, the primary site of visual processing is a midbrain structure known as the optic tectum. In the mammal, the homologue is the superior colliculus ("higher hill" in Latin; the inferior colliculus of the auditory system, *Figure 5.3*, is the "lower hill"), a structure that sits immediately caudal to the thalamus on the "roof" ("tectum") of the brainstem. There is a direct, monosynaptic pathway from the retina to the superior colliculus, which will receive more consideration further along in this chapter. The superior

colliculus projects to the **brainstem oculomotor nuclei**, which house the local circuitry and oculomotor neurons, analogous to the motor circuitry of the spinal cord, that process descending motor commands and send axons that innervate the extraocular muscles. Within cortex, regions contributing to oculomotor control can be divided into posterior and anterior systems, the posterior including, most prominently, V1, LIP, and V4, and the anterior comprising the frontal eye field (FEF) and the supplementary eye field (SEF, also sometimes referred to as the medial eye field [MEF] or dorsomedial frontal cortex [DMFC]). The primary distinction between the posterior and anterior systems is that the former regions project to the superior colliculus – superior colliculus lesions abolish the ability of electrical stimulation of posterior cortical areas to elicit saccades – whereas the anterior regions, although sending projections to the superior colliculus, also send projections directly to the brainstem oculomotor nuclei (*Figure 9.1*).

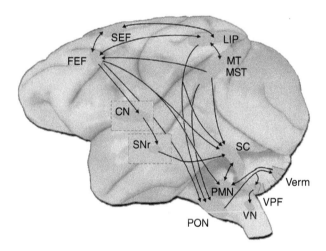

FIGURE 9.1 Important nodes and connections for oculomotor control. Note that among this diagram's simplifications are the absence of the V1 projection to SC, and the integration of LIP and FEF signals in the "oculomotor" cortico-basal ganglia-thalamic loop. CN = caudate nucleus; FEF = frontal eye field; LIP = lateral intraparietal area; MT = middle temporal area; MST = medial superior temporal area; PMN = brain stem (pre)motor nuclei (PPRF, riMLF, cMRF); PON = precerebellar pontine nuclei; SC = superior colliculus (intermediate and deep layers); SEF = supplementary eye field; SNr = substantia nigra pars reticulata; Verm = oculomotor vermis (cerebellum, lobules VI and VII); VN = vestibular nuclei; VPF = ventral paraflocculus (cerebellum). Source: From Krauzlis, Richard J. 2005. "The Control of Voluntary Eye Movements: New Perspectives." Neuroscientist 11 (2): 124–137. doi: 10.1177/1073858404271196. Reproduced with permission of SAGE Publications.

Let's now consider each of these major elements of the "oculomotor system" in turn. (Keep in mind that most, if not all, of the cortical areas to be discussed here can't be considered "specialized" for oculomotor control; the label "oculomotor system" is simply an economical way to say "regions in the brain that can contribute to oculomotor control.")

The superior colliculus

In primates, the superior colliculi (one on each side of the midline) are each about the size of a half-globe of a pea, and roughly divided into three layers: superficial, intermediate, and deep. It is the deep layers that send efferents to the brainstem oculomotor nuclei. The superficial layers receive direct input from the retina, via a specialized class of retinal ganglion cells called *w* cells. The deep layers are of greater interest, because they receive the cortical efferents illustrated in *Figure 9.1*, as well as sensory information from the visual, auditory, and tactile modalities. There are two interleaved populations of cells: sensory neurons that have multimodal receptive fields and represent locations in retinotopic coordinates; and saccade neurons whose bursting activity brings the center of gaze into the receptive field of adjacent sensory neurons. Thus, the superior

colliculus employs a vector code, in that activity in a certain area always produces a saccade of the same direction and magnitude.

At the rostral-most end of the superior colliculus, in the deep layers, are fixation neurons. In effect, in relation to the motor map illustrated in *Figure 9.2*, these command saccades of 0°. These neurons maintain a tonically elevated firing rate while one is fixating a stationary location. (Thus, at least from the reductionist perspective of activity in the superior colliculus, the term "passive fixation" that one frequently sees in the literature is a misnomer.)

At any moment in time, most of the motor units in the deep layers of the superior colliculus are under tonic inhibition from the substantia nigra pars reticulata, an arrangement analogous to the tonic inhibition of thalamus by the GPi that we detailed in *Chapter 8*.

The posterior system

Although neurons from layer 5 of V1 send projections to the superior colliculus, their contribution seems to be modulatory, rather than driving. This assessment comes from the fact that, while microstimulation of V1 can produce saccades, the currents required to do so are

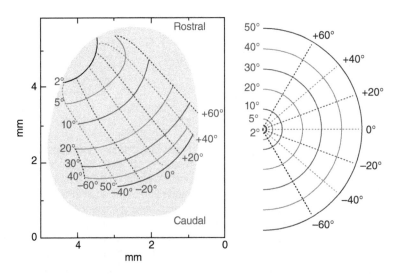

FIGURE 9.2 The functional organization of the superior colliculus. Left panel: motor map in deep layers of the left superior colliculus; right panel: corresponding regions of the right visual field. Green labels correspond to saccade amplitude and blue labels to degrees of polar angle from fixation. Thus, the left panel illustrates that motor neurons in the rostral portion of the SC produce small-amplitude saccades, and neurons in progressively more caudal areas produce progressively larger-amplitude saccades. Rostral and lateral to the 2° line in the left panel (i.e., in the upper-left corner of the image) are fixation neurons. Source: From Neeraj J. Gandhi and Husam A. Katnani, "Motor Functions of the Superior Colliculus Annual Review of Neuroscience", copyright 2011. Reproduced with permissions of Annual Reviews.

markedly higher than those required in other cortical areas, such as the FEF, or in the superior colliculus itself. However, cooling V1, a method that reversibly inactivates a region of cortex, has the effect of abolishing visual sensory responses in the deep layers. Also consistent with a modulatory function of V1 on the superior colliculus is the association of oscillatory state in occipital cortex with the generation of **express saccades**.

Express saccades have a latency that can be as much as one half that of standard saccade latency – in one recent study in humans, 90–135 ms after target onset for express saccades, 145–240 for standard saccades – and are often seen in response to the abrupt appearance of an unexpected stimulus. They are also readily produced experimentally with a "gap" paradigm, in which a gap of a few hundred milliseconds is inserted between the offset of the fixation point that began the trial and the onset of the target to which the subject must make a saccade. Although it would be tempting to postulate that express saccades are generated by a "short circuit" from the retina to superior colliculus to brainstem oculomotor nuclei, the anatomy would seem to argue against this, because, as far as we know, retinocollicular projections terminate exclusively in superficial layers of the superior colliculus, and these do not project directly to the deep layers. An alternative account is that "advanced motor preparation of saccadic programs" (Paré and Munoz, 1996, p. 3666) enables the cortical system to generate saccades more quickly than is typical. Consistent with this idea is the fact that express saccades are more likely to be generated on trials with elevated oscillatory synchrony in the low alpha-band range, measured at occipital EEG electrodes.

Consistent with their role in processing visual motion, the contributions of areas MT and MST to oculomotor control are particularly important for guiding smooth-pursuit movements.

As discussed in *Chapter 7* (section *Extracellular electrophysiological studies of monkeys*), the LIP is involved in "cognitive" aspects of eye-movement control – carrying out coordinate transformations and representing the spatial coordinates of potential targets (of eye movements) in the environment – functions related to this region's role in representing a priority map (section *Turning our attention to the future, Chapter 7*). Like area V1, microstimulation of neurons in LIP will drive the eyes to the location represented by the stimulated neuron's receptive field, although this may require stronger currents than other nodes in the oculomotor system.

The frontal eye field

As illustrated in *Figure 9.1*, the FEF in the monkey is located in the rostral bank of the arcuate sulcus, immediately anterior to premotor cortex. Among the first to study the FEF was David Ferrier, whose involvement in nineteenth-century research on the neural bases of vision made an appearance in *Chapter 1*. Ferrier (1875) found that electrical stimulation of this region produced "turning of the eyes and head to the opposite side" (p. 424), and that a unilateral lesion impaired the abilities of the animal to orient to stimuli presented contralateral to the lesion. Based on these and other observations, he stated most presciently in 1890 that "my hypothesis is that the power of attention is intimately related to the volitional movements of the head and eyes" (Ferrier, 1890, p. 151). Consistent with this idea has been the subsequent description, in monkeys and humans, of unilateral neglect after damage to the FEF.

As indicated earlier, a distinguishing characteristic of the FEF is its low threshold for saccade generation with microstimulation. Although its organization isn't strictly retinotopic, it's generally the case that small, contralaterally directed saccades are produced by microstimulation in the ventrolateral limb of the anterior bank of the arcuate sulcus, larger amplitude saccades by microstimulation in the dorsomedial limb. This region encodes vector saccades, in that repeated stimulation of the same site produces a saccade of the same metrics, regardless of the starting point of the eye (*Figure 9.3*). A sustained pulse of stimulation produces a series of discrete saccades, all of the same metric. This is an interesting phenomenon that illustrates the recoding that happens in the brainstem oculomotor nuclei, because sustained microstimulation of a single location in one of these nuclei produces a very different effect – a saccade whose magnitude is linearly dependent on the duration of the stimulation train. (The same is true, of course, for the extraocular muscles: they will only contract so long as ACh is being released into the neuromuscular junction.)

The supplementary eye field

Located dorsal and somewhat caudal to the FEF (*Figure 9.1*), the SEF lies just rostral to the SMA (*Chapter 8*). Its functions are associated with more "cognitive" aspects of oculomotor control. For example, when two targets are presented, and the monkey will receive a reward for saccading to either one (i.e., "free choice"), SEF, FEF, and LIP all show activity related to the impending choice, but that of the SEF starts earliest and is most robust. SEF activity is also prominent during the antisaccade task, when the rule is to make a saccade 180° opposite to the location of the target. The coding principle is also different between SEF and FEF, as is illustrated in *Figure 9.3*. While microstimulation of the FEF neuron produces a saccade with roughly the same metrics regardless of the starting point of the eyes, microstimulation of an SEF neuron always produces a saccade that lands in the same "termination zone." This is reminiscent of what we saw in the spinal cord in *Chapter 8*, where stimulation of one location will always bring the forepaw to the same location in space, regardless of its starting location. This also highlights a terminological ambiguity caused by differing conventions among scientists studying motor systems vs. sensory systems. A functional principle that has been ascribed to the SEF is that of employing a "place code," in that activity in one set of neurons always brings the center of gaze to the same place. In this context, "place coding" is comparable in meaning to "goal-directed." Recall that in the auditory system, in contrast, the *place* in place coding refers to the place along the length of the basilar membrane from which a fiber of the auditory nerve originates (section *Spectral decomposition performed by the basilar membrane, Chapter 5*).

THE CONTROL OF EYE MOVEMENTS, AND OF ATTENTION, IN HUMANS

Human oculomotor control

A classic task for assessing oculomotor control is the oculomotor delayed-response task, in which the subject begins by fixating a central location, the to-be-acquired location is cued by a flashed stimulus, but the subject must withhold the saccade to that location until the offset of the fixation point. This temporal separation allows for the dissociation of "sensory" activity related to presentation of the cue from motor activity prompted by fixation offset. (The interpretation of "delay-period activity" sometimes observed between cue offset and fixation offset is less straightforward, as we shall see further along in this chapter, and again in *Chapter 14*.) *Figure 9.4* illustrates this task, together with examples of activity measured from the monkey and the human. Note that in the human, there is robust delay-period activity in the intraparietal sulcus (IPS; one portion of which, IPS 2 [see *Figure 7.5*], is generally thought to be homologous to LIP in the monkey), as well

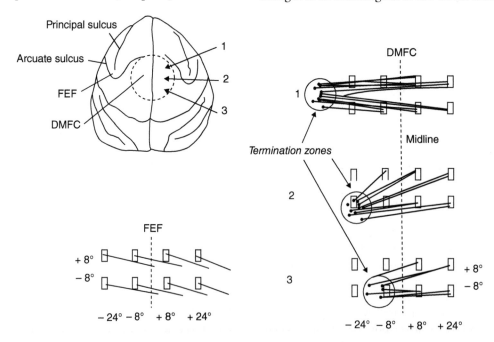

FIGURE 9.3 The effects of microstimulation of one neuron in FEF and of three neurons in SEF (referred to as DMFC by this author, the location of each indicated with an arrow, from each of eight starting fixation points). Source: From Schiller, Peter H. 1998. "The Neural Control of Visually Guided Eye Movements." In Cognitive Neuroscience of Attention: A Developmental Perspective, edited by John E. Richards, 3–50. Mahwah, NJ: Lawrence Erlbaum Associates. Reproduced with permission of Lawrence Erlbaum Associates.

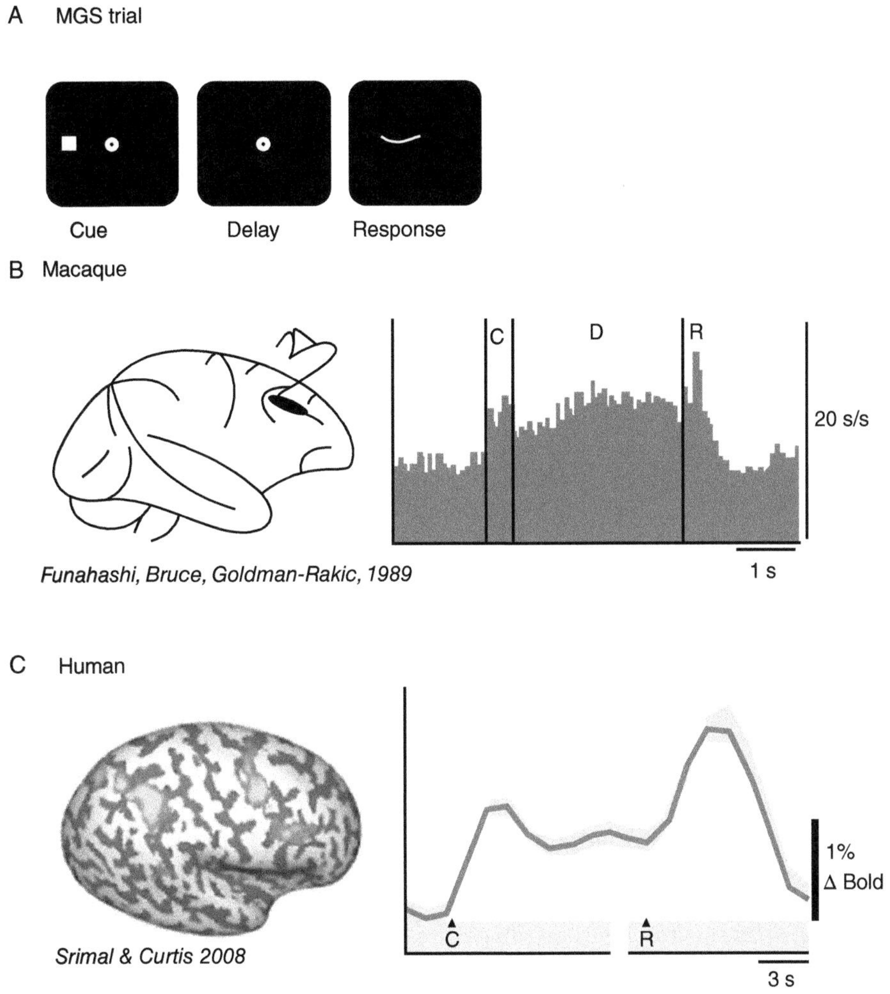

A   MGS trial

Cue   Delay   Response

B   Macaque

Funahashi, Bruce, Goldman-Rakic, 1989

20 s/s

1 s

C   Human

Srimal & Curtis 2008

1%
Δ Bold

3 s

FIGURE 9.4 The delayed-saccade task (also referred to as "memory-guided saccade [MGS]" or "oculomotor delayed response [ODR]"). Panel **A** illustrates a trial, as described in the text. Panel **B** illustrates the recording location in a monkey, in this instance at the caudal end of the principal sulcus, near the rostral boundary of the FEF, and activity recorded from a neuron at this location. Panel **C** illustrates regions with elevated delay-period activity in a human performing this task in an MRI scanner. The anatomical image has been "inflated," so that sulci (darker gray) are visible. Elevated activity is observed in most of the rostral IPS. The fMRI time-series data are interrupted between cue (C) and response (R) because the researchers used variable-length delay periods in this study. Note that this analysis was performed by involving a so-called "slow event-related design," in which individual trials are spaced out, typically by 10 seconds or more. To understand why, it might be helpful to turn back to the section of *Chapter 3*, on *Some considerations for experimental design*, and *Figure 3.20*. The data from the delayed-saccade task illustrated here (i.e., in *Figure 9.4*) needed to be handled in a manner most similar to the blocks featuring nine events occurring with a 15-second ISI (*Figure 3.20.A*), to minimize smearing of responses from one of the trial to others. A "linear deconvolution" analysis of the kind illustrated in *Figure 3.20.C* is not possible with a task like this delayed-saccade task, because the elements of the task can't be presented/performed in a randomized order; that is, the Cue must always precede the Delay, which must always precede the Response. Source: From Ikkai, Akiko, and Clayton E. Curtis. 2011. "Common Neural Mechanisms Supporting Spatial Working Memory." Neuropsychologia 49 (6): 1428–1434. doi:10.1016/j.neuropsychologia.2010.12.020. Reproduced with permission of Elsevier.

as at two distinct frontal foci (*Figure 9.4*, panel C). The human FEF has classically been associated with the more dorsal of the two, just rostral and ventrolateral to the point of intersection of the precentral sulcus and the superior frontal sulcus (sometimes referred to as the superior precentral sulcus [sPCS]). However, the more ventral focus of elevated activity in frontal cortex is also frequently observed. Note that although the frontal regions illustrated in *Figure 9.4* (panel C) show elevated activity during each epoch of the trial (*cue*, *delay*, *response*), these regions were identified statistically as having elevated activity during the delay period, regardless of whether or not they were differentially responsive during the *cue* and/or *response* epochs.

Human attentional control

Recall from *Chapter 7* that we restricted ourselves at that time to a detailed consideration of the *site* of the effects of attention, pushing off consideration of the *source* of attentional control to a later chapter. Well, *source*, your time has come.

From experimental psychology, we have long known that there are two general modes by which attention can be directed: the exogenously triggered *capture* of attention vs. the endogenously controlled *allocation* of attention. An example of an exogenously triggered capture of attention would be that you are sitting quietly, reading this book, when suddenly someone loudly, unexpectedly, bursts into the room. Your reflexive reaction is to turn your head in the direction of the noise. An example of the endogenous control of attention is when you are sitting quietly, pretending to be reading this book, but are, in fact, waiting for the imminent arrival of your roommate/boyfriend/girlfriend/spouse/whomever, and so are attending to the door, waiting for it to open at any moment.

The endogenous control of attention

Researchers at Washington University in St. Louis, led by neurologist Maurizio Corbetta and experimental psychologist Gordon Shulman, have used event-related fMRI to separate, for example, transient responses associated with a cue instructing the subject where to attend (i.e., to a location in the periphery) – or what to attend to (e.g., motion in a particular direction) – from the sustained activity associated with the top-down maintenance of attention at the cued location – or to the cued direction of motion. Importantly, on perceptually identical "passive viewing" trials, when subjects were told that there was no task to

perform, and that they merely had to watch the screen, the transient responses in posterior occipitotemporal regions were only slightly attenuated relative to what they had been in attention-demanding trials. This was consistent with what we have already seen at *sites* of the effects of attentional selection. Activity in rostral IPS and caudal superior frontal cortex, including the FEF, however, showed a very different pattern. It was markedly attenuated, almost to baseline levels, on passive trials. Also noteworthy was the fact that virtually identical patches of cortex in superior parietal and frontal regions showed elevated activity regardless of whether the sustained attention required by a task was location-based (e.g., *Detect the onset of a target at this location*) or feature-based (e.g., *Indicate when you see motion in this direction*). This led to the conclusion, consistent with results from many other studies performed in many other laboratories, that coordinated activity in this dorsal frontoparietal network acted as a *source* of top-down attentional control (*Figure 9.5*).

How might the dorsal frontoparietal system work? At the network level, projections from deep layers of cortex in these regions likely deliver the signals that manifest in sensory processing areas as baseline shifts (in the case of location-based attention) and/or as the biasing signal that influences the competition for representation of a behaviorally relevant stimulus vs. irrelevant items. (Revisit section *Effects of attention on neuronal activity*, *Chapter 7*, if the phenomena of the baseline shift and of biased competition aren't fluently springing to mind.) And at a functional level, many researchers have proposed that elevated activity in dorsal frontoparietal cortex (*Figure 9.5*) may reflect the operation of a priority map. The idea is that the brain maintains a representation of what's currently the most relevant for behavior, whether that be a representation of the location of potentially important stimuli in the environment, a representation of current behavioral priorities, or both.

Now, let's make this more concrete. Let's say that, I dunno, let's say you're new to Madison, WI, and are driving in a rental car from the airport, trying to navigate to this author's house where you are invited for dinner. As you drive along Old University Avenue, your eyes are, of course, on the road ahead of you. In your "mind" however, are additional representations: (1) you are searching for street signs at each intersection, so as to identify "Forest St." (which should be coming up on the left); (2) you are aware of a group of people standing near a crosswalk in front of you (*What's the law here? Do I have to stop and let them cross in*

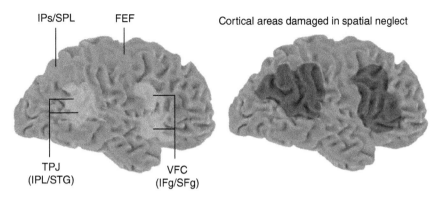

FIGURE 9.5 Two networks for the control of attention. At left, the networks associated with the endogenous (blue) vs. the exogenous (orange) control of attention. At right, regions whose damage is often associated with hemispatial neglect. Source: From Corbetta, Maurizio, and Gordon L. Shulman. 2002. "Control of Goal-Directed and Stimulus-Driven Attention in the Brain." Nature Reviews Neuroscience 3 (3): 201–215. doi: 10.1038/nrn755. Reproduced with permission of Nature Publishing Group.

front of me, or will they wait for a break in the traffic?); and, (3) what's up with the car behind you that is driving right up on your @&&? Thus, although your eyes are, indeed, on the road, your priority map will prioritize (1) the detection of green street signs; (2) movement into the crosswalk by a pedestrian; (3) evidence of trouble in the rearview mirror.

SUDDENLY, A BALL ROLLS OUT ONTO THE STREET IN BETWEEN TWO PARKED CARS JUST AHEAD OF YOU!

This unexpected event triggers what Corbetta and Shulman (2002) call the "circuit breaker," which overrides the influence of your current priority map on your behavior.

The exogenous control of attention

Exogenous control of attention occurs when something salient and unexpected occurs in the world around you, and the source is, therefore, *outside* of your brain. (The endogenous control of attention, in contrast, is considered more "cognitive," in the sense that your thoughts/plans/priorities/motivations are controlling your behavior. They are endogenous in the sense that they are generated within and maintained *inside* you, in your brain.) Corbetta and Shulman have documented that exogenous shifts of attention are associated with a right hemisphere-dominant network, ventral to the dorsal frontoparietal system, that includes the temporoparietal junction (TPJ) and the inferior frontal gyrus of the ventral prefrontal cortex (which they abbreviate as VFC). The ventral network is illustrated in orange in the brain image on

the left of *Figure 9.5*. Intriguingly, Corbetta and Shulman (2002) have noted that "the anatomy of neglect . . . matches the ventral TPJ–VFC system" (p. 212). (For one independent dataset generally consistent with this view [and generated several years later], revisit *Figure 7.4*.)

Next, we turn our attention to something that many readers will have already noticed, which is that the dorsal regions highlighted in panel C of *Figure 9.4*, which illustrates activity associated with a delayed-saccade task, show a remarkable overlap with the dorsal frontoparietal system illustrated in *Figure 9.5*.

THE CONTROL OF ATTENTION VIA THE OCULOMOTOR SYSTEM

Covert attention

There are many times in our lives when we want to attend to something, or someone, without directly looking at it (or him, or her). In high school math class there was that cute classmate on whom you had a crush, but at whom you didn't want to stare (never happened to me, but I've heard stories . . .). At the dinner table, one wants to convey to one's kids that one trusts them, and so doesn't want to appear to be closely monitoring their every move, yet one also wants to closely monitor their every move (this one I readily confess to . . .). These are examples of covert attention. (Overt attention is when you look directly at the object of your attention.) Sustained covert attention may be related to another attentional phenomenon, which is that there are considerable data from experimental psychology, and from neurophysiology, that the focus of attention jumps to the target of the next saccade tens of milliseconds prior to the initiation of the saccade itself. Therefore, might it be the case that the sustained allocation of covert attention to a location/object amounts to a pre-saccadic shift of attention, but one that does not culminate in the saccade itself?

Covert attention as motor preparation?

This last bit of speculation is what is posited by the **premotor theory of attention**, articulated in the 1980s by Giacomo Rizzolatti (of mirror neuron renown). It argues, in summary, that attention may fundamentally be "nothing more" than motor planning [my quotes]. Here's how Rizzolatti and Craighero (2010) have summarized it:

> Classically, spatial attention was thought of as a dedicated supramodal control mechanism, anatomically distinct from the circuits underlying sensorimotor processing . . . [However] there is no need to postulate two control mechanisms, one for action and one for attention. According to [the premotor theory,] spatial attention does not result from a dedicated control mechanism, but derives from a weaker activation of the same frontal-parietal circuits that, in other conditions, determine motor behavior toward specific spatial locations.

Empirical evidence for covert attention as motor preparation

Tirin Moore and his colleagues, first at Princeton University in the laboratory of Charlie Gross, and subsequently at his own lab at Stanford University, have carried out several experiments generating results that are broadly consistent with the premotor theory of attention. Here, we'll review one that he carried out at Princeton with Mazyar Fallah, that specifies in considerable detail how the oculomotor control system is involved in the endogenous control of spatial attention. The first step of the experiment by Moore and Fallah (2001) was to find a neuron in FEF whose microstimulation produced a repeatable saccade to the same location (e.g., *Figure 9.3*). This is referred to as the cell's motor field (by analogy to the sensory receptive field; panel A in *Figure 9.6*). Next, the layout of the covert attention task was arranged so that the target stimulus was located in that neuron's motor field (panel B

in *Figure 9.6*). In this task, the monkey had been trained to depress a lever to start the trial, and maintain central fixation while attending to the target. When the target dimmed, the monkey needed to release the lever in order to receive its reward. The dimming of the target was made difficult to detect by making the luminance change very small and very brief, by making the timing of this event unpredictable, and by including catch trials in which the target never dimmed. (On these the monkey was rewarded for holding the lever throughout; importantly, no reward was delivered on catch trials during which the animal committed false-alarm errors by releasing the lever.) Finally, as if that wasn't already enough, most trials also featured the concurrent presentation of flashing distractors throughout the trial. (Distractors flashed on and off, one at a time, in the locations illustrated with "sunbursts" in panel B in *Figure 9.6*)

Now for the really cool part. Recall that the target is positioned within the motor field of the FEF neuron. This means that, if the investigator wanted to, he could deliver a 100-ms train of ~ 50 μA of current and drive the eyes to the target location. Instead, they did what Tirin Moore sometimes characterizes as "just tickling" the neuron, by stimulating at half the amplitude that had been determined to produce a saccade 50% of the time. That is, the stimulation was subthreshold, in that it never produced an eye movement.

But would it – drumroll please – *produce a covert shift of attention?*

To assess this, the investigators varied the latency between when microstimulation was applied and when the target event occurred, a parameter called the "stimulation onset asynchrony." The findings suggested that electrically "tickling" an FEF neuron does, indeed, produce an effect equivalent to a covert shift of attention, and that it does so in a timing-dependent manner: the closer in time the microstimulation was to the target event, the stronger were its attention-like effects. These results, plotted in panel C of *Figure 9.6*, express the effect of microstimulation on the animal's "relative sensitivity," which is a measure of how much smaller the magnitude of the dimming needed to be made for the monkey to perform at the same level as it did on no-stimulation trials. Thus, in effect, what the plot shows is that delivering microstimulation with a stimulation onset asynchrony of 50 ms improved the animal's performance by roughly 17.5%. For comparison, the blue arrow shows that turning off the distracting stimuli improved the monkey's performance by ~ 21%.

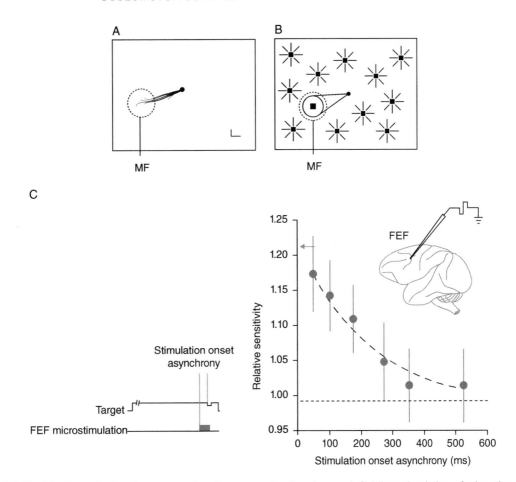

FIGURE 9.6 Electrically controlling the source of endogenous attentional control. **A**. Microstimulation of a location in the FEF of a monkey produces reliable vector saccades into a motor field (MF). **B**. The array of the visual search task is customized for that neuron, such that the target location falls within this MF. **C**. Microstimulation-related improvement in performance varied monotonically with the stimulation onset asynchrony (see text). Source: From Awh, Edward, Katherine M. Armstrong, and Tirin Moore. 2006. "Visual and Oculomotor Selection: Links, Causes and Implications for Spatial Attention." Trends in Cognitive Sciences 10 (3): 124–130. doi: 10.1016/j.tics.2006.01.001. Reproduced with permission of Elsevier Ltd.

In effect, instead of overtly cuing the animal about an impending target event, as is conventionally done with variants of the "Posner (1980) task," Moore and Fallah (2001) "reached into its brain" and directly shifted the animal's attention!

Because FEF microstimulation produced attention-like effects on behavior, the inference from Moore and Fallah (2001) was that subthreshold activity in the FEF acts as a source of attentional control. To fully understand how FEF might carry out this function, the next step was to show that FEF microstimulation produces attention-like effects in activity at a site of attentional control. To do this, Moore and then-graduate student Katherine Armstrong painstakingly found FEF neurons and V4 neurons whose motor

fields and receptive fields, respectively, overlapped. In this way, they were able to stimulate at a putative source of attentional control and simultaneously record from a site where attention influences visual processing. *Figure 9.7* illustrates that the effect of FEF stimulation was to boost the V4 response to a visually presented stimulus in a manner similar to the effects of selective attention that we considered in *Chapter 7*. (For example, compare the effect of FEF microstimulation in *Figure 9.7* to the effect of spatial attention in *Figure 7.11.B*.) Importantly, in this study *the monkey was not performing a task*, other than maintaining central fixation. Thus, together with the findings from the Moore and Fallah (2001) study, Moore and Armstrong (2003) offers a remarkable demonstration of the ability to

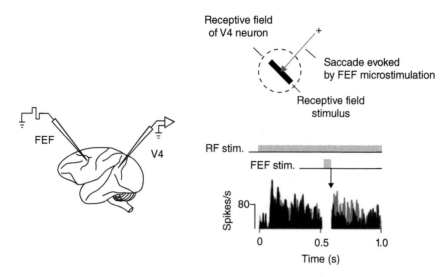

FIGURE 9.7 Generating the cause and measuring the effect of attentional control. Illustration at the top conveys the logic of the experiment: while the monkey is fixating the "+," apply subthreshold microstimulation to a location in the FEF where suprathreshold microstimulation would generate a saccade (red arrow) that would land in the receptive field of the V4 neuron that is simultaneously being recorded from (dashed circle), and evaluate the effect of this subthreshold microstimulation on the activity of this V4 neuron in response to a visual stimulus presented in that neuron's receptive field (the oriented black bar). Bottom panel is activity in the V4 neuron averaged across 10 trials when FEF microstimulation was delivered (red histogram), and 10 control trials when no FEF microstimulation was delivered (black histogram). Average activity on the two types of trials is identical from time 0 seconds, when the stimulus appears in the neuron's receptive field (RF stim.) until shortly after time .5 seconds, at which time the effect of FEF microstimulation is to boost the sensory-evoked response of this neuron. Source: From Awh, Edward, Katherine M. Armstrong, and Tirin Moore. 2006. "Visual and Oculomotor Selection: Links, Causes and Implications for Spatial Attention." Trends in Cognitive Sciences 10 (3): 124–130. doi: 10.1016/j.tics.2006.01.001. Reproduced with permission of Nature Publishing Group.

control the focus of attention via a causal manipulation of activity in the FEF, suggesting that covert eye-movement planning is an important source of the endogenous control of attention.

Where's the attentional controller?

Remember in *Chapter 8*, when a skeptic chimed up about "some little homuncular mathematician perched up on the precentral gyrus . . . calculating vector sums prior to each movement"? At this point they might chime in again, asking sarcastically if we've placed the spotlight operator (from *Chapter 7*'s *Day at the circus*) in the FEF. Stilted rhetorical devices aside, this is a legitimate concern, and one about which the cognitive neuroscientist and, particularly, the theoretician, needs always to be mindful. *Where does this control come from? Does assigning it to the dorsal frontoparietal system make it any less mystical, any more material? Have we accomplished anything more than assigning a more specific*

address to the ghost in the machine? The answer is that, in fact, in the domain of attentional control, and in many others, cognitive neuroscience has made important strides toward developing very explicit accounts of how control emerges as a consequence of network interactions. In these models, there is no one place from which control signals are generated de novo, no figurative "corner office" from which a neural executive spontaneously decides to issue commands. Such models are often most convincingly achieved via computational simulations. Although such simulations invariably include some idiosyncratic choices made by the modeler that may not account for every detail that we (think we) know about how the brain works, their value is in demonstrating that it is possible, in principle, to understand, to describe, and to measure every step, every connection, every causal factor in a system capable of producing complex behavior. For the question of the source of attentional control, we can consider a computational model by the German theoretician Fred Hamker.

The reentry hypothesis

The goal of the Hamker (2005) model was to simulate the findings of Chelazzi, Duncan, Miller, and Desimone (1998; *Figure 7.15*) that provide an important empirical foundation for the biased-competition model of visual attention. In particular, this model provides an explicit account for how the FEF and IT act as sources of the top-down attentional control that provide the biasing signal. Additionally, by specifying and explicitly modeling the activity of different classes of neurons in the FEF, it provides a plausible scenario for how a saccade plan might be prevented from actualization in a saccade, and, thus, could also serve as the basis for sustained covert attention. (Hint: it would involve the fixation neurons.)

Figure 9.8 illustrates the architecture of the model (panel B), and the anatomical connections on which it is based (panel A). One of the features of this model, which we haven't considered up to this point, is the heterogeneity of neuronal classes within FEF. Not unlike the deep layers of the superior colliculus, within FEF one can find visually responsive neurons, fixation neurons, visuomotor neurons, and movement neurons. Three of these are explicitly represented in the model (panel B in *Figure 9.8*), as are their temporal dynamics. Visuomovement cells in deep layers are active from stimulus onset until saccade execution. Typically their initial response does not distinguish between distractor or target, but the activity decays when

a distractor is in the receptive field. Movement cells are active prior to saccades and do not show any response to stimulus onset. Fixation cells decrease their activity before a planned eye movement (Hamker, 2005, pp. 433–434).

Now, how does this network implement attentional control? We can watch it happen, in *Figure 9.9*, on a (simulated) millisecond-by-millisecond basis. This figure can be thought of as simulating a visual search experiment like the one illustrated in *Figure 7.15*, in which the experimenters are simultaneously recording from the dendritic branches of a V4 neuron that is receiving signals from earlier stations in the visual system ("Input" plot), from the V4 neuron itself ("Gain modulated rate" plot), from a neuron in IT (plot with axis labeled "IT target firing rate"), and from an FEF movement neuron.

1. Not shown is the beginning of the trial, when initial presentation of the target object establishes the search template for that trial, which is retained in "working memory" in recurrent connections between PF and IT (between "Working memory" and "Target" populations) until the search array is presented. (For a cartoon of the task, see *Figure 7.15.A*.)

2. When one stimulus of the search array appears within the simulated receptive field, increased input elevates activity within the V4 neuron at roughly time 60 ms.

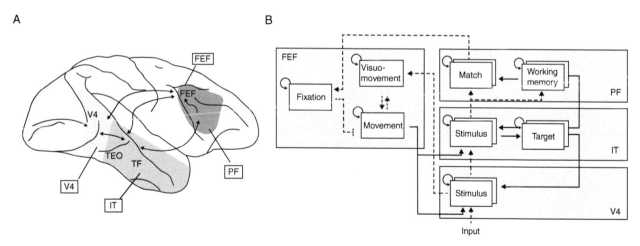

FIGURE 9.8 Illustrations from Hamker's simulation of the source of top-down control. Panel **A**, the known neural connections included in the model. Panel **B**, a schematic of the implementation of the model. Bottom-up connections are indicated in dotted arrows, feedback ("reentrant") connections in solid arrows. Each white box represents a population of neurons, "stacked" boxes indicating populations processing two stimulus features (color and form) in parallel. The "turning-back-on-itself" arrow at the upper-left corner of each of the boxes indicates that individual units within a population exert lateral inhibition on each other. Source: From Hamker, Fred H. 2005. "The Reentry Hypothesis: The Putative Interaction of the Frontal Eye Field, Ventrolateral Prefrontal Cortex, and Areas V4, IT for Attention and Eye Movement." Cerebral Cortex 15 (4): 431–447. doi: 10.1093/cercor/bhh146. Reproduced with permission of Oxford University Press.

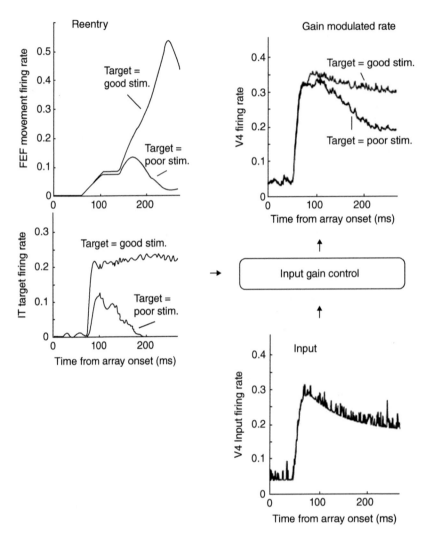

FIGURE 9.9 Multiple windows on the temporal evolution of activity occurring simultaneously at multiple stages of the simulated oculomotor system from *Figure 9.8*, for a V4 receptive field/FEF movement field processing information about a stimulus that is the V4 neuron's "good" stimulus on trials when this stimulus is the target of the visual search (Target = good stim.) vs. when it is not (Target = bad stim.; definitions in *Figure 7.15* and associated text). The "Input" box shows the stimulus-driven input to a V4 neuron from visual cortex (i.e., the aggregation of feedforward signals from V1, VP, and V3). The "Input gain control" implements the **convolution** of this Input signal with the "Reentry" signals from FEF and IT (i.e., it's the operation in the model that computes the effects of the reentrant signals on the Input) and the "Gain modulated rate" shows the resultant activity in V4 that feeds forward to IT. Source: From Hamker, Fred H. 2005. "The Reentry Hypothesis: The Putative Interaction of the Frontal Eye Field, Ventrolateral Prefrontal Cortex, and Areas V4, IT for Attention and Eye Movement." Cerebral Cortex 15 (4): 431–447. doi: 10.1093/cercor/bhh146. Reproduced with permission of Oxford University Press.

This V4 activity feeds forward to FEF, producing the initial ramp-up in the *FEF movement firing rate* that begins at ~ 75 ms into the trial. Note that, during the initial ramp-up of activity in V4, this activity doesn't yet differ as a function of whether or not the presented stimulus is the Target (i.e., if it is to be attended *Target = good stim.* or not *Target = bad stim.*).

3. Meanwhile, the neuron in IT does have information (from working memory) about the identity of that trial's Target, and so its early activity to the stimulus

does differentiate whether or not the stimulus is the Target.

4. [First wave of *reentry*] Feedback from IT to V4 produces an initial little bump in *Target = good stim.* activity in the "Gain modulated rate" of V4 that begins just prior to 100 ms.

5. Although the receptive field of any one V4 neuron is relatively large, the aggregate result of feedforward activity from a large population of them is that visuo-movement neurons in FEF that encode the region of the visual field containing the target will receive the strongest feedforward drive (a phenomenon reminiscent of population coding in M1). The visuomovement neurons in FEF receiving the strongest feedforward drive will, in turn, activate the FEF movement neurons whose movement field overlaps the location of the target. The accumulating effect of the small differential feedforward drive from V4 begins to emerge in FEF activity at ~ 130 ms.

6. [Second wave of *reentry*] Increasing activity in FEF movement neurons is experienced by V4 neurons as top-down feedback, and those representing features of the target, which already have a slightly elevated level of activity, benefit from this top-down boost to further increase their activity and, in turn, apply lateral inhibition to V4 neurons representing features of the nontarget stimulus.

7. These processes continue, in parallel, until a threshold is reached and a saccade to the target triggered.

The point of walking through this example has been to show how attention-related feedback from FEF can be generated without FEF "knowing" what the search target is. Thus, no "ghost" need live in the "machine" of the FEF for this region to act as a source of top-down attentional control. More broadly, it gives a very detailed example of how the dynamic interplay of regions within a recurrently connected network can give rise to sophisticated behavior. (This is at least schematically relevant, for example, for thinking about the recurrent activity that many believe is critical for visual object recognition, as considered in section *A critical role for feedback in the ventral visual processing stream, Chapter 6*.)

What we haven't done, however, is satisfactorily address the question of *Where is the attentional controller?* That is, were we to stop here, one could object that all that this exercise has done is to relocate the homunculus from the FEF to the box in *Figure 9.8* that's labeled "working memory," because it's working memory that enabled IT, in our step 3 of walking through *Figure 9.9*, to respond differentially on *Target = good stim.* vs. on *Target = bad stim.* trials. This was, after all, the event that provided the initial "nudge" of biasing the activity in the system toward ultimately producing a saccade to the Target on *Target = good stim.* trials. Fair enough. But we can't expect to do something so ambitious as banning the homunculus in just nine chapters! (We will have 11 more to go!) When we get to *Chapter 14, Working memory* and *Chapter 15, Cognitive control*, we'll return to this question, and we'll see that combining some of the properties of recurrent systems that we've considered here with mechanisms from reinforcement learning (section *Habits and reinforcement learning, Chapter 8*) might just enable us to ban the homunculus from working memory, too.

One final note about top-down signaling from the FEF before we move on. Although the empirical work from Moore and colleagues (2001, 2003) that we've reviewed here, and the modeling from Hamker (2005), only take into account firing rates, there is a growing body of electrophysiological evidence that dynamic changes in oscillatory phase synchrony between regions also play an important role in attentional control. For example, there's a considerable amount research suggesting that the efficacy of FEF influence on posterior regions may depend, in part, on long-range synchrony between these regions in the beta band. This is also something that we'll consider explicitly when we get to *Chapter 15, Cognitive control*.

ARE OCULOMOTOR CONTROL AND ATTENTIONAL CONTROL REALLY THE "SAME THING"?

So, we've seen evidence that FEF can act as a source of attentional control when it's electrically stimulated, and we've considered a model describing how it might come to serve this function. But is there evidence that this *is* how attention is controlled in humans? That is, can the endogenous control of attention truly be reduced to "nothing more" than covert oculomotor control? Although this is a difficult question to answer unequivocally, we'll conclude this chapter considering one way of addressing it.

The "method of visual inspection"

As noted previously, there is a striking similarity between the dorsal frontoparietal regions activated in the delayed-saccade task of Srimal and Curtis (2008; *Figure 9.4*) and the endogenous control network of Corbetta and Shulman (2002; *Figure 9.5*). Indeed, there have been many demonstrations in the literature that tasks engaging these two categories of behavior activate overlapping regions of the brain. However, there's only so far that one can get by, in effect, saying, *See? These two patterns of activity look the same.* The reason for this is that brain activity – whether considered at the level of a single neuron, a cortical column, a population of 100 or so neurons, a 3-mm^3 fMRI voxel containing tens of thousands of neurons, or a whole-brain fMRI volume comprising ~ 30,000 such voxels – is simply too complex to be interpretable by visual inspection. Thus, for example, one cannot compare the patterns of activity illustrated in *Figure 9.4* and *Figure 9.5* and conclude, with any certainty, that the brain is "doing the same thing" during the two tasks that produced these images. Why? First, one can find many examples of experiments in which these same regions are active, but subjects are not performing either a spatial attention task or an oculomotor task. The IPS, for example, also shows elevated activity during tests of mental arithmetic and numerosity, of learning sequences of finger movements, of "understanding the goals of others," of perceptual decision making, and so on. Second, what if activity that you can't see determines the function that IPS carries out? Perhaps there's different "unseen" activity happening in *Figure 9.4* vs. *Figure 9.5* (recall, from *Figure 3.19*, the uncolored parts aren't inert, their activity simply isn't exceeding a statistical threshold). Simply put, there is no brain area whose activity is specific for any single cognitive function.

Another way to think about this is that a limitation of the "visual inspection" approach is that it is inherently univariate. That is, we look at the region of elevated activity in IPS in *Figure 9.4* and can only watch it vary along one dimension: its signal intensity either increases or decreases. This can lead to thinking of it as one "activation" that is doing only one thing.

Finally, *these two patterns of activity look the same* isn't really what we care about. To address the question posed in the title of this section, what one would really want to be able to ask is: *Is the brain actually performing the same computations during the performance of these two tasks?* One approach that offers a way to get closer to being able to ask this question is to apply one of the relatively recently developed methods of multivariate pattern analysis (MVPA) to one's neuroimaging data. After first introducing some of the principles of MVPA, we'll wrap up this chapter by considering how one study has used it to ask this fundamental question.

MVPA of fMRI data

Let's consider the hypothetical example illustrated in *Figure 9.10*, in which there are two voxels in a reduced dataset, voxel *F* and voxel *P*. On 10 trials our subject attends to the left, and on 10 trials she attends to the right. The first analysis that we carry out is to "collapse across" all the voxels in our ROI. That is, to average their signal together, then compare the spatially averaged signal from the 10 rightward trials vs. the spatially averaged signal from the 10 leftward trials. The bar graph in panel A of *Figure 9.10* illustrates that one can detect no appreciable difference between the two conditions. If we consider the two voxels separately, i.e., we determine the average signal from voxel *F* separately for the 10 leftward trials and the rightward trials, and do the same for voxel *P*, we seem to do a little bit better: keeping in mind that these are made-up data, paired *t*-tests indicate that voxel *F* shows significantly greater activity for leftward than rightward trials, and voxel *P* the converse. Nonetheless, each voxel also shows robust activity for its "nonpreferred" direction. As a consequence, if we were presented with data from a 21st trial, we'd have difficulty classifying it as a leftward or a rightward trial if we only knew that voxel *F*'s activity on this trial was, say, 0.6. To illustrate this, consider panel C of *Figure 9.10*, which shows each voxel's response on each of our 20 trials (these are the data that produced plots A and B): where, on the voxel *F* plot, would you draw the line above which you'd be confident that a trial was a leftward trial?

The two "stacks" of data points plotted in panel C of *Figure 9.10* are 1-D values – they only vary along a line. Panel D of *Figure 9.10* illustrates how the same data used to generate the plots in A, B, and C can be redepicted as data that vary along *two* dimensions. To create the plot in panel D of *Figure 9.10*, all we've done is rotate the plot of voxel *P*'s activity – from panel C – from vertical to horizontal. That permits us to "project the data from each trial into a higher-dimensional space." (Sounds highfalutin, I know, but it *is* what we are doing; 2 is greater than 1.) Put more simply, it permits us to plot what each trial's values were at the two voxels simultaneously. As a result, it's easier to decide "where to draw the line": if the hypothetical 21st

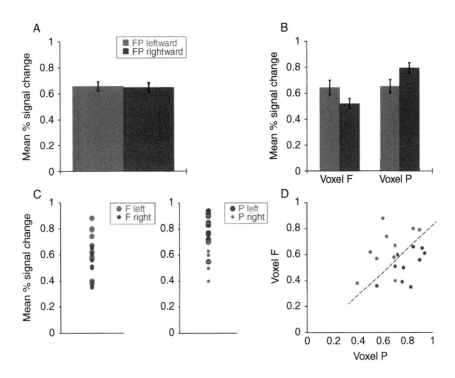

FIGURE 9.10 Improved sensitivity via MVPA. Panel **A** illustrates the data for the two conditions, averaged across voxels *F* and *P*. In panel **B**, the data are broken out by voxel. The plots in panel **C** illustrate the signal intensity from each trial, for each voxel. Panel **D** is a multivariate representation of the data from panel **C**, with a discriminant line added to illustrate how one would classify a 21st trial, depending on which side of the line it falls. Source: © Bradley R. Postle.

trial falls above and to the left of the discriminant line that cuts through the cloud of data points, it's probably a leftward trial.

With real fMRI datasets, this same principle is applied, but the assertion that one is "projecting the data into a high-dimensional space" no longer elicits snickers, because, with, say, 300 voxels in an ROI, one is working in 300-D space. Within this space, a machine learning algorithm would seek to find the 299-D hyperplane that, like our dotted line in panel D of *Figure 9.10*, would provide the best discrimination between conditions A and B. The higher sensitivity afforded by MVPA will be exploited in many experiments that we will consider from here on out.

"Prioritized maps of space in human frontoparietal cortex"

And so, *is the brain actually performing the same computations during the performance of tasks that require oculomotor control vs. visual attention* (and, for good measure) *vs. visual working memory?* A study from the laboratory of Clayton Curtis at

New York University addressed this question by scanning subjects with fMRI while they performed tests of "intention," "attention," and "memory." First, the authors carried out an fMRI scanning session to identify cortical regions supporting retinotopic maps of space, using a method similar to the one illustrated in *Figure 7.5.A*. The regions identified were largely overlapping with those shown in *Figure 9.4*, and they used the same heuristic used by Saygin and Sereno (2008) to identify boundaries between regions (section *The functional anatomy of the dorsal stream, Chapter 7; Figure 7.5.A*), which defined the ROIs to be used in the second part of the experiment: IPS0; IPS1; IPS2; IPS3; superior precentral sulcus (sPCS); and inferior precentral sulcus (iPCS). (Note that sPCS, IPS3, and IPS2 make up the dorsal attention network [DAN] as identified with resting state functional connectivity [*Chapter 3, Figure 3.18*].) In the second scanning session, subjects performed the three spatial cognition tasks. Trials for the test of motor intention began with the presentation of a central fixation point, followed by a target stimulus located anywhere from 5° to 15° of visual angle to the left or the right of fixation, and anywhere from 4° to 5° above or below the horizontal meridian. Both the target and the fixation point remained visible during a variable-length delay period, and subjects were to make a saccade to the target as soon as the fixation point disappeared. Trials for the test of spatial attention started with a fixation point flanked by one shape (a square or a circle) 5.5° to the left of fixation and the other shape to the right 5.5°, each one centered on the horizontal meridian. The fixation point was then replaced by either a square or a circle, which cued the subject to attend to the flanking shape that matched it, and count the number of times the cued flanking shape dimmed during the ensuing variable-length delay period. Trials for the test of spatial delayed recognition began with the presentation of a sample stimulus anywhere from 5° to 15° to the left or the right of fixation, and anywhere from 4° to 5° above or below the horizontal meridian. After a variable-length delay period, a probe stimulus appeared in a location that was either the same as or close to the location where the sample had appeared, and the subject indicated whether probe's location was a match or a nonmatch with that of the sample. To address the question of *is the brain actually doing the same thing during the performance of these* three *tasks?*, the authors used a clever application of MVPA and produced a remarkable result.

A common use of MVPA is as a tool for "decoding" brain activity. Consider the toy example illustrated in *Figure 9.10*. After plotting the results of 20 trials of leftward vs. rightward attention, we felt reasonably confident that we would be able to classify the identity of a 21st trial, based solely on the neural activity associated with this trial. That is, we felt reasonably confident that we could *decode* the neural activity so as to know what information it was representing. What Jerde et al. (2012) did with their data, as illustrated in *Figure 9.11*, is a step beyond "just" predicting the identity of a 21st trial drawn from the same task that produced the first 20 trials. Rather, what they did was first train a classifier to discriminate between *attend-left* and *attend-right* trials on the attention task, and then ask if this "attention classifier" could also successfully classify trials from the memory and the intention tasks.

To be more concrete, let's walk through an example from the sPCS ROI. Each subject performed 64 trials of each task (32 leftward and 32 rightward). Furthermore, let's say, for the purpose of this walk-through, that the sPCS ROI contained 200 voxels. Training the classifier would then begin by generating a cloud of 32 points in 200-D space that correspond to delay-period activity from each *attend-left* trial and a cloud of 31 points in this same

200-D space that correspond to delay-period activity from all but one of the *attend-right* trials and then by determining the 199-D hyperplane that separates the most *attend-left* points from *attend-right* points. The effectiveness of this "fold" of training would then be tested by plotting the point in this same 200-D space of the delay-period activity from the held-out trial and determining whether it falls on the correct or the incorrect side of the 199-D hyperplane separating *attend-left* from *attend-right*. If this trial was classified as an attend-right trial, the test of this fold would be scored as "correct." The process would then be repeated 63 more times, each time with a different trial held out from the cloud-plotting and hyperplane-fitting (a process referred to as "k-fold cross validation). After 64 iterations, the mean performance of the classifier (i.e., the number of correct classifications divided by 64) was calculated, and it was determined if this level of performance was above chance. (Note that, in practice, the cloud-plotting and hyperplane-fitting was carried out with logistic regression.) In *Figure 9.11*, for sPCS, the mean performance for classifiers trained and tested on attention trials, using the procedure described here, is indicated for each of the four subjects in this study as a blue dot inside the blue box. The clever application referred to a few paragraphs previously is what they did next, a procedure known as cross-task decoding. First, an attention classifier was trained by plotting the data from all 64 trials of the attention task into the 200-D space corresponding to the sPCS ROI, and determining the 199-D hyperplane that best discriminated *attend-left* from *attend-right*. Next, the 200-D point from each of the 64 memory trials was tested by determining for each whether, for example, a *remember-right* trial would be classified as belonging to the *attend-right* cloud of points (correct) or to the *attend-left* cloud of points (incorrect). In *Figure 9.11*, for sPCS, the mean performance for classifiers trained on attention trials and tested on memory trials is indicated for each of the four subjects as a green dot inside the blue box.

Now that we have a feel for the logic of the analysis of the Jerde et al. (2012) inspection of *Figure 9.11* reveals two general trends in the results. First, within-task decoding was only consistently successful in sPCS and IPS2. Second, cross-task decoding was successful only in sPCS and IPS2. This latter result is especially intriguing, because it provides compelling evidence for the idea that the brain may be performing the same computations for spatial attention (and for spatial working memory) as it performs for oculomotor control, actually doing the same thing during the performance of both of these tasks. Stated another way,

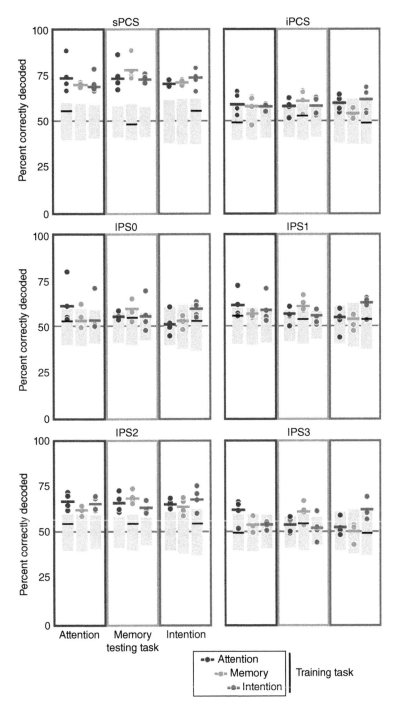

FIGURE 9.11 Classification results from Jerde et al. (2012) of data from tests of "attention," "memory," and "intention." Each dot corresponds to a subject. The color of each set of dots indicates the task used to train a classifier; the color of each rectangle around a set of dots indicates the task used to test the classifier. Thus, for example, the upper-left rectangle illustrates that, within sPCS, not only were *attention* trials decodable by the attention-trained classifier, but so, too, were *memory* trials and *intention* trials. Source: From Jerde, Trenton A., Elisha P. Merriam, Adam C. Riggall, James H. Hedges, and Clayton E. Curtis. 2012. "Prioritized Maps of Space in Human Frontoparietal Cortex." Journal of Neuroscience 32 (48): 17382–17390. doi: 10.1523/JNEUROSCI.3810-12.2012. Reproduced with permission of Society of Neuroscience.

"oculomotor control," "spatial attentional control," and "spatial working memory" may be three verbal labels that we use in different behavioral contexts but that, in fact, engage the same underlying neural activity.

OF LABELS AND MECHANISMS

As humans, we like (need?) to categorize things. I just traded in my son's "hybrid" bicycle (with properties of a "street bike" and of a mountain bike) for a (expensive!) mountain bike. To a nonexpert, the two would seem to be only superficially different from each other, but to my 11-year-old expert, they're as different as are a minivan from a sports car. To him, and to the biking community, they belong in different categories. To our intuitions, the act of moving one's eyes and of choosing to focus one's attention on object A vs. object B may feel like two different kinds of behavior. Certainly, up until quite recently,

psychology and neuroscience have treated these as two different domains. But what if, as the data that we reviewed in this chapter would suggest, it is the very same underlying neural systems and processes that produce these two putatively different kinds of behavior? Surely, at least on the neural level, they can no longer be assigned to different categories. This is one of the reasons why so much attention, and often debate, is focused on seemingly arcane questions such as *Are PPC neurons encoding* intention *or* attention?, or *Are the attention-like results from Moore and colleagues (2001, 2003) attributable solely to the activation of sensory neurons in the FEF, or do movement neurons also contribute?* This tension between different levels of description will reoccur often during the remainder of our time together. Just one example is manifested in *Figure 9.11*, which impels us to include spatial working memory in this discussion of neural mechanisms vs. psychological categories, even though we're not "supposed" to discuss it in detail until *Chapter 14.*

END-OF-CHAPTER QUESTIONS

1. What are some important differences between the posterior and anterior oculomotor control systems?

2. There are at least three different neural codes used to encode saccade metrics at different stations of the oculomotor system. Name the three codes, and the region(s) associated with each.

3. Define the exogenous vs. the endogenous control of attention. What cortical networks are associated with each?

4. How has the development of event-related methods for fMRI been critical for the study of oculomotor control and of attention? Describe one example from each domain.

5. Summarize the data that implicate the FEF as an important *source* of attentional control.

6. How does explicit computational modeling of the network interactions underlying visual search dispel the need to locate an attentional control homunculus in dorsal frontoparietal cortex?

7. How does MVPA differ fundamentally from neuroimaging data analysis methods that predominated during the 1990s and the 2000s? What advantages does MVPA offer?

8. Are "oculomotor control" and "attentional control" just two different labels that we have given to the same underlying neural mechanisms depending on the behavioral context in which they are used, or are there fundamental differences between the two? Defend your answer with data.

REFERENCES

Chelazzi, Leonardo, John Duncan, Earl K. Miller, and Robert Desimone. 1998. "Responses of Neurons in Inferior Temporal Cortex during Memory-Guided Visual Search." *Journal of Neurophysiology* 80 (6): 2918–2940.

Corbetta, Maurizio, and Gordon L. Shulman. 2002. "Control of Goal-Directed and Stimulus-Driven Attention in the Brain."

Nature Reviews Neuroscience 3 (3): 201–215. doi: 10.1038/nrn755.

Ferrier, David. 1875. "The Croonian Lecture: Experiments on the Brain of Monkeys (Second Series)." *Philosophical Transactions of the Royal Society of London.* 165: 433–488. doi: 10.1098/rstl.1875.0016.

Ferrier, David. 1890. *The Croonian Lectures on Cerebral Localisation.* London: Smith, Elder and Co.

Gandhi, Neeraj J., and Husam A. Katnani. 2011. "Motor Functions of the Superior Colliculus." *Annual Review of Neuroscience* 34: 205–231. doi: 10.1146/annurev-neuro-061010-113728.

Hamker, Fred H. 2005. "The Reentry Hypothesis: The Putative Interaction of the Frontal Eye Field, Ventrolateral Prefrontal Cortex, and Areas V4, IT for Attention and Eye Movement." *Cerebral Cortex* 15 (4): 431–447. doi: 10.1093/cercor/bhh146.

Ikkai Akiko, and Clayton E. Curtis. 2011. "Common Neural Mechanisms Supporting Spatial Working Memory." *Neuropsychologia* 49 (6): 1428–1434. doi: 10.1016/j.neuropsychologia. 2010.12.020.

Jerde, Trenton A., Elisha P. Merriam, Adam C. Riggall, James H. Hedges, and Clayton E. Curtis. 2012. "Prioritized Maps of Space in Human Frontoparietal Cortex." *Journal of Neuroscience* 32 (48): 17382–17390. doi: 10.1523/JNEUROSCI.3810-12.2012.

Krauzlis, Richard J. 2005. "The Control of Voluntary Eye Movements: New Perspectives." *Neuroscientist* 11 (2): 124–137. doi: 10.1177/1073858404271196.

Moore, Tirin, and Katherine M. Armstrong. 2003. "Selective Gating of Visual Signals by Microstimulation of Frontal Cortex." *Nature* 421 (6921): 370–373. doi: 10.1038/nature 01341.

Moore, Tirin, and Mazyar Fallah. 2001. "Control of Eye Movements and Spatial Attention." *Proceedings of the National Academy of Sciences of the United States of America* 98 (3): 1273–1276. doi: 10.1073/pnas.98.3.1273.

Paré, Martin, and Douglas P. Munoz. 1996. "Saccadic Reaction Time in the Monkey: Advanced Preparation of Oculomotor Programs Is Primarily Responsible for Express Saccade Occurrence." *Journal of Neurophysiology* 76 (6): 3666–3681.

Posner, Michael I. 1980. "Orienting of Attention." *Quarterly Journal of Psychology* 32 (1): 3–25. doi: 10.1080/00335558 008248231.

Rizzolatti, Giacomo, and Laila Craighero. 2010. "Premotor Theory of Attention." *Scholarpedia* 5 (1): 6311. doi: 10.4249/scholarpedia.6311.

Schiller, Peter H. 1998. "The Neural Control of Visually Guided Eye Movements." In *Cognitive Neuroscience of Attention: A Developmental Perspective*, edited by John E. Richards, 3–50. Mahwah, NJ: Lawrence Erlbaum Associates.

Sherman, S. Murray, and Ray W. Guillery. 2006. *Exploring the Thalamus and Its Role in Cortical Function.* Cambridge, MA: MIT Press.

Srimal, Riju, and Clayton E. Curtis. 2008. "Persistent Neural Activity during the Maintenance of Spatial Position in Working Memory." *NeuroImage* 39 (1): 455–468. doi: 10.1016/j.neuroimage.2007.08.040.

OTHER SOURCES USED

Awh, Edward, Katherine M. Armstrong, and Tirin Moore. 2006. "Visual and Oculomotor Selection: Links, Causes and Implications for Spatial Attention." *Trends in Cognitive Sciences* 10 (3): 124–130. doi: 10.1016/j.tics.2006.01.001.

Bastos, Andre, Conrado A. Bosman, Jan-Mathijs Schoffelen, Robert Oostenveld, and Pascal Fries. 2010. "Interareal Directed Interactions and Their Modulation by Selective Attention Assessed with High Density Electrocorticography in Monkey." Presented at the annual meeting for the Society for Neuroscience, San Diego, CA, November 13–17.

Bollimunta, Anil, Yonghong Chen, Charles E. Schroeder, and Mingzhou Ding. 2008. "Neuronal Mechanisms of Cortical Alpha Oscillations in Awake-Behaving Macaques." *Journal of Neuroscience* 28 (40): 9976–9988. doi: 10.1523/JNEUROSCI. 2699-08.2008.

Curtis, Clayton E., and Jason D. Connolly. 2008. "Saccade Preparation Signals in the Human Frontal and Parietal Cortices." *Journal of Neurophysiology* 99 (1): 133–145.

Fischer, Burkhart, and R. Boch. 1983. "Saccadic Eye Movements after Extremely Short Reaction Times in the Monkey." *Brain Research* 260 (1): 21–26. doi: 10.1016/0006-8993(83) 90760-6.

Funahashi, Shintaro, Charles J. Bruce, and Patricia S. Goldman-Rakic. 1989. "Mnemonic Coding of Visual Space in the Monkey's Dorsolateral Prefrontal Cortex." *Journal of Neurophysiology* 61 (2): 331–349.

Hamilton, Antonia F. de C., and Scott T. Grafton. 2006. "Goal Representation in Human Anterior Intraparietal Sulcus." *Journal of Neuroscience* 26 (4): 1133–1137. doi: 10.1523/JNEUROSCI.4551-05.2006.

Hamm, Jordan P., Kara A. Dyckman, Lauren E. Ethridge, Jennifer E. McDowell, and Brett A. Clementz. 2010. "Preparatory Activations across a Distributed Cortical Network Determine Production of Express Saccades in Humans." *Journal of Neuroscience* 30 (21): 7350–7357. doi: 10.1523/JNEUROSCI. 0785-10.2010.

Ikkai, Akiko, and Clayton E. Curtis. 2011. "Common Neural Mechanisms Supporting Spatial Working Memory." *Neuropsychologia* 49 (6): 1428–1434. doi:10.1016/j.neuropsychologia. 2010.12.020.

Javal, Louis Émile. 1878. "Essai sur la physiologie de la lecture." *Annales d'oculistique* 80: 61–73.

Kriegeskorte, Nikolaus, Rainer Goebel, and Peter Bandettini. 2006. "Information-Based Functional Brain Mapping." *Proceedings of the National Academy of Sciences of the United States of America* 103 (10): 3863–3868. doi: 10.1073/pnas.0600244103.

Moore, Tirin, and Mazyar Fallah. 2004. "Microstimulation of the Frontal Eye Field and Its Effects on Covert Spatial Attention."

Journal of Neurophysiology 91 (1): 152–162. doi: 10.1152/jn. 00741.

Norman, Kenneth A., Sean M. Polyn, Greg J. Detre, and James V. Haxby. 2006. "Beyond Mind-Reading: Multi-Voxel Pattern Analysis of fMRI Data." *Trends in Cognitive Sciences* 10 (9): 424–430. doi: 10.1016/j.tics.2006.07.005.

Pereira, Francisco, Tom Mitchell, and Matthew Botvinick. 2009. "Machine Learning Classifiers and fMRI: A Tutorial Overview." *NeuroImage* 45 (1–1): S199–S209. doi: 10.1016/j. neuroimage.2008.11.007.

Poldrack, Russell A. 2006. "Can Cognitive Processes Be Inferred from Neuroimaging Data?" *Trends in Cognitive Sciences* 10 (2): 59–63. doi: 10.1016/j.tics.2005.12.004.

Rizzolatti, Giacomo, Lucia Riggio, Isabella Drascola, and Carlo Umiltá. 1987. "Reorienting Attention across the Horizontal and Vertical Meridians: Evidence in Favor of a Premotor Theory of Attention." *Neuropsychologia* 25 (1–1): 31–40. doi: 10.1016/0028-3932(87)90041-8.

Shipp, Stewart. 2004. "The Brain Circuitry of Attention." *Trends in Cognitive Sciences* 8 (5): 223–230. doi: 10.1016/j.tics. 2004.03.004.

Squire, Ryan Fox, Nicholas A. Steinmetz, and Tirin Moore. 2012. "Frontal Eye Field." *Scholarpedia* 7 (10): 5341. doi: 10.4249/ scholarpedia.5341.

FURTHER READING

Aguirre, G. K., and M. D'Esposito. 1999. "Experimental Design for Brain fMRI." In *Functional MRI*, edited by C. T. W. Moonen and P. A. Bandettini, 369–380. Berlin: Springer-Verlag.
A comprehensive and accessible treatment of how and why the BOLD fMRI signal imposes constraints on experimental design and of various design strategies that take these constraints into account.

Awh, Edward, Katherine M. Armstrong, and Tirin Moore. 2006. "Visual and Oculomotor Selection: Links, Causes and Implications for Spatial Attention." *Trends in Cognitive Sciences* 10 (3): 124–130. doi: 10.1016/j.tics.2006.01.001.
An accessible, concise review of the cognitive neuroscience of attention with an emphasis on its relation to oculomotor control.

Luo, Liqun, Eugenio Rodriguez, Karim Jerbi, Jean-Philippe Lachaux, Jacques Martinerie, Maurizio Corbetta, . . . A. D. (Bud) Craig. 2010. "Ten Years of Nature Reviews Neuroscience: Insights from the Highly Cited." *Nature Reviews Neuroscience* 11 (10): 718–726. doi: 10.1038/nrn2912.
A "10 years on" commentary about subsequent developments in the two-attentional-pathways model from Corbetta and Shulman, whose 2002 review introducing this idea was the most highly cited paper to appear in Nature Reviews Neuroscience *that year.*

Norman, Kenneth A., Sean M. Polyn, Greg J. Detre, and James V. Haxby. 2006. "Beyond Mind-Reading: Multi-Voxel Pattern Analysis of fMRI Data." *Trends in Cognitive Sciences* 10 (9): 424–430. doi: 10.1016/j.tics.2006.07.005.
An accessible review of the logic underlying many implementations of MVPA and some applications to cognitive neuroscience questions.

Pereira, Francisco, Tom Mitchell, and Matthew Botvinick. 2009. "Machine Learning Classifiers and fMRI: A Tutorial Overview." *NeuroImage* 45 (1–1): S199–S209. doi: 10.1016/j.neuroimage.2008.11.007.
An authoritative presentation of the principles of statistical machine learning that underlie many implementations of MVPA.

Poldrack, Russell A. 2006. "Can Cognitive Processes Be Inferred from Neuroimaging Data?" *Trends in Cognitive Sciences* 10 (2): 59–63. doi: 10.1016/j.tics.2005.12.004.
A prescient and thorough consideration of the logic of "reverse inference" that underlies the interpretation of functional neuroimaging data, and thereby provides a useful context from which to consider the relative advantages and limitations of univariate vs. multivariate approaches to neuroimaging data analysis. Although explicitly addressing neuroimaging, much of its argumentation is equally applicable to other types of neurophysiological data.

Rizzolatti, Giacomo, and Laila Craighero. 2010. "Premotor Theory of Attention." *Scholarpedia* 5 (1): 6311. doi: 10.4249/ scholarpedia.6311.
A "23 years on" commentary about subsequent developments relevant to this theory, including the role of skeletomotor systems.

Sherman, S. Murray, and Ray W. Guillery. 2006. *Exploring the Thalamus and Its Role in Cortical Function.* Cambridge, MA: MIT Press.
Provocative text by two of the foremost authorities on the thalamus, proposing, among other ideas, that cortico-thalamo-cortical loops are the primary circuits for information transmission, with many cortico-cortical connections serving modulatory roles.

SECTION III
MENTAL REPRESENTATION

INTRODUCTION TO SECTION III
MENTAL REPRESENTATION

In *Section II* we covered "the inputs and the outputs." And as I've stated previously, some would contend that everything from here on out is an elaboration of the principles that we've already learned. Well . . . yes and no. One of the central features of the brain, its plasticity, means that it's constantly changing, constantly updating its models of the world. Consequently, no two instances of perceiving a stimulus are the same, because the circuits responsible for perception are always somewhat different, somewhat changed, from the previous instance. At another level, *Section III* will examine ways in which perception and action are necessarily dependent on the brain's interpretations of the *meaning* of the inputs and the outputs. This idea was summarized eloquently in a commentary by the neurophysiologist Joaquin Fuster, as follows:

> Perception is . . . based on long-term memory. Every percept of the world around us is an "interpretation" of current sensory data in the light of our experience with similar data stored in long-term memory. Thus, perception has been appropriately called the testing of hypotheses about the world. We not only remember what we perceive, but also perceive what we remember. Perceiving must essentially involve the matching of external gestalts to internalized representations in long-term memory. There cannot be an entirely new percept, because any conceivable new sensory configuration can resonate by associations of similarity or contiguity with others in long-term storage, which the new experience modifies and updates. In other words, the new experience modifies and updates previously established cortical networks. Intelligence and language, in all of their manifestations, operate in basically the same way, that is, by the activation of long-term memory networks towards the processing of intelligent behavior or speech. (Fuster, 2003, p. 737)

Long-term memory (LTM), the neural record of past experience that can influence future behavior, is the major focus of this section – how it's formed; how it's organized; how it influences cognition, perception, and action. Indeed, as an entrée into this topic, *Chapter 10* will, in effect, pick up where *Chapter 6* left off, considering how the sensation and perception of a visual object interact with preexisting representations of the category, and perhaps the particular token, of the object being perceived.

REFERENCE

Fuster, Joaquín M. 2003. "More Than Working Memory Rides on Long-Term Memory." *Behavioral and Brain Sciences* 26 (6): 737–737. doi:10.1017/S0140525X03300160.

CHAPTER 10
VISUAL OBJECT RECOGNITION AND KNOWLEDGE

KEY THEMES

- The neurological syndrome of visual agnosia raises many puzzling, and profound, questions about the relationship between perception and knowledge representation.

- Perhaps most intriguing is the seemingly selective inability of prosopagnosic patients to discriminate among individual faces.

- Despite the insight that can be gained from careful assessment of neurological deficits, models intended to explain "what must be broken" in these patients can sometimes posit cognitive mechanisms of questionable neural validity.

- Parallel distributed processing (PDP) models suggest that meaning (and, therefore, recognition) may reside within circuits that mediate perception and interactions between perception and action systems.

- fMRI evidence for a fusiform face area (FFA) that responds more strongly to faces than to other categories of stimuli has launched several vigorous debates about the neural and cognitive organization of the ventral stream.

- The FFA may also be sensitive to perceptual expertise, by which account its response profile would be better explained in terms of nurture than nature.

- MVPA analyses indicate that face information is also represented outside the FFA, and that the FFA can represent information about non-face categories, thereby challenging strong claims of the specificity of this region for face processing.

- Despite the previous point, direct electrical stimulation of the FFA, guided by electrocorticography (ECoG), selectively distorts the phenomenology of perceiving faces, thereby confirming a causal role of the FFA in face perception.

- Decomposing images of faces into principal components based on their image properties has revealed a principle underlying the brain's representation of faces, such that it provides a quantitative model with which to predict which synthetically generated face stimuli will drive neurons that, using conventional methods, had seemed to be highly specific to just one face.

- Face-related activity in the FFA can be shown to adhere to principles of predictive coding.

CONTENTS

Although, as promised, we won't begin this chapter with an overview of the anatomy underlying the function of interest in this chapter, we will begin, as we did for *Chapter 7*, by introducing and motivating our topic with a neurological syndrome.

VISUAL AGNOSIA

In *Chapter 6*, it was noted that "By the 1960s, decades of clinical observations in humans and lesion studies in experimental animals had established that the temporal lobes play a critical role in object recognition." Particularly germane for our present interests are the visual agnosias. This is a term for which understanding the etymology is more than a historical curiosity. *Gnosis* is ancient Greek for "knowledge," and the *a-* in front of it confers negation. Thus, visual agnosia is the absence of visual knowledge. There are two broad categories of visual agnosia, a cardinal feature of both being profoundly impaired visual perception despite relatively intact visual sensation. That is, these patients aren't blind. The symptoms that make them agnosic reflect damage to cortical networks that are downstream from V1.

Apperceptive agnosia

A case of Benson and Greenberg (1969), Mr. S., was selected by Martha Farah to introduce the syndrome in her influential (1990) book on the topic. After suffering accidental carbon monoxide (CO) poisoning, Mr. S. retained the ability to detect small differences in luminance, wavelength (i.e., color), and size. However, he was "unable to distinguish between two objects of the same luminance, wavelength, and area when the only difference between them was shape." Further, although failing to identify a safety pin, he reported that it was "silver and shiny like a watch or a nail clipper." Thus, some low-level functions of sensation and feature identification were spared, as was, indeed, an abstract understanding of the visual properties of real-world objects. Additionally, "Recent and remote memory, spontaneous speech, comprehension of spoken language and repetition were intact," indicating the absence of a general, nonspecific cognitive decline. And, most intriguingly, although "unable to name objects, pictures of objects, body parts, letters, numbers, or geometrical figures on visual confrontation, . . . he could readily identify and name objects from tactile, olfactory, or auditory cues" (Benson and Greenberg, 1969, pp. 83–85).

An interesting question to ask at this point – one that we won't try to answer right away – is whether this syndrome may arise from a disconnection between early visual processing and the knowledge of the world that's necessary to recognize objects. After all, the knowledge might still seem to be intact in this patient, in that he can access it through other sensory modalities. One caveat to accepting this interpretation, however, is that it assumes a property of the cognitive architecture that need not be true: it assumes that knowledge about the world is stored in an "amodal" format, one that is not tied to any one sensory modality. Could it not be the case that, at least to some degree, some long-term representations of knowledge remain tied to the sensory channel through which we first acquired this knowledge? In other words, might it not be the case that Mr. S.'s *visual* knowledge of objects was compromised, but his tactile knowledge, olfactory knowledge, and auditory knowledge remained intact?

Other interesting variants of apperceptive agnosia are manifested by patients who can recognize objects in optimal viewing conditions, but whose performance declines precipitously under more challenging conditions. For example, a patient may be able to recognize, say, a bicycle in a black and white line drawing, but fail at the same task if a piece of string is dropped onto the picture, or a handful of pencil shavings is scattered across it (i.e., if noise is introduced to the scene). Another classic deficit is the inability to recognize an object when it is presented from a noncanonical viewpoint. Thus, the patient may be able to recognize a wheelbarrow from a side view, but not a rear view (in which the handles are "pointing at" the viewer, perpendicular to the picture plane). Such deficits clearly implicate impairment of perceptual processes that we considered in *Chapter 6,* including viewpoint-dependent vs. viewpoint-independent object encoding, the disambiguating functions of top-down feedback, and, more generally, the hierarchical nature of object recognition.

Associative agnosia

This syndrome was famously characterized by (German-born, US-based) neuropsychologist Hans-Lukas Teuber (1916–1977) as "appear[ing] in its purest form as a normal percept that has somehow been stripped of meaning" (Milner and Teuber, 1968, p. 293). *Figure 10.1* shows an iconic example from a patient who was able to execute highly precise copies of line drawings, yet was unable to recognize what the drawings represented. Examples

Eagle

Guitar

Owl

Bee

FIGURE 10.1 Drawings by a patient with an associative agnosia. On the left are the models and on the right are the patient's copies. Despite reasonably good reproduction of all the visual features, the patient was unable to name the objects that they represent. He named the eagle "a cat sitting up," the guitar "some kind of machine, a press," the owl "a pattern," and the bee "an animal with horns and a tail . . . a rhino?" Source: From Farah, Martha J. 1990. Visual Agnosia. Cambridge, MA: MIT Press. Copyright 1990. Reproduced with permission of MIT Press.

such as this very strongly give the sense of a disconnection syndrome (i.e., of perception from knowledge), as considered previously. However, although this idea (and, indeed, Teuber's quote) remains prominent in the literature and textbooks, it should be noted that the perception of many of these patients is unlikely to be truly normal. For example, despite the high fidelity of the drawings in *Figure 10.1*, many patients producing copies of comparable quality have been described as doing so in a "slavish," "line-by-line" manner, and doing so "exceedingly slowly" (various authors, quoted by Farah, 1990, p. 60). Thus, the quality of the reproductions may owe more to "drawing strategies" than to truly intact visual perception.

It may be that the fundamental distinction between the two groups of patients (i.e., apperceptive vs. associative agnosic) isn't that visual perception is impaired in one and spared in the other, but that they differ with regard to where along the visual processing hierarchy their impairment is most pronounced. For example, whereas apperceptive agnosic patients try to identify objects based on low-level features (e.g., color, texture), associative agnosic patients typically try to rely on shape. Thus, for example, Farah (1990) notes that on four separate occasions she showed the associative agnosic patient L. H. a drawing of a baseball bat, and on each his incorrect guess was an object with a similar shape: a paddle, a knife, a baster, and a thermometer.

The most intriguing subset of associative agnosic patients are those whose primary deficit seems not to be with type identification, but with token individuation. And within this subset, the sub-subset whose impairment very saliently includes face recognition has claimed a disproportionate share of the attention of the visual cognition community.

Prosopagnosia

Following in the tradition of the German neurologist Heinrich Lissauer (1861–1891), who coined the term agnosia, it was also a German neurologist, Joachim Bodamer who, in 1947, welded on the prefix "prosopon" (from the ancient Greek for face) to characterize his patients with a selective agnosia for the recognition of faces. In the decades that followed, there have been many case studies of patients with this syndrome. In many cases the deficit is quite striking, with the patient, for example, not recognizing his wife as she walks past the hospital bed, and not even realizing that the strange person staring at him is actually a reflection of himself in a mirror.

What over a half-century of research hasn't resolved, however, is just how specific to faces this syndrome can be.

There is, for example, a report of a patient who could identify close relatives by voice, but not by face, but who had no difficulty recognizing, for example, his electric razor from among many similar-looking electric razors, his wallet from among similar wallets, and so on. Intriguing dissociations between patients have also been reported, such as one group with a face-recognition deficit but who have no difficulty recognizing facial expressions, and others showing the converse. Many other reports, however, indicate that within-category recognition of other categories can also be impaired, with animals and plants often compromised. Specific examples include an avid bird-watcher who, upon developing prosopagnosia, also lost the ability to discriminate between different species of birds. Perhaps most famous is the case of a farmer who, along with human faces, lost the ability to recognize each of the cows in his herd. This and other reports gave rise to the idea that prosopagnosia and other seemingly category-specific deficits of visual perception may reflect a more general loss of perceptual expertise, an idea that we'll take up further along in this chapter.

COMPUTATIONAL MODELS OF VISUAL OBJECT RECOGNITION

Two neuropsychological traditions

Interestingly, although much of the research summarized in the preceding sections occurred in parallel to the neurophysiological study of object recognition in the monkey, including studies of face processing (*Chapter 6*), the two traditions of research proceeded largely independent of each other. One reason is that they had largely different goals. Whereas systems neuroscience, including the subfield that I'll call experimental neuropsychology, prioritizes understanding how the brain works, the discipline of cognitive neuropsychology is primarily interested in how the mind works (i.e., in how the cognitive system is organized). Examples of experimental neuropsychology that we have already considered include the studies cited in *Chapter 6* as influences of Pohl (1973), and the research by Ramachandran on patients with phantom limb syndrome (*Chapter 5*). In relation to visual agnosia, see Milner and Teuber (1968) as an example of an experimental neuropsychology approach. An example of cognitive neuropsychology, in contrast, is the influential model of object recognition from Humphreys and Riddoch (1987), depicted in *Figure 10.2*.

Cognitive neuropsychology-influenced models

Note from *Figure 10.2* that the emphasis of this model is on stages of information processing, rather than stages of neural processing. From this perspective, although it is intended to account for object recognition deficits reported from a variety of brain-damaged patients, it addresses a level of explanation that is more closely related to cognitive psychology than to neuroscience. Thus, for example, each of the boxes in this model is inferred, rather than directly measured. They correspond to hypothesized stages of processing that are proposed in order to account for one or more findings from visual agnosic patients.

To summarize, Humphreys and Riddoch (1987) offer an analogy that's evidently a popular one – searching for a book at the library in the pre-internet age. (Because your author graduated from university in 1987, it is an analogy that he understands quite well.) When one wants to find a particular book at the library, one looks it up by title or by author in the card catalog, and the card corresponding to that book indicates the specific location in the stacks (the "address") where the book will be located. (In the United States, this address is encoded via the Dewey decimal system.) The library patron writes down the encoded address on a scrap of paper and then begins to search the stacks. In this analogy, the finding of the appropriate card in the card catalog corresponds to the transduction of visual information into a neural code, and the writing of the address onto the scrap of paper is early visual processing. The scrap of paper containing the link to the book is the "abstract episodic object description," which corresponds to a higher-level visual object representation. (In the analogy, this representation is *abstract* because the Dewey code doesn't literally say "5th floor, third stack from the north wall, fourth shelf, 437 books in from the end"; nor does it have any inherent meaning, it's just a code that will help you find your book – therefore, this level of perceptual representation is posited to be without inherent meaning.) The key step in the process is recognition, when one matches the code on one's scrap of paper with the code on the spine of the sought-after book. (The code on the spine of the book corresponds to "object form knowledge" in the model.) Recognition having successfully occurred, the patron now has access to the meaning of the card in the catalog – the contents of the book. Phenomenologically, this corresponds to recognizing the object that one is looking at, and knowing everything that there is to know about it. (For example, *That's my cat. Cats are quadrupedal domesticated mammals, probably bred from the lynx or ocelot or some smallish feline carnivore [etc.]. This cat's name is* Mahalia, *named after Mahalia Jackson*

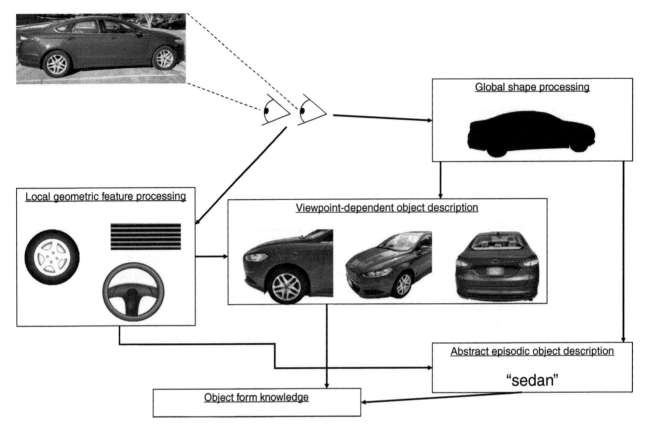

FIGURE 10.2 A cognitive neuropsychology-derived model of visual object processing.

because she's black, brown, and beige [the cat, that is], *the title of a Duke Ellington piece (on which M.J. sang) that we were into when we "rescued" her as a kitten from that crazy cat lady's house in Billerica, MA. She's really old* [again, referring to the cat].) Associative agnosia, by this account, is caused by damage to object form knowledge. (Note that this is the interface that Fuster [2003], in the *Introduction to Section III,* refers to as "the matching of external gestalts to internalized representations in long-term memory" that is central to perception.)

Critiques of cognitive neuropsychology-influenced models

Farah (1990) notes that this type of modular, "symbol-manipulation" model "has appealed to many [cognitive] neuropsychologists seeking an account of associative agnosia" because it provides a framework for the intuitions of observers of the syndrome. She summarizes these, drawing on literature spanning from 1914 through 1985, as "loss of the 'stored visual memories' or '**engrams**' of objects," and disruption of"'the matching of signal to trace' . . . [and the]

destruction of stored 'templates'" (pp. 96–97). However, there are at least two types of concern that one could have about such models. (And this is now me talking again, not Martha Farah). One is that they are not quantitative, and thus are difficult to disprove. With all models, and certainly with "verbal models" such as this, one must guard against post hoc elements that satisfy one's intuition about how a system might work. For our current discussion, make note of the fact that an explicit tenet of the Humphreys and Riddoch (1987) model, embodied in the *abstract episodic object description,* is that visual object recognition proceeds in two stages: (1) the processing of "structural" representations that have no inherent meaning, and (2) "visual semantics," the portion of long-term memory that represents meaning. A second concern about modular models is that they are often not constrained by/do not incorporate what we know about how the brain works. The model in *Figure 10.2,* for example, has no feedback from higher levels of visual processing to lower levels. Further, it contains no specification that a certain kind of neuron or area of the brain (such as those we reviewed in *Chapter 6*) corresponds to any of the processing stages in the model.

PDP-based models offer a neurally plausible alternative for modeling cognitive processes and architectures

At this juncture, a fair objection might be to remind me that the goal of a model such as that illustrated in *Figure 10.2* is to better understand behavior at a *cognitive* level of description, not at a *neural* level. And it is certainly true for such models that slapping "neuro-sounding" labels onto a model won't, in and of itself, confer any more legitimacy or plausibility to it. An alternative approach is offered by PDP, or "connectionist," modeling, which, although also often addressing questions at the cognitive level of description, is generally considered to generate models that are more compatible with many of the assumptions and practices of contemporary cognitive neuroscience (section *A PDP model of picture naming, Chapter 3*).

For this author, the really powerful implication of the model described in section *A PDP model of picture naming, Chapter 3*, is the way that it encourages a radical rethinking of how object recognition works. While writing *Chapter 6*, for example, I was feeling impelled to press "forward" (i.e., downstream) to "get to the place where recognition happens." Unbeknownst to me, as it were, I may have been describing *where*, if not *how*, object recognition happens while detailing feedforward and feedback connections from V1 to V4 and MT, from V4 to IT, and the response profiles of neurons all along these pathways. There may not, in fact, have been a finish line to cross.

The cognitive neuroscience revolution in visual cognition

With these ideas in mind, we now return to the physiology of the ventral pathway, this time (as opposed to how we did it in *Chapter 6*) as measured with fMRI, and in humans. This next section will illustrate the truly revolutionary influence that developments in cognitive neuroscience, in this case neuroimaging, have had on fields such as object recognition and visual cognition. One example is that the divide between the cognitive neuropsychology and the neurophysiology of object recognition, which we have just reviewed, has been steadily narrowing. Indeed, it is increasingly rare today to see models of object recognition and/or visual agnosia that don't explicitly incorporate constraints from functional neuroanatomy and/or neurophysiology. In other respects, however, some longstanding debates persist,

just reframed in terms of the newer technologies and the newer intellectual paradigms. Thus, for example, questions of localization-of-function vs. distributed processing remain front-and-center, and the even older dialogue between nature and nurture is also at play in this arena.

CATEGORY SPECIFICITY IN THE VENTRAL STREAM?

Are faces special?

"The Guns of August"

You'll recall, from *Chapter 6*, the puzzling properties of neurons of temporal cortex that seemed to respond to "the idea of a face" more so than to any individual face. In the 1990s, cognitive psychologists familiar with this literature, and with the puzzling questions raised by the prosopagnosic syndrome, saw in neuroimaging an opportunity to address these problems in a new way. Now I say "cognitive psychologists" – and Farah, in later editions of her 1990 book, chronicles many of the early contributors – but much of the intellectual and empirical energy driving these "early" cognitive neuroscience developments was one individual: Nancy Kanwisher. Kanwisher trained as a cognitive psychologist, and early in her career worked on a phenomenon known as repetition blindness, influentially relating it to how the brain handles the type vs. token distinction (e.g., why is it more difficult to individuate two different flowers presented sequentially in a serial stream of stimuli than, say, a flower and a face?) (Kanwisher, 1991). Perhaps this underlies her subsequent fascination with face processing. In any event, in 1997 she, together with then-postdoctoral fellow Marvin Chun and then-undergraduate research assistant Josh McDermott (now both of them highly accomplished cognitive neuroscientists in their own right) published a study that spurred a subfield of cognitive neuroscience that shows no sign of slowing down 20+ years later.

"The Fusiform Face Area: a module in human extrastriate cortex specialized for face perception"

The proclamation of this title of the Kanwisher, McDermott, and Chun (1997) study remains provocative and controversial to this day. The study employed a block design, in which subjects viewed the serial presentation of unfamiliar faces (ID photos of Harvard undergraduates) alternating with serial presentation of control stimuli.

Blocks were 30 seconds long; stimuli presented for 500 ms with an interstimulus interval of 150 ms. The experiment proceeded in three phases, the first an unapologetic "far and wide" search for regions that might respond with greater activity for faces vs. objects, the second and third more rigorous tests to replicate the phase 1 finding while ruling out potential confounds and further assessing the specificity of the regions identified in phase 1. The authors note that "Our technique of running multiple tests applied to the same region defined functionally within individual subjects provides a solution to two common problems in functional imaging: (1) the requirement to correct for multiple statistical comparisons and (2) the inevitable ambiguity in the interpretation of any study in which only two or three conditions are compared" (p. 4302).

In phase 1 of the study, 12 of 15 subjects showed significantly greater activity for faces than objects in the right fusiform gyrus, on the ventral surface of the temporal lobe, at the level of the occipitotemporal junction, and this region therefore became the focus of the subsequent scans. Phase 2, illustrated in *Figure 10.3*, panels 1B and 1C, was intended to rule out the potential confounding factors from the phase 1 study: differences in luminance between face and object stimuli, or the fact that the face, but not the object, stimuli were all exemplars drawn from the same category. Phase 3, illustrated in *Figure 10.3*, panels 2B and 2C, was intended to rule out the potential confound that the FFA was simply responding to any human body parts. Additionally, it established that FFA activity generalized to images of faces that were less easy to discriminate, and to a condition that

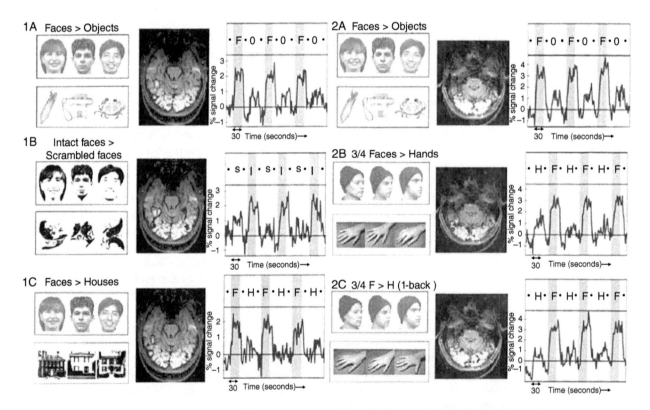

FIGURE 10.3 Examples of stimuli, fMRI statistical maps, and time series, from several assays of response properties of the FFA. The contrast illustrated in panels **1A** and **2A** (Faces > Objects) was used to define the FFA ROI, outlined in green in the axial images, each for a different individual subject. Panels **B** and **C** illustrate the statistical map generated by each of the contrasts, each for the same subject, with the green-outlined ROI from panel **A** superimposed. This procedure was repeated on the individual-subject data of five subjects, and each time series is the average from the FFA ROI of each. Source: From Kanwisher, Nancy, Josh McDermott, and Marvin M. Chun. 1997. "The Fusiform Face Area: A Module in Human Extrastriate Cortex Specialized for Face Perception." Journal of Neuroscience 17 (11): 4302–4311. http://www.jneurosci.org/content/17/11/4302.full.pdf+html. Reproduced with permission of the Society of Neuroscience.

imposed stronger attentional demands on the processing of stimuli, by requiring a decision of whether or not each stimulus matched the identity of the previous one.

As illustrated in *Figure 10.3*, the results were quite striking, showing considerably higher activity in FFA to faces than to any other stimuli. Thus, they were consistent with the idea that faces receive privileged processing in this region. But is this region *specialized* for face processing? Sometime in the late 1990s, I attended a presentation by Kanwisher at a colloquium at the University of Pennsylvania. The cognitive scientist Saul Sternberg asked what to make of the fact that the region also shows elevated activity (relative to baseline) to other stimuli. Kanwisher replied (and here I'm paraphrasing from memory) that "there must be something to this question, because I get it almost every time I give this talk," but that quantitative superiority of response to a particular class of stimuli has always been interpreted by physiologists as evidence for specificity of function. This stance will come under closer scrutiny in later portions of this chapter.

One reason that the discovery of the FFA generated so much interest was that it seemed to effect a neural test of an influential theory that had been advanced by the philosopher Jerry Fodor in his 1983 book *Modularity of Mind*. In a sentence, Fodor's idea was that particular cognitive operations were "encapsulated" and "cognitively impenetrable" once launched, and thus not susceptible to influence by other factors (attention, intention, behavioral priorities, distraction, etc.) until they had run their course. Visual object recognition, as we've been emphasizing here, and auditory perception (particularly speech perception) would be two such candidate processes. Not unlike some of the cognitive neuropsychological models that we considered earlier in this chapter, Fodor's idea gave a good account of many observations from studies of human behavior (in this case, from cognitive psychology), but, at the same time, also seemed contrary to much of what we know about how the brain works. Thus, one really intriguing aspect of the Kanwisher, McDermott, and Chun (1997) study was that it could be interpreted as a presentation of neural evidence in favor of a hypothesis that seemed contrary to how many thought that the brain works!

Questions raised by the Kanwisher and colleagues (1997) findings

Now, to be clear, no one questioned the veracity of the findings by Kanwisher, McDermott, and Chun (1997). Rather, the question was, *How do we make sense out of them?* Indeed, another reason for their impact was that they gave new urgency to the longstanding puzzle presented by

prosopagnosia: no one denied that the syndrome existed, but a widely accepted explanation had eluded scientists for at least 100 years. This puzzle was articulated by Hans-Lukas Teuber in an anecdote, from the "Perception" section of a chapter coauthored with Brenda Milner (Milner and Teuber, 1968). Stylistic and cultural aspects of this quote will be offensive to contemporary ears, but let's look past the dated language and focus on the substance:

> There is the familiar example of perceptual learning, on the part of a white visitor to China (or conversely, a Chinese visiting America). At first, most of the Chinese faces are indiscriminable, but, once one is fairly familiar with a few individual faces, something seems to have been learned that generalizes to all Chinese faces: "Knowing" a few makes all the others more distinguishable. Conceivably, an agnosia for faces could be due to a loss of such an acquired system of rules for classifying facial differences. Alternatively, and even more speculatively, we could assume that the initial basis for discrimination and identification of faces and of facial expressions involves some innate schemata onto which the subsequent, perceptual differentiations were merely imposed . . . This strong but as yet unproveable assumption would make a selective loss of capacity for identifying faces, following certain focal lesions in man, somewhat less paradoxical than it now appears to be (pp. 300–301).

The concluding sentence of Kanwisher, McDermott, and Chun (1997) echoes Teuber's reasoning: "Future studies can also evaluate whether extensive visual experience with any novel class of visual stimuli is sufficient for the development of a local region of cortex specialized for the analysis of that stimulus class, or whether cortical modules like area FF must be innately specified (Fodor, 1983)" (p. 4310).

Perceptual expertise

Many of the "future studies" foreshadowed by Kanwisher, McDermott, and Chun (1997) have been led by cognitive neuroscientists Isabel Gauthier and Michael Tarr. (Tarr, like Kanwisher, started his career as a cognitive psychologist and has become increasingly "neuro" as his career evolves.) These two, and their colleagues, have pursued the idea that there's "nothing special" about faces at the neural level – no genetically specified circuits responsible for their processing. Rather, faces are stimuli that we see thousands of times each day, every day, and being able to individuate them is vitally important for success in our highly social environment. Thus, for example, they have argued that it is possible that when Kanwisher, McDermott, and Chun (1997) showed different houses vs. different faces in the experiment illustrated in

Figure 10.3, panel 1C, the two types of stimuli may have been processed differently. The house stimuli may have been processed at a more categorical level, the faces at an individual level. (For example, the subjects may have been thinking "there's another house, and another one, and another one, ... ," whereas they may have been scrutinizing each face to determine if it was someone they knew.) By this account, the greater activity for faces than for buildings in that experiment may have had more to do with how stimuli were processed than with what the stimuli were.

This perspective is one that emphasizes the role of expertise in perception. At the neural level, it argues, FFA may be best construed as a "fusiform expertise area." (Although no one, to my knowledge, has ever proposed the moniker "FEA," Tarr and Gauthier [2000] did try rebranding it the "flexible face area." This never caught on.) Note that what we'll call the "expertise" account has a ready explanation for why some prosopagnosic patients also lose the ability to recognize other kinds of stimuli with which they have a high degree of expertise, such as the farmer who lost the ability to individuate his cows or the bird watcher who lost the ability to distinguish among species of bird. Gauthier and colleagues tested this idea with a clever set of studies that entailed training subjects (for some studies, healthy young individuals, for others, prosopagnosic patients) to become experts at recognizing individual tokens of a category of stimuli for which there could not exist a genetic template: Greebles.

Greeble expertise

Gauthier and Tarr (1997) developed Greebles as a class of stimuli that have many of the same properties as faces: (1) they are processed configurally – the relation between parts (for faces, distance between eyes, from nose to eyes, etc.) is critically important for recognition; (2) they show an inversion effect – it's disproportionally difficult, relative to other stimuli, to recognize them when they are upside down; (3) they tend to be classified at the individual level, rather than the category level. They can be categorized at the level of Greeble, family, gender, and individual. As illustrated in *Figure 10.4*, the fMRI data of Gauthier et al. (1999) suggest that, in contrast to Greeble novices, experts at Greeble recognition evince a statistically comparable stimulus-evoked activity for Greebles as for faces in the FFA. From this, they concluded that "The strongest interpretation suggested by our results . . . is that the face-selective area in the middle fusiform gyrus may be most appropriately described as a general substrate for subordinate-level discrimination that can be fine-tuned by experience with any object category" (p. 572).

Evidence for a high degree of specificity for many categories in ventral occipitotemporal cortex

Quickly on the heels of the 1997 publication from Kanwisher and colleagues came a flurry of publications suggesting a further modularization of inferior temporal cortex, some of them from this same group. A *parahippocampal place area* (PPA) was discovered that responds preferentially to viewing a scene (be it a room or a landscape); an *extrastriate body area* (EBA) that responds preferentially to viewing body parts; and a *visual word-form area*. A summary of the location of these areas is provided in *Figure 10.5*. These findings, too, have not been without their skeptics: some, à la Tarr and Gauthier, have pointed to potential confounds in the studies producing the results summarized in *Figure 10.5*; others, as we'll see in the next section, have called into question the validity of the assumptions underlying the methods that produce these results.

FIGURE 10.4 FFA activity in Greeble novices vs. Greeble experts. Source: Gauthier, Tarr, Anderson, Skudlarski, and Gore, 1999. Reproduced with permission of Nature Publishing Group.

A

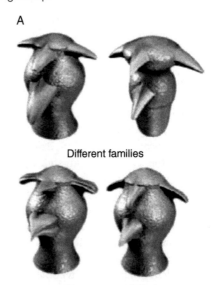

Different families

Different individuals

FIGURE 10.4.A Examples of Greebles. The two in the top row are from different "families," as defined by the large central part. The two at the bottom are two individuals from the same family, differing only in terms of the shapes of smaller parts. Subjects were determined to have achieved expert status with Greebles when they could distinguish two from the same family as quickly as they could two from different families.

B

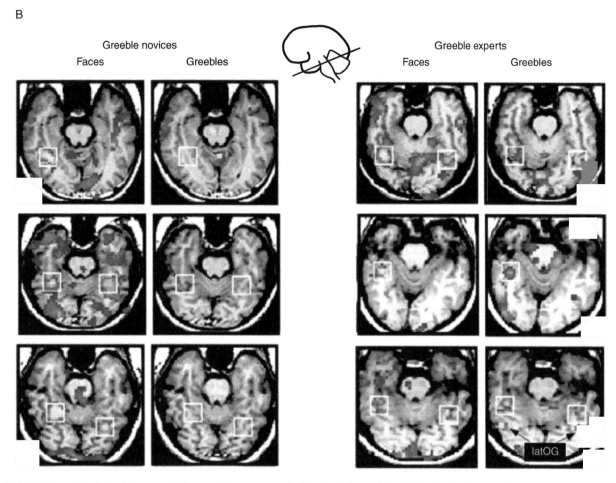

FIGURE 10.4.B Ventral temporal lobe activity, measured with fMRI, from six individuals, during passive viewing of faces vs. Greebles. White squares identify the FFA.

Evidence for highly distributed category representation in ventral occipitotemporal cortex

Four years after the publication of the Kanwisher, McDermott, and Chun (1997) report on face perception, James Haxby and colleagues published a study of object recognition that has also had a remarkable impact on cognitive neuroscience. In contrast to the modular models summarized in the previous paragraph, Haxby et al. (2001) proposed that "The representation of a face or object is reflected by a distinct pattern of response across a wide expanse of cortex in which both large- and small-amplitude responses carry information about object appearance," and, therefore, that "each category elicits a distinct pattern of response in ventral temporal cortex that

is also evident in the cortex that responds maximally to other categories" (p. 2425). To test these ideas, they acquired fMRI while subjects viewed two blocks, each of which serially presented stimuli drawn from one of each of eight categories: faces, houses, cats, bottles, small man-made tools, shoes, chairs, and smeary "nonsense images." Next, they compared the similarity of the pattern of activity for each category in its even-numbered block vs. in its odd-numbered block, by correlating the two – that is, they assessed the extent to which knowing the level of signal intensity in each voxel of ventral occipitotemporal cortex during one block predicted what that same voxel's signal intensity was in the other block. (Importantly, they did this regardless of whether any given voxel's value was large or small, positive or negative.) For comparison, they also correlated the pattern produced by the category in question

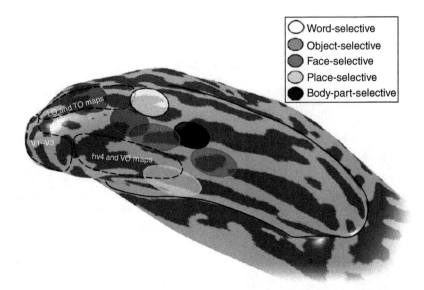

Word-selective
Object-selective
Face-selective
Place-selective
Body-part-selective

FIGURE 10.5 Juxtaposition of regions on the ventral surface of the occipital and temporal lobes that contain retinotopic maps (dashed lines and written labels) and those that have been identified as preferentially responsive (via univariate analyses) to visual stimuli from various categories. Source: From Wandell, Brian A., and Jonathan Winawer. 2011. "Imaging Retinotopic Maps in the Human Brain." Vision Research 51(7): 718–737. doi: 10.1016/j.visres.2010.08.004. Reproduced with permission of Elsevier.

against all of the other categories. (Thus, for each category, eight correlations were performed [one within-category and seven between-category], and for the experiment, eight within-category correlations and 56 between-category correlations.) This is illustrated in *Figure 10.6*, in which, for each row, the red bar represents the within-category correlation coefficient, and the dark blue the between-category coefficients. (Note that, because the correlations were computed by comparing the values of every voxel in ventral occipitotemporal cortex from one block vs. another, this was a multivariate analysis – indeed, it is considered the grandpappy of the MVPA methods introduced in *Chapter 9*.) The results indicated that coefficients from within-category correlations were higher than those from between-category correlations for 96% of the comparisons. The fact that this analysis was successful for each stimulus category provided support for the authors' proposal that the neural representation of visual categories is broadly distributed.

MVPA challenges models of modular category specificity

Two follow-up analyses went further and directly challenged the interpretation of studies supporting modular models of the neural organization of categories of information. First,

Haxby et al. (2001) redid the analyses described above, but after excluding from the dataset the voxels that responded maximally to the critical category. Thus, for faces, for example, they removed the voxels from the FFA before performing the within- and between-category correlations for face blocks. The outcome, as evidenced by comparison of the yellow vs. red bars in *Figure 10.6*, is that while the within-category correlation coefficients declined to varying degrees in the "excluded" datasets, they nonetheless remained higher than the between-category coefficients for 94% of comparisons. The implication of this finding is that, for faces, for example, face information is represented in cortex outside of the FFA. This, therefore, argues against the view that faces are processed in a highly modular manner.

But what about the FFA? Even if this region seems not to have "exclusive rights" to the processing of visually presented faces, it could still be "specialized" for face processing. But is it? This question hearkens back to the anecdote of Saul Sternberg's question to Nancy Kanwisher about what to make of the fact that FFA does respond, albeit with a smaller magnitude, to stimuli other than faces. To address it, Haxby et al. (2001) identified the FFA, the analogous voxels that responded maximally to houses, and so on. Then, using only FFA voxels, for example, they performed within- vs. between-category correlations for each of the other seven categories. That is, for example, within the FFA, does the pattern of activity evoked by a block of houses better predict the pattern evoked by the second block of houses than the pattern evoked by bottles, shoes, and so on? The answer was "yes," for 83% of comparisons. The same was true for six of the analogous ROIs defined with univariate methods as being selective for other categories: the house ROI supported within-category correlations (for non-house stimuli) at a rate of 93%; the tools ROI at 94%; and the cats ROI at 85%. Only the bottle ROI did not differentially represent non-bottle stimuli.

The implication of these findings is that the FFA cannot be characterized as being "specialized," or "selective," or "specific" for face processing. Is it, however, necessary?

Convincing as the Haxby et al. (2001) analyses were at establishing the distributed nature of object representation in ventral occipitotemporal cortex, it is important to keep in mind that the data were generated in an intact ventral occipitotemporal cortex. With regard to faces, for example, even though one of their analyses found evidence for face representation in cortex outside the FFA (i.e., when the FFA was excluded from the analysis), the tissue of the FFA was intact and functioning normally in the brains of their subjects when the scans were performed. Thus, the Haxby et al. (2001) findings cannot tell us whether non-FFA regions would be able to support a distributed representation of faces if the FFA were not influencing their activity when faces are being presented.

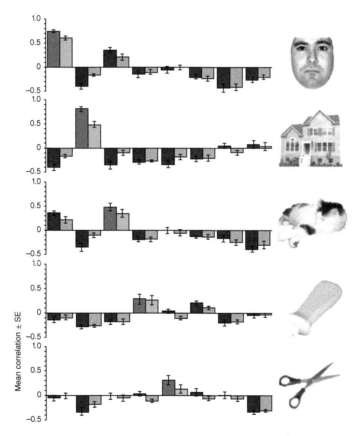

FIGURE 10.6 Correlation matrix showing within- and between-category correlations for the entirety of ventral occipitotemporal cortex that was analyzed, as well as for the same region, but with each category's maximally responsive voxels removed from the analysis. (The category *tools* is denoted by the scissors.) Source: From Haxby, James V., M. Ida Gobbini, Maura L. Furey, Alumit Ishai, Jennifer L. Schouten, and Pietro Pietrini. 2001. "Distributed and Overlapping Representations of Faces and Objects in Ventral Temporal Cortex." Science 293 (5539): 2425–2430. doi: 10.1126/science.1063736. Reproduced with permission of AAAS.

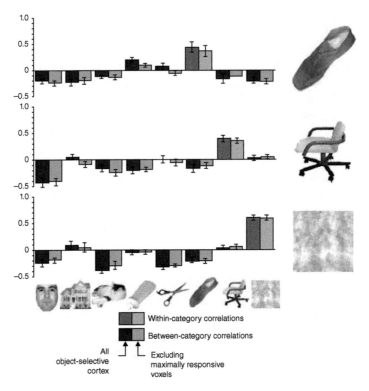

FIGURE 10.6 (*Continued*)

Demonstrating necessity

The caveat at the end of the previous paragraph is a replay of a principle that we have now encountered several times: one cannot know the necessity of a region's contribution to a function without perturbing it. Thus, the next study that we will consider in this chapter employed ECoG, as well as electrical stimulation via the ECoG grid, to effect just such a focused perturbation.

The study illustrated in *Figure 10.7* was led by neurologist Josef Parvizi, in collaboration with cognitive neuroscientist Kalanit Grill-Spector (a former postdoctoral trainee of Nancy Kanwisher). Parvizi and colleagues (2012; *Figure 10.7*) were interested in comparing findings from fMRI with those from ECoG, and so they focused on oscillatory power in the gamma band, which is believed to be monotonically related to the BOLD signal. (Although this has been demonstrated empirically in simultaneous fMRI and LFP measurements in the monkey [e.g., Logothetis et al., 2001], it can also be deduced from what we know about these two signals: the BOLD signal increases with increased metabolic demands driven by synaptic activity, and higher-frequency fluctuations in the LFP result from higher

rates of synaptic activity.) The image in panel B of *Figure 10.7* shows the location of two face-specific patches, produced with essentially the same methods as Kanwisher, McDermott, and Chun (1997). (Note from *Figure 10.3* that more than one face-selective patch was often identified in the right hemisphere of ventral occipitotemporal cortex, and this is also codified in the schematic in *Figure 10.5*) The fMRI scan that produced the image in panel B of *Figure 10.7* was acquired prior to the surgical implantation into this same individual of the ECoG array illustrated in panel A, and, although the investigators clearly hoped for such an outcome, it was "mere luck" that, for this patient, two electrodes from the ECoG array fell directly over these two face patches.

The pie charts in panel A of *Figure 10.7* represent the results of ECoG recording while the subject viewed stimuli from four categories. Note that these results could, in principle, be interpreted as being consistent with both modularist and distributed accounts of face processing. With regard to the former, electrodes over the fusiform gyrus clearly measure stronger gamma-band responses for faces than for non-face stimuli; with regard to the latter, there are clearly prominent responses to faces, as well as to non-faces, across a large expanse of posterior occipitotemporal cortex

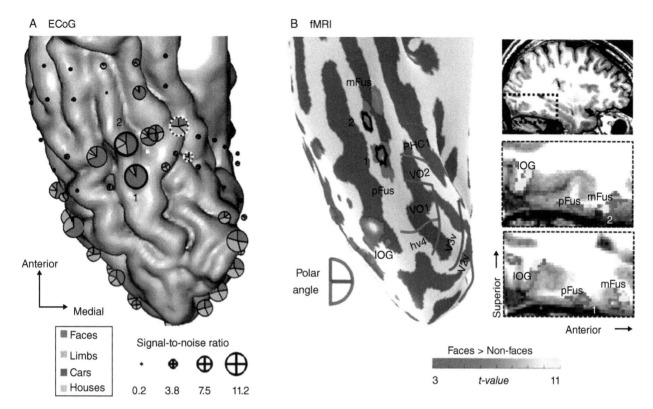

FIGURE 10.7 ECoG and electrical stimulation in the human brain. Panel **A** illustrates the layout of electrodes (black dots and pie charts) on the ventral surface of the patient's posterior right hemisphere. From their placement, one can surmise that two strips of electrodes, each containing two rows, were inserted lateral-to-medial along the ventral surface of the temporal lobe, the more posteriorly positioned one wrapping along the medial wall, and additional strips were positioned more superiorly, lateral and medial to the occipital pole. Each pie chart corresponds to an electrode that responded to one or more of the four categories of visual stimuli and is color-coded to indicate relative selectivity within these four. The size of each pie chart corresponds to the "signal-to-noise ratio," which can be understood as a normalized measure of the stimulus-specific increase in broadband power in the low- and high-gamma range (40–160 Hz) relative to the magnitude of fluctuations at this electrode at baseline. Panel **B** shows different views of regions whose univariate fMRI responses to faces exceeded responses to the other three categories, with colored lines on the inflated surface indicating retinotopically defined boundaries of the regions identified in black type. Superimposed on the inflated image in panel **B** is the location of the electrodes labeled "1" and "2" in panel **A**. Source: From Parvizi, Josef, Corentin Jacques, Brett L. Foster, Nathan Withoft, Vinitha Rangarajan, Kevin S. Weiner, and Kalanit Grill-Spector. 2012. "Electrical Stimulation of Human Fusiform Face-Selective Regions Distorts Face Perception." Journal of Neuroscience 32 (43): 14915–14920. doi: 10.1523/JNEUROSCI.2609-12.2012. Reproduced with permission of the Society for Neuroscience.

from which recordings were obtained. The additional step taken in this study that differentiates it from studies "limited" to ECoG is that the investigators stimulated the cortex by selectively passing current between pairs of electrodes in the array. The stimulation protocol (2–4 mA delivered in alternating square-wave pulses at 50 Hz) was intended to disrupt normal functioning in the tissue between the two electrodes. In this way, with the same experimental apparatus, the investigators could both localize activity that was correlated with a function of interest and assess the necessity of the regions so identified by disrupting their activity.

Although the published report contains considerable detail and illustration relating to the ECoG results, the truly novel and consequential finding from this study, produced by the focused electrical stimulation, was so clear and striking that no statistics, indeed, no quantification, were needed. It was sufficient to post a video clip of the patient's real-time report of what he was experiencing while looking at Parvizi while stimulation was delivered across electrodes 1 and 2: "'You just turned into somebody else. Your face metamorphosed.' When probed further, he reported that features appeared distorted: 'You almost look like somebody I've seen before, but somebody different. That was a trip. . . . It's almost like the shape of your face, your features drooped'" (Parvizi et al., 2012, p. 14918). The patient experienced no such perceptual metamorphoses when he attended to "Get Well Soon" helium balloons in the room, nor when he attended to the television. The video is astounding, and leaves no uncertainty about the answer to our question of whether the FFA plays a causal role in face processing.

The code for facial identity in the primate brain (!?!)

Tsao, Freiwald, Tootell, and Livinstone (2006) published a paper entitled "A cortical region consisting entirely of face-selective cells." They made this audacious claim based on the results of a study in which they first used a method similar to that of Kanwisher, McDermott, and Chun (1997) to identify face-selective activity in the brains of monkeys, with fMRI, then recorded with extracellular electrophysiology from the hotspots identified with fMRI (see *Figure 10.8* for a similar result from a different study). What they found when they recorded from this region is, well, stated in the title of their paper. In subsequent work, Doris Tsao and Winrich Freiwald, both in collaboration and in their own labs, identified a total of six face patches, along a caudal to rostral axis along the superior temporal sulcus and the inferior temporal gyrus. Previous work suggested a functional gradient along this axis, with neurons in the middle lateral (ML; "lateral" referring to the convexity of the inferior temporal gyrus) and middle fundus (MF; fundus of the superior temporal sulcus) patches

showing viewpoint-specific tuning, neurons in the anterior lateral (AL) patch achieving partial viewpoint invariance, and neurons in anterior medial (AM; "wrapping onto the ventral surface of ITG) achieving almost full viewpoint invariance. Indeed, in AM, characterized as "the final output stage of IT face processing," tuning had been observed to be "sparse" [i.e., highly specific], with neurons that "encode exemplars for specific individuals . . . respond[ing] to faces of only a few specific individuals, regardless of head orientation" (Chang and Tsao, 2017, p. 1013). Note that this is very different from the neurons that seem to be tuned for "the idea of a face," that we considered in *Chapter 6* (section *Broader implications of visual properties of temporal cortex neurons*). Indeed, this is much closer to the idea of the grandmother cell.

In a paper with a title perhaps as audacious as Tsao's 2006 paper, Chang and Tsao (2017) have presented a very sophisticated set of analyses and modeling that rule out the grandmother cell interpretation of neurons in face patch AM, thereby dramatically revising the interpretation that Tsao had given to her own previous work (e.g., Freiwald and Tsao, 2010). The key to the Chang and Tsao (2017) study was to approach the study of face processing as a quantitative problem, rather than one of engaging in a trial-and-error process of *let's-show-lots-of-stimuli-to-our-monkey-until-we-stumble-upon-one-that-this-neuron-responds-to*. Think back, for example, to the method of Tanaka and colleagues that led to the articulation of the principle of combination coding in IT (section *Combination coding, an alternative to the "grandmother cell," Chapter 6*). Although this approach led to an important conceptual breakthrough, a limitation is that it didn't provide an explanation for *why* neurons developed the seemingly arbitrary response properties that they did. For the neuron illustrated in *Figure 6.10*, for example, we don't know why it is that this neuron's preferred stimulus

FIGURE 10.8 Neuronal responses within face patches. Source: From Chang, L. and Tsao, D. Y. (2017). The code for facial identity in the primate brain. Cell 169: 1013–1028. Copyright 2017. Reproduced with permission of Elsevier.

A

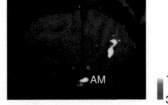

FIGURE 10.8.A fMRI activity in face patches in the temporal lobe. (Because the "shadow" of electrode tracks [caused by magnetic inhomogeneity] can be seen in ML and in AM, we can assume that these images are from a repeat scan, because images from an earlier scan would have been used for electrode targeting.)

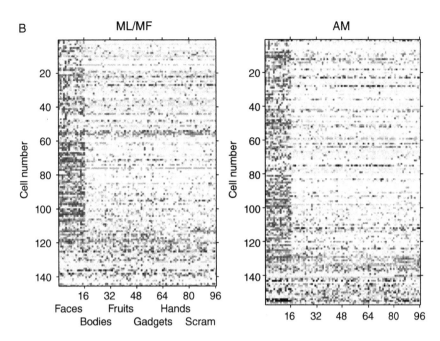

FIGURE 10.8.B Responses of neurons, averaged from 50 to 300 ms after the onset of each of 16 stimuli drawn from the categories listed along the horizontal axis of the figure. Each row corresponds to a single neuron, warmer colors indicating higher firing rate.

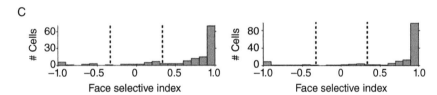

FIGURE 10.8.C Distribution of face selectivity of the neurons shown in *Figure 10.8.B*, as computed with the formula for a face selectivity index: [({mean response$_{faces}$} − {mean response$_{non-face\ objects}$})/({mean response$_{faces}$} + {mean response$_{non-face\ objects}$})]. (And so, at least for the animals in this study, these face patches didn't consist *entirely* of face-selective cells.)

seems to be for this configuration of a white square with two superimposed black rectangles, rather than, say, just the two black rectangles, or the square with black and white stripes with two superimposed black rectangles. Phrased another way, knowing that the reduction method revealed that this neuron is just as responsive to a white square with two superimposed black rectangles as it is to a top-down view of the head of a toy tiger doesn't provide any principled basis for predicting what this neuron's level of response would be for novel stimuli never before seen by this monkey. It doesn't provide information about what are the rules that determine this neuron's responsivity.

What Chang and Tsao (2017) did was to quantify each of 200 faces drawn from a database by identifying each of

58 landmarks on each of the faces and determining two values for each landmark: a location in a 2-D landmark-defined "face space" and a grayscale value (presumably from 0 to 255; *Figure 10.9*). With this method, I could hand you a piece of paper that had 116 numbers on it – a location value and a grayscale value for each of the 58 landmarks – and you would be able to use only those 116 numbers to recreate an image of the person that these numbers correspond to, an image that that person's mother would recognize as her own child. These two values make up what Chang and Tsao referred to as the "shape" dimension and the "appearance" dimension in their matrix of digitized values for the 200 faces in their stimulus set. Next, they applied two **Principal component analysis (PCA)** to

this matrix, one to identify the 25 largest principal components (PCs) in the shape dimension, the other to identify the 25 largest PCs in the appearance dimension, and these were used to construct a parameterized shape space. That is, by determining the 50 dimensions that are most discriminative among the 200 faces, they effected a "dimensionality reduction" that allowed for the conversion of each individual's location in 116-D "landmark space" to a 50-D location in "face PC space." The advantage of this step, in addition to the pragmatic benefit of reducing the size of each face file (i.e., fewer bytes are required to store a 50-D file on a computer than a 116-D file), is that it recoded face information out of the human-defined (and decidedly arbitrary) factors of the location and grayscale values of each of 58 landmarks, and into the "natural" dimensions along which each of the 200 faces actually varied in terms of their physical image properties. And, as it

turns out, knowing this face PC space gives one a remarkable ability to predict what never-before-seen faces will drive a neuron, including neurons that had previously been characterized as "sparse" and therefore assumed to be narrowly tuned for the face of just one or a few individuals.

The transformation from landmark space into the face PC space effected a recoding from a human-defined code into a code that reflected the statistical distribution of the image properties of the set of 200 faces. Most of the dimensions in face PC space were "holistic," in that they captured change along a dimension that involved multiple parts of the face. For example, moving along the first shape dimension produces simultaneous changes in hairline, face width, and height of eyes. Why are we going into this level of detail about this face PC space? (*Surely*, you may be thinking to yourself, *there must be some reason other than just to develop an appreciation for the fact that Chang and Tsao are*

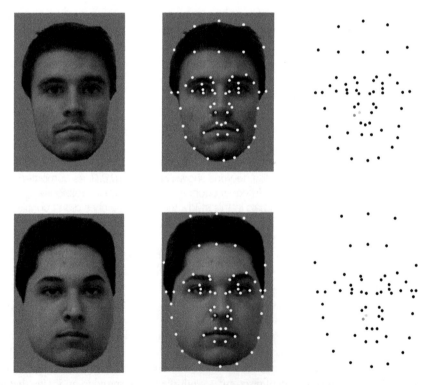

FIGURE 10.9 Quantifying what makes my face different from yours. Two faces taken from the FEI face database used by Chang and Tsao (2017) (database assembled by the Artificial Intelligence Laboratory of FEI in São Bernardo do Campo, São Paulo, Brazil). Each of the 200 faces was hand-labeled with 58 landmark points that corresponded to the same feature on each face. For illustrative purposes, the right-edge-of-nose landmark is colored blue, and the inside-edge-of-right-nostril landmark is colored yellow. Because the two model faces illustrated here have different shapes, the right-edge-of-nose landmark for these two will get a different location value, as well as a different grayscale value, and the same is true for each of the other 57 landmarks.

really good at math.) The reason is that the visual system of the monkey seems to use this face PC code for the perceptual analysis of faces! That is, if Chang and Tsao first used the trial-and-error method to find which of the 200 faces a neuron responds to, they could next look up where this face fell in the face PC space and then show this neuron different morphs of this "good" face that varied along one of the 50 axes of face PC space on which this good face could sit. Before too long they'd identify which was the best axis for that neuron, and could then demonstrate (i.e., they could predict) that any face that fell along this axis in face PC space would optimally drive this neuron, and that any face that didn't fall along this axis would produce a weaker response, the amount of weakening based on the distance from the preferred axis. Impressively, this meant that even for neurons that, by traditional definition, were highly selectively tuned to just one or a few face identities, Chang and Tsao could create never-before-seen faces that would drive this putatively "sparse" neuron. Furthermore, an implication of this model is that a single grandmother unit could not uniquely represent an individual face, because an infinite number of faces can be represented along any single axis in face PC space. Rather, theoretically, you'd need activity from at least 50 neurons, one corresponding to each dimension in the face PC space, to completely represent the identity of any one person's face. In practice, with their empirically collected data, Chang and Tsao (2017) found that they needed signals from a minimum of 200 neurons in MF/ML and AM to decode face identity with a high level of accuracy.

A heuristic take-home lesson from this study is that in brain regions where physiologists find neurons that seem to be highly selective for just a narrow set of variables, it is not necessarily the case that this reflects the true nature of the computations that this region carries out. Alternatively, these results may be due to the fact that the researchers simply do not know the principles underlying the activity in this brain area. A take-home for this chapter's topic of visual cognition is that the Chang and Tsao (2017) study offers a robust challenge to exemplar-based models, in which object recognition is assumed to be mediated by neurons tuned to specific objects.

VISUAL PERCEPTION AS PREDICTIVE CODING

Now that we've had three chapters devoted to visual object perception – covering the earliest stages from transduction through feature extraction, the anatomy and physiology

and some computational principles of the ventral visual stream, and now high-level object recognition – it is an apt time for us to give more thorough attention to models of visual perception as predictive coding. It gets very math-y very fast, but for our purposes I'm going to try to keep it in layman's terms. The premise of predictive coding is that the brain's modus operandi is to maintain a model (a set of predictions) about the current state of the outside world and to update that model when sensory evidence is inconsistent with it. This perspective flips the classical bottom-up perspective of perception on its head, construing the causal direction of information processing as one of top-down signals from higher-level areas informing lower-level areas of its predictions (i.e., of what kind of signals *should* be coming from the outside world) and lower-level areas signaling mismatches between these predictions and incoming evidence (i.e., errors in the prediction), with feedforward error signals used to update the internal model. Thus, in this framework, the feedforward signals that are traditionally understood as carrying sensory information about newly detected stimuli are reinterpreted as "after-the-fact" feedback about how good the internal prediction was. (In this text, we'll continue to use the terms "feedforward" and "feedback" with reference to the anatomically defined hierarchy.) (Suggestions for more formal treatments of predictive coding theory are listed in *Further Readings*.)

Key for models of the neural implementation of predictive coding is an account for how prediction signals and error signals are propagated and segregated. Most models hold that, within a cortical column in a generic region in the ventral stream (we can use V4), a representation of the brain's internal model (of the prediction) is maintained in deep layers, and feedback from layer V (of V4, for our example) to layers II/III (of V2) (via the connection illustrated in *Figure 2.4*) is the channel for conveying this prediction to a region closer to the sensory interface. In this feedback signal, the V4-level of stimulus representation is decomposed into the hierarchically "earlier" level of representation that V2 "can understand." In V2, input from V1 is then compared to the prediction that had been delivered by V4, and the error between prediction and feedforward signal is then propagated from V2 to V4 via the layer II/III-to-layer IV projection (*Figure 2.4*). This error signal also updates the predictive model encoded in V2's deep layers, and the resultant updated prediction from V2 is fed back onto V1. Laminar recordings of monkeys performing a variety of tasks relying on visual perception indicate that feedback signaling generated by layer V pyramidal neurons whose oscillations synchronize in the 10–30 Hz range (i.e., the alpha and beta bands) and feedforward signaling by layer II/III neurons whose oscillations synchronize in the 40–90 Hz range (the gamma band). This scheme for clocking a transmission within a (relatively) narrow frequency band is referred to as using frequency channels. One advantage of such a scheme is that, from the perspective of a cortical column in V2, it's less likely to confuse the ascending error signal from V1 and the descending prediction signal from V4 if the two are being carried in different frequency channels.

The experiment with which we'll conclude this chapter applied the ideas summarized in the two previous paragraphs to test the idea that face perception can be understood as a product of predictive coding. In it, a team led by Alla Brodski-Guerniero in the laboratory of Michael Wibral at Goethe University in Frankfurt, Germany, started with the assumption that if an individual knows that they are in a situation in which they might see a particular stimulus (e.g., if cued by a "Beware of the Dog" sign; *Figure 10.10.A*), they'll "preactivate" knowledge about this stimulus to facilitate its detection if it is encountered. For the experiment, Brodski-Guerniero et al. (2017) acquired MEG while subjects discriminated difficult-to-perceive black-and-white silhouette images of faces, houses, and scrambled versions of these (*Figure 10.10.B*).

The MEG data were analyzed to assess Active Information Storage (AIS), an **information theoretic** measure that indexes the amount of stimulus-related information in a signal (for our purposes, analogous to MVPA). An advantage of carrying out this experiment with MEG was that, because MEG

signals lend themselves to reasonably accurate spatial localization (section *Magnetoencephalography, Chapter 3*), it was capable of resolving frequency band–specific signaling localized to particular regions (*Figure 10.11*).

Of highest relevance for the predictive coding account, "directed information transfer" (assessed with the information theoretic measure called transfer entropy) indicated that, during the anticipatory period preceding each stimulus, face-related information flowed from aIT to FFA and from PPG to FFA. Thus, these findings were broadly consistent with the idea that face perception may be supported by the top-down signaling of predictions about faces.

PLAYING 20 QUESTIONS WITH THE BRAIN

Is anyone else dizzy? We've covered many profound questions, and in so doing have reviewed many cutting-edge techniques for modeling, measuring, and manipulating activity in the brain. One thing that comes to mind is the admonition delivered by AI pioneer Allen Newell (1973) almost 50 years ago: *You can't play 20 questions with nature and win.* Some questions about brain and behavior can never be satisfactorily addressed without formal computational modeling. This is particularly evident when one finds, as we have here, that seemingly incontrovertible evidence can be generated for each of two seemingly incompatible perspectives on how a function like face perception is carried out.

FIGURE 10.10 Rationale and stimuli for study of role of predictive coding in face recognition (Brodski-Guerniero et al., 2017). Brodski-Guerniero, A., G.-F. Paasch, P. Wollstadt, I. Özdemir, J. T. Lizier, and M. Wibral. 2017. "Information-Theoretic Evidence for Predictive Coding in the Face-Processing System." Journal of Neuroscience 37: 8273–8283. Licensed under Creative Commons Attribution 4.0 International License (CC-BY).

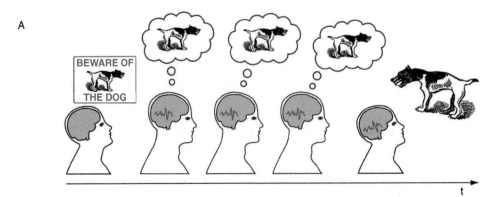

FIGURE 10.10.A Upon seeing an informative cue, an individual preactivates information from semantic memory (*Chapter 13*) and maintains an active representation of it (cartooned "brain signal") until the predicted object is perceived.

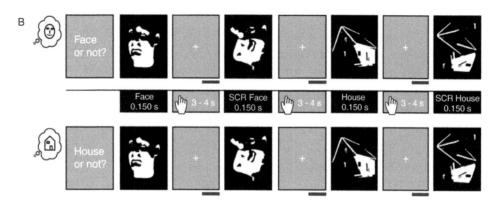

FIGURE 10.10.B Schematic illustration of blocks of trials in which faces, scrambled faces (SCR face), houses, and scrambled houses (SCR houses) were presented for discrimination. Note that stimuli of all four types were presented in both types of blocks, but all stimuli except faces were to be rejected during face blocks, and all stimuli except houses were to be rejected during house blocks. Hand icon indicates yes/no response; red bar indicates interval from which MEG data were analyzed.

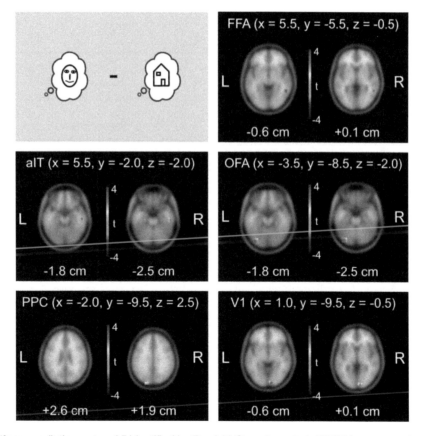

FIGURE 10.11 A "face-prediction network" identified by Brodski-Guerniero et al. (2017) in a comparison of face vs. house blocks. In each of these regions (anterior IT [aIT]; posterior parietal cortex [PPC]; fusiform face area [FFA]; occipital face area [OFA]; and v1), the strength of face information (indexed by AIS) correlated on a trial-by-trial basis with power in the alpha (8–14 Hz) and beta (14–32 Hz) bands. The correct source of these figures is Brodski-Guerniero, A., G.-F. Paasch, P. Wollstadt, I. Özdemir, J. T. Lizier, and M. Wibral. 2017. "Information-Theoretic Evidence for Predictive Coding in the Face-Processing System." *Journal of Neuroscience* 37: 8273–8283. Licensed under Creative Commons Attribution 4.0 International License (CC-BY).

END-OF-CHAPTER QUESTIONS

1. What factors call into question the idea that associative agnosias may reflect truly unimpaired perception "stripped of its meaning"?

2. How can prosopagnosic patients successfully navigate our modern, intensely social, society?

3. Do PDP models of visual object recognition "prove" that there is no distinction between "structural" and "semantic" levels of visual representation?

4. What was it about the Kanwisher, McDermott, and Chun (1997) results that some viewed as contrary to how they thought that the brain works?

5. What are the alternative explanations of the Kanwisher and colleagues (1997) results proffered by Gauthier and colleagues?

6. In what sense do the findings of Haxby et al. (2001) rule out a strong version of the argument that the FFA is specialized for face processing?

7. In what sense do the findings of Parvizi et al. (2012) establish that the FFA makes necessary contributions to face processing?

8. What is the premise of the face PC code proposed by Chang and Tsao (2017) and how does it obviate the need for grandmother cells?

9. How is activity in FFA related to face perception explained from a predictive coding framework?

REFERENCES

Benson, D. Frank, and John P. Greenberg. 1969. "Visual Form Agnosia: A Specific Defect in Visual Discrimination." *Archives of Neurology* 20 (1): 82–89. doi: 10.1001/archneur.1969.00480070092010.

Brodski-Guerniero, A., G.-F. Paasch, P. Wollstadt, I. Özdemir, J. T. Lizier, and M. Wibral. 2017. "Information-Theoretic Evidence for Predictive Coding in the Face-Processing System." *Journal of Neuroscience* 37: 8273–8283.

Chang, L., and D. Y. Tsao. 2017. "The Code for Facial Identity in the Primate Brain." *Cell* 169: 1013–1028.

Farah, Martha J. 1990. *Visual Agnosia*. Cambridge, MA: MIT Press.

Fodor, Jerry A. 1983. *The Modularity of Mind*. Cambridge, MA: MIT Press.

Freiwald, W. A., and D. Y. Tsao. 2010. "Functional Compartmentalization and Viewpoint Generalization within the Macaque Face-Processing System." *Science* 330: 845–851.

Fuster, Joaquín M. 2003. "More Than Working Memory Rides on Long-Term Memory." *Behavioral and Brain Sciences* 26 (6): 737–737. doi: 10.1017/S0140525X03300160.

Gauthier, Isabel, and Michael J. Tarr. 1997. "Becoming a 'Greeble' Expert: Exploring Mechanisms for Face Recognition." *Vision Research* 37 (12): 1673–1682. doi: 10.1016/S0042-6989(96)00286-6.

Gauthier, Isabel, Michael J. Tarr, Adam W. Anderson, Pawel Skudlarski, and John C. Gore. 1999. "Activation of the Middle Fusiform 'Face Area' Increases with Expertise in Recognizing Novel Objects." *Nature Neuroscience* 2 (6): 568–573. doi: 10.1038/9224.

Haxby, James V., M. Ida Gobbini, Maura L. Furey, Alumit Ishai, Jennifer L. Schouten, and Pietro Pietrini. 2001. "Distributed and Overlapping Representations of Faces and Objects in Ventral Temporal Cortex." *Science* 293 (5539): 2425–2430. doi: 10.1126/science.1063736.

Humphreys, Glyn W., and M. Jane Riddoch. 1987. "The Fractionation of Visual Agnosia." In *Visual Object Processing: A Cognitive Neuropsychological Approach*, edited by Glyn W. Humphreys and M. Jane Riddoch, 281–307. Hove, UK: Lawrence Erlbaum Associates.

Kanwisher, Nancy. 1991. "Repetition Blindness and Illusory Conjunctions: Errors in Binding Visual Types with Visual Tokens." *Journal of Experimental Psychology: Human Perception and Performance* 17 (2): 404–421. doi: 10.1037/0096-1523.17.2.404.

Kanwisher, Nancy, Josh McDermott, and Marvin M. Chun. 1997. "The Fusiform Face Area: A Module in Human Extrastriate Cortex Specialized for Face Perception." *Journal of Neuroscience* 17 (11): 4302–4311. http://www.jneurosci.org/content/17/11/4302.full.pdf+html.

Logothetis, Nikos K., Jon Pauls, Mark Augath, Torsten Trinath, and Axel Oeltermann. 2001. "Neurophysiological Investigation of the Basis of the fMRI Signal." *Nature* 412 (6843): 150–157. doi: 10.1038/35084005.

Milner, Brenda, and Hans-Lukas Teuber. 1968. "Alteration of Perception and Memory in Man: Reflections on Methods." In *Analysis of Behavioral Change*, edited by Lawrence Weiskrantz, 268–375. New York: Harper and Row.

Newell, A. 1973. "You Can't Play 20 Questions with Nature and Win: Projective Comments on Papers at This Symposium." In

W. G. Chase (ed.), *Visual Information Processing: Proceedings of the Eighth Annual Carnegie Symposium on Cognition.* Pittsburgh, PA Academic Press.

Parvizi, Josef, Corentin Jacques, Brett L. Foster, Nathan Withoft, Vinitha Rangarajan, Kevin S. Weiner, and Kalanit Grill-Spector. 2012. "Electrical Stimulation of Human Fusiform Face-Selective Regions Distorts Face Perception." *Journal of Neuroscience* 32 (43): 14915–14920. doi: 10.1523/JNEUROSCI.2609-12.2012.

Pohl, Walter. 1973. "Dissociation of Spatial Discrimination Deficits Following Frontal and Parietal Lesions in Monkeys."

Journal of Comparative and Physiological Psychology 82 (2): 227–239. doi: 10.1037/h0033922.

Tarr, Michael J., and Isabel Gauthier. 2000. "FFA: A Flexible Fusiform Area for Subordinate-Level Visual Processing Automatized by Expertise." *Nature Neuroscience* 8 (3): 764–769. doi: 10.1038/77666.

Tsao, D.Y., W. A. Freiwald, R. H. Tootell, and M. S. Livingstone. 2006. "A Cortical Region Consisting Entirely of Face-Selective Cells." *Science* 311: 670–674.

OTHER SOURCES USED

Bukach, Cindy M., and Jessie J. Peissig. 2010. "How Faces Become Special." In *Perceptual Expertise: Bridging Brain and Behavior*, edited by Isabel Gauthier, Michael J. Tarr, and Daniel Bub, 11–40. Oxford: Oxford University Press.

Ellis, Hadyn D. 1996. "Bodamer on Prosopagnosia." In *Classic Cases in Neuropsychology*, edited by Chris Code, Claus-W. Wallesch, Yves Joanette, and André Roch Lecours, 69–75. Hove, UK: Psychology Press.

Foster, Brett L., and Josef Parvizi. 2012. "Resting Oscillations and Cross-Frequency Coupling in the Human Posteromedial

Cortex." *NeuroImage* 60 (1): 384–391. doi: 10.1016/j.neuroimage.2011.12.019.

Gross, Charles G. 1994. "Hans-Lukas Teuber: A Tribute." *Cerebral Cortex* 4 (5): 451–454. doi: 10.1093/cercor/4.5.451.

Pribram, Karl H. 1977. "Hans-Lukas Teuber: 1916–1977." *American Journal of Psychology* 90 (4): 705–707. http://www.jstor.org/stable/1421744.

Riddoch, M. Jane, and Glyn W. Humphreys. 1987. "A Case of Integrative Visual Agnosia." *Brain* 110, 1431–1462.

Rubens, Alan B., and Frank Benson. 1971. "Associative Visual Agnosia." *Archives of Neurology* 24 (4): 305–316.

FURTHER READING

Bastos, A. M., W. M. Usrey, R. A. Adams, G. R. Mangun, P. Fries, and K. J. Friston. 2012. "Canonical Microcircuits for Predictive Coding." *Neuron* 76: 695–711.

A formal proposal for how cortical columns can be understood as implementing the simultaneous operations of receiving top-down signals that update the system's predictive model, bottom-up signals conveying the error computed at an earlier stage of processing, and the calculation and feedforward transmission of the error between these two incoming signals.

Çukur, Tolga, Shinji Nishimoto, Alexander G. Huth, and Jack L. Gallant. 2013. "Attention During Natural Vision Warps Semantic Representation across the Human Brain." *Nature Neuroscience* 16 (6): 763–770. doi: 10.1038/nn.3381.

A "mind-warping" demonstration of a state-of-the-art (as of this writing) fMRI investigation of the neural bases of visual knowledge representation.

Farah, Martha J. 1990. *Visual Agnosia*. Cambridge, MA: MIT Press.

An encyclopedic synthesis of 100 years of research on the topic; be sure to look at the most recent edition, as it will contain updates on developments in this fast-moving literature since the original 1990 publication.

Fodor, Jerry A. 1985. "Précis of *The Modularity of Mind*." *Behavioral and Brain Sciences* 8 (1): 1–5. doi: 10.1017/S0140525X0001921X.

A quicker read than the 1983 monograph, plus, it is supplemented by interesting critiques from others in the field, followed by Fodor's rejoinder.

Friston, K. 2018. "Does Predictive Coding Have a Future?" *Nature Neuroscience* 21: 1019–1021.

This concise commentary from a leading theoretician recounts the "strange inversion" in scientific thinking, whereby the brain has come to be seen as "an organ of inference, actively constructing explanations for what's going on 'out there', beyond its sensory epithelia."

Logothetis, Nikos K. 2003. "The Underpinnings of the BOLD Functional Magnetic Resonance Imaging Signal." *Journal of Neuroscience* 23 (10): 3963–3971.

A review summarizing the relation between the LFP and BOLD signals that trade accessibility for some of the technical details packed into Logothetis et al. (2001).

Wandell, Brian A., and Jonathan Winawer. 2011. "Imaging Retinotopic Maps in the Human Brain." *Vision Research* 51(7): 718–737. doi: 10.1016/j.visres.2010.08.004.

If you didn't look this one up after reading Chapter 2, here's another prompt! This paper presents a thorough summary of neuroimaging studies of the functional organization of the ventral stream, including category representation.

CHAPTER 11
NEURAL BASES OF MEMORY

KEY THEMES

- Foundational knowledge about the organization of memory, and the amnesic syndrome, has resulted from the pioneering research of Brenda Milner and her associates with the patient H.M.

- The model assembled from the global amnesic syndrome is one of dissociable systems supporting declarative vs. nondeclarative vs. working memory and of dissociable processes of encoding, consolidation, storage, and retrieval.

- Long-term memory (LTM) is underlain by the principle of association, and an influential model of a neurobiological and computational mechanism that might implement associative learning in the brain was proposed by D.O. Hebb in 1949.

- Hebbian learning seems to be embodied in the phenomenon of long-term potentiation (LTP) of synaptic strength, discovered by Lømo and Bliss in the 1960s and 1970s.

- A critical implementer of many types of LTP is the NMDA subtype of glutamate receptor, which acts as a synaptic coincidence detector.

- As of today, there are many proposals for "how the hippocampus works," but there does not yet exist a broad consensus on which one is likely to be correct.

- The phenomena of memory consolidation, and the more-recently characterized reconsolidation, pose important challenges for cognitive neuroscience, but also potential therapeutic opportunities.

CONTENTS

PLASTICITY, LEARNING, AND MEMORY

In several previous chapters, most notably when we considered use-dependent somatosensory reorganization, motor learning, and the simulation of learning by adjusting connection weights in a parallel distributed processing (PDP) model, we have touched on a fundamental property of the nervous system, which is its ability to modify itself in order to refine its representation of the outside world. Here, finally, we will address the phenomenon of *neural plasticity*, and its behavioral and theoretical counterparts *memory* and *learning*, head-on. We'll begin with a summary of an influential neurosurgical case that revealed many principles of human learning and memory, then consider computational and physiological models of cellular and molecular bases of these phenomena, and conclude at the systems level. This will provide a foundation for the subsequent three chapters' explorations of the kinds of memory that we use every day: episodic, semantic, and working.

THE CASE OF H.M.

Whether in terms of raw citation counts or subjective assessment of breadth and depth of legacy, H.M. may well be the single most influential case in the history of cognitive neuroscience. He spent his life in the vicinity of Hartford, CT, where, in 1953, at the age of 27, he underwent a neurosurgical procedure that truly shaped the course of scientific history.

Bilateral medial temporal lobectomy

In the back seat of his parents' car, on the day of his fifteenth birthday, Henry Molaison was stricken by his first grand mal epileptic seizure. Over the next few years, his neurological condition progressively worsened, to the point where he dropped out of high school, couldn't hold down a steady job, and, by his mid-20s, was effectively house-bound. For several years, he and his family had consulted with a neurosurgeon at Hartford Hospital, Dr. William Scoville, who, after exhausting other treatment strategies, proposed an experimental surgical intervention. Because EEG had failed to identify a clear focus of Henry's seizures, Scoville proposed the surgical removal of the hippocampus, known to be the most epileptogenic tissue in the brain. (A subsequent section summarizing the circuitry

of the hippocampus will provide some insight into why this may be.)

A diagram of the procedure, the surgeon's estimate of the extent of removal, and MR images acquired decades after the surgery, during the 1990s, are illustrated in *Figure 11.1*. Although the surgery was successful from the narrow perspective that it brought H.M.'s epilepsy under control, its unexpected consequences were profound and permanent: H.M. lost the ability to consciously remember anything that he experienced from that time forward for the remaining 50+ years of his life.

The global amnesic syndrome

The influence of H.M.'s case is due, in large part, to the pioneering research of Dr. Brenda Milner, a British-born and -trained neuropsychologist who has spent the majority of her scientific career at McGill University and the Montreal Neurological Institute, working with such luminaries as Wilder Penfield (whom we met in section *Somatotopy, Chapter 5*) and Donald Hebb (who we met in the section on *Hebbian plasticity, Chapter 8*). Shortly after H.M.'s surgery, Scoville came across a report from Penfield and Milner detailing rare instances of profound memory impairment resulting from unilateral removal of the medial temporal lobes (MTLs) and speculating that this may have resulted from the fact that the unoperated hemisphere was also, unbeknownst to them, dysfunctional. Scoville contacted Penfield to tell him about his case, and, soon thereafter, Milner traveled from Montreal to Hartford to initiate her historic series of studies. This research established several facts about the consequences of bilateral damage of the MTL:

- Within the realm of cognition, its effects are, by and large, restricted to memory. Thus, for example, H.M.'s ability to understand and produce language was unaffected by his surgery, as was his IQ, which actually increased marginally after his surgery, no doubt due to the reduction of epileptic symptoms.
- Within the realm of memory, its effects are primarily restricted to the acquisition of new memories, in that facts acquired and events occurring before the damage remain accessible. Thus, bilateral damage of the MTL produces an *anterograde amnesia* (loss of memory going forward in time).
- The comparatively less severe *retrograde amnesia* is limited to the few years immediately prior to the insult – remote memories are spared.

- Within the realm of memory acquisition, the deficit is limited to events and facts that can be consciously recalled; patients can continue to learn new skills and habits that don't require conscious recollection.
- Also spared by MTL damage is working memory (a.k.a. short-term memory), the ability to remember a telephone number while looking for your phone or, as considered in *Chapter 9*, to briefly remember the location previously occupied by a stimulus that is no longer there.

Some of these propositions have been refined and/or called into question in the ensuing years, but at this early juncture in the chapter it will be useful to summarize them as principles whose tenets, across this and the next three chapters, we can then test and revise as appropriate.

The organization of memory

The principles arising from the work of Milner and colleagues include the following:

1. There is a fundamental distinction between **declarative memory** (or **explicit memory**) and **nondeclarative memory** (or **implicit memory**). The former refers to consciously accessible records of facts and events and depends on the integrity of the MTL. The latter includes any example of experience-dependent plasticity that modifies a behavior in a manner that does not require conscious access for its expression and includes conditioning, motor learning, and perceptual learning.
2. Declarative memory and nondeclarative memory are both considered LTM and are distinct from working memory.

FIGURE 11.1 H.M.'s lesions.

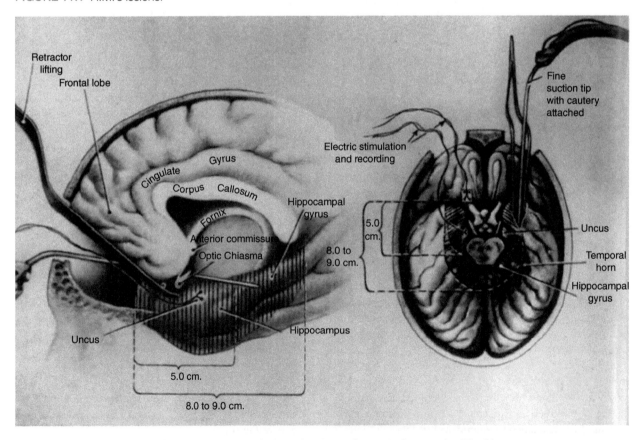

FIGURE 11.1.A Illustration of the surgical procedure, and estimated extent of removal, of the hippocampus.
Source: Scoville, William Beecher, and Brenda Milner. 1957. "Loss of Recent Memory after Bilateral Hippocampal Lesions." Journal of Neurology, Neurosurgery, and Psychiatry 20 (1): 11–21. doi: 10.1136/jnnp.20.1.11. Reproduced with permission of BMJ.

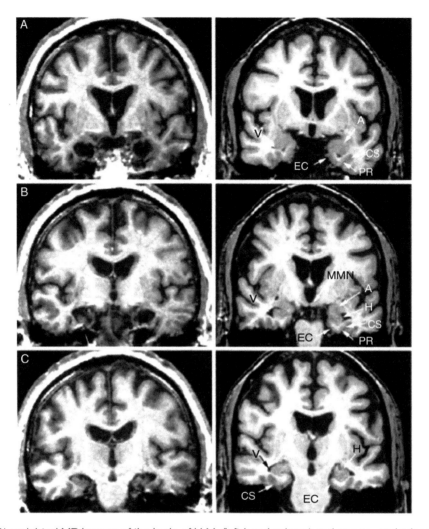

FIGURE 11.1.B T1-weighted MR images of the brain of H.M. (left-hand column) and an age- and education-matched control subject, at three progressively more caudal locations. Labels on the control images identify landmarks and regions that are damaged or missing in H.M. (A = amygdala; CS = collateral sulcus; EC = entorhinal cortex; PR = perirhinal cortex; MMN = medial mammillary nuclei; H = hippocampus; V = ventricle). At the caudal-most slice, atrophied remnants of hippocampal tissue that weren't removed by the surgery are visible. MMN, not directly affected by the surgery, are reduced in size in H.M., presumably due to the loss of afferent drive that they would typically receive from the hippocampus, via the **fornix** (see panel B). Source: Adapted from Corkin, Amaral, González, Johnson, and Hyman, 1997. Reproduced with permission of the Society of Neuroscience.

3. There is a neuroanatomical basis for dissociating the processes of encoding, storage, and retrieval in declarative memory. In the case of H.M., as presented so far, his deficit seems to be one of **encoding** – processing newly encountered information such that it can later be remembered. If either storage or retrieval were compromised, he wouldn't be able to retrieve memories from prior to his surgery. Thus, storage and retrieval would not seem to depend on the MTL.

4. Long-term declarative memories undergo a period of **consolidation**, to which the MTL contributes importantly, after which they become relatively immune to disruption by MTL damage.

The subjective experience of amnesia

While the cases of H.M. and many other patients have been invaluable for the scientific study of memory, they also raise profound questions that get at the core of what it means to be a sentient human being and confront us with heartbreakingly poignant conditions before which, at the present time, we are helpless.

Self-report from H.M. suggests that one phenomenological quality of anterograde amnesia is a pervasive anxiety about what may have happened just beyond the edge of the present moment:

> Right now, I'm wondering, have I done or said anything amiss? You see, at this moment everything looks clear to me, but what happened just before? . . . It's like waking from a dream; I just don't remember . . . Every day is alone, in itself. Whatever enjoyment I've had, whatever sorrow I've had. (Hilts, 1996, p. 138)

On another occasion, during an exchange between H.M. and researcher William Marslen-Wilson (relayed by Hilts, 1996), the patient confessed to worrying about giving the wrong answer, whether during formal testing or just in conversation:

> "It is a constant effort," Henry said. You must always "wonder how is it going to affect others? Is that the way to do it? Is it the right way?" . . . Asked if he worried about these things a lot, struggled with his thought to get right answers, he said yes, all the time. "But why?" "I don't know," said Henry. (p. 140)

The anxiety and frustration of amnesia is even more evident in Clive Wearing, a distinguished British musicologist, whose amnesia resulted from herpes encephalitis. This condition can produce severe damage to the hippocampus while leaving the rest of the brain relatively unscathed. In a clip from a television documentary, he and his wife are sitting in a city park:

WIFE: Do you know how we got here?
WEARING: No.
WIFE: You don't remember sitting down?
WEARING: No.
WIFE: I reckon we've been here about 10 minutes at least.
WEARING: Well, I've no knowledge of it. My eyes only started working now . . .
WIFE: And do you feel absolutely normal?

WEARING: Not absolutely normal, no. I'm completely confused.
WIFE: Confused?
WEARING (AGITATEDLY):   Yes. If you've never eaten anything, never tasted anything, never touched anything, never smelled something, what right have you to assume you're alive?
WIFE: Hmm. But you are.
WEARING: Apparently, yes. But I'd like to know what the hell's been going on!

(In making these pronouncements about his senses, it is clear that Wearing is speaking figuratively, not literally.) Thus, for Wearing, too, each waking moment feels as though he is just waking up from sleep. The journal that he keeps is filled with multiple entries that all contain variants of the same message. For example, directly under the entry "10:49 am I Am Totally Awake – First time," which appears on the first line of a page, is a second entry "11:05 am I Am Perfectly Awake – First time," and so on. When left alone in his room, the patient fills entire pages in this way with entries made at intervals ranging from 5 to 45 minutes.

These anecdotes capture an essential quality of the conscious phenomenology of the MTL amnesic patient, the near-continual impression of just having awakened from unconscious sleep. The plight of the MTL amnesic patient, then, is to be fully cognizant of, if not preoccupied by, the fact that one is not cognizant of the daily events of one's life.

Hippocampus vs. MTL?

At the time of the initial findings with H.M., and for most of the ensuing quarter century, most of the findings and principles summarized here were framed in terms of the hippocampus (e.g., "declarative memory is hippocampally dependent"). Subsequently, there has been an upswing in interest in the contributions of the neocortical regions of the MTL – the entorhinal, perirhinal, and parahippocampal cortex (PHC) – to memory functions and the extent to which each may make distinct contributions. These are questions about the organization of memory at a systems level – that is, at the level of circuits or regions of the brain and how they interact. To consider these systems-level questions in a substantive way, however, we'll first need a more detailed understanding of the cellular and molecular bases of memory formation and retention than what we've covered up to this point.

ASSOCIATION THROUGH SYNAPTIC MODIFICATION

Recall that in *Chapter 8*'s section on Synaptic plasticity we applied Hebb's postulate to the example from Pavlovian conditioning. What we said is that there was something special about the fact that metronome-related activity in the auditory system coincided with the activity in the digestive neurons. Thus, even though the auditory activity initially had no causal role in driving the digestive neurons, something about the coincidence of their activity led to a strengthening of the ability of auditory activity to drive salivary activity. But what we didn't get into at the time was the question of *why*. What was that special something? To address this question, we'll next turn to experiments initiated in Oslo in the mid-1960s.

Long-term potentiation

Hippocampal circuitry and its electrophysiological investigation

For our fanciful trip to Oslo to be most profitable, we'll need to be familiar with the circuitry of the hippocampus. As illustrated in *Figure 6.5*, the hippocampus can be construed as sitting atop the cortical hierarchy, receiving highly processed inputs from the highest levels of the dorsal stream (via feedforward projections from PPC to the PHC) and the ventral stream (via feedforward projections from IT and STP to perirhinal cortex [PRC]). PHC and PRC each funnel intoentorhinal cortex (EC), which then projects to hippocampus. The hippocampus is considered the top of the hierarchy because its projections don't go to yet another downstream area. Instead, after the incoming information is processed through its circuitry, the pattern of cortical convergence onto the hippocampus is inverted, and hippocampal outputs are fed back out to EC, then divergently to many regions of neocortex.

The hippocampus itself differs from six-layer neocortex in that it is, in effect, a rolled-up sheet of three-layer archicortex ("old" cortex). *Figure 11.2.A* shows the hippocampus in cross section and *Figure 11.2.B* how, within this 2-D slice, one can trace a complete circuit from the inputs from EC to the dentate gyrus (DG), through the CA fields, then back out to cortex. This circuitry lends itself to various experimental procedures that wouldn't be feasible in other brain regions. Now we can drop in on the University of Oslo in the mid-1960s, where the electrophysiologist Per Andersen had pioneered the measurement of the "population spike" in the DG. Due to the homogeneity and relative feedforward simplicity of the circuitry, if one delivers electrical stimulation to the axons projecting from EC to the DG (the "perforant path"), a field potential recorded from the granule cell body layer of the DG gives a straightforward measure of the aggregate strength of the connection between neurons in these two regions (*Figure 11.3*).

The breakthrough finding relating to Hebbian plasticity was first made by Andersen's graduate student Terje Lømo, and later confirmed and codified by Lømo in collaboration with visiting British neurophysiologist Timothy Bliss. The protocol of Bliss and Lømo (1973) was to deliver high-frequency (20 Hz) trains (15 seconds in duration) of electrical pulses ("tetani") to the perforant path of an anaesthetized rabbit and measure the resultant population spike from the granule cell-body layer of the DG.

Prior to delivery of the first tetanus, the researchers measured the baseline strength of the EC–DG connection by delivering single pulses. The waveforms at the top of *Figure 11.3* show the baseline response, a negative-going potential (here, negative is plotted up) with two bumps, that then returns to baseline. Why is negative plotted up? The negative-going potential corresponds to the depolarization of granule cells (intracellular depolarization causing extracellular hyperpolarization, which is what the extracellular electrode records), the "dimple" in the middle of the depolarization wave is the population spike (after discharge, a neuron rapidly hyperpolarizes, producing a positive shift in the extracellular potential). At baseline, very few DG neurons generate an action potential in response to single pulses to the perforant path, and so the population spike is very small. Each data point along the horizontal axis indicates the magnitude of the population spike produced by a single probing pulse (responses to the tetani themselves are not shown), and the baseline population spikes from time 0 to 0.75 hour (i.e., 45 minutes) are between 1 and 0 mV. At time 0.75 h, a tetanus is delivered to the perforant path in the "test" hemisphere, and the population spike to subsequent probing pulses in this hemisphere is briefly elevated (filled circles), whereas the control hemisphere is largely unchanged. At time 1.35 h, after the second tetanus, the response of the population spike is larger, and only weakens slightly, leveling off at 1 mV over the ensuing 30 minutes. At time 1.85 hours,

FIGURE 11.2 The hippocampus in cross section.

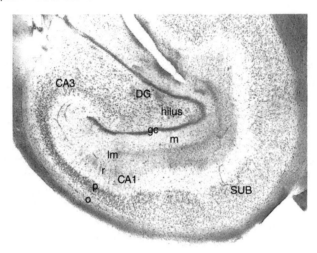

FIGURE 11.2.A Coronal section through the right hemisphere. DG = dentate gyrus, and within the DG, gc = granule cell layer and m = stratum moleculare. The layers of the CA fields are lm = stratum lacunosum moleculare; r = stratum radiatum; p = stratum pyramidale; o = stratum oriens. SUB = subiculum, an output station not included in *Figure 11.2.B*. Source: Rhythms of the Brain by Gyorgy Buszáki, 2006. Reproduced with permission of Oxford University Press, USA.

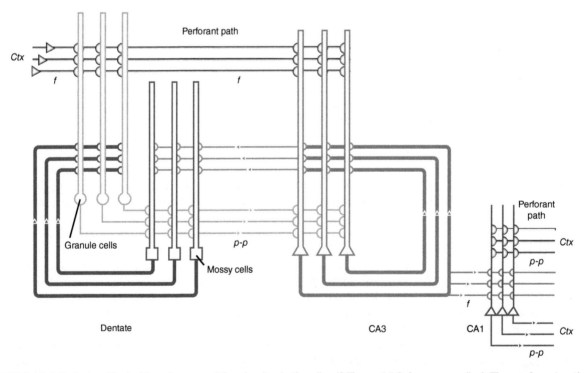

FIGURE 11.2.B A simplified wiring diagram of the circuitry in the slice if *Figure 11.2.A* were unrolled. The perforant path projects from entorhinal cortex to DG, CA3, and CA1. The projections from EC to DG are "fanning" (i.e., divergent), as are those from CA3 to itself (i.e., recurrent projections) and from CA3 to CA1, whereas those from EC to CA1 and from DG to CA3 are "point-to-point" (p-p).

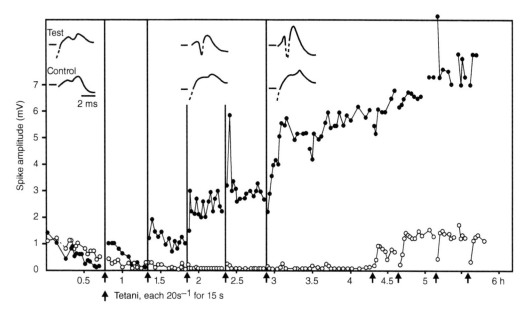

FIGURE 11.3 LTP in the DG of the hippocampus (Bliss and Lømo, 1973). Source: From Lomo, Terje. 2003. "The Discovery of Long-Term Potentiation." *Philosophical Transactions of The Royal Society of London, Series B: Biological Sciences* 358 (1432): 617–620. doi: 10.1098/rstb.2002.1226. Reproduced with permission of the Royal Society Publishing.

after the third tetanus, the population spike is further strengthened to 2.5 mV. (Representative waveforms from this epoch are illustrated at the top of the figure.) Finally, just before time 3 hours, a final tetanus is delivered. From this point forward, the population spike starts to grow "on its own," without any further tetani, from a magnitude of ∼ 3 mV to in excess of 7 mV. (The four tetani delivered after time 4 hours were delivered to the control hemisphere.) Critically, the potentiated state of the population spike was sustained for the next 3 hours, until the experiment was terminated. Ergo, "long-term potentiation." (After 6 hours of recording, preceded by several hours of setup and surgery, it was surely time to go to the pub!)

From these results it was inferred (and later confirmed) that this strengthening of the population spike reflected Hebbian plasticity, because it came about after the initial pulses of the tetanus had depolarized the granule cells, such that later pulses in the tetanus generated action potentials. (Delivery of the same number of single pulses, separated in time, would not have this effect.) Thus, it was the coincidence of (upstream) perforant path activity and (downstream) granule cell firing that resulted in the strengthening (the "potentiation") of the synapse.

There is near-universal consensus that LTP, as illustrated in *Figure 11.3*, is a core phenomenon underlying synaptic

plasticity in many regions of the brain, and that it is fundamental to learning and memory. However, although we've now seen how LTP can be induced, we still haven't addressed how it happens. How does a synapse strengthen itself? To do this, we have to move to the in vitro slice preparation.

The NMDA glutamate receptor as coincidence detector

Another convenient feature of hippocampal circuitry for neurophysiologists is that it is relatively straightforward to remove a cross-sectional "slice" of the hippocampus from a live animal (most typically a rodent) and keep it alive in vitro by perfusing it with tissue culture medium. Such a slice will contain intact circuitry as illustrated in *Figure 11.2*, and its electrophysiology and pharmacology can be studied with many techniques that are either not possible or certainly much more difficult, in an in vivo preparation. Indeed, much of what we know about the cellular and molecular bases of LTP has come from study of the synapse of CA3 axons on CA1 dendrites, a synapse that is much easier to study in the slice than in vivo. This research has revealed that there are many variants of LTP and many factors that influence it, only a few of which we can review here. Our focus will be on a prominent form of Hebbian

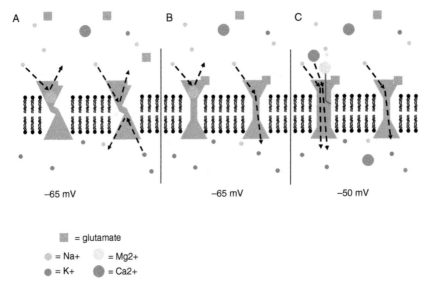

FIGURE 11.4 The two major classes of glutamate receptor. **A.** Cartoon of an NMDA receptor (blue) and an AMPA receptor (salmon), both membrane-spanning complexes of proteins that, when their binding site is not bound with a molecule of glutamate, are twisted such that ions in the extracellular and intracellular fluid cannot pass through them. Additionally, the NMDA receptor has an Mg^{2+} ion lodged in its pore. **B.** The binding of glutamate causes both of these receptors to untwist, thereby opening a pore that acts like a tube connecting the intra- and extracellular spaces. At a resting voltage of −65 mV, this allows for an influx of Na^+ ions impelled by both the electrical and chemical gradients, resulting in a depolarization of the local membrane potential. For the NMDA receptor, however, at this same resting voltage of −65 mV, the Mg^{2+} ion remains lodged in the receptor's pore (tugged by the electrical gradient). Thus, at the neuron's resting potential, binding with glutamate will not allow any ions to pass through it. **C.** As the local membrane potential begins to depolarize due to the influx of Na^+ ions through nearby AMPA channels, the net force exerted on the Mg^{2+} ion reverses, and its displacement from the channel's pore allows for the influx of positively charged ions. Importantly, because its pore is larger, the NMDA receptor also allows for the passage of Ca^{2+} ions.

LTP, one that is observed at the EC-DG synapse and also at the CA3-CA1 synapse: NMDA-dependent LTP.

As illustrated in *Figure 11.4*, there are two major classes of glutamate receptor, each named after a compound that is a selective agonist. (AMPA receptors received considerable treatment in section *The Neuron, Chapter 2*; if you're hazy on the details, now would be a good time to review them.) The NMDA receptor differs from the AMPA receptor in that its pore is occupied by a magnesium ion (Mg^{2+}). As a result, binding by glutamate, which untwists its pore, is not sufficient for the NMDA receptor to allow the influx of positively charged ions, because the Mg^{2+} ion remains in place, blocking the pore. To activate the NMDA receptor, there must be a coincidence of two events: (1) presynaptically released glutamate binds to its binding site; and (2) the postsynaptic terminal is already depolarized. A depolarized state of the postsynaptic terminal has the effect of repelling the positively charged Mg^{2+} ion,

thereby forcing it out of the pore, which allows for the influx of Na^+ and, more importantly, Ca^{2+}. As discussed previously, Ca^{2+} acts as a catalyst for many intracellular chemical cascades, and in the postsynaptic terminal, those cascades lead to LTP.

Before detailing the events triggered by the influx of Ca^{2+}, however, let's take another moment to emphasize the functional role of the NMDA receptor. It acts as a coincidence detector, in that it can only be activated by the coincidence of two events that implement Hebb's postulate: (1) release of glutamate by the presynaptic terminal of the synapse that is to be strengthened and (2) a prior state of depolarization of the postsynaptic terminal. To make this concrete, we can return once again to our Pavlovian conditioning vignette (from *Chapter 8*). The presentation of food activates olfactory neurons, which, in turn, activate salivary neurons. This provides the postsynaptic depolarization. If, during this time, the metronome clicks, glutamate released by auditory neurons onto their synapses with salivary neurons will activate NMDA receptors, due to the coincidence of olfactory activity and auditory activity. This activation of NMDA receptors, as we're about to see, can trigger LTP.

Multiple stages of synaptic strengthening

Finally, we can now explicitly consider the events that lead to the long-term potentiation of synapses. The influx of calcium ions through NMDA receptors has both short-term and long-term consequences. In the short term, it triggers the insertion of more AMPA receptors into the postsynaptic density. (As it turns out, postsynaptic terminals are equipped for rapid strengthening, with "racks" of AMPA receptors at the ready, to be inserted into the postsynaptic density as needed.) Because the binding of neurotransmitter with a receptor is a stochastic event, adding more AMPA receptors to the postsynaptic density will increase the probability that any one molecule of glutamate will bind with a postsynaptic receptor. The aggregate effect is that each action potential from the presynaptic neuron has a stronger influence on the postsynaptic neuron – the synapse will be potentiated.

In the long term, Ca^{2+}-initiated cascades of intracellular signaling initiate the synthesis of new proteins that are necessary for new synapses to be constructed. Electron-microscopic images have documented the appearance of new dendritic spines, presumed to be forming new synapses with efferent axons, a so-called structural form of LTP.

The necessity of NMDA channels for LTM formation

Since the pioneering work of Lømo and Bliss, the study of LTP has exploded into a subfield of its own. *What molecules are necessary for or influence its expression? Are there presynaptic as well as postsynaptic changes underlying it? Are there non-Hebbian, non-NMDA-dependent variants? What is the importance of the precise timing of depolarization of the postsynaptic cell and the firing of the presynaptic cell?* Of fundamental importance, however, was determining the necessity of NMDA-mediated neurotransmission for learning in the awake, behaving animal. One experiment designed to do this was carried out by British neuroscientist Richard Morris and colleagues and is perhaps the most famous experiment to have used the experimental apparatus of his invention, the "Morris Water Maze." The Morris Water Maze is a circular pool (~2 m in diameter) filled with water that has been

made opaque by the addition of milk. At one location is a small platform that is located just below the surface of the water. The water is kept chilly, to guarantee that when an already water-averse rat is placed in the pool, it will seek to climb up onto the platform ASAP. Because the platform cannot be seen, however, on the first trial of an experiment, the rat must swim around and around until bumping into it. On subsequent trials, it can draw on its memory for where the platform had been on previous trials, thereby escaping faster.

To assess the importance of NMDA receptors for spatial learning, Morris and colleagues (1986) administered to rats the NMDA antagonist AP5. Importantly, previous studies in the slice had established that although AP5 blocks LTP formation, it does not influence normal synaptic transmission. Together with two control groups, the rats were given 15 trials, across four days, to learn where the platform was located. On the critical test day, the platform was removed from the pool, and the rats were made to swim for 60 seconds. As illustrated in *Figure 11.5*, a rat in the control group that is caught in this diabolical quandary will swim back-and-forth over the location where it remembers the platform to have been. The AP5 animals, however, just swim around in circles, showing no bias for the location where the platform had been in its 15 previous trips to the pool. To quantify this, the investigators divided the pool into four imaginary quadrants, one centered on the location where the platform had been. The "time per quadrant" plots show that both the saline control animals and the animals treated with a chemically ineffective variant of AP5 (L-AP5) spent a disproportionate period of their 60 seconds in the remembered quadrant, whereas the group treated with the active variant of the compound, "D, L-AP5," did not.

As is often the case in science, the enthusiasm for the investigation of LTP has prompted a skeptical backlash from some who have questioned the importance of LTP in the grand scheme of understanding learning and memory. Some examples from this vibrant, dynamic literature are listed in the *Further Reading* section.

HOW MIGHT THE HIPPOCAMPUS WORK?

I had an argument with myself (as H.M. used to say) about whether to use the word "does" or "might" in this section heading. In the end, I've concluded that "does" would

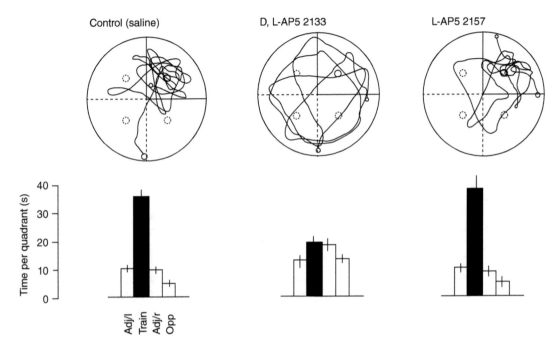

FIGURE 11.5 NMDA receptor blockade blocks spatial LTM in the rat. Top row shows the swimming trajectory on the test day of an animal in each of the three groups. Dotted circles indicate the location in each quadrant analogous to where the platform had been in the critical quadrant (top right for these three illustrations). Bottom row shows the group-average time spent in each of the four quadrants of the pool, labeled according to their position relative to the platform ("Train"). Source: From Morris, Richard G. M., E. Anderson, Gary S. Lynch, and Michael Baudry. 1986. "Selective Impairment of Learning and Blockade of Long-Term Potentiation by an N-Methyl-D-Aspartate Receptor Antagonist, AP5." Nature 319 (6056): 774–776. doi: 10.1038/319774a0. Reproduced with permission of Nature Publishing Group.

imply that this question is settled to a degree that it simply is not. What follows, however, are some of the more intriguing and influential possibilities.

Fast-encoding hippocampus vs. slow-encoding cortex

The "complementary learning systems" model was formulated by a collaboration between two PDP modelers, James McClelland and his then-student Randall O'Reilly, and hippocampal neurophysiologist Bruce McNaughton. Their model was motivated by two primary considerations, one empirical and one theoretical. The empirical consideration was the evidence for a time-limited role for the hippocampus in LTM storage. That is, the fact that retrograde memory loss is typically limited to memories acquired for the few years prior to the insult suggests either a transfer of memories from the hippocampus to cortex or a time-limited role for the hippocampus in the consolidation of cortically based long-term memories. The theoretical

consideration is as follows. There are two primary functions carried out by LTM: the retention of memories for specific episodes that we have experienced; and the extraction from life experiences of general patterns about how the world works. The former need to be highly detailed and discriminable from other episodes that may have some overlapping features. (For example, I only have one free half-hour in my entire day, and it's deadline day for submitting two letters of recommendation, one for a mentee, "A," who is applying for a faculty job, the other for a colleague, "B," who is applying for a research grant from a private foundation. Was it A or B who recently emailed to say that a one-day delay wouldn't jeopardize the evaluation of their application?) These are called **episodic memories**. The latter need to discard details that are idiosyncratic to any one specific episode and extract general principles. (For example, private foundation research grants tend to be smaller than research grants from government agencies, and they are often restricted to early-stage investigators.) These are called **semantic memories** and correspond to knowledge about the world that does not require reference to the episode(s) in which it was acquired.

Computational modeling has demonstrated that the formation of distinct episodic memories, an operation sometimes referred to as "pattern separation," is computationally incompatible with the abstraction across specific episodes to distill and/or update general principles, an operation sometimes referred to as "pattern completion." In particular, the rapid representation and retention of arbitrary associations, a hallmark of episodic memory, cannot be supported by a system that learns general principles by abstracting across multiple individual episodes, a necessarily slow process. The complementary learning systems model of McClelland, McNaughton, and O'Reilly (1995), therefore, proposes that the hippocampus serves as the special-purpose rapid learner of episodic memories, and that the way in which these memories are periodically reinstated (i.e., in full detail, or just the important principle that it contains) determines whether or not it will gradually migrate to the cortex as an episodic or a semantic memory.

Episodic memory for sequences

Figure 11.2 illustrates that the circuitry of the hippocampus includes two recurrent loops of circuitry: in addition to feeding forward to CA3, granule cells of the DG also project to mossy cells of the DG, which, in turn, project back onto the granule cells; and, in addition to feeding forward to CA1, CA3 neurons also feed back onto other CA3 neurons (as well as back onto mossy cells). Indeed, a critical detail that is not captured in this diagram is that individual CA3 axons vary in length from an astounding 150 to 400 mm, and each makes between 25,000 and 50,000 synapses with other neurons! This makes the CA3 layer a "giant random connection space, a requisite for combining arbitrary information" (Buszáki, 2006, p. 285). That is, it is an ideal substrate for the rapid formation of episodic memories invoked by McClelland, McNaughton, and O'Reilly (1995). The neurophysiologist and theoretician John Lisman has emphasized that this recurrent circuitry, combined with the direct EC-to-CA3 projection of the perforant path, enables the hippocampus to represent two kinds of information that are critical for episodic memory: the sequence in which events occurred and the context in which they took place.

Place cell precession

An empirical finding that underlies Lisman's (1999) idea comes from the phenomenon of **phase precession** of the firing of hippocampal place cells, discovered by John

O'Keefe, whom you'll remember from *Chapter 7* as the original discoverer of place cells (an accomplishment for which he was awarded the Nobel Prize). The phenomenon is cartooned in *Figure 11.6.B*. Lisman's explanation is that, because the ion channels of NMDA receptors in the CA3 and CA1 regions remain open for 100 ms or longer, the recurrent circuitry in this region can encode (via LTP) the temporal association of "location *B* comes after location *A*," and so on. Importantly, these associations are asymmetrical: although NMDA receptors at the *A*-to-*B* synapse can detect the "lagged coincidence" that A *was recently active, and* B *is now firing*, at the *B*-to-*A* synapse, the reverse cannot happen, because when *B* is firing, *A* is no longer depolarized. (Thus, although glutamate released from *B* might untwist the pores of NMDA channels in *A*, the absence of depolarization in *A* means that the Mg^{2+} blockade will remain in place.) Lisman further posits that if there is a buffer that can sustain activity in a representation over several seconds, either within the hippocampus or elsewhere, this same mechanism could also encode the sequential order of events spaced out by seconds, rather than tens of milliseconds (we'll consider data consistent with this idea from human fMRI studies in *Chapter 12*). The gist of this idea is that the mechanism observed in the phase precession of activity in hippocampal place cells (*Figure 11.6*) might underlie LTM for sequences of events.

Although not illustrated explicitly in *Figure 11.6.B*, the biophysics underlying the phase encoding of place cell activity is phase-amplitude coupling (PAC). Each of the five neurons illustrated in *Figure 11.6.B* is bursting at the trough of a gamma oscillation that is "nested" in the lower-frequency theta oscillation. If you revisit *Figure 3.11*, you'll note how the gamma-filtered component of the signal waxes and wanes as a function of the concurrent phase of the alpha cycle. Whereas *Figure 3.11* illustrates alpha-high gamma PAC recorded by an EcoG electrode on the cortical surface, the PAC in the hippocampus involves oscillatory gamma (~35–90 Hz) that is coordinated by the ongoing theta rhythm. Lisman's (1999) account of episodic memory recall is that each burst of activity in hippocampal PAC, rather than corresponding to a place cell as it does in *Figure 11.6.C*, corresponds to an ensemble of neurons that represent discrete propositional elements that make up the memory.

One example that Lisman (1999) uses to make concrete the connection between place cell precession in the rodent and human episodic memory retrieval is to invoke a memory of a visit to the zoo in which candy is purchased for a

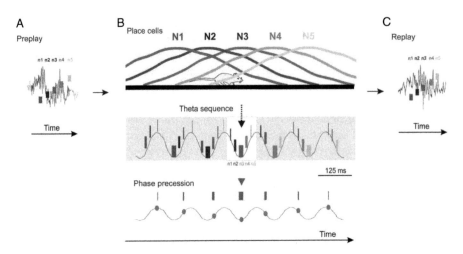

FIGURE 11.6 Phase precession of place cell firing. **A.** A sharp-wave ripple "preplaying" the sequence that will later come to represent the sequence of locations illustrated in panel B (see *Figure 11.7*, and text, for details about sharp-wave ripples). **B.** As a rat runs along a path, the hippocampal place cells representing five adjacent locations have overlapping place fields. At the moment in time captured in the top panel, when the rat is in the middle of the path, it's at the location for which neuron N3 is maximally active; it's leaving the location for which neuron N2 is maximally active, and N2's activity is beginning to decrease; and it will next be at the location for which neuron N4 is maximally active, and N4's activity is increasing. The middle panel shows how the activity of each of these five place cells (vertical bars) is not simultaneous, rather it occurs in an unfolding sequence of bursts, the order of which corresponds to the order in which the rat will pass through these locations. Furthermore, this sequence of activity is "compressed," in that the full sequence plays through in roughly 125 ms, which is much faster than the rat is actually moving. The degree of compression is dictated by frequency of the theta rhythm, which acts like a clock in the hippocampus. The area in white is the cycle of theta occurring while the rat is at the pictured location. Note that the "location" of each neuron's burst on the theta wave (more technically, the "phase angle") indicates where the rat is (the cell bursting at the trough), where it has been (the cells bursting on the falling slope of the cycle), and where it will be (the rising slope). This phenomenon is referred to as phase encoding. In this cartoon, the rat is moving at a rate of one place field per 125 ms. Thus, if you look across time (i.e., from the leftmost cycle to the rightmost), you'll see that the phase angle of neuron N3's bursting changes from cycle to cycle: when the rat was at location "N1," N3 fired at the peak of the cycle; as the rat moved forward N3's burst "advanced" to earlier points along the cycle, firing at the trough when the rat was at location "N3," then firing progressively earlier before the trough as the animal continued forward. This phenomenon is known as "precession" and is even easier to see in the bottom panel, which illustrates the activity of only N3, across the same five cycles of theta. **C.** A sharp-wave ripple "replaying" the sequence of locations illustrated in panel B (see *Figure 11.7*, and text, for details about sharp-wave ripples). Source: Adapted from Dragoi, G. (2013). Internal operations in the hippocampus: single cell and ensemble temporal coding. Frontiers in Systems Neuroscience 7. 10.3389/fnsys.2013.00046.

little boy named Jerry, but Jerry then drops the candy on the ground near the monkey cage, a monkey reaches through the bars of the cage to snatch it, and Jerry bursts into tears. (Extra credit to anyone who wants to dredge through John Lisman's biography to uncover the origin of this story.) If retrieval of this memory (i.e., thinking about it) were applied to *Figure* 11.6.*B*, N1 might correspond to an ensemble of cells representing "Jerry was given candy," N2 to "Jerry went to the monkey house," N3 to "Jerry dropped the candy," and so forth. Although these three

facts are seemingly activated simultaneously while one thinks about this event (and indeed, although they activate again-and-again over many sequential theta cycles), the order in which they occurred is encoded in the phase of the theta cycle at which each ensemble fires. To be more mechanistic, explicit memory recall, by this scheme, would play out as follows. First, neurons of EC that represent the context of "the zoo" depolarize CA3 neurons at a subthreshold level. Second, the subset of these CA3 neurons that also receive input from the DG representing *Jerry dropped his candy* would be depolarized to their firing threshold, and their activity would represent *Jerry dropped his candy at the zoo*. This activity, would, in turn, activate the CA3 representation of the puckish monkey who snatched away the candy, and so on. Note that regardless of the actual time elapsed between two sequentially occurring events, the representation of their ordinal relation would be compressed into a timescale clocked by theta-nested gamma-band oscillations in the hippocampus.

Sharp-wave ripples

To do justice to our current understanding of memory encoding and consolidation, and, in particular, how sleep relates to these processes, we need to drill down a little bit further into hippocampal neurophysiology. When a rodent is actively moving about, exploring, rearing, sniffing, the theta rhythm dominates activity in the CA fields of the hippocampus. Whenever the animal stops moving, however, and engages in behaviors such as eating, drinking, grooming, or sleeping, theta subsides, and in its place one sees irregular bursts of "sharp-wave ripples." (During sleep, sharp-wave ripples are seen during slow-wave sleep (SWS),

not rapid eye movement [REM] sleep.) As illustrated in *Figure 11.7*, sharp waves are large-amplitude negative-going deflections in the local field potential, and nested in the sharp waves are ripples – short-lived oscillatory patterns in the 100–200 Hz range. (Note that, as can be inferred from *Figure 11.7*, ripples are far too small in amplitude to be detected by ECoG methods for which power in this same frequency range is considered "high-gamma," and taken as a proxy for multiunit activity [MUA].) György Buzsáki (2015; who we first encountered at the end of *Chapter 2,* in the section entitled *Complicated and complex*) has characterized sharp-wave ripples as "the most synchronous events in the mammalian brain, associated with a robustly enhanced transient excitability in the hippocampus and its partner structures" (p. 1074). In effect, sharp-wave ripples are ideal for transferring information from the hippocampus to the neocortex, thereby, for example, providing a circuit-level mechanism for the processes considered in section *Fast-encoding hippocampus vs. slow-encoding cortex* in this chapter. An additional feature of sharp-wave ripples is that each deflection in a ripple corresponds to a spike from a

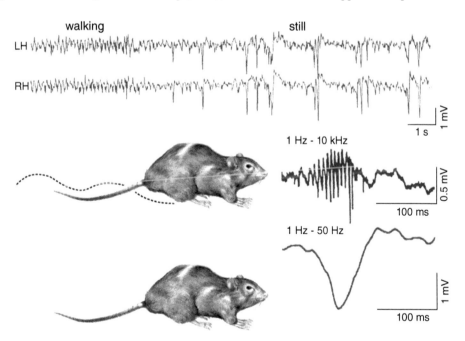

FIGURE 11.7 The hippocampal sharp-wave ripple. Green traces at top illustrate recordings made simultaneously in CA1 in the left and right hemisphere of a rat while walking, then standing still. Sharp waves are the large-amplitude deflections during the still epoch. Red trace is a blow-up of a single sharp wave, lowpass filtered to remove evidence of ripple. Blue trace is a ripple. Source: From Buzsaki, G. (2015). Hippocampal sharp-wave ripple: A cognitive biomarker for episodic memory and planning. Hippocampus 25: 1073–1188. Reproduced with permission of John Wiley & Sons.

different neuron. And the pattern of spiking is not arbitrary, but, rather, reflects repeating sequences of the same neurons always firing in the same sequence. Now we can return to *Figure 11.6* and explicate panels A and C, which we've ignored until now.

Starting with *Figure 11.6.C*, and consistent with an important role in consolidating memories by transferring information from the hippocampus to cortex, sharp-wave ripples are often seen to "replay" sequences of neural activity that occurred previously when the animal was in an active state. A classic example is illustrated in *Figure 11.6*, in that panel B shows a sequence of place cell activity that is recorded while the rat is exploring a novel environment (i.e., when it is "learning something new"). Later, when the animal is in SWS, the same sequence of N1-N2-N3-N4-N5 is observed within individual sharp-wave ripples. Because sharp-wave ripples are also characteristic of human neurophysiology (indeed, of every mammalian species in which they've been sought), they are believed to underlie the role of SWS in learning and memory. Finally, *Figure 11.6.A* – specific sequences of neuronal firing have been observed during sharp-wave ripples recorded during SWS that later turn out to be the same sequences of place cell firing that are recorded when the animal is exploring a never-before-encountered environment! The phenomenon has been termed "preplay" and is perhaps best accounted for by the proposal that the generation of self-organized sequences of activity is a natural consequence of the rapidly induced plasticity caused by sharp-wave ripples, themselves a consequence of the recurrent circuitry of CA3 and its connections with CA1. Perhaps the hippocampus is, in effect, always generating "templates" of sequential connections, and evolution has capitalized on this by learning to "harness" a "template" each time a novel sequence of events is experienced, thereby creating a mechanism for representing sequential order in long-term memories.

Episodic memory as an evolutionary elaboration of navigational processing

For the final model of hippocampal function that we'll consider here, it will be useful to bring to mind the place cell data from *Chapter 7*, as well as *Figure 7.10*'s illustration of the dorsal location of the hippocampus in the rodent. At the level of implementation it draws on many of the same mechanisms as Lisman's memory-for-sequences model, differing primarily in the emphasis on "how our memory came to work the way that it does." An influential account by György Buszáki uses as a starting point the fact that hippocampal place cells encode allocentric representations of space, and thus would be an effective substrate for one-dimensional (1-D) "dead reckoning" navigation (a.k.a. "route following"). 1-D maps, like that illustrated in *Figure 11.6.B*, would be sufficient to represent an episodic memory, such as Lisman's story of Jerry at the zoo (see this chapter's section on *Place cell precession*). If one adds together a sufficient number of 1-D routes, one can construct a 2-D map, and the CA3 field as a whole can be construed as a 2-D map, with distances encoded in synaptic weights. (That is, by the same logic that Lisman uses to argue that these cells can encode asymmetric sequences of events, one would also expect stronger synaptic links between place cells representing nearby locations than between those representing distant locations.) Indeed, Buszáki notes, CA3 probably holds many, many maps that are anatomically intermingled, thus obscuring any topography that may exist for any one map. Such 2-D maps would be used in "landmark-based" navigation (a.k.a. "map reading"), a more efficient form of navigation that has been observed in many animal species, as well as in humans. Such 2-D maps, Buszáki argues, may be the neural basis of explicit memories. Thus, the hippocampus, a region phylogenetically older than the neocortex, may have originally evolved to support navigation, foraging, and caching behaviors in our vertebrate ancestors. As the neocortex in early mammals acquired the ability to represent information other than just what have been called the "four Fs" – *feeding, fighting, fleeing, and . . . mating* – it may have coopted the substrate already in place in the hippocampus for the encoding and storage of information other than space.

Compatible with this "navigational origins" view is the "hippocampus as cognitive map" perspective championed by the cognitive neuroscientist Lynn Nadel, coauthor with John O'Keefe of the influential 1978 book of the same name. In a recent review, Nadel, Hoscheidt, and Ryan (2013) add an additional level of detail by drawing on the fact that the size of hippocampal place fields in the rodent changes along a gradient from small in the dorsal to relatively large in the ventral hippocampus. They infer from this that in the rodent, the dorsal hippocampus may be more important for fine-grained spatial processing and the ventral hippocampus for what they call "context coding" (i.e., information about what a particular locale is like). In a final inferential step, they review fMRI studies of memory retrieval in the human and propose that "Retrieving detailed spatial relational information preferentially activate[s] the posterior hippocampus, whereas retrieving information about locales (or contexts) preferentially activate[s] the anterior hippocampus" (p. 22).

WHAT ARE THE COGNITIVE FUNCTIONS OF THE HIPPOCAMPUS?

Some of the ideas reviewed in the two previous sections, like the discovery of Hebbian plasticity in the hippocampus and the dependence of performance of (at least some) tests of LTM on the NMDA receptor, fit well with the provisional principles that we listed in the earlier subsection *The organization of memory*. Others, like the putative role for the hippocampus in the retrieval of memories, would seem incompatible with the principles derived from initial studies of H.M. In this final section, we'll revisit those principles articulated near the beginning of this chapter in light of more recent theoretical and empirical developments.

Standard anatomical model

After some puzzling early failures to replicate H.M.'s amnesia in nonhuman primates, by the 1980s, teams led by Mortimer Mishkin (yes, *that* Mort Mishkin) and by Larry Squire and Stuart Zola (then Zola-Morgan) produced a series of findings indicating that surgical lesioning of the amygdala and hippocampus, bilaterally, produced a marked impairment of LTM. The standard task was the delayed nonmatch-to-sample (DNMS), and bilateral MTL animals began to show significant impairments with delays between sample and test stimuli as short as 15 seconds.

Notably, lesions targeting just selected portions of the MTL seemed to produce quantitatively, but not qualitatively, different patterns of impairment. Thus, for example, Squire (1992) summarizes that lesions restricted to the amygdala had no effect compared to unoperated controls, lesions of hippocampus plus the overlying parahippocampal gyrus (PHG) and caudal PRC ("H$^+$") produced a marked impairment at 15 seconds, and lesions of hippocampus plus PHC plus PRC and EC ("H^{++}") produced an even larger impairment at 15 seconds. When the delay was extended to 60 seconds, then to 10 minutes, the rate of decline for each group (including controls) was comparable. Note that, with the surgical aspiration methods that were used to produce these lesions, an "H" lesion that did not compromise any of the overlying tissue was not possible, due to its location deep in the MTL.

The model that emerged from results such as these was of the hierarchy as drawn in *Figure 6.5.B,* with regions feeding into PRC and PHG classified as "perceptual," and PHG, PRC, EC, and the hippocampus all part of the "MTL memory system."

Challenges to the standard anatomical model

Subsequent research using different, more recently developed lesioning methods and behavioral tasks, however, have called this hierarchical view into question.

Newer lesioning methods

With regard to lesions, Elisabeth Murray and colleagues at the NIMH's Laboratory of Neuropsychology have used injections of ibotenic acid to produce more circumscribed lesions. Ibotenic acid is a glutamate agonist that, through a phenomenon called excitotoxicity, destroys neurons whose dendrites it comes into contact with. Importantly, it spares so-called "fibers of passage," that is, axons traveling through the injection site, but that neither initiate nor terminate in it, because these fibers lack glutamate receptors. Mechanical lesions, in contrast, whether produced by surgical excision, heat coagulation, high electrical current, or aspiration, will destroy any fibers that pass through the lesioned area. With ibotenic acid injections, in contrast, Murray and colleagues could, for example, produce clean "H" lesions. Doing so produced results strikingly different from the earlier findings: H lesions produced no impairment on DNMS and other tests of LTM, whereas selective lesions of PRC and EC produced profound impairments.

Perceptually challenging behavioral tasks

Research by Murray, Oxford-based colleague David Gaffan, and collaborators has also evaluated MTL function from the perspective that PRC, being one synapse downstream from AIT, may subserve high-level object recognition functions in addition to LTM encoding functions. In addition to the anatomy, they were motivated by computational modeling suggesting that PRC might be particularly important for representing complex conjunctions of features. *Figure 11.8* illustrates one experiment in which monkeys first had to learn, through trial-and-error guessing, which one in each pair of stimuli was rewarded. For the test, these stimulus pairs were re-presented as they had appeared during training, along with two additional pairs per training pair that were constructed by partially morphing the two trained images together. Thus, as illustrated in *Figure 11.8.A*, the "feature ambiguity" pairs required discriminating between two stimuli that shared many features. The finding was that H- and PRC-lesioned animals did not differ when discriminating between trained stimuli, but the PRC-lesioned animals were selectively impaired on pairs with high feature ambiguity.

Importantly, Murray, Bussey, and Saksida (2007) note that PRC-lesioned animals are not impaired on just any difficult task. For example, they perform normally on tasks requiring difficult color or size discriminations. Thus, on the strength of these and other results, Murray and colleagues propose that "the perirhinal cortex participates in both perception and memory: it is important for perception because it contains mechanisms enabling the representation of complex conjunctions of features; it subserves memory because it serves as a storage site of complex stimulus representations in both the short- and long-term" (pp. 104–105).

The two previous subsections have summarized arguments that (1) the hippocampus is not needed for relatively simple object memory performance, whereas PRC and EC are; and (2) PRC performs important perceptual, in addition to mnemonic, functions. Both of these arguments represent serious challenges to the notion that the MTL can be considered a monolithic "memory system." Argument (1), in particular, may seem alarming, because it challenges the longstanding assumption that the

FIGURE 11.8 Memory vs. perception in the PRC.

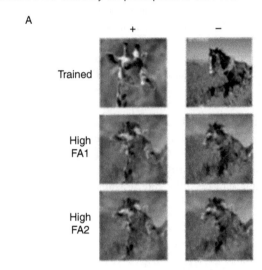

FIGURE 11.8.A On each trial the monkey saw a pair of stimuli and had to select one. During training the same images were always paired, and the reward contingencies were always the same. At test, the animals saw each of the training pairs, along with two variants of each pair that contained two levels of "feature ambiguity" (*High FA1* and *High FA2*).

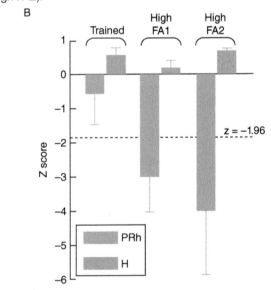

FIGURE 11.8.B Performance of animals with H vs. PRC (here, labeled "PRh") lesions on the three levels of perceptual difficulty. Source: From Murray, Elisabeth A., Timothy J. Bussey, and Lisa M. Saksida. 2007. "Visual Perception and Memory: A New View of Medial Temporal Lobe Function in Primates and Rodents." Annual Review of Neuroscience 30: 99–122. doi: 10.1146/annurev.neuro.29.051605.113046. Reproduced with permission of Annual Reviews.

hippocampus is critical for declarative memory encoding. The next subsection summarizes some more recent proposals for how to conceive of the role of the hippocampus in cognition.

"New roles" for the hippocampus?

One proposed reconceptualization of the hippocampus, championed by neurobiologist Howard Eichenbaum and cognitive neuroscientist Neal Cohen, is that its primary function is one of *relational binding*. In brief, this theory holds that the hippocampus affects the operation of representing and learning the relations between items in the environment. This might include the arbitrary rule for written English of "*i* before *e*, except after *c*," or the concrete spatial content of something like "Zidane struck the free kick from the left side of the field, lofting the ball over the heads of the Brazilian defenders and into the right side of the goal box, where Henry, running in unmarked, volleyed it into the back of the net." One implication of this idea is that the classical assumption that the hippocampus is not critical for STM may also need to be revisited. Indeed, recent studies have demonstrated that patients with hippocampal damage can be impaired on tests of STM for the spatial relations between items in a display, even with delays as brief as one second. Thus, from the relational binding perspective, the question isn't *How long must the information be held in memory?* but *What kind of information needs to be processed?*

A second, relatively recent, conceptualization of the function of the hippocampus is a role in what some call "mental time travel," the ability to anticipate a future event and imagine various outcomes, as well as the ability to imagine entirely made-up events. In one study, led by cognitive neuroscientist Eleanor Maguire, patients with bilateral hippocampal damage were asked to construct new imagined experiences, such as "Imagine you are lying on a white sandy beach in a beautiful tropical bay" and "Imagine that you are sitting in the main hall of a museum containing many exhibits." Their results indicated that the imagined experiences of the patients contained markedly less experiential richness than did those of healthy control subjects. A more detailed analysis also revealed lower "spatial coherence" (a measure of the contiguousness and spatial integrity of an imagined scene) in the performance of the patients. If the latter were at the root of the overall poor performance of the patients, these findings would be consistent with the hippocampus-as-cognitive-map framework.

A third development that postdates what we've called the standard model, that the hippocampus may be particularly critical for recollection-based, but not familiarity-based memory retrieval, will be considered in detail in *Chapter 12*.

Consolidation

The concept of consolidation has also become more complicated since the 1960s and 1970s. Consistent with the original idea of a process that yielded a time-limited role for the hippocampus in LTM, we have reviewed that McClelland, McNaughton, and O'Reilly (1995) proposed a computational basis for consolidation as the gradual integration of hippocampally based representations into the neocortex. A theoretical challenge to the idea of consolidation, however, has come in the form of the *multiple trace theory*, proposed by Lynn Nadel and neuropsychologist Morris Moscovitch. The multiple trace theory posits that each instance of memory retrieval also prompts the creation by the hippocampus of a new memory trace, such that over time a single memory comes to be stored as multiple traces. To the extent that elements of these traces overlap, this process leads to the development of semantic knowledge that is independent of the episodes in which the information was learned. So, for example, if learning about US presidents in primary school and taking a family trip to Washington DC both create traces representing the proposition that *Thomas Jefferson was the third President of the United States*, repeated iterations of this process create a representation that can be retrieved independently of any reference to any one of the contexts in which this information was encountered. In this way, the memory that *Thomas Jefferson was the third President of the United States* becomes a *semantic* memory. Should damage to the hippocampus be sustained several years after the learning took place, the patient would nonetheless be able to retrieve this knowledge. On this prediction, the multiple trace theory and the standard neuropsychological model are in accord. The two differ, however, in that the former holds that the specific memory of the visit to the Jefferson Memorial during the family trip to Washington DC remains an *autobiographical episodic* memory that is dependent on the hippocampus for the remainder of the subject's life. Thus, multiple trace theory would predict that access to this autobiographical episodic memory would be severely impoverished, if not completely impossible, after extensive damage to the MTL (particularly to the hippocampus). The standard neuropsychological model, in contrast, would hold that access to remote autobiographical episodic memories would be intact. This remains an area of vigorous and contentious debate.

Reconsolidation

The concept of consolidation got even *more* complicated in 2000, when the phenomenon of reconsolidation was discovered in the context of a fear conditioning paradigm, a variant of Pavlovian conditioning in which an initially neutral stimulus (often a tone, or sometimes the novel experimental chamber in which the conditioning occurs [referred to as the "context"]) is paired with an aversive stimulus, often an electric shock delivered through the floor. It differs from Pavlovian conditioning in that a single trial is sufficient for the animal to learn the conditioned response, which is to freeze the next time it hears the tone (or, is placed back into the chamber). Although we won't consider this in detail until *Chapter 18*, plasticity that is critical for the consolidation of a conditioned fear memory occurs in the amygdala. Evidence for this is that infusion of a protein synthesis inhibitor into the amygdala blocks the expression of the conditioned response when the animal is presented with the tone a day later. (Without the protein synthesis inhibitor, control animals freeze when they again hear the tone, and the duration of freezing is a dependent measure of how strong the memory is.)

Evidence for a reconsolidation process was produced by Karim Nader and colleagues in the laboratory of neurobiologist Joseph LeDoux when, instead of administering the protein synthesis inhibitor at the time of the learning episode, they administered it when the conditioned stimulus was re-presented, i.e., at the time of memory retrieval. The finding was that administering the protein synthesis inhibitor at the time of retrieval trial 1 did not disrupt the expression of the memory on retrieval trial 1; however, it did disrupt the expression of the memory on retrieval trial 2. This finding was controversial at first, but has since been replicated several times, in several species (including humans), and is now widely accepted. Although the mechanisms underlying reconsolidation aren't fully understood, the implication is that the act of retrieving a memory puts

it into a labile state that renders it vulnerable to disruption. (Here, it might be helpful to think of, e.g., Lisman's idea for how the activation of one item in a sequence can then trigger the activation of the subsequent items.) Part of the puzzle is that the LTP underlying the memory has already been through the (or, at least, a) structural phase. Is it reversed? Or somehow overwritten?

Even though the neurobiological and computational details of how and why reconsolidation happens remain uncertain, the finding of Nader, Schafe, and Le Doux (2000) very quickly initiated considerable excitement about potential clinical applications. As Schwabe and colleagues (2012) note: "Emotionally arousing experiences are usually better remembered than neutral experiences. Although generally adaptive to survival, this emotional memory enhancement may contribute to anxiety disorders, such as posttraumatic stress disorder (PTSD)" (p. 380). Therefore, could one leverage the phenomenon of reconsolidation to develop a "pill to forget" for treating victims of trauma? The study summarized in *Figure 11.9* represents an initial step in this direction, by demonstrating that a pharmacological intervention delivered at the time of retrieval of an emotional memory can, in effect, "reduce its emotionality" for subsequent retrieval episodes.

TO CONSOLIDATE

Memory is of central importance to human consciousness and cognition. At the conscious level, it creates our sense of identity, and provides the soothing, subjectively continuous internal narrative that enables us to go about our business without constantly stopping to ask "Why am I here? What was I doing 5 minutes ago?" At an unconscious level, it shapes our habits, our proclivities and dispositions, and makes us who we are. Thus, it's not surprising that different facets of memory are the focus of such intense research activity. Nor should it be a surprise when, if a large number of really smart, really driven people are working on a similar problem, multiple answers will emerge, not all of them mutually compatible.

The next three chapters will build on the themes that we have introduced here. Not all of the controversies will be resolved, but the reader will be equipped with a good sense of what some of the most interesting, most important problems are, and the ways that some of the leading scientists in the field are addressing them.

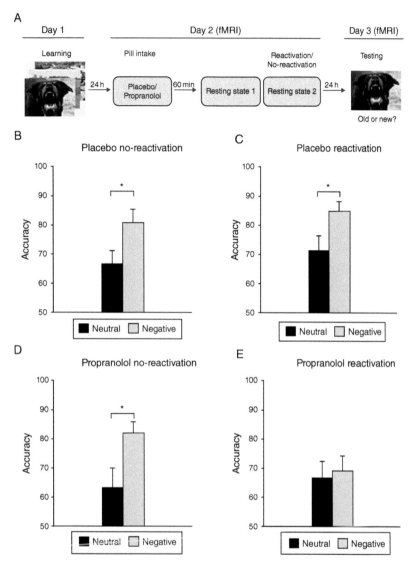

FIGURE 11.9 Weakening a consolidated emotional memory by disrupting reconsolidation. On Day 1, subjects viewed 25 emotionally negative and 25 emotionally neutral images. On Day 2, they either did or didn't receive a β-adrenergic antagonist ("beta-blocker") ("Propranolol"/"Placebo") and either were or weren't tested on their memory for the images ("Reactivation"/"No-reactivation"). (fMRI data were also acquired, but are not considered here.) On Day 3, memory for the images was tested in each of the four groups, and memory for emotionally negative images was impaired only in the "Propranolol reactivation" group. (In *Chapter 18*, we'll consider the role of norepinephrine in emotional processing.) Source: From Schwabe, Lars, Karim Nader, Oliver T. Wolf, Thomas Beaudry, and Jens C. Pruessner. 2012. "Neural Signature of Reconsolidation Impairments by Propranolol in Humans." Biological Psychiatry 71 (4): 380–386. doi: 10.1016/j.biopsych.2011.10.028. Reproduced with permission of Elsevier.

END-OF-CHAPTER QUESTIONS

1. What aspect of the global amnesic syndrome suggests a distinction between LTM encoding and LTM retrieval? Why?

2. What aspect of the global amnesic syndrome suggests a process of consolidation in LTM formation?

3. State Hebb's postulate. (Paraphrasing is okay, it's the idea that's important.)

4. In what sense does the NMDA receptor act as a coincidence detector? How is this important for LTP?

5. What principle is emphasized in the model of McClelland, McNaughton, and O'Reilly (1995)? Of Lisman (1999)? Of Buszáki (2006)?

6. Explain the phenomenon of phase precession in place cells. How might such a scheme also support the representation of temporal order in episodic memory?

7. What are sharp-wave ripples, and what is their relevance for understanding consolidation of LTM?

8. What is the principal difference between the standard model of consolidation and that of the multiple trace theory? What different predictions do the two make with respect to dependence on the hippocampus of autobiographical episodic memories?

9. Summarize the evidence for a process of reconsolidation.

10. In what respects are the experiments of Morris et al. (1986) and of Nader, Schafe, and Le Doux (2000) similar, and in what respects are they different?

REFERENCES

Bliss, Timothy Vivian Pelham, and Terje Lømo. 1973. "Long-Lasting Potentiation of Synaptic Transmission in the Dentate Area of the Anesthetized Rabbit Following Stimulation of the Perforant Path." *Journal of Physiology* 232 (2): 331–341. http://jp.physoc.org/content/232/2/331.full.pdf+html.

Buszáki, György. 2006. *Rhythms of the Brain*. Oxford: Oxford University Press.

Buszáki, György. 2015. "Hippocampal Sharp-Wave Ripple: A Cognitive Biomarker for Episodic Memory and Planning." *Hippocampus* 25: 1073–1188.

Dragoi, George. 2013. "Internal Operations in the Hippocampus: Single Cell and Ensemble Temporal Coding." *Frontiers in Systems Neuroscience* 7. doi: 10.3389/fnsys.2013.00046.

Corkin, Suzanne, David G. Amaral, R. Gilberto González, Keith A. Johnson, and Bradley T. Hyman. 1997. "H. M.'s Medial Temporal Lobe Lesion: Findings from Magnetic Resonance Imaging." *Journal of Neuroscience* 17 (10): 3964–3979. http://www.jneurosci.org/content/17/10/3964.full.pdf+html.

Hebb, Donald Olding. 1949. *The Organization of Behavior: A Neuropsychological Theory*. New York: Wiley and Sons.

Hilts, Philip J. 1996. *Memory's Ghost: The Nature of Memory and the Strange Tale of Mr. M.* Touchstone edition. New York: Simon and Schuster. First published 1995.

Lisman, John E. 1999. "Relating Hippocampal Circuitry to Function: Recall of Memory Sequences by Reciprocal Dentate–CA3 Interactions." *Neuron* 22: 233–242.

Lømo, Terje. 2003. "The Discovery of Long-Term Potentiation." *Philosophical Transactions of The Royal Society of London, Series B: Biological Sciences* 358 (1432): 617–620. doi: 10.1098/rstb.2002.1226.

Lüscher, Christian, and Robert C. Malenka. 2012. "NMDA Receptor-Dependent Long-Term Potentiation and Long-Term Depression (LTP/LTD)." *Cold Spring Harbor Perspectives in Biology* 4 (6): a005710. doi: 10.1101/cshperspect.a005710.

McClelland, James L., Bruce L. McNaughton, and Randall C. O'Reilly. 1995. "Why There Are Complementary Learning Systems in the Hippocampus and Neocortex: Insights from the Successes and Failures of Connectionist Models of Learning and Memory." *Psychological Review* 102 (3): 419–457. doi: 10.1037/0033-295X.102.3.419.

Morris, Richard G. M., E. Anderson, Gary S. Lynch, and Michael Baudry. 1986. "Selective Impairment of Learning and Blockade of Long-Term Potentiation by an N-Methyl-D-Aspartate Receptor Antagonist, AP5." *Nature* 319 (6056): 774–776. doi: 10.1038/319774a0.

Murray, Elisabeth A., Timothy J. Bussey, and Lisa M. Saksida. 2007. "Visual Perception and Memory: A New View of Medial Temporal Lobe Function in Primates and Rodents." *Annual Review of Neuroscience* 30: 99–122. doi: 10.1146/annurev.neuro.29.051605.113046.

Nadel, Lynn, Siobhan Hoscheidt, and Lee R. Ryan. 2013. "Spatial Cognition and the Hippocampus: The Anterior-Posterior Axis." *Journal of Cognitive Neuroscience* 25 (1): 22–28. doi: 10.1162/jocn_a_00313.

Nader, Karim, Glenn E. Schafe, and Joseph E. Le Doux. 2000. "Fear Memories Require Protein Synthesis in the Amygdala for Reconsolidation After Retrieval." *Nature* 406 (6797): 722–726. doi: 10.1038/35021052.

O'Keefe, John, and Lynn Nadel. 1978. *The Hippocampus as a Cognitive Map*. Oxford: Oxford University Press.

Pohl, Walter. 1973. "Dissociation of Spatial Discrimination Deficits Following Frontal and Parietal Lesions in Monkeys."

Journal of Comparative and Physiological Psychology 82 (2): 227–239. doi: 10.1037/h0033922.

Scoville, William Beecher, and Brenda Milner. 1957. "Loss of Recent Memory After Bilateral Hippocampal Lesions." *Journal of Neurology, Neurosurgery, and Psychiatry* 20 (1): 11–21. doi: 10.1136/jnnp.20.1.11.

Schwabe, Lars, Karim Nader, Oliver T. Wolf, Thomas Beaudry, and Jens C. Pruessner. 2012. "Neural Signature of Reconsolidation Impairments by Propranolol in Humans." *Biological Psychiatry* 71 (4): 380–386. doi: 10.1016/j.biopsych.2011.10.028.

Squire, Larry R. 1992. "Memory and the Hippocampus: A Synthesis from Findings with Rate, Monkeys, and Humans." *Psychological Review* 99 (2): 195–231. doi: 10.1037/0033-295X.99.3.582.

OTHER SOURCES USED

Baddeley, Alan D., and Elisabeth K. Warrington. 1970. "Amnesia and the Distinction between Long- and Short-Term Memory." *Journal of Verbal Learning and Verbal Behavior* 9 (2): 176–189. doi: 10.1016/S0022-5371(70)80048-2.

Cooper, Jack R., Floyd E. Bloom, and Robert H. Roth. 1991. *The Biochemical Basis of Neuropharmacology*, 6th ed. New York: Oxford University Press.

Corkin, Suzanne. 2013. *Permanent Present Tense: The Unforgettable Life of the Amnesic Patient, H.M.* New York: Basic Books.

Eichenbaum, Howard, and Neal J. Cohen. 2001. *From Conditioning to Conscious Recollection: Memory Systems of the Brain*. Oxford Psychology Series. Oxford: Oxford University Press.

Fuller, Leah, and Michael E. Dailey. 2007. "Preparation of Rodent Hippocampal Slice Cultures." *Cold Spring Harbor Protocols* 2007 (10): 4848–4852. doi: 10.1101/pdb.prot4848.

Gabrieli, John D. E., M. M. Keane, Ben Z. Stanger, Margaret Kjelgaard, Suzanne Corkin, and John H. Growdon. 1994. "Dissociations among Structural-Perceptual, Lexical-Semantic, and Event-Fact Memory in Amnesia, Alzheimer's, and Normal Subjects." *Cortex* 30: 75–103.

Jarrard, Leonard E. 1989. "On the Use of Ibotenic Acid to Lesion Selectively Different Components of the Hippocampal Formation." *Journal of Neuroscience Methods* 29 (3): 251–259. doi: 10.1016/0165-0270(89)90149-0.

Milner, Brenda, Suzanne Corkin, and H.-L. Teuber. 1968. "Further Analysis of the Hippocampal Amnesic Syndrome:

14-Year Follow-Up Study of H. M." *Neuropsychologia* 6 (3): 215–234. doi: 10.1016/0028-3932(68)90021-3.

Nadel, Lynn, and Morris Moscovitch. 1997. "Memory Consolidation, Retrograde Amnesia and the Hippocampal Complex." *Current Opinion in Neurobiology* 7 (2): 217–227. doi: 10.1016/S0959-4388(97)80010-4.

Nader, Karim, and Oliver Hardt. 2009. "A Single Standard for Memory: The Case for Reconsolidation." *Nature Reviews Neuroscience* 10 (3): 224–234. doi: 10.1038/nrn2590.

Pearce, John M. S. 2001. "Ammon's Horn and the Hippocampus." *Journal of Neurology, Neurosurgery, and Psychiatry* 71 (3): 351. doi: 10.1136/jnnp.71.3.351.

Schwabe, Lars, Karim Nader, and Jens C. Pruessner. 2013. "β-Adrenergic Blockade during Reactivation Reduces the Subjective Feeling of Remembering Associated with Emotional Episodic Memories." *Biological Psychology* 92 (2): 227–232. doi: 10.1016/j.biopsycho.2012.10.003.

Squire, Larry R., and Stuart Zola-Morgan. 1991. "The Medial Temporal Lobe Memory System." *Science* 253 (5026): 1380–1386. doi: 10.1126/science.1896849.

Squire, Larry R., Craig E. L. Stark, and Robert E. Clark. 2004. "The Medial Temporal Lobe." *Annual Review of Neuroscience* 27: 279–306. doi: 10.1146/annurev.neuro.27.070203.144130.

FURTHER READING

Clelland, Claire D., M. Choi, C. Romberg, G. D. Clemenson Jr., Alexandra Fragniere, P. Tyers . . . Tim J. Bussey. 2009. "A Functional Role for Adult Hippocampal Neurogenesis in Spatial Pattern Separation." *Science* 325 (5937): 210–213. doi: 10.1126/science.1173215.
The precise role(s) of neurogenesis – the "birth" of new neurons and their migration into the dentate gyrus – in LTM are still being worked out, but this paper captures much of the excitement and potential importance of the phenomenon.

Corkin, Suzanne. 2002. "What's New with the Amnesic Patient H. M.?" *Nature Reviews Neuroscience* 3 (2): 153–160. doi: 10.1038/nrn726.
From a student of Brenda Milner who has studied H. M. intensively from the 1960s up through the final years of his life.

Corkin, Suzanne. 2013. *Permanent Present Tense: The Unforgettable Life of the Amnesic Patient, H. M.* New York: Basic Books.
A book that with equal parts compassion and scientific acuity, alternates between the biography and tragedy of Henry Molaison, and the manifold discoveries about human memory that have resulted from research with this remarkable patient.

Dębiec, Jacek. 2012. "Memory Reconsolidation Processes and Posttraumatic Stress Disorder: Promises and Challenges of Translational Research." *Biological Psychiatry* 71 (4): 284–285. doi: 10.1016/j.biopsych.2011.12.009.
Commentary addressing the implications of the study presented in Research Spotlight 10.1.

Eichenbaum, Howard. 2013. "What H. M. Taught Us." *Journal of Cognitive Neuroscience* 25 (1): 14–21. doi: 10.1162/jocn_a_00285.
Review from a leading expert on LTM and the functions of the hippocampus.

Frankland, Paul W., Stefan Köhler, and Sheena A. Josselyn. 2013. "Hippocampal Neurogenesis and Forgetting." *Trends in Neuroscience* 36 (9): 497–503. doi: 10.1016/j.tins.2013.05.002.
MTL neurogenesis is highest early in life, and this review suggests that this may account for the phenomenon of "infantile amnesia" – our inability to remember events from the first few years of our lives.

Gaffan, David. 2002. "Against Memory Systems." *Philosophical Transactions of the Royal Society of London, Series B: Biological Sciences* 357 (1424): 1111–1121. doi: 10.1098/rstb.2002.1110.
A provocative set of arguments for a radical rethinking of LTM. I don't agree with all of it, but nonetheless always assign it to my grad seminar on the cognitive neuroscience of memory.

Luu, Paul, Orriana C. Sill, Lulu Gao, Suzanna Becker, Jan Martin Wojtowicz, and David M. Smith. 2012. "The Role of Adult Hippocampal Neurogenesis in Reducing Interference." *Behavioral Neuroscience* 126 (3): 381–391. doi: 10.1037/a0028252.
Whereas Frankland, Köhler, and Josselyn (2013) propose that MTL neurogenesis may be responsible for retroactive interference (in the very young), this paper presents a scheme whereby MTL neurogenesis in the adult may aid in the individuation of distinct episodic memories.

Murray, E. A., S. P. Wise, M. L. K. Baldwin, and K. S. Graham (2020). *The Evolutionary Road to Human Memory.* New York: Oxford University Press.
A highly readable and authoritative book, written for a general audience, that explains all the mnemonic phenomena that we've covered in this chapter, and more, from the perspective of the authors' insightful and provocative evolutionary perspective.

Nader, Karim, and Oliver Hardt. 2009. "A Single Standard for Memory: The Case for Reconsolidation." *Nature Reviews Neuroscience* 10 (3): 224–234. doi: 10.1038/nrn2590.
A vigorous defense of the phenomenon of reconsolidation that, in the process, provides a nice review of the principles of consolidation and reconsolidation in fear conditioning.

Oudiette, Delphine, and Ken A. Paller. 2013. "Upgrading the Sleeping Brain with Targeted Memory Reactivation." *Trends in Cognitive Sciences* 17 (3): 142–149. doi: 10.1016/j.tics.2013.01.006.
A role for sleep in memory consolidation is revealed when LTM for specific items can be improved when auditory or olfactory stimuli that are associated with those items are presented to subjects while they sleep.

Postle, B.R. (2016). The hippocampus, memory, and consciousness. In S. Laureys, O. Gosseries, and G. Tononi (Eds.) *Neurology of Consciousness*, 2nd Ed. Elsevier (Amsterdam), pp. 349–363.
A review chapter by Yours Truly that goes into more detail about the pathophysiology of amnesia and the cognitive neuroscience of long-term memory.

CHAPTER 12
DECLARATIVE LONG-TERM MEMORY

KEY THEMES

- Early neuroimaging studies of LTM encoding were difficult to design, and their results difficult to interpret, due to uncertainty over how to activate the hippocampus differentially in different conditions, and how to unconfound effects of novelty from effects of encoding per se.

- fMRI studies of the subsequent memory effect, made possible by the development of "fast" event-related techniques, have identified PFC and inferior temporal regions associated with level-of-processing manipulations as being important for successful memory encoding.

- Another important innovation in neuroimaging data analysis, particularly for studies of LTM, is methods for measuring functional connectivity between brain regions.

- Intensive memorization of a highly detailed and idiosyncratic 2-D map, followed by daily navigation through this environment (i.e., being a licensed taxi driver in London), leads to gross anatomical expansion of the posterior hippocampus, and commensurate reduction of the anterior hippocampus.

- Accurate interpretation of impaired vs. spared performance of anterograde amnesic patients on tests of LTM requires an understanding of the level of awareness of the learning episode that is required by the task.

- MVPA of fMRI data can be used to predict the category of information that a subject is about to recall, a feat that wouldn't be possible if contextual reinstatement was not a principle underlying recall from LTM.

- Models from cognitive psychology that posit a distinction between processes of familiarity and recollection in memory retrieval have received considerable support from cognitive neuroscience research, with the latter being explicitly linked to the hippocampus.

CONTENTS

THE *COGNITIVE* NEUROSCIENCE OF LTM

The previous chapter emphasized the neurobiology of memory. With that groundwork in place, this chapter will focus much more on cognitive-level questions, in many instances looking at how longstanding cognitive theories (e.g., levels of processing, contextual reinstatement) have been studied by cognitive neuroscientists. So let's jump in!

ENCODING

Neuroimaging the hippocampus

Neuroimaging studies of LTM from the 1980s and 1990s were particularly challenging to design, and their results challenging to interpret, due to the constraints imposed by block designs. To understand why, let's think back to how the Washington University group went about studying the visual perception of words (*Figure 3.19*). They presented words, serially, for a block of 40 seconds, and compared that with a 40-second block of fixation. Although one might (and some did) object that there was insufficient experimental control to support strong interpretation of what cognitive functions were engaged during the word blocks, there was little ambiguity about the control task: staring at an unchanging fixation point would produce minimal activation of the visual system relative to 1 Hz presentation of words. (And, indeed, the [words – fixation] subtraction revealed considerably higher levels of activity in ventral occipitotemporal cortex for the former than the latter.) It's not clear, however, that the same approach would work if one sought to "image the hippocampus," primarily because it's believed that this structure is always encoding. For example, had subjects in the Petersen et al. (1988) study been debriefed an hour, or a day, after their scanning session, they would surely have remembered having endured the blocks of fixation as well as the blocks of word viewing. Thus, there's no reason to posit that their hippocampi would be any less (or more) active during the control than the experimental task. Were they differently active during the two conditions? Sure, but MVPA (and, therefore, any hope of detecting such a difference with neuroimaging) was more than a decade away.

A variant of this problem will also crop up in the section of this chapter that focuses on retrieval from LTM, and is one that was foreshadowed in the previous chapter's section on reconsolidation: even the act of retrieving information from LTM will be encoded by the hippocampus. If the darned thing never shuts off, how can we measure its function with a subtractive methodology?

Novelty-driven "encoding"

The reasoning laid out in the previous section was invoked often in the early to mid-1990s, when a series of PET studies of encoding into LTM failed to show evidence for differential hippocampal activity during encoding vs. control blocks. One way that experimenters tried to address this problem was to reason, in effect, that *if we can't devise a condition when it is not active, perhaps we can at least devise one when it is* less *active.* To achieve this, two groups, one led by (University of Toronto) cognitive psychologist Endel Tulving and another by Chantal Stern and colleagues at the Massachusetts General Hospital (MGH) NMR Center, explicitly controlled novelty. Tulving et al. (1994) pre-exposed subjects to 40 color photos of complex scenes the day prior to the scanning session, and then scanned subjects on blocks that either featured the "old" items from the day before or 40 new items that they had never seen before. Although one can imagine that the "old" condition might have generated a considerable amount of memory retrieval, what Tulving et al. (1994) emphasized was that the "new" items had greater novelty, and thus were likely to emphasize encoding processes. Their findings included greater "new" than "old" activity in the hippocampus and parahippocampal gyrus.

Stern et al. (1996) also presented 40 novel images during the experimental block, and instructed subjects to encode them for a subsequent memory test. During the control block, they presented a single image (also novel) during the entire duration of the block. Similar to the Tulving et al. (1994) findings, this study also found greater activity for the varying-identity stimuli than the repeated-identity stimulus in the posterior hippocampus and parahippocampal gyrus, as well as more posteriorly in the lingual and fusiform gyri.

In retrospect, knowing what we do now about, for example, category-preferential responses in the ventral occipitotemporal cortex (*Chapter 10*), we can surmise that at least some of the activity reported in these studies related more to visual processing than to memory encoding per se. Additionally, both explicitly confounded stimulus novelty with the mental process that they sought to engage. Thus, even at the time that these studies appeared, more impactful than their conceptual contributions was the fact that they, and a handful of other contemporaneous studies, succeeded in "activating" the elusive hippocampus. This had the effect of accelerating the pace at which the memory research community worked to develop better methods that would produce more definitive results. Indeed, two years after the Stern et al. (1996) study, two groups simultaneously published breakthrough studies of memory encoding using novel event-related techniques.

Subsequent memory effects in the PFC and MTL

Since the early 1980s, ERP researchers had employed the following procedure to study encoding into LTM: (1) present subjects with several items, one at a time; (2) create a series of ERPs time locked to stimulus onset; (3) after the EEG recording session, administer a surprise memory test for the items that had been presented during the EEG session; (4) selectively average the ERPs according to whether each item was or was not remembered in the subsequent memory test. Any differences between the two resultant ERPs would identify signal associated with successful vs. unsuccessful memory encoding. The logic of this approach has much to recommend it. In particular, unlike the neuroimaging studies reviewed in the previous section, subjects are performing the same task on every trial, and so this avoids the problem of whether differences in physiological signal are due to the memory-encoding processes of interest or some other factor (novelty per se, boredom, level of vigilance, etc.). The problem with the effect in the ERP literature was that it was capricious – not every lab could produce it, and even within a lab the effect would come and go, depending on factors such as whether the subsequent memory test was a test of free recall, cued recall, or recognition. The latter was particularly puzzling because the ERP signal, of course, was collected prior to the time that the subsequent memory test would be administered. If subjects didn't even know that there was going to be a memory test, how could it be that the format of that future test would influence encoding-related activity? [*Hey, that'd be a good extra-credit question!*] The result of these problems was that interest in this approach had largely fizzled out. Until, that is, those fMRI people came along . . .

Randy Buckner earned his PhD at Washington University in St. Louis, under the tutelage of Petersen and colleagues. On the strength of his graduate training in PET imaging he was invited to the MGH NMR Center as a kind of postdoctoral "Minister without Portfolio." He

gravitated to Cambridge, MA-based memory luminaries Daniel Schacter and Suzanne Corkin, and together with Schacter, postdoctoral fellow Anthony Wagner, and others, adapted the subsequent memory paradigm for event-related fMRI. At the same time, back at the Stanford University laboratory where Wagner had recently earned his PhD, former Corkin student John Gabrieli and *his* student James Brewer were carrying out a very similar study. The two were published, back-to-back in *Science*, in 1998, and the principal results of the Wagner et al. (1998) study are presented in *Figure 12.1*. To understand it thoroughly, we'll need to review some of the details.

Wagner and colleagues (1998) designed their study within the levels-of-processing framework, the venerable finding from cognitive psychology that a stimulus is better remembered when one thinks about its meaning (i.e., when one processes it at a "deep" level) than when one thinks about a superficial property (i.e., processes it at a "shallow" level). Their first experiment did this explicitly, presenting individual words to subjects and instructing them to either judge whether each corresponded to a living or a nonliving thing (deep processing) or, during alternate blocks, whether each was printed in upper- or lowercase type. A surprise memory test after the fMRI scanning session confirmed that the manipulation worked, with subjects subsequently recognizing 85% of deeply processed items vs. 47% of shallowly processed items. In the brain, regions showing greater activity on deep vs. shallow blocks included left inferior frontal gyrus (IFG) of the ventrolateral PFC and left parahippocampal and fusiform gyri on the ventral surface of the temporal lobe. This set up predictions for the subsequent memory study: finding regions of overlap would be consistent with the idea that item-to-item fluctuations in how we think about information determine, at least in part, how effectively we encode it into LTM.

For the subsequent memory study, fMRI data were acquired while subjects viewed a total of 480 words across six blocks. Each block included 40 abstract words, 40 concrete words, and 40 fixation events, presented in a randomly determined order for 2 seconds each. Additionally, because the order in which words would subsequently be remembered or forgotten was expected to be random, the authors used a "fast" event-related analysis of the kind outlined in section *Some considerations for experimental design, Chapter 3*.

It is important to note that neither the Wagner nor the Brewer study identified a "memory network," per se.

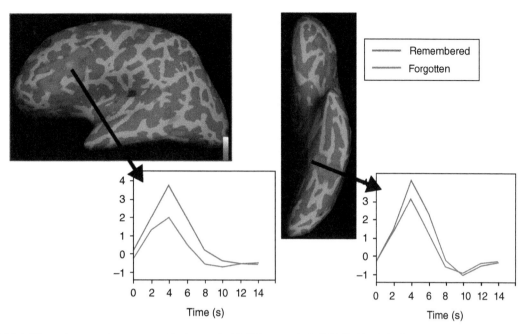

FIGURE 12.1 The subsequent memory effect reported in Wagner et al. (1998). Statistical maps, superimposed on two views of an inflated left hemisphere, show regions exhibiting higher activity for words that were later recognized with high confidence ("remembered") than for words that were later not recognized. Inset plots show the fMRI-evoked responses, estimated from the event-related analysis, from posterior IFG (left) and a region spanning parahippocampal and fusiform gyri (right: responses plotted in arbitrary units of BOLD signal intensity). Source: From Adapted from Paller, Ken A., and Anthony D. Wagner. 2002. "Observing the Transformation of Experience into Memory." Trends in Cognitive Sciences 6 (2): 93–102. doi: 10.1016/S1364-6613(00)01845-3. Reproduced with permission of Elsevier.

Wagner et al. (1998), in their study of memory for words, identified exclusively left-hemisphere regions. Brewer, Zhao, Desmond, Glover, and Gabrieli (1998), although using a similar subsequent-memory design and analysis, used photos of indoor and outdoor scenes, and the regions in which they found greater activity for high-confidence hits vs. misses were the parahippocampal gyrus, bilaterally, and right dorsolateral PFC. Thus, what both studies found was that activity in the very regions that are associated with verbal stimulus processing (in this case reading words) and visual stimulus processing (looking at pictures), respectively, is an important determinant of whether or not that information will later be remembered. (We'll see in *Chapter 19* that language processing is heavily biased to the left hemisphere. Additionally, other studies from Gabrieli's group had indicated that the left IFG, in particular, is recruited for tasks requiring semantic processing in the absence of any overt demands on episodic encoding or retrieval. And we've already seen that the PHG is important for representing information about

"places.") The fact that these studies both identified the magnitude of activity in their respective networks to be an important factor in subsequent memory retrieval is consistent with the idea that the strength of attentional focus may be an important determinant in LTM encoding. They are also consistent with the idea, recurring throughout this book, that ongoing fluctuations in brain state can influence the quality of cognition at any moment in time.

Uneven evidence for encoding-related activity in hippocampus

It is noteworthy that neither of these two studies, the two most sophisticated fMRI studies of memory encoding up to that time, identified differential activity in the hippocampus during the processing of items that would subsequently be remembered vs. subsequently forgotten. Importantly, there are two conclusions that CANNOT be drawn from this. At the physiological level, it would be a mistake to conclude that the hippocampus was somehow "not active" during these scans, for reasons that we've taken up previously. At the functional level, it would be a mistake to conclude from these studies, alone, that the hippocampus does not contribute to encoding into LTM. An interpretation from the perspective of the standard model, as outlined in the previous chapter, might be that the hippocampus is "doing its best" to encode everything that we experience, all the time, and that those thoughts and experiences that are somehow more salient, that are processed to a deeper level, are the ones that tend to get remembered. The primary function of the hippocampus, from this perspective, is to bind together the cortical representations whose association, after all, is what constitutes the memory.

Incidental encoding into LTM during a short-term memory task

Short-term memory (STM) will be the focus of *Chapter 14*, but one study of STM merits consideration here, for four reasons. First, it reinforces the idea that incidental encoding happens constantly, albeit with fluctuating efficacy, during the waking state, regardless of what behavior we're engaged in. (The didactic point being that this is true *even* when we are engaged in a task that is nominally testing something other than LTM.) The second reason for considering this study is because it illustrates the power of relating individual differences in physiological signal to individual differences in behavior. Third, it relates to a mechanism that we considered in the previous chapter's section on how the hippocampus might work. Finally, it provides an opportunity to introduce another important method for the analysis of functional neuroimaging data.

Signal intensity-based analysis of the subsequent memory effect

Each trial in this study, from the laboratory of cognitive neuroscientist Charan Ranganath at the University of California, Davis, began with the subject viewing a sample line drawing of a novel 3-D shape (multifaceted, blocky objects). After a delay period varying from 7 to 13 seconds in length, a memory probe appeared that matched the sample 50% of the time. After the scanning session, subjects were given a surprise test of recognition of the items that had been presented as sample stimuli during the STM task, and the fMRI data sorted as a function of subsequent LTM recognition. As illustrated in *Figure 12.2*, activity in

the left dorsolateral PFC and the left hippocampus during the early delay period of the STM task predicted subsequent LTM for each item. This illustrates the point that, in principle, one should be able to find a subsequent memory effect for *any* kind of behavior, not just in experiments designed to assess LTM encoding.

Another notable feature of this study is illustrated in *Figure 12.2.B*. Here, the authors show that, on a subject-by-subject basis, the magnitude of each subject's neural subsequent-memory effect is related to his or her subsequent LTM for items that had been presented during the test of STM. Stated another way, the magnitude of the subsequent memory effect in left hippocampus predicted how well the subject would later remember items from the test of STM. Such a demonstration of **individual differences** can be powerful, because it suggests that the mechanism under study might explain, at least in part, why some individuals remembered more items on the subsequent memory test and why others remembered fewer. Note that this is still (indeed, explicitly) a test of correlation, and therefore doesn't establish a causal role for, in this case, hippocampal activity. However, one could argue that such a relationship would be a necessary (though not sufficient) property of this system if it were causally related to LTM formation.

A short-term memory buffer to bridge the temporal gap between to-be-associated events

Recall from section *Episodic memory for sequences, Chapter 11*, that Lisman (1999) emphasized the fact that many events that we need to associate in LTM are separated in time by much longer than the ~100 ms activation time course of the CA3 NMDA receptor. Thus, although NMDA receptor dynamics could be the basis for associating location A with B and B with C as a rat runs along a path, many more than 100 ms would have elapsed, for example, between when the hapless Jerry dropped his candy and when the opportunistic monkey snatched it up. How might these two events be associated in LTM? It could be that the sustained hippocampal and prefrontal activity documented by Ranganath, Cohen, and Brozinsky (2005) serves this buffering function. (Interestingly, Hebb (1949; which featured prominently in *Chapter 11*) anticipated the need for such a temporary buffer, positing a "dual trace mechanism" by which transiently elevated, recurrent activity between the to-be-associated neural elements would persist until the second, structural trace (in his scheme, synaptic strengthening) would be established.)

FIGURE 12.2 Subsequent LTM effects in left dlPFC and left hippocampus during a test of visual STM. Source: Ranganath, Charan, Michael X. Cohen, and Craig J. Brozinsky. 2005. "Working Memory Maintenance Contributes to Long-Term Memory Formation: Neural and Behavioral Evidence." Journal of Cognitive Neuroscience 17 (7): 994–1010. doi: 10.1162/0898929054475118. Reproduced with permission of The MIT Press.

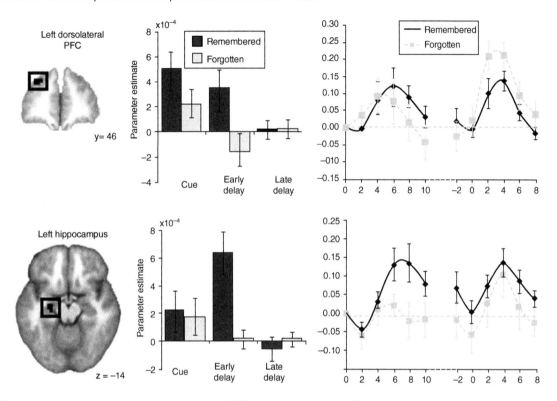

FIGURE 12.2.A Left-hand column: regions of left dlPFC (coronal view) and left hippocampus (axial view) showing the effect. Center column: bar graphs show magnitude of response during sample presentation ("Cue") and two portions of the delay period, expressed as parameter estimates from the GLM, as a function of whether they were subsequently remembered or forgotten on the surprise LTM test. Right-hand column: time series show activity (% signal change from baseline as a function of time), time locked to sample onset (first 0) and probe onset (second 0), also averaged as a function of subsequent LTM.

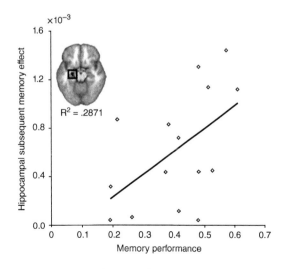

FIGURE 12.2.B Strength of early-delay hippocampal activity (vertical axis) predicts subsequent LTM performance.

Functional connectivity-based analysis of the subsequent memory effect

The final reason for devoting so much space to the Ranganath, Cohen, and Brozinsky (2005) study is a separate analysis that was applied to this dataset, and published as Ranganath et al. (2005): a functional connectivity-based analysis of the subsequent memory effect. All of the neuroimaging studies that we have considered up to this point in this chapter used the mass-univariate approach, effectively identifying voxels in which greater signal intensity related to stronger memory encoding. Thus, none of them directly assayed the element that is fundamental to all contemporary models of LTM – the standard model, the multiple trace model, reconsolidation models, etc. – which is the *interaction* between MTL regions and other regions of cortex, in the encoding, consolidation, and retrieval of LTM.

The very basis of LTM is assumed to be the association of neural representations, many of them supported in anatomically distal regions. Nonetheless, the massively parallel univariate analyses used in each of the studies reviewed in the first half of this chapter could neither prove nor disprove this assumption. One way to demonstrate these associations is to measure functional connectivity (section *Functional connectivity, Chapter 3*).

Note in *Figure 12.2.A* that the authors used a "slow" event-related design (see *Figure 9.4*) that partitioned variance in the fMRI signal into four epochs in the trial (probe covariate not shown in the figure). For the functional connectivity analysis, the loadings on each of these trial epoch-related covariates from the general linear model analysis (a.k.a. the parameter estimates, or the beta weights) were used as the data to determine which areas were functionally coupled across trials. Because the authors were specifically interested in whether there were areas for which functional connectivity with the hippocampus predicted stronger encoding into LTM, they implemented a "seeded analysis" by selecting the region of left hippocampus identified in *Figure 12.2.A*, extracting from it the early-delay parameter estimates from subsequently remembered trials, and correlating these against the early-delay parameter estimates from all other voxels in the brain. Then they did the same for subsequently forgotten trials. Finally, they determined the areas in which the correlation with the hippocampus was significantly greater on remembered than on forgotten trials. Their findings (*Figure 12.3*) provided direct evidence for something that most assumed, but that few, if any, had demonstrated directly: successful encoding was associated with enhanced functional connectivity

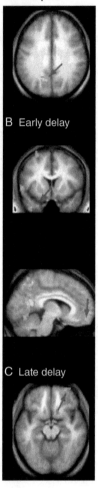

A Cue period

B Early delay

C Late delay

FIGURE 12.3 Seeded functional connectivity analysis, illustrating regions, by trial epoch, whose activity was more strongly correlated with the left hippocampus on trials presenting subsequently remembered vs. subsequently forgotten sample stimuli. Source: Ranganath, Heller, Cohen, Brozinsky, and Rissman, 2005. Reproduced with permission of Wiley.

between the hippocampus and many neocortical regions within and outside the MTL. Thus, they demonstrated that successful memory formation is associated with transient increases in cortico-hippocampal interaction.

The years since the mid-2000s have witnessed an explosion of studies of subsequent memory effects and of MTL-neocortical interactions during consolidation. What we've covered here provides a solid foundation from which to tackle this literature, and next we'll move on to consider memory retrieval. On the way, however, we'll take a detour to the streets of London.

THE HIPPOCAMPUS IN SPATIAL MEMORY EXPERTS

A fitting interregnum between the halves of this chapter focused on encoding-into or retrieval-from LTM are the studies by Eleanor Maguire and colleagues, which demonstrate how both of these processes can influence the macrostructure of the hippocampus. One account in the popular press sets the stage as follows:

> Manhattan's midtown streets are arranged in a user-friendly grid. In Paris, 20 administrative districts, or arrondissements, form a clockwise spiral around the Seine. But London? A map of its streets looks more like a tangle of yarn that a preschooler glued to construction paper than a metropolis designed with architectural foresight. Yet London's taxi drivers navigate the smoggy snarl with ease, instantaneously calculating the swiftest route between any two points (Jabr, 2011).

What does this have to do with the hippocampus? Maguire had noted, in addition to what we already know about experience-dependent synaptic change in the hippocampus, that the hippocampi of some species of food-catching birds undergo seasonal fluctuation in volume, expanding when demands on spatial abilities are greatest. She hypothesized, therefore, that a similar phenomenon might occur in humans whose job requires the memorization, and daily interrogation, of an incredibly detailed mental map of the streets of the city of London. Although several tests of this idea have now been published, perhaps the most tightly controlled of these is a comparison of the brains of London taxi drivers and London bus drivers. The two groups are well matched in terms of socioeconomic status, number of hours per day spent seated while pushing foot pedals and turning a wheel by hand, number of daily interactions with strangers, etc. Importantly, they are also well matched for specific perceptual and physiological factors known to affect the hippocampus, including self-induced movement (linked to the hippocampal theta rhythm), optic flow, and stress. Two salient differences, however, are that (1) taxi cab drivers have undergone intensive training to acquire "The Knowledge" – the layout of 25,000 streets in the city and the locations of thousands of places of interest – and (2) the daily activity of taxi drivers entails planning and navigating in reference to this 2-D

mental map, whereas bus drivers follow 1-D routes, often the same one(s) over and over.

Maguire et al. (2000) compared the brains of bus and taxi drivers with voxel-based morphometry (VBM; section *Voxel-based morphometry, Chapter 3*) and reported the following:

> Examination of [MRI] scans using whole-brain [VBM] . . . showed greater gray matter volume in the posterior hippocampi, and reduced anterior hippocampal gray matter volume in taxi drivers when compared with an age-matched control group. In addition, the longer taxi drivers navigated in London, the greater the posterior hippocampal gray matter volume, and the more decreased the anterior gray matter volume. (p. 1091)

Mindful of the caveats that accompany VBM analyses, Maguire et al. (2000) corroborated their VBM finding with a separate "pixel counting" analysis in which an observer literally measured the surface area of hippocampus in each slice, and volumes were computed from these measurements.

Note from *Figure 12.4* that taxi drivers had greater gray matter volume than bus drivers in mid-posterior hippocampus, and the reverse in anterior hippocampus. (Nadel, Hoscheidt, and Ryan (2013) drew on this when emphasizing the rostrocaudal gradient in hippocampal functioning that was considered in section *Episodic memory as an evolutionary elaboration of navigational processing, Chapter 11*.) The latter can be interpreted in terms of the caudal-to-rostral gradient of dorsal-stream input into the hippocampus, in that it would be expected that parietal and parahippocampal inputs drive the formation of highly detailed spatial maps in the hippocampus. The anterior hippocampus effects are not easy to interpret. They are the result of smaller VBM values in the taxi group, not larger values in the bus group. Paralleling this neuroanatomical finding was the fact that the taxi group was significantly impaired on a test of delayed recall for an abstract, complex figure, a task that is sensitive to right anterior hippocampal damage. Thus, one possible interpretation is that the unique circumstances of being a London taxi driver produce a trade-off of regional hippocampal processing power.

RETRIEVAL

In cognitive psychology, an influential framework for organizing different kinds of memory retrieval has been Tulving's (1985) distinction between *autonoetic*, *noetic*, and *anoetic* states

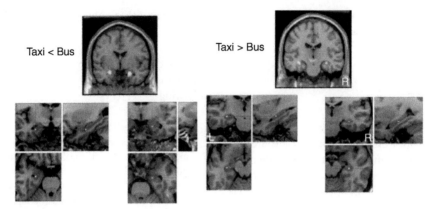

FIGURE 12.4 Differences in gray matter volume in the hippocampi of taxi cab drivers vs. bus drivers. The top image in both the Taxi > Bus and Taxi < Bus panels displays the effects on a full coronal slice from an atlas, and the images below show, in coronal, sagittal, and axial views, on the higher-resolution structural image of a single subject from the study, the location of the voxel from each hemisphere showing the largest VBM difference at the group level. Source: Maguire, Woollett, and Spiers, 2006. Reproduced with permission of Wiley.

of awareness with respect to memory retrieval. Derived from a Greek word appropriated by philosophers to refer to "mind" or "intellect," the term "noetic" in this context roughly corresponds to "knowing." Thus, autonoetic (or "self-knowing") awareness refers to an instance of memory retrieval that "is not only an objective account of what has happened or what has been seen or heard . . . [but also] necessarily involves the feeling that the present recollection is a reexperience of something that has happened before" (Wheeler, 2000, p. 597). (Note that the concept of "recollection" will take on a technical definition as the parameter in a model of recognition memory that we will consider in some detail further along in this chapter.) Noetic ("knowing") awareness can occur in two contexts. One is when retrieving information from semantic memory, for which there is not an associated "something that has happened before," the other is having a "feeling of familiarity" when encountering something or someone, but not being able to retrieve the episode in which this thing or person may have previously been encountered. ("Familiarity," too, will come up in the context of formal models of recognition memory.) Finally, the concept of anoetic ("not knowing") retrieval captures the fact that nondeclarative memory can be expressed without the individual's awareness that his or her performance is being influenced by a prior experience. In this section, we will consider each of these types of memory retrieval, beginning with how a misattribution of memory-test performance to noetic vs. anoetic memory retrieval led to erroneous conclusions about the bases of the amnesic syndrome.

Retrieval without awareness

In the modern era, the initial demonstrations of spared learning in amnesia were generated by Milner and her students and colleagues. First, in 1962, came the report of H.M.'s performance on a test of mirror tracing. This task required the subject to trace out the shape of a star on a piece of paper, while seated at an apparatus that required the paper to be viewed via its reflection off a mirror, because a barrier occluded its direct observation. Although it sounds simple, it's maddeningly difficult at first, as one repeatedly watches one's own hand pushing the pencil in directions that seem to bear no relation to one's volition. Over time, however, one learns the novel sensorimotor mapping imposed by this apparatus (a kind of sensorimotor adaptation largely mediated by the cerebellum), and the number of errors declines with each trial. Impressively, H.M. also showed a steady decline in errors across trials, and, astoundingly, his first trial on day 2 of testing had many fewer errors than he had made at the beginning of day 1! By day 3, his performance on the task was near error-free, this despite the complete absence of any conscious memory of ever having performed the task. Soon thereafter, it was confirmed that H.M. demonstrated preserved motor learning on many motor performance-based tasks.

H.M. also demonstrated learning on what came to be called repetition priming tasks. For example, a set of 20 line drawings of common objects was modified to create four incomplete versions of each, with version 1 resembling a series of arbitrary dashes and marks on the page – very difficult to recognize – and each subsequent version containing progressively more details. Subjects were first shown version 1 of each image, and encouraged to name it as quickly as possible. If they didn't get all 20 correct (and no individual did), they were then shown all the images from version 2, and so on, until they correctly named each drawing. H.M. correctly identified 6 from version 1, 12 from version 2, 19 from version 3, and all 20 from version 4. An hour later, he was retested on the same procedure, and this time he correctly identified 14 from version 1, 15 from version 2, and all 20 from version 3, overall making 48% fewer errors. Although this was a marked learning effect, it was characterized by Milner, Corkin, and Teuber (1968) as "vestigial," and was smaller than that of age- and education-matched control subjects, whose improvement ranged from 52% to 93%. "All in all," they concluded, "we remain impressed with the central role of the learning impairment in these patients" (p. 233). Around this time, however, another group was interpreting very similar findings in a radically different way.

Logically, the phenomenon of relatively spared retrograde memory after bilateral damage to the MTL does not, alone, permit one to distinguish between a "consolidation block" (i.e., impaired encoding) account versus a disordered storage or a disordered retrieval account of the consequence of MTL damage. This was illustrated most famously by British neuropsychologist Elizabeth Warrington and British neuroscientist Lawrence Weiskrantz, with a series of papers in the late 1960s and early 1970s. A 1974 study, for example, highlighted performance on two memory tests administered 10 minutes after the presentation of 16 words. Amnesic patients performed as well as control subjects on what the authors termed a "cued recall" task, in which the first three letters of a studied word (so-called "word stems") were presented and the subject was "required to identify the stimulus word" (p. 420). The patients were markedly impaired, however, on a test of Yes/No recognition. These results, together with evidence of disproportionate sensitivity to interference from items presented prior to or after the critical information, led Warrington and Weiskrantz (1974) to argue that MTL amnesia was better characterized by "altered control of information in storage" (p. 419), rather than a consolidation block. That is, they argued that new information "gets in there" (my quotes), but the amnesic syndrome makes it more difficult to access this information.

Subsequent research has established that the flaw in Warrington and Weiskrantz's reasoning was their failure to appreciate the distinction between noetic and anoetic states

of memory retrieval. With word-stem completion, for example, it has subsequently been shown that the performance of amnesic subjects relative to control subjects depends on the precise phrasing of instructions about how to process the three-letter stems: when subjects are instructed to complete the three-letter stems to "the first word that comes to mind," and no reference is made to the prior study episode, amnesic patients often generate target words at a level that is comparable to that of control subjects. When, in contrast, they are instructed to use the three-letter stem as a cue with which to retrieve an item from the studied list, amnesic patients are typically impaired. The former procedure, which more closely resembles that used by Warrington and Weiskrantz (1974), has come to be known as word-stem completion *priming*, the latter as word-stem *cued recall*. Thus, the intact performance of amnesic patients in the Warrington and Weiskrantz (1974) study has come to be reinterpreted as an early demonstration of intact performance by amnesic subjects on a priming task, a phenomenon that falls under the rubric of *nondeclarative memory*. (See Squire, 2006, *Further Reading*.)

The term "priming" is a metaphor referring to priming a pump – filling it with fluid so that when subsequently turned on (or activated via its handle) it will start producing fluid more quickly than it would have had it not been primed. With the psychological phenomenon of repetition priming, for example, the initial visual presentation of a stimulus makes it easier for the visual system to process that stimulus on subsequent repeat presentations. This comes about as a result of plasticity in the cortical systems that process these stimuli, regardless of whether or not there is an intact MTL to make a record of this processing that can later be explicitly recalled.

Documenting contextual reinstatement in the brain

"Our results ground Tulving's speculations about mental time travel in neural fact" (Polyn, Natu, Cohen, and Norman, 2005, p. 1966).

Mic drop.

The study to which we now turn speaks directly to the neural instantiation of autonoetic retrieval. It was motivated by the idea from cognitive psychology that the British psychologist Sir Frederick Bartlett (1886–1969) had articulated in 1932, and that Tulving had codified in 1973 as "contextual reinstatement": when asked to recall a particular memory, people first activate knowledge about the general properties of the event in question, and then use this general knowledge to constrain the search for the memory of the target event. To assess the neurobiological plausibility of this "speculation," Sean Polyn, his then-graduate mentor Kenneth Norman (himself a mentee of Schacter and O'Reilly), and their colleagues devised an innovative application of multivariate pattern analysis (MVPA). The resultant publication (Polyn et al., 2005) was one among a handful of seminal publications, all published within a year or so of each other, that brought to the attention of the broader cognitive neuroscience community just how powerful this technique could be.

The dynamics of multivariate state changes underlying behavior

The experiment of Polyn et al. (2005) proceeded in three phases. The first was the study phase, during which subjects were scanned while viewing 90 stimuli, 30 drawn from each of three categories: famous people; famous locations; and common household objects. This was an incidental encoding task in that, for each stimulus, subjects were requested to indicate (a) how much they liked the famous person; (b) how much they would like to travel to the famous location; or (c) how often they use the common object. ("*But wait,*"

you ask, "*I thought this section was about memory retrieval. So why this study in which they scan during encoding?*" "*A perceptive question,*" I respond, "*we'll see why in a moment.*")

The second phase of the experiment was to train a neural network pattern classifier to discriminate brain activity from each of the three categories. The process of training the classifier was not different, in principle, from how Rogers et al. (2003; section *A PDP model of picture naming, Chapter 3*) trained their PDP model to recognize verbal and visual objects: they presented it with the vector of signal intensity values produced by a given stimulus item, evaluated the network's guess as to which category the stimulus belonged to, and, when appropriate, propagated an error signal through the network that adjusted its weights such that it would be more likely to correctly classify that pattern of signal intensity values in the future. Once the classifier was fully "trained up," it could be fed the data from phase 3.

Phase 3 was a free recall session, performed in the scanner, in which subjects were simply instructed to recall as many items from the study phase as possible, by speaking into an MR-compatible microphone. Data from one subject are shown in *Figure 12.5*, and the first thing you may notice is that items were recalled in clusters: first, the

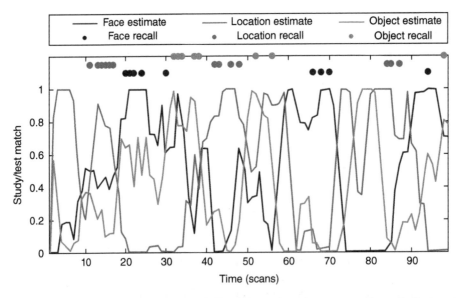

FIGURE 12.5 Contextual reinstatement during free recall. The blue, red, and green waveforms indicate, at each point in time during a scan of a single subject, the extent to which neural activity matches the pattern of neural activity independently determined to correspond to that stimulus category. Each dot indicates the recall of an item (larger dots for the recall of multiple items within the same 1.8-second volume acquisition; dots have been shifted forward in time by three TRs (i.e., 5.4 seconds) to account for the hemodynamic lag). Note how each instance of recall is preceded by brain activity taking on that category's pattern. Source: From Polyn, Sean M., Vaidehi S. Natu, Jonathan D. Cohen, and Kenneth A. Norman. 2005 "Category-Specific Cortical Activity Precedes Retrieval During Memory Search." Science 310 (5756): 1963–1966. doi: 10.1126/science.1117645. Reproduced with permission of AAAS.

subject recalled ~ 8 locations; then ~ 6 celebrities ("faces"); then ~ 6 objects; and so on. This kind of clustering in recall has been known to psychologists for decades and, indeed, is one of the behavioral phenomena consistent with a contextual reinstatement mechanism. What's really amazing about this figure is the relation that it reveals between this distinctive pattern of free recall performance and the concurrent fluctuations in neural activity: just prior to each "recall cluster," one can see a reconfiguration of brain activity into the pattern corresponding to that cluster's category! That is, we see the "neural *context*" associated with thinking about, let's say, famous locales, getting *reinstated* in the subject's brain just before they come out with "Taj Mahal . . . Eiffel Tower . . . Machu Pichu." Thus, an important component of autonoetic awareness in retrieval is the reconfiguration of brain activity into the state that it was in when encoding the to-be-remembered information.

Implications for "mind reading" . . . well, "brain reading"

Aside from what it tells us about memory retrieval, the Polyn et al. (2005) study garnered considerable excitement because it demonstrated a primitive form of "brain reading." Particularly intriguing was that, unlike the examples of MVPA that we've considered up to this point, in this study the experimenters were not demonstrating that a classifier could correctly guess what was the independent variable on a particular trial (e.g., "on this trial the stimulus was a picture of a cat"). Rather, their classifier was tracking the evolution of a mental *dependent* variable – memory recall – that was unaccompanied by any external cues or special instructions. I qualify it as primitive, because we have to keep in mind that this demonstration occurred within a highly constrained environment in which the classifier only "knew about" three states in the world – celebrities, famous locales, and common objects. Thus, if Polyn's subjects ever thought about climate change, or that evening's social plans, or how physically uncomfortable they felt while lying in the scanner, the classifier would only be able to assess whether the corresponding pattern of activity was more celebrity-like, locale-like, or object-like.

Familiarity vs. recollection

Dual-process models

We'll conclude this chapter with a consideration of whether the theoretical distinction between autonoetic and noetic retrieval corresponds to qualitatively, or just

quantitatively, different neural mechanisms. There are numerous, by and large quite similar, theoretical models from cognitive psychology that can be classified as dual-process models. These posit that free recall depends on a process called **recollection**, and that recognition memory can be supported by recollection and/or by a process called **familiarity**. According to these theories, recollection entails the retrieval of qualitative information associated with the learning episode. For example, when I recollect what I ate for dinner last night (it was my twentieth wedding anniversary! [really!]), I remember not only the scallops that I ate, but also where we sat at the restaurant, what the weather was like outside, the dress that my lovely dinner companion was wearing, etc. On a more mundane level, subjects in the Polyn et al. (2005) study, for example, would be assumed to have recollected contextual information about the moments when each of the images that they recalled had been presented at study. (For example, perhaps, when recalling having seen an image of the actor John Wayne, they also recollected how they had rated John Wayne during the study phase.) Recollection is characterized as a relatively slow process.

Familiarity, in contrast, is a fast, quantitative memory signal that only conveys the information that one has or has not previously encountered the item in question. Let's return again to the author's personal life, and engage in some mental time travel. Let's say that later today, when I need to take a break from writing, I will go jogging in my neighborhood and will encounter, walking along the sidewalk, a woman and, few minutes later, a man. As I run past the woman I have a strong sense that I've seen her somewhere recently, but where was it? Being the omniscient narrator as well as the protagonist in this imagined anecdote, I can tell you that this woman was the waitress from last night's dinner, but that I didn't recognize her in this very different context. Subsequently, as I run past the man, I have a vague sense, but less strong, that he, too, "looks familiar." He, it turns out, drove the taxi cab that took us to the restaurant. (Or was he the driver from the ride home? . . .) Thus, familiarity can vary along the dimension of strength, whereas recollection is assumed to be an all-or-none phenomenon. (Recollected memories can, of course, contain more or less detail, but this has more to do with the stored memory than the process of retrieval, per se.)

And these two processes are assumed to work in tandem. As it turns out, I actually did go out for a run this afternoon, and although I encountered neither the waitress nor the cabbie, I did run past someone worthy of note. He was coming the other direction along the Lakeshore Path, and at first looked like just another afternoon jogger. Then, I thought, as we were passing, "that guy looks familiar." After another three or four strides it came back to me: *His name is Charlie, he's in the Business School (Marketing Department?), and he and I used to serve together on the Graduate School Fall Competition committee.* By the dual-processes account, the visual processing of Charlie coming toward me along the path first generated a familiarity signal, which, after a few seconds of cogitation, gave way to a recollection.

As stated previously, there are many dual-process models. There have also been many, many behavioral, ERP, and fMRI studies testing different facets of these models. A general pattern that has emerged from this literature is that recollection depends on the hippocampus proper, whereas familiarity is supported by neocortical regions of the MTL, particularly the perirhinal cortex. We'll conclude this chapter by illustrating just one narrow slice from this literature, highlighting the methods underlying one particular model – the **receiver operating characteristic (ROC)**-based model – and a study that addresses the role of the hippocampus in these two hypothesized processes.

Dual processes of memory retrieval in the rodent

One of the simplest forms of recognition test is "yes/no" recognition, in which subjects indicate whether or not they have seen the memory probe previously. In two-alternative forced choice, an old and a new item are presented together, and the subject must indicate which is old (or, in the case of delayed nonmatch-to-sample [DNMS], which is new). Although both of these procedures assess memory performance, neither addresses what underlying processes may contribute to this performance. Several procedures for estimating the relative contribution of familiarity- vs. recollection-based processes to recognition performance do so empirically. For example, in the "remember/know" procedure, instead of responding "yes" or "no" to the probe, subjects indicate whether they "remember" having seen it, in which case they are assumed to have used recollection, or whether they only "know" that they have seen it, but can't retrieve a memory of having done so, in which case they are assumed to have used familiarity. The Wagner et al. (1998) and Brewer et al. (1998) studies with which we opened this chapter each effectively used this procedure, by asking subjects to rate their confidence for each positive response, and only using high-confidence hits in their subsequent-memory analyses.

An alternative procedure is to derive the contribution of the two processes to recognition performance by estimating ROC curves from one's subjects. This is the approach taken by Fortin, Wright, and Eichenbaum (2004) in the study with which we'll conclude this chapter.

The receiver operating characteristic (ROC) gets its arcane name from the fact that it was first developed during World War II to understand the performance of radar operators whose difficult (and consequential) task was to observe a noisy screen on the radar receiver and to detect signals (e.g., blips corresponding to enemy aircraft) from among the noise. They faced the classic trade-off that was later formalized in signal detection theory: by setting too lenient a criterion the operator risked issuing "false alarms" – announcing that he had detected enemy aircraft when, in fact, it was just noise – whereas, by setting too strict a criterion, he risked missing actual targets. Note that, in this scenario, the radar operator can change his performance by "setting his criterion" to be liberal or to be strict, but none of this changes the inherent signal-to-noise ratio (SNR) of the radar receiver.

Applied to recognition memory, all items that one encounters are assumed to generate some recognition signal, and one's "mnemonic SNR" is the extent to which *old* items generate a stronger signal than *new* items. Independent of one's SNR is where one chooses to set one's criterion. The trade-off associated with different criterion settings is implemented in an ROC-based recognition test by requiring subjects to rate each recognition probe on a scale from 1 to, say, 5, with "5" corresponding to "I'm absolutely certain that I saw this previously" and "1" corresponding to "I'm certain that I did not see this previously." After collecting the data, one can then derive an ROC curve in the following way. First, set up a pair of axes like those illustrated in *Figure 12.6* with the probability of hits along the vertical axis and the probability of false alarms along the horizontal axis. Next, score the subject's data as though only their "5" responses were "yes"es, and all their other responses were "no"s. This estimates their performance at its most conservative: their hit rate will be the lowest, but so, too, will their false-alarm rate. Next, rescore the data as though responses of "5" and "4" were "yes"es, and the others "no"s. Because the subject had assigned 4s to probes about which they had relatively high confidence, the "5s + 4s" hit rate will be higher than was the "5s only" hit rate. But so, too, will the false alarm rate. Thus, the "5s + 4s" data point will be higher along both the hits axis and the false alarms axis. The progression from most conservative to most liberal can be seen in panel D of *Figure 12.6.*

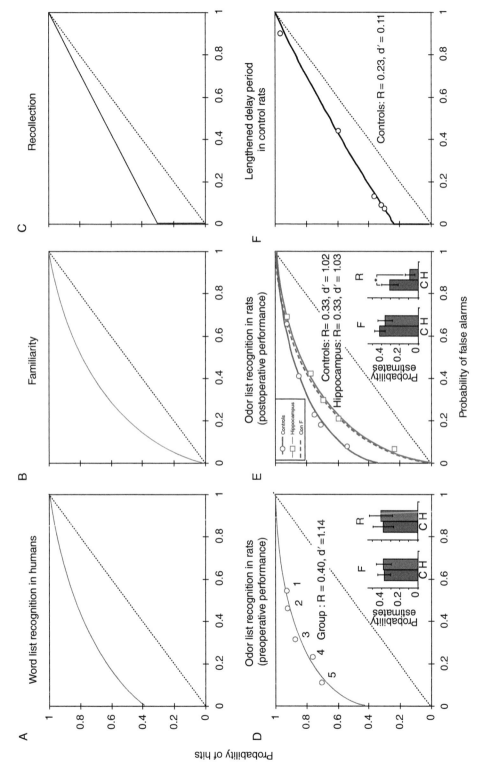

FIGURE 12.6 ROC-based dual-process model of recognition, applied to humans and rats. Source: From Fortin, Norbert J., Sean P. Wright, and Howard Eichenbaum. 2004. "Recollection-Like Memory Retrieval in Rats Is Dependent on the Hippocampus." Nature 431 (7005): 188–191. doi: 10.1038/nature02853. Reproduced with permission of Nature Publishing Group.

Panel A of *Figure 12.6* is an idealized plot of ROC-analyzed data from a human subject in a word-list recognition experiment. When this "asymmetric ROC" is decomposed, as in panels B and C, it can be understood as being made up of independent factors represented by each of these two: the degree of curvilinearity of the function labeled "familiarity" and the y-intercept of the function labeled "recollection" (Yonelinas, 2001). Note that the overall memory performance is captured by the area under the curve, with the diagonal equivalent to chance performance.

How did Fortin and colleagues manage to get rats to give a range of five levels of confidence judgments? Let's start with their basic procedure. Rats were trained to retrieve rewards (pieces of the breakfast cereal Cheerios) buried in cups of sand that was scented with one of 30 distinctive household scents. Next, they learned the DNMS procedure by which, if a test scent was a new scent, they were to dig in that sand to retrieve the reward. If, on the other hand, it was an old scent, the cup would not contain a reward, but the rat would be rewarded for refraining from digging and, instead, moving to the back of the cage to an alternate, empty cup to which a reward would be delivered. Different levels of bias were then created by, in effect, varying how hard the rat would have to work to receive its reward for correctly choosing "new" vs. how much food it would receive for correctly choosing "old." A liberal bias was created by making it more onerous to respond "new," by placing the same amount of sand in a taller cup and reducing the amount of reward that it would contain. Thus, if the animal "wasn't sure" about whether a new scent was old or new, it would tend to favor an "old" response, because a "new" response required too much effort. (If, however, it was certain that the test scent was new, then the only choice was to climb and dig.) In tandem, the amount of reward delivered for correct "old" responses was increased, thereby further incentivizing this choice.

Now let's look at the results, in the bottom three panels of *Figure 12.6*. Panel D shows the preoperative performance of the animals that were to receive hippocampal lesions, and of the control animals. The combined ROC of the two groups shows the standard asymmetric shape, and the model estimates (inset bar graphs) indicate comparably strong familiarity and recollection estimates for both groups. Panel E illustrates the key result: after hippocampal lesions, the estimated contribution to performance of recollection is devastated, but that of familiarity is intact. The dotted line labeled "Con F" shows the estimated performance of the control group when the contribution of recollection is algebraically removed. Finally, panel F shows the effect, in the control rats, of lengthening the delay period from 30 to 75 minutes. This manipulation removes the contribution of familiarity to their performance but leaves recollection relatively intact. Not shown in this figure is the effect of the hippocampal lesion on overall performance: the control animals at the 30-minute delay performed 73% correct, the hippocampal animals 66% correct (i.e., a modest impairment), and the control animals at the 75-minute delay performed 64% correct. (Note that while lesioning the hippocampus removes the estimated contribution of recollection to performance, increasing the delay by and large removes the estimated contribution of familiarity to performance.)

This suggests a possible solution to why the monkeys with H lesions described in the Murray, Bussey, and Saksida (2007) review (*Figure 11.8*) were not impaired on tests of DNMS: perhaps they could solve the tasks based on a familiarity signal.

KNOWLEDGE

That's the topic of our next chapter. Having considered the principles of how information gets into, and gets retrieved from, the brain, we next turn to how it is organized "while it sits there," waiting to be accessed and thought about.

END-OF-CHAPTER QUESTIONS

1. Why can one not "just trial-average" stimulus-evoked fMRI signal as a function of subsequent memory performance, as can be done with ERP data?

2. Is it possible to learn anything about LTM encoding from a neuroimaging study of STM?

3. What principle of LTM formation is better captured with functional connectivity analyses than with signal intensity-based analyses?

4. What is it about the experience/expertise of London taxi drivers, and the organization of cortical projections

to the hippocampus, that accounts for the pattern of structural change observed in these individuals?

5. Define autonoetic, noetic, and anoetic memory retrieval, and describe a task that would be well suited to assess each.

6. Would it be possible to document contextual reinstatement in the brain with univariate analyses of fMRI data? Explain how, or why not.

7. Is there neural evidence for a distinction between familiarity and recollection in memory retrieval? To what kind of task is it assumed that only one of these can contribute, and to what can both contribute?

REFERENCES

Bartlett, Frederic C. 1932. *Remembering: A Study in Experimental and Social Psychology*. Cambridge, MA: Cambridge University Press.

Brewer, James B., Zuo Zhao, John E. Desmond, Gary H. Glover, and John D. E. Gabrieli. 1998. "Making Memories: Brain Activity that Predicts How Well Visual Experience Will Be Remembered." *Science* 281 (5380): 1185–1187. doi: 10.1126/science.281.5380.1185.

Fortin, Norbert J., Sean P. Wright, and Howard Eichenbaum. 2004. "Recollection-Like Memory Retrieval in Rats Is Dependent on the Hippocampus." *Nature* 431 (7005): 188–191. doi: 10.1038/nature02853.

Hebb, Donald Olding. 1949. *The Organization of Behavior: A Neuropsychological Theory*. New York: Wiley and Sons.

Jabr, Ferris. 2011. "Cache Cab: Taxi Drivers' Brains Grow to Navigate London's Streets." *Scientific American*, December 8. http://www.scientificamerican.com/article.cfm?id=london-taxi-memory.

Lisman, John E. 1999. "Relating Hippocampal Circuitry to Function: Recall of Memory Sequences by Reciprocal Dentate–CA3 Interactions." *Neuron* 22: 233–242. doi: 10.1016/S0896-6273(00)81085-5.

Maguire, Eleanor A., David G. Gadian, Ingrid S. Johnsrude, Catriona D. Good, John Ashburner, Richard S. J. Frackowiak, and Christopher D. Frith. 2000. "Navigation-Related Structural Change in the Hippocampi of Taxi Drivers." *Proceedings of the National Academy of Sciences of the United States of America* 97 (8): 4398–4403. doi: 10.1073/pnas.070039597.

Maguire, Eleanor A., Katherine Woollett, and Hugo J. Spiers. 2006. "London Taxi Drivers and Bus Drivers: A Structural MRI and Neuropsychological Analysis." *Hippocampus* 16 (12): 1091–1101. doi: 10.1002/hipo.20233.

Milner, Brenda, Suzanne Corkin, and H.-L. Teuber. 1968. "Further Analysis of the Hippocampal Amnesic Syndrome: 14-Year Follow-Up Study of H. M." *Neuropsychologia* 6 (3): 215–234. doi: 10.1016/0028-3932(68)90021-3.

Murray, Elisabeth A., Timothy J. Bussey, and Lisa M. Saksida. 2007. "Visual Perception and Memory: A New View of Medial Temporal Lobe Function in Primates and Rodents." *Annual Review of Neuroscience* 30: 99–122. doi: 10.1146/annurev.neuro.29.051605.113046.

Nadel, Lynn, Siobhan Hoscheidt, and Lee R. Ryan. 2013. "Spatial Cognition and the Hippocampus: The Anterior-Posterior Axis." *Journal of Cognitive Neuroscience* 25 (1): 22–28. doi: 10.1162/jocn_a_00313.

Petersen, Steven E., Peter T. Fox, Michael I. Posner, Mark Mintun, and Marcus E. Raichle. 1988. "Positron Emission Tomographic Studies of the Cortical Anatomy of Single-Word Processing." *Nature* 331 (6157): 585–589. doi: 10.1038/331585a0.

Polyn, Sean M., Vaidehi S. Natu, Jonathan D. Cohen, and Kenneth A. Norman. 2005 "Category-Specific Cortical Activity Precedes Retrieval during Memory Search." *Science* 310 (5756): 1963–1966. doi: 10.1126/science.1117645.

Ranganath, Charan, Michael X. Cohen, and Craig J. Brozinsky. 2005. "Working Memory Maintenance Contributes to Long-Term Memory Formation: Neural and Behavioral Evidence." *Journal of Cognitive Neuroscience* 17 (7): 994–1010. doi: 10.1162/0898929054475118.

Ranganath, Charan, Aaron Heller, Michael X. Cohen, Craig J. Brozinsky, and Jesse Rissman. 2005. "Functional Connectivity with the Hippocampus during Successful Memory Formation." *Hippocampus* 15 (8): 997–1005. doi: 10.1002/hipo.20141.

Rogers, Timothy T., John R. Hodges, Matthew A. Lambon Ralph, and Karalyn Patterson. 2003. "Object Recognition under Semantic Impairment: The Effects of Conceptual Regularities on Perceptual Decisions." *Language and Cognitive Processes* 18 (5–6): 625–662. doi:10.1080/01690960344000053.

Stern, Chantal E., Suzanne Corkin, R. Gilberto González, Alexander R. Guimaraes, John R. Baker, Peggy J. Jennings, . . . Bruce R. Rosen. 1996. "The Hippocampal Formation Participates in Novel Picture Encoding: Evidence from Functional Magnetic Resonance Imaging." *Proceedings of the National Academy of Sciences of the United States of America* 93 (16): 8660–8665.

Tulving, Endel. 1985. "Memory and Consciousness." *Canadian Psychology* 26 (1): 1–12. doi: 10.1037/h0080017.

Tulving, Endel, Hans J. Markowitsch, Shitij Kapur, Reza Habib, and Sylvain Houle. 1994. "Novelty Encoding Networks in the Human Brain: Positron Emission Tomography Data." *NeuroReport* 5 (18): 2525–2528. doi: 10.1097/00001756-199412000-00030.

Wagner, Anthony D., Daniel L. Schacter, Michael Rotte, Wilma Koutstaal, Anat Maril, Anders M. Dale, Bruce R. Rosen, and Randy L. Buckner. 1998. "Building Memories: Remembering and Forgetting of Verbal Experiences as Predicted by Brain Activity." *Science* 281 (5380): 1188–1191. doi: 10.1126/science.281.5380.1188.

Warrington, Elizabeth K., and Lawrence Weiskrantz. 1974. "The Effect of Prior Learning on Subsequent Retention in Amnesic Patients." *Neuropsychologia* 12 (4): 419–428. doi: 10.1016/0028-3932(74)90072-4.

Wheeler, Mark A. 2000. "Episodic Memory and Autonoetic Awareness." In *The Oxford Handbook of Memory*, edited by Endel Tulving and Fergus I. M. Craik, 597–608. New York: Oxford University Press.

Yonelinas, Anderew P. 2001. "Components of Episodic Memory: The Contribution of Recollection and Familiarity." *Philosophical Transactions of the Royal Society of London, Series B: Biological Sciences* 356 (1413): 1363–1374. doi: 10.1098/rstb.2001.0939.

OTHER SOURCES USED

Gabrieli, John D. E., Margaret M. Keane, Ben Z. Stanger, Margaret M. Kjelgaard, Suzanne Corkin, and John H. Growdon. 1994. "Dissociations among Structural-Perceptual, Lexical-Semantic, and Event-Fact Memory Systems in Amnesia, Alzheimer's Disease, and Normal Subjects. *Cortex* 30 (1): 75–103. doi: 10.1016/S0010-9452(13)80325-5.

Holyoak, Keith J. 1999. "Psychology." In *The MIT Encyclopedia of the Cognitive Sciences*, edited by Robert A. Wilson and Frank C. Keil, xxxix–xlix. Cambridge, MA: MIT Press.

Lee, Diane W., Troy G. Smith, Anthony D. Tramontin, Kiran K. Soma, Eliot A. Brenowitz, and Nicola S. Clayton. 2001. "Hippocampal Volume Does Not Change Seasonally in a Non Food-Storing Songbird." *NeuroReport* 12 (9): 1925–1928. doi: 10.1097/00001756-200107030-00031.

Moscovitch, Morris. 2000. "Theories of Memory and Consciousness." In *The Oxford Handbook of Memory*, edited by Endel Tulving and Fergus I. M. Craik, 609–625. New York: Oxford University Press.

Rissman, Jesse, Adam Gazzaley, and Mark D'Esposito. 2004. "Measuring Functional Connectivity during Distinct Stages of a Cognitive Task." *NeuroImage* 23 (2): 752–763. doi: 10.1016/j.neuroimage.2004.06.035.

Rugg, Michael D., and Kevin Allan. 2000. "Event-Related Potential Studies of Memory." In *The Oxford Handbook of Memory*, edited by Endel Tulving and Fergus I. M. Craik, 521–538. New York: Oxford University Press.

Smulders, Tom V., A. D. Sasson, and Timothy J. DeVoogd. 1995. "Seasonal Variation in Hippocampal Volume in a Food-Storing Bird, the Black-Capped Chickadee." *Journal of Neurobiology* 27 (1): 15–25. doi: 10.1002/neu.480270103.

Warrington, Elizabeth K., and Lawrence Weiskrantz. 1968. "New Method of Testing Long-Term Retention with Special Reference to Amnesic Patients." *Nature* 217 (5132): 972–974. doi: 10.1038/217972a0.

Warrington, Elizabeth K., and Lawrence Weiskrantz. 1970. "The Amnesic Syndrome: Consolidation or Retrieval?" *Nature* 228 (5272): 628–630, doi: 10.1038/228628a0.

Williams, Pepper, and Michael J. Tarr. 1997. "Structural Processing and Implicit Memory for Possible and Impossible Figures." *Journal of Experimental Psychology: Learning, Memory, and Cognition* 23 (6): 1344–1361. doi: 10.1037/0278-7393.23.6.1344.

Williams, Pepper, and Michael J. Tarr. 1999. "Orientation-Specific Possibility Priming for Novel Three-Dimensional Objects." *Perception and Psychophysics* 61 (5): 963–976. doi: 10.3758/BF03206910.

Zola, Stuart M., Larry R. Squire, Edmond Teng, Lisa Stefanacci, Elizabeth A. Buffalo, and Robert E. Clark. 2000. "Impaired Recognition Memory in Monkeys after Damage Limited to the Hippocampal Region." *Journal of Neuroscience* 20 (1): 451–463.

FURTHER READING

Ashburner, John, and Karl J. Friston. 2001. "Why Voxel-Based Morphometry Should Be Used." *NeuroImage* 14 (6): 1238–1243. doi: 10.1006/nimg.2001.0961.
A response to the critique from Bookstein (2001).

Bookstein, Fred L. 2001. "'Voxel-Based Morphometry' Should Not Be Used with Imperfectly Registered Images." *NeuroImage* 14 (6): 1454–1462. doi: 10.1006/nimg.2001.0770.
One critique of VBM, suggesting circumstances in which face-value interpretation of its results may lead to flawed inference.

Cowan, Emily, Anli Liu, Simon Henin, Sanjeev Kothare, Orrin Devinsky, and Lila Davachi. 2020. "Sleep spindles promote the restructuring of memory representations in ventromedial prefrontal cortex through enhanced hippocampal-cortical functional connectivity." *Journal of Neuroscience*. doi: 10.1523/JNEUROSCI.1946-19.2020.
Whereas subsequent memory designs assess item-by-item encoding processes, this study assesses consolidation processes by relating density of sleep spindles, a substrate for sharp-wave ripples, to hippocampal-neocortical connectivity at retrieval.

Norman, Kenneth A., Sean M. Polyn, Greg J. Detre, and James V. Haxby. 2006. "Beyond Mind-Reading: Multi-Voxel Pattern Analysis of fMRI Data." *Trends in Cognitive Sciences* 10 (9): 424–430. doi: 10.1016/j.tics.2006.07.005.
An accessible tutorial review on applications of MVPA decoding models in neuroimaging research.

Paller, Ken A., and Anthony D. Wagner. 2002. "Observing the Transformation of Experience into Memory." *Trends in Cognitive Sciences* 6 (2): 93–102. doi: 10.1016/S1364-6613(00)01845-3.
Review of the theoretical and ERP literatures on the subsequent memory effect that laid the groundwork for the fMRI studies detailed in this chapter.

Rissman, Jesse, and Anthony D. Wagner. 2012. "Distributed Representations in Memory: Insights from Functional Brain Imaging." *Annual Review of Psychology* 63: 101–128. doi: 10.1146/annurev-psych-120710-100344.
Review of MVPA analyses of declarative memory.

Rugg, Michael D., and Andrew P. Yonelinas. 2003. "Human Recognition Memory: A Cognitive Neuroscience Perspective." *Trends in Cognitive Sciences* 7 (7): 313–319. doi: 10.1016/S1364-6613(03)00131-1.
A good place to start if contemplating a plunge into the massive literature on the cognitive neuroscience of dual-process models of retrieval.

Serences, John T., and Sameer Saproo. 2012. "Computational Advances towards Linking BOLD and Behavior." *Neuropsychologia* 50 (4): 435–446. doi: 10.1016/j.neuropsychologia.2011.07.013.
A tutorial review that this author found to be very helpful for understanding the principles of multivariate encoding models.

Squire, Larry R. 2006. "Lost Forever or Temporarily Misplaced? The Long Debate about the Nature of Memory Impairment." *Learning and Memory* 13 (5): 522–529. doi: 10.1101/lm.310306.
A detailed, authoritative review of consolidation block vs. retrieval-based accounts of anterograde amnesia.

Yonelinas, Anderew P. 2001. "Components of Episodic Memory: The Contribution of Recollection and Familiarity." *Philosophical Transactions of the Royal Society of London, Series B: Biological Sciences* 356 (1413): 1363–1374. doi: 10.1098/rstb.2001.0939.
Detailed theoretical and quantitative background for the ROC-based dual-process model of retrieval.

CHAPTER 13
SEMANTIC LONG-TERM MEMORY

KEY THEMES

- An intact MTL is not necessary for the preservation and retrieval of semantic memory.

- The study of patients showing (doubly dissociable) selective impairments at naming or identifying living things vs. nonliving things has given rise to three classes of theory: (1) knowledge is organized according to taxonomic categories (living vs. nonliving being the most salient); (2) knowledge is organized according to the functional modalities (e.g., visual perception of color and shape, motor sequences for manipulating tools) from which meaning derives; (3) lexical representation and access (i.e., words per se, not necessarily knowledge) is organized according to taxonomic categories.

- Important, and difficult, questions have been raised about the logical limits of what can be inferred from the results from single neuropsychological cases, as well as from demonstrations of double dissociations of function.

- The dependence on language of many experimental procedures used for testing semantic memory can complicate the interpretation of whether a particular set of neuroimaging findings reflects the processing of knowledge, per se, or reflects the language functions (often, lexical retrieval) engaged by the behavioral task.

- Attributing neuroimaging signal to knowledge vs. visual perception and/or mental imagery can also be tricky.

- fMRI adaptation, which takes advantage of the phenomenon of repetition suppression, can be an effective technique for deducing key functional properties of a targeted region of cortex.

- Research on patients with different variants of frontotemporal dementia has been characterized by debates on whether the core deficit underlying systematic patterns of cognitive decline is semantic or language based.

- Neuroimaging (both voxel-based morphometry [VBM] and activation studies) and computational modeling indicate an important role for the anterior temporal lobe in the control, if not the storage itself, of semantic memory.

CONTENTS

KNOWLEDGE IN THE BRAIN

Questions about how we represent knowledge have been at the core of formal inquiry into human cognition, and the human condition, for as far back in history as we have written records. From the perspective of cognitive neuroscience, it's one of the domains in which there is considerable contact with scholarship in the humanities and the social sciences, including philosophy, linguistics, and cognitive development. It is also noteworthy for its relative lack of overlap with other branches of neuroscience, both because it is so difficult to study in nonhuman species and because it deals with mental faculties that other animals simply may not possess.

As we have done in other chapters, we'll outline some of the key questions by considering findings from neuropsychology. This time, however, instead of starting with an exotic "new" neurological syndrome, we'll begin with our old friend H.M.

DEFINITIONS AND BASIC FACTS

Semantic memory is defined as memories for facts whose retrieval is not accompanied by information pertaining to how, when, or where that information was acquired – its retrieval is noetic. One of the striking findings with H.M. was that his memory for facts that he had learned prior to his surgery remained intact. So, too, did his vocabulary. Indeed, our knowledge of the meanings of words, often referred to as the mental lexicon, is an important subset of our semantic memory. Although it was initially believed that H.M.'s episodic memory for the distant past was intact, more recent interpretations, many of them prompted by the multiple trace theory, argue that even these were impoverished, "semanticized," and lacking the feeling of "reexperienc[ing] . . . something that has happened before." There is, however, no comparable debate about his semantic memory. For example, in 1953, H.M. was administered the Wechsler Adult Intelligence Scale (WAIS, i.e., an IQ test) as a part of preoperative cognitive testing in the run-up to his surgery. His performance on the subtests that measure lexical memory was within the normal range on that occasion, and it remained unchanged across 19 different testing sessions that were performed during the ensuing decades, through 2000. Thus, an intact MTL is not necessary for the preservation and retrieval of semantic memory.

How are semantic memories acquired? One formal account that we reviewed in *Chapter 11* is the slow cortical component of the complementary learning system models of McClelland, McNaughton, and O'Reilly (1995). It is generally accepted that all memories begin as episodic, and gradually transition to becoming semantic. (This view, however, is not universally held, as we shall see in just a bit.) A concrete example familiar to many readers might be one's knowledge of Barack Obama. The first time that a typical American (okay, let's pick on me again) may have heard of this individual may have been while Obama was campaigning to be the president of the United States. Let's say that I first encountered the name while reading a newspaper article about the Democratic Primary race during the early months of 2008. (He was elected President in November 2008.) The next time I thought about Barack Obama may have been when a friend asked, "Which Democrat do you think will win the party's nomination?" To answer this question, I would have retrieved a memory of (the episode of) having read that newspaper article. Let's say that the third time I heard the name was while watching a televised debate between the candidates. At that point, would I think back to the newspaper article, or the discussion with the friend, each time the camera turned to Obama? Or would the knowledge *Black man, junior senator from Illinois, graduate of Harvard Law School, or Democratic candidate for President* just pop into my head noetically? If not by this third episode, shortly thereafter repeated iterations of this process would have created a representation that can be retrieved independently of any reference to any one of the episodes in which this information was encountered. In this way, the knowledge of "Barack Obama" becomes a *semantic* memory. Should I sustain bilateral damage to my MTL in, say, 2021 (pause here for a moment while the author knocks on wood to negate this premonition), I would still be able to retrieve this knowledge after that time.

Another core fact about semantic memory is conveyed by its name: a semantic memory represents meaning. Further, we know that it is organized in relation to its meaning. In the case of Barack Obama, for example, numerous experiments have demonstrated that if one is thinking about him, one can more quickly access information about John F. Kennedy (the US President from the 1960s) than if one was thinking of, say, Pelé, or John Lennon. Precisely how knowledge is organized, however, remains uncertain, and will be the question underlying much of the research that we'll review in this chapter.

And *now* we can turn to some exotic cases from cognitive neuropsychology.

CATEGORY-SPECIFIC DEFICITS FOLLOWING BRAIN DAMAGE

Animacy, or function?

Two seminal publications that have greatly influenced the contemporary study of the neural bases of semantics came from British neuropsychologist Elizabeth Warrington and her colleagues in the mid-1980s. First, Warrington and McCarthy (1983) described a patient (V.E.R.) whose large left hemisphere stroke produced a massive impairment in the ability to speak or to understand spoken language, including the following of simple instructions. Nevertheless, they were able to demonstrate some spared comprehension, in that, in response to a spoken word, she could point to a picture corresponding to the word. Her performance showed a noteworthy pattern, however, in that she produced many more correct responses when the words named flowers, animals, or foods than when they named manmade objects. Thus, she seemed to have a relative impairment of knowledge for nonliving things. The next year, Warrington and Shallice (1984) reported four patients recovering from herpes encephalitis who showed the opposite pattern: all four demonstrated worse performance with living than with nonliving words. Further, two of these patients had sufficiently spared language abilities that they could be tested on picture naming, and here the pattern was the same. One of them (J.B.R.) named 90% of the nonliving objects he was shown, but only 6% of the living objects. The second (S.B.Y.) named 75% vs. 0%. Examples of S.B.Y.'s definitions given to spoken words, in *Table 13.1*, are particularly compelling, in that they illustrate that the deficit is not a deficit of speech production, or general verbal fluency.

Together, these two studies produced a double dissociation, thereby seeming to rule out any trivial reasons that one category might be inherently more difficult than the other, and therefore more vulnerable to brain trauma of any kind. What remained, of course, was to explain this intriguing pattern of findings. One interpretation is that they provide evidence that knowledge is represented according to taxonomic categories in the brain. This view has been championed perhaps most vigorously by cognitive neuropsychologist Alfonso Caramazza. In one formulation of a "domain-specific knowledge hypothesis," for example, Caramazza and Shelton (1998) propose that "evolutionary pressures have resulted in specialized mechanisms for perceptually *and* conceptually distinguishing animate and inanimate kinds . . . leading to a *categorical* organization of this knowledge in the brain." We will return to this hypothesis in a bit, but first let's consider an alternative, because one of the factors that makes the domain-specific knowledge hypothesis provocative is that many investigators, including Warrington herself, have *not* interpreted these patterns of category-specific impairment as evidence for categorical organization in the brain.

Warrington and Shallice (1984) noted that patient J.B.R., for example, also showed impaired performance for several categories of nonliving objects: metals, types of cloth, musical instruments, and precious stones. Thus, perhaps the (relatively) selective impairments of knowledge for living vs. nonliving things in these patients were by-products of two different underlying factors with which the categories of living and nonliving are highly correlated. Warrington and Shallice proposed that these factors might be two cardinal functions of the brain, perception and action. We primarily experience living things, they reasoned, by vision, whereas we experience objects by how we use them. Thus, patterns of relative impairment or sparing of knowledge about living or nonliving things may result from the damage to brain areas primarily involved in

TABLE 13.1 Example definitions to spoken words by patient S. B. Y.

Duck: an animal	*Wheelbarrow:* object used by people to take material about
Wasp: bird that flies	*Towel:* material used to dry people
Crocus: rubbish material	*Pram:* used to carry people, with wheels and a thing to sit on
Spider: a person looking for things, he was a spider for his nation or country	*Submarine:* ship that goes underneath the sea

Source: Adapted from Warrington, Elizabeth K., and Tim Shallice. 1984. "Category Specific Semantic Impairments." Brain 107 (3): 829–853. doi:10.1093/brain/107.3.829. Oxford University Press.

visual perception or in action planning. This alternative can be called a modality-specific scheme, because it posits that the organizing factor for semantics is the functional modality (e.g., vision, motor control) that a given piece of knowledge relies on.

A PDP model of modality specificity

Martha Farah (who we met in *Chapter 10*) and James McClelland (*Chapter 11*) sought to capture this intuition with a PDP model similar to the one illustrated in *Figure 3.21*, although with details of the architecture carefully selected to address their question. First, to quantify the idea that living things are more strongly associated with visual representations and nonliving things with their function, they obtained dictionary definitions for each of the living and nonliving words used by Warrington and Shallice (1984), and then asked volunteers who were naïve to their hypotheses to score these definitions. Twenty-one individuals read through each definition and underlined "all occurrences of words describing any aspect of the visual appearance of an item," and 21 different individuals read through each and underlined "all occurrences of words describing what the item does or what it is for" (Farah & McClelland, 1991; p. 342). The results identified an average of 2.68 visual descriptors for living things and 1.57 for nonliving things, and an average of 0.35 functional descriptors for living things and 1.11 for nonliving things. Thus, a visual:functional ratio of 7.7:1 for living things and 1.4:1 for nonliving things. (Note that the ratio for nonliving things was not <1, as Warrington and Shallice's explanation had assumed.) These ratios were then used to determine the proportion of units in the hidden layer of the PDP model (the "semantic layer") that were labeled "visual" vs. "functional."

The model represented 10 living and 10 nonliving things, each as a distinct, randomly determined pattern of +1 and −1, with each item represented by the full set of 24 visual input units and 24 verbal input units, but each living thing represented by an average of only 16.1 visual and 2.1 functional units in the semantic layer, and each nonliving thing by an average of only 9.4 visual and 6.7 functional units in the semantic layer. This captured the ratios derived from the dictionary definitions. Note that nowhere in the model was "livingness" or "nonlivingness" explicitly represented.

Next, the model was trained, by iteratively adjusting weights in the manner described in *Chapter 3* (section *A PDP model of picture naming*) until, upon being given each

visual item, it produced the correct pattern in the semantic and verbal layers, and upon being given each verbal item, it produced the correct pattern in the semantic and visual layers. Finally, it was time to simulate brain damage. As illustrated in *Figure 13.1*, this was done by progressively turning off 20%, 40%, 60%, 80%, or 100% of the units in the visual partition of the semantic layer (left plot) or in the functional partition of the semantic layer. The results produced the double dissociation that has been observed across cases, in that progressively greater damage to visual semantic memory preferentially impaired performance with living things, and progressively greater damage to functional semantic memory preferentially impaired performance with nonliving things. Like any other computational simulation, these results of the modeling of Farah and McClelland (1991) can't prove that knowledge is organized in a modality-specific rather than a category-specific manner. It does, however, provide a quantitatively explicit proof of concept for an alternative that many would argue is more neurally plausible. For example, in a subsequent paper, Farah, Meyer, and McMullen (1996) opined that a drawback of category-related models is that they "represent a significant departure from everything else we know about brain organization, according to which subsystems are delineated by function or modality but not by semantic content" (p. 143).

The domain-specific knowledge hypothesis

To defend the category-based interpretation of category-specific deficits, Caramazza and colleagues rely heavily on logical argumentation, and on additional case studies. From the latter, for example, Caramazza and Shelton (1998) observe that the categories of animals, fruits and vegetables, and artifacts can be impaired in isolation across patients. From this observation they have proposed that "these, and only these, three categories form the basis for the organization of conceptual knowledge" (p. 19). The basis for these categories would be "evolutionary pressures [that] led to specific adaptations for recognizing and responding to animal and plant life" (p. 20). The argument is plausible. One can certainly see how the ability to quickly recognize and categorize an animal as potential predator or prey, or a plant as potential food or medicine or poison, might raise the odds of survival and, therefore, to evolutionary selection of those traits. However, although such ideas about evolutionary factors can be used to assemble a plausible, if not intriguing, "back story" for a

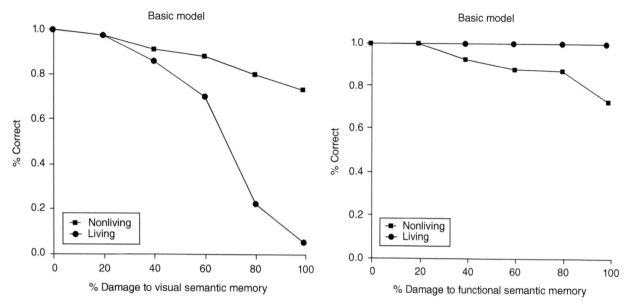

FIGURE 13.1 Performance of the model as a function of the percentage of damage to visual semantic memory (left) or to functional semantic memory (right). Source: From Farah, Martha J., and James L. McClelland. 1991. "A Computational Model of Semantic Memory Impairment: Modality Specificity and Emergent Category Specificity." Journal of Experimental Psychology: General 120 (4): 339–357. Reproduced with permission of APA.

model, such ideas are impossible to test, and so offer limited explanatory power. Further, because they are necessarily post hoc, one quickly finds that analogous, and equally plausible, evolutionary accounts can be marshaled for alternative models. Thus, for example, an equally plausible story can be told about the evolution of sensory and motor systems, and how increased computing power of the evolving primate brain led to the abstraction and elaboration of portions of these two core systems into the semantic systems that humans enjoy today. As there is no direct way to test which of these evolutionary stories may be a closer approximation of the truth, neither can contribute directly to a mechanistic understanding of how this aspect of cognition is carried out by the brain.

How definitive is a single case study? A double dissociation?

The ideas considered in the previous paragraph illustrate one complication of working with neuropsychological data, particularly with single case studies. Without question, neuropsychology is an invaluable tool for understanding human behavior, which is complex, variable, and noisy. Systematic deviations from typical patterns of behavior, particularly when they can be associated with specific patterns of brain damage, can provide important insight

into the organization and principles of a particular aspect of cognition. To draw on just one familiar example, the case of H.M. provided evidence for a fundamental distinction between processes associated with LTM encoding vs. with LTM storage and retrieval. In retrospect, however, the enduring contributions from this research are those facts that have been replicated in other patients, confirmed via complementary studies with other species, refined via computational and neuroimaging research, etc. (Indeed, in many cases the initial "facts" deduced from patterns in H.M.'s behavior [such as the idea of completely spared episodic memory for events experienced prior to the penumbra of retrograde amnesia] have been revised in light of subsequent findings and theoretical developments [for example, multiple trace theory and associated research].)

An example of a case study that might have turned out to have been "the H.M. of semantic memory" is the patient K.C., who has been described in several studies as having a selective deficit in the encoding of episodic, but not semantic, LTM. A face-value acceptance of the findings from K.C. would require revision of the idea that all memories begin as episodic memories, and then gradually, through the process of consolidation, transition to becoming semantic. But although Tulving – who unarguably occupies the highest echelon of respected and influential theoreticians of memory – has argued for this perspective very cogently (for review, see Rosenbaum et al., 2005), the absence, in the years since the initial descriptions of K.C. in the 1980s, of any additional patients presenting a similar behavioral profile, may explain the fact that this case has had negligible influence on most current views of LTM encoding. Whether or not the same is (or, will be) true of single cases in the literature on the organization of semantic memory remains a matter of vigorous debate (with, e.g., Mahon and Carramazza (2009) arguing for the importance of single cases, and Rogers et al. (2006) calling their importance into question).

Often tied up in these debates is the more technical question of *How important is the double dissociation formed by comparing case A with case B?* This question has arisen from neural network simulations demonstrating that double dissociations can be procured from unitary computational architectures. Although the details of these arguments are beyond the scope of what we can cover in this textbook, a feel for the essence of some of these arguments is captured in a chapter from Rogers and Plaut (2002):

> apparent double dissociations even in . . . homogeneous network[s] with no assumed neuroanatomic specialisation . . . [and] double-dissociations . . . aris[ing] from damage to anatomically distinct areas that are in no way specialised to subserve the cognitive functions dissociated . . . call into question the conclusion . . . that the dissociated functions must be subserved by modules that may be damaged independently. (p. 252)

Counterarguments have pointed to simulations such as Farah and McClelland (1991), in which meaningful structure can exist, and can therefore be selectively impaired, within computational architectures that started out homogeneous before training. In view of these complications associated with cognitive neuropsychology, it is not surprising that semantic memory researchers have turned to functional neuroimaging as an important source of data. So too, now, will we.

THE NEUROIMAGING OF KNOWLEDGE

The meaning, and processing, of words

Many pioneering neuroimaging studies addressing the questions that we have reviewed up to this point have come from Alex Martin and his collaborators at the National Institutes of Health (NIH). Martin trained as a neuropsychologist and spent the early part of his career studying the breakdown of language and memory functions in patients with Alzheimer's disease and HIV–AIDS. In 1989, he moved to the National Institute of Mental Health (NIMH), where he quickly fell in with Haxby, Ungerleider, and colleagues and soon began adapting their neuroimaging methods for the study of semantic memory. As a starting point, Martin considered the PET study of Petersen, Fox, Posner, Mintun, and Raichle (1988; *Figure 3.19.A* in this book), specifically the contrast that isolated verb generation and is entitled "Thinking about words." The fact that this contrast revealed greater activity in left inferior PFC (centered in BA 47) for verb generation than for either reading visually presented words or repeating auditorily presented words suggested a role for left PFC in semantic processing, independent of input modality. That is, this finding made left inferior PFC a candidate locus for an amodal semantic store. However, Martin doubted that semantics would be localizable to this particular region of the brain, for a number of reasons. First, theories from cognitive psychology and from linguistics posited that "the meaning of a concrete noun is not unitary, but rather is composed of parts – specifically, knowledge about the physical and functional properties of the object." One would therefore predict word meanings to be anatomically distributed. Complementary to this perspective were the computational and neural canons that we have established in the previous two chapters, that long-term memories are composed of the association of the neural representations of many sensations, actions, emotions, and so on. Additionally, he reasoned that "even if semantic networks were confined to a single region, the neuropsychological evidence suggested that the critical area would be the left temporal, not the left frontal, lobe" (Martin, Ungerleider, and Haxby, 2000, pp. 1023–1024).

An aside about the role of language in semantics and the study of semantics

Note that, for each type of study that we've considered thus far in this chapter, language has been central to the theoretical motivation and/or the experimental procedures employed. Warrington and McCarthy's 1983 paper is entitled "Category Specific Access Dysphasia," **dysphasia** being an impairment in the ability to communicate following brain damage. Thus, their title appeals to the directly observable clinical presentation of the patient, rather than to the psychological construct of semantic memory. Indeed, experiment 1 from this paper reported the phenomenon that they presumably set out to study in this patient, a dissociation between her relatively spared ability to match pictures of common objects to exemplars of the objects themselves versus a relative impairment in matching spoken words to the same objects. The subsequent several experiments then examined factors such as how the rate of presentation affected her performance, the effect of using spoken vs. written cues, and so on. It was only at experiment 10 (sic!) that the issue that this paper is famous for arose: "An incidental observation in a pilot study . . . suggested that V.E.R.'s comprehension deficits might to some extent be more severe for some taxonomic categories than others" (p. 870.). The Farah and McClelland (1991) model, intended to emulate the Warrington studies, was trained, and tested, by mapping verbal labels to pictures, and vice versa. Peterson et al. (1988) studied the generation of a verb from a visually or auditorily presented noun, and many, many subsequent neuroimaging studies have also used linguistic material to study semantics. Perhaps this is a necessary association? Whether language is fundamental to the mental representation of knowledge, or at least to its acquisition, is a question that has occupied philosophers for centuries, and is certainly not one that we can hope to resolve in this textbook. What is important for our purposes is to be aware of this association while evaluating the research that we'll review here, and of the competing claims about the centrality of language to the impairments seen in various neurological syndromes, and to patterns of activity observed in brain imaging studies.

PET scanning of object knowledge

Alex Martin's analysis of the neuropsychological literature drew heavily on the visual agnosias, which could produce selective impairment of particular stimulus attributes, such as cortical color blindness (achromatopsia) and motion blindness (akinetopsia). Thus, his first PET study, published in 1995, was designed to isolate the contributions to object knowledge of the cortical representation of the features *color* and *action*. It presented subjects with black-and-white line drawings of common objects, and in one scan asked subjects to name them, in a second, to name a color associated with that object, and in a third, to name an action that could be performed with that object. With a procedure comparable to that used by Petersen et al. (1988), the data were analyzed by subtracting the object-naming images from both the color- and action-naming images, thereby emphasizing the processing of those two object attributes. The results, presented in *Figure 13.2*, indicated that color naming was selectively associated with activity in a region of ventral temporal cortex, just anterior to a region of ventral occipital cortex associated with color perception, and

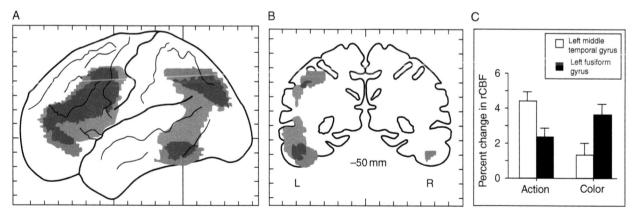

FIGURE 13.2 Superimposed subtraction images of object attribute naming from black-and-white line drawings. Green = object color generation (minus object naming); gray = object action generation (minus object naming); and blue indicates the overlap. **A**. Lateral view, with vertical line six tick marks in the from the right side of the image indicating the plane of section illustrated in **B**. **B**. Coronal slice is 50 mm posterior to the "0 mm" location, along the rostrocaudal axis of the anterior commissure. **C**. Peak magnitude of these effects in two temporal lobe regions. Source: From Martin, Alex, James V. Haxby, François M. Lalonde, Cheri L. Wiggs, and Leslie G. Ungerleider. 1995. "Discrete Cortical Regions Associated with Knowledge of Color and Knowledge of Action." Science 270 (5233): 102–105. doi: 10.1126/science.270.5233.102. Reproduced with permission of AAAS.

that action naming preferentially activated a broad expanse of left lateral cortex encompassing posterior middle temporal gyrus (MTG), superior temporal gyrus (STG), and posterior inferior parietal lobule (IPL). In addition to some overlap in these posterior regions, both tasks also preferentially activated a large region of left PFC and premotor cortex (PMC), with the action-generation task preferentially engaging posterior inferior frontal gyrus, including Broca's area. In a second experiment, subjects performed the same tasks, but in response to written words instead of line drawings, and the results were, by and large, the same.

Martin et al. (1995) interpreted their results as evidence that "knowledge of colors and of actions is represented in discrete cortical areas . . . [that are] proximal to color and motion perceptual-processing areas" (p. 104). They further argued for a framework in which the representation of knowledge about objects is distributed among the representation of the features that make up an object, and tied to the modality that characterizes these features (e.g., the perception of chrominance and the perception of motion). Note the similarities between this perspective and the modality-specific models of Warrington and colleagues and of Farah and McClelland (1991). Indeed, consistent with the modality-specific view were the results of a second study that contrasted the silent naming of animals with the silent naming of tools (Martin, Wiggs, Ungerleider, and Haxby, 1996), effectively a PET version of the Warrington studies. The findings from this study were also consistent with a modality-specific model: naming animals was associated with greater activity in visual cortex, and naming tools with greater activity in regions that largely overlapped those identified with action generation in the 1995 study – posterior middle temporal gyrus and premotor cortex. In interpreting these results, and in several subsequent studies, Martin and colleagues have argued that the tool-related activity in the posterior middle temporal gyrus relates to the processing of artifact motion, and that the tool-related activity in premotor cortex corresponds to movement plans associated with object use. The involvement of premotor cortex, in particular, has prompted the suggestion of a role for mirror neurons in the representation of tool semantics. There are also clear points of contact with the rapidly growing "embodied cognition" perspective in psychology.

To just review the results from these two PET studies as we have done up to this point, and to then move on to the next topic, would be to misrepresent this field of study.

There are many conceptual nuances, and pragmatic challenges, with which one must grapple when studying a cognitive faculty as abstract as *knowledge* with a method as concrete as functional neuroimaging. Publishing in this arena often sparks vociferous debate. To get a feel for some of this nuance and challenge, let's follow the trajectory, as it were, of just one claim from the Martin et al. (1995) study, considering some of the scrutiny it has received and how it has, in turn, been critiqued and defended. The claim that we'll follow is that activity in left fusiform gyrus is associated with the storage of color-related knowledge about objects (*Figure 13.2*). We'll begin by returning to the Martin et al. (1995) paper itself, in whose pages we can already find evidence for objections to this claim.

Knowledge retrieval or lexical access?

In the discussion of their findings, Martin and colleagues (1995) include a footnote to reinforce the interpretation of their findings in terms of "knowledge," per se, rather than of activation of the mental lexicon.

> Findings from studies of patients with focal brain lesions argue against alternative interpretations of these activations as the sites for storage of the color and action words themselves, or of word-specific retrieval mechanisms located at sites distant from where information is stored. A patient with a color anomia can produce color words. What the patient cannot do is correctly name the colors seen or correctly answer questions about object-associated colors (the patient will respond with an inappropriate color word). To show areas of activity by functional brain imaging technologies, it is necessary for subjects to be engaged in performing a task. Thus, it might be formally more accurate to state that the activations in the temporal lobes show the sites of storage during the act of retrieval. In this sense, however, storage and retrieval cannot be distinguished. (p. 105)

Although one can't know for sure (and I am deliberately choosing not to contact Alex Martin to confirm or deny my speculation), it is not infrequently the case that a footnote such as this gets added to a manuscript after it has gone through peer review. For example, if there's a particular point about which the author and a reviewer disagree, sometimes an editor will say something like this to the authors: "Although you disagree with this argument raised by the reviewer, you need to explicitly raise it in your manuscript. In addition, if you like, you can also give your rationale for disagreeing with it." Whether or not a similar scenario played out in this instance, what is clear is that this footnote is intended to refute an alternative interpretation with which the authors strongly disagree. There are several instructive aspects to this footnote. First, it highlights the fact that, particularly in this research area, the design and interpretation of neuroimaging studies is heavily influenced by patient studies, and vice versa. Second, it highlights one of the pragmatic challenges raised by the tight coupling of language and semantics. Finally, it highlights a second alternative perspective (in addition to the domain-specificity model) to the modality-specific model: the true driver of patterns of neural activity in the left temporal lobe, such as those described here, and, by extension, of category-specific deficits in patients with lesions to these areas, may be lexical retrieval.

The lexical-retrieval view has been advocated by the husband-and-wife team of neurologists Hanna Damasio and Antonio Damasio. They have argued, from patient and neuroimaging data presented in a 1996 paper, that lateral temporal-lobe regions, although not classical language areas, make important contributions to the retrieval of words, and do so in a domain-segregated way, with lexical access to the names of people mediated by the temporal pole, to the names of animals by mid-inferior temporal gyrus wrapping onto the ventral surface of the temporal lobe, and to the names of tools by posterior middle temporal gyrus. They have interpreted the core deficit of patients with damage to each of these regions to be one of lexical retrieval (i.e., word finding), because the naming deficits of their patients were demonstrated after it had been determined that they understood the concept of the item in the picture. (This specific claim, incidentally, is one with which Caramazza and Shelton (1998) take issue.) Thus, Damasio, Grabowski, Tranel, Hichwa, and Damasio (1996) have argued for a different functional interpretation of the ventral temporal-lobe activity that Martin et al. (1995) attribute to color knowledge (as well as for a different functional interpretation of the area that Martin et al. (1995) attribute to action generation).

Recall that we previously noted that, in composing their title, Warrington and McCarthy (1983) emphasized the directly observable clinical presentation of their patient (dysphasia) rather than that psychological construct that the patient's clinical presentation implicated. It may be that the emphasis that the Damasios have placed on a lexical explanation for their data also derives, at least in part, from their experience in the clinic. Both are behavioral neurologists whose contributions to cognitive neuroscience have been strongly anchored by their experience and expertise in working with patients. Further along in this chapter we'll also see that a naming interpretation has been favored by many neurologists in relation to a neurodegenerative disease that they have referred to as "primary progressive aphasia" (aphasia being an inability to use language), whereas many approaching the same disease from a psychological perspective call it "semantic dementia."

Repetition effects and fMRI adaptation

Recall, from *Chapters 11* and *12*, that repetition priming is a type of nondeclarative memory in which previous exposure to a stimulus biases or facilitates the processing of that stimulus (or fragments thereof) when presented on subsequent occasions. At the neural level, a correlate of repetition priming is the repetition suppression effect, whereby neurons that responded to the initial presentation of the stimulus with an increase in firing rate respond with a smaller increase to the second and subsequent presentations. With neuroimaging, an analogous effect was shown by Buckner and his mentors at Washington University with PET studies of word-stem completion, and later, more definitively, by Buckner, Dale, and colleagues at the MGH NMR Center with their rapid event-related fMRI technique.

A methodological innovation that takes advantage of this phenomenon, called fMRI adaptation, was developed by the Israeli team of Kalanit Grill-Spector and her mentor Rafael Malach. To deduce the functional properties of a region of visual cortex, one first presents a stimulus several times to produce a repetition-suppression effect. Then, one begins varying the stimulus properties along a single dimension. If the region of interest is invariant to changes along that dimension, its response will remain suppressed. If, however, it is sensitive to changes along this dimension, any change will result in a "release from adaptation" and a return to its initial level of response.

Illustrated in *Figure 13.3* is a study that employed the logic of fMRI adaptation. Subjects were presented with images of familiar objects (both living and nonliving) and asked to indicate for each, with a button press, whether or not it would fit into a shoe box. During the study phase, each of 32 items was presented on three occasions. During the test phase (the transition from study to test occurring without the subjects' knowledge), each of 64 items was presented once. Of these, 32 were repeats of the 32 study

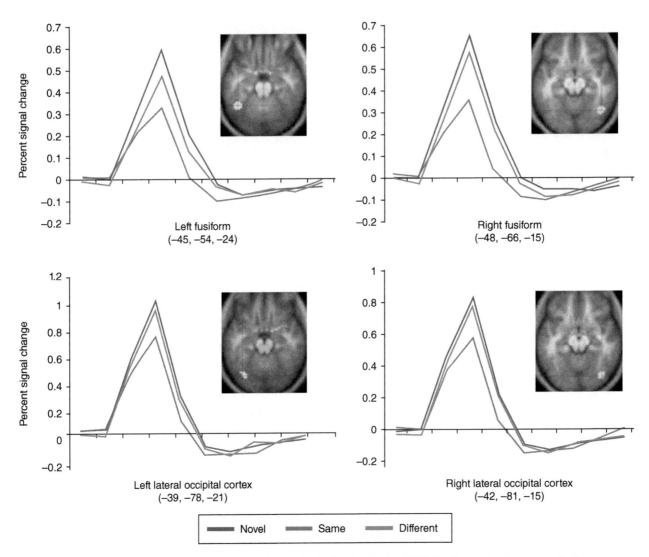

FIGURE 13.3 Repetition effects in four posterior regions. Each plot shows the fMRI stimulus-evoked response to the presentation of a picture that is being seen for the first time (Novel), is a repeat presentation of a stimulus that had already been presented on three previous occasions during the same scan (Same), or is a different exemplar of one of the already-thrice-presented items (Different). Source: From Simons, Jon S., Wilma Koutstaal, Steve Prince, Anthony D. Wagner, and Daniel L. Schacter. 2003. "Neural Mechanisms of Visual Object Priming: Evidence for Perceptual and Semantic Distinctions in Fusiform Cortex." NeuroImage 19 (3): 613–626. doi:10.1016/S1053-8119(03)00096-X. Reproduced with permission of Elsevier.

items, 16 were different exemplars of the same kind of item as the 32 (Different), and 16 were items that had not appeared among the 32 studied items (Novel).

As illustrated in *Figure 13.3*, three of the regions of interest in this study showed an identical response profile: suppressed activity for a repeated item, but a release from adaptation for a same-type-but-different-token (i.e., "Different") item. These regions, therefore, were assumed

to encode the visual characteristics of stimuli, but not more abstract properties. (For example, for a study picture of an octopus, if a region's response to a physically different picture of an octopus is not suppressed, it must be that this region doesn't register that the two physically different images correspond to the same kind of thing. The interpretation is that this region doesn't encode the meaning of "octopus-ness.") The left fusiform gyrus ROI, in contrast, showed only a partial release from adaptation. (That is, its response to a Different stimulus showed a level of adaptation intermediate between a repeated image and a Novel image.) Thus, this region's activity registered the semantic similarity between the two images, even though their physical properties differed.

Although repetition effects have been fairly well characterized at the neural level, it is uncertain how fMRI repetition/adaptation effects are brought about. A paper from Grill-Spector, Henson, and Martin (2006) considers three possible mechanisms: fatigue, sharpening, and facilitation.

The study by Simons et al. (2003) that is illustrated in *Figure 13.3* is one of several senior authored by Daniel Schacter, a cognitive psychologist who trained with Endel Tulving and, during the 1980s and 1990s, came to be known as arguably the world's foremost authority on repetition priming. Initially in collaboration with Randy Buckner, Schacter and his mentees performed many of the first fMRI studies of this phenomenon. Among them, this study by then-visiting scholar Jon Simons and colleagues (2003) was one to which Martin could refer as evidence against the possibility that the left fusiform gyrus activity produced by his color-naming condition corresponded to "just" visual imagery that did not "do any conceptual work."

This concludes our deeper look at claims about the proposed role for left fusiform gyrus in the storage of color-related knowledge about objects (Figure 13.2). The realm of nuance and challenge to which we will devote the remaining portion of this chapter is that of a neurodegenerative disease whose cognitive impairments manifest prominently in language and semantic memory.

THE PROGRESSIVE LOSS OF KNOWLEDGE

Unlike the acute-onset neurological syndromes that we have considered up until now (stroke, viral encephalitis, and, indeed, surgical excision), neurodegenerative diseases are characterized by progressive degeneration of the integrity of the affected neural systems, and in the behaviors that depend on those systems. *Dementia* refers to the progressive impairment of intellect and behavior that, over time, dramatically alters and restricts normal activities of daily living. Most dementias are of "insidious onset," medicalese for "subtle and gradual." The most common form of dementia is Alzheimer's disease, which typically initially

manifests as a progressively worsening anterograde amnesia, and gradually compromises other mnemonic and other cognitive functions. Atypical dementias are a class that first manifest in domains of cognition other than episodic memory. Of interest for our present purposes is a subtype of frontotemporal dementia that has variously been called *primary progressive aphasia* or *semantic dementia*.

Primary Progressive Aphasia or Semantic Dementia, what's in a name?

As alluded to previously, there was a period during which one's preferred label for this disease seemed to reflect, in part, the scholarly tradition from which one approached it. In the 1970s, neuropsychologists Elizabeth Warrington in the United Kingdom and Myrna Schwartz and her colleagues in Philadelphia both described cases of dementia with anomia (naming difficulty) that seemed to reveal a selective impairment of semantic memory. Subsequently, neurologists who saw similar patients in clinic labeled the condition semantic dementia. In parallel, in 1982, the behavioral neurologist M.-Marsel Mesulam wrote a highly influential paper describing patients with a condition he labeled "primary progressive aphasia," and that he characterized it as follows:

> Patients with Alzheimer's disease come to medical attention because of forgetfulness, usually accompanied by apathy. Misplacing personal objects, repeating questions, and forgetting recent events are among the presenting symptoms. Although the patient may forget people's names, word-finding during conversation is usually not a major problem. In contrast, patients with primary progressive aphasia come to medical attention because of the onset of word-finding difficulties, abnormal speech patterns, and prominent spelling errors. Primary progressive aphasia is diagnosed when other mental faculties, such as memory of daily events, visual and spatial skills (assessed by tests of drawing and face recognition), and behavior (assessed by a history obtained from a third party), remain relatively intact; when language is the only area of prominent dysfunction for at least the first two years of the disease; and when structural brain imaging studies do not reveal a specific lesion, other than atrophy, that can account for the language deficit. (Mesulam, 2003, pp. 1535–1536).

In this and a contemporaneous review, Mesulam and a formidable lineup of colleagues asserted (via the title of their paper) that the language impairment is the core deficit in all forms of primary progressive aphasia, and poor word comprehension a peripheral "halo" in one subtype that they termed the fluent subtype of primary progressive aphasia. Alternatively, they argued that there is a second distinct neurodegenerative syndrome characterized by two independent factors: an aphasia due to pathology in left hemisphere perisylvian language cortex and an associative agnosia due to pathology in the inferior temporal cortex.

In the remainder of this chapter we will consider a series of studies advocating an alternative view, whereby the core deficit of the putatively fluent variant of primary progressive aphasia is, in fact, semantic memory. The pedagogical strategy here is analogous to what we did with the color-naming finding from Martin et al. (1995) – by zooming in for a fine-grained consideration of this particular debate, we will encounter experimental methods and lines of argumentation that illustrate broader themes about how scientists think about the neural bases of semantic memory. Much of this work that we will consider has been led by neurologist John Hodges and neuropsychologist Karalyn Patterson, both based in Cambridge, UK at the time of this work, and their colleagues and trainees, who have taken the position that semantic dementia and the fluent variant of primary progressive aphasia may be "two sides of the same coin." They have framed the debate as follows: are there "subtle nonverbal semantic deficits . . . in fPPA [(fluent primary progressive aphasia)][that are] fundamental to the anomia and reflect, therefore, damage to an amodal semantic system associated with bilateral, though often asymmetric, anterior temporal lobe (ATL) atrophy?" Alternatively, does semantic dementia "result from a disease process that encompasses two separate neurocognitive networks – a left hemisphere language network and a bilateral fusiform network for face and object recognition – which, though jointly compromised in SD [i.e., "semantic dementia"], can also be damaged independently?" (Adlam et al., 2006, p. 3067).

Nonverbal deficits in fluent primary progressive aphasia?

To test between the two alternatives laid out in the previous paragraph, Adlam et al. (2006) recruited seven patients who met the diagnostic criteria of early-stage fluent primary progressive aphasia. Along with standard neuropsychological tests of language, memory, and other cognitive functions, they also administered tasks intended to probe their earlier findings that semantic dementia progressively restricts the conceptual knowledge of patients to general and typical features. Examples of these tasks are shown in *Figure 13.4*. The results of the behavioral testing indicated

that, on the standardized tests, patients demonstrated a more marked verbal than nonverbal impairment. This is an oft-replicated finding, and follows from the fact that these patients had been selected for meeting diagnostic criteria for fluent primary progressive aphasia. Additionally, however, results from the tests designed to assess the selective loss of specific knowledge contradicted Mesulam's assertion that "language is the only area of prominent dysfunction for at least the first two years of the disease." On the Levels of Specificity and Typicality (LOST) word-matching task (*Figure 13.4.A,* and *B*), for example, the patients were selectively impaired at matching low-typical targets among close distractors. On the object-color decision task (*Figure 13.4.C*), patients were only impaired on NR > R trials, that is, when the foil was a more domain-typical color than the target. The authors interpreted these findings as evidence that fluent primary progressive aphasia/ semantic dementia impairs conceptual knowledge, in addition to language functions, at early stages of the disease:

> In summary of this issue: how are we to classify patients like the ones studied here, who present with a prominent fluent, anomic aphasia and a less prominent but still significant agnosia that is modulated by concept familiarity/typicality? These patients qualify for a diagnosis of fPPA on the basis of their preserved everyday function and the predominance of their aphasia; but (i) when tested with appropriate materials, they have clear non-verbal impairments, even at this early stage, and (ii) in our (by now fairly extensive) experience, they invariably progress to a pattern that every FTD researcher would call SD. It seems more logical and more useful to label them as having mild or early-stage SD (p. 3077).

The locus of damage in fluent primary progressive aphasia?

The anatomical prediction of Adlam et al. (2006), that fluent primary progressive aphasia would be characterized by degeneration focused on the anterior temporal lobe, stemmed from prior neuropathology and neuroimaging studies of patients who they had defined as having semantic dementia. Additionally, a PDP model by Rogers and McClelland (2004) suggests the existence of "a set of shared representation units that tie together all of an object's properties across different information types." After damage to or degeneration in these units, "distinctions between items that are very similar semantically tend to be lost . . . while distinctions between highly dissimilar objects are maintained" (pp. 378–379). (Note that this is

FIGURE 13.4 Testing semantic impairments in fluent primary progressive aphasia/semantic dementia.

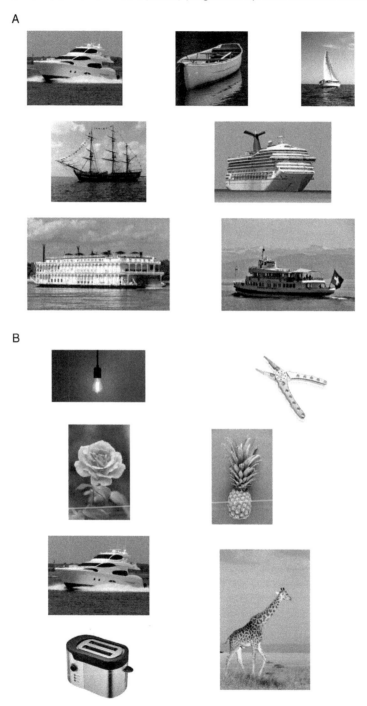

FIGURE 13.4.**A. AND B.** Two trials similar to the Levels of Specificity and Typicality (LOST) word-matching task, illustrating the manipulation of specificity, in that subjects are asked to point to the yacht (a low-typicality target from the category "boats") from among a set of close semantic distractors (**A**) and from among a set of unrelated semantic distractors (**B**). An example of a high-typicality target boat would be a sailboat. Source: Bradley R Postle.

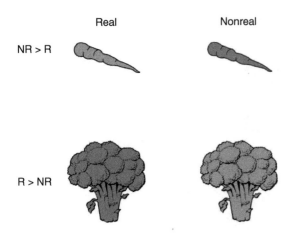

FIGURE 13.4.C. Two sets of stimuli from the object-color decision task from Adlam et al. (2006). "NR > R" denotes a trial when the nonreal item has a color that is more typical of the category than the real item (i.e., vegetables are more typically green than orange; however, carrots are orange), and R > NR a trial when the real item has a color that is more typical of the category than the nonreal item (i.e., vegetables are more typically green than orange, and broccoli is green). Source: Adapted from Adlam, Anna-Lynne R., Karalyn Patterson, Timothy T. Rogers, Peter Nestor, Claire H. Salmond, Julio Acosta-Cabronero, and John R. Hodges. 2006. "Semantic Dementia and Fluent Primary Progressive Aphasia: Two Sides of the Same Coin?" Brain 129 (11): 3066–3080. doi:10.1093/brain/awl285. Oxford University Press.

the pattern of deficit observed in the patients of Adlam et al., 2006.) In contrast, most prior research explicitly performed on patients diagnosed with fluent primary progressive aphasia had assumed that the language impairments in these patients resulted from atrophy in the left hemisphere language network, particularly "perisylvian" regions of posterior superior temporal gyrus/inferior parietal lobule and inferior prefrontal cortex of the left hemisphere. Further, Mesulam (2003) had argued that fluent primary progressive aphasia accompanied by associative agnosia would be expected to exhibit degeneration in inferior occipitotemporal cortex to account for the latter.

Results from the VBM analysis performed by Adlam et al. (2006), illustrated in *Figure 13.5*, show clear evidence for bilateral degeneration in the anterior temporal lobes of these patients. Further, performance of patients on both verbal and nonverbal tasks was predicted by gray-matter density values in the left inferior temporal region marked with cross hairs in *Figure 13.5.C*, suggesting that the integrity of this region may be crucial for performance on their behavioral tests of semantic memory.

How are we to make sense of the absence of any overlap between the predictions of Mesulam (2003) on the one hand and, on the other, those of Adlam et al. (2006) with respect to which areas are expected to be compromised by fluent primary progressive aphasia? One possible answer comes from an earlier PET study of semantic dementia (Mummery et al., 1999) whose results suggested that activity in regions presumably spared the direct effects of degeneration nonetheless showed abnormal patterns of activity.

Distal effects of neurodegeneration

The study of Mummery et al. (1999) compared patients diagnosed with semantic dementia with age- and education-matched control subjects. Results from a VBM analysis were highly similar to those of Adlam et al. (2006; *Figure 13.5* in this book), indicating a degenerative focus in anterior temporal lobe and, importantly, a sparing of the gray-matter integrity of the ventral surface of temporo-occipital cortex. This finding was particularly noteworthy in light of the PET results, which indicated that activity related to word matching and to picture matching was significantly reduced in patients, relative to control subjects, in lateral and inferior temporo-occipital cortex (*Figure 13.6*). This finding suggests a way to reconcile the predictions from Mesulam (2003), on the one hand, and Adlam et al. (2006), on the other: neurodegeneration associated with fluent primary progressive aphasia/semantic dementia may directly impact the integrity of anterior temporal cortex, and this structural degeneration may lead to functional abnormalities in the inferior and lateral temporo-occipital cortex, the normal functioning of which is known to be critical for high-level visual perception and, by many accounts, the perceptual features that contribute to the semantic representation of objects.

Systems-level accounts of how disruption of ATL can lead to abnormal functioning in the visual ventral stream include the designation of this region as a critical "hub" for semantic representation (Rogers et al., 2006), and the related idea that it may be a source of "top–down modulatory input necessary for successful retrieval of semantic representations stored in posterior regions" (Martin and Chao, 2001, p. 198).

Entente cordiale

This chapter concludes with an all-too-rare consensus resolution of a hotly debated topic in cognitive neuroscience. An important factor pushing for this outcome was the fact that the stakes in the debate about the neural and cognitive

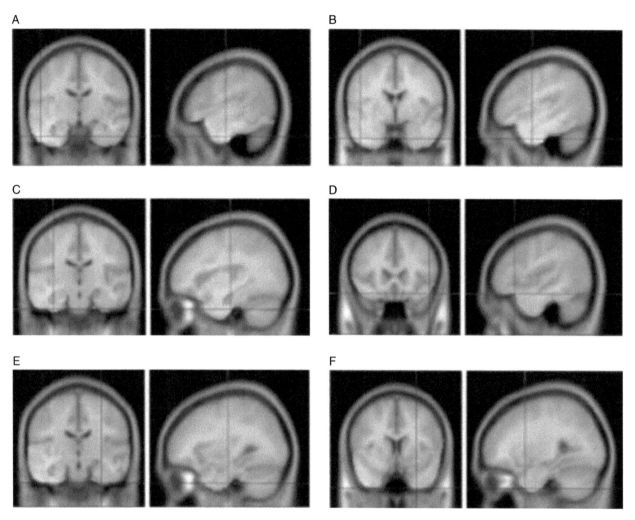

FIGURE 13.5 Results from the whole-brain VBM analysis from Adlam et al. (2006), showing extensive gray matter reduction (relative to control subjects) in each of six patients with fluent primary progressive aphasia. The cross hairs on each image indicate the peak areas of atrophy. Note that these occur in both hemispheres, and exclusively in anterior temporal lobe. Images are shown in neurological display convention (i.e., right side of coronal image corresponds to right hemisphere). Source: Adlam, Patterson, Rogers, Nestor, Salmond, Acosta-Cabronero, and Hodges, 2006. Reproduced with permission of Oxford University Press.

impairments of fluent primary progressive aphasia/semantic dementia have very important implications for the diagnosis and treatment of patients who are suffering from the disease. To achieve it, an international "consensus group" including all of the neurologists whose work we have highlighted here met on three different occasions. Rather than just "getting together to argue," however, they followed a systematic procedure:

The group reviewed video presentations of 12 PPA cases from different sites. A list of 17 salient speech and

language features was provided and each clinician rated whether specific language features were present or not. Videos included a component of spontaneous speech and various portions of formal language evaluations. The analysis of the responses revealed a high level of agreement. There were 15 clinical features in which experts agreed over 80% of the time. Thirteen of these features were included in an operationalized classification scheme listing the main features of the three most commonly recognized clinical presentations of PPA (Gorno-Tempini et al., 2011).

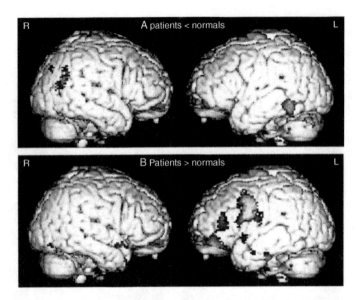

FIGURE 13.6 Difference images from PET scans comparing semantic task-related activity in semantic dementia patients vs. control subjects (Normals). Source: Mummery, Patterson, Price, Vandenbergh, Price, and Hodges, 1999. Reproduced with permission of Oxford University Press.

The three subtypes of primary progressive aphasia identified by the consensus group are nonfluent/agrammatic variant; semantic variant; and logopenic variant. It is on the existence and characteristics of the semantic variant that we have focused in this chapter, and the consensus diagnostic and classification criteria are very similar to what one would expect from the findings of Mummery et al. (1999) and Adlam et al. (2006).

NUANCE AND CHALLENGES

I told you it would be like this! As cognitive functions become more abstract, it becomes increasingly difficult to design studies that unequivocally measure the construct of primary interest. The tradeoff, of course, is that what one gives up in reductive clarity, one gains in being able to grapple with questions that lie at the very heart of what it means to be human (from the perspective of "Je pense, donc je suis"). Additionally, although we haven't given it much explicit consideration here, semantic memory is one of the domains in which cognitive neuroscience makes considerable contact with the cognitive psychology school of embodied cognition, a perspective that is very prominent in the zeitgeist, and that many feel might unify the study of cognition from psychological, cognitive neuroscience, and philosophical traditions.

END-OF-CHAPTER QUESTIONS

1. What are three interpretations of the fact that patients have been characterized with a selective deficit in the ability to name nonliving things and others with a selective deficit in the ability to name living things?

2. Summarize the arguments for and against placing a high level of importance on the findings from a single case

study, and on the findings from two case studies that together form a double dissociation.

3. Name at least two possible explanations when a neuroimaging study finds selectively elevated activity for the naming of the color of an object relative to the naming of its shape, or of an action that one could take with that

object, or some other property. What additional evidence can be drawn on to constrain the interpretation (e.g., to rule out alternative explanations)?

4. How has fMRI adaptation been used to argue for a functionally different role for left mid-fusiform gyrus relative to its right hemisphere homolog and to areas of lateral occipital cortex?

5. How do neurodegenerative disorders, such as variants of frontotemporal dementia, differ from acute-episode neurological insults, such as stroke or viral encephalitis?

6. What pattern(s) of behavioral performance has (have) been pointed to as evidence for an inherently conceptual,

as opposed to lexical, deficit in the semantic variant of primary progressive aphasia?

7. What brain area is selectively compromised in early stages of the semantic variant of primary progressive aphasia? What function, relevant to semantic dementia, is this region believed to support?

8. What are the salient differences between the consensus classifications of the nonfluent/agrammatical variant of primary progressive aphasia and the semantic variant of primary progressive aphasia? Despite these differences, how do the two differ in more fundamental ways from Alzheimer's disease?

REFERENCES

Adlam, Anna-Lynne R., Karalyn Patterson, Timothy T. Rogers, Peter Nestor, Claire H. Salmond, Julio Acosta-Cabronero, and John R. Hodges. 2006. "Semantic Dementia and Fluent Primary Progressive Aphasia: Two Sides of the Same Coin?" *Brain* 129 (11): 3066–3080. doi:10.1093/brain/awl285.

Caramazza, Alfonso, and Jennifer R. Shelton. 1998. "Domain-Specific Knowledge Systems in the Brain: The Animate–Inanimate Distinction." *Journal of Cognitive Neuroscience* 10 (1): 1–34. doi:10.1093/neucas/4.4.399-k.

Damasio, Hanna, Thomas J. Grabowski, Daniel Tranel, Richard D. Hichwa, and Antonio R. Damasio. 1996. "A Neural Basis for Lexical Retrieval." *Nature* 380 (6574): 499–505. doi:10.1038/380499a0.

Farah, Martha J., and James L. McClelland. 1991. "A Computational Model of Semantic Memory Impairment: Modality Specificity and Emergent Category Specificity." *Journal of Experimental Psychology: General* 120 (4): 339–357.

Farah, Martha J., Michael M. Meyer, and Patricia A. McMullen. 1996. "The Living/Nonliving Dissociation is Not an Artifact: Giving an A Priori Implausible Hypothesis a Strong Test." *Cognitive Neuropsychology* 13 (1): 137–154. doi:10.1080/026432996382097.

Gorno-Tempini, Maria L., Argye E. Hillis, Sandra Weintraub, Andrew Kertesz, Mario F. Mendez, Stefano F. Cappa, . . . Murray Grossman. 2011. "Classification of Primary Progressive Aphasia and Its Variants." *Neurology* 76 (11): 1006–1014. doi:10.1212/WNL.0b013e31821103e6.

Grill-Spector, Kalanit, Richard Henson, and Alex Martin. 2006. "Repetition and the Brain: Neural Models of Stimulus-Specific Effects." *Trends in Cognitive Sciences* 10 (1): 14–23. doi:10.1016/j.tics.2005.11.006.

Mahon, Bradford Z., and Alfonso Caramazza. 2009. "Concepts and Categories: A Cognitive Neuropsychological Perspective." *Annual Review of Psychology* 60: 27–51. doi:10.1146/annurev.psych.60.110707.163532.

Martin, Alex, and Linda L. Chao. 2001. "Semantic Memory and the Brain: Structure and Processes." *Current Opinion in Neurobiology* 11 (2): 194–201. doi:10.1016/S0959-4388(00)00196-3.

Martin, Alex, James V. Haxby, François M. Lalonde, Cheri L. Wiggs, and Leslie G. Ungerleider. 1995. "Discrete Cortical Regions Associated with Knowledge of Color and Knowledge of Action." *Science* 270 (5233): 102–105. doi: 10.1126/science.270.5233.102.

Martin, Alex, Leslie G. Ungerleider, and James V. Haxby. 2000. "Category Specificity and the Brain: The Sensory/Motor Model of Semantic Representations of Objects." In *The New Cognitive Neurosciences*, 2nd ed., edited by Michael S. Gazzaniga (pp. 1023–1036). Cambridge, MA: MIT Press.

Martin, Alex, Cheri L. Wiggs, Leslie G. Ungerleider, and James V. Haxby. 1996. "Neural Correlates of Category-Specific Knowledge." *Nature* 379 (6566): 646–652. doi:10.1038/379649a0.

McClelland, James L., Bruce L. McNaughton, and Randall C. O'Reilly. 1995. "Why There Are Complementary Learning Systems in the Hippocampus and Neocortex: Insights from the Successes and Failures of Connectionist Models of Learning and Memory." *Psychological Review* 102 (3): 419–457. doi:10.1037/0033-295X.102.3.419.

Mesulam, M.-Marsel. 1982. "Slowly Progressive Aphasia without Generalized Dementia." *Annals of Neurology* 11 (6): 592–598. doi:10.1002/ana.410110607.

Mesulam, M.-Marsel. 2003. "Primary Progressive Aphasia – A Language-Based Dementia." *New England Journal of Medicine* 346 (16): 1535–1542. doi:10.1056/NEJMra022435.

Mummery, Catherine J., Karalyn Patterson, Richard J. Price, R. Vandenbergh, Cathy J. Price, and John R. Hodges. 1999.

"Disrupted Temporal Lobe Connections in Semantic Dementia." *Brain* 122 (1): 61–73. doi:10.1093/brain/122.1.61.

Petersen, Steven E., Peter T. Fox, Michael I. Posner, Mark Mintun, and Marcus E. Raichle. 1988. "Positron Emission Tomographic Studies of the Cortical Anatomy of Single-Word Processing." *Nature* 331 (6157): 585–589. doi:10.1038/331585a0.

Rogers, Timothy T., Julia Hocking, Uta Noppeney, Andrea Mechelli, Maria Luisa Gorno-Tempini, Karalyn Patterson, and Cathy J. Price. 2006. "Anterior Temporal Cortex and Semantic Memory: Reconciling findings from Neuropsychology and Functional Imaging." *Cognitive, Affective, and Behavioral Neuroscience* 6 (3): 201–213. doi:10.3758/CABN.6.3.201.

Rogers, Timothy T., and James L. McClelland. 2004. *Semantic Cognition: A Parallel Distributed Processing Approach.* Cambridge, MA: MIT Press.

Rogers, Timothy T., and David C. Plaut. 2002. "Connectionist Perspectives on Category-Specific Deficits." In *Category Specificity in Brain and Mind*, edited by Emer M. E. Forde and Glyn W. Humphreys (pp. 251–289). Hove: Psychology Press.

Rosenbaum, R. Shayna, Stefan Köhler, Daniel L. Schacter, Morris Moscovitch, Robyn Westmacott, Sandra E. Black, . . . Endel Tulving. 2005. "The Case of K. C.: Contributions of a Memory-Impaired Person to Memory Theory." *Neuropsychologia* 43 (7): 989–1021. doi:10.1016/j.neuropsychologia.2004.10.007.

Simons, Jon S., Wilma Koutstaal, Steve Prince, Anthony D. Wagner, and Daniel L. Schacter. 2003. "Neural Mechanisms of Visual Object Priming: Evidence for Perceptual and Semantic Distinctions in Fusiform Cortex." *NeuroImage* 19 (3): 613–626. doi:10.1016/S1053-8119(03)00096-X.

Warrington, Elizabeth K., and Rosaleen McCarthy. 1983. "Category Specific Access Dysphasia." *Brain* 106 (4): 859–878. doi:10.1093/brain/106.4.859.

Warrington, Elizabeth K., and Tim Shallice. 1984. "Category Specific Semantic Impairments." *Brain* 107 (3): 829–853. doi:10.1093/brain/107.3.829.

OTHER SOURCES USED

Buckner, Randy L., Julie Goodman, Marc Burock, Michael Rotte, Wilma Koutstaal, Daniel L. Schacter, . . . Anders M. Dale. 1998. "Functional-Anatomic Correlates of Object Priming in Humans Revealed by Rapid Presentation Event-Related fMRI." *Neuron* 20 (2): 285–296. doi:10.1016/S0896-6273(00)80456-0.

Buckner, Randy L., Steven E. Petersen, Jeffrey G. Ojemann, Francis M. Miezin, Larry R. Squire, and Marcus E. Raichle. 1995. "Functional Anatomical Studies of Explicit and Implicit Memory Retrieval Tasks." *Journal of Neuroscience* 15 (1): 12–29.

Chao, Linda L., James V. Haxby, and Alex Martin. 1999. "Attribute-Based Neural Substrates in Temporal Cortex for Perceiving and Knowing about Objects." *Nature Neuroscience* 2 (10): 913–919. doi:10.1038/13217.

Kensinger, Elizabeth A., Michael T. Ullman, and Suzanne Corkin. 2001. "Bilateral Medial Temporal Lobe Damage Does Not Affect Lexical or Grammatical Processing: Evidence from Amnesic Patient H. M." *Hippocampus* 11 (4): 347–360. doi:10.1002/hipo.1049.

Martin, Alex. 2009. "Circuits in Mind: The Neural Foundations for Object Concepts." In *The Cognitive Neurosciences*, 4th ed., edited by Michael S. Gazzaniga (pp. 1031–1045). Cambridge, MA: MIT Press.

McCarthy, Rosaleen A., and Elizabeth K. Warrington. 1988. "Evidence for Modality-Specific Meaning Systems in the Brain." *Nature* 334 (6181): 428–430. doi:10.1038/334428a0.

Mummery, Catherine J., Karalyn Patterson, Cathy J. Price, John Ashburner, Richard S. J. Frackowiak, and John R. Hodges. 2000. "A Voxel-Based Morphometry Study of Semantic Dementia: Relationship between Temporal Lobe Atrophy and Semantic Memory." *Annals of Neurology* 47 (1): 36–45. doi:10.1002/1531-8249(200001)47:1<36::AID-ANA8>3.0.CO;2-L.

Paulesu, Eraldo, John Harrison, Simon Baron-Cohen, J. D. G. Watson, L. Goldstein, J. Heather, . . . Chris D. Frith. 1995. "The Physiology of Coloured Hearing: A PET Activation Study of Colour–Word Synaesthesia." *Brain* 118 (3): 661–676. doi:10.1093/brain/118.3.661.

Plaut, David C. 1995. "Double Dissociation without Modularity: Evidence from Connectionist Neuropsychology." *Journal of Clinical and Experimental Neuropsychology* 17 (2): 291–321. doi:10.1080/01688639508405124.

Schacter, Daniel L., and Randy L. Buckner. 1998. "Priming and the Brain." *Neuron* 20 (2): 185–195. doi:10.1016/S0896-6273(00)80448-1.

Schwartz, Myrna F., Oscar S. M. Marin, and Eleanor M. Saffran. 1979. "Dissociations of Language Function in Dementia: A Case Study." *Brain and Language* 7 (3): 277–306. doi:10.1016/0093-934X(79)90024-5.

Simner, Julia, Catherine Mulvenna, Noam Sagiv, Elias Tsakanikos, Sarah A. Witherby, Christine Fraser, Kirsten Scott, and Jamie Ward. 2006. "Synaesthesia: The Prevalence of Atypical Cross-Modal Experiences." *Perception* 35 (8): 1024–1033. doi:10.1068/p5469.

Warrington, Elizabeth K. 1975. "Selective Impairment of Semantic Memory." *Quarterly Journal of Experimental Psychology* 27 (4): 635–657. doi: 10.1080/14640747508400525.

FURTHER READING

Gorno-Tempini, Maria L., Argye E. Hillis, Sandra Weintraub, Andrew Kertesz, Mario F. Mendez, Stefano F. Cappa, . . . Murray Grossman. 2011. "Classification of Primary Progressive Aphasia and Its Variants." *Neurology* 76 (11): 1006–1014. doi:10.1212/WNL.0b013e31821103e6.
The consensus paper on PPA and semantic dementia.

Grill-Spector, Kalanit, Richard Henson, and Alex Martin. 2006. "Repetition and the Brain: Neural Models of Stimulus-Specific Effects." *Trends in Cognitive Sciences* 10 (1): 14–23. doi:10.1016/j.tics.2005.11.006.
No consensus, but nonetheless constructive; three scientists with different explanations for repetition effects seen with fMRI describe their models and articulate a way forward for adjudicating between them.

Lewis, Penelope A., and Simon J. Durrant. 2011. "Overlapping Memory Replay during Sleep Builds Cognitive Schemata." *Trends in Cognitive Sciences* 15 (8): 343–351. doi:10.1016/j.tics.2011.06.004.
Proposal for how information processing during sleep may contribute to the semanticization of long-term memories.

Mahon, Bradford Z., and Alfonso Caramazza. 2009. "Concepts and Categories: A Cognitive Neuropsychological Perspective." *Annual Review of Psychology* 60: 27–51. doi:10.1146/annurev.psych.60.110707.163532.
The perspective of two prominent advocates of the category-specific models of semantic memory.

Martin, Alex. 2007. "The Representation of Object Concepts in the Brain." *Annual Review of Psychology* 58: 25–45. doi:10.1146/annurev.psych.57.102904.190143.
Comprehensive review from one of the leading cognitive neuroscientists of semantic memory.

Rogers, Timothy T., and James L. McClelland. 2008. "*Précis of Semantic Cognition: A Parallel Distributed Processing Approach.*" *Behavioral and Brain Sciences* 31 (6): 689–749. doi:10.1017/S0140525X0800589X.
Précis of Rogers and McClelland's PDP model of semantic cognition, followed by many critiques by others working in the field, followed by a rejoinder from R & M.

CHAPTER 14
WORKING MEMORY

KEY THEMES

- The terms "short-term memory" (STM) and "working memory" refer to the span of information (e.g., a telephone number, an array of objects) that one can hold in mind without assistance from the MTL. Although this information is quickly lost when attention is directed elsewhere, it can, in principle, be maintained indefinitely.

- "Working memory" is the more general term, because it also refers to the mental manipulation of information being held "in mind," such as performing mental arithmetic or reordering items into numerical or alphabetical order.

- From the 1940s through the 1960s, lesion findings indicated that the PFC was important for performance on tasks that required guiding behavior with working memory, but not for the short-term retention per se, of recently encoded information.

- Despite the previous point, the 1971 extracellular electrophysiology studies reporting elevated delay-period-spanning activity of PFC neurons was interpreted by some (although not the authors themselves) as a neural correlate of the short-term retention of information.

- The conceptual integration of working memory theory from cognitive psychology with neurophysiological findings from PFC led to the development of an influential model working memory theory, which holds the idea that PFC houses domain-segregated STM buffers.

- The superior sensitivity of MVPA over univariate analysis methods has provided compelling evidence that the short-term retention of information is supported in regions also responsible for its perception, including V1 for visual working memory.

- The superior specificity of MVPA over univariate analysis methods suggests that elevated delay-period activity, including in the PFC, does not always correspond to stimulus representation per se.

- The retention of information in working memory may not always depend on elevated activity in neurons, but in some situations might be supported by a "structural trace," a pattern of modified synaptic weights created by short-term synaptic plasticity.

CONTENTS

"PROLONGED PERCEPTION" OR "ACTIVATED LTM?"

Working memory refers to the ability to hold in mind information that's needed to achieve some immediate goal, like remembering the digits of an unfamiliar phone number until you get a chance to dial it, or remembering from the road sign that you just passed the ordering of the roads leaving the roundabout that you're about to enter. Although the order of chapters in this book by and large follows what one finds in most textbooks on cognitive psychology and cognitive neuroscience, the placement of this one is at variance with its typical placement at the transition point between a just-completed section on sensation/perception/attention and an ensuing section on LTM. This scheme, in effect, portrays working memory as "prolonged perception." My rationale for positioning the chapter that you are currently reading here, instead of there, is twofold. First, it is increasingly the case that working memory is understood as being supported by the activation of semantic, as well as perceptual, representations. (Additionally, the idea of a role for the hippocampus in some types of short-term memory is gaining prominence in the literature.) Thus, having an understanding of the organization and neural substrates of LTM will be important for what's to follow in this chapter. Second, short-term and working memory are closely intertwined with cognitive control and other aspects of high-level cognition that will be the focus of the final section of this book, and so this chapter will serve as a nice transition to this final section.

DEFINITIONS

One thing to sort out before we go any further is what's meant by "immediate memory" vs. "short-term memory" vs. "working memory." Different people use these terms to mean different things, sometimes in different contexts. *Immediate memory* and *short-term memory* (STM) have a long history going back at least as far as William James and Ebbinghaus. To experimental psychologists, STM is expressed when subjects are asked to recognize or recall a small number of items more-or-less immediately after these items were presented. (In most contexts, the terms "immediate memory" and STM are synonymous, so we'll focus on the latter.) Psychologists and neuropsychologists measure the capacity of verbal STM with variants of the

digit span test, in which the subject first hears the digits "5 . . . 2" and must repeat them back to the experimenter. Assuming success, the subject next hears "7 . . . 9 . . . 4" and repeats these back. This process continues, with the span of digits increasing by one item, and the subject being given two attempts at each span, until they make an error on two consecutive attempts at the same length. The largest number of items that the subject is able to recall is referred to as their "digit span." The psychologist George Miller (1920–2012), often credited as one of the progenitors of cognitive neuroscience as a formal discipline, famously noted the high probability that this subject's span would fall within that of "the magical number 7 +/− 2."

This seems fairly straightforward, but what if after hearing each series of digits, the subject was asked to wait 10 seconds before reciting them back? What about 10 minutes? 10 hours? At what point would their performance stop reflecting STM? Alternatively, what if they were asked to recite the digits back immediately, but to do so in reverse order ("backward digit span")? Or to recite them back in their cardinal order, from smallest to largest (i.e., "7 . . . 9 . . . 4" becomes "4, 7, 9")? Or to mentally calculate the sum of the remembered digits and respond with that? We can begin to address some of these questions with a vignette featuring this author's go-to patient for explaining all things mnemonic:

> One of the privileges afforded graduate students at MIT's Behavioral Neuroscience Laboratory has been the opportunity to transport . . . H.M. from and to his home, a handful of hours distant, for his roughly semiannual research visit . . . The convention for H.M.'s transport was for the designated graduate student to travel with a companion, and there was never a shortage of volunteers (typically a fellow student or a postdoc) eager to take a road trip with this famous patient. On one such trip the two scientists-in-training and their charge sought to pass the time by playing a game in which each player selects a color – on this occasion, green, blue, and white – and accumulates points for each car painted in his or her color that passes in the opposite direction on the highway. Each player counts aloud, and the gaps between passing cars are typically filled with cheering and good-natured banter. H.M. participated fully in the game, selecting his color, accurately keeping track of his running total, and participating in the debate about whether a teal-colored car should be scored as blue, green, or neither. Indeed, on this occasion H.M. won, accruing a score of 20 first. A round of congratulations was exchanged, followed by a lull as the car rolled through the undulating central Massachusetts countryside. A few minutes later, the guest traveler, eager to maximize his once-in-a-lifetime opportunity

to gain first-hand insight from this famous patient, asked "Henry, what are you thinking about right now?" H.M. replied that his count of white cars had now increased to 36.

The driver and guest were both impressed that this patient, famously incapable of remembering virtually anything that had occurred in his life since his 1953 surgery, had accurately maintained and updated a running count of arbitrarily selected "target stimuli" across a span of several minutes, with no evident source of external support or reinforcement. The three travelers commented on this before the guest traveler redirected the conversation to a line of questions that was typical of these trips: "Do you know what today's date is?" "Do you know who the current President is?" "Do you know who we are; where we're going today?" H.M. complied with good-natured responses, as always, clearly enjoying the interaction with and attention from these young, engaged travelling companions. Very quickly, however, H.M. initiated another typical element in the driving-with-H.M. script, by steering the conversation towards a reminiscence from his youth, the portion of his life still mentally accessible after his surgery. (The story, about riding the Silver Meteor passenger train on a multi-day trip to visit an aunt in Florida, had already been told several times at that point in the trip, a product of the teller not remembering the previous tellings.) At that point, sensing a "teachable moment," the driver of the car interjected with a question of his own: "Henry, do you remember what game we were playing a few minutes ago?" No, he didn't. "It involved counting cars of different colors; do you remember what your color was?" No. "The three colors were green, blue, and white; do you remember which was yours?" No, the cues didn't help. "Do you remember who won the game?" No recollection even of the triumph that had, only a few minutes before, produced in H.M. a modest chuckle and satisfied smile. (Postle, 2009, pp. 326–327)

The main point here is that the "short" in short-term memory does not relate to the passage of time, per se. So long as H.M. continued thinking about the car-counting game, he could have maintained a memory for the number 36 (or, continued adding to it) indefinitely. *Continuous thought*, however, does seem to be a key factor for STM, as illustrated in the fact that just a few minutes of thinking about other things removed any trace of memory of even the circumstances that gave rise to the number 36, not to mention the number itself. Indeed, this anecdote also illustrates that the "long" in long-term memory can't be taken at face value: The fact that both of the neurologically healthy individuals in the car *were* able to effortlessly recall all the details from the game played only a few minutes previously implies that processing in their intact MTL was

critical for this outcome. One conclusion from these considerations is that the informal use of "short-term memory" to refer to memory for things that happened earlier today, or earlier this week, is neurobiologically inaccurate. (It's okay, and understandable, when friends or TV newscasters misuse "STM" but, maddeningly, many neurologists and psychiatrists are also guilty of using it in this imprecise way. Confusingly, many neurobiologists studying synaptic plasticity also use "STM" to refer to the early, pre-protein-synthesis-dependent component of LTP.)

Now what about "STM" vs. "working memory"? By some definitions, tasks that only require the short-term retention of a small amount of information, like the digit span task, are tests of STM, whereas tasks that require any kind of manipulation of that information, such as reordering it or performing mental arithmetic on it, are tests of working memory. A complicating factor with this definition, however, is that it implies that every test of working memory has an STM component. Additionally, some take the view that "STM" is best used only as a label for a class of tasks, but not to refer to a discrete cognitive or neural function. What we'll do is adopt the practice that is increasingly common in cognitive neuroscience, just using the term "working memory" to refer to all explicit tests of memory that can be performed without requiring a contribution from the MTL. When we want to be more specific, for example, to refer to the short-term retention of information, we'll specify "working memory storage," or "working memory maintenance."

The reason for taking the time to give explicit considerations to these labels, and what they can refer to, will be evident as we start to consider the history of working memory research.

WORKING MEMORY AND THE PFC? THE ROOTS OF A LONG AND FRAUGHT ASSOCIATION

By the 1930s, the functions of most regions of cerebral cortex had been determined, at least to a first order of approximation, with the scientific methods that we introduced way back in *Chapter 1*. An exception to this was the prefrontal cortex, the region that, in the primate, is rostral to those in which damage produced paralysis, or other disorders of motor control. To investigate its function, Carlyle Jacobsen, at Yale University, trained two monkeys to perform a delayed-response task in an apparatus quite similar to the

WGTA (Jacobsen, 1936; *Figure 6.2* in this book). After watching while one of two covered food wells was baited, the monkey waited for several seconds during which the lowered screen blocked it from seeing or reaching the wells. When the screen was raised, the animal was given one reach with which to displace the cover and, on a correct response, retrieve the food. Pre-lesion, the animals learned to perform the task almost perfectly. After recovery from bilateral surgical removal of the frontal cortex anterior to the arcuate sulcus (i.e., these were MASSIVE PFC lesions), the animals' performance never deviated from chance (50% correct). That is, they were reduced to guessing. Importantly, Jacobsen included trials on which no delay was imposed between well-baiting and food retrieval. On these, performance improved dramatically, meaning that poor performance on delay trials could be ascribed to neither visual nor motoric nor motivational problems. Thus, a reasonable description of the findings might seem to be that, on delay trials, monkeys with bilateral PFC lesions could not remember which well had been baited. Jacobsen concluded that the prefrontal cortex is responsible for "immediate memory."

The idea that Jacobsen had *localized* immediate memory to the PFC was soon called into question, however, by a subsequent study performed at the same laboratory. Robert Malmo (1942) replicated the basic procedure from Jacobsen's experiment, but added the additional experimental factor of turning off the lights in the lab on one half of the trials. (The laboratory had remained illuminated throughout Jacobsen's experiment.) Remarkably, this simple manipulation had the effect of "rescuing" the performance of the PFC-lesioned monkeys, in that they performed correctly on roughly 85% of lights-off trials, despite still getting only 50% correct on lights-on trials.

Thus, even though the delayed-response task is a test of working memory, it was evidently incorrect to assume that the ability to briefly hold information in mind, per se, had been disrupted by damage to the PFC. (If it had been, then turning off the lights wouldn't have restored performance in the Malmo [1942] experiment.) Malmo (1942) attributed his findings to an increased susceptibility to interference in the monkeys after bilateral PFC removal. That is, whereas the ability to retain information over a short period of time remained intact, the ability to guide behavior with this information was compromised when other stimuli competed for attention, as would be more likely to happen in the light than in the dark. Twenty years later, in the chapter that introduced the concept and the term "working memory" to neuroscience, Pribram, Ahumada, Hartog, and Roos

(1964) reinforced this view, asserting that "aspects other than the trace of memory [i.e., working memory storage] were involved by the frontal procedure [i.e., were compromised by PFC lesions]: action at the time of stimulus presentation and distractability were found to be important" (p. 29).

Early focus on role of PFC in the control of STM

In the wake of the Malmo (1942) study, the majority of research on behavior mediated by the PFC concentrated on the control of contents of working memory, and the control of behavior that is guided by working memory. The latter was shown to be compromised, for example, when monkeys with PFC lesions were shown to be impaired on tests of delayed alternation. Delayed alternation differs from delayed response in that no to-be-remembered information is presented at the beginning of the trial. Instead, what the animal must remember is what its response was on the previous trial, whether or not that response was correct (i.e., was rewarded), and what the rules of the task are. On the first trial of a test of delayed alternation, the monkey sees two covered food wells and reaches to one of them (we'll call it A). Unbeknownst to the monkey, both food wells are empty on this trial (thereby guaranteeing that the choice on this first trial will be incorrect). During the ensuing delay (during which the barrier is lowered), the well that the animal has just selected is baited. This is alternation because, from the animal's point of view, it's as though the food was in well B on the first trial, but it selected A. Now, on the second trial, the food will be in A. In the third trial, it will be in B, and so on. Once the task is underway, the animal needs not only to retain a memory of where the food had been on the previous trial, but also to overcome the natural tendency to return to the place where it had most recently been rewarded, a level of control of behavior that we will consider in more depth in *Chapter 15*. (In rodents, delayed alternation is often tested in a "T maze," a maze in which the animal begins each trial at the "bottom" of the T and has to run to the other end of the start arm and then to decide whether to turn left or right.)

In humans, Lila Ghent, Mort Mishkin, and Hans-Lukas Teuber (1962) found no impairment of forward digit span performance in a group of 24 patients with PFC lesions, although they did find an impairment in a group of 20 patients with lesions in left temporoparietal cortex (more on this in *Chapter 19*). Two years later, another study revealed echoes in the human of what the Malmo (1942) study had shown in the monkey. In it, Brenda Milner (1964) reported on the performance of PFC-lesioned patients on tests of 60-second delayed recognition. When tested with novel nonsense shapes, with no reuse of stimuli across trials (i.e., an "open set" of stimuli), performance of these patients was intact. With other types of stimuli, however, the patients were tested on closed sets, and their performance was impaired on those tests. A "closed set" refers to a stimulus pool made up of a finite number of items, such that they are reused over the course of testing. (For example, the digit span test uses a closed set of the digits 0–9; because the test begins with a span of two, and each length is tested twice, the first four-item list in the series is necessarily composed of items that have already appeared on earlier trials of the test.) From experimental psychology, we know that the use of closed sets increases the influence of proactive interference (often abbreviated PI), a phenomenon whereby retrieval from memory "gets interfered with" by the residual memory traces of no-longer-relevant items from previous trials. On tests of recognition, PI can produce false-alarm responses, such that the subject incorrectly endorses a memory probe that had appeared on a previous trial, but not the current one. On tests of recall, it can produce intrusions, such that an item from a previous trial is recalled as having appeared on the current one. Because patients with PFC lesions were impaired when tested with closed sets, but not with open sets, Milner (1964) attributed their impairment on the former to the control of PI, rather than to working memory storage per se.

Thus, in the 1960s, the role of the PFC in working memory was generally understood to be one of the control of the contents of working memory, and/or the control of behavior that must be guided by the contents of working memory, rather than of the short-term retention, per se, of information. "This all changed" (well, the thinking of many researchers changed) when extracellular electrophysiological techniques were brought to the study of delay-task performance.

Single-unit delay-period activity in PFC and thalamus

In 1971, the groups of Joaquin Fuster and Garrett Alexander at UCLA, and of Kisou Kubota and Hiroaki Niki at Kyoto University, published the results of recordings from neurons in the PFC while monkeys performed delayed recognition and delayed alternation. The cortical recording sites for both groups were the banks of the caudal half of the principal sulcus (the area labeled "DL" in *Figure 14.1.B.*; many studies subsequent to those of Jacobsen and Malmo had indicated that damage restricted to just this region was sufficient to impair delay-task performance). Additionally, Fuster and Alexander (1971) also recorded in the mediodorsal (MD) nucleus of the thalamus, the thalamic station of the lateral frontal cortico-basal ganglia-thalamic circuits illustrated in *Figure 8.3*.

During delayed-response performance, Fuster and Alexander found that many neurons in both PFC and MD displayed elevated firing rates that spanned the duration of the delay period, which varied in length, unpredictably, within a range of 15–65 seconds. During delayed-alternation performance, Kubota and Niki (1971) observed two classes of task-related activity: neurons with elevated activity during the delay ("D") and neurons that became active just prior to, and during, the response period ("E"; presumably because their activity predicted the onset of activity in the electromyogram [EMG] that was recording from muscles in the animal's arm). Interestingly, the response profile of many E cells changed quantitatively when the animals performed a simple alternation task with no delay period interposed between responses – pre-response bursts were of lower intensity (i.e., slower firing rate), and they preceded the motor response by a smaller period of time.

Both sets of authors provided interpretations of their data that, at the time of the writing of this book, fit well with current thinking about the PFC.

Fuster and Alexander (1971):

> The temporal pattern of firing frequency observed in prefrontal and thalamic units during cue and delay periods suggest the participation of these units in the acquisition and temporary storage of sensory information which are implicated in delay response performance. Their function, however, does not seem to be the neural coding of information contained in the test cues, at least according to a frequency code, for we have not found any unit showing differential reactions to the two positions of the reward.
>
> It is during the transition from cue to delay that apparently the greatest number of prefrontal units discharge at firing levels higher than the intertrial baseline. This may be the basis of the d-c negative potential shift that has been reported to occur at that time in the surface of the prefrontal cortex . . . We believe that the excitatory reactions of neurons in MD and granular frontal cortex during delayed response trials are specifically related to the focusing of attention by the animal on information that is being or has been placed in temporary memory storage for prospective utilization (p. 654).

(Two points of clarification: First, we'll define "d-c negative potential shift" further along in this chapter; second, "granular" refers to the fact that this area of PFC has a prominent layer IV, the neurons of which have large cell bodies that appear granular under a microscope [c.f. *Figure 2.3*].)

Kubota and Niki (1971):

in the delay task, periprincipal E units may be causally coupled with the initiation of the voluntary lever pressing. During delay task the E unit activity may be coupled with the initiation and sustaining of . . . lever pressing rather than [a] memory retrieval process.

   . . . D units are hardly correlated with the memory storage (1) or retrieval of the memory for the lever pressing (2). Frequency of D unit during delay phases is not apparently different between right and left lever pressings. However, this interpretation does not exclude the possibility that the activities of neurons in the prefrontal cortex represent a memory function related to the choice of the correct performance on the basis of immediate past experience, i.e., remembering the spatially directed response on the preceding trial. (p. 346)

The implications, and influence, of these first studies of PFC electrophysiology

Perhaps to a greater extent than the details of the findings, the very fact of the publication of these studies was compelling, evocative, and highly influential. At the most general level, they represented the first direct studies of brain activity underlying a behavior that was unequivocally *cognitive*. For over a century, of course, scientists had been systematically studying the effect of lesions on various aspects of cognition, but there's somehow something more compelling, more seductive, about "watching the workings of the mind" compared to inferring them from patterns of behavioral impairment. It was also true that electrophysiologists had been studying brain function for decades. But, prior to 1971, even the most ground-shattering, Nobel Prize–winning studies were being performed on the "machine-like" functions of perception and motor control – often the animals under investigation were anaesthetized and, even when not, the reports of these studies contained little to no consideration of the implications of the research for *thinking*. The two papers in question, on the other hand, didn't have to explicitly invoke "thinking"; thinking was, quite simply, what they were about.

    With regard to working memory, what seems to have made these studies so compelling was, paradoxically, an interpretation that the authors themselves explicitly

discounted: some saw in them a physiological correlate of the first of two memory mechanisms that Hebb had postulated in 1949. His theory was, in fact, a dual-trace theory with the first being a "reverberatory" mechanism for "a transient 'memory' of [a] stimulus" (p. 61) that would sustain a representation until it could be encoded into a more permanent state via synaptic strengthening. The attribution of explicitly mnemonic interpretations to the 1971 findings became more prevalent when, a few years later, a highly influential model of working memory was introduced in the cognitive psychology literature.

The multiple-component model of working memory

The model of working memory proposed in 1974 by the British experimental psychologists Alan Baddeley and Graham Hitch was a response to two factors. First, in their view, contemporaneous models of STM didn't capture the fact that mental operations performed on information in conscious awareness can be carried out independent of interaction with, or influence on, LTM. (One implication from levels-of-processing theory [section *Subsequent memory effects in the PFC and MTL, Chapter 12*], for example, was that indefinite "maintenance rehearsal" of information in STM did not predict encoding into LTM. Contemporary models, in the view of Baddeley and Hitch [1974], could not accommodate this fact.) Second, they had observed that performance on each of two tasks under dual-task conditions could approach levels of performance under single-task conditions if the two engaged different domains of information, specifically verbal and visuospatial. Thus, early versions of their model called for two storage buffers (dubbed the "phonological loop" and the "visuospatial sketchpad," respectively) that could operate independently of each other and independently of LTM, although both under control of a Central Executive. This model has proven to be enormously influential, spawning a prodigious amount of research that continues through to this day.

Extracellular electrophysiology inspired by the multiple-component model

The next major development in the cognitive neuroscience of working memory was the proposal, by the neuroscientist Patricia Goldman-Rakic (1937–2003), that the sustained delay-period activity in the PFC of the monkey (reported by, e.g., Fuster and Alexander [1971] and Kubota and Niki [1971]) and the storage buffers of the multiple-component model of working memory from cognitive psychology were cross-species manifestations of the same fundamental mental phenomenon.

Earlier in her career, Goldman-Rakic had trained and collaborated with such luminaries as neuropsychologists Haldor Rosvold and Mortimer Mishkin and neuroanatomist Walle Nauta. When she moved to Yale University in 1979, she was a leading authority on the anatomy, pharmacology, and functional anatomy (i.e., the effects of lesions) of the PFC. Motivated by the ongoing neurophysiological research coming out of the UCLA and Kyoto groups, on the one hand, and the growing influence of the cognitive model of Baddeley and Hitch, on the other hand, she

FIGURE 14.1 Studying the organization of visual working memory with ODR tasks.

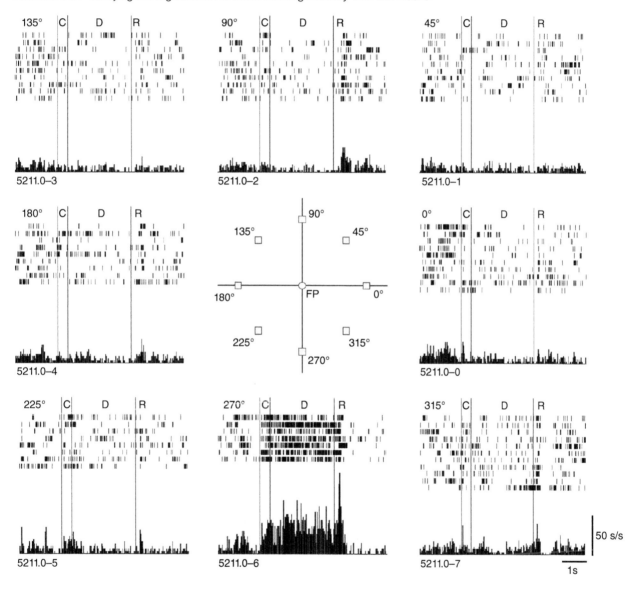

FIGURE 14.1.A The center illustrates the fixation point (FP) and the location of the eight targets, each 13° of visual angle from FP (and only visible to the animal when flashed briefly to cue a location as the target for an ODR). The plots illustrate firing rate (rows of rasters) for several trials at each location, with a histogram (below) summing up the activity. Source: From Funahashi, Shintaro, Charles J. Bruce, and Patricia S. Goldman-Rakic. 1989. "Mnemonic Coding of Visual Space in the Monkey's Dorsolateral Prefrontal Cortex." Journal of Neurophysiology 61 (2): 331–349. Reproduced with permission of The American Physiological Society (APS).

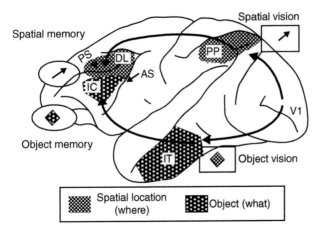

FIGURE 14.1.B Diagram, assembled from the results of many studies from Goldman-Rakic and colleagues, proposing how visual working memory may be supported by extensions into PFC of the where and what pathways. In frontal cortex: PS = principal sulcus; DL = dorsolateral; AS = arcuate sulcus; IC = inferior convexity (the gyrus ventral to the PS). Source: From Wilson, Fraser A. W., Séamas P. Ó Scalaidhe, and Patricia S. Goldman-Rakic. 1993. "Dissociation of Object and Spatial Processing Domains in Primate Prefrontal Cortex." Science 260 (5116): 1955–1958. Reproduced with permission of AAAS.

recruited a former student of Kubota, Shintaro Funahashi, to help her establish an electrophysiology laboratory.

The task that Funahashi, Goldman-Rakic, and colleagues used in many of their studies was the oculomotor delayed-response (ODR) task (as introduced and illustrated in *Figure 9.4*). *Figure 14.1.A* illustrates activity from a single left-hemisphere PFC neuron, recorded during several trials of ODR to each of eight locations. Note that this neuron responds selectively during trials for which a saccade to the downward (270°) target will be required. (Note that a suggestion that it may also be conveying information during trials to other locations comes from the fact that its delay-period activity is suppressed below baseline levels on trials requiring saccades to the three upper visual field locations.) One may ask why Funahashi and colleagues found such selectivity in PFC neurons, when neither Fuster and Alexander (1971) nor Kubota and Niki (1971) did. This may have been due, in part, to changes in experimental procedures, including the finer grain of spatial discrimination required by, the ODR task. It could also have to do with how the monkeys were trained, a factor that we'll consider in more depth further along.

One important contribution of Goldman-Rakic's theorizing was to suggest that the organization of the visual system into two parallel pathways (recall from *Chapter 6*, that the idea of a *what* vs. *where* organization had been introduced in the early 1980s) may extend to visual working memory, such that working memory for object identity may be dissociable from visual working memory for object locations. This idea was confirmed in subsequent experimental psychology, neuroimaging, and neuropsychological studies and led to a "fractionation" of the visuospatial sketchpad, by Baddeley and colleagues, into a "visual cache" (representing *what*) and an "inner scribe" (representing *where/how*) – a clear instance of neuroscience influencing cognitive theory.

In terms of brain organization, Goldman-Rakic's proposal is captured in *Figure 14.1.B*. There are three important points to make about this diagram. First, the heavy arrows representing the flow of information, although schematized cartoons, do reflect the anatomical bias of fibers projecting from posterior regions into PFC, many of these pathways discovered and characterized by Goldman-Rakic and colleagues. (We should note, however, the inaccuracy of suggesting unidirectional patterns of connectivity.) Second, the paper presenting this diagram did also report data from neurons in the inferior convexity of the PFC that showed elevated activity that was preferential for colors or faces over locations in space. Third – and this may well be the only instance in the entire book for which I have to say this – many researchers today wouldn't endorse the main point of the model illustrated in *Figure 14.1.B*. Reasons are to follow, but this has simply been such a highly influential model in the cognitive neuroscience of working memory that it couldn't not be included here.

Alternative accounts of delay-period activity in the PFC

How is one to reconcile the seemingly incompatible findings between the electrophysiological findings depicted in *Figure 14.1* and the results of the lesion studies that we reviewed at the beginning of this chapter? Starting from the position of strongest inference, we have to accept that the lesion studies made a strong case that performance on simple working memory tasks is by-and-large insensitive to the integrity of the PFC. This implies that the PFC does not play a necessary role in supporting working memory storage buffers. But if that's true, what might have been the function of the stimulus-selective delay-period activity in

PFC that we just reviewed? Next we'll consider two studies that have sought to answer this question.

A study by Lebedev, Messinger, Kralik, and Wise (2004) used a variant of ODR that required the animal to keep two locations "in mind" – a remembered location and an attended one. Each trial of their task began with the cuing of a "remembered location" (at one of four possible positions). Next, the stimulus began moving along the diameter of an imaginary circle centered on the fixation point, later stopping at one of the three remaining locations (not the remembered location), thereby defining the "attended location." Careful attention to the moving stimulus was required, because where it would stop was unpredictable and because as soon as it stopped the subsequent "trigger signal" was very subtle: a dimming of the stimulus instructed a saccade to the stimulus itself (i.e., to the attended location), whereas a brightening instructed a saccade to the remembered location (*Figure 14.2.A*). In this way, Lebedev, Messinger, Kralik, and Wise (2004) effectively unconfounded the focus of attention from the contents of working memory, and their results indicated that the majority of neurons from which they recorded in the PFC tracked the focus of attention, not the remembered location (*Figure 14.2.B*). Thus, the results from Lebedev, Messinger, Kralik, and Wise (2004) offer an alternative explanation to the results presented in *Figure 14.1.A* that many of the PFC neurons recorded by Funahashi, Bruce, and Goldman-Rakic (1989) may have been supporting attentional, rather than mnemonic, functions.

A second study that was designed to test an alternative interpretation to the working memory buffer account of PFC delay-period activity addressed the proposed dorsal-ventral segregation of working memory function in PFC (Wilson, Ó Scalaidhe, and Goldman-Rakic, 1993; *Figure 14.1.B* in this chapter). In this study from Rao, Rainer, and Miller (1997), monkeys were trained to perform a two-step What-then-Where ODR task (*Figure 14.3.A*). They reasoned that perhaps in the earlier studies, the serial training of monkeys to perform STM for one type of information, then for another, had led their PFCs to organize themselves according to the tasks that they were trained to perform. (Other studies had shown, for example, that intensive training on a color-based visual search task results in color selectivity in the frontal eye field [FEF], which is not a typical property of this region.) The results indicated that over half of the PFC neurons displaying elevated delay-period activity showed tuning for both object identity and location, and that these were topographically intermingled with those that responded only to "What" or only to "Where," with no evident dorsal-ventral segregation. This finding cast doubt on the labeled line-like organization of function in PFC illustrated in *Figure 14.1.B*.

We'll return to what these alternative accounts might mean for understanding PFC contributions to working memory after we review some more recent developments from EEG and fMRI studies of visual working memory in humans.

FIGURE 14.2 The representation of attention vs. memory in PFC. Source: From Carmena JM, Lebedev MA, Crist RE, O'Doherty JE, Santucci DM, Dimitrov DF, et al. (2003) Learning to Control a Brain–Machine Interface for Reaching and Grasping by Primates. PLoS Biol 1(2): e42. https://doi.org/10.1371/journal.pbio.0000042. Public Domain.

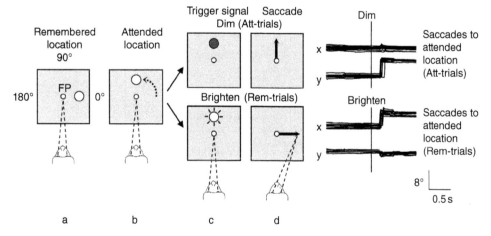

FIGURE 14.2.A Timeline of the task, with panels on the right showing eye-position traces corresponding to upward and rightward saccades.

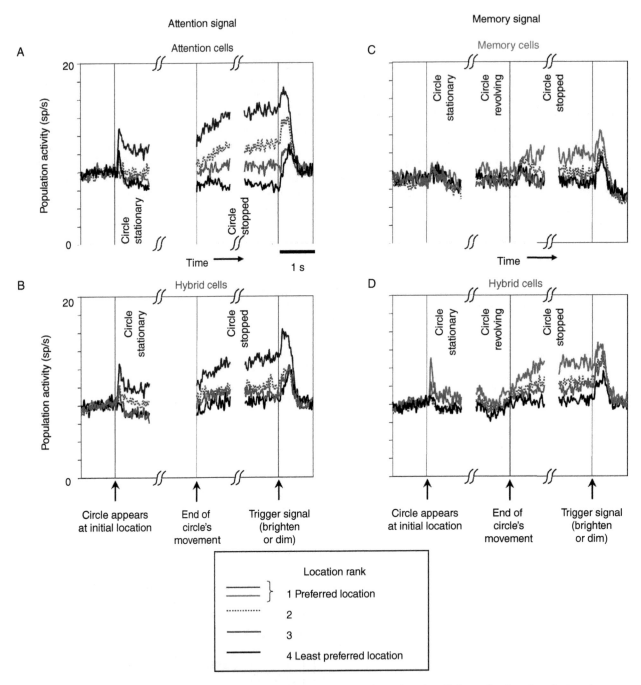

FIGURE 14.2.B Left- and right-hand columns illustrate responses on "Attention signal" (i.e., stimulus dimming) and "Memory signal" (i.e., stimulus brightening) trials, respectively. Across two monkeys, 61% of delay-active PFC neurons were sensitive to only the attention demands of the task, 16% to only the memory demands, and the remaining 23% to both ("Hybrid"). Plots A–D illustrate activity from these three types of neurons on the two types of trials.

FIGURE 14.3 Integration of *what* and *where* in primate prefrontal cortex.

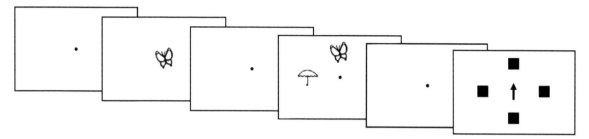

FIGURE 14.3.A An example behavioral trial of *what*-then-*where*. The trial began with the central presentation of a sample stimulus that the animal had to hold in memory across the first delay until its re-presentation, together with a second stimulus, each in one of four possible locations. At this point, the animal had to encode the location occupied by the sample stimulus, hold this information across the second delay, and then, when the central fixation point was replaced by four target locations, make a saccade to the location that had been occupied by the sample. Source: Rao, Rainer, and Miller. 1997. Reproduced with permission of AAAS.

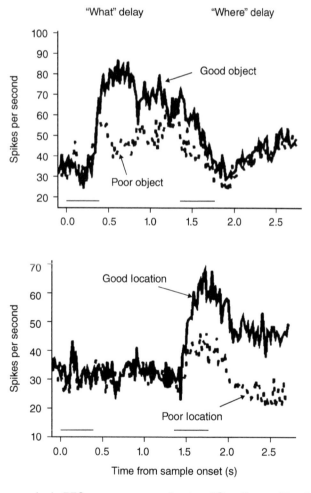

FIGURE 14.3.B Four traces from a single PFC neuron, responding to a "Good" vs. a "Poor" object (same convention as *Figure 7.15.C*), and to a "Good" vs. a "Poor" location. Source: From Rao, S. Chenchal, Gregor Rainer, and Earl K. Miller. 1997. "Integration of What and Where in the Primate Prefrontal Cortex." Science 276 (5313): 821–824. doi: 10.1126/science.276.5313.821. Reproduced with permission of AAAS.

WORKING MEMORY CAPACITY AND CONTRALATERAL DELAY ACTIVITY

One aspect of working memory that is a focus of intensive inquiry, and debate, is the nature of its capacity limitations. We already noted at the beginning of this chapter that, for 95% of healthy young adults, digit span will be between five and nine items. Furthermore, this turns out to be a value that is inflated by our ability to speak quickly (even if only covertly, in our "mind's ear"), because when verbalizing is prevented (e.g., by requiring subjects to repeatedly say a word aloud) digit span drops to 4 +/− 1. An important development in working memory research occurred in the late 1990s when Steve Luck and his then-graduate student Ed Vogel published a series of studies in which subjects were shown a sample array of objects (colored squares in the canonical version of the task [*Figure 14.4*]), then, after a brief delay, a test array that was either identical to the sample or differed by one feature of one of the objects (e.g., the color of one of the squares was different). (In other variants only one test object from the array is presented, and it either is or isn't different from how it had appeared in the sample array.) In the jargon of the field, this has come to be known as a "change detection" task. There were two critical factors that were varied. The first is that arrays of many different sizes were used, varying from 1 item up to 12 or more. As *Figure 14.4.B* illustrates (and as you surely could have guessed!), performance drops off with increasing set size. Second, and more controversially, Luck and Vogel found the same pattern of performance regardless of whether the individual stimuli in the test array had only one dimension (e.g., color) or many (e.g., bars of different color, length, orientation, and size of a gap in their middle). This gave rise to the proposal that working memory capacity is constrained at the level of objects, rather than at the level of features. (That is, because performance with arrays of, say, four items, was comparable for "4-dimension" items and for "1-dimension" items, the fact that the 4-item 4-dimension arrays contained 16 discrete features, whereas the 4-item 1-dimension arrays contained only 4, didn't seem to matter.) We'll come back to this idea of the object as the fundamental unit of visual working memory, but first we need to flesh out the idea of a capacity limitation.

Luck and Vogel (1997) noted that another way to characterize data like those shown in *Figure 14.4.B* was to transform them from this description of accuracy at different loads into an estimate of the number of items that the subject was holding in mind at each load. To do this,

they used a simple formula first introduced by the psychologist Hal Pashler that entails subtracting the false-alarm rate (FAR) from the hit rate (HR) and multiplying the result by the number of items that had been in the memory array. Let's try it by eyeballing the data from *Figure 14.4.B*. The variable used to represent capacity is k.

at load 1, $k = [(\mathrm{HR}\ (1.0) - \mathrm{FAR}\ (0)) \star 1] = 1$;
at load 2, $k = [(\mathrm{HR}\ (.98) - \mathrm{FAR}\ (.02)) \star 2] = 1.92$;
at load 3, $k = [(\mathrm{HR}\ (.97) - \mathrm{FAR}\ (.03)) \star 3] = 2.82$;
at load 4, $k = [(\mathrm{HR}\ (.90) - \mathrm{FAR}\ (.1)) \star 4] = 3.2$;
at load 5, $k = [(\mathrm{HR}\ (.86) - \mathrm{FAR}\ (.14)) \star 5] = 3.6$;
at load 6, $k = [(\mathrm{HR}\ (.81) - \mathrm{FAR}\ (.19)) \star 6] = 3.72$;
at load 7, $k = [(\mathrm{HR}\ (.73) - \mathrm{FAR}\ (.27)) \star 7] = 3.22$;
at load 8, $k = [(\mathrm{HR}\ (.67) - \mathrm{FAR}\ (.33)) \star 8] = 2.72$.

FIGURE 14.4 Visual working memory capacity. Source: Hakim and Vogel, 2018. Source: From Hakim N, Vogel EK (2018) Phase-coding memories in mind. PLoS Biol 16(8): e3000012. https://doi.org/10.1371/journal.pbio.3000012 Published: August 29, 2018. Public Domain.

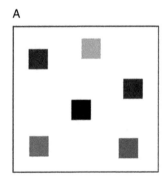

A

FIGURE 14.4.A A memory array at set-size 6.

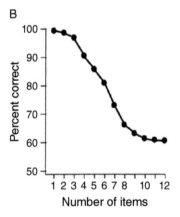

B

FIGURE 14.4.B Performance of a subject at multiple set sizes.

C

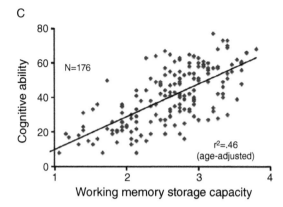

FIGURE 14.4.C Correlation of visual working memory capacity with a composite measure of cognitive ability made up of tests of processing speed, working memory, verbal learning, visual learning, attention/vigilance, reasoning/problem solving, and social cognition. This sample includes neurologically and psychiatrically healthy adults and patients with schizophrenia. The correlations were very similar when carried out separately on the data from each of the groups, justifying their aggregation here (Johnson et al., 2013).

So the intuition is that at low loads, one is holding in mind as many items as were presented in the array, but as the set size continues to grow one reaches the point where no more items can be held in working memory, and so k asymptotes (or sometimes starts to decline, as is the case from this quick-and-dirty set of calculations – a common convention for determining someone's working memory capacity is to take the largest value from the range of set sizes). Estimates of working memory capacity are stable across repeated measurements, and so are considered by many to be a cognitive trait. Furthermore, as *Figure 14.4.C* illustrates, there is considerable variability across individuals. All this has led to one popular model that holds that working memory capacity limitations are due to the cognitive system having a finite number of hypothetical "slots," the number of slots that one has determining the number of items that one can hold in working memory at any one moment in time.

The fact that working memory capacity is correlated with many laboratory measures of cognitive ability (e.g., *Figure 14.4.C*), as well as with real-world outcomes like reading proficiency and scholastic achievement, further reinforces the idea that it is a critical contributor to many aspects of high-level cognition. This, in turn, further

motivates the importance of understanding the neural bases of working memory capacity limitations.

The electrophysiology of visual working memory capacity

Soon after Vogel started his own lab, he and graduate student Maro Machizawa described an ERP component that closely tracked estimates of k. They named this component the "contralateral delay activity" (CDA) for reasons made clear in *Figure 14.5*. The discovery of the CDA has generated a great deal of interest because it seems to track what's happening "inside people's heads" rather than on the surface of their retinas. That is, if an individual's visual STM capacity is estimated to be 3 items, the CDA will increase in magnitude, monotonically, as the memory array increases from 1 to 2 to 3 items (*Figure 14.5*) but then it saturates, and will remain at that magnitude for arrays of 4, 8, or 12 items, presumably because the subject can only select and remember 3 items out of an array, regardless of whether it contains 3 or 12 items.

At the circuit level, the neuronal bases of the CDA remain poorly understood. One observation is that a sustained shift in the voltage of a signal, as is observed with the CDA, can be thought of as a "DC shift," by analogy to direct current, a constant (i.e., not alternating) current source. When the magnitude of current flow changes to a new, constant value, this is a DC shift. Thus, the load-related changes in the amplitude of the CDA may correspond to sustained increases in current flow in the neurons engaged by the task. In this regard, it's intriguing to speculate that the ECoG-measured DC shift referred to in the extended quote from Fuster and Alexander (1971; earlier in this chapter) may relate to the CDA. Regardless of this speculative observation, the CDA is considered by many to be an analogue in humans to the elevated, sustained delay-period activity recorded from neurons in the PFC of monkeys, as described earlier in this chapter. It is important to keep in mind, however, that while the neural activity reviewed early in this chapter is recorded in the PFC, the CDA is recorded from scalp electrodes positioned over parietal and occipital cortex, and is generally believed to originate from activity in these regions.

The CDA in visual cognition

The CDA has become an important tool in research on visual working memory and visual cognition. In particular, it has been used in several different contexts to demonstrate tight links between visual working memory and behaviors

FIGURE 14.5 Contralateral delay activity (CDA), an ERP component sensitive to STM load. Source: From Vogel, Edward K., and Maro G. Machizawa. 2004. "Neural Activity Predicts Individual Differences in Visual Working Memory Capacity." Nature 428 (6984): 679–782. doi: 10.1038/nature02447. Reproduced with permission of Nature Publishing Group.

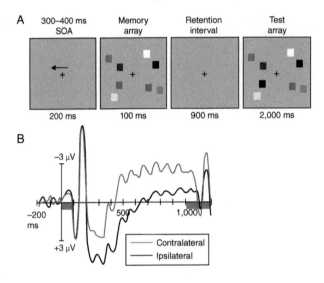

FIGURE 14.5.A Panel **A** illustrates the task, a variant of the color "change detection" task introduced in *Figure 14.4.A*. Critically, at the beginning of each trial a cue indicates whether the array in the left or right hemisphere is to be remembered. Panel **B** shows the grand average (12 subjects) ERP recorded at parietal and occipital electrodes, first averaged within each subject across electrodes from just the right and just the left hemisphere, then averaged across trials as a function of whether the attended array was in the visual field ipsilateral or contralateral to the hemisphere. To generate the CDA (*Figure 14.5.B*), this ipsilateral signal is subtracted from the contralateral signal. The rationale is that, because subjects are fixating centrally, visual stimulation for both hemispheres is comparable, and the difference in the two waveforms should reflect the effect of the additional processing of the array in the cued hemifield (i.e., encoding it, then holding it in working memory). (Note that, for both waveforms, one can count about five to six cycles of an oscillation between time 500 and 1000 ms, which would correspond to 10–12 Hz (i.e., the alpha band of the EEG). It's evidently the case that the phase reset triggered by the memory array is so potent that alpha-band oscillations aren't fully obscured by trial averaging (section *Event-related analyses, Chapter 3*).

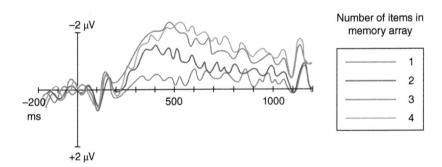

FIGURE 14.5.B The CDA at four memory loads. Beginning with load 4, the green waveform is the subtraction of the two illustrated in *Figure 14.5.A* (contralateral – ipsilateral). Note that this removes the early stimulus-evoked deflections, because they are comparable for the two hemispheres. The remaining CDAs illustrated here are for arrays of load 1, 2, and 3. An important detail to note here is that the magnitude of the CDA on load 4 trials is no different than on load 3 trials. This fits with the fact that the group-mean visual working memory capacity of this group of 12 subjects is 2.8 items.

that, on the surface, one may not have thought of as "memory tasks." One example is visual search. For decades, cognitive psychologists have studied visual search as a way of understanding how we are able to pick out and analyze small amounts of information in the face of the huge amount of information that is present in most visual scenes. (The anecdote from section *Synchronous oscillation, Chapter 2,* about the walk in the woods, and survival depending on attending to the bear rather than to the bird or the clouds, is just one example of such a situation. A more mundane one is the anecdote at the end of *Chapter 9, The endogenous control of attention,* about driving in an unfamiliar city and looking for a particular street.) Psychologists distinguish between "pop-out" search tasks, which assess the exogenous capture of attention, typically by an array in which one item differs from all the others by a distinguishing feature, and "conjunction search," in which all the elements in the array often have all the same features, and the target is either configured or oriented in a unique way. The visual search task that we considered in some depth in *Chapter 7, Figure 7.15.B,* can be thought of as a (very) simplified version of a conjunction search, in that there's nothing more inherently attention-demanding for the monkey about the butterfly vs. the wheelbarrow in the search array, it's only the instruction given at the beginning of the trial, in the form of the search target, that determines which of these two the animal should select when they are presented in the search array. In experimental psychology, the search target is often given to the subject verbally. In *Figure 14.6.A,* it would be *search for the upright letter T.* In this kind of task, interpreting patterns in the CDA from the concurrently measured EEG has provided important insights into the role of working memory in visual search.

For conjunction search, as illustrated in *Figure 14.6.A,* it's well established that the average search time is monotonically related to the number of distractors in the array. The interpretation of this is that, because all the stimuli look very similar, subjects must serially inspect each one until they come upon the target. Stephen Emrich and colleagues (2009), then at the University of Toronto, recorded the EEG while subjects performed a conjunction search task and a visual working memory "change detection" task (*Figure 14.6.A*). Their search task was modified from the standard procedure in that, like the working memory task, a cue that preceded the stimulus display indicated which of the two arrays, one in each visual field, was relevant to that trial. Their EEG results

(*Figure 14.6.B*) showed that when the data from the visual search task were processed in the same way as those from the working memory task, the resultant ERP between the two tasks was almost identical, with the exception that the "contralateral *search* activity" (CSA), as they dubbed it, started ramping up slightly later than the CDA from the working memory task. They interpreted these results as evidence for working memory being engaged in this task: as subjects begin to search through the array, they hold on to a memory of where their attention has recently been, so as to prevent returning to an already-visited item during the search. The progressive increase in the amplitude of the CSA was explained as reflecting the accumulation of items in working memory as the subject progressively searched through the array. Importantly, the CSA was observed to plateau at a level corresponding to each individual's working memory capacity, regardless of the number of distractors in the array, and, therefore, independent of the search time. Furthermore, if the CSA were sensitive to the total number of items that needed to be inspected during the search, on target-absent trials it would be expected to increase in a manner corresponding to the 10 items in the search array. However, this was not observed, the CSA plateauing at the same level for target-absent as for target-present trials, despite the evidence that subjects did, indeed, search through more items on target-absent (mean RT = 1826 ms) than on target-present (mean RT = 1301 ms) trials.

Just to be clear, Emrich, and colleagues (2009) concluded from their data that the CSA they recorded during the visual search task and the CDA they recorded during the working memory task reflected the operation of the same stimulus-retention operation during both tasks. They used a different label for the ERP to be clear that one was recorded while a search array was on the screen and one was recorded during a delay period.

Another way in which working memory is involved in visual search was made explicit is the computational model that we considered in section *The reentry hypothesis, Chapter 9.* Recall that in this model the search target is held in working memory, and it is this representation of the target that results in IT responding differentially to the same stimulus on *Target = good stim.* vs. on *Target = bad stim.* trials. Indeed, independent of the search-related activity that we just considered in *Figure 14.6,* a CDA has been observed during the delay between the presentation of the present trial's search target and the onset of the array.

FIGURE 14.6 Visual working memory in visual search. Source: From Emrich, Stephen M., Naseem Al-Aidroos, Jay Pratt, and Suzanne Ferber. 2009. "Visual Search Elicits the Electrophysiological Marker of Visual Working Memory." PLoS One 4: e8042

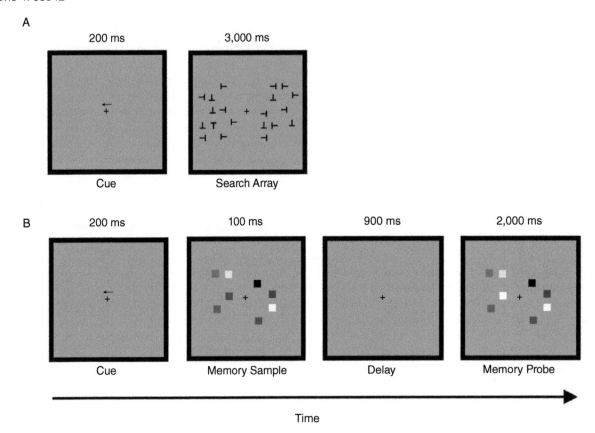

FIGURE 14.6.A Diagrams of the conjunction search task (**A**) and the visual working memory task ("color change detection;" **B**).

This working memory representation of the search target is often referred to as the "search template," and measurement of this template-related CDA has provided insight into what had been a puzzling observation from the cognitive psychology literature. The puzzle was from dual-task experiments in which subjects are first given information to hold in working memory, then presented a target for a visual search, then presented the search array, and then finally tested on their memory for the information from the beginning of the trial. The consistent finding is that search performance is markedly impaired relative to that of trials with no concurrent working memory task

(another piece of evidence for overlap between attention and working memory). However, if the search target is the same on several consecutive trials, performance on the search task gets better, until it is no longer impaired by the concurrent working memory task. An explanation is suggested by the fact that, across successive trials with a recurring search target, the template-related CDA is seen to decrease in amplitude. This has been interpreted as evidence for a "handoff of the attentional template from visual working memory to long-term memory as subjects search for the same target object across runs of trials" (Carlisle, Arita, Pardo, and Woodman, 2011, p. 9320).

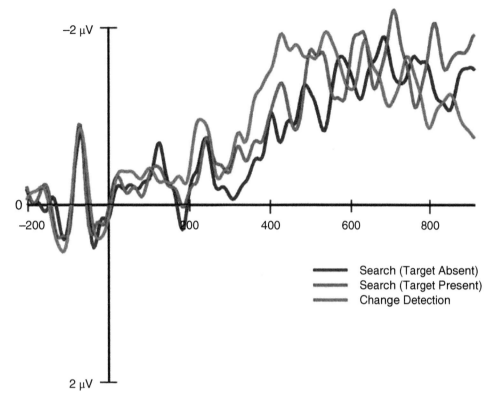

FIGURE 14.6.B The group-average ERP data superimposed from three types of trials: visual search when there was no target present in the array (blue); visual search when there was a target in the array (red); and the CDA from the working memory task ("change detection;" green).

A follow-up study replicated the effect that consecutive-trial repetitions of the search target were associated with a decline in the template-related CDA, and also showed that a different ERP component, the P170, increased with each repetition. Because the P170 indexes the accumulation of information that supports recognition from long-term memory, this provided further evidence for an interplay between working memory and long-term memory during visual search.

Another challenging attentional task that isn't traditionally thought of as a working memory task is multiple-object tracking (MOT). A trial of this task begins with a display of several circles on a screen, all the same color. One briefly changes color, then changes back. Next, the circles all begin moving around the screen in a random manner, as though the circles were a colony of swarming ants, each one following a different zigzag path. Finally, the circles all stop moving, one of

them changes color again, and the subject's job is to indicate whether it's the same as the one that had changed color at the beginning of the trial. The "multiple" part of MOT, as you may have surmised, is that the number of color-changing items varies from trial to trial, such that on some trials the subject has to (try to) track as many as five or more. The maximum number of items that someone can track turns out to be very similar to their visual working capacity, and, furthermore, (you guessed it) MOT performance generates a CDA with all the same properties that we've already considered. One conclusion to all these findings may be that, in keeping with Emrich and colleagues (2009), visual working memory is engaged by many different tasks that require selective attention, even tasks that don't seem to have an overt memory component. An alternative way of thinking about this is suggested by Luck and Vogel in a more recent review from them: "visual working memory may not be a memory

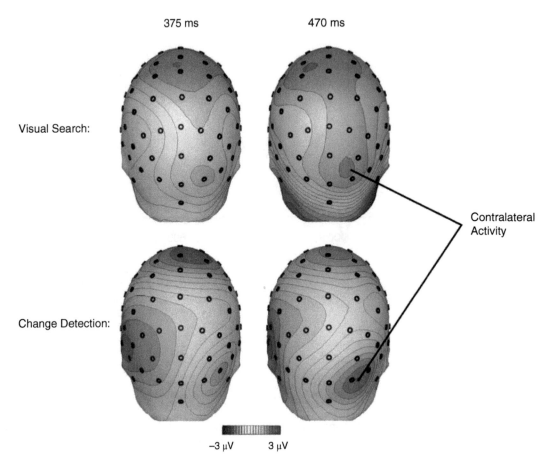

FIGURE 14.6.C Two "snapshots in time" of scalp topography of voltages recorded during the visual search and visual working memory ("change detection") tasks, from left-cued trials. The times listed at the top of the two columns of voltage maps correspond to the timeline of the waveforms from *Figure 14.6.B*. The similarity between the topographies from the two tasks, including their similar evolution through time, led the authors to conclude that "the same neural resources were recruited during both visual search and change detection" (p. 6).

system per se, but may instead be a general-purpose visual representation system that can, when necessary, maintain information over short delays" (Luck and Vogel, 2013, p. 394).

Before we leave this section, it needs to be noted that, although the CDA is most often considered to be evidence for the slot model of visual working memory, this model is not universally accepted. Many point out, for example, that the estimated precision of the mental representation of information begins to decline as soon as more than one item needs to be held simultaneously. This is seen by many as evidence that capacity limitations are the result of some

factor like a limited resource (maybe attention?) that gets spread ever more thinly as more items need to be held in working memory, or that having more items in working memory results in more interference, and a consequent decline in performance. Regardless of what theoretical model of working memory one prefers, it is objectively the case that the precision of recall does decline with each additional item held in visual working memory. The fact that the CDA does not show any evidence of this load-related decline in precision indicates that it may not capture "all there is to know" about the neural bases of visual working memory.

NOVEL INSIGHTS FROM MULTIVARIATE DATA ANALYSIS

Your humble and devoted author has been an active researcher in the cognitive neuroscience of working memory for over 20 years (yikes!), and so one might think that this would be the easiest chapter for him to write. I certainly did. However, I'm now realizing that my insider status makes what I'm about to do all the more difficult, because it means leaving out so much important and influential work, by so many scientists whom I really admire. (To say nothing of older work from my own career.) But here goes . . .

The tradition of univariate analyses

There was a period of time, during the 1990s and 2000s, when one could be forgiven for characterizing the functional-neuroimaging-of-human-working-memory literature as a reactive one that parroted the monkey elec-trophysiology literature with a lag of 3–5 years. Thus, for example, a few years after the studies from Funahashi, Bruce, and Goldman-Rakic (1989) and Wilson, Ó Scalaidhe, and Goldman-Rakic (1993), neuroimaging studies appeared to confirm the gist of these findings in humans; then a few years after the Rao, Rainer, and Miller (1997) study that challenged the conclusion of Wilson, Ó Scalaidhe, and Goldman-Rakic (1993), neuroimaging studies came out challenging the conclusions of the first set of neuroim-aging studies; and so on. Now, of course, this would be a cynical and overly facile summary. Some examples of neu-roimaging studies of working memory making important, original contributions have already been highlighted in, for example, *Figure 9.4* and *Figure 12.2*. Nonetheless, more recent developments in the field have shown that much of the neuroimaging literature based on univariate analyses may need to be reinterpreted, for two principal reasons, both deriving from advances in multivariate pattern analy-sis (MVPA) of fMRI datasets. The first relates to sensitivity: MVPA from several groups has now established that regions that do not show elevated levels of regionally aggregated signal intensity during the delay period of a working mem-ory task can nonetheless be shown to be actively represent-ing the to-be-remembered information. The second relates to specificity: MVPA has also shown that elevated delay-period signal, even when it shows properties, such as load sensitivity, that have been considered hallmarks of working memory storage, doesn't always represent the information that researchers have assumed that it does.

MVPA of fMRI

Increased sensitivity reveals delay-period storage in "unactivated" areas

In 2009, two groups published striking demonstrations that the active retention of information in STM can occur in regions that do not show elevated activity. One of them, from John Serences at the University of California, San Diego, and his long-time collaborators who were then at University of Oregon (Serences, Ester, Vogel, and Awh, 2009), was designed to test the "sensory recruitment hypothesis," the idea that the short-term retention of information may be supported, at least in part, by the very same neural systems that were responsible for their percep-tual processing. (Note how this differs from the scheme, illustrated in *Figure 14.1.B*, whereby the "back of the brain" does vision and the "front of the brain" does mem-ory.) To test it, they scanned subjects performing 10-second delayed recognition in which each sample stimulus com-prised a set of black oriented bars (referred to as a grating) superimposed on a colored square, and the memory probe matched the sample along both dimensions on half the trials, differing subtly by color or by orientation in the other half. (Trials were blocked by condition.)

Panel A of *Figure 14.7* shows the trial-averaged fMRI signal from the 62 most responsive voxels in V1, as defined with a (univariate) GLM analysis (analogous to *Figure 9.10*, which shows the spatially averaged response collapsed across the two voxels in that simulated dataset). Two things are noteworthy. First, the signal intensity drops to baseline during the delay period of both conditions. (This is what the label "unactivated" in the title of this subsection is referring to.) Second, it's clear by mentally superimposing the two that one wouldn't be able to discriminate orienta-tion from color trials with these data. Panel B of this figure, however, shows that these 62 voxels evidently carried infor-mation about the sample, because Serences, Ester, Vogel, and Awh (2009) were able to discriminate 45°-oriented from 135°-oriented gratings on Remember Orientation trials and red from green on Remember Color trials.

Demonstrating stimulus-specific patterns of delay-period activity in a putatively "unactivated" part of the brain was an impressive technical feat. The real conceptual advance from this study, however, was the insightful way in which the authors tested a specific prediction of the sen-sory recruitment hypothesis: they trained a classifier to dis-criminate 45°-oriented from 135°-oriented gratings (and red from green displays) from "perception" trials – signal

FIGURE 14.7 MVPA reveals delay-period stimulus representation in V1. Source: From Serences, John T., Edward F. Ester, Edward K. Vogel, and Edward Awh. 2009. "Stimulus-Specific Delay Activity in Human Primary Visual Cortex." Psychological Science 20 (2): 207–214. doi: 10.1111/j.1467-9280.2009.02276.x. Reproduced with permission of SAGE Publications.

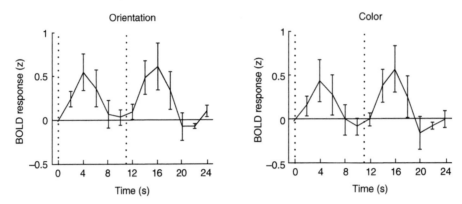

FIGURE 14.7.A Trial-averaged fMRI activity from V1 during orientation and color blocks; vertical lines indicating the onset of sample and probe stimuli. Note robust stimulus-evoked responses, but that activity drops back to baseline during the delay period.

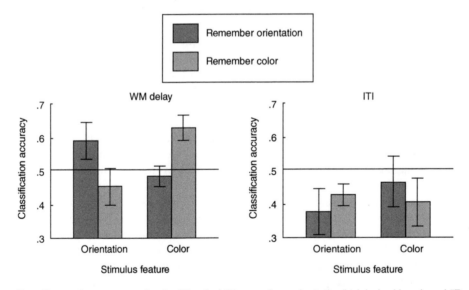

FIGURE 14.7.B Classifier performance on (or, for ITI, after) "Remember orientation" trials (red bars) and "Remember color" trials, as a function of stimulus feature being decoded. Note that the WM delay and ITI epochs were identical in length and with respect to the lag after the most recently presented stimulus (delay decoding performed on data from 4 to 10 seconds after presentation of the sample; ITI decoding from 4 to 10 seconds after presentation of the probe).

corresponding to the runs used to derive the ROIs, during which subjects just viewed stimuli – then used these classifiers to decode data from working memory trials. Successful decoding would necessarily mean that patterns of activity during the working memory trials were largely similar to patterns of activity measured during the perception trials. Thus, working memory for oriented gratings

(or color) could be said to engage the same neural representation as the visual perception of oriented gratings (or color). Serences, Ester, Vogel, and Awh (2009) summarized it as follows: "the sustained activation patterns observed during the delay period were similar to patterns evoked by the continuous presentation of identical sensory stimuli, which suggests that early feature-selective visual areas are recruited to maintain a 'copy' of remembered stimulus attributes, as opposed to a more abstract or categorical representation" (p. 213).

Depending on whether you were a glass-half-full or a glass-half-empty kind of person, the results from this study (and a similar one from Stephenie Harrison and Frank Tong, 2009) were either very exciting or very worrying. They were perhaps worrying from the perspective that of the hundreds (thousands?) of neuroimaging studies of working memory that had been published in the years prior to these two, almost all of them employed univariate analysis methods. How many of them may have failed to identify stimulus-related delay-period activity due to the relatively poor sensitivity of these univariate methods? On the exciting side, multivariate methods have provided researchers with novel tools to address not just *where* in the brain one can find evidence for working memory storage but also *how* these processes might work.

Increased specificity raises questions about the interpretation of elevated delay-period activity

In this section, the author will immodestly summarize some research from his own group, emphasizing a study led by former postdoc Steve Emrich and former graduate student Adam Riggall. (Emrich had been a graduate student when he did the work presented in *Figure 14.6*.) Emrich, Riggall, LaRocque, and Postle (2013) tested working memory for the direction of motion with three serially presented apertures, each containing a field of dots, colored red in one aperture, green in another, and blue in the third. On some trials only one field of dots was moving, on some two, and on some all three. Thus, memory load was manipulated, and, indeed, results of the GLM-based univariate analysis indicated that regions of rostral intraparietal sulcus (IPS) and several regions of PFC – caudal superior frontal gyrus and middle frontal gyrus, and more rostral regions of middle and inferior frontal gyri – demonstrated load sensitivity. Several years earlier, in the same issue of the journal *Nature* in which Vogel and Machizawa (2004) had published the first description of the CDA (*Figure 14.5*), Jay Todd and René Marois (2004) described an fMRI analogue: a region of IPS in which signal intensity increased with memory load up until the number of items corresponding to the subject's working memory capacity, but never exceeded this level at higher loads. I, myself, just a few years earlier, had written that load sensitivity is accepted as the "gold-standard" hallmark of a region's involvement in working memory storage of

stimulus information. MVPA, however, has called this assumption into question.

The working memory task used by Emrich, Riggall, LaRocque, and Postle (2013) tested recall by presenting a dial at the end of each trial and asking the subject to rotate the needle until it pointed in the remembered direction of motion. (Sometimes this procedure is referred to as "delayed estimation.") The color of the dial indicated whether direction of the red, the blue, or the green sample was to be recalled. The reason for using recall rather than recognition (recognition was used with the "change detection" tasks that we considered previously, for example, *Figure 14.4, Figure 14.5,* and *Figure 14.6*) is that recall provides a continuous measure of error, and this can be interpreted as a proxy for the precision of the representation being held in working memory. See *Figure 14.8* for an illustration.

An important point to emphasize about the recall vs. recognition distinction is that because recognition responses are binary (*Yes/No; Match/Nonmatch; Change/No Change*), one has little information from the plot in *Figure 14.4.B*, for example, about why performance is declining with increasing load. In contrast, the fit of the three-factor mixture model to the behavioral performance of subjects while being scanned by Emrich, Riggall, LaRocque, and Postle (2013) indicated that the precision of responses declined monotonically as memory load increased from 1 to 2 to 3. (Additionally, P_T dropped slightly, mirrored by an increase in P_{NT}, with increasing memory load.) For comparison, there would be no way of knowing from our back-of-the-envelope conversion of performance from *Figure 14.4.B* into estimates of capacity, whether the estimates of a capacity of 1 for a one-item array, a capacity of 1.92 for a two-item array, and a capacity of 2.82 for a three-item array were accompanied by progressive declines in the quality of the working memory representations. And so one can make a case that, in certain situations, the recall/estimation procedure provides more details about the workings of working memory. It is nonetheless also important to note that, for all the conceptual and mathematical sophistication of the three-factor mixture model, it's still restricted to evaluating the subject's responses. That's why we keep saying that these measures are only a *proxy* for what we want to know, which is the nature of the mental representation(s) while they are being held in working memory – that is, during the delay period.

Among the most intriguing results from the Emrich, Riggall, LaRocque, and Postle (2013) study was how MVPA decoding performance in occipital cortex tracked

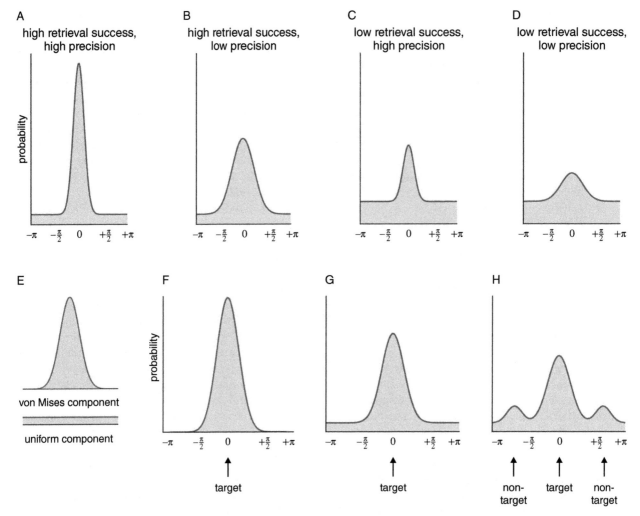

FIGURE 14.8 Modeling the precision of recall. Over the course of a testing session, a subject's responses will center on each trial's remembered direction of motion, and when s/he makes an error, small errors are more frequent than large errors. When each trial is shifted to 0°, to facilitate assessing overall performance at a glance, the distribution of responses is roughly Gaussian. **A.** A plot of the data from a hypothetical subject whose responses are very precise, as captured by the narrowness of the distribution of errors. **B.** A hypothetical subject whose responses are less precise than those of the subject in **A**, as indicated by the broader distribution of errors. **C.** A hypothetical subject whose responses that are centered on the target are just as precise as those of the subject in **A**, but who has also completely forgotten the direction of motion on several trials, and on these trials is just guessing. A true guess is equally likely to land anywhere along the possible range of 360°, so a high rate of guessing (i.e., of forgetting) manifests as an increase in the probability of a response at every possible value. **D.** This hypothetical subject performing poorly – their responses, on trials when they remembered the probed item, were imprecise, and their guess rate was also high. **E.** The tendencies displayed in panels **A–D** are descriptive. For example, one way to estimate the precision of subject A's memory vs. that of subject B is to compute 1/SD of their responses. But what about subject A in comparison to subject C? It wouldn't be accurate to say that their performance was comparable just because the precision of their responses to the probed item was comparable, because this wouldn't take into account the higher rate of forgetting in subject C. To quantify everyone's performance, so that "fair comparisons" can be made across all of them, the data from each subject can be construed as a mixture of three

the load-related decline in the precision of responses. Consider the subject whose data are portrayed in red in panel E of *Figure 14.9*. For load-of-1 trials, the precision of this subject's behavioral responses was high (a concentration parameter estimate of nearly 80), and MVPA decoding of the fMRI data from these trials was at 80% correct. For load-of-2 trials, the precision of this subject's behavioral responses declined to a concentration value of approximately 50, and MVPA decoding of the fMRI data from these trials declined to around 70% correct. For load-of-3 trials, the precision of this subject's behavioral responses declined even further, to a concentration value of about 43, and MVPA decoding of the fMRI data from these trials also declined further, to 65% correct. (*Concentration parameter* is defined in legend for panel E of *Figure 14.8*.) Although we don't understand exactly what MVPA decoding at 80% vs. 70% vs. 65% actually means in terms of the properties of a neural representation of a stimulus, we can take it as a proxy (there's that word again!) for the fidelity of the neural representation. With the caveat that this is a correlation, an interpretation of these results is that, with increasing load, the fidelity of delay-period neural representations of information declines, resulting in noisier "read-out" via end-of-trial recall.

Thus, the results from Emrich, Riggall, LaRocque, and Postle (2013) indicate that not only does working memory storage of visual perceptual information take place in occipital cortex, the fidelity of these occipital representations may also influence the precision of these memories. But what about IPS and PFC? Interestingly, in this study MVPA decoding of stimulus information was unsuccessful in IPS and posterior PFC, the regions that showed elevated,

load-sensitive delay-period activity. (Although more recent studies have shown that it can be possible to decode delay-period stimulus information from these areas, the decoding is generally weaker, and has never been shown with memory loads greater than 2.) If not the representation, per se, of stimulus information, what delay-period functions might account for the elevated activity in these regions? One possibility is the function of context binding. Recall that the testing procedure for this task was that the color of the response dial indicated which of the sample items was to be recalled. It's conceivable that, at the end of a trial, one could have a highly accurate memory for each of the three sample directions of motion, but have forgotten which one had been shown in which color. In this example, the color of the moving dots is considered "context," because the task isn't a color-memory task, it's a direction-of-motion task. Nonetheless, forgetting the association, the *binding*, between each direction and its color would produce an error if it were a test of recognition, or a response to a non-target (*Figure 14.8,* panel E) if it were a test of recall. (For the "change detection" tasks that we have considered previously [*Figure 14.4, Figure 14.5,* and *Figure 14.6*], each item's location on the screen was its critical context.) This author's group has tested the context-binding hypothesis with experiments that have varied memory load between one and three items, but also varied the domain of stimuli in the memory set. In one experiment, for example, subjects had to remember the orientation of one bar (a load-of-1 trial), the orientations of three bars (a load-of-3 "homogeneous" trial), or the orientation of one bar, the chrominance of one color patch, and the brightness of one luminance patch

---

FIGURE 14.8 (*Continued*) kinds of responses: responses corresponding to the subject's memory of the direction of the probed item (called "probability of a response to the target" [P_T]), guesses (P_G), and responses mistakenly made to one of the items in the memory set that was *not* probed (responses to a non-target [P_{NT}]). (Responses to a non-target are illustrated in panel **H**.) What panel **E** illustrates are the parameter that fits the distribution of responses to targets and responses to non-targets (the "von Mises component"), and the parameter that fits the distribution of guesses (the "uniform component"). A von Mises distribution is the analogue of a Gaussian for stimuli that exist in a circular space, such as the 360° of possible directions of motion on a 2-D plane. The width of the fitted von Mises provides the estimate of the precision of responses, and is referred to as the "concentration parameter." In the Emrich, Riggall, LaRocque, and Postle (2013) study, the partitioning of variance between P_T, P_{NT} and P_G – that is, the "mixture modeling" – was implemented with maximum likelihood estimation. **F.** Illustration of a model that only fits a von Mises function to the data. **G.** Illustration of a model that fits the data as a mixture of responses to the target, fit with a von Mises function, and of guesses, fit as a uniform distribution. **H.** Illustration of a model that fits the data as a mixture of responses to the target, of guesses, and of responses to non-targets. Source: From Franziska R Richter, Rose A Cooper, Paul M Bays, Jon S Simons,. "Distinct neural mechanisms underlie the success, precision, and vividness of episodic memory", University of Cambridge, United Kingdom, Oct 25, 2016. eLife Sciences Publications Ltd. Public Domain.

FIGURE 14.9 MVPA of delayed recognition of direction of motion.

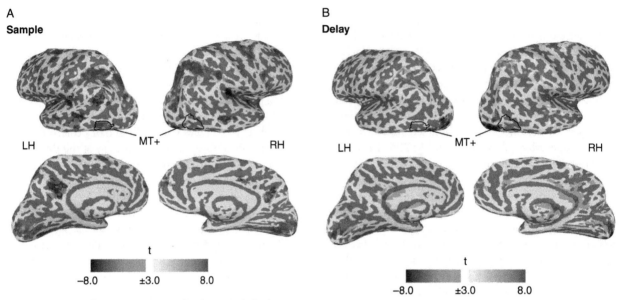

FIGURE 14.9.A Group-average univariate statistical maps of changes in fMRI activity, relative to baseline, when subjects are viewing and encoding into memory the direction of motion of a sample stimulus on a load-of-1 trial, from a study by Riggall and Postle (2012). On the lateral surface one can see elevated activity in occipital cortex, along the IPS, and in two patches of posterior PFC (note the similarity of these to the regions described in section *Human oculomotor control, Chapter 9,* and illustrated in *Figure 9.4*). Also note elevated activity is stronger in the left than in the right hemisphere in this sample, in MT+, which was identified in a separate localizer scan that alternated between coherently moving dots and stationary dots. On the medial surfaces one can see strong elevated activity in calcarine and pericalcarine cortex, and in a medial frontal area that may correspond to the supplementary eye field (SEF). Finally, among the areas showing a significant decrease relative to baseline are regions associated with the Default Mode Network: inferior parietal lobule, posterior cingulate/precuneus, and medial frontal cortex. Source: From "The Relationship between Working Memory Storage and Elevated Activity as Measured with Functional Magnetic Resonance Imaging" by Adam C. Riggall and Bradley R. Postle, The Journal of Neuroscience, September 19, 2012. Reproduced with permission of Society for Neuroscience.

FIGURE 14.9.B Group-average univariate statistical maps of changes in fMRI activity, relative to baseline, during the delay period, when subjects are remembering the item whose encoding is illustrated in load-of-1 trial, *Figure 14.9.A*. There are two overall impressions: First, activity remains elevated in the IPS and in posterior PFC, albeit with lower statistical values; and second, activity in occipital cortex, including calcarine cortex and MT+, has declined to baseline or lower-than-baseline levels. Source: From "The Relationship between Working Memory Storage and Elevated Activity as Measured with Functional Magnetic Resonance Imaging" by Adam C. Riggall and Bradley R. Postle, The Journal of Neuroscience, September 19, 2012. Reproduced with permission of Society for Neuroscience.

(a load-of-3 "heterogeneous" trial). The logic was that, if at the end of a load-of-3 "heterogeneous" trial, the subject remembers each of the three stimulus values but has forgotten where on the screen each was presented, they would still perform fine if their memory for the color patch was probed by a color wheel response dial. The finding from this and related studies is that delay-period signal in IPS and posterior PFC is highest for load-of-3 "homogeneous" trials, and markedly lower for load-of-3 "heterogeneous"

C

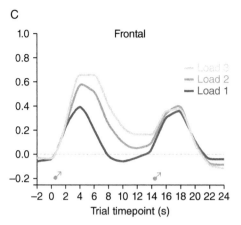

FIGURE 14.9.C Group- and trial-averaged activity from a different study – Emrich, Riggall, LaRocque, and Postle the Emrich and colleagues (2013) study that is described in the text – but from posterior PFC regions comparable to those shown in Figure 14.9.B. Note that encoding-related (4–6 seconds) and delay-related (10–14 seconds) activity shows a load effect, with higher signal intensity for higher loads. Arrow icons along the time axis indicate when moving-dot stimuli were on the screen. Source: From Emrich, Stephen M., Adam C. Riggall, Joshua J. LaRocque, and Bradley R. Postle. 2013. "Distributed Patterns of Activity in Sensory Cortex Reflect the Precision of Multiple Items Maintained in Visual Short-Term Memory." Journal of Neuroscience 33 (15): 6516–6523. doi: 10.1523/JNEUROSCI. 5732-12.2013.

D

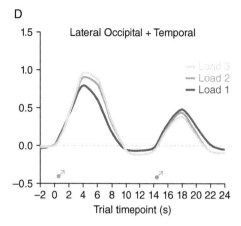

FIGURE 14.9.D Group- and trial-averaged activity from lateral occipital regions comparable to those shown in Figure 14.9.A, but also from the Emrich, Riggall, LaRocque, and Postle (2013) study that is described in the text. Note that although the encoding-related response is higher than that in posterior PFC, activity in these regions drops to just below baseline levels during the delay period.

E

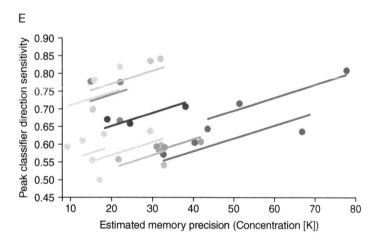

FIGURE 14.9.E Results from an analysis of covariance (ANCOVA) relating the behavioral precision of each individual subject to the fidelity of MVPA decoding of data from lateral occipital cortex at time 12 seconds (i.e., late in the delay period) from Emrich, Riggall, et al. (2013). Data points are color-coded by subject, with each subject's left-most point corresponding to load 3, the middle point to load 2, and the right-most point to load 1. For example, the behavioral precision (as plotted on the horizontal axis) of the subject shown in indigo is among the highest in this sample, but MVPA decoder performance for this subject starts at 65% correct and declines from there. The behavioral precision of the subject portrayed in the light violet, in contrast, is markedly lower, yet this subject's MVPA decoder performance is the highest for this sample. Despite this intersubject variability, the regression lines illustrate the common group-level effect that the load-related decline in MVPA performance relates to the load-related decline in behavioral precision with an r^2 = .35. Chance-level MVPA decoding for this analysis is 33.3%; behavioral precision is displayed as values of the concentration parameter of the three-factor mixture model illustrated in panel H of Figure 14.8.

trials, sometimes no different from load-of-1 trials. Therefore, a pattern of activity that had previously been interpreted as supporting a working memory storage function (e.g., Todd and Marois, 2004) may correspond, at least in part, to a different cognitive function: context binding.

The neural bases of visual working memory

The MVPA analyses from studies of fMRI data from Serences, Ester, Vogel, and Awh (2009; *Figure 14.7*), Harrison and Tong (2009), and Emrich, Riggall, LaRocque, and Postle (2013; *Figure 14.9*) are all consistent with the sensory-recruitment hypothesis, whereby the active retention, per se, of information in visual working memory occurs in the networks responsible for the perception of these stimuli and the features that make them up. And so this is consistent with the first of Hebb's dual traces, with the caveat that this process occurs at a grain of anatomical detail that is not detectible in bulk changes of signal intensity (as assessed with univariate methods). There are data from lesion studies and from extracellular neurophysiology that are consistent with this account, and are referenced as *Further Reading*.

But what about the sustained, elevated activity that many studies, in monkeys and in humans, have documented in PPC and PFC? For spatial working memory tasks, much of this likely reflects engagement of the dorsal frontoparietal network implicated in the endogenous control of attention (as reviewed in *Chapter 9*), and in functions such as context binding. Additionally, for feature- and object-based working memory, some insight comes from the MVPA of extracellular activity; read on!

Retrospective MVPA of single-unit extracellular recordings

In principle, the same caveats that we have considered in relation to the interpretation of univariate analyses of neuroimaging data also apply to single-unit electrophysiological data. That is, face-value interpretation of elevated activity in a neuron that has been identified as "coding for" a specific kind of information can also lead to flawed inference. There's a sense in which we saw this in *Chapter 10, The code for facial identity in the primate brain (!?!)*, in that Chang and Tsao (2017) showed with a PCA-based analysis (also a multivariate method) that neurons whose activity had previously been interpreted as responding to the identities of individual faces were in fact doing something quite different. Next, we'll see a similar demonstration of reevaluating the interpretation of the results from earlier univariate analyses, but with MVPA.

The traditional approach to in vivo extracellular electrophysiology is to begin each day's recording session by inserting an electrode into the microdrive, positioning it over the region of cortex that is of interest, and lowering it into the brain until a neuron is isolated. (It's painstaking work, and, when training the animal(s) is taken into account, it can take considerably longer than a year [sometimes many years] to acquire a complete set of data. In terms of the investment of blood, sweat, and tears – not to mention the PI's research funding – these datasets are priceless.) Seemingly at odds with this "single unit" approach, the rationale behind multivariate analyses is that the system under study (i.e., the brain) experiences the simultaneous activity of many processing elements all at the same time (e.g., the 62 voxels in the Serences, Ester, Vogel, and Awh [2009] V1 ROI, or the hundreds of "sensory" vs. "delay" voxels in the Emrich, Riggall, LaRocque, and Postle [2013] study), and that the overall influence of these patterns of activity is very likely to be different from what one could deduce by studying any single processing element in isolation. What is a neurophysiologist to do?

One approach that has yielded important insights is to retrospectively assemble all of the records of single neurons recorded while the animal performed identical trials and analyze them all together as though all had been recorded simultaneously during a single "virtual trial." Pioneers in this approach have been Ethan Meyers and his mentor Tomaso Poggio at MIT. In one study, they reanalyzed a dataset from Earl Miller's laboratory in which monkeys had viewed images of 3-D computer-generated animals and judged whether each was more cat-like or dog-like. The original study had compared neuronal activity in PFC vs. IT, and found that IT seemed "more involved in the analysis of currently viewed shapes, whereas the PFC showed stronger category signals, memory effects, and a greater tendency to encode information in terms of its behavioral meaning" (Freedman, Riesenhuber, Poggio, and Miller, 2003, p. 5235). The multivariate reanalysis, although generally confirming these broader patterns, uncovered novel, surprising information about the interpretation of sustained, elevated firing rates: the representation of stimulus category information in a delayed-match-to-category task "is coded by a nonstationary pattern of activity that changes over the course of a trial with individual neurons . . . containing information on much shorter time scales than the population as a whole" (Meyers et al., p. 1407). That is, this information was not carried for extended periods in time by any individual neuron, regardless of whether or not its activity level was elevated at a sustained level across the delay period.

Whoa! This observation, if true, would have major implications for how one interprets neuronal recordings. And lest one think that this may have been a spurious finding attributable to the retrospective nature of the analyses, a group led by (former Goldman-Rakic trainee) Matthew Chafee at the University of Minnesota reported an analogous phenomenon from simultaneous (i.e., "prospective") recordings acquired with multiunit arrays recording from posterior parietal cortex, illustrating that a *stable* mental representation of a part of an object is sustained by a *dynamic* succession of activity patterns that change on a 100 ms-by-100 ms timescale.

ACTIVITY? WHO NEEDS ACTIVITY?

This chapter has been based around the assumption that working memory requires that an active representation of the to-be-remembered stimulus needs to be held – somewhere and somehow – by the brain. This is seen in the studies from the 1990s to 2000s of delay-period activity in the PFC, in the studies of the CDA, and in MVPA studies of fMRI signals and of retrospectively assembled single-neuron signals. But we also know that other examples of memory, habits laid down by reinforcement learning (*Chapter 8*), MTL-dependent episodic long-term memories (*Chapter 11* and *Chapter 12*) and semantic memories (*Chapter 13*) don't depend on elevated activity. Rather, these "live" in our heads as passive, "silent" representations that are encoded in patterns of synaptic weights. Could the same be true for working memory? Certainly, there's no principled reason that I know of why a weight-based storage scheme that can hold a semantic memory for 20 years or longer wouldn't also be able to store information about a visual array, or a string of digits, for 20 seconds, or for 2 seconds. All that one would need would be a mechanism for rapid adjustment of synaptic weights.

Such mechanisms do exist in many varieties. More-or-less instantaneous adjustment is seen, for example, as long-term depression at synapses between parallel fibers and Purkinje cells in the cerebellum (section *Hebbian plasticity, Chapter 8*), and PFC-striatal synapses implicated in acquisition and refinement of habitual behaviors (section *Habits and reinforcement learning, Chapter 8*). Importantly, however, these mechanisms entail "adjustments," the gradual sculpting of motor behaviors that are performed thousands of times across the lifetime. But if I need to remember how many left turns I still need to make in order to find the rental-car returns counter in the massive parking structure of an unfamiliar airport (and I'm running late and

worried about missing my flight!), a gradual, "refining" process won't work. There would need to be a mechanism for creating robust changes in broadly distributed networks of synapses that would last for as long as needed, but no longer than needed. (This is why working memory is so useful, I'll never again in my life need to remember the specific route from the entrance gate to the drop-off booth of that rental car agency at that particular airport, so why clog my LTM with an enduring representation of it?) Well, we're in luck, because mechanisms like this also exist!

The general phenomenon is called short-term synaptic plasticity (STSP). One specific example is "spike time-dependent plasticity" (STDP) which, for our purposes, we can think of it as a quasi-Hebbian phenomenon whereby a synapse will be strengthened if the action potential propagated by the presynaptic neuron precedes an action potential fired by the postsynaptic neuron by a few milliseconds to tens of milliseconds, and it weakens if the action potential propagated by the presynaptic neuron lags behind the action potential of the postsynaptic neuron within a similarly narrow time window. Importantly, these effects can be relatively transient, and so synapses modified by STDP can return to their baseline state after a relatively short period of time. The details of how/when/where/and under what conditions STDP works are beyond the scope of this book, but the point for now is to know that there is a biological basis for building STSP into computational simulations.

Motivated in part by results from recent fMRI studies, a team led by David Freedman at the University of Chicago and Xiao-Jing Wang and New York University developed a recurrent neural network (RNN) to model performance on tests of working memory for the direction of motion, similar to what we've reviewed in this chapter's section on *Increased specificity raises questions about the interpretation of elevated delay-period activity*. The network by Masse et al. (2019) can be thought of conceptually, if not in implementational detail, as a hybrid with some properties similar to the PDP network that we considered in depth in *Chapter 3* (section *A PDP model of picture naming; Figure 3.21.A*), and some properties similar to the DNN that we considered in *Chapter 6* (section *Using deep neural networks to understand computations underlying object recognition, Figure 6.11.A*). It's similar to the PDP model in that it has three layers (input, hidden, and output), and each simulated trial is initiated by presenting a "hand-coded" sample stimulus to the input layer, that is, the input layer is made up of 24 units, each corresponding to one of the 24 possible directions of motion that would be used in the simulation. (The input layer of the DNN from *Chapter 6*, in contrast, acted like a retina, and getting to the point of being able to "perceive" what direction was contained in the

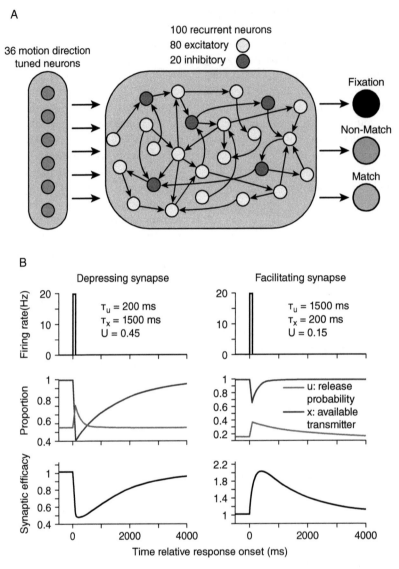

FIGURE 14.10 An RNN model of visual working memory. **A.** The architecture of the model. Of the 80 excitatory neurons and 20 inhibitory neurons in the hidden layer, half from each population had depressing synapses and half had facilitating synapses. **B.** For depressing synapses (left-side column), a brief pulse of incoming activity (i.e., activation from another unit in the model; top panel) produced a transient-then-rapidly-decaying increase in neurotransmitter utilization (red trace), accompanied by a large and slow-to-recover decrease in available neurotransmitter (blue trace; middle panel), which resulted in an abrupt drop and slow recovery of synaptic efficacy (bottom panel). For facilitating synapses (right-hand column), a brief pulse of incoming activity (i.e., activation from another unit in the model; top panel) produced a long-lasting increase in neurotransmitter utilization (red trace), accompanied by a small and rapidly recovering decrease in available neurotransmitter (blue trace, middle panel), which resulted in a rapid increase in synaptic efficacy followed by a slow return to baseline (bottom panel). The parameters determining these properties were T_u, STSP neurotransmitter utilization; T_x, STSP neurotransmitter time constant; U, STSP neurotransmitter increment. Source: From Masse, N. Y., Yang, G. R., Song, H. F., Wang, X.-J. and Freedman, D. J. (2019). Circuit mechanisms for the maintenance and manipulation of information in working memory. Nature Neuroscience 22: 1159–1167. Reproduced with permission of Nature Publishing Group.

sample would be the whole point of a simulation with that kind of DNN.) It's similar to the DNN in that it has built-in elements intended to simulate specific neuronal properties (processing steps like convolution and pooling for the DNN; different subtypes of units that show depressing vs. facilitating plasticity for this RNN), and in that multivariate decoding of values from the network is used to assess the state of stimulus representation (*Figure 14.10*).

For the delayed-recognition task, Masse et al. (2019) trained 20 models (to account for the stochastic nature of the outcome of any one simulation) and, once each was trained and performing the task at a high level of accuracy, evaluated *how* they achieved this level of performance in two ways. First, they used multivariate decoding to assess the representation of stimulus information in (a) the (100-D)

patterns of activity of neurons in the hidden layer, and (b) the (100-D) patterns of synaptic "efficacies". (To equate the number of dimensions for these two properties of the network, the authors summarized across the weights of every connection to any given neuron to compute a single summary value for that neuron's "synaptic efficacy.") Second, they causally tested the necessity of activity patterns and of synaptic patterns by shuffling the activation values but leaving synaptic values intact just before output to the output layer, or shuffling synaptic values and leaving activation values intact (*Figure 14.11*). The results, remarkably, indicated that stimulus information encoded in synaptic weights was more important for successful performance of the working memory task than was stimulus information encoded in patterns of activity!

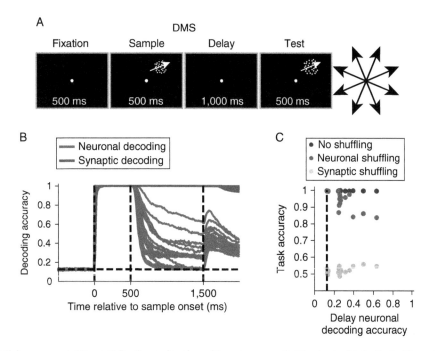

FIGURE 14.11 Activity vs. synaptic contributions visual working memory. **A.** The task. (Note that although the authors call this a "DMS" task (**delayed match-to-sample**), this is incorrect. DMS would be if the sample plus a foil are presented at test, and the subject has to, well, match to the sample. Rather, this is **delayed recognition**. (*Is the probe the same as the sample? Yes or No.*)) **B.** Time course of moment-by-moment decoding accuracy, plotted for all 20 iterations of training, indicating that decoding accuracy of "activity-silent" patterns of synaptic weights (pink) allows for perfect discrimination of each of the 24 possible directions of motion, at any point during the trial. The same is not true for patterns of activity: although decoding accuracy (green) is at 100% while the sample is on the screen, it drops off during the delay, ranging anywhere from roughly 40% for the best-performing simulation all the way down to chance-level performance for several of the poorest-performing simulations. **C.** Task accuracy. This panel displays mean proportion correct performance, comparable to what is displayed, for example, in *Figure 14.4.B*, plotted against neuronal decoding accuracy, for each of the 20 simulations. Indigo dots show that performance was near perfect for each of the simulations when no perturbation was applied. Red dots show that the effect of randomly shuffling activity values among the 100 neurons in the hidden layer had only a negligible effect on performance accuracy. Cyan dots, in contrast, show that shuffling synaptic efficacies drops performance to chance (50% correct). Source: Adapted from Masse, N. Y., Yang, G. R., Song, H. F., Wang, X.-J. and Freedman, D. J. (2019). Circuit mechanisms for the maintenance and manipulation of information in working memory. Nature Neuroscience 22: 1159–1167. Nature Publishing Group.

Before leaving this study, it's important to note that Masse et al. (2019) simulated many different working memory tasks, this one-item delayed-recognition task being simplest. The gist from the totality of their findings is that task demands have an important influence on determining whether patterns of synaptic weights or patterns of activity are relatively more important for task performance. STSP is sufficient to support performance on tasks that only require the simple retention of sample stimulus information across of brief delay (such as illustrated in Figure 14.10), whereas elevated activity may not be. As tasks require increasing levels of stimulus transformation and/or control to execute the correct response, however, the importance of STSP declines, and that of elevated activity increases. Examples of tasks that require elevated activity include tasks for which a correct recognition probe will be one whose direction of motion is rotated by 90° relative to the direction of the sample pattern, or for which two items are presented, and only during the delay period will a "retrodictive" cue (opposite of "predictive") indicate which of the two will be tested. So it's a complicated state of affairs, but the Masse et al. (2019) simulations provide yet another demonstration that *you can't play 20 questions with the brain and win.*

FOUR-SCORE AND A HANDFUL OF YEARS (AND COUNTING)

We've covered research spanning from 1936 to 2019, from massive bilateral PFC lesions to sophisticated computational simulations. Over the course of those 80+ years, we have learned a great deal about how people (and monkeys) hold information in mind for short periods of time in order to guide their behavior. One of the themes that we've emphasized – that there's an important difference between "just" holding information in mind (which may not depend on the PFC, and may not even depend on elevated activity) and performing various mental operations with, or on, that information – offers a nice segue into our next chapter, because working memory is a critical contributor to cognitive control.

END-OF-CHAPTER QUESTIONS

1. What does the "short" in short-term memory refer to?

2. Were the interpretations that Fuster and Alexander (1971) and that Kubota and Niki (1971) gave to their electrophysiological findings consistent with the lesion data that had come before? How, or how not?

3. How did the development of the multicomponent model of working memory, from cognitive psychology, influence some people's interpretation of the findings from Question 2?

4. Summarize the findings from Funahashi, Bruce, and Goldman-Rakic (1989) and Wilson, Ó Scalaidhe, and Goldman-Rakic (1993), and how subsequent studies have been designed to evaluate and, in some cases, reinterpret them.

5. What is the CDA? What distinctive property makes it particularly interesting to students of visual working memory?

6. In what ways have recent MVPA analyses of fMRI studies of working memory raised questions about the interpretation of earlier studies analyzed with univariate methods? (Hint: sensitivity; specificity.)

7. If an fMRI scan shows that the level of activity in region X does not differ from baseline during the delay period of a working memory task, does this mean that the region is not involved in working memory storage?

8. Is elevated activity at the level of networks of neurons necessary for working memory storage? If not, what other mechanism could carry out this function?

REFERENCES

Baddeley, Alan D., and Graham Hitch. 1974. "Working Memory." In *The Psychology of Learning and Motivation: Advances in Research and Theory*, vol. 8, edited by Gordon H. Bower, 47–89. New York: Academic Press.

Carlisle, N. B., J. T. Arita, D. Pardo, and G. F. Woodman. 2011. "Attentional Templates in Visual Working Memory." *The Journal of Neuroscience* 31: 9315–9322.

Emrich, Stephen M., Adam C. Riggall, Joshua J. LaRocque, and Bradley R. Postle. 2013. "Distributed Patterns of Activity in Sensory Cortex Reflect the Precision of Multiple Items Maintained in Visual Short-Term Memory." *Journal of Neuroscience* 33 (15): 6516–6523. doi: 10.1523/JNEUROSCI.5732-12.2013.

Emrich, S. M., N. Al-Aidroos, J. Pratt, and S. Ferber. 2009. "Visual Search Elicits the Electrophysiological Marker of Visual Working Memory." *PLoS One* 4: e8042.

Freedman, David J., Maximilian Riesenhuber, Tomaso Poggio, and Earl K. Miller. 2003. "A Comparison of Primate Prefrontal and Inferior Temporal Cortices during Visual Categorization." *Journal of Neuroscience* 23 (12): 5235–5246.

Funahashi, Shintaro, Charles J. Bruce, and Patricia S. Goldman-Rakic. 1989. "Mnemonic Coding of Visual Space in the Monkey's Dorsolateral Prefrontal Cortex." *Journal of Neurophysiology* 61 (2): 331–349.

Fuster, Joaquin M., and Garrett E. Alexander. 1971. "Neuron Activity Related to Short-Term Memory." *Science* 173 (3997): 652–654. doi: 10.1126/science.173.3997.652.

Ghent, Lila, Mortimer Mishkin, and Hans-Lukas Teuber. 1962. "Short-term Memory after Frontal-Lobe Injury in Man." *Journal of Comparative and Physiological Psychology* 55 (5): 705–709. doi: 10.1037/h0047520.

Hakim, N., and E. K. Vogel. 2018. "Phase-Coding Memories in Mind." *PLoS Biology* 16: e3000012.

Harrison, Stephenie A., and Frank Tong. 2009. "Decoding Reveals the Contents of Visual Working Memory in Early Visual Areas." *Nature* 458 (7238): 632–635. doi: 10.1038/nature07832.

Hebb, Donald Olding. 1949. *The Organization of Behavior: A Neuropsychological Theory*. New York: Wiley.

Jacobsen, C. F. 1936. "The Functions of the Frontal Association Areas in Monkeys." *Comparative Psychology Monographs* 13: 1–60.

Johnson, M. K., R. P. McMahon, B. M. Robinson, A. N. Harvey, B. Hahn, C. J. Leonard, S. J. Luck, and J. M. Gold. 2013. "The Relationship between Working Memory Capacity and Broad Measures of Cognitive Ability in Healthy Adults and People with Schizophrenia." *Neuropsychology* 27: 220–229.

Kubota, Kisou, and Hiroaki Niki. 1971. "Prefrontal Cortical Unit Activity and Delayed Alternation Performance in Monkeys." *Journal of Neurophysiology* 34 (3): 337–347.

Lebedev, Mikhail A., Adam Messinger, Jerald D. Kralik, and Steven P. Wise. 2004. "Representation of Attended versus Remembered Locations in Prefrontal Cortex." *PLoS Biology* 2 (11): e365. doi: 10.1371/journal.pbio.0020365.

Luck, S. J., and E. K. Vogel. 1997. "The Capacity of Visual Working Memory for Features and Conjunctions." *Nature* 390: 279–281.

Luck, S. J. and E. K. Vogel. 2013. "Visual Working Memory Capacity: From Psychophysics and Neurobiology to Individual Differences." *Trends in Cognitive Sciences* 17: 391–400.

Malmo, Robert B. 1942. "Interference Factors in Delayed Response in Monkeys after Removal of Frontal Lobes." *Journal of Neurophysiology* 5 (4): 295–308.

Masse, N. Y., G. R. Yang, H. F. Song, X.-J. Wang, and D. J. Freedman. 2019. "Circuit Mechanisms for the Maintenance and Manipulation of Information in Working Memory." *Nature Neuroscience* 22: 1159–1167. https://doi.org/10.1038/s41593-019-0414-3.

Meyers, Ethan M., David J. Freedman, Gabriel Kreiman, Earl K. Miller, and Tomaso Poggio. 2008. "Dynamic Population Coding of Category Information in Inferior Temporal and Prefrontal Cortex." *Journal of Neurophysiology* 100 (3): 1407–1419. doi: 10.1152/jn.90248.2008.

Milner, Brenda. 1964. "Some Effects of Frontal Lobectomy in Man." In *The Frontal Granular Cortex and Behavior*, edited by John M. Warren and Konrad Akert, 313–334. New York: McGraw-Hill.

Miller, George Armitage. 1956. "The Magical Number Seven, Plus or Minus Two: Some Limits on Our Capacity for Processing Information." *Psychological Review* 63 (2): 81–97. doi: 10.1037/h0043158.

Postle, Bradley R. 2009. "The Hippocampus, Memory, and Consciousness." In *Neurology of Consciousness*, edited by Steven Laureys and Giulio Tononi, 326–338. Amsterdam: Elsevier.

Pribram, Karl H., Albert J. Ahumada Jr., J. Hartog, and L. Roos. 1964. "A Progress Report on the Neurological Processes Distributed by Frontal Lesions in Primates." In *The Frontal Granular Cortex and Behavior*, edited by John M. Warren and Konrad Akert, 28–55. New York: McGraw-Hill.

Rao, S. Chenchal, Gregor Rainer, and Earl K. Miller. 1997. "Integration of What and Where in the Primate Prefrontal Cortex." *Science* 276 (5313): 821–824. doi: 10.1126/science.276.5313.821.

Richter, F. R., R. A. Cooper, P. M. Bays, and J. S. Simons. 2016. "Distinct Neural Mechanisms Underlie the Success, Precision, and Vividness of Episodic Memory." *eLife* 5: e18260. doi: 10.7554/eLife.18260.

Riggall, Adam C., and Bradley R. Postle. 2012. "The Relationship between Working Memory Storage and Elevated Activity as Measured with Functional Magnetic Resonance Imaging." *Journal of Neuroscience* 32 (38): 12990–12998. doi: 10.1523/JNEUROSCI.1892-12.2012.

Serences, John T., Edward F. Ester, Edward K. Vogel, and Edward Awh. 2009. "Stimulus-Specific Delay Activity in Human Primary Visual Cortex." *Psychological Science* 20 (2): 207–214. doi: 10.1111/j.1467-9280.2009.02276.x.

Vogel, Edward K., and Maro G. Machizawa. 2004. "Neural Activity Predicts Individual Differences in Visual Working Memory Capacity." *Nature* 428 (6984): 679–782. doi: 10.1038/nature02447.

Wilson, Fraser A. W., Séamas P. Ó Scalaidhe, and Patricia S. Goldman-Rakic. 1993. "Dissociation of Object and Spatial Processing Domains in Primate Prefrontal Cortex." *Science* 260 (5116): 1955–1958.

OTHER SOURCES USED

Balaban, H., and R. Luria. 2019. "Using the Contralateral Delay Activity to Study Online Processing of Items Still within View." *Neuromethods*. 10.1007/7657_2019_22.

Crowe, David A., Bruno B. Averbeck, and Matthew V. Chafee. 2010. "Rapid Sequences of Population Activity Patterns Dynamically Encode Task-Critical Spatial Information in Parietal Cortex." *Journal of Neuroscience* 30 (35): 11640–11653. doi: 10.1523/JNEUROSCI.0954-10.2010.

Erickson, Martha A., Lauren A. Maramara, and John Lisman. 2010. "A Single Brief Burst Induces GluR1-Dependent Associative Short-Term Potentiation: A Potential Mechanism for Short-Term Memory." *Journal of Cognitive Neuroscience* 22 (11): 2530–2540. doi: 10.1162/jocn.2009.21375.

Funahashi, Shintaro, Charles J. Bruce, and Patricia S. Goldman-Rakic. 1990. "Visuospatial Coding in Primate Prefrontal Neurons Revealed by Oculomotor Paradigms." *Journal of Neurophysiology* 63 (4): 814–831.

Funahashi, Shintaro, Matthew V. Chafee, and Patricia S. Goldman-Rakic. 1993. "Prefrontal Neuronal Activity in Rhesus Monkeys Performing a Delayed Anti-Saccade Task." *Nature* 365 (6448): 753–756. doi: 10.1038/365753a0.

Jensen, Ole, Jack Gelfand, John Kounios, and John E. Lisman. 2002. "Oscillations in the Alpha Band (9–12 Hz) Increase with Memory Load during Retention in a Short-Term Memory Task." *Cerebral Cortex* 12 (8): 877–882. doi: 10.1093/cercor/12.8.877.

Jerde, Trenton A., Elisha P. Merriam, Adam C. Riggall, James H. Hedges, and Clayton E. Curtis. 2012. "Prioritized Maps of Space in Human Frontoparietal Cortex." *Journal of Neuroscience* 32 (48): 17382–17390. doi: 10.1523/JNEUROSCI.3810-12.2012.

Miller, George Armitage, Eugene Galanter, and Karl H. Pribram. 1960. *Plans and the Structure of Behavior*. New York: Holt.

Miller, Earl K., Cynthia A. Erickson, and Robert Desimone. 1996. "Neural Mechanisms of Visual Working Memory in Prefrontal Cortex of the Macaque." *Journal of Neuroscience* 16 (16): 5154–5167.

Miller, Earl K., Lin Li, and Robert Desimone. 1991. "A Neural Mechanism for Working and Recognition Memory in Inferior Temporal Cortex." *Science* 254 (5036): 1377–1379. doi: 10.1126/science.1962197.

Miller, Earl K., and Robert Desimone. 1994. "Parallel Neuronal Mechanisms for Short-Term Memory." *Science* 263 (5146): 630–522. doi: 10.1126/science.8290960.

Mongillo, Gianluigi, Omri Barak, and Misha Tsodyks. 2008. "Synaptic Theory of Working Memory." *Science* 319 (5869): 1543–1546. doi: 10.1126/science.1150769.

Nakamura, Katsuki, and Kisou Kubota. 1995. "Mnemonic Firing of Neurons in the Monkey Temporal Pole during a Visual Recognition Memory Task." *Journal of Neurophysiology* 74 (1): 162–178.

Niki, Hiroaki. 1974. "Differential Activity of Prefrontal Units during Right and Left Delayed Response Trials." *Brain Research* 70 (2): 346–349. doi: 10.1016/0006-8993(74)90324-2.

Niki, Hiroaki, and Masataka Watanabe. 1976. "Prefrontal Unit Activity and Delayed Response: Relation to Cue Location *versus* Direction of Response." *Brain Research* 105 (1): 79–88. doi: 10.1016/0006-8993(76)90924-0.

Reinhart, R. M., and Woodman, G. F. 2014. "High Stakes Trigger the Use of Multiple Memories to Enhance the Control of Attention." *Cerebral Cortex* 24: 2022–2035.

Rushworth, Matthew F. S., Philip D. Nixon, Madeline J. Eacott, and Richard E. Passingham. 1997. "Ventral Prefrontal Cortex Is Not Essential for Working Memory." *Journal of Neuroscience* 17 (12): 4829–4838.

Suzuki, Wendy A., Earl K. Miller, and Robert Desimone. 1997. "Object and Place Memory in the Macaque Entorhinal Cortex." *Journal of Neurophysiology* 78 (2): 1062–1081.

Warren, John M., and Konrad Akert, eds. 1964. *The Frontal Granular Cortex and Behavior*. New York: McGraw-Hill.

FURTHER READING

Meyers, Ethan, and Gabriel Kreiman. 2011. "Tutorial on Pattern Classification in Cell Recording." In *Visual Population Codes: Toward a Common Multivariate Framework for Cell Recording and Functional Imaging,* edited by Nikolaus Kriegeskorte and Gabriel Kreiman, 517–538. Cambridge, MA: MIT Press.
Details on Meyers' and Kreiman's method for retrospective MVPA of single-unit datasets.

Pasternak, Tatiana, and Mark W. Greenlee. 2005. "Working Memory in Primate Sensory Systems." *Nature Reviews Neuroscience* 6 (2): 97–107. doi: 10.1038/nrn1603.
Extensive review of evidence, from monkey neurophysiology and human psychophysics, for the sensory-recruitment hypothesis of STM.

Postle, Bradley R. 2006. "Working Memory as an Emergent Property of the Mind and Brain." *Journal of Neuroscience* 139 (1): 23–38. doi: 10.1016/j.neuroscience.2005.06.005.
Review covering much of the univariate analysis–based neuroimaging literature on human working memory that this chapter skips over.

SECTION IV
HIGH-LEVEL COGNITION

INTRODUCTION TO SECTION IV
HIGH-LEVEL COGNITION

Who's to say what's "high-level cognition" vs. "mental representation"? Isn't sorting out whether it was Brazil or Uruguay who won the first World Cup, or whether it was Graham Greene or Evelyn Waugh who wrote *Our Man in Havana*, high-level cognition? Sure. But they're also examples of semantic cognition. Isn't the endogenous control of attention a form of cognitive control? Of course it is. So, as we have noted, is working memory. The labels aren't so important. The rationale behind this book's organization is that *Section II* provided a necessary foundation for exploring different facets of mental representation (i.e., *Section III*) in a substantive way, and *Section II* and *Section III* have provided the background and framework needed for us to do the same with the topics to be covered here. None of them won't require a good understanding of perception, of attention, of motor control, and of memory. (Nor, for that matter, of oscillatory coherence, of fMRI, of neuropsychology, of computational modeling, and so on.) Further, the chapters in this section will build on each other: principles from cognitive control will apply to decision-making, emotions, social behavior, and language; principles from decision-making will apply to emotions and social behavior; and so on.

It's THE NEURAL BASES OF HIGH-LEVEL COGNITION, people! Let's go!

CHAPTER 15
COGNITIVE CONTROL

KEY THEMES

- Cognitive control refers to the ability to guide behavior and prioritize information processing in accordance with internal goals, regardless of the salience of goal-irrelevant stimuli in the environment or of goal-irrelevant behavioral choices.

- The environmental-dependency syndrome resulting from lateral PFC damage produces behavioral disinhibition that is a dramatic illustration of failures of cognitive control.

- A cardinal laboratory-based operationalization of impaired cognitive control is perseverative responding on the Wisconsin Card-Sorting Test.

- ERPs from patients with lateral PFC lesions who are performing a working memory task reveal two distinctive patterns: an abnormally reduced N1 to the target stimulus, suggesting impaired attentional selection/encoding; and abnormally large medium-latency AEPs to delay-period distracting stimuli, suggesting impaired filtering of task-irrelevant information.

- A computational simulation of the A-not-B error, a perseverative response in infants attributed to an immature PFC, suggests that it may arise from ineffective top-down selection of the weaker, but behaviorally appropriate, response, rather than

a failure to inhibit the stronger, but inappropriate response.

- The PFC has unique properties that may underlie its central role in cognitive control: the ability to transition between an "up" state that sustains an active pattern of activity and a "down" state during which it can update to a new pattern; heavy innervation by midbrain dopamine neurons that carry reward-related information; and extensive anatomical connectivity through which it can implement control.

- The dynamics of ascending dopamine signals, including the reward prediction error (RPE) signal, are of central importance for triggering state transitions in PFC and providing performance-related information to performance-monitoring circuits in the dorsomedial frontal cortex (dmFC).

- Rather than contributing to behavior directly, dmFC seems to track the current level of performance and acts as a governor on some aspects of PFC activity, including whether to stay in a more effortful proactive mode of control, or to revert to a reactive mode.

- As a consequence of its unique connectivity, combined with the effects of dopamine-mediated reinforcement learning signals, the dorsolateral PFC can support such "meta" cognitive functions as learning to learn.

CONTENTS

Cognitive control governs much of our lives, much of the time, often in seemingly trivial ways that nonetheless add up to us conducting our daily lives much more efficiently and efficaciously. Often we are not aware of its operation until it fails. A classic example for this writer is stepping onto the elevator in the Psychology building, lifting his eyes off the document he's reading long enough to press "5," and then going back to the document. The elevator rises, stops, the doors open, and he steps out. Only to find himself on the fourth floor. Here is how two leading researchers (one of them Earl Miller, from the previous chapter) introduced a very influential paper on the topic:

> One of the fundamental mysteries of neuroscience is how coordinated, purposeful behavior arises from the distributed activity of billions of neurons in the brain. Simple behaviors can rely on relatively straightforward interactions between the brain's input and output systems. Animals with fewer than a hundred thousand neurons (in the human brain there are 100 billion or more neurons) can approach food and avoid predators. For animals with larger brains, behavior is more flexible. But flexibility carries a cost: although our elaborate sensory and motor systems provide detailed information about the external world and make available a large repertoire of actions, this introduces greater potential for interference and confusion. The richer information we have about the world and the greater number of options for behavior require appropriate attentional, decision-making, and coordinative functions, lest uncertainty prevail. To deal with this multitude of possibilities and to curtail confusion, we have evolved mechanisms that coordinate lower-level sensory and motor processes along a common theme, an internal goal. (Miller and Cohen, 2001, pp. 167–168)

In the elevator anecdote, the internal goal of our protagonist was to get to his office (located on the fifth floor), the detailed sensory information that was driving his behavior was being displayed on the electronic tablet in his hand. (The ubiquity of "smartphones" and other mobile gadgets in current society has made the value of cognitive control in everyday life all the more apparent.) As this anecdote suggests, cognitive control often implicates the idea of inhibition – the absent-minded professor's attention focused elsewhere, he failed to inhibit the prepotent reflex of walking through the just-opened doors. Whether or not control truly engages inhibition, at a mechanistic level, will be a theme that will recur here, and in subsequent chapters.

As with many other domains of human behavior, behavioral neurology has provided a rich set of cases, and

theories, that provide a backdrop for contemporary thinking about cognitive control.

THE LATERAL FRONTAL-LOBE SYNDROME

In reproducing the quote from Miller and Cohen (2001), above, I left out the following, because I want to be careful not to advocate a "one-brain-area = one-kind-of-behavior" style of thinking. To exclude it altogether, however, would be to mischaracterize the received wisdom on this subject: "This ability for cognitive control no doubt involves neural circuitry that extends over much of the brain, but it is commonly held that the prefrontal cortex (PFC) is particularly important" (p. 168). This emphasis on the PFC derives, in part, from a large body of undeniably compelling evidence from behavioral neurology. The French neurologist François Lhermitte (1921–1998) captured this succinctly in companion papers entitled *Human Autonomy and the Frontal Lobes*. "The patients' behavior was striking," he writes, "as though implicit in the environment was an order to respond to the situation in which they found themselves" (Lhermitte, Pillon, and Serdaru, 1986, p. 335). *"Human autonomy"?!? I thought this book was about neuroscience. This isn't a trick to lure us into reading a treatise on existentialism, is it?* Well, no. On the other hand, from here on out there's not a topic that we'll consider that doesn't have clear and profound implications for what it is that makes us uniquely human.

Environmental-dependency syndrome

Building on the well-characterized phenomenon of **utilization behavior** – an exaggerated tendency for one's behavior to be determined by the external environment – Lhermitte, Pillon, and Serdaru (1986) carried out a study in which the experimenter simply sat opposite each subject and, with a "completely neutral and indifferent" demeanor, performed a series of gestures: common body gestures (e.g., crossing legs, tapping fingers); symbolic gestures (e.g., thumbing one's nose, military salute); manipulating objects (folding a piece of paper and putting it into an envelope, combing hair); and so on. No advance explanation or instructions were provided. The description of the response of neurologically intact individuals is not only hilarious but also provides context that helps one appreciate just how striking, and profound, are the impairments of the patients. And why Lhermitte invokes autonomy in relation to the frontal-lobe syndrome:

> Normal subjects never imitated the examiner. They were unconcerned but surprised, without otherwise making the slightest remark. . . . When asked why the examiner had performed these gestures, they hesitated and replied: "To test me"; "I don't know – perhaps to see my reactions"; and so on. The answers varied little with personality or age. When the examiner asked them if it had crossed their mind to imitate him, their answer was: "No, not at all." Boys and girls between the ages of 12 and 16 reacted by laughing and calling the examiner a clown. Children between the ages of 5 and 6 . . . later [said] "The doctor was very nice, but it's funny how bad-mannered he is; he thumbed his nose at all of us." Children 2 to 4 years old sometimes took an object – a ball, for instance – and threw like the examiner [had thrown], but they were merely playing with him. (Lhermitte, Pillon, and Serdaru, 1986, p. 329)

In contrast, almost all patients with frontal-lobe damage began imitating the experimenter with his first gesture. They did this seemingly without surprise and with careful attention to detail. For example, when pretending to light a candle, no imitating patient forgot to "blow out the match." Later, when interviewed about the experiment, all patients could remember all of the gestures. Strikingly, "After being told not to imitate, most patients [nonetheless continued to] display the same IB" (p. 329). (With this study, Lhermitte, Pillon, and Serdaru were describing a "new" variant of utilization behavior: "imitation behavior" [IB].) IB was found in 28 of 29 patients with frontal lobe lesions in the study's sample. The 29th, recently diagnosed with a right frontal glioma (a tumor originating in glial cells), began to spontaneously exhibit IB 3 weeks later. Overall intellectual functioning of these patients was numerically, but not statistically, lower than that of control subjects. (In the 1940s, Hebb had reported that patients with PFC lesions can have normal IQ scores.) The frontal patients had varied damage, with inferior frontal gyrus involved in the majority of the patients. In contrast, IB was only seen in 1 of 21 non-frontal lesion control patients (and the factors behind this one patient's performance were characterized as "doubtful").

Lhermitte suggested the term "environmental-dependency syndrome" for even more exotic behavior by patients that he observed outside of the laboratory. One, who had undergone a large left-hemisphere frontal lobectomy to remove a glioma, was invited to the doctor's apartment. Upon walking past the bedroom and seeing the bed, the patient undressed, climbed into bed, and prepared to sleep! Subsequently, other neurologists have adopted "environmental-dependency syndrome" as a blanket term encompassing all these types of behavior. In summarizing them, two other behavioral neurologists, Robert Knight and Mark D'Esposito (2003), have noted that:

> It has been postulated that the frontal lobes may promote distance from the environment and the parietal lobes foster approach toward one's environment. Therefore, the loss of frontal inhibition may result in overactivity of the parietal lobes. Without the PFC, our autonomy in our environment would not be possible. A given stimulus would automatically call up a predetermined response regardless of context. (p. 262)

Next, let's consider how inhibition has also been implicated in a classic laboratory finding in lateral frontal-lobe patients – perseveration on the Wisconsin Card-Sorting Task (WCST).

Perseveration

In the late 1950s, decades of research on the effects of various brain lesions on cognitive flexibility had yielded contradictory and equivocal results, with some studies implicating the frontal lobes, some posterior regions, and some one hemisphere over another. In 1963, this debate was resolved, and a new avenue for research opened up, by a single study. Brenda Milner (of course) had the sagacity to realize that tasks that had been used in previous studies had the dual drawbacks of lacking sensitivity and lacking clearly defined psychological underpinnings. She had the insight to select a task that was likely to yield clearer results. And she had the benefit of working at the Montreal Neurological Institute (MNI), where she could test neuro-surgical patients pre- and postoperatively – a repeated-measures design, of course, affording much greater statistical power.

The WCST was designed by (University of Wisconsin-based) psychologist David Grant (1916–1977) and his graduate student Esta Berg, who modeled it in part on a task by the German psychologist Egon Weigl and in part on problem-solving tasks developed for monkeys, and shown to be sensitive to frontal-lobe damage, by Grant's colleague Harry Harlow (1905–1981; developer of the Wisconsin General Testing Apparatus [WGTA; *Figure 6.2.A*]). As illustrated in *Figure 15.1*, the WCST consists of a stack of cards for which subjects must first learn, and then apply, a sorting rule. Milner tested 53 patients pre- and post-op: 18 with

FIGURE 15.1 The WSCT, as the subject sees it. They are given the stack of 128 response cards and instructed to sort them by placing each, one at a time, under the appropriate "stimulus card" ("wherever you think it should go"), but with no further instruction. Each response card varies along three dimensions: color, shape, and number; each level of each dimension being represented across the four stimulus cards. In the example shown here, the two red crosses on the response card at the top of the stack match the *color* of the stimulus card on the left, the *number* of the next stimulus card to the right, and the *shape* of the next stimulus card to the right. The first sorting dimension is color, which the subject must learn by trial and error. The only feedback they receive throughout the task is whether each sort is "right" or "wrong." After 10 correct consecutive trials (i.e., after they have learned to sort to color), the sorting dimension is switched (without warning) to shape. After 10 correct consecutive sorts to shape, it is switched to number. The procedure continues through another full cycle of the three sorting dimensions, or until all 128 cards have been sorted. There are three dependent measures: number of categories achieved (i.e., number of dimension-changes + 1), number of perseverative errors, and number of nonperseverative errors. The former is a choice that would have been correct prior to the most recent switch. Source: Photograph by Michael Starrett, in Bradley R Postle's laboratory.

excisions invading dorsolateral PFC, 7 with excisions of orbitofrontal and temporal cortex, and the remainder with excisions of temporal, parietal, or temporal-parietal-occipital cortex. The average number of categories achieved decreased for the dorsolateral PFC group from pre-op (3.3) to post-op (1.4), whereas this measure was effectively unchanged for all other groups (overall mean, 4.6–4.7; for

the orbitofrontal + temporal group, a slight decrease from 5.3 to 4.9). The culprit was perseveration (see caption to *Figure 15.1*; the term derives from Old French, and before that Latin, for *persistence* and *stubbornness*). For dorsolateral PFC patients, perseverative errors increased, from 39.5 pre-op to 51.5 post-op, whereas the average for other groups declined from 20.0 to 12.2; for the orbitofrontal + temporal group they increased slightly from 9.3 to 12.0. Note that, when considering pre-op performance, we need to keep in mind that the surgeries for these patients targeted tissue that was epileptogenic, and so likely already contributing in only an impaired fashion, if at all, to task performance.

A perseverative error would seem to be one of a failure of inhibition, in that the patient "can't stop themself" from repeating a choice already known to be incorrect. Indeed, as Milner (1963) notes, "perhaps the most interesting finding after frontal-lobe injury has been the dissociation between the subject's actual responses to the stimulus-objects and his ability to verbalize what those responses should be." Many patients were "able themselves to say 'right' or 'wrong' as they placed the cards, and yet did not modify their responses accordingly" (p. 99). That is, it was literally the case for some patients that their words indicated (well, suggested) that they knew what they were supposed to do, but that they couldn't control their skeletomotor responses accordingly.

When considering the effects of damage to subregions of the PFC, Milner (1963) wrote that:

> removal of the inferior and orbital portions of one frontal lobe causes no significant change, even if combined with a temporal lobectomy. One patient in seven showed a deficit in performance after such a removal, and this proportion does not differ from that showing loss after posterior excisions. In contrast to these negative results, all removals which encroached upon the superior frontal region were associated with poor performance on the sorting test. The critical lesion cannot be more precisely defined, although area 9 of Brodmann was implicated in most cases. (p. 97)

It will be worth remembering, when we get to emotions and social behavior in later chapters, that the bias in the sample of Lhermitte, Pillon, and Serdaru (1986), who showed impairments on tasks that entailed direct social interaction, was toward damage in inferior regions of PFC, whereas in the Milner (1963) study with the WCST, the procedure minimized social factors and emphasized "cold reasoning." And so, the story thus far is that a defining function of the PFC, and of cognitive control, is the overriding of prepotent responses (often summarized as "inhibition"), often triggered by environmental stimuli. Can we find physiological evidence for this?

Electrophysiology of the frontal-lobe syndrome

Robert Knight is a behavioral neurologist who got his scientific training with EEG electrophysiologists Steven Hillyard and Robert Galambos, senior authors on the auditory-evoked potential (AEP) studies that we considered in *Chapter 5*. Among Knight's noteworthy contributions to cognitive neuroscience have been studies of ERPs in patients with PFC damage. Particularly relevant for us here is work that Knight carried out in the 1990s with then-graduate student Linda Chao. Chao and Knight (1995) reported that patients with lateral PFC lesions were disproportionately impaired on (5-second) delayed yes/no recognition of environmental sounds when distracting tone pips were played during the delay period. Going back to Malmo (1942), this is an oft-replicated consequence of lateral PFC damage. For a follow-up study, Chao and Knight (1998) repeated the experimental procedure, but while recording the EEG. To fully appreciate their study, it's worth taking a moment to flip back to *Figure 5.4* and refresh your memory about the organization of the AEP into early-, middle-, and long-latency components. What we're about to consider here is a masterful application of the logic, mentioned in that chapter, that "sensory evoked responses . . . are stereotyped, stable responses that are effectively identical in all neurologically healthy individuals[, making] them very effective 'tools' for studying the neural bases of many different behavioral and physiological states."

For the follow-up ERP study, Chao and Knight (1998) obtained ERPs during task performance from seven patients with unilateral damage to lateral PFC and from matched control subjects. The previous behavioral finding of exaggerated sensitivity to distraction in the PFC group was replicated. To understand why, they examined the middle-latency components of the AEP (MAEP) to the distracting tone pips, because the MAEP is known to reflect the initial cortical processing of the auditory signal. From *Figure 15.2* one can see that several peaks in the MAEP are larger for patients than for control subjects. The proximal conclusion that can be drawn from this finding is that the patients' memory for the sample stimulus may have been disrupted by the abnormally high level of perceptual processing of the distracting tone pips. The inference about mechanism that follows from this is that one function of

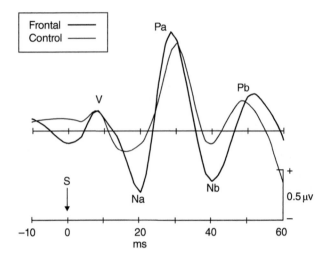

FIGURE 15.2 Auditory-evoked responses in PFC-lesioned patients performing auditory delayed recognition. The middle-latency components – MAEP – to tone pips played during the 5-second delay period of an auditory delayed-recognition task, for patients with lateral PFC lesions and control subjects. Source: From Chao, Linda L. and Robert T. Knight 1998. "Contribution of Human Prefrontal Cortex to Delay Performance." Journal of Cognitive Neuroscience 10 (2): 167–177. Reproduced with permission of The MIT Press.

the PFC is to control the processing of incoming sensory stimuli, such that this processing doesn't interfere with behaviorally relevant internal representations. Further evidence for this interpretation comes from the fact that, in the patients, the magnitude of the Pa component of the MAEP predicted the magnitude of their behavioral deficit. The term often used for this hypothesized mechanism is "filtering," as though the PFC imposes a filter that selectively passes only some incoming sensory information. Indeed, there's good evidence from the cat and the monkey that PFC literally controls just such a filter via its projections to the thalamic reticular nucleus (TRN; see *Chapter 4* and *Chapter 5*). In the cat, these projections have been shown to regulate the magnitude of the AEP.

A second finding from this study was seen in a long-latency component of the AEP (LAEP) to sample stimuli: The N1 was smaller across the scalp in the PFC-lesioned patients. The auditory N1 is believed to arise from auditory association cortex, and previous research from Knight and colleagues has shown it to be disrupted in PFC patients in conjunction with impaired performance on tests of selective attention. Thus, Chao and Knight (1998) suggested that the reduced N1 in this study may have reflected an impairment in the ability of their patients "to focus attention on task-relevant stimuli" (p. 173).

Integration?

The two main findings from the study described by Chao and Knight (1998) highlight two distinct processes whose disruption may contribute to the lateral frontal-lobe syndrome: the filtering of irrelevant information and the focusing of attention on behaviorally relevant information. Do either of these mechanisms help us understand the environmental-dependency syndrome, or the phenomenon of perseveration? It's not immediately apparent to me that they do. Whereas the Chao and Knight (1998) ERP findings relate to perceptual processing, the earlier findings that we reviewed related to the control of behavior. The environmental-dependency syndrome is characterized by a remarkable influence of external stimuli and contexts on behavior, one that supersedes the autonomous control that is central to the human condition. Perseveration on the WCST occurred after a behavioral set was learned and repeated (reinforced), then needed to be changed when reward contingencies changed unexpectedly. Is it possible to unify these disparate findings under a single framework, or is cognitive control just a term that actually corresponds to a collection of task-specific proficiencies that don't share any meaningful underlying relation, apart from a common address in the lateral PFC? This is precisely the kind of question that computational modeling is good at addressing, and so it's to there that we now turn our attention.

MODELS OF COGNITIVE CONTROL

Developmental cognitive neuroscience

The phenomenon of perseveration isn't restricted to behavioral neurology. Neurologically healthy adults will generate prepotent or habitual responses in inappropriate circumstances when their working memory is at capacity (such as in a dual-task paradigm), or when otherwise under stress (or when intently reading something while riding on the elevator). Perseveration is also a classic feature of cognitive development. The Swiss developmental psychologist Jean Piaget (1896–1980) codified this as the "A-not-B" error, seen in infants younger than 12 months of age.

The A-not-B error
To demonstrate the A-not-B error, a toy is placed under one of two boxes (box A) that are within reach of an infant,

and she displaces the box to retrieve the toy. This is repeated several times. On the critical trial, the infant watches while the toy is hidden under box B, she nonetheless reaches for it under box A. Sometimes on the critical trial, in a behavior reminiscent of the dissociation between verbal report and card placement described by Milner (1963), the infant will look at box B (where the toy is hidden) yet reach for box A. Piaget ascribed this error to an immature sense of "object permanence," the understanding that objects continue to exist after they disappear from view. More recently, in 1989, the developmental cognitive neuroscientist Adele Diamond and Patricia Goldman-Rakic, with whom Diamond did a postdoctoral fellowship, documented the phenomenon in young rhesus macaques, replicated it qualitatively in adult monkeys with dorsolateral PFC lesions, and showed the absence of the effect in adult monkeys with PPC lesions. It is known that the primate PFC is among the last brain areas to mature, and that, in the human, it (along with other brain regions) undergoes rapid development during the first few years of life. (From an "anthropology of scientists" standpoint, these findings provide an interesting example of how one's scholarly perspective can influence how one interprets data. Some have seen the PFC lesion findings as evidence that the theoretical construct of object permanence may not exist, and that the A-not-B error in young children is attributable to immature circuitry that supports a cognitive function like working memory, inhibition, or attention. Others have simply concluded that knowledge of object permanence must be supported by the PFC.)

Is the A-not-B error a failure of working memory?
The association of the A-not-B error with the PFC led to proposals that success on the task depends on working memory, and that underdeveloped working memory abilities underlie the error in infants. (Note that the Diamond and Goldman-Rakic findings were published at the same time as the latter's neurophysiological recordings from PFC, as summarized in section *Extracellular electrophysiology inspired by the multiple-component model, Chapter 14*.) The working memory view was soon challenged, however, by studies with human infants using the procedure illustrated in *Figure 15.3.A*, in which all the information needed to solve the task was always visible to the infant (i.e., no overt demands on working memory were made), yet infants still produced the A-not-B error. What *Figure 15.3.B* illustrates is a schematic of a neural network simulation by developmental cognitive neuroscientist Yuko Munakata and students. In offering an explanation for the A-not-B error, it addresses several other questions that have arisen in this

FIGURE 15.3 Modeling perseveration in the A-not-B error.

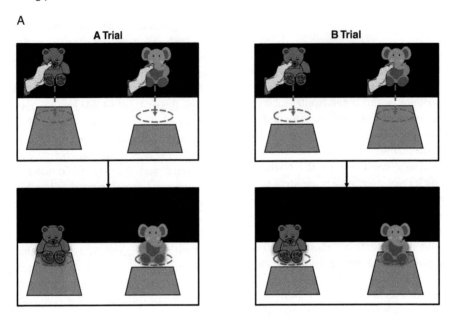

FIGURE 15.3.A An example of an experimental procedure that produces the A-not-B error in infants, even though there is no overt working memory component. Infants sit on a parent's lap and watch as two toys are placed side by side, out of reach, but one on a piece of cloth and one not on a piece of cloth. (The parent is listening to music on headphones so that they won't bias the child's performance.) The closer ends of the cloths are within reach, and the infant knows to pull on the cloth on which the toy sits, in order to retrieve it. This is repeated several times. On the B trial, the cloths are switched, such that it is now the cloth on the right that should be pulled to retrieve the toy. Infants under the age of 12 months, however, commit the A-not-B error.

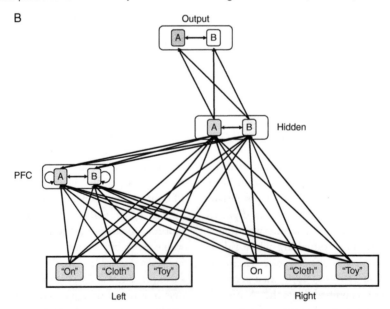

FIGURE 15.3.B Simulation of an A trial, in which stronger input on the left results in stronger activation of the A unit in the hidden layer, and, consequently, stronger activation of the A unit in the output layer. (Although the authors do not state this explicitly, the input layer can be construed as extrastriate visual areas, the hidden layer as the parietal reach area [*Chapter 7* and *Chapter 8*], and the output layer as motor cortex. The layer labeled as the PFC is what's known as a "context layer." A context layer is "off to the side," not in the direct pathway of input ➔ hidden ➔ output, but its inputs to the hidden layer can bias the activity of the latter.)

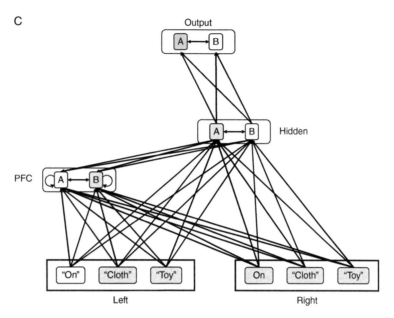

FIGURE 15.3.C Simulation of a first B trial after a series of A trials, with an immature PFC. Although the input values are higher on the right, the stronger connections between left-side inputs and the hidden layer result in stronger activation of the *A* unit in the hidden layer, and, consequently, stronger activation of the *A* unit in the output layer. The immature PFC cannot effectively bias the competition in the hidden layer.

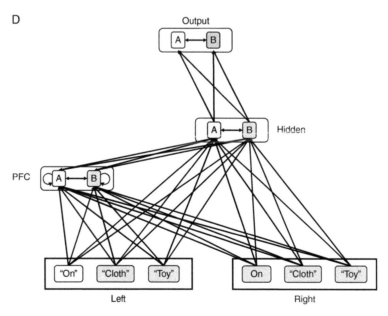

FIGURE 15.3.D Simulation of a first B trial after a series of A trials, with a matured PFC. Now, despite the stronger input-to-hidden drive from the left-side inputs, the PFC with stronger recurrent connections is able to bias the competition in the hidden layer to produce stronger activation of the *B* unit in the hidden layer, and, consequently, stronger activation of the *B* unit in the output layer.

chapter: it provides an explicit account for why perse-veration happens, and what might be the PFC mechanism that controls it; it proposes a mechanism that would unify the ERP findings from Chao and Knight (1998) with the observations of perseveration and of environmental-dependency syndrome; and it offers one answer to the question of whether the lateral frontal-lobe syndrome is fundamentally one of a loss of inhibitory control.

The simulation of Stedron, Sahni, and Munakata (2005) used a neural network simulation environment created by O'Reilly (of McClelland, McNaughton, and O'Reilly renown) and Munakata (2000) that is grounded in the principles of the PDP framework, and that explicitly simu-lates the properties of biological networks by modeling membrane potential within each unit and determining the propagation of activity from one layer to another via an equation that converts the membrane potential into an activation value (a proxy for the rate and number of action potentials). Another important feature of the model is the Hebbian strengthening of connections, which occurs when anatomically connected units are simultaneously active.

On A trials in the A-not-B task, the attributes of *Toy*, *Cloth*, and *On* are activated for the left side (i.e., A side) of the display, and the attributes of *Toy* and *Cloth* are activated for the right side (i.e., B side). As a result, units representing choice A in the hidden layer and in the PFC layer receive more activation than do those representing choice B, because 3 > 2. Thus, choice A receives more activation in the output layer, and the correct choice is made. As this process is repeated across several trials, the connections between the left-sided input layer and the representation of A in the hidden layer, and between the representation of A in hidden layer and the representation of A in the output layer, get strengthened by the Hebbian learning rule. This models, in a very concrete way, prepotency. A consequence of this prepotency is that, on the critical B trial, even though 2 < 3, the strength of the drive received by the *A* unit in the output layer is greater than that received by the *B* unit in the output layer. This is because multiplying the input value of 2 by larger numbers (due to stronger connections) can yield a larger result than does multiply-ing a starting value of 3 by smaller numbers. End result? The model will choose output A – it will perseverate. How can this problem be overcome? By top-down input from the PFC.

Unlike with the classic A-not-B task, where the toy is hidden from view under a box, success on this always-visible variant comes earlier in development: 9-month-olds

perseverate with this procedure, but 11-month-olds do not. Stedron and colleagues simulated the maturation of PFC from 9 to 11 months by progressively strengthening recurrent weights between units within the PFC layer, thereby allowing it to progressively enhance its ability to sustain a representation of the input side that had a higher value, regardless of the strength of learned connections from input to hidden to output layers. Their finding was that at a certain level of PFC "maturity," the ability of this layer to sustain a representation of the correct response provided a sufficiently strong top-down boost of the hid-den layer that output B was selected, despite the structural advantage of the representation of A. It is noteworthy that this simulation refutes the logic that was employed by the group that thought that, by creating a version of the task that lacked an overt memory component, it could be shown that working memory does not contribute. (And this is 8 years before Luck and Vogel [2013] wrote that "visual work-ing memory may not be a memory system per se, but may instead be a general-purpose visual representation system . . ." and here I'll add *to implement cognitive control*.)

Generalizing beyond development

Implied in the Stedron, Sahni, and Munakata (2005) results (*Figure 15.3*) is something that was also implied in my chapter-opening elevator anecdote, and stated explicitly by Miller and Cohen (2001): lurking within all of us is a machinery whose default mode of operation is to respond to the most salient stimulus in the environment, or act in the most habitual, stereotyped way. Explicit in their model is the idea that cognitive control may not be implemented by inhibiting inappropriate responses but, rather, by select-ing the most appropriate response and activating it to the point that it will win out over other alternatives. This is equivalent to the idea of biased competition (*Chapter 7*). Can the ERP findings of Chao and Knight (1998; *Figure 15.2*) be understood in this framework? Recall that, at a descriptive level, they suggested impairments of "fil-tering" and of "focusing." The former, that PFC-lesioned patients showed abnormalities in the MAEP, can be explained as a dysfunction at the *site* of cognitive (or atten-tional) control – a failure of sensory filtering at the level of the thalamus due to a weakened top-down signal to the TRN. The latter, manifesting as a diminished N1, could be the result of a weakened top-down selection signal. If one were striving for a unifying account of PFC function, one might posit that these could be the same signals that, when directed to the appropriate networks in parietal and/or motor cortex, would select the appropriate response over the prepotent but inappropriate one.

But can the "unitary mechanism" simulated in the Stedron, Sahni, and Munakata (2005) model really account for the lateral frontal-lobe syndrome in adults? There are at least a couple of arguments that one might raise. First, the A-not-B error is a behavior that infants grow out of before they can even walk. Second, that simulation is limited to a drastically impoverished and constrained environment (only 3 × 2 inputs; only two choices of response). How applicable can it really be to understanding the richness and complexity of adult cognition and neurological syndromes? And then there's that pesky concern that's been lurking at least since we considered the Hamker (2005) model back in *Chapter 9 – Who is controlling the controller?* For all of the computational models that we've considered, "who" is keeping track of the rules governing the task that's being performed? For a task like the WCST, how does simply receiving the feedback "wrong," as every subject does each time the sorting dimension is surreptitiously changed, prompt us to try a different strategy? The facetious answer is that there's that "little man" inside our head, the homunculus, who knows what different parts of the brain are supposed to do in different situations, and tells them how, and when, to do it. Now of course, it's been hundreds of years since any serious scientist or philosopher may have proposed such a thing. On the other hand, the majority of contemporary models that seek to understand one aspect of cognition depend, in effect, on what Rougier et al. (2005) refer to as "some unexplained source of intelligence" (p. 7338) to do the things that the model can't. The model that we'll consider next develops the capacity for error correction and cognitive control without resorting to a homunculus, and without leaving this question unspecified.

What makes the PFC special?

I said earlier that I didn't want to endorse a "PFC = cognitive control" perspective, and yet we're more than halfway through the chapter and what is the only brain area that we have discussed? Consideration of this neural network simulation by Rougier et al. (2005) will accomplish two goals in this regard. First, it will consider some aspects of the PFC that justify this bias, by calling explicit attention to unique properties of this region and the neurons that make it up, relative to other parts of the brain. Second, it explicitly considers how the PFC must interact with other

brain systems in order to accomplish cognitive control. (That is, it can't literally do it all by itself.)

Properties of the PFC

The Rougier et al. (2005) model of PFC includes three properties that distinguish it from other areas of cortex:

- Sustained activity that is robust against potentially interfering signals: PFC pyramidal neurons have the property of bistability – the ability to operate in an "up" state or a "down" state. This is partly due to the density and distribution of D1-type dopamine receptors along their dendrites, and so the state that they occupy at any point in time is influenced by dopamine levels. When in the up state, a high density of NMDA-type glutamate receptors and recurrent connections with other pyramidal cells can support sustained, elevated levels of activity, such as those illustrated in many of the figures in the opening section of *Chapter 14*.

- Capacity for rapid updating of activity patterns via bistability, that is, switching from an active maintenance mode, when one pattern of activity is sustained so as to prioritize one type of behavior, into an "updating mode," when the previous pattern is destabilized and a new pattern of activity can be established. This is effected by dopamine signals and by input from the basal ganglia, the same elements that are central to reinforcement learning (section *Habits and reinforcement learning, Chapter 8*), and turns out to be key for circumventing the homunculus problem, in that it helps "tell the PFC what to do."

- Capacity for PFC modulation of other cortical areas (as we have already seen in this chapter with Chao and Knight [1998], and in *Chapter 9* with FEF implementation of the endogenous control of attention).

Before returning to neural network models of cognitive control, let's get a foreshadowing of what's to come in *Chapter 16* with a snippet of an admonition from Wolfram Schultz about how to think about reward and (though he doesn't say it explicitly here) the RPE signal:

Primary sensory systems have dedicated physical and chemical receptors that translate environmental energy and information into neural language. Thus, [their] functions . . . are governed by the laws of mechanics, optics, acoustics, and receptor binding. By contrast, there are no dedicated receptors for reward . . . [and so] the investigation of reward functions requires behavioral theories that can conceptualize the different effects of rewards on behaviors . . . Animal learning theory and microeconomics are two prominent examples of such behavioral theories. (Schultz, 2006, p. 91)

Reward, uncertainty, contingency, probability, expected value, these don't physically exist in the world, but must be constructed and represented neurally, in what Weigl (1941) called *Abstraktionsprozesse*.

Influence of the DA reward signal on the functions of PFC

Turning now to *Figure 15.4*, we can see that the Rougier et al. (2005) model is, for the most part, just a more-finely-carved-up variant of the model from Stedron, Sahni, and Munakata (2005; *Figure 15.3* in this chapter.) The finer grain of details in the input layer, for example, will enable the simulation of tasks like the WCST, in which there are multiple stimulus dimensions, and multiple levels within each dimension. For our purposes, the key qualitative difference is the little red box in the upper right, labeled "AG," for adaptive gating. This carries dopaminergic RPE signals.

Not unlike PDP models that we've considered previously, the Rougier et al. (2005) model started from scratch, knowing nothing about its world. It was trained by learning to discriminate and make decisions about stimuli via the four training tasks (see the legend to *Figure 15.4.B*). A critical fact about this process, however, is that instead of the teaching/learning process being implemented "deus ex

FIGURE 15.4 PFC and flexible cognitive control: rule-guided behavior without a homunculus. Source: From Rougier, Nicolas P., David C. Noelle, Todd S. Braver, Jonathan D. Cohen, and Randall C. O'Reilly. 2005. "Prefrontal Cortex and Flexible Cognitive Control: Rules without Symbols." Proceedings of the National Academy of Sciences of the United States of America 102 (20): 7338–7343. Reproduced with permission of PNAS.

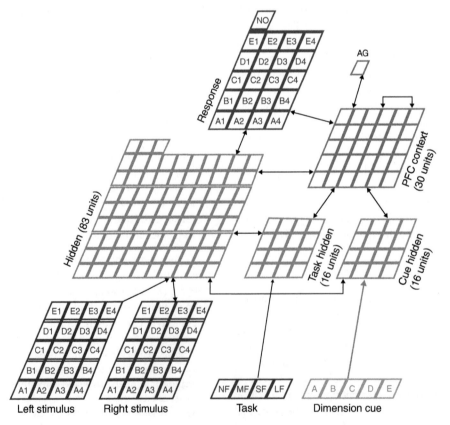

FIGURE 15.4.A The architecture of the model. "AG" is the "adaptive gating" unit, which corresponds to the ascending dopaminergic RPE signal.

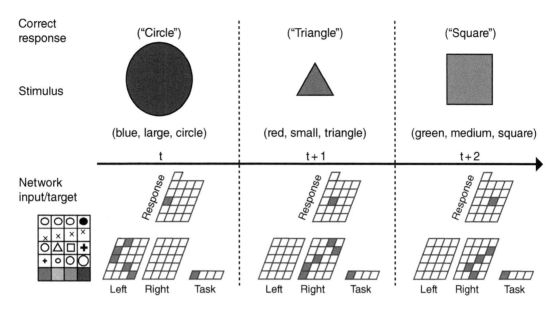

FIGURE 15.4.B Illustration of how high-dimensional inputs were fed into the model. The key in the lower left illustrates how one stimulus dimension is represented in each of the five rows of the input array – color, size, shape, location, and texture – and the columns correspond to different levels of each dimension. From left to right is an illustration of how the presentation of three consecutive stimuli (top row) would be represented and processed by the network. The "task" layer tells the model which of four tasks to perform: naming the feature in a particular stimulus dimension; determining whether two features match according to a particular stimulus dimension; indicating which item is larger; and indicating which item is smaller.

machina," by a scientist delivering a backpropagated error signal each time they observed that the model got an answer wrong, it was implemented by a DA-mediated RPE in a manner that emulated what the authors knew about how this system works. Think back to the example of Pavlovian conditioning that we considered in *Chapter 8*. What Schultz and colleagues did in some of their studies was literally to monitor the responses of midbrain DA neurons throughout the entire process of learning a conditioned stimulus (CS)–unconditioned stimulus (US) pairing. As a result, we know, for example, that in pairings where the CS precedes the US in time (as is illustrated in *Figure 8.11*), there is initially a phasic increase in DA associated with reward delivery. (This has been interpreted by some models as signaling that "the current state is better than expected" [Montague, Dayan, and Sejnowski, 1996, p. 1944].) Over time, as the pairing is learned, the US ceases to elicit this burst of DA. (By definition, once the pairing has been learned, the arrival of the US is now expected. Thus, its delivery can no longer be characterized as being "better than expected.")

The architecture of this network incorporated the fact that PFC is much more densely innervated by DA fibers than are posterior cortical regions, and as a consequence, the hidden layer corresponding to temporal and parietal cortex developed differently than did the PFC (also a context layer, as it was in the model of Stedron, Sahni, and Munakata, 2005). The hidden layer developed to represent stimulus features, and thus the ability to discriminate between features. The PFC, in contrast, developed representations of the rules of the tasks that the network was trained on: how to select the correct name when presented with two objects that both have names; or how to decide if a medium-small-sized, yellow, vertically striped square located in the lower left of the screen is the same as or different from a small, green, vertically striped square located in the upper right if the matching dimension is size. This PFC-based representation of abstract rules (there it is again, the *Abstraktionsprozesse*!) facilitated task performance by providing top-down excitation to the representations of the relevant stimulus dimension in the hidden layer of the network. We have already seen that this is one of the functions of the PFC. The new insight for us is the mechanism by which it influenced the network's performance on the WCST.

Combining an RPE signal with a flexible PFC to overcome perseveration

At the end of the training process, when the model was mature, Rougier et al. (2005) introduced the WCST. The task was simulated by the feature-naming task, but minus any input to the "task" layer or to the "dimension cue" layer that would indicate which dimension was to be named. Thus, it had to rely on trial-and-error learning, when presented with a medium-small, yellow, vertically striped square located in the lower left of the screen, to decide whether to give the response "medium-small," "yellow," "vertical stripes," "square," or "lower left." Feedback was conveyed by the DA RPE, operationalized by the AG unit, following the dynamics described above. When, in the course of trial-and-error guessing, the network stumbled upon the correct dimension by naming the color of a stimulus, the PFC received a phasic burst of DA. And here's where the dopaminergic control of PFC bistability comes in. The response of the model's PFC to this burst of DA was to switch into its up state, in effect "locking in" its active representation of the dimension *color*. (There is considerable biophysical, neuropharmacological, and neurophysiological evidence that a phasic increase of DA can trigger such a state change in PFC pyramidal neurons.) The sustained representation of the category of *color* in the PFC, in turn, biased the competition within the hidden layer of the model in favor of representing the color of each ensuing stimulus relative to their other properties. This state of affairs persisted across the next several consecutive trials. On the switch trial, when (unbeknownst to the network) the sorting rule was changed to *shape*, the network blithely named the color of the stimulus, just as it had in the previous several trials. This time, however, it was given the feedback "incorrect." The failure to receive the expected positive feedback produced a negative RPE – a pause in the release of DA (see *Figure 8.11*), which had the effect (again, based on experimental evidence) of shunting the PFC out of its active maintenance mode and into its updating mode. This destabilized the PFC activity pattern corresponding to the dimension of *color*, thereby permitting a different activity pattern, corresponding to a different stimulus dimension, to emerge.

One take-home from all this is that the role of the PFC in enabling flexible adaptation to unexpected changes in the environment isn't mysterious; isn't the result of "some unexplained source of intelligence." Rather, it results from a combination of the unique biophysical and connectivity properties of its neurons, its uniquely dense innervation by midbrain DA neurons, and the principles of reinforcement learning that are believed to be implemented by these DA neurons. It is also noteworthy that the rapid, flexible adaptation of PFC, such as emphasized in the Rougier et al. (2005) model, has more recently been characterized empirically by the application of dynamic network neuroscience analyses to fMRI data, as summarized in section *Network science and graph theory, Chapter 3*.

NEURAL ACTIVITY RELATING TO COGNITIVE CONTROL

Having considered what can happen when control goes awry, and some computational principles underlying its implementation, we'll conclude this chapter by considering some neurophysiological correlates of cognitive control. This is a fascinating area of research, and one in which, as we've seen elsewhere, new findings can sometimes raise more questions than they answer.

Error monitoring

The ERN/Ne

In the early 1990s, two research groups, one in Dortmund, Germany, and one in Ann Arbor, Michigan, reported an ERP component that corresponded to making an error on speeded reaction-time tasks. The Dortmund group, led by Michael Falkenstein, called it the Ne (error negativity) and the Ann Arbor group, led by William Gehring, called it the ERN (error-related negativity). The fact that it started out with, and, to some extent, still goes by, two names, is emblematic of the enduring lack of consensus that there has been for what process(es) it is likely to index. Here, for simplicity, we will refer to it as the ERN. As illustrated in *Figure 15.5*, the onset of the ERN occurs shortly before, or at, the moment of the erroneous button press, and peaks roughly 100 ms later. The reason for the intense interest in the ERN is clear, as summarized by Gehring, Liu, Orr, and Carp (2012):

> Cognitive control functions include processes that detect when control is needed – as when performance breaks down – and processes that implement control through changes in attentional focus and other strategic adjustments. Because an error is a salient marker that performance has broken down, the ERN is generally thought to reflect a process involved in evaluating the need for, or in implementing, control. (p. 233)

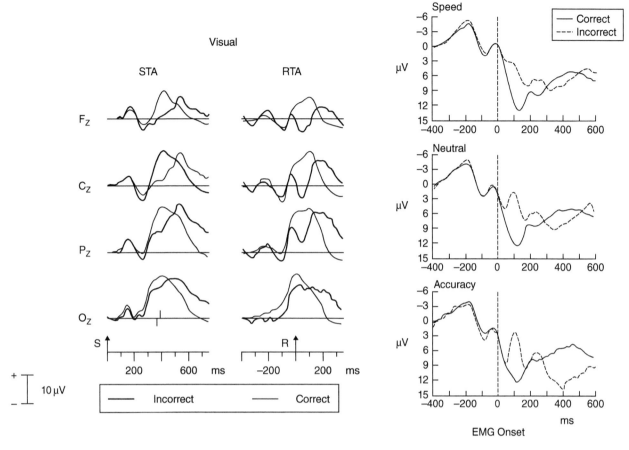

FIGURE 15.5 Two exemplars of the ERN. On the left, based on data from Falkenstein, Hohnsbein, Hoormann, and Blanke (1991), are ERP waveforms aligned to stimulus onset (STA; marked by the arrow on the timeline) or to the response (RTA) in a visual discrimination task. Note that the negative-going component on incorrect trials is only detectable for the latter. (Note, too, that negative is plotted down.) On the right, based on data from Gehring et al. (1993), are ERPs from three conditions of a flanker task, time-locked to the onset of the electromyogram (EMG) of the muscles controlling the finger. (In these plots, negative is plotted up.) Source: From Gehring, William J., Yanni Liu, Joseph M. Orr, and Joshua Carp. 2012. "The Error-Related Negativity (ERN/Ne)." In The Oxford Handbook of Event-Related Potential Components, edited by Steven J. Luck and Emily S. Kappenman, 231–291. New York: Oxford University Press. Reproduced with permission of Oxford University Press.

Subsequent research indicates that an ERN can be generated in at least two circumstances. One, illustrated in *Figure 15.5*, is the "response ERN." The second is the "feedback ERN," which peaks 250–300 ms after the delivery of negative feedback. Some consider these to be two manifestations of activity in the same error-processing system. The response ERN, however, differs from the feedback ERN in that its timing raises the fascinating question about whether it is accompanied by awareness. We've all had the experience of making a mistake, doing something wrong, and not realizing that it was a mistake until minutes, hours, days later. For example, "I thought that this new hairstyle was a good idea, until the barber finished and I saw the result in the

mirror for the first time." Similarly, in the WCST, one can't know on a switch trial that one has made an error until being told that the previous action was "wrong." Here, in contrast, the question is whether, on a moment-to-moment basis, there are circumstances in which our nervous system knows that we are making a mistake even while the action is being carried out, tens if not hundreds of milliseconds before we will become aware of the error. One example is what just happened to me: in the previous sentence I typed "errro," but didn't realize it until I saw the jumble of letters on my computer screen. This fascinating literature, full of nuance, is reviewed in the *Further Reading* from Wessel (2012). (Just did it again! I typed "Wellel.")

Performance monitoring and the dmFC

Considerable evidence, from source localization of EEG and MEG data, from recordings from nonhuman primates, and from fMRI scans of similar tasks, indicate that the ERN is likely generated in frontal midline areas generally referred to as the dorsomedial frontal cortex, a term that encompasses the posterior anterior cingulate cortex (ACC) and the pre-supplementary motor area. (Note that, in this context, the area in question is much more extensive than the "DMFC" whose oculomotor functions were illustrated in *Figure 9.3*.) Indeed, as illustrated in *Figure 15.6*, and as we will review here, this region has been implicated in many types of performance-monitoring situations.

Traditionally, the dmFC has been thought of as part of the motor system, as exemplified by the organizational scheme of Picard and Strick (1996), illustrated in panel A of *Figure 15.6*. It distinguishes between a rostral cingulate zone (RCZ) activated in relation to complex motor tasks and a smaller caudal cingulate zone (CCZ) activated during simpler tasks. The near-complete overlap between these two regions and the sites of activity identified by the many studies contributing to panel B in *Figure 15.6* illustrates very concretely the idea that we have contemplated elsewhere, which is that many aspects of high-level cognition can be construed as elaborations on more basic principles of perception and action.

Panel B of *Figure 15.6* shows the results of a meta-analysis, led by the Dutch cognitive neuroscientist Richard Ridderinkhof, of the foci of activity in neuroimaging studies that were designed to isolate one of four types of performance-monitoring phenomena:

Pre-response conflict

Response conflict occurs when a task concurrently activates more than one response tendency, a classic example being the Stroop task, in which it's more difficult to name the color of ink in which a word is printed if the two aren't congruent. (That is, there's more response conflict, indexed by longer reaction times, more errors, and greater activity in dmFC, when red ink is used to spell out the word "BLUE" than when it is used to spell out the word "RED" [a congruent stimulus] or the word "BOOK" [a neutral stimulus].) Response conflict processing can theoretically also occur after an incorrect response, as a residual representation of the correct response conflicts with the activated representation of the executed response. The canonical statement of the response conflict theory is presented in Botvinick et al. (2001).

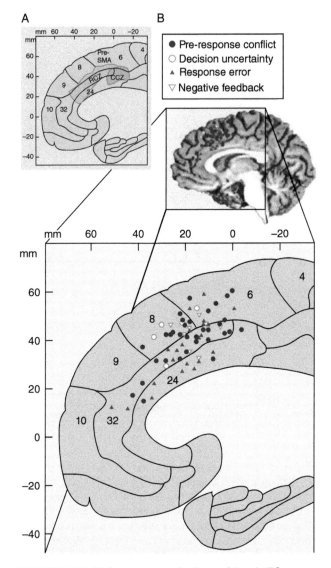

FIGURE 15.6 Performance monitoring and the dmFC. **A** illustrates a functional distinction proposed by Picard and Strick (1996), and **B** a meta-analysis from Ridderinkhof and colleagues (2004). Source: From Ridderinkhof, K. Richard, Markus Ullsperger, Eveline A. Crone, and Sander Nieuwenhuis. 2004. "The Role of the Medial Frontal Cortex in Cognitive Control." Science 306 (5695): 443–447. Reproduced with permission of AAAS.

Decision uncertainty

Also referred to as "underdetermined responding," this is operationalized by conditions requiring a choice from a set of responses, none of which is more compelling than the others.

Response error

This category includes the response ERN and the commission of errors on two-alternative forced-choice tasks. It is often associated with premature responses. One theory motivated by these effects draws explicitly on the dopaminergic RPE. It posits that the RCZ processes these signals in accordance with the principles of reinforcement learning to improve task performance. A place to learn more about this perspective is Holroyd and Coles (2002).

Negative feedback

The canonical event for this category from panel B of *Figure 15.6* is one that would produce a feedback ERN. Included in this category are trials in economic games when an unexpectedly low monetary reward is given, or when a more abstract form of negative feedback is given (like being told that a response on the WCST is "wrong"). An idea that unifies these types of trials is that they entail a comparison of expected vs. actual outcomes. Ridderinkhof, Ullsperger, Crone, and Nieuwenhuis (2004) also note that "Similar parts of the pMFC are activated by primary reinforcers such as pain affect and pleasant tastes, suggesting that the pMFC plays a general role in coding the motivational value of external events."

dmFC–PFC interactions

What does the brain do with all these performance- and error-monitoring signals that we've just cataloged? There's a general consensus that these represent another source of biasing and context signal for the PFC, together with DA-ergic RPEs that we've detailed earlier in this chapter, and "action-refinement" signals from the cerebellum (section *Cerebellum: motor learning, balance, . . . and mental representation?, Chapter 8*). Indeed, dmFC input may play a role in prompting, and maintaining, the increased involvement of activity-based stimulus representation on working memory tasks that require transformations between the first-order sensorimotor associations afforded by sample stimuli and more complicated contextually appropriate control of behavior. Recall, for example, that one of the tasks simulated in the recurrent neural network (RNN) stimulations of Masse et al. (2019) that we considered in *Chapter 14*, section *Activity? who needs activity?*, was the delayed recognition of a rotation of the sample. In that task, a test patch with motion in the same direction of the sample required a negative response to be correct, and a test patch that was rotated by 90° required a positive response. Here, we'll consider dmPF influence on dorsolateral (dl) PFC in a task with formal similarities to that task from Masse et al., (2019): the antisaccade task (section *The supplementary eye field, Chapter 9*), in which the onset of a peripheral target requires a speeded eye movement *away* from the target.

In their study, Liya Ma and colleagues (2019), in the laboratory of Stephan Everling at the University of Western Ontario, studied the influence of the ACC of the dmFC on dlPFC activity during alternating blocks of prosaccade and antisaccade performance. As illustrated in *Figure 15.7*,

FIGURE 15.7 Antisaccade, rule changing, and cortical cooling. Source: From Ma, L., Chan, J. L., Johnston, K., Lomber, S. G. and Everling, S. (2019). Macaque anterior cingulate cortex deactivation impairs performance and alterns lateral prefrontal oscillatory activities in a rule-switching task. PLoS Biology. Public Domain.

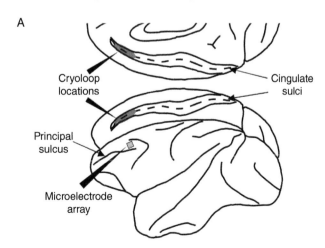

FIGURE 15.7.A Positioning of the cryoloops at the rostral end of the cingulate sulcus, bilaterally, and of the 100-contact "Utah array" in caudal dlPFC, just superior to the principal sulcus.

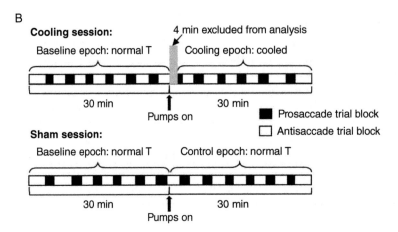

FIGURE 15.7.B The timeline of the experimental procedure.

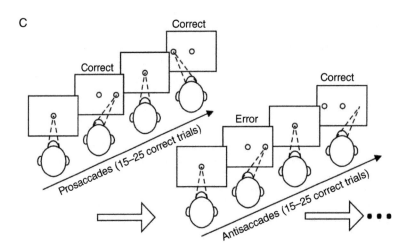

FIGURE 15.7.C Illustration of prosaccade and antisaccade trials.

their task required multiple switches between runs of consecutive prosaccade and antisaccade trials, with the rule switches happening unpredictably, and with no signal, after anywhere from 15 to 25 correct trials. Therefore, this task contained elements of many tests of working memory and of cognitive control that we've already seen in this chapter and the previous chapter: the need to transform the first-order stimulus–response association inherent in the visual stimulus; the frequent (but unpredictable) and uncued change of rules, shortly after the subject had acclimated to the current rule; frequent errors (necessitated by the uncued change of rules). *And* their experiment included cryogenic inactivation (a.k.a. "cortical cooling") of ACC,

thereby effecting a test of the necessity of this region's *performance-monitoring* activity for skilled performance of this task, as well as an assessment of its influence on task-related activity of the dlPFC, a presumed target of these performance-monitoring signals. It's as though they published this study just for us!

The behavioral effects of cooling the ACC were marked, although perhaps different from what one might have expected having just read the preceding section on *Performance monitoring and the dmFC*. As illustrated in *Figure 15.8*, cooling had a selective effect on antisaccade accuracy, in that performance was lower during antisaccade blocks. Notably, however, there was no evidence of an

FIGURE 15.8 Effects of ACC cooling on pro- and antisaccade performance. Source: From Ma, L., Chan, J. L., Johnston, K., Lomber, S. G. and Everling, S. (2019). Macaque anterior cingulate cortex deactivation impairs performance and alterns lateral prefrontal oscillatory activities in a rule-switching task. PLoS Biology. Public Domain.

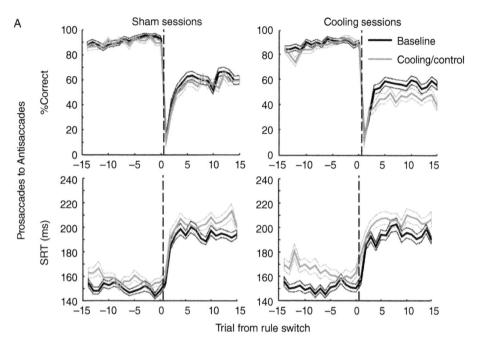

FIGURE 15.8.A Performance on the trials leading up to, and then following, a switch from the prosaccade rule to the antisaccade rule. Accuracy is plotted in the top row, and saccadic reaction time (SRT) in the bottom row.

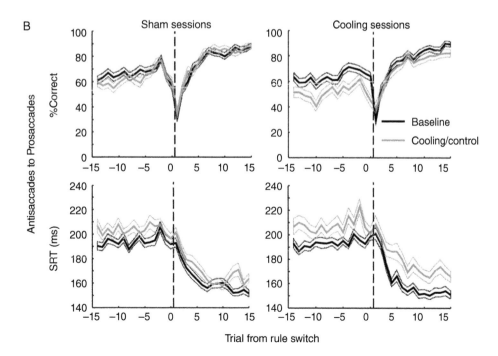

FIGURE 15.8.B Performance on the trials leading up to, and then following, a switch from the antisaccade rule to the prosaccade rule.

exaggerated effect on the trials immediately following a switch from one rule to the other. Rather, the effect was comparable throughout the run of antisaccade trials. In particular, there was no obvious absence of a phasic signal relating to conflict, or error, or feedback. For RT, cooling slowed saccadic RTs in both conditions, and these effects were also across-the-board, with no exaggerated effects around the time of the switch. More generally, the cooling-related impairment of performance did not show the hallmarks classically seen with damage to the dlPFC, such as perseveration, or impulsivity. Rather, the authors summarized it as one of "weakened representation of the current rule or the ability to use the rule to guide behavior" (p. 9). In effect, they reasoned that because making a speeded saccade to a target is such a prepotent response, continued "effort" is required to maintain a mode of antisaccade responding.

Cognitive neuroscientist Todd Braver has proposed that distinct frontal circuitry underlies whether one is in a state of "proactive control," actively anticipating the need to counteract reflexive or otherwise prepotent tendencies, vs. one of "reactive control," effectively waiting until response conflict, or an error, arises before engaging top-down control (e.g., Braver, 2012). Here, Ma et al. (2019) are effectively saying that cooling has the effect of decreasing the animals' ability to sustain a state of proactive control. Indeed, more intriguingly, Ma et al. (2019) speculate that "[the] antisaccade deficit may be related to a loss in the animals' ability or stamina to exert cognitive effort" (p. 23). Although this might seem, upon first hearing, to be uncharacteristically, well, "not very science-y," there's actually a precedent for it. In 2013 Josef Parvizi and colleagues published a report entitled "The will to persevere induced by electrical stimulation of the human cingulate gyrus." Not unlike the approach described in the report of the patient for whom electrical stimulation of the fusiform gyrus disrupted face perception (section *Demonstrating necessity, Chapter 10*), in the 2013 paper that we are featuring here, Parvizi and colleagues demonstrate the effects of stimulation of the ACC with the patient's verbal reports. Here's an example, just to give you a flavor:

> It's . . . like this thing of trying to figure out your way out of, how you're going to get through something. . . . Let's say . . . if you knew you were driving your car and it was . . . one of the tires was half flat and you're only halfway there and you have no other way to turn around and go back,

you have to keep going forward. That type of a, you know, feeling you have. You're like, you're like (pats chest) am I gonna, am I gonna to get through this?

> Am I gonna get through this? . . . I'd say . . . it was . . . a positive thing like . . . push harder, push harder, push harder to try and get through this. (Parvizi et al., 2013, p. 1361)

For me, even though I knew about the Braver model of two modes of cognitive control, and about Parvizi's "will to persevere paper," it was a surprise to see the absence of any clear effect of ACC cooling in the immediate aftermath of switch trials. Do the electrophysiological results from dlPFC help make sense of these findings? As illustrated in *Figure 15.7.A*, Ma and colleagues (2019) recorded neural signals from dlPFC with a Utah array, which is a dense array of 100 microelectrodes embedded in a 10×10 grid. Recall that in the previous chapter, it was observed that "The traditional approach to in vivo extracellular electrophysiology is to begin each day's recording session by inserting an electrode into the microdrive, positioning it over the region of cortex that is of interest, and lowering it into the brain until a neuron is isolated." The next step is then to find a stimulus, or a movement, that selectively drives this neuron. The Utah array, however, is "not your mother's" traditional electrode. It is implanted surgically into the region of cortex that is of interest to the researcher, but it does not afford the ability to isolate individual neurons during its implantation, nor the ability to raise or lower it, once implanted. Rather, it trades the high SNR (signal-to-noise ratio) that one can get from recordings with individual penetrating electrodes, and the precision of mapping out response properties (e.g., tuning curves) of individual neurons, for the extensive coverage that is provided by the Utah array's 100 recording channels. (Stable recordings from just a portion of the channels in a Utah array can support MVPA, an approach that, by definition, is not possible with the [univariate] signal recorded by a single electrode.) In some cases, researchers can isolate individual units from signals recorded from a Utah array, more commonly they record and report MUA and/or LFPs. Ma et al. (2019) reported results from LFPs.

Recordings from dlPFC suggested an important role for elevated oscillatory power in the beta band (Figure 15.9). During fixation, when the animal was waiting for the next saccade target, power in the beta band was elevated for both trial types, and during the entirety of each run of consecutive trials of the same type (Figure 15.9.A). This suggested to

Ma et al. (2019) a role for beta-band oscillations in the representation and application of the currently relevant rule. (By contrast, the only noteworthy effect in lower frequency bands was that increases in alpha-band power, which was otherwise not different from baseline, predicted errors, a pattern consistent with nonspecific lapses in task engagement (Figure 15.9.A).) ACC cooling had effects on power in all frequency bands, but in beta they seemed to be most consequential, in that they reduced power to baseline levels (Figure 15.9.B).

So earlier I wrote that I was surprised by the pattern of effects produced by cooling of the ACC, but should I have been? After all, the PFC was not directly affected by the experimental intervention, nor would have been the midbrain-originating dopaminergic RPEs that were likely generated after the response on the first trial after each

switch. Recall that we said that this element of the task gave it a property comparable to switches on the WCST, and Rougier et al. (2005) demonstrated that a negative RPE was sufficient to push the PFC into its down state. Furthermore, fronto-basal ganglia-thalamic circuitry plays an important role in response selection and Ma et al. (2019) summarize it as such: "In short, our findings support a role of prefrontal beta activities in information gating and interference inhibition that are critical in rule maintenance. We speculate that this suggested function of prefrontal beta rhythm is potentially linked to the role of dACC in allocating cognitive effort. That is, dACC may be responsible for detecting the need for cognitive effort such as interference inhibition, which then took place in the dlPFC upon receiving communication from the dACC."

FIGURE 15.9 Effects of ACC cooling on dlPFC electrophysiology during fixation (relative to ITI). Source: From Ma, L., Chan, J. L., Johnston, K., Lomber, S. G. and Everling, S. (2019). Macaque anterior cingulate cortex deactivation impairs performance and alterns lateral prefrontal oscillatory activities in a rule-switching task. PLoS Biology. Public Domain.

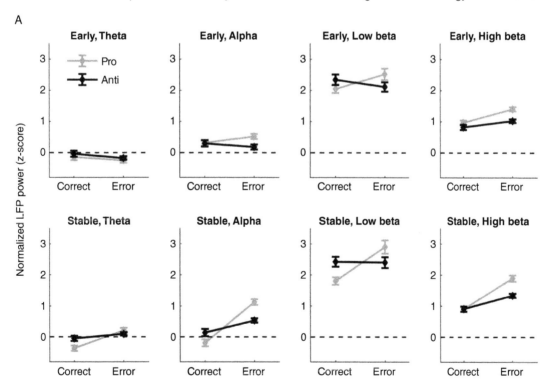

FIGURE 15.9.A LFP power in dlPFC by task epoch (*Early*, *Stable*), by performance (correct, error), and by frequency band, during baseline trials (i.e., prior to cooling or sham). The first four trials following each rule switch are categorized as *Early*, and trials 5–12 following each rule switch are categorized as *Stable*. These show that LFP power in the "low-beta" band (13–20 Hz) and the "high-beta" band (21–20 Hz) is engaged by task performance.

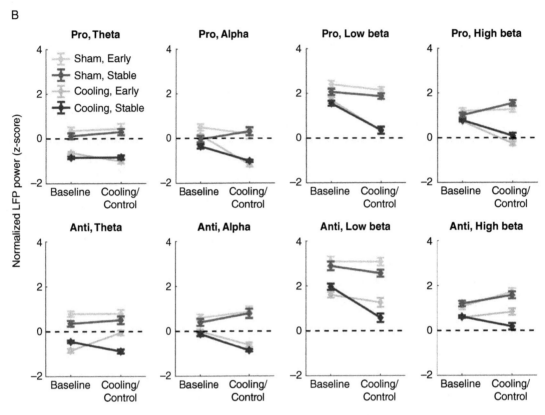

FIGURE 15.9.B Effects of cooling shown only for correct trials, so that interpretation of changes in oscillatory power isn't complicated by deviations of performance unrelated to the cooling intervention. Cooling had an across-the-board effect of reducing LFP power, and this effect was stronger for *Stable* than *Early* trials in the theta, low-beta, and high-beta frequency bands.

GOING META

One area of cognitive neuroscience research that has really exploded in the past decade is the application of the principles of reinforcement learning to cognitive control. This approach emphasizes the important role that the cortico-basal ganglia-thalamic system plays in gating cortical activity, and the fact that cognitive control is often enlisted to govern the selection of one from among many different possible habitual actions. Another important factor, however, is the fact that in-the-moment decision making and response selection can be seen to be governed by some of the same factors – reward probability, choice history, object value, and prediction error – that are fundamental to reinforcement learning. (Some of these constructs, particularly reward probability and object value, will be explored in depth in *Chapter 16*'s treatment of economic decision making.)

As was noted in *Chapter 14* (section *Activity? Who needs activity?*), however, fronto-striatal reinforcement learning is a slow process of gradually refining stereotyped motor behaviors. It's reinforcement learning, after all, that created those "many different possible habitual actions." Using reinforcement learning to select among actions that were themselves created via reinforcement learning would be kind of *meta*, wouldn't it?

A recent development at the intersection of machine learning, deep neural network modeling, and reinforcement learning has been the idea that the PFC exhibits the property of "meta reinforcement learning." That is, dynamic adaptations of behavior that follow the principles of reinforcement learning can be implemented on a trial-by-trial basis by patterns of activity in the PFC. The gist is that unique physiological properties and anatomical connectivity of the PFC allow for DA-based reinforcement learning to train this region, over time, to be able to operate as a "learning system" that implements principles of reinforcement learning in patterns of activity. Thus, although conventional reinforcement learning is slow, based, as it is, on incremental changes in synaptic weights, meta reinforcement learning can change behavior on a moment-by-moment basis because the constructs that we listed in the previous paragraph – reward probability, choice history, object value, and RPEs – can be represented dynamically in distributed patterns of activity, rather than in patterns of weights that bias connection strengths between different neurons or, in the case of RPEs, rather than with a midbrain DA signal.

One demonstration of meta reinforcement learning comes from Jane X. Wang, Zeb Kurth-Nelson, and a team of collaborators (2018) at DeepMind in London, UK, led by Matthew Botvinick. Their computational model treats PFC-basal ganglia-thalamic circuitry as a recurrent network, conceptually similar to the hidden layer in the model of Masse et al. (2019), with which we concluded the previous chapter (*Figure 14.10*). Additionally, in a manner conceptually similar to the model of Rougier et al. (2005), the model is trained "from scratch" with simulated dopaminergic RPEs. Thus, there is no homunculus involved in either its training or its trained performance (see section *Influence of the DA reward signal on the functions of PFC*, earlier in this chapter). The aspect of this model's performance that we'll consider here is a simulation of the task that gave rise to the concept of "learning to learn." In the original experiment of Harlow (1949), a monkey in a WGTA was presented with two unfamiliar objects, one covering a baited food well and one covering an empty food well, and could choose one, retrieving the food reward if it had guessed correctly. This sequence was repeated with the same two objects for five more trials, with left–right position determined randomly on each trial. After the sixth trial, the two objects were replaced with two new objects and the procedure continued. Each six-trial set of trials with the same two objects was called an "episode." *Figure 15.10.B* shows that the animals tested with this procedure were slow to fully learn the rule, but by the 250th episode they performed at 100% correct on the final five trials of each episode regardless of whether their initial choice had been correct or incorrect.

To simulate the Harlow (1949) study, Wang et al. (2018) added a deep neural network similar to the one illustrated in *Figure 6.11* to the front end of their PFC-basal ganglia-thalamic model, so that they could present to the model

FIGURE 15.10 Teaching the PFC to learn to learn.

A

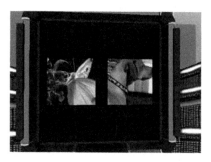

FIGURE 15.10.A A trial from Wang et al. (2018). After initial fixation (top image), the network was shown two images drawn at random from the ImageNet database (middle image), and selected one of the two by "saccading to it" (bottom image). On this trial the network selected the dog, indicated by the dog occupying the network's central vision. Prefrontal Cortex as a Meta-Reinforcement Learning System by Jane X. Wang, Zeb Kurth-Nelson, Dharshan Kumaran, Dhruva Tirumala, Hubert Soyer, Joel Z. Leibo, Demis Hassabis & Matthew Botvinick. Nature Neuroscience volume 21, pages 860–868 (2018).

B

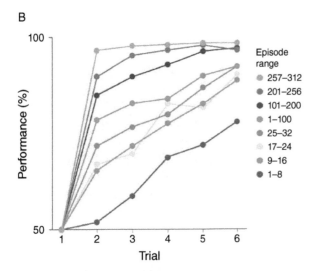

C

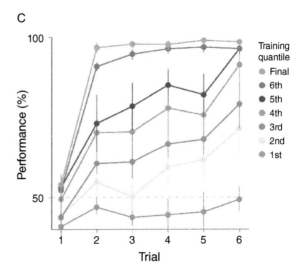

FIGURE 15.10.B Accuracy per trial in each six-trial episode of the monkeys from the Harlow (1949) study, averaging across different ranges of episodes. (For example, although the reddish orange, yellowish orange, yellow, and green symbols present the first 32 trials of testing, the cyan symbol includes these 32 trials in its representation of the first 100 trials of testing.) From Wang, J.X., Kurth-Nelson, Z., Kumaran, D. et al. Prefrontal cortex as a meta-reinforcement learning system. Nat Neurosci 21, 860–868 (2018). Reproduced with permission of Springer Nature.)

FIGURE 15.10.C Accuracy per trial of the neural network, across seven successive stages of training. For each stage, weights within the network were fixed, to assess the level of performance that could be accomplished solely by activity in PFC units. (Note that performance of the network is qualitatively different from that of Harlow's [1949] monkeys during the early stages of training. This is because this simulation required not only learning abstract rule structure but also learning to process visually complex images [i.e., the process described in section *Using deep neural networks to understand computations underlying object recognition, Chapter 6*], learning to stay oriented toward the simulated computer screen, learning to maintain central fixation, and learning to respond with a saccade to the selected image. As a result, much of the initial training time was spent at chance levels of performance.) From Wang, J.X., Kurth-Nelson, Z., Kumaran, D. et al. Prefrontal cortex as a meta-reinforcement learning system. Nat Neurosci 21, 860–868 (2018). Reproduced with permission of Springer Nature.

two novel images from the ImageNet database (section *Deep neural networks, Chapter 3*) for each episode. After training (from scratch) was complete, weights in the recurrent network were fixed, to ensure that the reinforcement learning algorithm used to train the network did not contribute to task performance. Results showed that, after training, with model weights fixed, the model, just like Harlow's (1949) monkeys, learned how to respond to each stimulus pair after the first trial of the episode (*Figure 15.10*). It had learned to learn.

WHERE IS THE CONTROLLER?

In this chapter we have seen that the dorsolateral PFC plays a key role in the control of cognition. Damage to this structure leads to a dramatic and debilitating behavioral disinhibition characteristic of the lateral frontal-lobe syndrome. On closer inspection, we saw that the implementation of cognitive control involves many stages, and many "players." The activity of midbrain dopamine neurons plays two roles

simultaneously. Its "long game" is the shaping of habitual behaviors, while its in-the-moment role is to "nudge" the dorsolateral PFC to maintain or switch behavioral strategies. The dmFC monitors performance, processes information about conflict and errors, and likely acts as a governor on some aspects of PFC activity, including whether to stay in a more effortful proactive mode of control, or to revert to a reactive mode. The unique architecture and connectivity of the PFC-basal ganglia-thalamic system enables it to apply the principles of reinforcement learning so as to optimize the outcome of behavior in novel situations requiring in-the-moment decisions.

END-OF-CHAPTER QUESTIONS

1. Name two or three examples from your daily life that illustrate failures of cognitive control. In each case what was it about the behavior that you *did* produce that resulted in its being expressed rather than the behavior that you should have produced? (Need a hint to get you started? Have you ever done anything "klutzy" while walking and texting on your mobile phone?)

2. What brain region is most strongly associated with cognitive control? Might there be subregional organization such that different aspects of control are preferentially carried out by different subregions?

3. Define a perseverative error on the WCST. What factors explain this distinctive failure of cognitive control? What are two possible mechanisms whereby perseverative errors are prevented in the normally functioning nervous system?

4. The study of PFC-lesioned patients with ERP has revealed two distinct loci of impaired neuronal functioning during an STM task – what are they?

5. How can models of PFC-based cognitive control rule out the logical problem of a PFC-based homunculus who decides when it's time to apply cognitive control?

6. Summarize three explanations for performance-monitoring-related activity in dorsomedial PFC.

7. In what way are principles from reinforcement learning relevant for understanding cognitive control?

REFERENCES

Braver, T. S. 2012. "The Variable Nature of Cognitive Control: A Dual Mechanisms Framework." *Trends in Cognitive Sciences* 16: 106–113.

Botvinick, Matthew M., Todd S. Braver, Deanna M. Barch, Cameron S. Carter, and Jonathan D. Cohen. 2001. "Conflict Monitoring and Cognitive Control." *Psychological Review* 108 (3): 624–652. doi: 10.1037/0033-295X.108.3.624.

Chao, Linda L., and Robert T. Knight. 1995. "Human Prefrontal Lesions Increase Distractibility to Irrelevant Sensory Inputs." *Neuroreport* 6 (12): 1605–1610.

Chao, Linda L., and Robert T. Knight. 1998. "Contribution of Human Prefrontal Cortex to Delay Performance." *Journal of Cognitive Neuroscience* 10 (2): 167–177. doi: 10.1162–08989299 8562636.

Falkenstein, Michael, Joachim Hohnsbein, Jörg Hoormann, and L. Blanke. 1991. "Effects of Cross-Modal Divided Attention on Late ERP Components: II. Error Processing in Choice Reaction Tasks." *Electroencephalography and Clinical Neurophysiology* 78 (6): 447–455.

Gehring, William J., Brian Goss, Michael G. H. Coles, David E. Meyer, and Emanuel Donchin. 1993. "A Neural System for Error Detection and Compensation." *Psychological Science* 4 (6): 385–390. doi: 10.1111/j.1467-9280.1993.tb00586.x.

Gehring, William J., Yanni Liu, Joseph M. Orr, and Joshua Carp. 2012. "The Error-Related Negativity (ERN/Ne)." In *The Oxford Handbook of Event-Related Potential Components*, edited by Steven J. Luck and Emily S. Kappenman, 231–291. New York: Oxford University Press. doi: 10.1093/oxfordhb/9780195374148.013.0120.

Hamker, Fred H. 2005. "The Reentry Hypothesis: The Putative Interaction of the Frontal Eye Field, Ventrolateral Prefrontal Cortex, and Areas V4, IT for Attention and Eye Movement." *Cerebral Cortex* 15 (4): 431–447. doi: 10.1093/cercor/bhh146.

Harlow, H. F. 1949. "The Formation of Learning Sets." *Psychological Review* 56: 51–65.

Holroyd, Clay B., and Michael G. H. Coles. 2002. "The Neural Basis of Human Error Processing: Reinforcement Learning, Dopamine, and the Error-Related Negativity." *Psychological Review* 109 (4): 679–709. doi: 10.1037/0033-295X.109.4.679.

Knight, Robert T., and Mark D'Esposito. 2003. "Lateral Prefrontal Syndrome: A Disorder of Executive Control." In *Foundations of Cognitive Neuroscience* edited by Mark D'Esposito, 259–280. Cambridge, MA: MIT Press.

Lhermitte, François, Bernard Pillon, and M. Serdaru. 1986. "Human Autonomy and the Frontal Lobes. Part I: Imitation and Utilization Behavior: A Neuropsychological Study of 75 Patients." *Annals of Neurology* 19 (4): 326–334. doi: 10.1002/ana.410190404.

Malmo, Robert B. 1942. "Interference Factors in Delayed Response in Monkeys after Removal of Frontal Lobes." *Journal of Neurophysiology* 5 (4): 295–308.

Ma, L., J. L. Chan, K. Johnston, S. G. Lomber, and S. Everling 2019. "Macaque Anterior Cingulate Cortex Deactivation Impairs Performance and Alters Lateral Prefrontal Oscillatory Activities in a Rule-Switching Task." *PLoS Biology* 17: e3000045. doi.org/10.1371/journal.pbio.3000045.

Miller, Earl K., and Jonathan D. Cohen. 2001. "An Integrative Theory of Prefrontal Cortex Function." *Annual Review of Neuroscience* 24: 167–202.

Milner, Brenda. 1963. "Effects of Different Brain Lesions on Card Sorting: The Role of the Frontal Lobes." *Archives of Neurology* 9 (1): 90–100. doi: 10.1001/archneur.1963.00460070100010.

O'Reilly, Randall C., and Yuko Munakata. 2000. *Computational Explorations in Cognitive Neuroscience: Understanding the Mind by Simulating the Brain.* Cambridge, MA, and London: MIT Press.

Parvizi, J., V. Rangarajan, W. R. Shirer, N. Desai, and M. D. Greicius. 2013. "The Will to Persevere Induced by Electrical Stimulation of the Human Cingulate Gyrus." *Neuron* 80: 1359–1367.

Picard, Nathalie, and Peter L. Strick. 1996. "Motor Areas of the Medial Wall: A Review of Their Location and Functional Activation." *Cerebral Cortex* 6: 342–353. doi: 10.1093/cercor/6.3.342.

Ridderinkhof, K. Richard, Markus Ullsperger, Eveline A. Crone, and Sander Nieuwenhuis. 2004. "The Role of the Medial Frontal Cortex in Cognitive Control." *Science* 306 (5695): 443–447. doi: 10.1126/science.1100301.

Rougier, Nicolas P., David C. Noelle, Todd S. Braver, Jonathan D. Cohen, and Randall C. O'Reilly. 2005. "Prefrontal Cortex and Flexible Cognitive Control: Rules without Symbols." *Proceedings of the National Academy of Sciences of the United States of America* 102 (20): 7338–7343. doi: 10.1073/pnas.0502455102.

Schultz, Wolfram. 2006. "Behavioral Theories and the Neurophysiology of Reward." *Annual Review of Psychology* 57: 87–115. doi: 10.1146/annurev.psych.56.091103.070229.

Stedron, Jennifer Merva, Sarah Devi Sahni, and Yuko Munakata. 2005. "Common Mechanisms for Working Memory and Attention: The Case of Perseveration with Visible Solutions." *Journal of Cognitive Neuroscience* 17 (4): 623–631. doi: 10.1162/0898929053467622.

Wang, Jane X., Zeb Kurth-Nelson, Dharshan Kumaran, Dhruva Tirumala, Hubert Soyer, Joel Z. Leibo, Demis Hassabis, and Matthew Botvinick. 2018. "Prefrontal Cortex as a Meta-reinforcement Learning System." *Nature Neuroscience* 21: 860–868.

Weigl, Egon. 1941. "On the Psychology of So-Called Processes of Abstraction." Translated by Margaret J. Rioch. *Journal of Abnormal and Social Psychology* 36 (1): 3–33. doi: 10.1037/h0055544. Originally published in 1927 as "Zur psychologie sogenannter Abstraktionsprozesse." *Zeitschrift Fur Psychologie* 103: 2–45.

OTHER SOURCES USED

Diamond, Adele, and Patricia S. Goldman-Rakic. 1989. "Comparison of Human Infants and Rhesus Monkeys on Piaget's A-Not-B Task: Evidence for Dependence on Dorsolateral Prefrontal Cortex." *Experimental Brain Research* 74 (1): 24–40. http://www.devcogneuro.com/Publications/Diamond&Goldman-RakicAB.pdf.

Durstewitz, Daniel, Jeremy K. Seamans, and Terrence J. Sejnowski. 2000. "Dopamine-Mediated Stabilization of Delay-Period Activity in a Network Model of Prefrontal Cortex." *Journal of Neurophysiology* 83 (3): 1733–1750.

Eling, Paul, Kristianne Derckx, and Roald Maes. 2008. "On the Historical and Conceptual Background of the Wisconsin Card Sorting Test." *Brain and Cognition* 67 (3): 247–253. doi: 10.1016/j.bandc.2008.01.006.

Falkenstein, Michael, Joachim Hohnsbein, and Jörg Hoormann. 1990. "Effects of Errors in Choice Reaction Tasks on the ERP under Focused and Divided Attention." In *Physiological Brain Research*, vol. 1, edited by Cornelis Henri Marie Brunia, Anthony W. K. Gaillard, and A. Kok, 192–195. Tilburg, NL: Tilburg University Press.

Grant, David A., and Esta Berg. 1948. "A Behavioral Analysis of Degree of Reinforcement and Ease of Shifting to New Responses in a Weigl-Type Card-Sorting Problem." *Journal of Experimental Psychology* 38 (4): 404–411. doi: 10.1037/h0059831.

Harlow, Harry F., and J. Dagnon. 1943. "Problem Solution by Monkeys Following Bilateral Removal of the Prefrontal Areas. I. The Discrimination and Discrimination-Reversal Problems." *Journal of Experimental Psychology* 32 (4): 351–356. doi: 10.1037/h0058649.

Lhermitte, François. 1986. "Human Autonomy and the Frontal Lobes. Part II: Patient Behavior in Complex and Social Situations: The 'Environmental Dependency Syndrome.'" *Annals of Neurology* 19 (4): 335–343. doi: 10.1002/ana.410190405.

Lisman, John E., Jean-Marc Fellous, and Xiao-Jing Wang. 1998. "A Role for NMDA-Receptor Channels in Working Memory." *Nature Neuroscience* 1 (4): 273–275. doi: 10.1038/1086.

MacLeod, Colin M. 1991. "Half a Century of Research on the Stroop Effect: An Integrative Review." *Psychological Bulletin* 109 (2): 163–203. doi: 10.1037/0033-2909.109.2.163.

Miller, Earl K., and Timothy J. Buschman. 2013. "Cortical Circuits for the Control of Attention." *Current Opinion in Neurobiology* 23 (2): 216–222. doi: 10.1016/j.conb.2012.11.011.

Montague, P. Read, Peter Dayan, and Terrence J. Sejnowski. 1996. "A Framework for Mesencephalic Dopamine Systems Based on Predictive Hebbian Learning." *Journal of Neuroscience* 16 (5): 1936–1947.

Settlage, Paul, Myra Zable, and Harry F. Harlow. 1948. "Problem Solution by Monkeys Following Bilateral Removal of the Prefrontal Areas: VI. Performance on Tests Requiring

Contradictory Reactions to Similar and to Identical Stimuli." *Journal of Experimental Psychology* 38 (1): 50–65. doi: 10.1037/h0054430.

Skinner, James E., and Charles D. Yingling. 1976. "Regulation of Slow Potential Shifts in Nucleus Reticularis Thalami by the Mesencephalic Reticular Formation and the Frontal Granular Cortex." *Electroencephalography and Clinical Neurophysiology* 40 (3): 288–296. doi: 10.1016/0013-4694(76)90152-8.

Yingling, Charles D., and James E. Skinner. 1976. "Selective Regulation of Thalamic Sensory Relay Nuclei by Nucleus Reticularis Thalami." *Electroencephalography and Clinical Neurophysiology* 41 (5): 476–482. doi: 10.1016/0013-4694(76)90059-6.

Zable, Myra, and Harry F. Harlow. 1946. "The Performance of Rhesus Monkeys on Series of Object-Quality and Positional Discriminations and Discrimination Reversals." *Journal of Comparative Psychology* 39 (1): 13–23. doi: 10.1037/h0056082.

FURTHER READING

Knight, Robert T., and Mark D'Esposito. 2003. "Lateral Prefrontal Syndrome: A Disorder of Executive Control." In *Neurological Foundations of Cognitive Neuroscience*, edited by Mark D'Esposito, 259–280. Cambridge, MA: MIT Press.
An authoritative review, from two behavioral neurologists, of the clinical presentation, and theoretical interpretations, of patients with damage to the lateral PFC.

Miller, Earl K., and Jonathan D. Cohen. 2001. "An Integrative Theory of Prefrontal Cortex Function." *Annual Review of Neuroscience* 24: 167–202.
Paper that really set the agenda for cognitive control research for the first decade of the new millennium.

Miller, Earl K., and Timothy J. Buschman. 2013. "Cortical Circuits for the Control of Attention." *Current Opinion in Neurobiology* 23 (2): 216–222. doi: 10.1016/j.conb.2012.11.011.
Summary of evidence, from human and monkey research, of the key role for interareal oscillatory synchrony as a mechanism for cognitive control.

Passingham, Richard E., and Steven P. Wise. 2012. *The Neurobiology of the Prefrontal Cortex*. Oxford: Oxford University Press.
A systematic, focused, theory of frontal-lobe function, including a unique and provocative evolutionary focus, from two leading students of its anatomy, physiology, and function.

Stuss, Donald T., and Robert T. Knight, eds. 2013. *Principles of Frontal Lobe Function*, 2nd ed. New York: Oxford University Press.
An exhaustive collection of chapters from many of the world's authorities on the myriad functions of the PFC.

Wessel, Jan R. 2012. "Error Awareness and the Error-Related Negativity: Evaluating the First Decade of Evidence." *Frontiers in Human Neuroscience* 6: 88. doi: 10.3389/fnhum.2012.00088.
Review of the nuanced and fascinating literature on the role of awareness in performance monitoring and the ERN.

CHAPTER 16
DECISION MAKING

KEY THEMES

- Three classes of decisions that are each studied differently, and may be differently implemented by the brain, are perceptual decisions, value-based decisions (the focus of neuroeconomics), and foraging decisions.

- Microstimulation of MT influences the decisions that monkeys make about the visually perceived direction of motion.

- Signal detection theory and drift-diffusion modeling interpret the activity of LIP neurons as representing the buildup of evidence in a decision variable.

- If the response is to be made manually, the buildup in the decision variable is seen in skeletomotor circuits.

- In a value-based oculomotor decision task, LIP neurons can represent the value of a saccade, raising the possibility that the representation of a motor value map, rather than a salience map, may be the fundamental function of the oculomotor system.

- Extracellular electrophysiology in the monkey and fMRI in the human implicate orbital and ventromedial PFC in the representation of the "common currency" that underlies behavioral economic theory.

- In contrast to explicit two-alternative forced choice decision, foraging entails the weighing of different factors -- the value of exploiting versus the value of exploring conditioned by the cost of exploring -- to decide how to guide behavior.

- Some aspects of adolescent behavior can be understood as reflecting differing rates of maturation of neural systems that guide decision-making.

CONTENTS

BETWEEN PERCEPTION AND ACTION

The simplest sensorimotor event is the reflex – the patellar tendon of your knee is tapped by the doctor's mallet and your leg kicks out. This response to a stimulus involves just one synapse in the CNS: the stretch receptor in your quadriceps muscle makes a synaptic connection, in the spinal cord, directly onto a motor neuron which, in turn, commands a contraction of the quadriceps muscle. This circuit is called a reflex arc, and its action does not entail any kind of decision. Once initiated, its outcome is known with certainty. It is a deterministic system.

For behaviors that qualify as cognitive, on the other hand, all actions can be said to be the result of a decision. Most are not deterministic: given the same stimulus inputs, an individual may respond to it in different ways, depending on the context, the current state of any of many systems in his brain, and recent history with the same or similar types of stimuli.

There are three domains of decision making that we will consider here. **Perceptual decisions** are closely related to the processes of recognition – "snap" decisions by which we classify what kind of thing we are seeing (or hearing, or feeling, or . . .) and then take the appropriate action – *If that pedestrian is walking out into the street, I will apply my brakes; if she is standing still, I will continue driving past the crosswalk.* **Value-based decisions** are the purview of **neuroeconomics** – *Out of the (seemingly) hundreds of choices on the shelves, which bottle of red wine will I select to purchase and bring home to enjoy with dinner?* The third domain, **foraging decisions**, evokes (for me, at least) that timeless question posed by the British punk rock band The Clash: *Should I stay or should I go?* That is, *Do I remain where I am to* exploit *the resources currently before me, or do I set out to* explore *new opportunities?* (This is yet another question that is at the heart of reinforcement learning.)

We begin with perceptual decision making – *was the overall direction of motion within that cloud of dots moving up or down?* We'll begin, as we have in other domains of cognitive neuroscience, by considering the "laws of . . . optics . . . and receptor binding." From this starting point, how far will we get in our understanding of this behavior, and its neural underpinnings, before, as Wolfram Schultz admonished in the previous chapter, we need to turn to the behavioral sciences, particularly psychology and economics, for models that will help us understand even this simplest of decision-making behaviors?

PERCEPTUAL DECISION MAKING

Judging the direction of motion

Microstimulation of MT

Decisions about sensory stimuli are a scientifically tractable place to start because one can systematically vary properties of the inputs, and determine how this variation influences the decision, thereby gaining important knowledge about how the decision-making process works. A highly influential example of such an experiment, performed in the laboratory of neurophysiologist William Newsome, is illustrated in *Figure 16.1*. Newsome, Britten, and Movshon (1989) had previously studied the relationship between the activity of neurons in MT and a monkey's decision about whether the overall, integrated motion signal from a cloud of randomly moving dots (referred to as the "global" motion of the cloud, to distinguish it from the "local" motion of any one dot) was in one direction or another. Intriguingly, they found that the stimulus-related activity of some individual MT neurons was as good as, and in some cases even better than, the performance of the monkey! As a next step, to establish the causal role of MT neurons tuned to a specific direction of motion in the animal's decisions about motion in that direction, they performed the experiment illustrated in *Figure 16.1*.

The results were striking. Let's first consider just one data point in panel A of *Figure 16.2* – the *no stimulation* trials when the RDM stimulus contained 0% correlated motion. On these trials, there was no global motion, and the animal decided *Pref* on roughly half the trials (to my eye it looks like 55% of these trials). Now look at its performance on the perceptually identical trials when microstimulation was delivered to an MT neuron tuned to detect motion in the *Pref* direction: on *stimulation* trials, the monkey decided *Pref* much more often, somewhere between 65% and 70% of the time. The same is true, qualitatively, at every level of stimulus signal strength, when there was true global motion in one of the two directions.

When thinking about the effect of microstimulation for one stimulus type (e.g., 0% correlation), it's easier for me to think about it in terms of "How does microstimulation influence the monkey's decision?" Answer: it biases it to give more *Pref* responses. But this leads to additional questions whose answers are less straightforward. For example, does this mean that microstimulation makes the monkey *perceive* more of the stimuli as having motion in the preferred direction? Of course, we can't know this

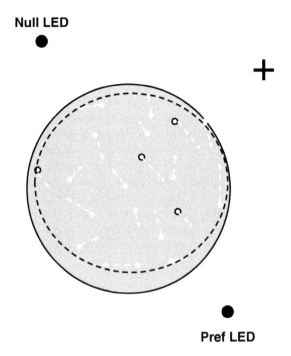

Null LED

Pref LED

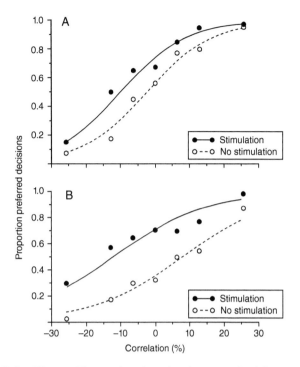

FIGURE 16.1 Influencing perceptual decisions with microstimulation. Diagram of the task, in which the directional tuning of an MT neuron was first determined (in this illustration, down and to the right), and then the monkey performed several trials in which random dot motion (RDM) was presented in a stimulus aperture that covered the neuron's receptive field (dotted circle). For each, the animal made a forced choice as to whether the global direction of motion was toward the location marked "Pref LED," or toward the "Null LED," and indicated its decision with a saccade made to that LED. The strength of the motion signal was controlled by varying the percentage of dots all moving in the same direction, such that a 25% "correlation" meant that 25% of the dots were moving in the neuron's preferred (Pref) direction, and a –8% correlation that 8% of the dots were moving in the neuron's anti-preferred (Null) direction. (This cartoon shows a display in which the four highlighted dots are producing a 21% correlation.) On a randomly determined subset of trials, microstimulation was delivered to the neuron.

FIGURE 16.2 The results of varying the strength of the motion signal, and of microstimulation, at two stimulation sites within MT. The data are plotted as the animal's choice (expressed as the proportion of saccades made to the Pref LED) as a function of motion signal strength (Correlation). Source: From Salzman, C. Daniel, Kenneth H. Britten, and William T. Newsome. 1990. "Cortical Microstimulation Influences Perceptual Judgements of Motion Direction." Nature 346 (6280): 174–177. Reproduced with permission of Springer Nature.

with any certainty, and so this makes characterization of the effects of microstimulation in terms of the animal's performance seem somehow subjective and less than satisfying. For this reason, even though on a stimulus-by-stimulus basis the effect of microstimulation is to move each data point "up" relative to the ordinate, a more useful

way to quantify its effects is in terms of how it shifts the animal's psychometric curve along the abscissa. The reason is that this allows the effects of microstimulation to be quantified in terms of the physical properties of the stimulus that was varied in order to derive this curve. Thus, for the stimulation site depicted in the upper panel of *Figure 16.2*, microstimulation was equivalent to adding 7.7% dots moving in the preferred direction to each stimulus display. For the site in the lower panel, the effect was much larger, corresponding to a shift of 20.1% correlated dots in the preferred direction. (Note that, for both sites illustrated, particularly the one in the lower panel, there also seems to be some distortion to the shape of the psychometric function. What might explain this?)

Part of the importance of the Salzman, Britten, and Newsome (1990) study was that it prescribed a well-understood circuit in which perceptual decision making could be studied in greater depth: first, the microstimulation results indicated that MT is, indeed, causally involved in perceptual decision making (about motion); second, the task was, in effect, a visually guided eye-movement task. "This link," as Gold and Shadlen (2007, p. 545) point out, "enables investigators to treat the decision as a problem of movement selection." Because the eye-movement command would have to pass through FEF and/or SC, the logical next area to investigate in the perceptual decision-making process, one that links MT with these two oculomotor control centers, was LIP.

LIP

Because LIP is two synapses downstream from MT, the question is whether its contribution to the perceptual decision task reflects perceptual processing, decision making, or motor planning and execution. In keeping with the idea of treating the decision as a problem of movement selection, the procedure for studying neurons in LIP is to place one of the two saccade targets inside the neuron's motor field, (i.e., not the aperture displaying the RDM). With this procedure, it turns out to be fairly straightforward to rule out a perceptual role for LIP, simply by examining trials that present RDM containing 0% correlated motion. On roughly half of these trials, a saccade will be made into the motor field of the neuron being recorded from, and on the other half it will be made away from it. If the neuron's function were perceptual in nature, its response would be the same for both decisions, because on all trials the stimulus is identical. However, this is not what is observed. Instead, for the former, the neuron shows a gradual increase in firing rate, until the time of the saccade, and for the latter, it shows a suppressed level of activity until after the saccade. Interpretation of the ramp-up observed on trials when the saccade will come into the neuron's motor field, however, is more complicated. One possibility is that it corresponds to an accumulation of information related to the decision. The alternative is that it represents motor preparation, the decision having already been made elsewhere. Is it possible to rule out one of these two accounts?

Initial evidence that activity in LIP may correspond to the buildup of decision-related information comes from

the rate of increase in activity that is observed as a function of stimulus strength: on trials in which the RDM contains a larger percentage of correlated motion, the increase in firing rate is steeper than it is on trials presenting weaker directional information. Does this steeper increase in firing rate result in the animal reaching its decision more quickly? It turns out that it does, as can be determined by making a change to the experimental procedure: rather than having the animal make a response after a fixed-length, longish delay, allow it to respond as quickly as possible, incentivizing speed by providing larger rewards for quicker responses. The first thing one notices on the speeded variant of the task is that response time (RT) is slower for stimuli that provide weaker directional information. This pattern in the behavior is also seen in the activity of LIP neurons (*Figure 16.3*).

Modeling perceptual decision making

A psychological model (signal detection theory)

In a framework proposed by Gold and Shadlen (2007), perceptual decision-making can be formalized in terms of signal detection theory and the related theory of sequential analysis; we'll abbreviate these as SDT. "A decision," they write, "is a deliberative process that results in the commitment to a categorical proposition. An apt analogy is a judge or jury that must take time to weigh evidence for alternative interpretations and/or possible ramifications before settling on a verdict" (p. 536). The key factors to understand, from this perspective, are how the deliberative process works, and how it is translated into the commitment. In *Figure 16.3*, the neural correlate of the deliberative process is captured by the evolution of any two same-colored PSTHs, which are interpreted as reflecting the temporal evolution of evidence for a *Select T_{in}* vs. a *Select T_{out}* decision. The sum of these two processes, running in parallel, is the decision variable, which Gold and Shadlen (2007) note "is not tied to the (possibly fleeting) appearance of stimuli but spans the time from the first pieces of relevant information to the final choice" (p. 538). In this regard, the concept of the decision variable can be understood in relation to working memory, as can the sustained, elevated activity in LIP illustrated in *Figure 16.3*. (Recall that we introduced SDT in section *Dual processes of memory retrieval in the rodent, Chapter 12*. From *this* perspective, our present topic suggests similarities to the buildup of evidence from a familiarity-based signal.)

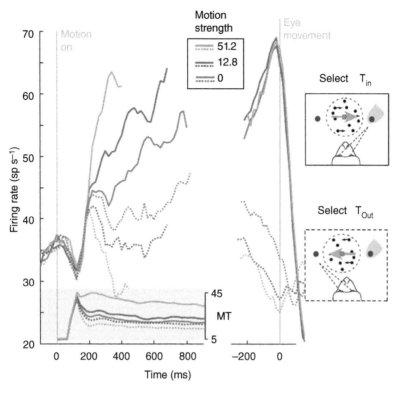

FIGURE 16.3 Activity in LIP, displayed as peristimulus time histograms (PSTHs), in speeded RDM direction decision task, as a function of motion strength and whether the direction selected corresponds to a saccade made into the neuron's motor field ("Select T$_{in}$"; solid lines) or away from the neuron's motor field ("Select T$_{out}$"; dotted lines). Data on the left are aligned to trial onset, and on right are aligned to saccade onset. Inset, in shaded box in lower left, is activity in an MT neuron tuned for motion in the Select T$_{in}$ direction. Source: From Gold, Joshua I., and Michael N. Shadlen. 2007. "The Neural Basis of Decision Making." Annual Review of Neuroscience 30: 535–574. doi: 10.1146/annurev.neuro.29.051605. 113038. Reproduced with permission of Annual Reviews.

Regardless of stimulus strength, the theory predicts that the decision variable must achieve a certain value before a decision is made, and this would seem to correspond to when LIP neurons representing the decision (i.e., "Select T$_{in}$" from *Figure 16.3*) reach a firing rate of 70 Hz. Because LIP neurons achieve this level of activity with RDM displays containing no global direction information, they (and the hypothesized **decision variable**) must be receiving additional information apart from what is fed forward from MT. SDT holds that this additional information corresponds to expectations that have been built up over past experience ("prior probabilities" in terms of Bayesian statistics) and information about the reward value of the choice. More generally, one might think of it in terms of the predictive coding framework introduced at the end of section *Where does sensation end? Where does perception begin?*,

Chapter 4, – where there is insufficient sensory evidence, behavior will be governed by the brain's current model of the environment.

A further detail in the framework of Gold and Shadlen (2007) is the distinction between "momentary evidence" and "accumulated evidence." Momentary evidence is what is generated by "the (possibly fleeting) appearance of stimuli" referred to earlier in this section. At this point, it will be useful to introduce a second type of model that can be useful for translating the abstract constructs from SDT into somewhat less abstract constructs that lend themselves better to mapping onto neural processes: drift-diffusion models (DDM; sometimes also referred to as "diffusion to bound" models). Developed by experimental psychologists to account for behavior in speeded decision tasks, DDMs are "mathematical models in which sensory information is integrated stochastically over time, resulting in a **random walk** of an abstract variable toward a preset decision threshold. A decision is made when the random walker reaches the threshold." (Lo and Wang, 2006, p. 956)

O.k., now back to the momentary vs. accumulated distinction. To assess the neural bases of these hypothesized processes, Hanks, Ditterich, and Shadlen (2006) took a page from the book of Shadlen's postdoctoral mentor, William Newsome, and compared the effects of micro-stimulation delivered to MT vs. to LIP. They fit a DDM to the behavioral data from the monkeys, and found that microstimulation of MT had the effect of steepening the drift rate in the model, thereby biasing decisions in favor of the direction preferred by the stimulated neurons. (Look at the cartoon of a drift-diffusion process in *Figure 16.4.B*; the drift rates are the slopes of the trajectories of the two

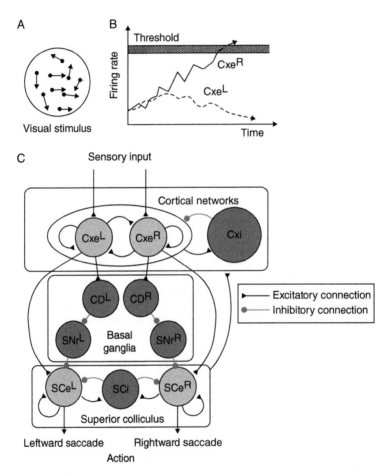

FIGURE 16.4 Corticobasal ganglia circuit mechanism for a decision threshold in reaction time tasks. **A.** Schematic of the RDM display. **B.** A schematic representation of the random walk in a DDM followed by two cortical populations (Cxe^L and Cxe^R), which we can think of as LIP neurons representing Select T$_{in}$ and Select T$_{out}$ from *Figure 16.3*. Commitment occurs (i.e., a saccade is generated) as soon as one of the population firing rates reaches the threshold. **C.** Schematic model architecture. Neural pools in the cortical network integrate sensory information and also compete against each other. They project to both the superior colliculus (SCe^L and SCe^R) and the caudate nucleus (CD^L and CD^R). Blue, populations of excitatory neurons; red, populations of inhibitory neurons. Each population is simulated by noisy spiking leaky integrate-and-fire neurons. Source: From Lo, Chung-Chuan, and Xiao-Jing Wang. 2006. "Cortico–Basal Ganglia Circuit Mechanism for a Decision Threshold in Reaction Time Tasks." Nature Neuroscience 9 (7): 956–963. Reproduced with permission of Nature Publishing Group..

decision variables.) Quantitatively, the model indicated that MT microstimulation had an effect comparable to adding 9.3% correlated motion to the RDM display. Microstimulation of LIP, in contrast, had the effect of moving the starting point of the decision variable corresponding to the site of microstimulation closer to the decision threshold. (Imagine "grabbing" the solid-line trajectory in *Figure 16.4.B* and sliding it up along the vertical axis, such that it would start off closer to the decision threshold.) For LIP, microstimulation had an effect comparable to handicapping the "race" between *choose left* and *choose right*, by moving the decision variable corresponding to the location of microstimulation 16.9% closer to the decision boundary. Other studies applying microstimulation to the frontal eye fields (FEFs) have similarly produced results consistent with a role for FEF in evidence accumulation.

A circuit model of spiking neurons

The model from Lo and Wang (2006) is a biophysically inspired spiking network model (*Figure 16.4*). That is, for even the most explicitly *neural* of the neural network models that we encountered in the previous two chapters (i.e., Stedron et al., 2005; Rougier et al., 2005; Masse et al., 2019), activation is passed from node to node as varying levels of an analog signal. Spiking network models, in contrast, explicitly simulate the analog-to-digital conversion that happens in biological neurons – the summation of postsynaptic currents that depolarize the neuron to the point of triggering the firing of one or more action potentials – then the "D to A" conversion at the synapse. When circuits in such models are recurrently connected with excitatory connections, they can achieve self-sustained stable states of activity within that circuit or, in this model from Lo and Wang (2006), evidence accumulation via "slow reverberatory network dynamics mediated by NMDA receptors" (p. 956) within and between the two modeled area LIPs (right and left hemisphere, commanding a leftward or rightward saccade, respectively). Ultimately, the decision is executed by burst neurons in the superior colliculus (SC). The commitment is achieved through a dynamic interplay between direct cortical input onto the SC and a basal ganglia-mediated input from cortex to caudate nucleus to substantia nigra pars reticulata (SNpr) to SC. The location of the threshold (i.e., how high the firing rate in LIP needs to get) is determined by the strength of synapses from cortex onto the caudate nucleus. This is because although LIP has a direct input to the SC,

the excitatory influence of this input is counteracted by the tonic inhibition from the SNpr. (Here's a sentence from section *The superior colliculus, Chapter 9,* to refresh our memory about this detail from the oculomotor control system: *at any moment in time, most of the motor units in the deep layers of the superior colliculus are under tonic inhibition from the substantia nigra pars reticulata, an arrangement analogous to the tonic inhibition of thalamus by the GPi that we detailed in Chapter 8.*)

Thus, a major conceptual contribution of the Lo and Wang (2006) model was to map the construct of the velocity of the decision variable from the DDM (i.e., its speed and its direction) to the dynamic interplay of activity in LIP, and of the construct from the DDM of starting distance from the threshold for commitment to a decision to the strength of corticostriatal synapses.

Controversy and complications

Is anyone surprised that in this realm of cognitive neuroscience with important implications for questions like how we drive, or how we make choices as consumers – how we carry out mental deliberation – there might be some controversy? No, I didn't think so. The controversy arose when research groups followed up on these studies from Shadlen and colleagues by doing the opposite of microstimulation – pharmacologically inactivating in LIP during perceptual decision making. To the surprise of many, these studies, first carried out in monkeys, subsequently in rodents, failed to find evidence that inactivating LIP (in the rodent, posterior parietal cortex [PPC]), or FEF (in the rodent, "Frontal Orienting Field" [FOF]), had any appreciable effect on decision making performance. One group, led by Princeton-based neuroscientist Carlos Brody, reasoned that the anterior dorsal striatum (ADS) of the rat (homologous to the caudate nucleus in primates) might be a candidate alternative to the PPC or the FOF, because it receives, and integrates, inputs from each of these two cortical areas, and because ramping activity reminiscent of evidence accumulation had previously been reported in this region. "The ADS," they reasoned, "is thus ideally positioned to participate in evidence accumulation as part of its established role in action selection" (Yartsev, Hanks, Yoon, and Brody [2018, p. 2 of 24]).

As illustrated in *Figure 16.5.A*, each trial began with the onset of a central light, which prompted the rat to begin engaging in "nose fixation" in the central port of a three-port testing apparatus. This triggered the delivery of two different trains of randomly timed acoustic clicks, one

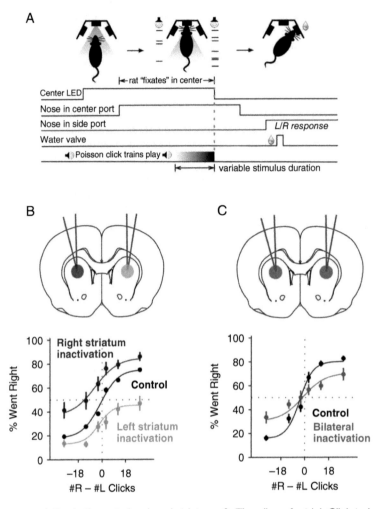

FIGURE 16.5 Evidence accumulation in the anterior dorsal striatum. **A**. Time line of a trial. Click trains were Poisson distributed, always presented at a combined rate of 40 Hz, with a lateralized difference ranging from 39:1 clicks/second (the easiest discrimination) to 26:14 clicks/second (the most difficult discrimination). **B**. Effects of unilateral muscimol infusion into the ADS. **C**. Effects of bilateral muscimol infusion into the ADS. In both **B** and **C**, control data were acquired the day before drug infusion sessions. Source: From Yartsev, M. M., Hanks, T. D., Yoon, A. M. and Brody, C. D. (2018). Causal contribution and dynamical encoding in the striatum during evidence accumulation. eLife Sciences Publications, Ltd. Public Domain.

from a speaker positioned to the left of fixation, and another from a speaker positioned to the right. Upon termination of the click trains, the central light extinguished, which cued the animal to release from the central port and enter the port on the side from which a greater number of clicks had been played. ADS activity was suppressed in one or both hemispheres via the infusion of muscimol, a GABAergic agonist. As shown in *Figure 16.5.B*, inactivation of the right ADS skewed performance dramatically,

resulting in many more "right" choices, and inactivation of the left ADS had the opposite effect. Bilateral inactivation made the psychometric curve flatter, indicating that discrimination was worse across the board.

So, do the results from Yartsev et al. (2018), together with those from the studies that failed to show effects of inactivating LIP/PPC and FEF/FOF, mean that the earlier work that we went over in so much detail has been invalidated? At the time of this writing, this remains a hotly debated question. However, the most recent development has been one that has pointed out (yet again!) that *it's not so simple*. Recall that Gold and Shadlen (2007) called attention to the fact that their procedure, of putting one of the two saccade targets within the motor field of the LIP neuron from which they were recording, "enables investigators to treat the decision as a problem of movement selection." Similarly, the findings of Yartsev et al. (2018) are best understood in the context of the "established role [of the ADS] in action selection." And so, all of the tasks that we've considered up to this point (including those from Shadlen's group) emphasized "deciding where to move." But what if, instead, LIP was assessed in a test of "deciding what you are looking at" (Zhou and Freedman 2019; p. 180)?

In addition to overseeing the study by Masse et al. (2019) that we considered near the end of *Chapter 14*, David Freedman has carried out several experiments assessing the involvement of different brain areas, including MT and LIP, in a variety of tests of visual cognition. Among these have been categorization, such as making a circular set of synthetic animals that progressively morph between a canonical dog and a canonical cat, and then training monkeys to learn which arbitrarily determined category each of the animals belongs to – that is, when shown an exemplar, the animal had to *decide* whether this exemplar belonged to category A or category B. In this experiment, Zhou and Freedman (2019) trained their animals to perform a motion-direction categorization task – five directions in one 180° arc, and five in the other – and a motion-direction discrimination task, comparable to the ones illustrated in *Figures 16.1* and *16.3*. Then they infused muscimol into LIP in one hemisphere to determine the effect of inactivating LIP on task performance. An important difference from the previous tasks, however, was that they varied whether it was the saccade with which the monkey indicated its decision that would be directed into the affected visual field, or the stimulus about which the monkey was making its decision that would be presented in the affected visual field.

The results from Zhou and Freedman (2019) indicated that inactivation of LIP had a much larger effect on performance on trials when the stimulus was presented in the impaired visual field than when the decision-guided movement had to made into the bad visual field. Thus, they concluded "LIP plays an important role in visually based perceptual and categorical decisions, with preferential involvement in stimulus evaluation compared with motor-planning aspects of such decisions (p. 181)."

Perceptual decision making in humans

Perceptual decision making has also been studied extensively in humans, for decades with behavioral studies, and also with many neurophysiological measures. The study summarized here was selected because it entailed the by-now familiar decision about RDM, but required manual button-press responses, rather than eye movements. It was performed by a group of German scientists based at institutions in Germany and in the Netherlands. The two senior authors, Andreas Engel and Pascal Fries, have both made pioneering contributions to our understanding of the functional and cognitive roles of oscillatory dynamics in the brain. Not surprisingly, then, oscillatory power is the principal dependent measure that they used.

The task required subjects to indicate, on each trial, whether or not there was coherent motion embedded in the RDM display. Magnetoencephalographic (MEG) measurements were acquired during task performance, and the analyses focused on spectral transforms of the data. Responses were lateralized – one hand for *yes*, and the other for *no* (counterbalanced across subjects) – to allow for clear isolation of *yes* hand-related vs. *no* hand-related preparatory and response activity. The researchers focused on two broad frequency bands, beta (which they operationalized as 12–36 Hz) and gamma (64–100 Hz). Prior research had indicated that limb movements are accompanied by suppression activity in the beta band and enhancement of activity in the gamma band, both typically stronger in the motor cortex contralateral than ipsilateral to the side of the movement (see *Figure 8.6*). These dynamics allowed for the computation of **choice probability** – a statistical estimate of how well a hypothetical ideal observer could predict a subject's impending choice based on differences in signal between the two hemispheres. (Conceptually, not unlike how MVPA classifier performance is evaluated.)

The findings, summarized in *Figure 16.6*, were consistent with the conclusion that had been drawn from the initial studies from Shadlen and colleagues, which is that the evolution of the decision variable can be observed in the premovement activity of the system controlling the motor effector that will make the response. In both the beta and the gamma bands, and in both ROIs, choice

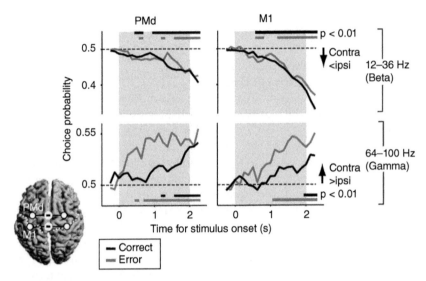

FIGURE 16.6 Buildup of choice-predictive activity in human motor cortex during perceptual decision making. Superimposed on the top-down view of a brain are the cortical regions of interest in the hand area of M1 and in premotor cortex (PMd). (The white dotted line traces the central sulcus.) The plots show choice probability (see text), estimated from two ROIs and from two frequency bands, ramping toward a decision in each cell. The shaded region indicates stimulus presentation; bars along the top and bottom indicate periods during which choice probability estimates are statistically significant. Source: From Donner, Tobias H., Markus Siegel, Pascal Fries, and Andreas K. Engel. 2009. "Buildup of Choice-Predictive Activity in Human Motor Cortex during Perceptual Decision Making." Current Biology 19 (18): 1581–1585. Reproduced with permission of Elsevier.

probability estimates began to ramp up as early as 1.5 seconds prior to the button press. In a further analysis, the authors leveraged a previous finding of theirs, in which they had found that gamma-band power in MT scales monotonically with motion coherence. They reasoned that, if the motor cortex oscillatory dynamics that they had observed were truly reflective of the evolution of the decision variable, trial-by-trial choice-predictive activity from these ROIs should relate to trial-to-trial fluctuations of the temporal integral of gamma-band activity in MT (i.e., to the strength of the momentary evidence). This prediction was borne out, thus reinforcing their interpretation of activity from precentral ROIs as reflecting the integration of sensory evidence provided by MT.

Up until now in this chapter, and really in the entire book, we've given little-to-no consideration to what is the reason that human and nonhuman research subjects even bother to perform hundreds (for humans) or thousands of trials in these cognitive neuroscience experiments that, no matter how clever and insightful in their motivation and

design, are invariably quite repetitive and boring for the subject. They do it, of course, for the rewards that they receive for diligent compliance with instructions/rules, and for accurate performance. Let's turn our attention to that.

VALUE-BASED DECISION MAKING

Most decision theories – from expected utility theory in economics (von Neumann and Morgenstern, 1944) to prospect theory in psychology (Kahneman and Tversky, 1979) to reinforcement learning theories in computer science (Sutton and Barto, 1998) – share a core conclusion. Decision makers integrate the various dimensions of an option into a single measure of its idiosyncratic subjective value and then choose the option that is most valuable. Comparisons between different kinds of options rely on this abstract measure of subjective value, a kind of "common currency" for choice. That humans can in fact compare apples to oranges when they buy fruit is evidence for this abstract common scale. (Kable and Glimcher, 2009, p. 734)

Thus, the quest for the common currency will be an important theme underlying this section of the chapter.

From the outset, value-based decision making would seem to be different in kind from the perceptual decision making that we just covered. Whereas the latter entails a decision about the objective state of the world – either the dots were drifting up or they were drifting down – the former is inherently subjective – if you randomly select two people and present them with the same choice, one can prefer the apple and one can prefer the orange, and neither of these choices can be said to be wrong. This is also called economic choice, as the quote from Kable and Glimcher indicates, and economists have studied this aspect of behavior for decades. In the context of this text-book, what we'll be discussing has come to be referred to as neuroeconomics. To get started, we'll begin with an experimental procedure that shares many surface charac-teristics with the perceptual decision making studies that we have just reviewed, and that measures neural activity in familiar territory.

The influence of expected value on activity in LIP

This study by Michael Platt and Paul Glimcher (1999), two prominent figures in neuroeconomics research, was designed from the perspective that "nearly all theories require the decision-maker to have some knowledge of two environmental variables: the gain expected to result from an action and the probability that the expected gain will be realized." In expected-value theory, for example, "a rational decision-maker should multiply expected gain by the probability of gain to establish an expected value for each course of action, and then should choose the option with the highest expected value" (p. 233). To evaluate this idea, they trained monkeys to fixate centrally while two target stimuli were presented in the periphery, one always in the movement field of the LIP neuron being recorded. Upon offset of the fixation point, the monkey was rewarded for making a saccade to either target (i.e., it was a "free choice"). For the first block of trials, the amount of juice reward associated with each of the targets (i.e., the gain) was identical. Subsequently, the gain associated with each target was modified, such that selection of one yielded 0.15 ml of juice, and the other 0.05 ml. These gains were switched between the targets, on a block-by-block basis,

such that across several blocks of trials each LIP neuron was recorded under perceptually identical conditions, but when the expected gain of selecting the target in its move-ment field was relatively high or relatively low.

The results, illustrated in *Figure 16.7.A*, revealed a marked difference in response profile of LIP neurons for high-gain vs. low-gain decisions, with the former showing elevated activity even prior to target onset. (One can spec-ulate that, even prior to target onset, the animal had "already made up its mind.") Note, however, that this was no longer true at the time of response. It is also important to note that the monkeys did not simply always choose the target with the highest expected gain. Rather, they fol-lowed the "matching law," a tendency for animals to allo-cate their choices in direct proportion to the rewards provided by each choice. (That is, they chose 0.15 ml on 75% of the trials, and 0.05 ml. on 25% of the trials.) As a result, Platt and Glimcher (1999) were able to compute an estimate of the animal's valuation of each response on a trial-by-trial basis (i.e., that response's expected value) by applying a formula from a formal theory of behavioral decision making – melioration theory – that took into account the animal's choices over the 10 preceding trials. *Figure 16.7.B* shows that activity in this LIP neuron related to expected value during all epochs of the trial, including the two that preceded target onset, with the exception of the "postmovement" epoch (which was actually time-locked to the initiation of the saccade).

Decision variables emergent from motor plans

Thus, to recap what we've learned in LIP, Shadlen and col-leagues have shown that neurons in this region participate in the accumulation of evidence – the deliberation – toward a perceptual decision (*Figure 16.3*), with Zhou and Freedman (2019) clarifying that this activity is more related to sensory evaluation than to motor planning. In the study just summarized, Platt and Glimcher (1999) have shown that, in a free-choice task that varies gains over time, LIP neurons can also represent the value of eye movements. One consequence of the latter, and related, findings is that some have proposed a reframed conception of the frontoparietal oculomotor complex. Rather than this system representing a salience map (*Chapter 9*) – which is a fundamentally attention-based construal of brain function – perhaps it represents a "saccade value" map – a fundamentally neuroeconomic construal of brain function.

FIGURE 16.7 Activity of a single LIP neuron (identified as CH980304) from kinematically identical saccades made into its movement field, but at two levels of expected gain (0.15 ml vs. 0.05 ml of juice). Source: From Platt, Michael L., and Paul W. Glimcher. 1999. "Neural Correlates of Decision Variables in Parietal Cortex." Nature 400 (6741): 233–238. Reproduced with permission of Springer Nature.

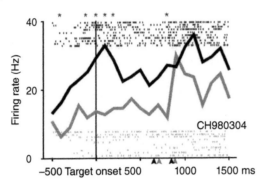

FIGURE 16.7.A Average activity, time-locked to target onset, for high (black) and low (gray) expected-value trials. Arrows along the timeline indicate, sequentially, and for each trial type, the average time of fixation offset and of saccade initiation. Asterisks indicate 100-ms bins in which firing rate differed significantly at the two levels of expected gain. Note that fixation onset preceded target onset by 500 ms.

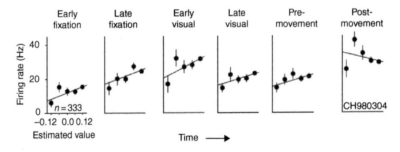

FIGURE 16.7.B Activity from the same neuron, divided into 200-ms samples, and regressed against expected value, for movements into the neuron's response field.

In-depth consideration of the relative merits of these two perspectives is beyond the scope of this book. However, there is one additional conclusion from this literature that will help us wrap up where we've been and push off toward our next destination. Studies of value-based decision making in parietal cortex have established that the representations of value found in LIP are limited to the value of decisions about where to move the eyes. Thus, by analogy to what we saw with perceptual decision making, value can be represented in the networks that will effect the value-based action. A further implication of this fact is that LIP does not seem to represent the common currency that is, in some sense, a Holy Grail of neuroeconomics. For that, neuroeconomic Sir Galahads have been more successful in the orbital and ventromedial PFC (sometimes summarized as "orbitomedial PFC" [omPFC]).

Common currency in the omPFC

It is well established in behavioral neurology that lesions of the omPFC impair choice behavior in many domains, including eating disorders, abnormal gambling, and, more generally, erratic decision making. In the monkey, extracellular recordings in omPFC have revealed neurons that encode particular flavors in a manner that is modulated by satiety. That is, a neuron may "prefer" raisins to pieces of apple, but its responses decrease in intensity as the animal gets closer to having eaten its fill of raisins. Thus, omPFC is a logical place to look for the representation of a common currency. Before we can start looking, though, we need to establish some definitions.

Transitivity, value transitivity, and menu invariance

A key property conferred by a common currency is *transitivity*. Let's return to my wine selecting dilemma from the beginning of the chapter, but simplify it, such that on one trip to my neighborhood wine shop (enigmatically named "Mike's," even though it's owned by Pat) there are only two varieties available, and I select bottle A over bottle B. Then, on a subsequent visit, the shop has sold out of A and has replaced it with C. Given this choice, I select B over C. On a third visit, if I find only A and C on the shelf, I will choose A. (To do otherwise would be to show "erratic decision making.") Another property conferred by a common currency is that the value of one item can be expressed in terms of a quantity of any other, what economists call *value transitivity*. Thus, let's also say that my wine preferences are similar in proportion to those of the monkey whose juice preferences are illustrated in *Figure 16.8.C*. If Mike's started selling wine B in bottles holding 1.3 times the volume (but at the same price), I'd be equally happy with a 975-ml bottle of B as with a 750-ml bottle of A. And for the same reason, I'd be equally happy, for the same price, leaving the shop with four bottles of C as with one bottle of A. In economics, the term "menu" refers to a set of choices. Thus, the A vs. B choice is a menu, and the B vs. C choice is a different menu. For a neural implementation of a common currency to be effective, it needs to display "menu invariance," which means that, assuming equal levels of satiety, the level of activity representing B will be the same whether it is on the A–B menu or on the B–C menu. (One can imagine that the economist(s) who coined this naming convention may have done so over a meal, at a restaurant.)

Now we're ready for the next two courses on our menu.

Evidence from the nonhuman primate

Figure 16.8.D summarizes evidence, from a study by Camillo Padoa-Schioppa and John Assad (2008), that OFC represents a common currency. The neuron illustrated in the top row represents one type of menu invariance, in that its representation of the offer value for C is the same whether it's being offered against A or against B. The neuron illustrated in the middle row encodes the value of the choice that the animal made, regardless of which two juices were "on the menu." The neuron illustrated in the third row represents a different type of menu invariance, taste B, regardless of what else is "on the menu."

There are two details about the experimental procedure employed by Padoa-Schioppa and Assad (2008)

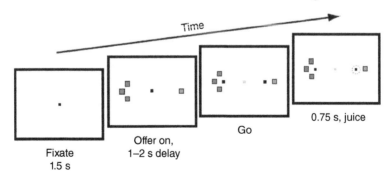

FIGURE 16.8 The representation of economic value in the OFC is invariant for changes of menu.

FIGURE 16.8.A On each trial, the monkey is made an offer that varies in terms of quantity and quality; in this case, three drops of agua frescas-flavored juice or one drop of peppermint tea. It indicated its choice by making a saccade to one of the two targets. Source: From Padoa-Schioppa, Camillo, and John A. Assad. 2008. "The Representation of Economic Value in the Orbitofrontal Cortex Is Invariant for Changes of Menu." Nature Neuroscience 11 (1): 95–102. Reproduced with permission of Nature Publishing Group.

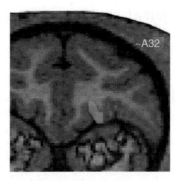

FIGURE 16.8.B Coronal slice of an MRI, with yellow shading indicating the location of area 13 m from which recordings were made. Note the orbits of the eyes, immediately ventral to this portion of cortex. Source: Padoa-Schioppa and Assad, 2008. Reproduced with permission of Nature Publishing Group.

A = grape, B = fruit punch, C= 1/2 apple

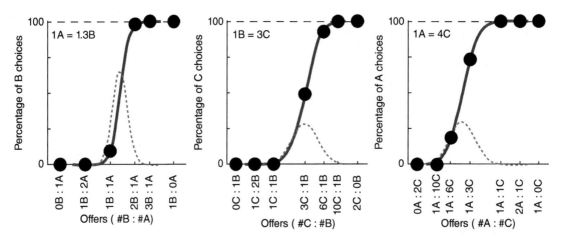

FIGURE 16.8.C Example of choice patterns from a session, featuring the three juices listed at top. This monkey liked grape juice better than fruit punch, and fruit punch better than apple juice diluted with equal parts of water (thus, "1/2 apple"). Estimation from the sigmoid fit indicates that the monkey valued 1 unit of A the same as 1.3 units of B, 1 unit of B the same as 3 units of C, and 1 unit of A the same as 4 units of C. The dotted red lines illustrate the underlying normal distributions, their width indicating the variability in these estimates. Source: From Padoa-Schioppa, Camillo, and John A. Assad. 2008. "The Representation of Economic Value in the Orbitofrontal Cortex Is Invariant for Changes of Menu." Nature Neuroscience 11 (1): 95–102. Reproduced with permission of Nature Publishing Group.

that are worth highlighting before we move on. First, the spatial layout of offers was randomized across each experiment, such that the responses illustrated in *Figure 16.8.D* were invariant with regard to the retino-topic location of the representation of an offer and to the metrics of the saccade required to make a choice. In this way, the activity in OFC seems to differ from that in LIP, or in M1 and PMd, in that, for the latter, "value modulates responses that are sensory or motor in nature . . . [whereas] neurons in the OFC encode

economic value per se, independently of visuomotor contingencies" (Padoa-Schioppa and Assad, 2008; p. 101). The second detail is that, over the course of their experiment, they used 11 different juices, recombined into 23 different juice triplets. Thus, their findings are unlikely to be stimulus specific.

Why the OFC? In interpreting their findings, Padoa-Schioppa and Assad (2008) note that "Current anatomical maps divide the medial and orbital prefrontal cortices into 22 distinct brain areas, organized in two mostly segregated

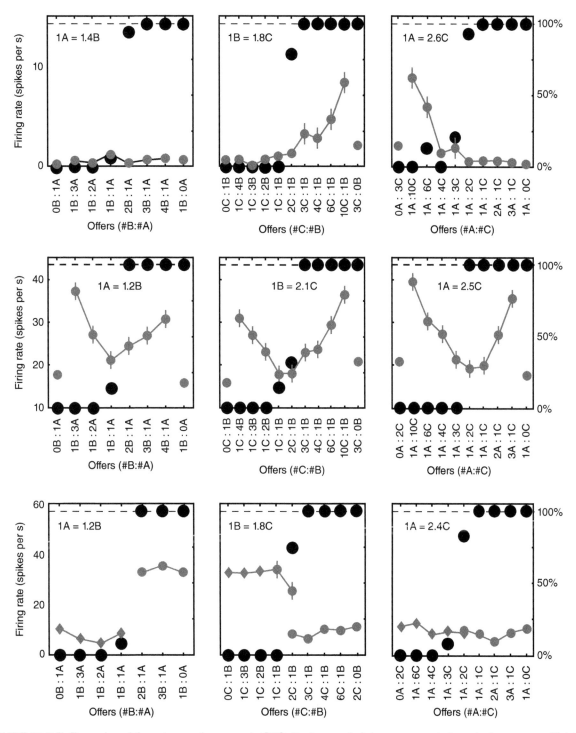

FIGURE 16.8.D Examples of three types of neurons in OFC. Each row of plots corresponds to a single neuron, with black dots indicating choices, with same conventions as *Figure 16.8.C*, except the ordinate for choices is on right side of each plot. Red dots represent firing rate. Neuron in top row encodes offer value C. Neuron in middle row encodes the "chosen value" (i.e., the difference in value between the chosen and unchosen offers, independent of the juice pair. Neuron in the bottom row encodes taste B, independent of pairing. Source: From Padoa-Schioppa, Camillo, and John A. Assad. 2008. "The Representation of Economic Value in the Orbitofrontal Cortex Is Invariant for Changes of Menu." Nature Neuroscience 11 (1): 95–102. Reproduced with permission of Nature Publishing Group.

networks. In the orbital network (including 13m and the surrounding areas [*Figure 16.8.B*]), anatomical input from all sensory modalities converges with anatomical input from limbic areas including the amygdala. The orbital network thus seems well placed to compute a quantity such as subjective value" (p. 100).

Still, *low-sugar tamarind Kool-Aid? Peppermint tea?* If these findings are to apply to human cognition, we wanna see 'em play out with factors that drive the behavior, and occupy the minds of modern humans, right? How about some money?!? How about some sex!?! (We'll get to drugs in an ensuing chapter. Not sure about the rock and roll, apart from my periodic gratuitous asides.)

Evidence from humans

The study we'll next describe, by Smith et al. (2010), has been identified as the first to provide direct evidence for a common-currency representation of value in the human. It comes from a Duke University team that included Michael Platt, whose group had previously performed a clever study of decision making with stimuli that were socially relevant for their male monkeys (Deaner,

Khera, and Platt, 2005). Not unlike the Padoa-Schioppa and Assad (2008) study, it determined value transitivity in its subjects, doing so in a "pay-per-view" scheme in which male monkeys evaluated offers pitting juice against the opportunity to view images of the backsides of female monkeys in estrus, or the opportunity to view images of male monkeys from the same colony and with a known dominance status relative to the animal in the study. In a follow-up to the monkey pay-per-view study, Platt's group established (to no one's surprise) that heterosexual university-aged human males will trade small amounts of money or expend effort to view photographs of attractive females.

Building on this context, Smith and colleagues (2010) carried out a two-part study on 26 self-reported heterosexual males, ranging in age from 18 to 28. First, during fMRI, subjects passively viewed faces of young adult women varying in attractiveness from "very attractive" to "very unattractive," and of images of paper money that varied in value from $5 to −$5 (*Figure 16.9.A*). (Attractiveness ratings were made by an independent group of raters on > 2000 images, rumored to have been

FIGURE 16.9 Common currency for money and social reward in human vmPFC. Source: Smith, Hayden, Truong, Song, Platt, and Huettel, 2010. Reproduced with permission of the Society of Neuroscience.

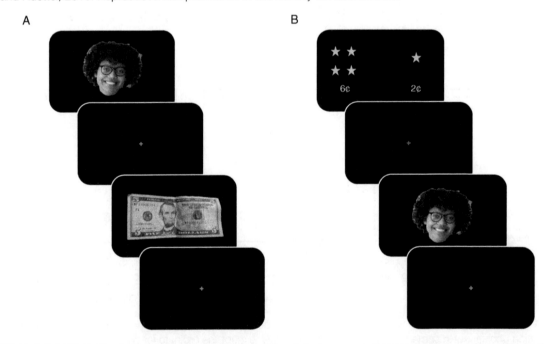

FIGURE 16.9.A AND B The first portion of the fMRI study (A) entailed no overt decision making, just passive viewing of images of faces that varied in attractiveness, and images of currency that varied in magnitude. The second portion (B) was an economic exchange task in which subjects chose, for each offer, whether to spend a larger sum of money or a smaller sum of money to view a more or less attractive face (indicated by the number of stars), respectively.

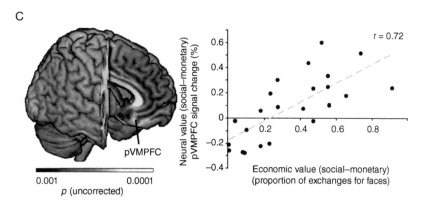

FIGURE 16.9.C Activity in posterior vmPFC, during passive viewing, predicts individual differences in willingness to exchange money to view attractive faces.

downloaded from the website www.hotornot.com.) The rationale for this first stage was to determine whether neural representations of decision values are computed even when no overt decision is being made. Immediately thereafter, while still in the scanner, subjects participated in an economic exchange task formally identical to that of Padua-Schioppa and Assad (2008), but instead of choosing between less volume of a more-preferred juice vs. more volume of a less-preferred juice, on each trial subjects chose between spending more money to view a "more-preferred" face vs. spending less money to view a "less-preferred" face. (In *Figure 16.9.B*, four stars correspond to "most attractive," and one star to "least attractive.")

Results from the economic exchange task revealed considerable individual variability in each subject's value transitivity scale – each person's "willingness to pay" (WTP) to view a more attractive face ranged from 0 to 6.4 cents/star. The authors also computed a related measure, the proportion of offers on which the subject was willing to exchange more money to view a more attractive face. This is the measure represented on the abscissa of the plot in *Figure 16.9.C*. On the ordinate is a measure of neural activity derived from the passive viewing task, computed to operationalize the extent to which brain activity "preferred" faces over money, or vice versa, with the contrast ([4-star face – 1-star face) – (large-monetary gains – large-monetary losses]). As illustrated in *Figure 16.9.C*, the one region in the brain in which this neural measure of social value predicted individual differences in value transitivity scale was located in posterior vmPFC.

Has neuroeconomics taught us anything about the economics of decision making?

Almost every paper that has been cited in this section on value-based decision making has cited decades-old principles and laws from behavioral economics: von Neumann and Morgenstern (1944); Samuelson (1937; 1947); Nash (1950a; 1950b); and so on. And so, cognitive neuroscientists, using the most cutting-edge neurophysiological measurement and analysis techniques available, are, in effect, looking for evidence that 75-year-old ideas might be implemented in the brain? Certainly, none of the work that was reviewed here discovered any new laws of economics. To address this question, let's return to the same review with which we kicked off this section:

We should be clear about the relationship between . . . neurophysiological model[s] of choice and the very similar theoretical models in economics from which [they are] derived. Traditional economic models aim only to predict (or explain) an individual's observable choices. They do not seek to explain the (putatively unobservable) process by which those choices are generated. In the famous terminology of Milton Friedman, traditional economic models are conceived of as being "as if" models (Friedman, 1953). Classic proofs in utility theory (i.e., Samuelson, 1937), for example, demonstrate that any decision-maker who chooses in a mathematically consistent fashion behaves as if they had first constructed and stored a single list of the all possible options ordered from best to worst, and then in a second step had selected the highest ordered of those available options. Friedman and nearly all of the neoclassical economists who followed him were explicit that the concept of utility was not meant to apply to anything about [its possible instantiation in the brain]. For Friedman, and for many contemporary economists (i.e., Gul and Pesendorfer, 2008), whether or not there are "neurophysiological correlates of utility" is, by construction, irrelevant (Kable and Glimcher, 2009, pp. 733–734).

The title of the Gul and Pesendorfer (2008) paper, for example, is "The Case for Mindless Economics." (Get it?) Such skepticism toward the study of the brain bases of what one studies certainly isn't limited to economists. It was noted in the *Introduction to Section I* of this book, for example, that there has been a considerable amount of push-back against cognitive neuroscience from cognitive psychology.

But, as my formerly 10-year-old daughter might have said: "What' everrrr."

First, just to state the obvious, unless our economist friends are dyed-in-the-wool dualists, they have to concede that these laws that do such an exquisite job of explaining human behavior must arise from the workings of the human brain. At a more pragmatic level, it's not their job to figure out how their laws are implemented by the brain. That's fine. Another branch of their discipline has ideas about how people who are good at one thing should focus on that one thing, and let other people do other things. Those who study neuroeconomics are not trying to derive new laws of choice. They are seeking to understand the neural mechanisms of an aspect of human behavior that influences our every waking moment, one whose dysfunction leads to misery and, in some cases, tragedy, which plays out thousands of times every day. Addiction, depression, obesity, and psychopathy are just some of the health problems that neuroeconomics research may help us to better understand, and to palliate. (Finally, for what it's worth there are many formally trained economists who have gravitated to neuroeconomics as one way to pursue their scholarly interests. I'm waiting for that email from Gul or from Pesendorfer.)

And so, *has neuroeconomics taught us anything about the economics of decision-making?*

Wrong question.

FORAGING

And now for the third category of decision-making invoked at the beginning of this chapter. There's a great deal of overlap between behavioral economics, psychological decision theory, and studies of group dynamics and optimal foraging behavior carried out under such disciplinary rubrics as anthropology, ecology, zoology, and dynamical systems. For example, the latter contributed importantly to many of the reviews by Glimcher and colleagues that we drew on in the previous section. Here, we'll consider one perspective that will give a feel for how this approach differs from the two that we have considered up to this point. Its consideration will also (once again) emphasize the overlap between the domains of "decision making" and "cognitive control."

For much of his career, Oxford University-based cognitive neuroscientist Matthew Rushworth has explored the functions of the dorsomedial frontal cortex (dmFC), including the anterior cingulate cortex (ACC). Based on his training and many of his early studies, it's fair to say that he has approached this work from the perspective that this region is important for high-level motor control and response selection. (For context, see the discussion of Picard and Strick (1996) in relation to *Figure 15.6*) In more recent work, Rushworth and colleagues have emphasized that while the majority of behavioral economics and neuroeconomics studies entail choices between two or more options, across evolutionary history primates have often been confronted with a different kind of decision. In particular, an organism often has to choose between remaining in the place where it is and exploiting the resources that are directly accessible, versus moving on to explore the environment for new, potentially more rewarding, opportunities. This foraging perspective, they argue, has implications for how one interprets activity in the dmFC.

In binary-choice decision-making tasks, dmFC signals are most often inversely related to the value difference between the chosen and unchosen options. This is what we saw in *Chapter 15, Figure 15.6*, in which difficult, or high-conflict, choices were associated with strong responses from dmFC. The alternative account from the foraging perspective, however, is that the error/difficulty/conflict accounts of this region's function may be artifacts of the two-alternative forced-choice tasks with which the majority of decision-making research is carried out. When assessed in foraging situations, dmFC signal strength is seen not to represent the unchosen option (and, therefore, a conflict), but, instead, to represent the relative benefit of exploring. One consequence of this view is that there may, in fact, be more than one common currency that guides decision making. By the foraging account, the research that we reviewed in the previous section that points to common currency representation in vmPFC/OFC is valid, and its conclusions generally solid. In addition, however, it argues that qualitatively different information may govern foraging decisions. In a foraging context there are three critical factors: the value of exploiting the option that one has encountered (encounter value); the potential value of exploring and finding something better (search value); and the energetic cost of embarking on the search (search cost). These three factors do not seem to be represented in the vmPFC/OFC, but are represented in the dmFC. Can a foraging-influenced model account for presumed functions of the dmFC that have, in other experimental contexts, been framed in terms of error, difficulty, and/or conflict? Developments in this literature will be among the more exciting to watch over the next several years. (Indeed, back in 2014, during the few months separating when I first wrote these words and when I subsequently corrected the final proofs for this book's first edition, a challenge to the foraging interpretation of dmFC had appeared from Shenhav, Straccia, Cohen, and Botvinick (2014, *Further Reading*). And between then and now (mid-2019), the Shedhav et al. (2014) paper has accrued 134 citations, not all of them in accord with its conclusions . . .)

Adolescence

Before we conclude this chapter, let's end with just a taste of one line of research that addresses a practical, social, and public-health application of decision making.

Boys being boys

So I'm sitting out on the patio in front of my house, writing a chapter about decision making. (True story.) I live in a leafy, residential neighborhood that is aptly named "University Heights." It's midmorning on a summer day and over the hill comes a group of six or so high-school-aged boys. Boisterous, jocular; you know, adolescent boys. They're obviously middle class, and in this neighborhood, the odds are that at least one of their mothers or fathers is on the faculty at the university where I teach. As they cross the street, walking through the neighbor's landscaping, one teases another sarcastically: "Oh nice, step on the flowers." Another retorts "Only step on the ones that are in bloom!" None of them actually do this, but when one notices me, he nonetheless shouts "Go Memorial," in a half-hearted attempt to fool me into thinking that they attend a different, rival high school, rather than the school in this neighborhood where they are almost certainly enrolled. The last one in the group makes eye contact with me, then, running to catch up with his buddies, deliberately kicks an oak tree sapling growing in my side yard.

The stakes were low, the consequences minor, but that kid just made a bad decision. Once I decided not to chase them down the street (*I didn't have shoes on . . . they hadn't actually done anything wrong . . . and what would I do once I caught up to them? . . .*), I was reminded of a presentation I had recently attended by developmental cognitive neuroscientist B. J. Casey. The take-home: had any of these boys been walking down the street by himself, he very likely would not have walked through the bed of flowers, nor maliciously kicked my tree.

Peer pressure

"Adolescence," as Casey, Jones, and Somerville (2011) have noted, "is a developmental period often characterized as a time of impulsive and risky choices leading to increased incidence of unintentional injuries and violence, alcohol and drug abuse, unintended pregnancy and sexually transmitted diseases" (p. 21). Casey's assessment, however, is that this is more of a social phenomenon than an indictment of the decision-making capabilities of any individual adolescent: it's not that their decision making is inherently bad, she argues, but that it's disproportionally susceptible to influence by what is popularly known as "peer pressure."

To illustrate this point, Casey's then-postdoctoral fellow Leah Somerville and colleagues (2013) scanned 69 subjects between the ages of 8 and 23. Subjects were misled into believing that there was a camera embedded in the MRI scanner, that during "video on" portions of the scan, their face would be viewed in real time by a same-sex age-matched peer. In addition to fMRI, measures of self-reported embarrassment and skin conductance (i.e., sweatiness) were collected. Model fits indicated that peak embarrassment associated with "being watched" occurred at age 17.2, peak skin conductance at age 14.4, and peak activity in a medial PFC region associated with social cognition and emotional valuation processes at age 15.8. The authors proposed that basal ganglia systems implicated in appetitive behavior are influenced by such social pressure, and that underdeveloped dorsal PFC regions that implement cognitive control can't counteract this susceptibility.

NEXT STOP

The South Shore of Long Island, a suburb of New York City, is so densely populated that what were once discrete villages have long-since merged into mile upon mile of, to an outsider's eye, unbroken suburban landscape. Although different towns have their own distinct character, without signs along the road (or announcements by the Long Island Rail Road conductor) one wouldn't know when one were leaving one town and entering another. The same might be said of *Chapters 16, 17,* and *18* – each will have a different emphasis, but the transitions between them are arbitrary, and wouldn't be noticeable without explicit chapter breaks. Prominent in the landscape of each are the estimation and representation of value, the influence of social interactions, and interactions between medial frontal cortex and subcortical structures. So, keep the facts and principles that we've just covered close at hand!

END-OF-CHAPTER QUESTIONS

1. What are the three different types of decisions that we need to make, and what distinguishes each?

2. How has microstimulation informed our understanding of perceptual decision making about the direction of motion in RDM displays?

3. How does the behavior of neurons in LIP relate to drift-diffusion models of perceptual decision making?

4. Why is the activity of LIP neurons so different on two-alternative forced-choice trials, requiring identical saccades, when the task is an RDM perceptual decision task vs. when it is a free choice between different levels of expected gain?

5. What are the currently proposed roles for the neostriatum in perceptual decision making? Of LIP?

6. How do the concepts of transitivity, value transitivity, and menu invariance relate to the idea of a common currency? In what brain areas are all these properties observed?

7. Has neuroeconomics taught us anything about the economics of decision making?

8. How do foraging decisions differ fundamentally from perceptual and value-based decisions? What brain area is implicated in the control of foraging?

REFERENCES

Casey, B. J., Rebecca M. Jones, and Leah H. Somerville. 2011. "Braking and Accelerating of the Adolescent Brain." *Journal of Research on Adolescence* 21 (1): 21–33. doi: 10.1111/j.1532-7795.2010.00712.x.

Deaner, Robert O., Amit V. Khera, and Michael L. Platt. 2005. "Monkeys Pay Per View: Adaptive Valuation of Social Images by Rhesus Macaques." *Current Biology* 15 (6): 543–548. doi: 10.1016/j.cub.2005.01.044.

Donner, Tobias H., Markus Siegel, Pascal Fries, and Andreas K. Engel. 2009. "Buildup of Choice-Predictive Activity in Human Motor Cortex during Perceptual Decision Making." *Current Biology* 19 (18): 1581–1585. doi: 10.1016/j.cub.2009.07.066.

Friedman, Milton. 1953. *Essays in Positive Economics.* Chicago, IL: University of Chicago Press.

Gold, Joshua I., and Michael N. Shadlen. 2007. "The Neural Basis of Decision Making." *Annual Review of Neuroscience* 30: 535–574. doi: 10.1146/annurev.neuro.29.051605.113038.

Gul, Faruk, and Wolfgang Pesendorfer. 2008. "The Case for Mindless Economics." In *The Foundations of Positive and Normative Economics: A Handbook,* edited by Andrew Caplin and Andrew Schotter (pp. 3–41). New York: Oxford University Press.

Hanks, Timothy D., Jochen Ditterich, and Michael N. Shadlen. 2006. "Microstimulation of Macaque Area LIP Affects Decision-Making in a Motion Discrimination Task." *Nature Neuroscience* 9 (5): 682–689. doi: 10.1038/nn1683.

Kable, Joseph W., and Paul W. Glimcher. 2009. "The Neurobiology of Decision: Consensus and Controversy." *Neuron* 63 (6): 733–745. doi: 10.1016/j.neuron.2009.09.003.

Kahneman, Daniel, and Amos Tversky. 1979. "Prospect Theory: An Analysis of Decision Under Risk." *Econometrica* 47 (2): 263–291. doi: 10.2307/1914185.

Lo, Chung-Chuan, and Xiao-Jing Wang. 2006. "Cortico–Basal Ganglia Circuit Mechanism for a Decision Threshold in Reaction Time Tasks." *Nature Neuroscience* 9 (7): 956–963. doi: 10.1038/nn1722.

Masse, N. Y., Yang, G. R., Song, H. F., Wang, X.-J. and Freedman, D. J. 2019. "Circuit Mechanisms for the Maintenance and Manipulation of Information in Working Memory". *Nature Neuroscience* 22: 1159–1167. https://doi.org/10.1038/s41593-019-0414-3

Nash, John R. 1950a. "The Bargaining Problem." *Econometrica: Journal of the Econometric Society* 18: 155–162.

Nash, John R. 1950b. "Equilibrium Points in n-Person Games." *Proceedings of the National Academy of Sciences of the United States of America* 36: 48–49.

Newsome, William T., Kenneth H. Britten, and Anthony Movshon. 1989. "Neuronal Correlates of a Perceptual Decision." *Nature* 341 (6237): 52–54. doi: 10.1038/341052a0.

Padoa-Schioppa, Camillo, and John A. Assad. 2008. "The Representation of Economic Value in the Orbitofrontal Cortex Is Invariant for Changes of Menu." *Nature Neuroscience* 11 (1): 95–102. doi: 10.1038/nn2020.

Picard, Nathalie, and Peter L. Strick. 1996. "Motor Areas of The Medial Wall: A Review of their Location and Functional Activation." *Cerebral Cortex* 6: 342–353. doi: 10.1093/cercor/6.3.342.

Platt, Michael L., and Paul W. Glimcher. 1999. "Neural Correlates of Decision Variables in Parietal Cortex." *Nature* 400 (6741): 233–238. doi: 10.1038/22268.

Rougier, Nicolas P., David C. Noelle, Todd S. Braver, Jonathan D. Cohen, and Randall C. O'Reilly. 2005. "Prefrontal Cortex and Flexible Cognitive Control: Rules without Symbols." *Proceedings of the National Academy of Sciences of the United States of America* 102 (20): 7338–7343. doi: 10.1073/pnas.0502455102.

Salzman, C. Daniel, Kenneth H. Britten, and William T. Newsome. 1990. "Cortical Microstimulation Influences Perceptual Judgements of Motion Direction." *Nature* 346 (6280): 174–177. doi: 10.1038/346174a0.

Samuelson, Paul A. 1937. "A Note on the Measurement of Utility." *Review of Economic Studies* 4 (2): 155–161. doi: 10.2307/2967612.

Samuelson, Paul A. 1947. *Foundations of Economic Analysis.* Vol. 80 of *Harvard Economic Studies.* Cambridge, MA: Harvard University Press.

Smith, David V., Benjamin Y. Hayden, Trong-Kha Truong, Allen W. Song, Michael L. Platt, and Scott A. Huettel. 2010. "Distinct Value Signals in Anterior and Posterior Ventromedial Prefrontal Cortex." *Journal of Neuroscience* 30 (7): 2490–2495. doi: 10.1523/JNEUROSCI.3319-09.2010.

Somerville, Leah H., Rebecca M. Jones, Erika J. Ruberry, Jonathan P. Dyke, Gary Glover, and B. J. Casey. 2013. "The Medial Prefrontal Cortex and the Emergence of Self-Conscious Emotion in Adolescence." *Psychological Science* 24(8): 1554–1562. doi: 10.1177/0956797613475633.

Stedron, Jennifer Merva, Sarah Devi Sahni, and Yuko Munakata. 2005. "Common Mechanisms for Working Memory and Attention: The Case of Perseveration with Visible Solutions." *Journal of Cognitive Neuroscience* 17 (4): 623–631. doi: 10.1162/0898929053467622.

Sutton, Richard S., and Andrew G. Barto. 1998. *Reinforcement Learning: An Introduction.* Cambridge, MA: MIT Press.

von Neumann, John, and Oskar Morgenstern. 1944. *Theory of Games and Economic Behavior.* Princeton, NJ: Princeton University Press.

Yartsev, M. M., Hanks, T. D., Yoon, A. M. and Brody, C. D. 2018. "Causal Contribution and Dynamical Encoding in the Striatum During Evidence Accumulation." *eLife* 7: e34929. doi.org/10.7554/eLife.34929

Zhou, Y. and Freedman, D. J. 2019. "Posterior Parietal Cortex Plays a Causal Role in Perceptual and Categorical Decisions." *Science* 365: 180–185.

OTHER SOURCES USED

Gold, Joshua I., and Michael N. Shadlen. 2000. "Representation of a Perceptual Decision in Developing Oculomotor Commands." *Nature* 404 (6776): 390–394. doi: 10.1038/35006062.

Gold, Joshua I., and Michael N. Shadlen. 2001. "Neural Computations that Underlie Decisions about Sensory Stimuli." *Trends in Cognitive Sciences* 5 (1): 10–16. doi: 10.1016/S1364-6613(00)01567-9.

Kolling, Nils, Timothy E. J. Behrens, Rogier B. Mars, and Matthew F. S. Rushworth. 2012. "Neural Mechanisms of Foraging." *Science* 336 (6077): 95–98. doi: 10.1126/science.1216930.

Öngür, Dost, and Joel L. Price. 2000. "The Organization of Networks within the Orbital and Medial Prefrontal Cortex

of Rats, Monkeys and Humans." *Cerebral Cortex* 10 (3): 206–219. doi: 10.1093/cercor/10.3.206.

Padoa-Schioppa, Camillo, and John A. Assad. 2006. "Neurons in the Orbitofrontal Cortex Encode Economic Value." *Nature* 441 (7090): 223–226. doi: 10.1038/nature04676.

Ratcliff, Roger, Anil Cherian, and Mark Segraves. 2003. "A Comparison of Macaque Behavior and Superior Colliculus Neuronal Activity to Predictions from Models of Two-Choice Decisions." *Journal of Neurophysiology* 90 (3): 1392–1407. doi: 10.1152/jn.01049.2002.

Ratcliff, Roger, and Jeffrey N. Rouder. 1998. "Modeling Response Times for Two-Choice Decisions." *Psychological Science* 9 (5): 347–356. doi: 10.1111/1467-9280.00067.

Rushworth, Matthew F. S., Nils Kolling, Jérôme Sallet, and Rogier B. Mars. 2012. "Valuation and Decision-Making in Frontal Cortex: One or Many Serial or Parallel Systems?" *Current Opinion in Neurobiology* 22 (6): 946–955. doi: 10.1016/j.conb.2012.04.011.

Shadlen, Michael N., and William T. Newsome. 1996. "Motion Perception: Seeing and Deciding." *Proceedings of the National Academy of Sciences of the United States of America* 93 (2): 628–633. doi: 10.1073/pnas.93.2.628.

Shadlen, Michael N., and William T. Newsome. 2001. "Neural Basis of a Perceptual Decision in the Parietal Cortex (Area LIP) of the Rhesus Monkey." *Journal of Neurophysiology* 86 (4): 1916–1936.

Wang, X.-J. 2008. Decision Making in Recurrent Neuronal Circuits. *Neuron* 60: 215–234.

FURTHER READING

Ebitz, R. Becket, and Benjamin Hayden. 2016. Dorsal Anterior Cingulate: a Rorschach test for cognitive neuroscience. Nature Neuroscience 19: 1278–1279. doi: 10.1038/nn.4387.
A "News and Views" overview accompanying two more recent articulations of the foraging vs. conflict interpretations of ACC function.

Glimcher, Paul W. 2010. *Foundations of Neuroeconomic Analysis*. Oxford: Oxford University Press.
As of this writing, the definitive statement of the development, current status, and future potential of neuroeconomics.

Gold, Joshua I., and Michael N. Shadlen. 2007. "The Neural Basis of Decision Making." *Annual Review of Neuroscience* 30: 535–574. doi: 10.1146/annurev.neuro.29.051605.113038.
Authoritative review of visual perceptual decision making, including thorough treatment of application of SDT.

Kable, Joseph W., and Paul W. Glimcher. 2009. "The Neurobiology of Decision: Consensus and Controversy." *Neuron* 63 (6): 733–745. doi: 10.1016/j.neuron.2009.09.003.
A broad-ranging and synthetic overview of neuroeconomics.

Rushworth, Matthew F. S., Nils Kolling, Jérôme Sallet, and Rogier B. Mars. 2012. "Valuation and Decision-Making in Frontal Cortex: One or Many Serial or Parallel Systems?" *Current Opinion in Neurobiology* 22 (6): 946–955. doi: 10.1016/j.conb.2012.04.011.
A comprehensive summary of the foraging perspective of the role of dmFC in control and decision making.

Schultz, Wolfram. 2013. "Updating Dopamine Reward Signals." *Current Opinion in Neurobiology* 23 (2): 229–238. doi: 10.1016/j.conb.2012.11.012.
Introduced in the previous chapter, but central to decision making, the most up-to-date statement, at the time of this writing, of our understanding of dopamine reward signals.

Shenhav, Amitai, Mark A. Straccia, Jonathan D. Cohen, and Matthew M. Botvinick. 2014. "Anterior Cingulate Engagement in a Foraging Context Reflects Choice Difficulty, Not Foraging Value." *Nature Neuroscience* 17: 1249–1254. doi: 10.1038/nn.3771.
A response to the foraging interpretation of cognitive control-related functions of dmFC.

CHAPTER 17
SOCIAL BEHAVIOR

KEY THEMES

- The trustworthiness of an actor in an economic exchange influences one's delay-discounting function – a variable from neuroeconomics – highlighting the fact that social context can influence many, if not all, of the behavioral and cognitive processes that we have studied to date.

- The case of Phineas Gage, whose medial PFC was destroyed in a freak rock-blasting accident, and whose personality was dramatically altered as a consequence, highlights the role of this region in controlling social behavior.

- Neural abnormalities in the vmPFC and its connections to medial temporal regions may be a factor in the "psychiatric" condition of psychopathy.

- Neuroscientific study of "theory of mind" (ToM), the ability to represent the thoughts and motivations of others, is an informative lens through which to observe how scientists go about studying the neural bases of social behavior – the methods, the theoretical perspectives, the complications, and the controversies.

- One hotly contested conceptual battleground has been whether the functions of the right temporoparietal junction (RTPJ) are better characterized as being specific to ToM and moral decision-making, or more generally recruited by tasks requiring attentional reorienting and, perhaps, the comparison of internal models of the world with external sensory evidence.

- Observing the behavior of others – an inherently social act – can be an important way to learn new information, even if we only observe others' actions but not the outcome(s) of those actions. Model-based analysis of fMRI provides evidence that different kinds of error prediction signals, processed by different brain areas at different times during learning, may underlie this otherwise perplexing phenomenon.

CONTENTS

TRUSTWORTHINESS: A PREAMBLE

As highlighted by the anecdote of boys kicking flowers (section *Boys being boys, Chapter 16*), people often behave differently when others are present than when they are alone. Because all behavior derives from the workings of the CNS, it must be that such social influences on our behavior are, themselves, also mediated by the brain. This should be reason enough to want to study and understand them. But there's more to it than that. This brief preamble illustrates just one example of how those who wish to assemble a comprehensive model of the neural bases of human cognition need to incorporate social factors. Otherwise, at best, theirs will be an incomplete model.

Delaying gratification: a social influence on a "frontal" class of behaviors

We've seen many examples over the past few chapters of failures of cognitive control being associated with the integrity of the PFC. One hallmark class of behaviors that we haven't considered explicitly is delaying gratification – being able to withhold the impulse to take an immediate reward in favor of achieving a longer term goal. One's ability to delay gratification seems to be an important trait that predicts success in many aspects of life. For example, during the preschool years, a period when the PFC is underdeveloped relative to other neural systems, it has been shown that the ability to resist a desirable immediate treat in favor of a larger delayed one predicts higher levels of success later in life, as indexed, for example, by higher scores on standardized tests and better social competence in adolescence, and lower rates of obesity and of substance abuse in adulthood.

In behavioral economics, and, therefore, more recently, in neuroeconomics, delaying gratification has been operationalized with **intertemporal choice** tasks. These measure the phenomenon of *delay discounting*, demonstrated by the fact that people tend to prefer a smaller amount of money (or some other reward) immediately over the promise of a larger amount of money at a later date. (It's "as if" they are discounting the value of the larger amount, because of the delay before they'd be able to access it.) Intertemporal choice behavior has been used by many to index behavioral impulsivity, a factor that is sensitive to the integrity of frontostriatal systems. Further, because delay discounting is also observed in nonhuman animals, including rodents and monkeys, it is at a rich nexus of overlapping research domains, including behavioral neuroscience, behavioral neuropharmacology, behavioral genetics, and neuroeconomics. To list just a few examples, in rats, it can

be studied via manipulation of neurotransmitter systems, such as serotonin, and in combination with drugs, such as methylphenidate (commercial names include Ritalin and Adderall), that act on noradrenergic and dopaminergic systems; in the monkey, in relation to the activity of midbrain dopaminergic neurons; in humans, as a function of variation in genotype that relates to dopamine processing; and so on. It is seen as an important tool for studying factors underlying **attention deficit hyperactivity disorder** (ADHD).

What does this all have to do with social behavior? A team lead by Yuko Munakata (who we previously encountered in section *Is the A-not-B error a failure of working memory?, Chapter 15*) offers one demonstration by pointing out that social cognition adds another dimension to the phenomenon:

> delaying gratification also relies on the fundamental assumption that a future reward will be delivered as promised . . . [and so] is not simply about choosing "more later" over "some now," but rather, requires choosing "*maybe* more later" over "some now," when there is doubt about whether those promising the future reward would come through [i.e., would honor their commitment to deliver the reward in the future]. (Michaelson, de la Vega, Chatham, and Munakata [2013; p. 1])

To show this, they designed a study in which, on an intertemporal choice task, they manipulated the perceived trustworthiness of one party in the exchange. Adult subjects read three vignettes, depicting a person who was trustworthy, untrustworthy, or neutral, and then judged delay gratification situations involving these individuals. (For example, "If [individual x] offered you $40 now or $65 in 70 days, which would you choose?") Their findings show that the probability that one will delay gratification depends, in part, on the perceived trustworthiness of the other actor in the exchange. And because one of their studies manipulated perceived trustworthiness as a within-subject factor, the influence of this factor must be, at least in part, independent of frontal-lobe functioning, neurotransmitter levels, genotype, and other explicitly neurobiological factors that are also known to influence intertemporal choice.

Now, let's think all the way back to section *Construct validity in models of cognition, Chapter 1*. The preamble to this current chapter directly invokes the admonition that not all phenomena that have a psychological reality ("conscientiousness" in *Chapter 1*, and "trustworthiness" here) are necessarily amenable to direct study with the methods of cognitive neuroscience. That still seems reasonable to me. But now we find ourselves in a situation where, if we want to fully understand this aspect of behavior – delay discounting — we will have to start thinking about incorporating the neural processing of information about the trustworthiness of other parties in the exchange. Just walking away from the problem because "it would be very difficult to study with the methods of cognitive neuroscience" isn't an option. As noted previously, for example, this line of research on delay discounting has made important contributions to understanding many aspects of human health. And so what this means is that a problem that we're interested in has (yet again) gotten more complicated.

Just a moment's reflection indicates that social factors are likely to influence many, if not all, of the behaviors and cognitive processes whose neural bases we have considered up to this point in this book. (For example, action selection/ motor control? The word "social" didn't appear once in *Chapter 8*, yet the influence of social context on the skeletomotor control of that tree-kicking cretin called out at the end of *Chapter 16* is undeniable.) Thus, we need to develop an understanding of how to study aspects of behavior, and influences on behavior, that are inherently social.

THE ROLE OF vmPFC IN THE CONTROL OF SOCIAL COGNITION

We'll start our consideration of social neuroscience with this brain region that featured prominently in the previous chapter's treatment of value-based decision-making.

Phineas Gage

If there's a neurological patient who might challenge H.M.'s status as most cited case in the brain sciences, it's Phineas Gage. The fantastic nature of his accident and injury, and its profound behavioral consequences, ensures that he appears in every textbook, and in more than his fair share of introductory slides in scientific presentations.

Horrible accident

In 1848, Gage was the foreman of a crew that was blasting rock to create a route through the mountains of south-central Vermont for a new line for the Rutland and Burlington Railroad. At that time, this entailed drilling a deep, narrow hole into the rock, filling it with blasting powder, then a fuse, then sand, compressing the charge with a metal tamping iron (of, say, 3′7″ in length and 1.25″ in diameter; and did we say fashioned out of *metal*?), then presumably walking a long distance away, and yelling something like "Fire in the hole!" before lighting the fuse. An article from the *Boston Post* picks up the story from here:

Horrible Accident — As Phineas P. Gage, a foreman on the railroad in Cavendish, was yesterday engaged in tamping for a blast, the power exploded, carrying an iron instrument through his head an inch and a fourth in circumference and three feet and eight inches [*sic*] in length, which he was using at the time. The iron entered on the side of this face, shattering the upper jaw [*sic*], and passing back of the left eye, and out at the top of the head.

The most singular circumstance connected with the melancholy affair is that he was alive at two o'clock this afternoon [i.e., on the following day], and in full possession of his reason, and free from pain — Ludlow, Vt., Union. (September 21, 1848)

(Everything qualified with a "*sic*" is not strictly accurate; but what's an inch here or there?) Most singular indeed. And the fact that he was to live for not just a day after the accident, but for nearly 12 additional years, and that a detailed assessment was made of the lasting consequences of his injuries, means that contemporary scientists have gone to great lengths to reconstruct the damage likely incurred by his brain (*Figure 17.1* and *Figure 17.2*).

FIGURE 17.1 Daguerrotype of Phineas Gage, taken years after his accident, holding the tamping iron that was "his constant companion for the remainder of his life." Source: Originally from the collection of Jack and Beverly Wilgus, and now in the Warren Anatomical Museum, Harvard Medical School. https://commons.wikimedia.org/ wiki/File:Phineas_Gage_Cased_Daguerreotype_ WilgusPhoto2008-12-19_Unretouched_Color.jpg. Licensed under CC BY-SA 3.0.

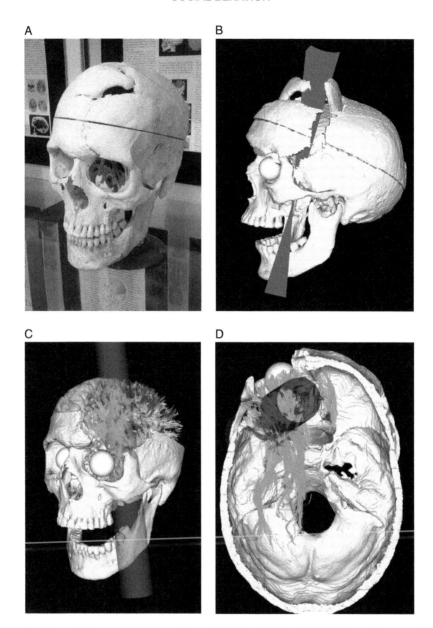

FIGURE 17.2 Reconstruction of damage to Phineas Gages's brain. **A.** The skull of Phineas Gage on display at the Warren Anatomical Museum at Harvard Medical School. **B.** The set of possible rod trajectory centroids, estimated from CT scans of the skull. **C.** A rendering of the Gage skull with the best fit rod trajectory and example fiber pathways in the left hemisphere intersected by the rod. **D.** Same data as **C**, but rendered in radiological conventions, and portrayed as though viewed from the underside, with a reconstruction of the ventral surface of the brain. Gray shading is the estimate of frank lesion to the left frontal lobe (note that circle within this region indicates the actual diameter of the tamping iron). Fiber pathways estimated to have been severed by the path of the tamping iron are illustrated with colors representing the orientation of the tract (e.g., green is parallel to the Y dimension, yellow-red to X, and blue to Z). Source: Van Horn, Irimia, Torgerson, Chambers, Kikinis, and Toga, 2012. Reproduced with permission of the authors.

Sequelae

Why all the fuss? Here's the most-quoted section from the first thorough account in the medical literature of the behavioral consequences of Gage's injury, published eight years after his death:

> The equilibrium or balance, so to speak, between his intellectual faculties and animal propensities, seems to have been destroyed. He is fitful, irreverent, indulging at times in the grossest profanity (which was not previously his custom), manifesting but little deference for his fellows, impatient of restraint or advice when it conflicts with his desires, at times pertinaciously obstinate, yet capricious and vacillating, and devising many plans of future operations, which are no sooner arranged than they are abandoned in turn for others appearing more feasible. A child in his intellectual capacity and manifestations, he has the animal passions of a strong man. Previous to his injury, although untrained in the schools, he possessed a well-balanced mind, and was looked upon by those who knew him as a shrewd, smart businessman, very energetic and persistent in executing all his plans of operation. In this regard his mind was radically changed, so decidedly that his friends and acquaintances said he was 'no longer Gage.' (Harlow, 1868, pp. 339–340)

The extent to which North American scientific culture was underdeveloped at that time, relative to that of Europe, is evident in the fact that it was decades before the scientific community (a) first learned of this case, then (b) fully accepted the veracity of its details, and finally (c) incorporated it into models of the behavioral functions of the medial PFC. (In 1877, David Ferrier [section *The localization of visual perception, Chapter 1*] wrote to the chair of Harvard's department of psychology to inquire about reports he had heard about this case. His letter included the following observation, too rich to not share: "In investigating reports on diseases and injuries of the brain, I am constantly amazed at the inexactitude and distortion to which they are subject by men who have some pet theory to support. The facts suffer so frightfully that I feel obliged always to go to the fountainhead – dirty and muddy though this frequently turns out" [Macmillan, 2000, p. 464]. *Why can't I have been the one to have said this? . . .*)

Contemporary behavioral neurology

Mesulam (1985) has summarized the frontal cortex as being organized into three divisions: motor–premotor, paralimbic, and heteromodal. It is the paralimbic that has been most strongly associated with social behavior. Comprising anterior cingulate and caudal orbitofrontal regions, it is characterized by a cytoarchitecture that is transitional from evolutionarily older archicortex of the hippocampus and piriform cortex of the olfactory system, at one boundary, and six-layer "granular" neocortex at the other. (Note that the caudal orbitofrontal component of paralimbic cortex corresponds to the ventromedial [vm] PFC that featured prominently in section *Common currency in the omPFC, Chapter 16*.) In an editorial that accompanied the publication of the two reports from Lhermitte (1986) that we considered in section *The lateral frontal-lobe syndrome, Chapter 14*, Mesulam (1986) notes that input from regions processing internal physiological states and "extensively preprocessed sensory information . . . suggests that [paralimbic] areas should be of crucial importance for channeling drive and emotion to appropriate targets in the environment . . ." Then, with explicit reference to social behavior, he continues:

> Experimental lesions in the paralimbic component of the frontal lobe also interfere with social interactions. In social animals such as monkeys, group bonds heavily rely on specific aggressive and submissive displays, grooming behavior, and vocalizations. Success depends on directing the proper behavior to the proper individual in the proper context. Animals with orbitofrontal lesions show a severe disruption of these conspecific affective and affiliative behaviors and eventually experience social isolation . . . These experiments provide a model for the socially maladaptive and emotionally inappropriate behaviors seen after frontal lobe damage and support the contention that these behaviors emerge after involvement of the paralimbic component in the frontal lobe (pp. 322–323).

Linking neurological and psychiatric approaches to social behavior

Consistent with the perspective represented in the preceding quote is work from Michael Koenigs and his colleagues at the University of Wisconsin–Madison, who have studied both patients with lesions of the vmPFC, and incarcerated inmates who have been diagnosed with psychopathy. Psychopathy is a psychiatric condition associated with callous and impulsive antisocial behavior, and the rationale behind some of Koenigs' research is that "For decades, neurologists have noted that the personality changes accompanying vmPFC damage (e.g., lack of empathy, irresponsibility, and poor decision-making) bear striking resemblance to hallmark psychopathic personality traits," to the extent that "the personality changes associated with vmPFC damage have been dubbed 'pseudopsychopathy' and 'acquired sociopathy'" (Motzkin, Newman, Kiehl, and Koenigs, 2012, p. 17348). Prompted by this similarity, Koenigs teamed up with clinical psychologist Joseph Newman to administer tests of economic decision-making to incarcerated criminals with diagnoses of psychopathy. Their tasks, like many we considered in *Chapter 16*, comprised a series of offers. However these offers, like the "prisoner's dilemma" and other assessments from game theory that some readers may be familiar with, were designed to assess the extent to which people's behavior in social situations can deviate from what a rational choice should be. Their findings (Koenigs, Kruepke, and Newman, 2010) indicated that psychopaths, like patients with vmPFC lesions, were disproportionately sensitive to offers that are perceived as unfair – that is, how fairly one thinks one is being treated by another person,

and/or one's assessment of the motives of that person, influences how one reasons about money. The next step was to see if their brains showed evidence of abnormality in the vmPFC. Enter Kent Kiehl.

Kiehl (profiled in Seabrook, 2008, *Further Reading*), a professor at the University of New Mexico, has a mobile MRI scanner, built into a trailer that can be towed around the continent, and well-worked-out specifications for how the local contractor is to pour the concrete slab in the prison yard to provide the needed stability for the scanner. On the strength of their behavioral findings, Koenigs and Newman teamed up with Kiehl to set up his scanner on the grounds of a prison and enroll incarcerated psychopaths. The first of their findings, illustrated in *Figure 17.3*,

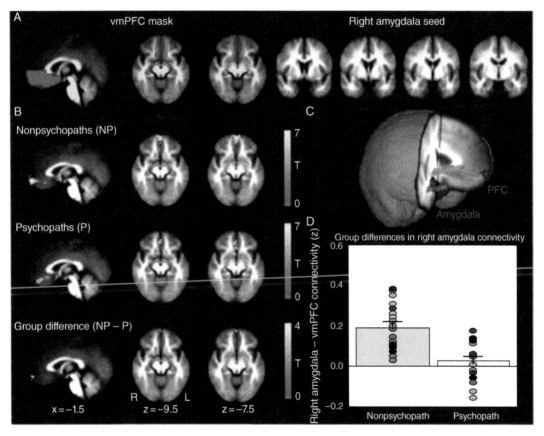

FIGURE 17.3 Amygdala–vmPFC connectivity differs between psychopathic and nonpsychopathic incarcerated inmates. **A.** Illustrates the amygdala seed and vmPFC mask used for resting-state connectivity analysis. **B.** Illustrates regions of significant resting-state functional correlation in the two groups, and the difference map. Similarly decreased functional connectivity was found between medial parietal cortex and vmPFC. **C.** Illustrates the path of the uncinate fasciculus, the primary white-matter pathway connecting vmPFC with the anterior medial temporal lobe, shown by DTI to be compromised in the psychopath group. **D.** Shows individual subject values that went into **B**. Source: From Motzkin, Julian C., Joseph P. Newman, Kent A. Kiehl, and Michael Koenigs. 2012. "Reduced Prefrontal Connectivity in Psychopathy." Journal of Neuroscience 31 (48): 17348 –17357. Reproduced with permission of the Society of Neuroscience.

emphasizes the message also evident in *Figure 17.2*, which is that we can't lose sight of the fact that no region of the brain does anything in isolation – understanding connections is central to understanding function. Here's how Motzkin et al. (2012) set it up:

> Both the amygdala and precuneus/PCC are densely and reciprocally connected with vmPFC . . . and both areas have been associated with reduced activity in psychopathy . . . Moreover, the interactions between vmPFC and these areas are thought to subserve key functions related to psychopathy. vmPFC–amygdala interactions are thought to underlie aspects of emotion regulation, aggression, and stimulus-reinforcement associations [we'll get to these in *Chapter 18*], while vmPFC–precuneus/PCC interactions are thought to underlie aspects of self-reflective processing (p. 17348).

For the study, they recruited prison inmates who did and did not have a diagnosis of psychopathy, and acquired T1-weighted MRI, diffusion tensor imaging (DTI), and resting-state fMRI data (section *Diffusion tensor imaging – in vivo "tract tracing"; T1-weighted imaging; Resting state functional correlations, Chapter 3*). *Figure 17.3* illustrates the connectivity results, which do, indeed, implicate some of the tracts also estimated to have been compromised by Phineas Gage's injury. Another analysis, by Martina Ly et al. (2012; not shown here), found evidence for cortical thinning in many areas, although none of them in vmPFC.

There are many, many fascinating questions being addressed by social neuroscience. An excellent and thorough summary of the breadth of this field can be found in Adolphs (2010, *Further Reading*). Here, as has been the practice in previous chapters, we will focus on just a few specific topics, so as to be able to drill down to a level of detail that allows us to engage at a substantive level with its key themes and problems.

THEORY OF MIND

One concept that is central to many accounts of cognitive development, and of adult social cognition, is the theory of mind (ToM). As described by Gallagher and Frith (2003), "It underpins our ability to deceive, cooperate and empathize, and to read others' body language. It also enables us to accurately anticipate other people's behaviour, almost as if we had read their minds" (p. 77). The ToM literature is considered in this section because the core questions underlying many claims, and debates, within this subfield

of social neuroscience have relevance to other subfields of social neuroscience and, more broadly, to other areas of cognitive neuroscience. An additional reason that this literature is of interest to many is the widely (although not universally) held view that impairments of ToM may underlie one or more variants of **autism spectrum disorder** (ASD).

The ToM network

ToM reasoning has been theorized to require two representational steps. First, one must represent another person as an individual. Second, one must represent that person's mental states. This second stage – engaging in mentation about the thoughts, motivations, and knowledge of another – is referred to as "mentalizing." The British cognitive neuroscientist Christopher Frith was the senior author on one of the first neuroimaging studies to explicitly study ToM mentalizing (Fletcher et al., 1995), and has subsequently made many influential contributions to the field. This first study, and many subsequent neuroimaging studies of ToM, used a story comprehension procedure in which subjects read brief vignettes, answering a question about each, and the contrast of interest compared activity associated with stories that required ToM mentalization vs. stories that did not. In their 2003 review, Gallagher and Frith emphasize three regions, each contributing differently to performance on theory-of-mind tasks. Two of these are in the temporal lobes, and the authors suggest that these aren't involved in mentalizing per se, but perhaps in functions that are necessary precursor steps. To the right posterior superior temporal sulcus they ascribe "understanding the meaning of stories and cartoons involving people, with or without the requirement to mentalize . . . understanding causality and intentionality . . . the attribution of intentions to the movements of geometric shapes . . . and taking the self-perspective" (pp. 80–81). They note that this region has also been associated with the perception of biological motion, including movements of the hands, body, lips, and eyes. To the temporal poles they ascribe retrieval of semantic and episodic memory that can be critical for succeeding on a ToM task. The anterior paracingulate cortex, in contrast, is argued to be recruited for tasks in which subjects take "an 'intentional stance', which is to treat a system as an agent, attributing to it beliefs and goals" (p. 79). This is seen, for example, when the subject is playing the game "rock, paper, scissors" (in the UK, "stone, paper, scissors") in two conditions, one in which they believe that the opponent is a computer (in which the anterior paracingulate cortex is not engaged), and the second in which they believe that the opponent is a human being outside the scanner (in which case the anterior paracingulate cortex is engaged). (This anterior paracingulate region corresponds closely to the "medial PFC" (MPFC) region that another group routinely identifies with its "ToM localizer" task, and which will be illustrated in *Figure 17.6*).

Two asides about anterior paracingulate cortex

First, the anterior paracingulate region implicated in ToM mentalizing by Frith and colleagues is in a similar location to the one identified by Somerville et al. (2013) when their adolescent subjects thought they were being watched while in the scanner (section *Peer pressure, Chapter 16*).

Second, there is a fascinating variant of frontotemporal dementia (FTD), the same category of disease that produces primary progressive aphasia of the semantic subtype (a.k.a. semantic dementia; section *The progressive loss of knowledge, Chapter 13*), which prominently affects the medial frontal cortex, including vmPFC and anterior paracingulate cortex, in addition to a focal region of the anterior insular cortex, deep within the Sylvian fissure. It is referred to as the behavioral variant of FTD (bvFTD), and manifests as a progressive decline in control over emotion, social behavior, personal conduct, and decision-making (see Seeley, Zhou, and Kim, 2012, *Further Reading*).

Now we move on to the posterior superior temporal sulcus.

The temporoparietal junction (TPJ)

In an influential, and controversial, paper, Rebecca Saxe and her then-PhD advisor Nancy Kanwisher challenged the localization conclusions of Frith and colleagues with regard to ToM mentalizing (Saxe and Kanwisher, 2003). In this and subsequent papers, Saxe (now a faculty member at MIT) and colleagues have argued, instead, for a critical role for the temporoparietal junction (TPJ; the same region that Gallagher and Frith (2003) refer to as the posterior superior temporal sulcus) in ToM mentalizing. Saxe and Kanwisher (2003) noted that previous studies, including some by Frith and colleagues, relied on the practice from developmental psychology of using "false belief" stories as the gold standard for assessing ToM. An example of a false belief scenario, adapted from Gallagher and Frith (2003), is

as follows: *(1) Chris knows that the chocolates have been moved from the drawer to the cupboard. (2) He also knows that Helen does not know that they were moved. (3) Chris knows that Helen wants a chocolate. (4) Chris predicts Helen will look for a chocolate in the* _____. Children younger than age 4, as well as patients with developmental disabilities, including some patients with ASD, will mistakenly answer that Chris thinks that Helen will look in the cupboard, which is interpreted as an inability to comprehend that Helen has a false belief about the location of the chocolate. That is, an inability to understand the world from Helen's perspective. With regard to the TPJ, what Gallagher and Frith (2003) have argued, in effect, is that TPJ plays a role in proposition 3, understanding that Helen wants a chocolate, but not in understanding proposition 4.

Now back to Saxe and Kanwisher (2003). They questioned what they characterized as an overreliance on false belief scenarios to engage ToM mentalizing, and developed a procedure for an fMRI study that used many different contrasts. Experiment 1 contrasted ToM stories vs. mechanical inference stories and vs. human action stories. A group analysis indicated greater activity for ToM than mechanical inference stories in five regions: left and right TPJ; left and right anterior superior temporal sulcus; and precuneus. At the single subject level, activity in TPJ was greater for ToM stories vs. mechanical inference stories in 22 of 25 subjects (bilaterally in 14, left in 5, and right in 3). Importantly, for 14 subjects the authors also performed a localizer scan for the extrastriate body area (EBA; see *Figure 10.5*), and found that although this region was anatomically adjacent to the TPJ, it did not respond to ToM stories. Thus, they reasoned that TPJ does not merely respond to the physical presence of a person. Experiment 2 from Saxe and Kanwisher (2003) contrasted activity associated with (1) false belief stories, (2) false photograph stories, (3) desires, (4) inanimate descriptions, and (5) physical people. Desire stories were designed to describe the character's goals or intentions, and thereby to also rely on ToM mentalizing. The findings from Experiment 2 largely mirrored those from Experiment 1, with the additional fact that desire stories also produced elevated activity in TPJ.

In a follow-up study, Saxe and Powell (2006) note in the Introduction that:

Research suggests that infants and young children understand mental-state concepts like desires, goals, perceptions, and feelings . . . However, not until the age of 4 do children seem to understand concepts like belief – concepts that require understanding that mental representations of the world may differ from the way the world really is. Developmental psychologists have therefore hypothesized that the later-developing system for representing the specific (representational) contents of mental states, such as beliefs, is a distinct component of human understanding of other minds (p. 692).

Thus, for this study, they hypothesized that if the late-developing component of ToM – what we've been calling "ToM mentalizing" – remains distinct in adulthood, then they would identify one or more brain regions recruited for ToM mentalizing stories, but not for stories invoking mental-state concepts such as those listed in the previous quote. Their findings indicated that TPJ responded specifically to ToM mentalizing stories, whereas more posterior regions, in the supramarginal gyrus of the inferior parietal lobule, responded preferentially to stories invoking mental state concepts.

Challenge to the specificity of TPJ for ToM

Why, at the beginning of this section, did I characterize this work as controversial? One reason relates to the point that has been made about the constructs of *conscientiousness*, and, earlier in this chapter, *trustworthiness* – many feel that such complex and nuanced mental constructs are very unlikely to be localizable to just one or a small handful of regions. At a psychological level, Decety and Lamm (2007) have noted that "Most if not all of the concepts used in social psychology emanate from our folk psychology; we are prisoners of words. Theory of mind, perspective-taking, empathy – these are all complex psychological constructs that cannot be directly mapped to unique single computational mechanisms, and the functions that they attempt to describe [must be] underpinned by a network of areas" (p. 584). From a different perspective, that of systems neuroscience, the portrayal of TPJ as specific for ToM mentalizing smacks of a strongly modular view of the brain that's almost phrenological.

What's at stake in this debate isn't just whether ToM mentalizing "activates" area X or area Y in the brain, but a deeper question that relates to the age-old question of "nature vs. nurture." From the "nature" perspective, evidence that the TPJ may be specialized for ToM mentalizing is taken by some to be consistent with nativist theories in cognitive development, and the idea that ToM may be a trait with which we are born, and which matures to take its place in the mind at around 4 years of age. From the "nurture" perspective, psychological abilities such as ToM mentalizing are portrayed as emerging, through learning, from a general-purpose architecture that's neither specialized for, nor specific to, any one function. This latter perspective is also articulated by Decety and Lamm (2007): "activation in the TPJ during social cognition may . . . rely on a lower-level computational mechanism involved in generating, testing, and correcting internal predictions about external sensory events (p. 583)." More specifically, Decety and Lamm have proposed that "social cognition has gradually arisen from general pervasive perception-action coupling mechanisms" (p. 580) and that a function of the TPJ may be "compar[ing] internal predictions with actual external events" (p. 581).

Note that, in many ways, these objections are reminiscent of some that we considered in relation to strong claims that the fusiform face area (FFA) is specialized for face processing (*Chapter 10*). Indeed, a 2008 paper from "social cognitive and affective neuroscientist" Jason Mitchell is clearly motivated by his concerns about strong claims of specificity for activity in TPJ. In the paper's Introduction, Mitchell notes that

> The preferential response of this region during belief stories has prompted claims that "temporo-parietal junction is selectively involved in reasoning about the contents of other people's minds" (Saxe and Kanwisher 2003, p. 1835); "the response of the RTPJ is highly specific to the attribution of mental states" (Saxe and Wexler 2005, p. 1397); and "the BOLD response in the RTPJ is associated with a highly specific cognitive function . . . the ability to attribute thoughts to another person" (Saxe and Powell 2006, p. 697).

He then goes on to say:

> Although this research has done an exhaustive job of demonstrating that RTPJ contributes to social cognition only when perceivers must infer the beliefs of another person, it has, importantly, neglected substantial evidence that this region may also subserve a set of attentional processes that are not specific to social contexts. In a literature that has developed in parallel with the work of Saxe and colleagues, researchers have repeatedly observed increased RTPJ activity when perceivers must break their current attentional set to reorient to task relevant stimuli (p. 262).

What Mitchell is referring to here is the "circuit breaker" in the Corbetta and Shulman (2002) model that we reviewed in some detail in *Chapter 9*. In this regard, Mitchell (2008), Decety and Lamm (2007), and others have noted that RTPJ is a key region whose damage often produces hemispatial neglect (see, e.g., *Figure 7.4*).

To investigate this overlap, Mitchell (2008) closely followed procedures used previously by Saxe and colleagues, most notably using Saxe's "ToM localizer" task (which she shared with him for this purpose), and then interrogating the so-identified "ToM regions" with an attention cuing task. The ToM localizer contrasts story comprehension for false belief scenarios that require ToM mentalization with story comprehension for "false photograph" scenarios, the latter requiring subjects to represent the false content of a physical representation, such as a photograph or map, that doesn't contain any "person" content. The spatial cuing task is a speeded target detection task in which a cue that precedes the target by a variable period of time informs subjects where the target will appear. On a small proportion of trials, the cue is invalid, thereby requiring a reorientation of attention to the unexpected target location (Posner, 1980).

Mitchell's finding was that the RTPJ, but not other regions identified by the ToM localizer (medial PFC, precuneus, superior frontal gyrus), showed elevated responses on all trials of the attentional cuing task, and disproportionately large responses on invalidly cued trials. His conclusion was unflinching:

> These results not only undermine the assertion that RTPJ is selective for inferring the beliefs of other people, but are also incompatible with any strong claim that this region participates selectively in social cognition, broadly construed. Greater activation in RTPJ was observed on a simple attentional cueing task in which no other people were relevant to the task: participants merely pressed 1 of 2 keys to indicate the location in which a target asterisk appeared. In so demonstrating that RTPJ activity is not selective for social cognition, the current results suggest that theory of mind and attentional reorienting may both require a solution to the same computational problem. That is, the observation that the same RTPJ region is engaged by these two tasks suggests the recruitment of a cognitive process that contributes to both belief attribution and attentional reorienting. Although the current study does not specify the exact nature of this shared process, it does suggest the need to develop a better conceptual account of what computational problems might jointly be faced by belief attribution and attentional reorienting and, subsequently, to test hypotheses about the processes that are deployed by the human mind to solve this particular cognitive challenge (p. 268).

False beliefs (?) about Rebecca Saxe's mind

Not surprisingly, Saxe objected to this interpretation. Although this author does not have any direct knowledge about the reasoning and motivation behind all the studies that have subsequently come out of her lab, at this point it will serve a useful didactic purpose for us to engage in theorizing about Rebecca Saxe's mind – even if it means entertaining a false belief or two ;-). Specifically, it will be instructive for us to consider three studies published since 2009 in terms of how effectively they support her argument that the RTPJ plays a privileged role in ToM mentalization, in the context of the Mitchell (2008) results. The first is a replication of the Mitchell (2008) study, but at a higher spatial resolution. The second is a study employing repetitive transcranial magnetic stimulation (rTMS) to the RTPJ while subjects perform a moral judgment task. The third is a study applying multivariate pattern analyses (MVPA) to fMRI data from neurotypical adults and adults with ASD performing a moral judgment task. We will consider each in turn, because they are instructive with regard to inferences about specificity that can be supported by different experimental approaches and different kinds of data.

fMRI with higher spatial resolution

This first of the three studies actually doesn't require any mentalizing on our part, because Scholz, Triantafyllou, Whitfield-Gabrieli, Brown, and Saxe (2009) explicitly frame it as a replication of the methods of Mitchell (2008), but in a scanner featuring a higher magnetic field (3 tesla, as opposed to the 1.5 tesla scanners that Mitchell and Saxe's group had previously used). The higher magnetic field strength allowed Scholz et al. (2009) to acquire higher resolution data (1.6 × 1.6 × 2.4 mm voxels; as contrasted with the 3.75 × 3.75 × 6 mm of Mitchell (2008), and of previous studies from Saxe's group). The empirical results, some of which are illustrated in *Figure 17.4*, were equivocal. First, Scholz et al. (2009) replicated Mitchell's (2008) finding that ToM-defined regions were sensitive to the valid–invalid contrast in the spatial attention task, and that spatial attention-defined regions were sensitive to the ToM contrast. In a second analysis, they determined that there was only a small amount of overlap (<10%) in voxels identified in the ToM localizer and the attention reorientation

A

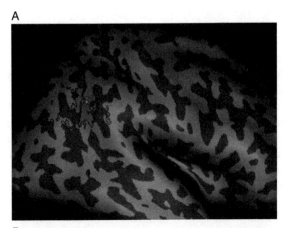

B

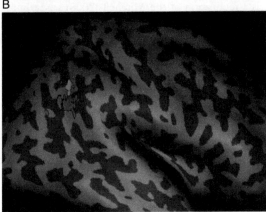

FIGURE 17.4 Activity associated with processing false belief stories (i.e., ToM mentalizing; red voxels), and with invalid spatial attention cues (i.e., attentional reorienting; green voxels). **A.** Shows group analysis results, with the same threshold applied to the statistical maps for each. **B.** Shows results from the same statistical contrasts, but with the threshold for the attentional orienting lowered, such that there would be roughly the same number of voxels for each contrast (82 for ToM mentalizing; 72 for attentional reorienting). Source: Scholz, Triantafyllou, Whitfield-Gabrieli, Brown, and Saxe, 2009. Reproduced with permission of the authors.

localizer. In a third analysis, they found no evidence of correlation of voxel activity across the two tasks. Finally, they evaluated whether there was reliable spatial separation between the "peak activations" for the two tasks, and determined that there was.

Let's take these findings in (roughly) reverse order. To a certain extent, the second, third, and fourth each depend on there being high test–retest reliability with fMRI. This is because interpretation of a finding that condition A and condition B do not activate precisely the same voxels during a scanning session is only meaningful if one knows that condition A at time 2 would activate the same voxels as had condition A at time 1, and that condition B at time 2 would activate the same voxels as did condition B at time 1. If, however, the fMRI signal is not highly reliable with regard to the specific voxels that are activated from session to session, then showing that condition A and condition B both activate the same region, but different specific voxels within that region, does not constitute strong support for the conclusion that the two conditions activate strictly nonoverlapping regions, because the nonoverlap resulting from these two scans could just be due to chance. To this author's knowledge, not many such assessments of test–retest reliability have been performed with fMRI. In one study that did, his then-postdoctoral fellow Eva Feredoes (now at the University of Reading [UK]) compared activation maps from a button-pushing task and a verbal working-memory task (Feredoes and Postle, 2007). She found a range from 15.2% to 6.5% of the same voxels appearing in thresholded statistical maps, depending on the particular contrast, when subjects performed the same tasks in different sessions. This suggests that the precise spatial stability of univariate fMRI maps may be relatively low, and therefore Scholz et al. (2009) would have needed to have empirically tested the overlap of $ToM_{Time\ 1}$ vs. $ToM_{Time\ 2}$ and of Spatial $Attention_{Time\ 1}$ vs. Spatial $Attention_{Time\ 2}$ for the results of their analyses to have supported a strong interpretation.

Turning to the first analysis, it replicated the findings of Mitchell (2008), and was, therefore, consistent with his conclusion that the ToM-defined region located in the vicinity of RTPJ is not specific for ToM.

rTMS of RTPJ during moral judgments

Moral judgment is another domain through which one can study social cognition. Koster-Hale, Saxe, Dungan, and Young (2013) have described it as such:

The distinction between intentional and accidental acts is particularly salient in the case of moral cognition. Adults typically judge the same harmful act (e.g., putting poison in a drink, failing to help someone who is hurt, making an insensitive remark) to be more morally wrong and more deserving of punishment when committed intentionally vs. accidentally . . . These moral judgments depend on individuals' ability to consider another person's beliefs, intentions, and knowledge, and emerge relatively late in childhood, around age 6–7 years (p. 5648).

In a 2010 study led by Saxe's then-postdoctoral fellow Liane Young (now at Boston College), and including TMS pioneer Alvaro Pascual-Leone, Saxe's group used TMS applied via two different protocols to test for evidence that RTPJ plays a causal role in moral reasoning. The first stage in the procedure of Young et al. (2010) was to identify the brain regions that they wanted to target with TMS by scanning subjects with fMRI while they performed the ToM localizer. Next, on a subsequent day, the subjects came to the TMS laboratory and were fitted with infrared markers that allowed for the coregistration of their skull with the computer's digital image of the skull. As a result, when the experimenters touched a part of a subject's skull with an infrared-detectable wand, the computer image of the subject's skull would also be "touched" by an image of a virtual wand. Next, in the computer's 3-D image of the head and all that was inside it, they could virtually "peel away" the skull to visualize where the wand was located relative to the subject's brain. Thus, with knowledge of the shape of the magnetic field produced by the "wand" (i.e., by the TMS stimulator), the researchers could position the TMS coil on the scalp such that its pulses would stimulate the area that they wanted to target. This approach is sometimes referred to as "frameless stereotaxy," in that it uses the same logic as stereotaxic surgery. In stereotaxic surgery, a patient is scanned while wearing a cage-like frame over their head, and the subsequent surgery is also performed while they are wearing the cage, so that the location of unseen brain regions can be calculated in reference to the invariant spatial relation between the brain and the frame.

Young et al. (2010) used two TMS protocols, which they referred to as "offline" and "online." The logic of offline TMS was to produce a virtual lesion with a prolonged train of TMS pulses delivered at 1 Hz, in their case for 25 minutes (see section "*Virtual lesion*" *TMS protocols, Chapter 3*). This approach is "offline" in the sense that low-frequency TMS is delivered first, and then the subject performs the task during the ensuing 25 minutes – TMS is not delivered simultaneous with task performance.

The logic of "online" TMS, in contrast, was analogous to that of microstimulation protocols that we have considered previously: it is delivered simultaneous with task performance and, importantly, can target specific epochs of a trial, leaving the others unaffected (section *High-frequency repetitive (r)TMS, Chapter 3*).

The study (Young et al., 2010) crossed three factors in each of two experiments: intention of the protagonist (malicious, neutral); outcome suffered by the recipient of the action (harm, no harm); TMS (RTPJ, control area). In both experiments, the story that each subject heard included four sections – "background," "foreshadow," "intention," and "action" – followed by a judgment in which the experimental subject rated the protagonist's actions on a permissible-to-forbidden rating scale. The inclusion of an "active stimulation" control region, which for this study was a region of parietal cortex located 5 cm posterior to the RTPJ, allowed the authors to evaluate whether any result that they might find by applying rTMS to RTPJ could be explained as a nonspecific result of simply applying magnetic stimulation to the brain. (Another type of control that is sometimes used in TMS studies, sham stimulation that produces the sounds and tapping sensation of TMS, is not as effective a control, because it does not deliver any stimulation to the brain.)

Both studies (offline and online) found the same result, which is that TMS delivered to the RTPJ selectively altered judgments on the malicious-intent stories for which the outcome was no harm. For these stories, rTMS to the RTPJ, but not the control area, had the effect of making subjects judge the action of the protagonist as being more "permissible" than they did when TMS was delivered to the control area. In the online study, a 500-ms train of 10 Hz TMS was delivered simultaneous to the onset of the moral judgment question, meaning that this effect could be isolated to the judgment itself, not to some other factor, like comprehension of the protagonist's intention.

These findings, therefore, revealed a necessary contribution of RTPJ and, possibly, other regions functionally coupled with it (i.e., having elevated functional connectivity with RTPJ), for moral reasoning. It is particularly striking that the TMS effects were seen in just this one cell of the experimental design, meaning, among other things, that rTMS did not also affect judgments for scenarios when the protagonist had ill intent *and* something bad happened. This suggests that RTPJ contributes most importantly to assessing morally ambiguous situations, in which there is a mismatch between intent and outcome.

Note, however, that the results of Young et al. (2010) do not speak to the possible specificity of this region's function in relation to moral judgment versus other aspects of cognition. In particular, the authors did not assess the effects of rTMS on attentional reorienting. Indeed, there have been several studies of spatial attention (that, in turn, did not test ToM) that have shown clear effects on attention when TMS is delivered to RTPJ. Thus, it's a reasonable assumption that, had Young et al. (2010) included a test of attentional reorienting, they would have found effects on this task, too. (This observation, in turn, must be qualified by the fact that the volume of tissue affected by TMS is sufficiently large – of the order of a cubic centimeter or so – that if the true organization of the brain were such that ToM functions and attentional reorienting functions were segregated in a strict but anatomically adjacent way (as portrayed, e.g., in *Figure 17.4*), the coarse resolution of rTMS would make it impossible to target just one of these regions.)

Moral judgments, TPJ, and autism spectrum disorder

The third of the three post-Mitchell (2008) studies that we will consider here is a study scanning ASD individuals while performing a moral judgment task, and comparing their performance and brain activity to those of neurotypical (NT) individuals. With regard to moral reasoning, the behavioral profile of high-functioning individuals with ASD has been summarized as follows by Jorie Koster-Hale and colleagues (2013):

> Individuals with ASD do not have impairments in moral judgment as a whole: children with ASD make typical distinctions between moral and conventional transgressions . . . and between good and bad actions . . . However, they are delayed in using information about innocent intentions to forgive accidents . . . Furthermore, even very high-functioning adults with ASD who pass traditional tests of understanding (false) beliefs neglect beliefs and intentions in their moral judgments compared with NT adults (p. 5651).

Koster-Hale et al. (2013) assessed reasoning about beliefs and intentions with stories similar to those used in the TMS study of Young et al. (2010), with the exception that information in this study was provided in the order of "background," "action," "outcome," "intent." High-functioning individuals with ASD were recruited via advertisements placed with a regional Asperger's Association. They were rated as scoring higher on the Autism Quotient questionnaire than NT subjects, and received a diagnosis of ASD from a clinician.

Behaviorally, subjects in both the ASD and NT groups judged intentional harms to be more blameworthy than accidental harms, and judged both of these to be worse than neutral acts. The two groups differed, however, in that ASD subjects assigned more blame for accidental harms than did NT adults, and they assigned less blame for intentional actions.

In both groups, univariate analyses of fMRI signal intensity, averaged over the whole trial, revealed higher responses for harmful than neutral actions in all four ROIs illustrated in *Figure 17.5*, panel A. However, the two groups differed when just the final 8 seconds of the trial were considered. This was the point in the trial when the intention of the protagonist was revealed to be malicious or neutral. During this portion of the trial, the response in the RTPJ of NT adults was higher for accidental than intentional harms, but there was no such difference in the ASD group. Findings from MVPA largely paralleled those from the univariate analyses, indicating that for both groups and in all ROIs, patterns of activity discriminated between harmful vs. neutral acts. Also mirroring the univariate analyses, it was only in NT subjects that patterns of activity in RTPJ discriminated reliably between accidental and intentional harms. Further analyses, illustrated in *Figure 17.5*, panel C, indicated that for these NT subjects the strength of the multivariate discriminability of intentional vs. accidental harms in RTPJ activity predicted how strong was the behavioral distinction that that subject would make between these types of harms. That is, individuals whose RTPJ evinced more distinct patterns between intentional and accidental harms also drew a larger distinction between intentional and accidental harms in their moral judgments.

Although one might be tempted to see parallels between the TMS study of Young et al. (2010) and the fMRI findings of Koster-Hale et al. (2013), they differ importantly in that TMS of TPJ had the effect of making subjects judge failed attempts to do harm more leniently. In the present study, however, MVPA of TPJ did not distinguish between neutral acts and malicious acts that failed to harm.

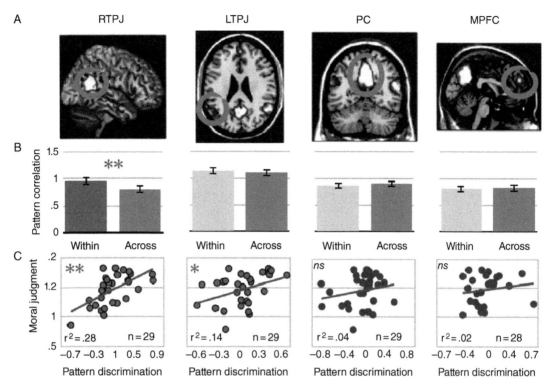

FIGURE 17.5 fMRI of judgment of accidental and deliberate harms. **A.** Shows the four regions that are routinely identified with Saxe's ToM localizer task. **B.** Illustrates, for each ROI, MVPA results. These analyses were performed as the correlation of multivariate patterns within each condition (i.e., correlation of the pattern from half of the *accidental harm* trials vs. the other half of the *accidental harm* trials, and of half of the *intentional harm* trials vs. the other half of the *intentional harm* trials) vs. across the two conditions (i.e., correlation of *accidental harm* trials vs. *intentional harm* trials). **C.** Illustrates the relation between the magnitude of the *Within* vs. *Across* difference from the MVPA (Pattern discrimination) and how differently that person assigned blame on accidental vs. intentional harm trials (Moral judgment). Source: From Koster-Hale, Jorie, Rebecca Saxe, James Dungan, and Liane L. Young. 2013. "Decoding Moral Judgments from Neural Representations of Intentions." Proceedings of the National Academy of Sciences of the United States of America 110 (14): 5648–5653. Reproduced with permission of PNAS.

A final assessment of the role of RTPJ in ToM mentalization

The TMS study of Young et al. (2010) and the fMRI study of Koster-Hale et al. (2013) each provide strong evidence that right R is involved in making moral judgments. This is not equivalent, however, to saying that they have shown it to be a "moral judgment area." Indeed, the findings of Mitchell (2008) and Scholz et al. (2009) both suggest that voxels of RTPJ recruited for ToM mentalization are also recruited for the reorienting of spatial attention. The findings from these two individual studies are consistent with the results of a meta-analysis of neuroimaging studies by Decety and Lamm (2007). The term "meta-analysis" refers

to an analysis of the results from many other studies. It's "meta" because one isn't collecting data from individual subjects and analyzing these data, but, instead, assembling the results of many individual experiments, and, in effect, treating each experiment as a data point in an analysis of analyses. For neuroimaging data, the goal is often to obtain a measurement that is less susceptible to sampling error in the precise localization of effects that would be characteristic of any one fMRI study. (That is, the problem of low test–retest reliability that was raised earlier in this section.) The meta-analysis of Decety and Lamm (2007) included 18 studies of spatial reorienting, 15 studies of agency (the feeling of being the cause of one's own thoughts and actions), 13 studies of empathy, and 24 studies of ToM. The finding was that, in the region of the RTPJ, results from the four kinds of studies largely overlapped. Thus, it seems to be that although the RTPJ contributes importantly to moral judgment and perhaps also to ToM mentalization, it is not specialized for these functions. Therefore, the question about how these aspects of social behavior are produced by the brain remains uncertain.

OBSERVATIONAL LEARNING

In the final section of this chapter we'll consider another aspect of social behavior, which is learning by observing the behavior of others. Although we learn many things about how the world works through direct trial and error, it can often be more efficient, and sometimes safer, to watch how others approach a novel situation and, as the expression has it, "learn from their mistakes." Formally, these two types of learning are called "experiential" and "observational," and the latter is inherently social. On an intuitive level, it makes sense that we can learn from observing the experiences of others, and each of us can immediately think of several examples. (Here's a simple one. You are hungry and go to the lobby in your building where the vending machines are located. There are two machines, and as you arrive you see that someone else has put money into machine A, and is now cursing and kicking the machine because it took her money but didn't deliver the item that she selected, nor will it return her money. As a result of this observation of the behavior of another, which vending machine are you going to use, A or B?) As we shall now see, however, the way that observational learning works can be nonintuitive.

In this chapter's final study that we will consider in detail, Burke, Tobler, Baddeley, and Schultz (2010) scanned an individual while they performed a probabilistic learning task. On each block of trials, three pairs of randomly selected fractal stimuli were presented, and the subject had to learn, for each pair, which of the two was associated with a positive outcome, and which with a negative outcome. (On "gain" blocks, the positive outcome was gaining 10 points for selecting the correct stimulus and gaining 0 points for the incorrect one; on "loss" blocks, the positive outcome was losing 0 points for selecting the correct stimulus and losing 10 points for selecting the incorrect one.) To capture some of the "noise" and variation that exist in the natural world, these reward contingencies were probabilistic, rather than absolute: during *gain* blocks, the "good" stimulus delivered the reward on 80% of trials, but the

contingencies were reversed on the remaining 20% of trials; during *loss* blocks, the same probabilities applied. Trial types were blocked such that the subject performed a series of *gain* trials during one several minute-long scan, and then a series of *loss* trials during the next.

Here's the social/observational part. On the day of the scan, it was arranged that a few minutes after the true research subject arrived to be scanned, a second individual who was actually a part of the research team (a "confederate") arrived and also sat in the waiting room, pretending to be a second research subject. A second experimenter then came into the room (playing the role of the experimenter) and explained to both that they would be playing the same casino-style gambling-machine game – trying to win the most points and lose the least. It was also explained that, while playing the game, one of the two would be in the MRI scanner (always the true research subject) and the other would be in a testing room. They would take turns, and each would be able to observe how the other was playing. Then a digital photo of each was taken, and it was further explained that the photo of the person currently taking a turn would be displayed on the screen.

The sequence of a trial is illustrated in *Figure 17.6.A–C* In addition to the blocking of *gain* and *loss* trials, the critical experimental factor was the amount of social information available to the research subject: none (*Individual learning*); some (*Action only*, in which the subject could observe the confederate's choice, but not the outcome of that choice); and all (*Action + outcome*). One might expect (at least I did) that observing only another's choice, but not the outcome of that choice, would be of little-to-no use to an individual. How could one "learn from another's mistakes" without the knowledge of whether or not that person's choice turned out to be a mistake? (In the vending machine example, if you watched the person put her money in the machine, but then stepped outside to make a phone call and, when you returned to the lobby, she was no longer there, how could having seen her put her money into the machine, but not having seen her lose it, possibly help you make a better decision?)

Predicting the outcome of someone else's actions

Although there is no obvious, intuitive reason why a subject should learn more quickly when they can observe the choices of another, but not the outcomes of those choices, there are reinforcement learning algorithms that can explain it. The senior author of the Burke et al. (2010) study is

FIGURE 17.6 Experiential and observational learning. Source: From Burke, Christopher J., Philippe N. Tobler, Michelle Baddeley, and Wolfram Schultz. 2010. "Neural Mechanisms of Observational Learning." Proceedings of the National Academy of Sciences of the United States of America 107 (32): 14431–14436. Reproduced with permission of PNAS Group.

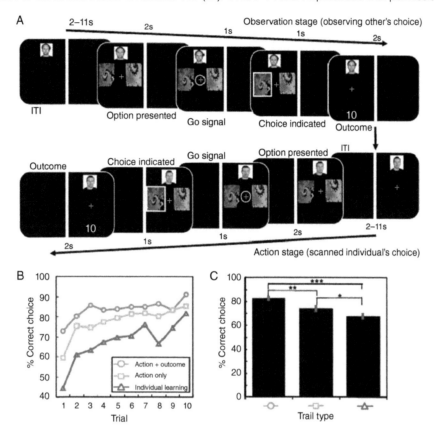

FIGURE 17.6.A–C Panel **A** illustrates the series of events on a single trial. First, in the *Observation stage*, the subject watched as the confederate's options were presented, and then as the confederate received the go signal, made his choice, and was informed of the outcome (in this illustration he won 10 points). Next, in the *Action stage*, the subject went through the same sequence. On *Individual learning* trials, the subject only experienced the *Action stage*; on *Action only* trials, the subject observed all the sequences of the confederate's trial (i.e., the *Observation stage*), with the exception that the outcome was not presented. During each session, the same pair of stimuli was presented for each *Individual learning* trial, a different pair for each *Action only* trial, and still a different pair for each *Action + outcome* trial. Panel **B** illustrates, with a single subject's performance, that even beginning with the first trial, having knowledge of another's action, even without knowing what the outcome of the action was, aided performance markedly. Panel **C** shows aggregated group performance in the three learning conditions.

Wolfram Schultz, whose discovery and characterization of the dopaminergic reward prediction error (RPE) signal, with Ranulfo Romo, was detailed in section *Habits and reinforcement learning, Chapter 8*. It was true to form, then, that Burke et al. (2010) proposed that human observational learning may be achieved by applying two newly hypothesized variants of prediction error signals. They implemented this idea in a computational model of probabilistic reward learning, the "Q" learning algorithm. The mathematical details of this model are beyond the scope of this book, but the gist of Q learning is that it solves the problem of learning which actions are the best to take in different circumstances when the outcome of the action (and, therefore, its rewarding value) won't be known until some later point in time.

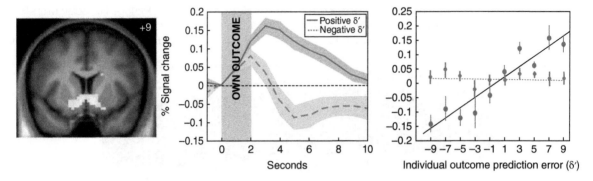

FIGURE 17.6.D The region correlating with RPEs from the model was the nucleus accumbens of the ventral striatum. The center panel shows activity from this region, time locked to the *Outcome* epoch of the *Action stage* from panel A, sorted into trials estimated by the model to produce a positive RPE or a negative RPE. The panel on the right illustrates the correlation between the model's estimates of individual outcome prediction error on each of the 10 trials of the session and the fMRI response from the corresponding trials, in the nucleus accumbens (red symbols). Note that, because all trial types had an *Outcome* stage, this analysis included trials from all three trial types. To demonstrate that social and nonsocial learning are integrated in this outcome prediction error in the nucleus accumbens, the gray symbols show the same regression with expected outcome prediction errors, but when social information was removed from the model on *Action only* and *Action + outcome* trials.

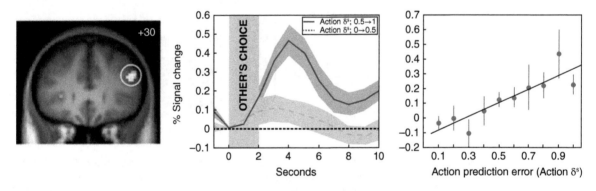

FIGURE 17.6.E The region correlating with the *action prediction error* signal from the model was the dorsolateral PFC. The center panel shows activity from this region, time locked to the *Choice indicated* epoch of the *Observation stage* from panel A, sorted into trials estimated by the model to produce a large *action prediction error* and a small *action prediction error*. The panel on the right illustrates the correlation between the model's estimate of *action prediction errors* and the fMRI response from the corresponding trials in the dorsolateral PFC.

When considering observational learning, the authors noted that the behavior of subjects differs as a function of whether or not the outcome of the confederate's choice is visible. If we consider *Figure 17.6.A–C, for example, on Action only* trials subjects much more frequently imitated the confederate's behavior than they did on *Action + outcome* trials. This confirmed that subjects "adaptively used available information in the different conditions," and led the authors to reason that a process of observing another's actions and combining this knowledge with one's own experience "can be thought of as a two-stage learning process; first, observing the action of another player biases the observer to imitate that action (at least on early trials), and second, the outcome of the observer's own action refines the values associated with each stimulus" (p. 14432). On *Action + outcome* trials, the authors posited an analogous two-stage process, but with the first stage corresponding to observing the outcome of the confederate's choice.

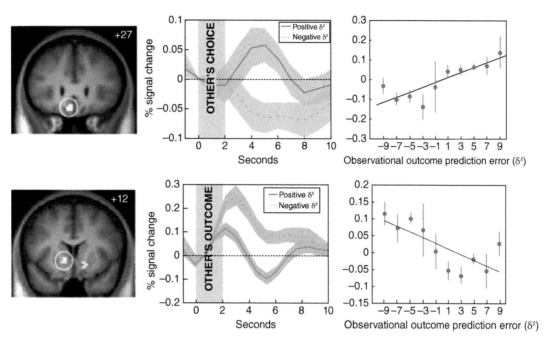

FIGURE 17.6.F The regions correlating with *observational-outcome prediction errors* from the model were the ventromedial PFC (top row) and the nucleus accumbens (bottom row), the two responding in opposite directions. The center panel shows activity from the two regions, time locked to the *Outcome* epoch of the *Observation stage* of from panel A, sorted into trials estimated by the model to produce a positive *observational-outcome prediction error* and a negative *observational-outcome prediction error*. The panel on the right illustrates the correlation between the model's estimate of *observational-outcome prediction errors* and the fMRI response from the corresponding trials in the two regions.

The mechanisms that Burke et al. (2010) proposed to implement the first stage of learning in these two two-stage processes were previously undescribed variants of RPEs. On *Action only* trials they hypothesized that an "action prediction error" would signal the extent to which the confederate's choice differed from what the observer expected the confederate to do. Because, unlike rewarded choices, simple motor actions do not have inherently rewarding or punishing values, they posited that action prediction errors could vary in magnitude, but not in sign. On *Action + outcome* trials, they hypothesized that an "observational outcome prediction error" would have the same properties as the RPE that we have already learned about, but would be based on "vicarious" rewards, in that they pertained to the confederate, and could never have been obtained by the observer. (This is related to the concept of "fictive learning.")

Studying observational learning with model-based fMRI

To assess the neurobiological reality of these hypothesized RPEs, Burke et al. (2010) first fit their model to the behavioral data from the subjects who were scanned while performing the task illustrated in *Figure 17.6.A*. Next, they used the resultant parameter estimates, corresponding to *reward prediction errors*, *action prediction errors*, and *observational-outcome prediction errors*, as regressors with which to analyze the fMRI data. This approach to fMRI data analysis is known as "model-based" fMRI. The rationale behind model-based fMRI is that if there's no clear intuitive understanding of a certain pattern of behavior, conventional analyses of fMRI scans of this poorly understood behavior wouldn't provide any new insight about it. All they would show is where in the brain there are changes in activity associated with something

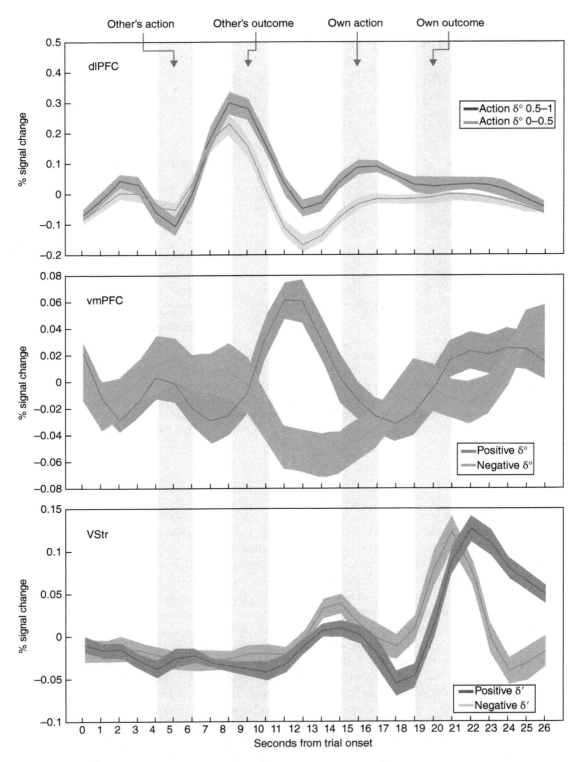

FIGURE 17.6.G The full-trial group- and trial-averaged fMRI time series from the three ROIs illustrated earlier in this figure. This panel provides a simultaneous presentation of the timing of each of the error prediction–related effects in the context of the full length of the trial, and of each of the ROIs implicated in these different facets of observation learning. ("VStr" = ventral striatum, the gross-anatomical location of the nucleus accumbens.)

that we don't understand. If, on the other hand, this behavioral phenomenon can be understood via a formal model, with parameters that we can understand, determining that there are areas of the brain whose activity covaries with the parameter estimates from the model can help us understand the neural implementation of the mechanisms described by the model. This is how Burke et al. (2010) analyzed the fMRI data from their study, as illustrated in *Figure 17.6*.

These results from the *observational-outcome prediction errors*, that two different regions showed opposite effects to the same prediction error signal, allow us to make a broader point. Although the results from Figure 17.6.D–F show results in different regions, it's not believed to be the case that, for example, RPE signals were delivered in a targeted manner to the nucleus accumbens, whereas *action prediction errors* were delivered in a targeted manner to dorsolateral PFC. Rather, each of these prediction errors is assumed to have been delivered via the same nonspecific "broadcast" from midbrain dopaminergic neurons. The differential patterns of activity result from the fact that brain areas that are processing information about reward, or about another's actions, or the outcome of another's actions, will respond differentially to this dopaminergic signal when they are processing information about the task. (For example, if dorsolateral PFC "doesn't care" about reward, nor about the rewarding consequences for someone else of the outcome of that person's actions, the dopaminergic prediction error won't change its activity. Recall that DA's effects are modulatory, not driving.)

A fun and speculative note to end on is the interpretation of the opposing effects that the *observational-outcome prediction error* had on vmPFC vs. on nucleus accumbens. In particular, the sign reversal of the signal in nucleus accumbens means that a positive *observational-outcome prediction error* occurs when viewing another person lose. Although the Burke et al. (2010) model did not have an a priori prediction about this finding, the authors speculate that this may suggest that "it [is] . . . rewarding to view the misfortunes of others" (p. 14435). Schadenfreude.

TRUSTWORTHINESS, REVISITED

We have just reviewed the fact that neural systems in the vmPFC and paralimbic regions contribute importantly to the control of social comportment. We have also considered research aimed at understanding the neural underpinnings of ToM. Finally, we have reviewed neural evidence for an explicit computational model of social learning. Having done so, we now have a better sense of how one might go about incorporating a representation of trustworthiness into neural models of intertemporal choice. We have a better sense of what are the important questions to ask, of what caveats and complications may arise. It won't be easy, or quick. It's unlikely to be uncontroversial. But now, at the end of this chapter, we have a better sense of how one might go about incorporating into our cognitive neuroscience models the very real influences of social cognition on brain and behavior.

END-OF-CHAPTER QUESTIONS

1. What are potential problems, or, at least, complications, with trying to study the neural bases of a social construct like *trustworthiness*? Why is it nonetheless important to confront these problems/complications?

2. In what ways do findings from contemporary behavioral neurology support the conclusions drawn from the case of Phineas Gage?

3. Does the vmPFC act alone in the control of social behavior? What evidence supports your answer?

4. What brain regions have been implicated as part of a "ToM network"? What function(s) have been ascribed to different components of this network?

5. What are the two differing accounts of the function(s) supported by the right TJP in relation to social behavior?

What fundamental theoretical perspectives underlie these two accounts?

6. What are the three different experimental protocols for using TMS to study neural function? (Note: only two are considered in this chapter.)

7. What is the circumstance when a model-based analysis of fMRI may be preferred over the univariate or multivariate approaches that we have emphasized up until this point in the book?

8. How might it be that learning is improved when one can observe the actions of others, but not the outcomes of these actions? Be specific about the proposed neurobiological mechanism(s).

REFERENCES

Burke, Christopher J., Philippe N. Tobler, Michelle Baddeley, and Wolfram Schultz. 2010. "Neural Mechanisms of Observational Learning." *Proceedings of the National Academy of Sciences of the United States of America* 107 (32): 14431–14436. doi: 10.1073/pnas.1003111107.

Corbetta, Maurizio, and Gordon L. Shulman. 2002. "Control of Goal-Directed and Stimulus-Driven Attention in the Brain." *Nature Reviews Neuroscience* 3 (3): 201–215. doi: 10.1038/nrn755.

Decety, Jean, and Claus Lamm. 2007. "The Role of the Right Temporoparietal Junction in Social Interaction: How Low-Level Computational Processes Contribute to Meta-Cognition." *Neuroscientist* 13 (6): 580–593. doi: 10.1177/1073858407304654.

Feredoes, Eva, and Bradley R. Postle. 2007. "Localization of Load Sensitivity of Working Memory Storage: Quantitatively and Qualitatively Discrepant Results Yielded by Single-Subject and Group-Averaged Approaches to fMRI Group Analysis." *NeuroImage* 35 (2): 881–903. doi: 10.1016/j.neuroimage.2006.12.029.

Fletcher, Paul C., Francesca G. Happé, Uta Frith, S. C. Baker, Ray J. Dolan, Richard S. J. Frackowiak, and Christopher D. Frith. 1995. "Other Minds in the Brain: A Functional Imaging Study of 'Theory of Mind' in Story Comprehension." *Cognition* 57 (2): 109–128. doi: 10.1016/0010-0277(95)00692-R.

Gallagher, Helen L., and Christopher D. Frith. 2003. "Functional Imaging of 'Theory of Mind'." *Trends in Cognitive Sciences* 7 (2): 77–83. doi: 10.1016/S1364-6613(02)00025-6.

Harlow, John Martyn. 1868. "Recovery from the Passage of an Iron Bar through the Head." *Bulletin of the Massachusetts Medical Society* 2: 327–347.

Koenigs, Michael, Michael Kruepke, and Joseph P. Newman. 2010. "Economic Decision-Making in Psychopathy: A Comparison with Ventromedial Prefrontal Lesion Patients." *Neuropsychologia* 48 (7): 2198–2204. doi: 10.1016/j.neuropsychologia.2010.04.012.

Koster-Hale, Jorie, Rebecca Saxe, James Dungan, and Liane L. Young. 2013. "Decoding Moral Judgments from Neural Representations of Intentions." *Proceedings of the National Academy of Sciences of the United States of America* 110 (14): 5648–5653. doi: 10.1073/pnas.1207992110.

Lhermitte, François, Bernard Pillon, and M. Serdaru. 1986. "Human Autonomy and the Frontal Lobes. Part I: Imitation and Utilization Behavior: A Neuropsychological Study of 75 Patients." *Annals of Neurology* 19 (4): 326–334. doi: 10.1002/ana.410190404.

Ly, M., J. C. Motzkin, C. L. Philippi, G. R. Kirk, J. P. Newman, K. A. Kiehl, and M. Koenigs. 2012. "Cortical thinning in Psychopathy." *American Journal of Psychiatry* 169 (7): 743–749. doi: 10.1176/appi.ajp.2012.11111627.

Macmillan, Malcolm. 2000. *An Odd Kind of Fame: Stories of Phineas Gage.* Cambridge, MA: MIT Press.

Mesulam, M.-Marsel. 1985. "Patterns in Behavioral Neuroanatomy: Association Areas, the Limbic System, and Hemispheric Specialization." In *Principles of Behavioral Neurology*, edited by M.-Marsel Mesulam, Vol. 26 of *Contemporary Neurology Series*, 1–70. Philadelphia, PA: F.A. Davis.

Mesulam, M.-Marsel. 1986. "Frontal Cortex and Behavior." *Annals of Neurology* 19 (4): 320–325. doi: 10.1002/ana.410190403.

Michaelson, Laura, Alejandro de la Vega, Christopher H. Chatham, and Yuko Munakata. 2013. "Delaying Gratification Depends on Social Trust." *Frontiers in Psychology* 4: 355. doi: 10.3389/fpsyg.2013.00355.

Mitchell, Jason P. 2008. "Activity in Right Temporo-Parietal Junction Is Not Selective for Theory-of-Mind." *Cerebral Cortex* 18 (2): 262–271. doi: 10.1093/cercor/bhm051.

Motzkin, Julian C., Joseph P. Newman, Kent A. Kiehl, and Michael Koenigs. 2012. "Reduced Prefrontal Connectivity in Psychopathy." *Journal of Neuroscience* 31 (48): 17348–17357.

Posner, Michael I. 1980. "Orienting of Attention." *Quarterly Journal of Experimental Psychology* 32 (1): 3–25. doi: 10.1080/00335558008248231.

Saxe, Rebecca, and Nancy Kanwisher. 2003. "People Thinking about Thinking People: The Role of the Temporo-Parietal Junction in 'Theory of Mind.'" *NeuroImage* 19 (4): 1835–1842. doi: 10.1016/S1053-8119(03)00230-1.

Saxe, Rebecca, and Lindsey J. Powell. 2006. "It's the Thought That Counts: Specific Brain Regions for One Component of Theory of Mind." *Psychological Science* 17 (8): 692–699. doi: 10.1111/j.1467-9280.2006.01768.x.

Scholz, Jonathan, Christina Triantafyllou, Susan Whitfield-Gabrieli, Emery N. Brown, and Rebecca Saxe. 2009. "Distinct Regions of Right Temporo-Parietal Junction Are Selective for Theory of Mind and Exogenous Attention." *PLoS ONE* 4 (3): e4869. doi: 10.1371/journal.pone.0004869.

Somerville, Leah H., Rebecca M. Jones, Erika J. Ruberry, Jonathan P. Dyke, Gary Glover, and B. J. Casey. 2013. "The Medial Prefrontal Cortex and the Emergence of Self-Conscious Emotion in Adolescence." *Psychological Science* 24(8): 1554–1562. doi: 10.1177/0956797613475633.

Van Horn, John Darrell, Andrei Irimia, Carinna M. Torgerson, Micah C. Chambers, Ron Kikinis, and Arthur W. Toga. 2012. "Mapping Connectivity Damage in the Case of Phineas Gage." *PLoS ONE* 7 (5): e37454. doi: 10.1371/journal.pone.0037454.

Young, Liane, Joan Albert Camprodon, Marc Hauser, Alvaro Pascual-Leone, and Rebecca Saxe. 2010. "Disruption of the Right Temporoparietal Junction with Transcranial Magnetic Stimulation Reduces the Role of Beliefs in Moral Judgments." *Proceedings of the National Academy of Sciences of the United States of America* 107 (15): 6753–6758. doi: 10.1073/pnas.0914826107.

OTHER SOURCES USED

Dunne, Simon, and John P. O'Doherty. 2013. "Insights from the Application of Computational Neuroimaging to Social Neuroscience." *Current Opinion in Neurobiology* 23 (3): 387–392. doi: 10.1016/j.conb.2013.02.007.

Kobayashi, Shunsuke, and Wolfram Schultz. 2008. "Influence of Reward Delays on Responses of Dopamine Neurons." *Journal of Neuroscience* 28 (31): 7837–7846. doi: 10.1523/JNEUROSCI.1600-08.2008.

Passingham, Richard E., and Steven P. Wise. 2012. *The Neurobiology of the Prefrontal Cortex: Anatomy, Evolution, and the Origin of Insight*. Oxford: Oxford University Press.

Smith, Christopher T., and Charlotte A. Boettiger. 2012. "Age Modulates the Effect of COMT Genotype on Delay Discounting Behavior." *Psychopharmacology* 222 (4): 609–617. doi: 10.1007/s00213-012-2653-9.

Winstanley, Catharine A., Jeffrey W. Dalley, David E. H. Theobald, and Trevor W. Robbins. 2003. "Global 5-HT Depletion Attenuates the Ability of Amphetamine to Decrease Impulsive Choice on a Delay-Discounting Task in Rats." *Psychopharmacology* 170 (3): 320–331. doi: 10.1007/s00213-003-1546-3.

FURTHER READING

Adolphs, Ralph. 2010. "Conceptual Challenges and Directions for Social Neuroscience." *Neuron* 65 (6): 752–767. doi: 10.1016/j.neuron.2010.03.006.

One of the leading figures in social neuroscience considers the challenges confronting, and promises awaiting, this field.

Amodio, David M. 2010. "Can Neuroscience Advance Social Psychological Theory? Social Neuroscience for the Behavioral Social Psychologist." *Social Cognition* 28 (6): 695–716. doi: 10.1521/soco.2010.28.6.695.

A perspective on the value, or otherwise, of bringing neuroscience methods into the traditionally behavioral field of social psychology.

Kennedy, Daniel P., and Ralph Adolphs. 2012. "The Social Brain in Psychiatric and Neurological Disorders." *Trends in Cognitive Sciences* 16 (11): 559–572. doi: 10.1016/j.tics.2012.09.006.

Psychiatric and neurological disorders have informed the concept of the social brain, and the authors argue that the social brain, and its dysfunction and recovery, must be understood not in terms of specific structures, but rather in terms of their interaction in large-scale networks.

Macmillan, Malcolm. 2000. *An Odd Kind of Fame: Stories of Phineas Gage*. Cambridge, MA: MIT Press.

Engrossing, thoughtful account of all things Gage, covering Gage's personal history, cultural history, history of science, and implications of the case for cognitive neuroscience, including an emphasis on the localization–mass action debate that reigned over the decades surrounding the turn of the twentieth century.

Seabrook, John. 2008. "Suffering Souls: The Search for the Roots of Psychopathy." *New Yorker*, November 10. http://www.newyorker.com/reporting/2008/11/10/081110fa_fact_seabrook.

Interesting profile of cognitive neuroscientist and psychopathy researcher Kent Kiehl.

Seeley, William W., Juan Zhou, and Eun-Joo Kim. 2012. "Frontotemporal Dementia: What Can the Behavioral Variant Teach Us about Human Brain Organization?" *Neuroscientist* 18 (4): 373–385. doi: 10.1177/1073858411410354.

This review provides entrée into a fascinating (ongoing) scientific detective story about the causes of this insidious disease, which prominently compromises social behavior, personal conduct, and decision making.

CHAPTER 18
EMOTION

KEY THEMES

- Neuropsychological and neuroimaging data suggest that the amygdala performs a kind of implicit, quick-and-dirty triage of visual stimuli, signaling, for example, whether or not faces look trustworthy, whether they belong to a certain race, or whether they are expressing an emotion. The evidence further suggests that the amygdala's "decision rules" are learned, and can therefore change with experience.

- Scientific interest in the amygdala was triggered by the drastic and bizarre changes in behavior associated with surgically produced Klüver–Bucy syndrome.

- Intensive study of Pavlovian fear conditioning has yielded detailed knowledge, at the circuit and molecular level, of the plastic change in the amygdala that underlies acquisition and expression of conditioned fear.

- The emotional component of declarative memories also seems to depend on the amygdala and, in particular, its influence on norepinephrine levels via its outputs to the locus coeruleus.

- From a neurobiological perspective, emotions may derive from the activity of systems related to survival, and perhaps to the underlying assessment that a stimulus or situation is safe to approach, or best to avoid.

- The extinction of a fear memory depends on new learning in the amygdala, but also on top-down input from ventromedial PFC (infralimbic [IL] cortex in the rat).

- Unlike conditioned memories, which are implicit and long-lasting, our current emotional state is amenable to real-time volitional control. The mechanisms by which we implement this control are the same as those used for the control of other domains of cognition.

- Many types of psychopathology (post-traumatic stress disorder, depression, anxiety disorder) are characterized by abnormal neural correlates of emotion regulation.

- One's awareness of the factors that are influencing one's emotional state may be an important determinant of one's ability to regulate their effects.

CONTENTS

WHAT IS AN **EMOTION?**

The Oxford English Dictionary defines *emotion* as "any strong mental or instinctive feeling, as pleasure, grief, hope, fear, etc., deriving esp. from one's circumstances, mood, or relationship with others . . ." and goes on to specify that it refers to "instinctive feeling as distinguished from reasoning or knowledge." Not very helpful if we want to study it scientifically. If we strip away from this definition examples of different kinds of emotions (grief, hope, fear), we're left with what it's not (reasoning, knowledge) and a few alternative words (feeling, mood) that would, themselves, need definition. From one perspective, this illustrates the principle that knowledge, fundamentally, may arise from "nothing more" than associations between the sensory and the motor (c.f., section *A PDP model of picture naming, Chapter 3* and *Chapters 10 and 13*), but let's not get distracted. How can we define and operationalize this aspect of cognition so that it can be studied with the methods of cognitive neuroscience?

Let's consider how two prominent groups have done so.

Approach/withdrawal

One perspective from which we can start to operationalize emotion comes from psychology, and the distinction between approach vs. withdrawal behaviors. As framed by Richard Davidson and then-graduate student Wil Irwin (1999):

> The approach system facilitates appetitive behavior and generates certain types of positive affect that are approach-related, for example, enthusiasm, pride, etc. . . . This form of positive affect is usually generated in the context of moving toward a desired goal . . . a second neural system [is] concerned with the implementation of withdrawal. This system facilitates the withdrawal of an individual from sources of aversive stimulation and generates certain forms of negative affect that are withdrawal-related. For example, both fear and disgust are associated with increasing the distance between the organism and a source of aversive stimulation (pp. 12–13).

From "feeling words" to neural systems

An alternative, though not necessarily incompatible, way to operationalize emotion is to acknowledge the linkage between the subjective experience of emotions and survival-related systems in the brain, but then to only study the latter. Thus, Joseph LeDoux (2012b) argues that the clearest way to make meaningful progress in understanding emotions and their regulation is to be explicit that the relevant systems whose neurobiological bases can be studied are "circuits involved in defense, maintenance of energy and nutritional supplies, fluid balance, thermoregulation, and reproduction. These survival circuits and their adaptive functions are conserved to a significant degree . . . across mammalian species, including humans" (p. 654). He further argues that it is important that we not conflate these circuits and functions with the "feeling words" that we use to describe the subjective experience of emotion: fear, anger, love, sadness, jealousy, and so on.

> By focusing on survival functions instantiated in conserved circuits, key phenomena relevant to emotions and feelings are discussed with the natural direction of brain evolution in mind (by asking to what extent are functions and circuits that are present in other mammals also present in humans) rather than by looking backward, and anthropomorphically, into evolutionary history (by asking whether human emotions/feelings have counterparts in other animals) (p. 654).

At the nexus of perception and social cognition

As we make our way through this chapter, we'll see that these two perspectives inform several themes that will be important as we consider the neural bases of emotion. One is the influence of social context on many aspects of cognition that we've already considered in depth, including perception, memory encoding and retrieval, and decision-making. Another is how we control *emotional* aspects of our thinking. A third, and related, theme is the influence of the implicit vs. the explicit on our emotional state and our ability to control it. Let's start with the interaction of the social with the perceptual.

TRUSTWORTHINESS REVISITED – AGAIN

So it turns out that, almost as if conjured on command to meet the needs of this book's author, *trustworthiness* has previously been studied in the context of affective neuroscience. And because the studies in question engage this social construct with which we're already familiar, as well as a cognitive process – face perception – that we have also considered in depth, they offer a tailor-made opportunity for us to introduce a neural structure that will be front-and-center in this chapter: the amygdala.

A role for the amygdala in the processing of trustworthiness

Neuropsychological evidence

A seminal study linking the amygdala to judgments of trustworthiness came from the team of Ralph Adolphs, Daniel Tranel, and Antonio Damasio (1998), who reported on judgments of approachability and trustworthiness in three patients with bilateral amygdala lesions. The findings, illustrated in *Figure 18.1*, showed that the patients tended to judge all faces to be approachable and trustworthy, even

those belonging to the 50 most negative faces, as rated by an independent group of individuals. Recall from Chapter *11* that the amygdala is located in the medial temporal lobe, just rostral, medial, and superior to the rostral end of the hippocampus. Although it is in close anatomical proximity to ventral anterior IT cortex, it has not typically been considered a part of the visual system (a point that is germane to the findings of this study, and many others to follow). Strikingly, the patients described by Adolphs, Tranel, and Damasio (1998) did not differ from healthy control subjects when making the same judgments based on *verbal* descriptions of different individuals. Thus, the impairment seems tied to the visual modality. Based on this finding, the authors proposed that "The amygdala appears necessary to trigger the retrieval of information on the basis of prior social experience or innate bias in regard

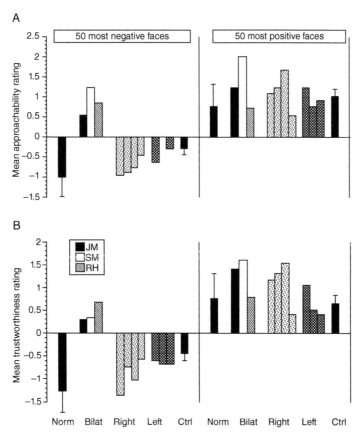

FIGURE 18.1 Performance of three patients with bilateral amygdala lesions (Bilat) – relative to neurologically healthy control subjects (Norm), unilateral right or left amygdala lesion patients, and non-amygdala lesion control subjects (Ctrl) – on rating the approachability and trustworthiness of 50 visually presented faces rated "most negative," and 50 "most positive," by an independent group. Source: From Adolphs, Ralph, Daniel Tranel, and Antonio R. Damasio. 1998. "The Human Amygdala in Social Judgment." Nature 393: 470–474. Reproduced with permission of Springer Nature.

to certain classes of faces" (p. 472). By this account, the social information itself is stored elsewhere, and accessible via other channels, such as spoken language.

fMRI evidence

Several fMRI studies prompted by the Adolphs, Tranel, and Damasio (1998) finding have shown that the magnitude of amygdala responses to faces corresponds to judgments of the trustworthiness of each face. Remarkably, one such study has shown that the correlation is stronger between the amygdala of an individual and the group-average ratings of trustworthiness made by 129 unscanned subjects than it is for that individual's own ratings of trustworthiness (Engell, Haxby, and Todorov, 2007; *Figure 18.2*)! I know; it also took me a while to wrap my brain around this finding. How could a person's brain show a response that corresponds more closely to a group-average judgment than to their own judgment? The answer to this question requires us to keep in mind that this finding only pertained to one region within this person's brain. As Engell, Haxby, and Todorov (2007) note, their finding is "consistent with the notion that the amygdala is involved in the initial automatic assessments of trustworthiness based on facial features, but that other neural systems, which are more closely linked to idiosyncratic perception, modulate the amygdala's influence on behavior" (p. 1516).

A final note about this study is that Engell and colleagues (2007) told their subjects that they were participating in a face memory experiment, and made no mention of trustworthiness prior to the scanning session. Then, with the same logic as a subsequent-memory study (section *Subsequent memory effects in the PFC and MTL, Chapter 12*), the researchers measured behavioral correlates of the actual mental construct of interest, in this case trustworthiness judgments, after the scanning session. This means that the activity measured in their scans corresponded to what could be characterized as "implicit trustworthiness judgments." That is, the subjects presumably weren't thinking to themselves *Is this person trustworthy?* as they viewed each face; they were just trying to memorize them. Nonetheless, a part of their brain was responding to each face in a way that very strongly mirrored how they would later judge trustworthiness when asked to do so explicitly.

If we relate the findings and interpretation of the Engell, Haxby, and Todorov (2007) study to those of Adolphs, Tranel, and Damasio (1998), a decidedly visual role for the amygdala emerges. Recall that the patients with bilateral amygdala lesions were not impaired at making approachability and trustworthiness judgments when they listened to brief spoken biographies of hypothetical individuals. Thus, despite what we earlier said about the amygdala not being considered a part of the visual system,

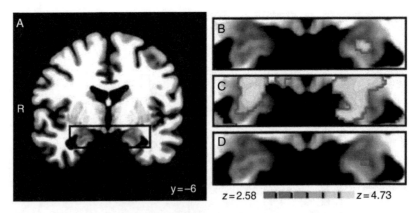

FIGURE 18.2 Amygdala responses to visually presented faces correlate with "consensus" judgments of trustworthiness. **A.** Coronal section with a rectangle indicating the area that is magnified in **B–D**. The black regions immediately lateral to each amygdala are the temporal horns of the lateral ventricles. **B.** The correlation of individual judgments of trustworthiness (acquired in the post-scanning test) with amygdala activity. **C.** Correlation of "consensus" judgments of trustworthiness, the average across 129 subjects who were not scanned, with amygdala activity. **D.** Correlation with individual judgments after partialling out variance shared between individual and consensus judgments. Source: Engell, Haxby, and Todorov, 2007. Reproduced with permission of The MIT Press.

the operation performed here seems to be a fundamentally visuoperceptual one: it may be that the amygdala detects visually perceived configurations of physical properties that can be used to classify other people according to any number of categories, including their moral character, and the emotion that they are currently experiencing. By this view, the amygdala may carry out fast, unconscious, "triage" operations on incoming sensory information, upon which this initial classification, this "early-warning signal," is fed to relevant cortical areas that can then engage in more deliberative processing.

Implicit information processing by the amygdala

One question for us to keep in mind as we proceed through this chapter is whether this function of the amygdala is limited to vision. For example, what if one presented to subjects the voices of different individuals, all speaking the same sentence, such that physical properties like timbre and emotional inflection, but not the semantic content of the speech, would vary as a function of moral character? Would the amygdala also implicitly classify this auditory information as a function of trustworthiness? We'll get a sense of the modality specificity of the amygdala in the next section, when we consider its functional anatomy in detail. But before leaving our friend *trustworthiness*, another important question to ask is whether this phenomenon of implicit social "judgment," of unconscious classification of people that we encounter, generalizes to other dimensions along which one can evaluate other people.

Neural bases of racial stereotyping

The implicit association test (IAT) is widely used by social psychologists to measure implicit social judgments, such as the judgments of trustworthiness illustrated in *Figure 18.1* and *Figure 18.2*. A study of racial bias by affective neuroscientist Elizabeth Phelps and colleagues (2000) used a design similar to that of Engell, Haxby, and Todorov (2007) – first scan, then obtain behavioral measures of the subjects' evaluation of the stimuli and see if the latter relate to the former – with the important difference that the post-scan behavioral measures included a test of implicit cognition, in addition to a test of explicit cognition.

Research subjects in experiment 1 of Phelps et al. (2000) were White Americans who were told they were participating in a memory study. During scanning, they viewed photographs of nine White male faces and nine Black male faces. After the scan, subjects were first administered an IAT during which they made interleaved speeded judgments of whether a face was Black or White (race classification) and whether a word was good or bad (word classification). The logic of the IAT is to vary the classification-to-response-button mapping, such that on some blocks the race classification maps Black to left and White to right, and on others the mapping is reversed. Independently, the word-classification mapping to the same two buttons is also varied, such that one can then compare the RT for endorsing a good word as "good" when the same button was required for classifying a face as Black, vs. the RT for endorsing a good word as "good" when that button was required for classifying a face as White. The oft-replicated finding is that White Americans show an implicit bias, in that they are slower to endorse good words as "good" when the same button is associated with Black faces. Notably, this effect is statistically independent of one's professed, explicit stance with regard to racial equality.

Experiment 1 of Phelps et al. (2000) did, indeed, replicate the standard finding of an implicit negative association between Black faces and positive words. This was despite the fact that the subjects also were rated with a mean of 1.89 on the Modern Racism Scale, a test of explicit racial attitudes that was administered after the IAT. The scale ranges from 1 (strongly pro-Black) to 6 (strongly anti-Black). Comparison of the behavioral vs. fMRI data showed that the magnitude of implicit anti-Black bias in an individual's IAT score was positively correlated with the extent to which the amygdala response in that subject was greater for Black than White faces. There was no evidence for a similar relationship between performance on the Modern Racism Scale and amygdala responsivity.

In experiment 2, Phelps et al. (2000) repeated their experimental procedure, but with the faces of famous and positively regarded Black and White men. For these stimuli, the IAT revealed diminished evidence for differential association of Black vs. White faces with positive words, and these scores no longer correlated with amygdala responses. Because experiment 1 used unfamiliar faces and experiment 2 famous faces, these findings were interpreted as evidence that the fast, implicit social categorization performed by the amygdala is learned, rather than hardwired.

We will return to the study of amygdala processing of implicit bias later in this chapter, when we consider the volitional control of emotion. At present, however, we will pursue a different theme, which is that the principles of plasticity and learning in the amygdala seem to play a prominent role in this region's functions. Indeed, we've reached a point where, having linked a handful of intriguing findings to the activity of this neural structure, we really need to learn more about the structure itself – its afferents, its internal anatomy and physiology, and its efferents.

THE AMYGDALA

Klüver–Bucy syndrome

The idea that the amygdala plays a central role in emotion processing has its roots in the work of the German-born, US-based experimental psychologist Heinrich Klüver (1897–1979) and the American neurosurgeon Paul Bucy (1904–1992). Klüver had pursued the first 3 years of his doctoral research at the University of Hamburg under the mentorship of Max Wertheimer (1880–1943), a founder of the tradition of **gestalt psychology**. After immigrating to the United States in 1923, he received his PhD from Stanford University in 1924 for research on visual imagery ("eidetic vision") in children (presumably getting credit for work done in Hamburg). Soon thereafter he began to compare his findings in children with subjective experiences produced by consuming the dried tops of the cactus *Lophophora williamsii* (a.k.a. peyote), which contains the psychoactive compound mescaline. These early studies of self-experimentation led to subsequent studies of the effects of mescaline on the behavior of monkeys, and this led Klüver to Paul Bucy.

Of particular interest to Klüver was the observation that mescaline produced chewing and licking behaviors, as well as seizures, in monkeys, and these suggested a type of seizure that the neurologist John Hughlings Jackson had named "uncinate fits." These seizures, implicating the temporal lobe terminations of the uncinate fasciculus (*Figure 17.3*), are characterized by hallucinations of taste and odor, no doubt a consequence of the involvement of the vmPFC (connected via the uncinate fasciculus with anterior medial temporal lobe) in olfactory processing. Klüver's goal, therefore, was to determine whether lesions of the medial temporal lobes would produce comparable effects, thereby localizing the circuitry responsible for mescaline's psychoactive effects. He enlisted Bucy, a neurosurgeon, to perform the surgical lesions. Although Klüver wanted

Bucy to restrict the region to the uncus (adjacent to the piriform cortex, see *Chapter 17*), Bucy "felt he couldn't perform such an operation and decided to take out the whole temporal lobe" (Nahm and Pribram, 1998, p. 9). The syndrome that resulted, which was exacerbated by an equivalent surgical removal in the other hemisphere, was dramatic, and characterized by five cardinal symptoms: (1) "psychic blindness" (later to be known as associative agnosia [Chapter *10*]); (2) "tameness" (i.e., docility); (3) hyperorality, characterized by putting nearly everything in the mouth to "explore" it (this has been reported to include such obviously hazardous items as lighted cigarettes); (4) dietary changes, including hyperphagia (an extreme drive to consume food); and (5) hypersexuality, including the mounting of animals from different species. Anecdotal reports also indicated that the monkeys lost their fear of snakes and of humans, the former of which is believed to be innate.

In the 1950s, research by Lawrence Weiskrantz (later to mentor Charles Gross, and of "Warrington and Weiskrantz [1974]" renown [section *Retrieval without awareness, Chapter 12*]) localized most of the symptoms of the Klüver–Bucy syndrome to the amygdala and established the idea that this structure was important for processing the rewarding or punishing value of stimuli. This set the stage for enormously influential breakthroughs from studies of the role of the amygdala in Pavlovian fear conditioning.

Pavlovian fear conditioning

Induction

Recall from section *Synaptic plasticity, Chapter 8,* that Pavlovian conditioning entails pairing an unconditioned stimulus (US) with a conditioned stimulus (CS), in temporal contiguity, until an association between the two is learned, at which time the CS, alone, will produce the conditioned response. In the classic fear conditioning protocol, the CS is an auditory tone and the unconditioned stimulus is a footshock. ("Footshock" is such a commonly used protocol that this formulation has become an accepted word in the literature.) Decades of intensive research, beginning in the 1980s, have established in exquisite detail the mechanisms underlying fear conditioning in mammals. For the following summary to make sense, you should first familiarize yourself with the circuitry of the amygdala, and of its afferents and efferents, by studying *Figure 18.3*.

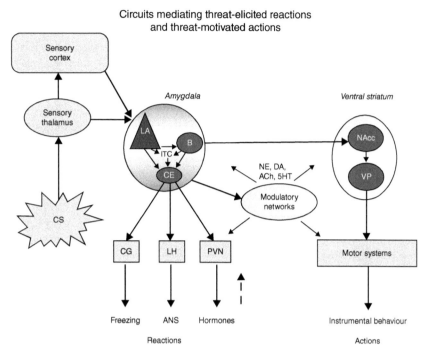

Circuits mediating threat-elicited reactions and threat-motivated actions

FIGURE 18.3 A schematic diagram of what is sometimes called the "fear circuit," which LeDoux (2012a) frames here as the threat-processing circuit. The CS, which is most often a tone or a light, is first registered in the sensory thalamus (e.g., the MGN or LGN) and then conveyed to the lateral nucleus of the amygdala (LA) via two routes, the putative direct "low road" and the cortical "high road." (The former presumably carries only low-resolution, magnocellular-dominated signals, but is central to many models of amygdala function; some, however, dispute its existence.) Within the amygdala complex are the lateral (LA), basal (B), and central (CE) nuclei, as well as the GABAergic intercalated cells (ITCs). The CE sends outputs to regions that produce the body's internal reactions to fear: the central gray (CG); the lateral hypothalamus (LH; which, in turn, activates the autonomic nervous systems [ANS]); and the paraventricular nucleus (PVN) of the hypothalamus. The basal nucleus sends outputs to the nucleus accumbens (NAcc) and ventral pallidum (VP) of the basal ganglia, thereby influencing motor actions. "Modulatory networks" refer to subcortical nuclei that innervate the brain to deliver the modulatory neurotransmitters listed. Source: From LeDoux, Joseph E. 2012a. "Evolution of Human Emotion: A View through Fear." In Evolution of the Primate Brain: From Neuron to Behavior, edited by Michael A. Hofman and Dean Falk, vol. 195 of Progress in Brain Research, edited by Stephen G. Waxman, Donald G. Stein, Dick F. Swaab, and Howard L. Fields, 431–442. Amsterdam: Elsevier. Reproduced with permission of Elsevier.

Okay, now that you've familiarized yourself with LeDoux's diagram, here's his summary of fear conditioning. Fasten your seatbelt:

CS and US convergence occurs in the lateral nucleus of the amygdala (LA), and specifically in the dorsal subregion of the LA. This convergence leads to synaptic plasticity and the formation of a CS–US association. Damage to LA, inactivation of LA, or manipulation of a variety of molecular pathways in LA prevents fear conditioning. A second important region is the central nucleus of the amygdala (CE). Manipulations of this region also disrupt conditioning. LA and CE are connected directly and by way of various intra-amygdala pathways. Once the CS–US association is formed, later exposure to the CS results in the retrieval of the learned association formed by CS–US convergence during conditioning. Information then flows from LA to CE, which then connects to hypothalamic and brainstem areas that control behavioral, autonomic, and hormonal responses that help the organism cope with the threat. Plasticity also occurs in CE, and in CS processing regions and motor control regions (LeDoux, 2012a, p. 435).

As noted by LeDoux, this is a simplified description that omits many details. Should one want more details, two places to begin are LeDoux (2000) and Davis and Whalen (2001), both listed in the *Further Reading* section. Here, we will limit ourselves to just some key concepts.

First, the necessary locus of LTP, without which fear conditioning does not occur, is the LA nucleus. Among the manipulations of LA that block fear conditioning is the administration of protein synthesis inhibitors, which block the formation of new synapses within the LA and between the LA and other amygdala nuclei. This is an evolutionarily ancient form of learning, observable in many invertebrate organisms with much simpler nervous systems. (These animals have neither amygdalae nor LA nuclei. However, synaptic plasticity underlying Pavlovian conditioning in these animals results from very similar, if not identical, intracellular molecular cascades that are triggered by the network-level convergence of signals relating to the CS and the unconditioned stimulus.)

Second, downstream from synaptic changes in LA, (at least) two different kinds of learning are supported by (at least) two pathways in the amygdala. Most of our attention in this chapter will focus on Pavlovian fear conditioning, which is mediated by circuits connecting LA to CE, and expressed via CE's outputs to systems that implement the innate "Reactions" (bottom left of *Figure 18.3*). A second pathway, from LA to the basal nucleus, implements a different kind of learning, **avoidance learning**.

Distinguishing avoidance learning from fear conditioning

Avoidance learning refers to context-dependent skeletomotor behaviors that, when engaged, enable the organism to avoid the adverse outcome predicted by the CS. One laboratory example of an avoidance behavior would be if the floor of the testing chamber has two different textures, and the rat learns to avoid a shock by, each time a tone is played, moving to the "safe zone" in the chamber. An avoidance behavior from everyday life is illustrated during my daily walk home from work. Having been unpleasantly startled more than once by the aggressive dog who is sometimes (fenced in) outdoors in front of a particular house, I have learned to turn left at the intersection preceding this house so as to avoid the unpleasant encounter.

There are three points to emphasize about the amygdala-mediated "instrumental behavior" (*Figure 18.3*, lower right), of which avoidance learning is a subtype. First is that it entails engaging the motor system via the basal nucleus-to-ventral striatum efferent pathway of the amygdala (i.e., a different pathway from LA to CE). Second is its context dependence. If I'm on a sailboat on Lake Mendota and suddenly I see a large motor boat bearing down on me, the notion of *turning left at the intersection of Kendall St. and Prospect Ave. to avoid this threat* doesn't even make sense.

That learned behavior is only relevant in the context of avoiding the dog on the walk home from work. Third is that the behaviors acquired via avoidance learning aren't emotional per se. Returning again to my pedestrian commute, if, on my way home from work, I need to pick up my son at his friend Alex's house I would also turn left at that same intersection, because that's the way to get to Alex's house. This is in contradistinction to fear conditioning, in which the CS triggers the same innate, hardwired defense mechanisms as does an unconditioned stimulus. Thus, for example, when a dog suddenly barks and lunges (an unconditioned stimulus), one's first responses are to freeze and to activate the autonomic nervous system (i.e., the "fight or flight" reflex). Similarly, when the sailor suddenly sees the large motor boat bearing down on her (also an unconditioned stimulus), her first response is also to freeze and to activate the autonomic nervous system. (It is only "a split second" after this initial stereotyped response that she engages the contextually appropriate response of grabbing the tiller to steer the sailboat out of the path of the oncoming motor boat.)

Emotional content in declarative memories

Pharmacological interference with norepinephrine

The two types of memory summarized in the preceding section, fear conditioning and avoidance learning, are nondeclarative – they do not depend on the integrity of the hippocampus or rhinal cortices. However, declarative memories can also have an emotional component. Is the amygdala involved in these? Before answering this question, we'll first revisit the study of pharmacological interventions with emotional declarative memory. Recall from section *Reconsolidation, Chapter 11*, that administration of the β-adrenergic antagonist propranolol at the time of retrieval of an emotional memory "knocks down" the emotional content of the memory on subsequent retrievals. This finding provided evidence that reconsolidation under normal circumstances entails the activation of the subcortical "modulatory networks" illustrated in *Figure 18.3*. Those reconsolidation studies considered in *Chapter 11* built on the earlier work performed by neurobiologists Larry Cahill, and colleagues, as summarized here. In one celebrated study, Cahill, Prins, Weber, and McGaugh (1994) used 12 images to construct a story with two variants, one emotionally arousing and one emotionally neutral. Both involved a mother and son leaving home, a hospital emergency room, and hailing a taxi. In the neutral story, the mother and son walk to the hospital where the father works, passing automobile wreckage in a junkyard along the way. They observe an emergency preparedness drill (which includes a brain scan and makeup artists creating simulated injuries), and the mother phones the day care facility where her younger child is enrolled, then hails a taxi to go pick up this child. The arousing story begins the same way, but then the boy is critically injured by a collision between cars while he and the mother are crossing the street, the brain scan and gore at the hospital are described as medical procedures undertaken to save the boy's life, and the phone call and taxi are the same.

For their experiment, Cahill et al. (1994) divided their subjects into four groups. An hour prior to hearing the story, subjects in two of the groups were administered propranolol, and those in the other two were administered a placebo. Subsequently, subjects in one of the two propranolol groups and in one of the two placebo groups heard the arousing version of the story, and those in the remaining two groups heard the neutral version. On the surprise memory test administered 1 week later, the *arousing-story/placebo* group (A/P in *Figure 18.4*) showed the expected benefit that emotional content gives to memory performance. Such emotion-related enhancement was not present, however, for the *arousing-story/propranolol* (A/BB) group, whose memory performance for the middle part of the story didn't differ from that of the two neutral-story groups (*Figure 18.4.A*). Notably, however, the A/BB did not differ from the A/P group with regard to how emotional they found the story to be, when asked to rate it immediately after having heard it. This latter finding suggests that the effect of propranolol was specific to emotional memory encoding and had minimal, if any, effect on the experience of emotion at the time that the story was presented.

Emotional declarative memory in patients with amygdala lesions

Now, you may be wondering why the description of the Cahill et al. (1994) experiment is presented in this section of this chapter that is focused on the amygdala. After all, propranolol administration in this study was systemic, and so antagonism of β-adrenergic receptors was likely to have occurred throughout the brain, as well as in the periphery. The reason for including it in this section is because a subsequent study used the same experimental materials and procedure to test the emotional memory of patients with bilateral amygdala lesions. In it, a team led by Ralph Adolphs, and including Larry Cahill, found that the

performance of amygdala-lesioned patients was very similar to that of healthy individuals who had been administered propranolol: first, just like the A/BB subjects in *Figure 18.4.A*, their declarative memory was selectively impaired for emotional portions of the story; second, like the A/BB subjects in *Figure 18.4.B*, their immediate rating of the emotionality of the story was comparable to that of control subjects(!). This certainly came as a surprise to me. Does this mean that amygdala isn't necessary for the experience of emotion?!? Anecdotally, Adolphs, Cahill, Schul, and Babinsky (1997) note that one of the two patients

> explained that she had found the slide with the surgically repaired legs quite emotional and unpleasant, a comment she also made spontaneously when she first saw the slides. She has made similarly appropriate comments in regard to her experience of a large number of emotionally charged stimuli we have shown her, as well as in regard to emotional events from her own life.

Then, they go on to conclude that

> It may thus be that the experience of emotional events does not depend essentially on the amygdala. The amygdala's function in humans appears to be rather more subtle, serving to mediate between the evaluation and experience of emotional events, on the one hand, and their acquisition and consolidation into long-term memory, on the other (p. 298).

Other findings, from other groups, are consistent with this idea of a distinction between the experience of emotion and the incorporation of emotional content into memory. For example, a PET study by Cahill and colleagues (1996) suggested that activity in the amygdala correlated more strongly with memory for emotionally arousing stimuli than with subjects' subjective reports of their state of emotional arousal while in the scanner. Thus, we may need to look beyond the amygdala to get a fuller picture of how emotion is represented while we experience it "in real time," and how it influences ongoing cognition and behavior. In doing so, we need to keep in mind, however, the dictum of which we've periodically been reminded, which is that one's introspection is a notoriously poor indicator of what processes are actually unfolding between the ears. For example, earlier in this chapter we've reviewed evidence of the unconscious influence of the amygdala on decisions related to trustworthiness and to racial bias. Thus, one caveat to keep in mind is whether, despite their reports on a single subjective measure of

FIGURE 18.4 Effect of propranolol on emotional memory encoding. Source: From Cahill, Larry, Bruce Prins, Michael Weber, and James L. McGaugh. 1994. "β-Adrenergic Activation and Memory for Emotional Events." Nature 371 (6499): 702–704. Reproduced with permission of Springer Nature.

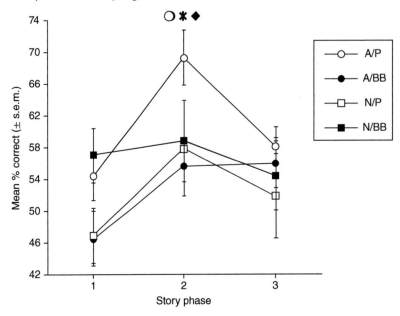

FIGURE 18.4.A Performance of four groups of subjects on a surprise memory test administered 1 week after hearing either the emotionally arousing (A) or neutral (N) version of a story, prior to which they had been administered either a placebo (P) or the drug propranolol (a "beta blocker" [BB]). These factors were fully crossed in a 2 × 2 between-subjects design. Story phases 1 and 3 were identical. The A/P group, but not the A/BB group, demonstrated superior memory performance for story phase 2, which contained the emotionally arousing content. The A/BB group's performance did not differ from that of the groups that heard the emotionally neutral version of the story.

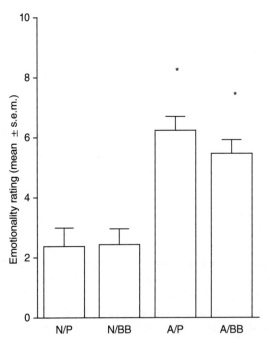

FIGURE 18.4.B Immediately after hearing the story, subjects were asked to rate on a 1–10 scale how emotional they found the story. Both groups that had heard the emotionally arousing version of the story rated it to be more emotional than did the groups that had heard the emotionally neutral version.

emotionality, patients with bilateral amygdala damage are truly experiencing the full richness of the emotional response to an evocative story that control subjects are experiencing. The same holds for subjects with propranolol in their systems.

And so, to recap, our understanding at this point in the chapter is that the amygdala can influence our decisions, and possibly our emotions, implicitly. Additionally, however, there's reason to think that our explicit, subjective experience of emotions "in the moment" is supported, at least in part, by systems outside of the amygdala proper.

The amygdala's influence on other brain systems

As a transition toward considering the processing of emotion outside of the amygdala, we can start by considering how the amygdala influences other brain systems. For example, one possible explanation of the findings from the emotional memory studies that we have just reviewed (Cahill, Prins, Weber, and McGaugh, 1994, and Adolphs, Cahill, Schul, and Babinsky, 1997) is that the amygdala's activation of the locus coeruleus (one of the "Modulatory

networks" in *Figure 18.3*), and the consequent increase in norepinephrine levels in many parts of the brain, results in enhanced memory consolidation. If β-adrenergic receptors are blocked by a pharmacological agent, then the emotional memory effect is blocked. Similarly, if there is no amygdala to trigger the increased release of norepinephrine, then enhanced consolidation will not take place. Note that, unlike the midbrain dopaminergic nuclei A8–A10, whose output is focused on the basal ganglia and PFC, the locus coeruleus sends projections diffusely throughout the cortical mantle and to several subcortical targets. Further, β-adrenergic receptors are found throughout the brain, most densely in hippocampus, but also in cerebellum, basal ganglia, thalamus, midbrain, and cerebral cortex. Thus, it is difficult to know how anatomically localized or diffuse are the effects of propranolol that blocks the consolidation of memory for the emotional components of a story.

Consistent with the caveat about the interpretation of self-reports of the subjective experience of emotion, there is fMRI evidence that damage to the amygdala does, indeed, influence the processing of emotional stimuli. As illustrated in *Figure 18.5*, the fMRI response of the

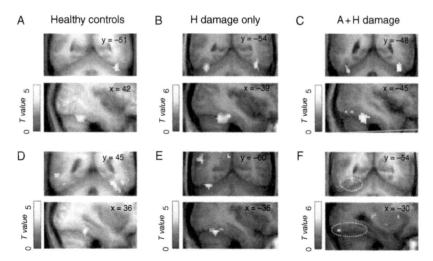

FIGURE 18.5 Activity evoked by attention to faces vs. houses, in control subjects, patients with medial temporal lobe damage encompassing the hippocampus, but not the amygdala ("H damage only"), and in patients with damage encompassing the hippocampus and amygdala ("A + H damage"). On each trial, two faces and two houses were presented, and, depending on the attentional cue, subjects had to decide whether either the two faces or the two houses were the same or different. Orthogonal to the factor of attention was the fact that half of the trials presented neutral faces and half presented fearful faces. **A–C** show that the response in fusiform gyrus for attention to faces was comparable across the three groups. **D–F**, in contrast, show a selective reduction of activity to fearful vs. neutral faces, regardless of whether subjects were attending to faces or houses, for the A + H group. Source: Vuilleumier, Richardson, Armony, Driver, and Dolan, 2004. Reproduced with permission of Nature Publishing Group

midfusiform gyrus to the visual presentation of faces, relative to houses, is no different in patients with amygdala damage relative to control subjects. Where these patients differ, however, is in the comparison of the response to fearful vs. neutral faces. In the inferior temporal and occipital regions shown in *Figure 18.5*, as well as in primary visual cortex and superior temporal sulcus (not shown), patients with amygdala lesions do not show the robust increase in activity for fearful vs. neutral faces that is seen in healthy control subjects, as well as in patients whose lesions compromise the hippocampus but spare the amygdala. And so our understanding of the functions of the amygdala does, indeed, need to be nuanced.

THE CONTROL OF EMOTIONS

Recall from the beginning of this chapter that Davidson and Irwin (1999) have suggested that positive emotions involve appetitive and approach-related behaviors, whereas negative emotions involve aversive and withdrawal-related behaviors. Davidson and colleagues have associated positive, approach-related emotions with elevated activity in the left PFC and negative, withdrawal-related emotions to elevated activity in the right PFC. This is seen in mood-induction studies, in which subjects who are made to feel happy or sad reliably produce greater activity in the left or right PFC, respectively. Further, in the resting EEG, when subjects are sitting quietly and not performing a task, alpha-band power over the PFC is relatively higher over the left than the right hemisphere in patients with depression, suggesting a relative hyperactivation of the right PFC. Finally, damage to the left frontal cortex is more likely to produce depression than a frank lesion of any other part of the brain.

The fact that one can carry out a "mood induction study" is a demonstration of something that each of us has experienced several times throughout our lives: external circumstances can *make* us feel a certain way, thereby influencing our emotional state. In this section, we will consider how, and under what circumstances, we exert (internal) control over our emotions. As you make your way through this section, keep in mind that the questions it addresses are of keen interest to those who study psychopathology and mental health.

Extinction

As important as it can be to an organism to learn that a particular stimulus, or context, represents a threat, and should be avoided, it can be equally important to extinguish this association once the threat no longer exists. There is, regrettably, a real-world example of this fact that will be familiar to almost any reader of this book, regardless of where on the planet they live. For a soldier who is in an active combat zone, it is highly adaptive to take evasive action when there is an unexpected loud noise. Often, this is produced by an explosion or by gunfire, and the soldier needs to protect themself and/or do what's necessary to repel the attackers. Once decommissioned and back in civilian life, however, it is no longer adaptive for the ex-soldier to respond in the same way to an unexpected loud sound. The sound is highly likely to be the result of something innocuous, like a car backfiring, or the crash of a garbage can knocked over by a dog. "Hitting the dirt" or running for cover in the civilian context can interfere with family and other social relations, can make it difficult to obtain or hold a job, and so on. Further along, we'll see that at pathological levels, an inability to extinguish no-longer-relevant conditioned fear can result in post-traumatic stress disorder (PTSD), and possibly other anxiety-related disorders.

Research in the rodent

Fear extinction refers to the gradual decrement in conditioned fear responses that occur with repeated presentation of a CS without the accompanying US. Extinction was originally thought by Pavlov and his contemporaries to be an inhibitory phenomenon, akin to "unlearning," or forgetting, the CS–US association. Research from physiological psychology has long-since established, however, that extinction of conditioned fear is better understood as the result of learning a new association with the CS, one that competes with the previously learned CS–US association. Behavioral phenomena such as spontaneous recovery and rapid relearning provide evidence against an "unlearning" mechanism. Consistent with the "new learning" account of extinction is the finding that fear extinction requires the participation of NMDA receptors in the basal nucleus of the amygdala. The principles and mechanisms that govern fear extinction have now been well worked out in the rodent, and translating these to the human, including into potential treatment strategies, are currently the focus of intensive research activity.

The roles of midline frontal cortex

As indicated above, the basal nucleus of the amygdala has been identified as an important locus for learning a new association for the CS that was previously associated with a threatening US. Learning the new association presumably entails the pairing of sensory information representing the CS with a representation of the context in which extinction trials are administered. (Extinction is more quickly learned if extinction trials – i.e., trials presenting the CS in the absence of the threat-related US – are carried out in a different context from the one in which the fear memory was acquired.) Additionally, however, there are top-down control signals that adjudicate between the older CS–fear association and the newer CS–neutral association. The sources of this top-down control are the midline frontal regions illustrated in *Figure 18.6*. Interestingly, the two seem to operate in antagonistic, "push me-pull you" fashion.

The role for the infralimbic (IL) cortex of the rodent in prioritizing the newer CS–neutral association (often referred to as the "extinction memory") has been demonstrated in a series of elegant studies by neurobiologist Gregory Quirk and his colleagues at the University of Puerto Rico. The first experiments in the series demonstrated that whereas lesions of IL did not interfere with extinction on day 1 of the experiment, they did interfere with retrieval of this newer CS–neutral association on day 2. That is, the animal's behavior on day 2 was similar to what it would have been had it not gone through extinction learning on day 1. At face value this could either reflect a deficit in the consolidation of the newly learned

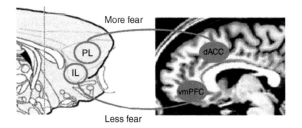

FIGURE 18.6 Anteromedial regions in the rat and the human believed to play homologous roles in the control of conditioned fear memories. PL = paralimbic; IL = infralimbic. Source: From Milad, Mohammed R., and Gregory J. Quirk. 2012. "Fear Extinction As a Model for Translational Neuroscience: Ten Years of Progress." Annual Review of Psychology 63: 129–151. Reproduced with permission of Annual Reviews.

CS–neutral association or an impairment in the newer memory's retrieval. Evidence in favor of the latter account came from recording studies indicating that although IL neurons responded to the CS (a tone) during neither the initial acquisition of the fear memory nor during its extinction, they did respond to the CS during retrieval of the new CS–neutral memory on day 2 of extinction. Furthermore, the magnitude of these "extinction memory" responses was inversely correlated with the amount of freezing at the retrieval test, suggesting that the strength of this IL signal was determining the influence of the newer CS–neutral memory on the animal's behavior. Causal evidence for this interpretation was produced by microstimulation of the IL neurons that represented the CS during retrieval of the extinction memory – microstimulation enhanced the expression of the newer CS–neutral memory over the older CS–US fear memory.

As alluded to earlier, a second region along the frontal midline, the prelimbic (PL) cortex (*Figure 18.6*), seems to play the opposite role. First, inactivation of PL has been shown to reduce the expression of conditioned fear (i.e., the duration of time that the animal freezes upon hearing the tone). Second, neurons in PL show sustained increases in firing rate in response to presentation of the CS, and the time course of this activity closely mirrors the duration of the freezing behavior itself. This sustained activity seems to be the product of activity in a recurrent network linking PL with the amygdala, in that the onset of the CS is first detected by the amygdala, which sends a transient "fear signal" to PL. PL then converts this into a sustained signal that feeds back onto the amygdala, resulting in sustained amygdala output to the systems that implement the freezing response (*Figure 18.3*). Finally, as they have done with the IL circuit, Quirk's group has demonstrated that microstimulation of PL neurons has the effect of increasing freezing responses to the CS and of interfering with extinction.

Fear conditioning and extinction in the human

As suggested previously in the text, and diagrammed explicitly in *Figure 18.6*, there seem to be important parallels from the rodent work that we have just reviewed, and findings with related behaviors in humans. In particular, activity in human vmPFC seems to support functions that are analogous to those of the rodent IL: during fear conditioning, activity in vmPFC is suppressed relative to baseline; during extinction learning, activity in vmPFC increased; and during retrieval of the extinction memory, the magnitude of vmPFC activity predicts the subject's success at inhibiting conditioned responding. In addition to these functional studies, analysis of structural MRI data suggests that individual differences in the thickness of cortex in the vmPFC (as measured with voxel-based morphometry [VBM; section *Voxel-based morphometry*, *Chapter 3*]) is positively correlated with the strength of extinction learning.

With regard to the establishment and maintenance of conditioned fear memories, there's good evidence that the human dorsal ACC supports functions analogous to those of the rodent PL (*Figure 18.6*): activity in this region correlates with behavioral and skin-conductance correlates of fear conditioning, and there's also a large body of evidence indicating that this region is highly responsive to aversive unconditioned stimuli (e.g., the delivery of pain, or an unexpected and loud noise). (Skin conductance measures increased activity in sweat glands in the skin, an [automatic] autonomic response to fear, stress, or anxiety.) Interestingly, the omission of an expected US also produces increased activity in this region, an effect that likely reflects a role for error prediction signals in fear conditioning.

The volitional regulation of emotion

Any particular emotion can influence us for varying lengths of time. Davidson and Begley (2013) summarize it as such: "The smallest, most fleeting unit of emotion is an emotional *state* . . . a feeling that . . . persist[s] over minutes or hours or even days, is a *mood* . . . And a feeling that characterizes you . . . for years is an emotional *trait*" (p. xi). In contrast to acquisition and extinction of conditioned fear, processes of which we are typically unaware, and which contribute implicitly to our emotional and personality traits, emotional states and moods are amenable to volitional control. One model of the cognitive control of emotion, which social cognitive neuroscientist Kevin Ochsner and colleagues have been building and refining for over a decade, is illustrated in *Figure 18.7*. In effect, the model posits that many of the same principles and mechanisms that we use for cognitive control are also used for emotion regulation.

In the model of emotion regulation illustrated in *Figure 18.7*, individual mechanisms that implement emotion regulation each act on one of the steps in emotion generation that are illustrated in panel A of *Figure 18.7*. "Situation selection" and "situation modification" are strategies for avoiding an encounter with a potentially aversive situation. "Attentional selection" and its complement, "attentional filtering," control what information is gated into, or out of, the emotion generation process. "Cognitive change" refers to a collection of processes and strategies that act on the process of emotional appraisal. The most widely studied of this is "cognitive reappraisal," which entails reinterpreting the meaning or significance of the triggering event, so as to change one's emotional response to it. Finally, the processes that act on the expression of emotion ("Response" in *Figure 18.7*) fall under the rubric of "response modulation." The most commonly studied of these is expressive suppression – controlling the muscles of one's face to disguise from observers what emotion one is experiencing (e.g., the "poker face").

Meditation as mental training in emotion regulation

Our emotional states, moods, traits, are arguably the most important factors in determining our quality of life. A growing body of promising findings indicates that through intensive, scientifically informed training, one can become more skilled at regulating one's emotional states, and reap collateral cognitive benefits, to boot. One example, which we'll consider here, is meditation. With an introduction seemingly more fitting for this book's *Chapter 9* than *Chapter 18*, here's how cognitive neuroscientist Heleen Slagter and colleagues (2007; including senior author Richard Davidson) set up their meditation study:

> A major limitation in human information processing arises from the time required to consciously identify and consolidate a visual stimulus in short-term memory . . . This process can take more than a half-second before it is free for a second stimulus, as is revealed by the attentional-blink paradigm: If the second target stimulus (T2) of two target stimuli is presented within 500 ms of the first one (T1) in a rapid sequence of distracters, it is often not detected . . . This deficit . . . is believed to result from competition between the two targets for limited attentional resources . . . : When many attentional resources are devoted to the processing of T1, too few may be available for T2 . . . Yet, the attentional blink does not reflect a general, immutable bottleneck, because most individuals are able to identify T2 on at least a portion of trials . . . suggest[ing] that some control . . . over the allocation of attentional resources is possible (p. 1).

In the study by Slagter et al. (2007), subjects performed the attentional-blink task, with concurrent EEG, immediately prior to beginning, and immediately upon completing, a 3-month retreat during which they meditated for 10–12 hours per day with Vipassana meditation. With this

A Strategies and processes

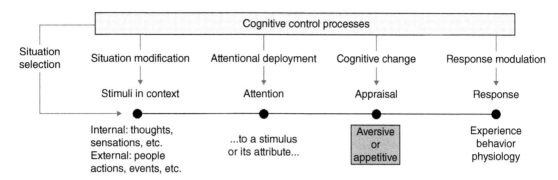

B Neural systems

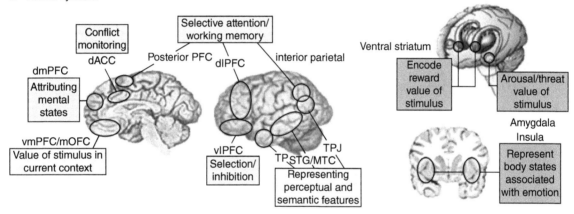

FIGURE 18.7 Ochsner's model of the cognitive control of emotion. In **A**, the four black dots represent the stages involved in generating an emotional state. First, a triggering event occurs, be it a spontaneous thought or an external event. Second, attention is drawn to all or just a part of the representation of the triggering event. Third, a process of appraisal happens, whereby a determination is made about the valence (positive or negative) of this event. There are many theories about the processes that may underlie emotional appraisal, and the neural bases of some processes that may be important appear in blue in panel **B**; here Ochsner, Silvers, and Buhle (2012) have simplified the appraisal step into a binary classification of "aversive or appetitive." (Note that this has parallels with Davidson's "avoid/approach" distinction.) Finally, the "Response" stage corresponds to "translating these appraisals into changes in experience, emotion-expressive behavior, and autonomic physiology" (p. e3). The control processes that can influence each of these stages are elaborated in the text. Panel **B** summarizes mental processes and neural systems proposed to be involved in appraisal (blue), generating responses (pink), and systems with "an undefined or intermediary role in reappraisal" (p. e3; yellow). Source: From Ochsner, Kevin N., Jennifer A. Silvers, and Jason T. Buhle. 2012. "Functional Imaging Studies of Emotion Regulation: A Synthetic Review and Evolving Model of the Cognitive Control of Emotion." Annals of the New York Academy of Sciences 1251: e1–e24. Reproduced with permission of John Wiley & Sons.

technique one starts by focusing or stabilizing concentration on an object such as breath, and subsequently broadens one's focus, cultivating a "nonreactive" form of sensory awareness, meaning that one does not respond emotionally to the foci of attention. Control subjects were individuals who were interested in learning about meditation; they were given a 1-hour class, and meditated for 20 minutes daily for the week prior to each EEG session.

Behavioral results showed a greater reduction of the attentional blink post- vs. pre-training for the meditation group relative to the control group. Notably, however, both groups improved slightly at T1 detection on session 2 vs. session 1, and the control group also showed a modest (though significantly smaller) improvement at T2 detection. This highlights the importance of including a control group in training studies. Without one, one cannot rule out effects of learning the task, or some other factor that may have nothing to do with the training procedure itself. The ERP data, illustrated in *Figure 18.8*, also showed a selective effect of meditation training.

Although other studies have shown that intensive performance of a task can produce similar reductions in the

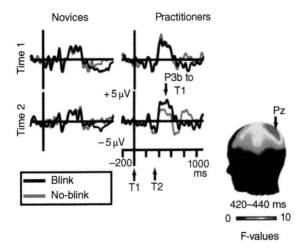

FIGURE 18.8 Comparison of the P3b ERP component evoked by the T1 stimulus as a function of group (Novices [i.e., controls]; Practitioners [i.e., meditation training group]); pre- vs. post-training session ("Time 1" and "Time 2," respectively) and, in each plot, whether the following T2 stimulus was detected ("No-blink") or missed ("Blink"). The P3b is associated with the allocation of attentional resources. The finding of a lower P3b to "No-blink" T1 stimuli in the post-training session for Practitioners, but not Novices, therefore, is interpreted as evidence that, as a result of training, they were able to devote fewer attentional resources to T1, thereby leaving more resources available for detection of T2. Also consistent with this interpretation was the fact that the individuals who showed the largest Time 1-to-Time 2 reductions in P3b were those who showed the largest reductions in attentional blink from Time 1 to Time 2. Source: From Slagter, Heleen A., Antoine Lutz, Lawrence L. Greischar, Andrew D. Francis, Sander Nieuwenhuis, James M. Davis, and Richard J. Davidson. 2007. "Mental Training Affects Distribution of Limited Brain Resources." PLoS Biology 5 (6): e138. Public Domain.

attentional blink, noteworthy about the present finding from Slagter et al. (2007) was that the benefits of training resulted from *purely mental* activity – subjects weren't pointing a video-game gun, controlling a joystick, or performing any other overtly sensorimotor task; they were "just thinking." ("Just thinking," of course, in a particularly disciplined, focused way.) Subsequent studies from Davidson and colleagues have demonstrated impressive effects with other types of mental training. One example is that 2 weeks of compassion training (relative to a reappraisal control group) alters performance on a neuroeconomic game entailing the altruistic redistribution of funds, a behavioral change accompanied by changes in task-related functional connectivity between dorsolateral PFC and the nucleus accumbens (Weng et al., 2013).

The role of awareness

One interpretation of the results from Slagter et al. (2007) is that intensive training on increasing awareness of one's body and one's thoughts can help one become more aware of events in the outside world that one might otherwise have missed. Here we'll look at the idea that the volitional control of one's emotional state may depend on awareness of the event that put one in that state. The study was led by Regina Lapate, now a professor at University of California Santa Barbara, at the time a graduate student with Richard Davidson at the University of Wisconsin–Madison. Lapate et al. (2016) used an affect misattribution task in which an emotionally evocative stimulus can bias the assessment of a subsequently encountered neutrally valenced stimulus. This is typically demonstrated in the laboratory by presenting the emotional stimulus under perceptually challenging conditions, and the typical finding is that the misattribution effect is strongest when subjects aren't aware of having perceived the emotional stimulus. For this study, Lapate et al. (2016) scanned subjects with fMRI while presenting fearful faces under conditions of "continuous flash suppression" (CFS), a technique whereby a stimulus can be presented to a subject, but not detected, if it is presented at low contrast to the nondominant eye while a series of high-contrast colorful stimuli (here, "Mondrian" patterns) are simultaneously presented to the dominant eye. Six-to-nine seconds later a neutral face was presented and subjects rated "How much do you like this person?" on a 1–4 scale. Neutral priming stimuli were images of flowers, and both fearful faces and flowers were also shown in blocks without CFS, during which they were visible.

Behavioral results indicated that there was considerable intersubject variability in the size of the emotion misattribution effect (indexed by subtracting neutral-face ratings on flower trials from neutral-face ratings on fearful-face trials). Ignoring these individual differences, univariate analyses showed two primary effects of interest: first, BOLD signal was greater on *aware* (CFS-absent) than *unaware* (CFS-present) trials, collapsing over fearful faces and flowers, in dorsolateral PFC (areas 9/46/45), dorsomedial PFC (areas 6/8), and fusiform and lateral occipital cortex (*Figure 18.9*); second, the central nucleus of the amygdala in the right hemisphere (identified a priori as an ROI) responded more strongly to fearful faces than to flowers on both *aware* and *unaware* trials. Notably, when individual differences in the emotion misattribution effect were taken into account, it was only in the unaware condition that the intensity of the amygdala response to fearful faces predicted the size of an individual's emotion misattribution effect (*Figure 18.10*). "Therefore," the authors noted, "when unchecked by conscious awareness, greater amygdala responses to fear cues were associated with a negative bias towards later-shown neutral stimuli—i.e., greater affective coloring" (p. 3).

Now onto the question of primary interest to us (and to the authors): the neural mechanisms associated with the control of emotional misattribution. Panel A in *Figure 18.11* shows the regions in left dorsolateral PFC and medial PFC for which a decrease in the value of functional connectivity with the right amygdala predicted a smaller emotion misattribution effect, effects consistent with the idea that PFC-based control signals play an important role in emotion regulation. This may need some unpacking. If two regions are involved in a process, let's say face perception, it makes sense to think that increased efficacy of that process would be associated with an increase in functional connectivity between these regions. If we consider V1 and the FFA, for example, the presentation of high-contrast faces might be expected to produce greater functional connectivity between V1 and FFA than would the presentation of low-contrast faces. That is, the correlation between these two regions would be higher. But what if we consider another circuit in the brain where one region has an inhibitory effect on the other – let's say the substantia nigra pars reticulata (SNpr) and the superior colliculus (SC; section *The superior colliculus, Chapter 9*). Because activity in these two structures is negatively correlated, the functional connectivity between them will be negative. Using this same logic, the more negative is the functional connectivity between the PFC and the amygdala, the more effective is the inhibitory control of the latter by the former.

HOW DOES THAT MAKE YOU FEEL?

Emotion processing – at least negative emotion processing, which is what we focused on in this chapter – is strongly associated with the amygdala. Although the amygdala can play a role in the top-down modulation of visual processing, it doesn't seem to be necessary for real-time assessment of the emotional content of a situation. It does seem to play a key role, however, in long-term memory for such

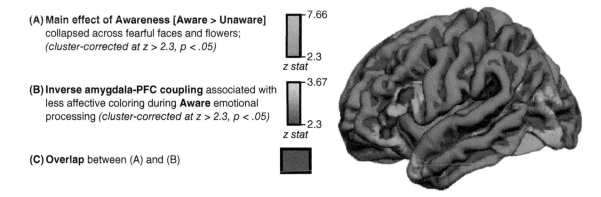

FIGURE 18.9 Lateral left-hemisphere view of regions showing a greater response to the 1-second binocular presentation of faces and flowers than to continuous flash suppression (1-second presentation of face/flower to nondominant eye plus concurrent 1.5-second presentation of 10-Hz Mondrian patterns to dominant eye). Source: From Lapate, R. C., Rokers, B., Tromp, D. P. M., Orfali, N. S., Oler, J. A., Doran, S. T., Adluru, N., Alexander, A. L. and Davidson, R. J. (2016). Awareness of emotional stimuli determines the behavioral consequences of amygdala activation and amygdala-prefrontal connectivity. Scientific Reports 6: 25826. 10.1038/srep25826. Reproduced with permission of Nature Publishing Group.

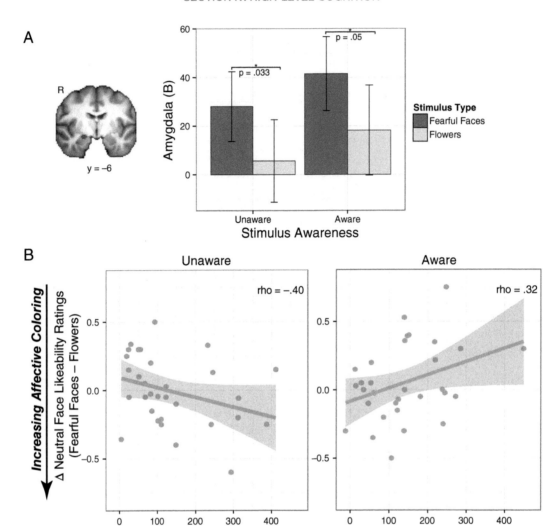

FIGURE 18.10 Amygdala responses to perceived vs. not-perceived fearful faces (and flowers), and their relation to emotion misattribution. **A.** Right-hemisphere central amygdala ROI, and its mean BOLD response to priming stimuli as a function of awareness condition. (BOLD values are in units of beta values from the GLM analysis.) **B.** The differential response of the amygdala to fearful faces vs. flowers (abscissa) plotted against the emotion misattribution effect ("affective coloring", ordinate). In each plot, each dot corresponds to 1 of the 31 subjects in the study. The plot on the left shows the significant trend for *unaware* trials, whereby greater amygdala activity for fearful faces relative to flowers predicts a greater emotion misattribution effect. The plot on the right shows a trend in the opposite direction for *aware* trials, although this effect did not exceed the threshold for significance. (The *unaware* vs. *aware* regression lines are significantly different from each other.) Source: From Lapate, R. C., Rokers, B., Tromp, D. P. M., Orfali, N. S., Oler, J. A., Doran, S. T., Adluru, N., Alexander, A. L. and Davidson, R. J. (2016). Awareness of emotional stimuli determines the behavioral consequences of amygdala activation and amygdala-prefrontal connectivity. Scientific Reports 6: 25826. 10.1038/srep25826. Reproduced with permission of Nature Publishing Group.

information. Research indicates that learning to control your emotions can make you "more aware," and that, conversely, being aware of what's affecting your mood might be important for being able to control it. With mental health being such a large, and growing, concern for our society, further expansion of our understanding of the neural bases of emotion and its regulation can't happen too soon.

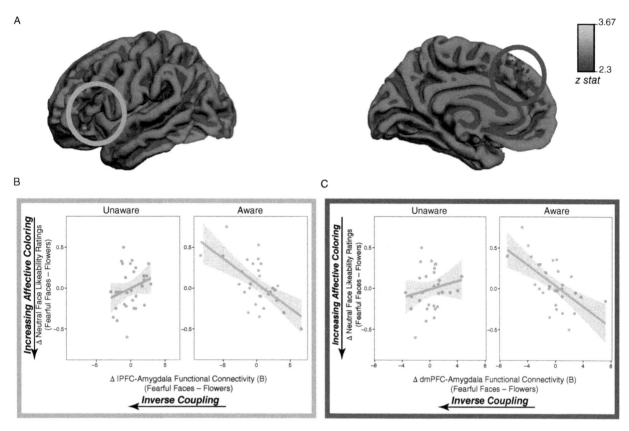

FIGURE 18.11 Emotion regulation by PFC control of amygdala responses to fearful faces. **A.** The cortical regions in the left hemisphere that show an inverse coupling with amygdala. (Dorsomedial PFC in the right hemisphere also showed this effect.) **B** and **C** illustrate that, for both of these regions, the effect is only seen on *aware* trials. Source: From Lapate, R. C., Rokers, B., Tromp, D. P. M., Orfali, N. S., Oler, J. A., Doran, S. T., Adluru, N., Alexander, A. L. and Davidson, R. J. (2016). Awareness of emotional stimuli determines the behavioral consequences of amygdala activation and amygdala-prefrontal connectivity. Scientific Reports 6: 25826. 10.1038/srep25826. Reproduced with permission of Nature Publishing Group.

END-OF-CHAPTER QUESTIONS

1. In what way(s) does it make sense to think of faces and the information that they convey – about emotion, trustworthiness, race – as conditioned stimuli?

2. Trace the circuitry that underlies fear conditioning. Is this same or different from the circuitry underlying avoidance learning?

3. Fear conditioning and avoidance learning are both types of nondeclarative memory. Is the amygdala also involved in processing the emotional component of declarative memories? If so, through what mechanism(s)?

4. How, and under what circumstances, does the amygdala influence visual processing?

5. What is an emotion? How are emotions different, neurobiologically, from the domains of cognition that we have studied up through this point in the book?

6. Is the role of the vmPFC in fear extinction best understood as learning the new CS, or retrieving it after it is learned? Contrast this with the role of the ACC.

7. How do the processes of emotion regulation differ from extinction? What brain systems and mechanisms are associated with emotion regulation?

8. How is emotion regulation related to psychopathology?

9. What's the role of awareness in emotion regulation?

REFERENCES

Adolphs, Ralph, Larry Cahill, Rina Schul, and Ralf Babinsky. 1997. "Impaired Declarative Memory for Emotional Material Following Bilateral Amygdala Damage in Humans." *Learning & Memory* 4 (3): 291–300. doi: 10.1101/lm.4.3.291.

Adolphs, Ralph, Daniel Tranel, and Antonio R. Damasio. 1998. "The Human Amygdala in Social Judgment." *Nature* 393: 470–474. doi: 10.1038/30982.

Cahill, Larry, Richard J. Haier, James Fallon, Michael T. Alkire, Cheuk Tang, David Keator, . . . James L. McGaugh. 1996. "Amygdala Activity at Encoding Correlated with Long-Term, Free Recall of Emotional Information." *Proceedings of the National Academy of Sciences of the United States of America* 93 (15): 8016–8021. doi: 10.1073/pnas.93.15.8016.

Cahill, Larry, Bruce Prins, Michael Weber, and James L. McGaugh. 1994. "β-Adrenergic Activation and Memory for Emotional Events." *Nature* 371 (6499): 702–704. doi: 10.1038/371702a0.

Davidson, Richard J., and Sharon Begley. 2013. *The Emotional Life of Your Brain: How Its Unique Patterns Affect the Way You Think, Feel, and Live – and How You Can Change Them.* New York: Plume.

Davidson, Richard J., and William Irwin. 1999. "The Functional Neuroanatomy of Emotion and Affective Style." *Trends in Cognitive Sciences* 3 (1): 11–21. doi: 10.1016/S1364-6613(98)01265-0.

Engell, Andrew D., James V. Haxby, and Alexander Todorov. 2007. "Implicit Trustworthiness Decisions: Automatic Coding of Face Properties in the Human Amygdala." *Journal of Cognitive Neuroscience* 19 (9): 1508–1519.

Hariri, Ahmad R., and Paul J. Whalen. 2011. "The Amygdala: Inside and Out." *F1000 Biology Reports* 3: 2. doi: 10.3410/B3-2.

Klüver, Heinrich. 1926. "Mescal Visions and Eidetic Vision." *American Journal of Psychology* 37 (4): 502–515.

Klüver, Heinrich. 1928. *Mescal: The "Divine" Plant and Its Psychological Effects.* London: Kegan Paul, Trench, Trubner.

Lapate, R. C., B. Rokers, D. P. M. Tromp, N. S. Orfali, J. A. Oler, S. T. Doran, N. Adluru, A. L. Alexander, and R. J. Davidson. 2016. "Awareness of Emotional Stimuli Determines the Behavioral Consequences of Amygdala Activation and Amygdala-Prefrontal Connectivity." *Scientific Reports* 6: 25826. doi: 10.1038/srep25826.

LeDoux, Joseph E. 2012a. "Evolution of Human Emotion: A View through Fear." In *Evolution of the Primate Brain: From Neuron to Behavior*, edited by Michael A. Hofman and Dean Falk, vol. 195 of Progress in Brain Research, edited by Stephen G. Waxman, Donald G. Stein, Dick F. Swaab, and Howard L. Fields, 431–442. Amsterdam: Elsevier. doi: 10.1016/B978-0-444-53860-4.00021-0.

LeDoux, Joseph E. 2012b. "Rethinking the Emotional Brain." *Neuron* 73 (4): 653–676. doi: 10.1016/j.neuron.2012.02.004.

Milad, Mohammed R., and Gregory J. Quirk. 2012. "Fear Extinction as a Model for Translational Neuroscience: Ten Years of Progress." *Annual Review of Psychology* 63: 129–151. doi: 10.1146/annurev.psych.121208.131631.

Nahm, Frederick K. D., and Karl H. Pribram. 1998. "Heinrich Klüver 1897–1979." In *Biographical Memoirs* 73: 1–19. Washington, DC: National Academies Press.

Ochsner, Kevin N., Jennifer A. Silvers, and Jason T. Buhle. 2012. "Functional Imaging Studies of Emotion Regulation: A Synthetic Review and Evolving Model of the Cognitive Control of Emotion." *Annals of the New York Academy of Sciences* 1251: e1–e24. doi: 10.1111/j.1749-6632.2012.06751.x.

Phelps, Elizabeth A., Kevin J. O'Connor, William A. Cunningham, E. Sumie Funayama, J. Christopher Gatenby, John C. Gore, and Mahzarin R. Banaji. 2000. "Performance on Indirect Measures of Race Evaluation Predicts Amygdala Activation." *Journal of Cognitive Neuroscience* 12 (5): 729–738. doi: 10.1162/089892900562552.

Slagter, Heleen A., Antoine Lutz, Lawrence L. Greischar, Andrew D. Francis, Sander Nieuwenhuis, James M. Davis, and Richard J. Davidson. 2007. "Mental Training Affects Distribution of Limited Brain Resources." *PLoS Biology* 5 (6): e138. doi: 10.1371/journal.pbio.0050138.

Vuilleumier, Patrik, Mark P. Richardson, Jorge L. Armony, Jon Driver, and Raymond J. Dolan. 2004. "Distant Influences of Amygdala Lesion on Visual Cortical Activation during Emotional Face Processing." *Nature Neuroscience* 7 (11): 1271–1278. doi: 10.1038/nn1341.

Weng, Helen Y., Andrew S. Fox, Alexander J. Shackman, Diane E. Stodola, Jessica Z. K. Caldwell, Matthew C. Olson, . . . Richard J. Davidson. 2013. "Compassion Training Alters Altruism and Neural Responses to Suffering." *Psychological Science* 24 (7): 1171–1180. doi: 10.1177/0956797612469537.

OTHER SOURCES USED

Adolphs, Ralph, Daniel Tranel, Hanna Damasio, and Antonio R. Damasio. 1994. "Impaired Recognition of Emotion in Facial Expressions following Bilateral Damage to the Human Amygdala." *Nature* 372 (6507): 669–672. doi: 10.1038/372669a0.

Cunningham, William A., Marcia K. Johnson, Carol L. Raye, J. Chris Gatenby, John C. Gore, and Mahzarin R. Banaji. 2004. "Separable Neural Components in the Processing of Black and White Faces." *Psychological Science* 15 (12): 806–813. doi: 10.1111/j.0956-7976.2004.00760.x.

Davidson, Richard J. 1998. "Affective Style and Affective Disorders: Perspectives from Affective Neuroscience." *Cognition and Emotion* 12 (3): 307–330. doi: 10.1080/026999398379628.

Demos, Kathryn E., William M. Kelley, S. L. Ryan, and Paul J. Whalen. 2008. "Human Amygdala Sensitivity to the Pupil

Size of Others." *Cerebral Cortex* 18 (12): 2729–2734. doi: 10.1093/cercor/bhn034.

Heller, Aaron S., Tom Johnstone, Alexander J. Shackman, Sharee N. Light, Michael J. Peterson, Gregory G. Kolden, Ned H. Kalin, and Richard J. Davidson. 2009. "Reduced Capacity to Sustain Positive Emotion in Major Depression Reflects Diminished Maintenance of Fronto-Striatal Brain Activation." *Proceedings of the National Academy of Sciences of the United States of America* 106 (52): 22445–22450. doi: 10.1073/pnas.0910651106.

Johnstone, Tom, Carien M. van Reekum, Heather L. Urry, Ned H. Kalin, and Richard J. Davidson. 2007. "Failure to Regulate: Counterproductive Recruitment of Top-Down Prefrontal-Subcortical Circuitry in Major Depression." *Journal of Neuroscience* 27 (33): 8877–8884. doi: 10.1523/JNEUROSCI.2063-07.2007.

Klüver, Heinrich, and Paul C. Bucy. 1937. "'Psychic Blindness' and Other Symptoms following Bilateral Temporal Lobectomy in Rhesus Monkeys." *American Journal of Physiology* 119 (2): 352–353.

LeDoux, Joseph E. 2013. "The Slippery Slope of Fear." *Trends in Cognitive Sciences* 17 (4): 155–156. doi: 10.1016/j.tics.2013.02.004.

Nitschke, Jack B., Issidoros Sarinopoulos, Desmond J. Oathes, Tom Johnstone, Paul J. Whalen, Richard J. Davidson, and Ned H. Kalin. 2009. "Anticipatory Activation in the Amygdala and Anterior Cingulate in Generalized Anxiety Disorder and Prediction of Treatment Response." *American Journal of Psychiatry* 166 (3): 302–310. doi: 10.1176/appi.ajp.2008.07101682.

OED Online. 2019. "emotion, n." Oxford University Press, September 2019, www.oed.com/view/Entry/61249. Accessed December 2, 2019.

Phan, K. Luan, Mike Angstadt, Jamie Golden, Ikechukwu Onyewuenyi, Ana Popovska, and Harriet de Wit. 2008. "Cannabinoid Modulation of Amygdala Reactivity to Social Signals of Threat in Humans." *Journal of Neuroscience* 28 (10): 2313–2319. doi: 10.1523/JNEUROSCI.5603-07.2008.

Reznikoff, Glen A., Scott Manaker, C. Harker Rhodes, Andrew Winokur, and Thomas C. Rainbow. 1986. "Localization and Quantification of Beta-Adrenergic Receptors in Human Brain." *Neurology* 36 (8): 1067–1073. doi: 10.1212/WNL.36.8.1067.

Vuilleumier, Patrik, Jorge L. Armony, Jon Driver, and Raymond J. Dolan. 2001. "Effects of Attention and Emotion on Face Processing in the Human Brain: An Event-Related fMRI Study." *Neuron* 30 (3): 829–841. doi: 10.1016/S0896-6273(01)00328-2.

Vuilleumier, Patrik. 2005. "How Brains Beware: Neural Mechanisms of Emotional Attention." *Trends in Cognitive Sciences* 9 (12): 585–594. doi: 10.1016/j.tics.2005.10.011.

FURTHER READING

Bavelier, Daphne, and Richard J. Davidson. 2013. "Brain Training: Games to Do You Good." *Nature* 494 (7438): 425–426. doi: 10.1038/494425a.
Leading experts in the training of attention and emotion, respectively, consider the question of "Because gaming is clearly here to stay, . . . how to channel people's love of screen time towards positive effects on the brain and behavior."

Davidson, Richard J., and Sharon Begley. 2013. *The Emotional Life of Your Brain: How Its Unique Patterns Affect the Way You Think, Feel, and Live – and How You Can Change Them*. New York: Plume.
An engrossing, authoritative "users' manual" for understanding and controlling one's emotional style, interwoven with facts from affective neuroscience and anecdotes from the career of one of its leading scholars.

Davis, Michael, and Paul J. Whalen. 2001. "The Amygdala: Vigilance and Emotion." *Molecular Psychiatry* 6 (1): 13–34.
A detailed review of the anatomy, physiology, and pharmacology of conditioning in the amygdala.

Greenwald, Anthony G., and Mahzarin R. Banaji. 1995. "Implicit Social Cognition: Attitudes, Self-Esteem, and Stereotypes." *Psychological Review* 102 (1): 4–27. doi: 10.1037/0033-295X.102.1.4.
This paper lays out the theoretical background for understanding implicit bias and stereotyping from a social psychology perspective.

Kubota, Jennifer, T., Mahzarin R. Banaji, and Elizabeth A. Phelps. 2012. "The Neuroscience of Race." *Nature Neuroscience* 15 (7): 940–948. doi: 10.1038/nn.3136.
A review of exciting, vitally important research that has revealed "overlap in the neural circuitry of race, emotion and decision-making."

LeDoux, Joseph E. 2000. "Emotion Circuits in the Brain." *Annual Review of Neuroscience* 23: 155–184.
Provocative argument by one of the leading scientists of fear conditioning for reconceptualizing emotion research as research on survival circuits.

Milad, Mohammed R., and Gregory J. Quirk. 2012. "Fear Extinction as a Model for Translational Neuroscience: Ten Years of Progress." *Annual Review of Psychology* 63: 129–151. doi: 10.1146/annurev.psych.121208.131631.
A review of the truly pioneering research, much of it performed by the authors, on the neural bases of extinction learning.

Ochsner, Kevin N., and Matthew D. Lieberman. 2007. "The Emergence of Social Cognitive Neuroscience." *American Psychologist* 56 (9): 717–734. doi: 10.1037//0003-066X.56.9.717.
As told by two of the field's leading practitioners.

CHAPTER 19
LANGUAGE

KEY THEMES

- Language, with its ubiquity in our lives, and its uniqueness to our species, is the focus of strongly debated theories that get at the core of what it means to be human.

- Nineteenth-century neurological models posited a network comprising posterior temporal cortex (Wernicke's area), temporoparietal regions, posterior inferior frontal gyrus (Broca's area), and, critically, the connections between them, that still influences contemporary ideas of the neural organization of language.

- EEG indicates that the earliest cortical generators of the AEP begin segregating speech from nonspeech sounds, probably bilaterally.

- The idea that there may exist a single "grammar gene" has prompted considerable excitement and research, but is probably "too good to be true."

- More than 150 years after Broca's first report of language impairment following damage to inferior PFC, the function(s) attributable to "his" area are still the focus of contentious debate.

- EEG and MEG suggest that predictive coding is an important principle governing language processing, both its comprehension and its production.

CONTENTS

A SYSTEM OF REMARKABLE COMPLEXITY

A human language is a system of remarkable complexity. To come to know a human language would be an extraordinary intellectual achievement for a creature not specifically designed to accomplish this task. A normal child acquires this knowledge on relatively slight exposure and without specific training. He can then quite effortlessly make use of an intricate structure of specific rules and guiding principles to convey his thoughts and feelings to others, arousing in them novel ideas and subtle perceptions and judgments. (Chomsky, 1975, p. 4)

Language is, indeed, a most remarkable element of human cognition. It is the medium through which culture is transmitted, technology propagated, and knowledge about topics as rich and complex as, say, *cognitive neuroscience*, is taught, updated, debated, refined, and so on. It also seems to be a uniquely human faculty. Although some other species have sophisticated means of communication, and some can be taught some rudimentary elements of human language, none has a system of communication whereby a finite number of elements can be recombined in an infinite number of ways, to generate an infinite number of messages. This is referred to as the generative property of human language.

Perhaps it is because language is so uniquely human, and so intertwined with virtually everything that we do as individuals, and as societies, that it is the most intensively studied domain of human cognition. One result of this intense interest is that there are many theoretical debates about, for example, how language is learned, the rules of grammar (indeed, whether there even exist rules of grammar), and the extent to which language depends on neural structures that are unique to our species. Although these debates can be rich and fascinating, and often get at core questions about human nature, many of them are simply outside the remit of this book. Nonetheless, when appropriate, theoretical motivations and/or implications of particular studies that we will consider will be flagged.

WERNICKE–LICHTHEIM: THE CLASSICAL CORE LANGUAGE NETWORK

The aphasias

In *Chapter 1* we introduced the work of Broca, including with his famous patient "Tan," as an important demonstration, alongside roughly contemporaneous research on motor control and vision, of localization of function in the brain. In his famous and influential paper from 1970, the neurologist Norman Geschwind (1926–1984) pointed out another important feature of this work: "The aphasias were . . . the first demonstrations of the fact that selective damage to the brain could affect one class of *learned behavior* while sparing other classes" (p. 940, emphasis added). That is, the first evidence that an unequivocally cognitive function (as contrasted with motor control and vision) could be localized to a particular part of the brain. But "the aphasias"? Broca only described one. What Geschwind was also referring to was the work of the Germany-based neurologist Carl Wernicke (1848–1904) who, in 1874, proposed a model for the neural bases of two classes of aphasia. Here's how Geschwind (1970) summarized them:

> The aphasic of the Broca's type characteristically produces little speech, which is emitted slowly, with great effort, and with poor articulation . . . Characteristically the small grammatical words and endings are omitted . . . These patients invariably show a comparable disorder in their written output, but they may comprehend spoken and written language normally. In striking contrast to these performances, the patient may retain his musical capacities. It is a common but most dramatic finding to observe a patient who produces single substantive words with great effort and poor articulation and yet sings a melody correctly and even elegantly.

This contrasts with Geschwind's summary of the aphasia first described by Wernicke in 1874:

> The Wernicke's aphasi[a] . . . patient usually has no paralysis of the opposite side, a fact which reflects the difference in the anatomical localization of his lesion. The speech output can be rapid and effortless, and in many cases the rate of production of words exceeds the normal. The output has the rhythm and melody of normal speech, but it is remarkably empty and conveys little or no information. The patient uses many filler words, and the speech is filled with circumlocutions. There may be many errors in word usage, which are called paraphasias . . . The Wernicke's aphasic may, in writing, produce well-formed letters, but the output exhibits the same linguistic defects which are observed in the patient's speech. He shows a profound failure to understand both spoken and written language, although he suffers from no elementary impairment of hearing or sight (pp. 940–941).

The loci of the regions described by Broca and by Wernicke, and redescribed by Geschwind (1970), are illustrated in *Figure 19.1*. Broca's area is located at the posterior end of the inferior frontal gyrus of the left hemisphere of the PFC, immediately anterior to premotor cortical

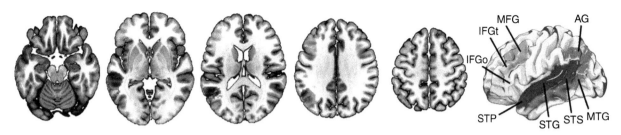

FIGURE 19.1 Cortical regions of the left hemisphere involved in language functions. Broca's area is traditionally localized to the IFGo and the IFGt, and Wernicke's area to the caudal portion of STG. AG = angular gyrus; MFG = middle frontal gyrus; IFGt = inferior frontal gyrus pars triangularis; IFGo = IFG pars opercularis; STP = superior temporal pole; STG = superior temporal gyrus; STS = superior temporal sulcus; MTG = middle temporal gyrus. Source: From John Del Gaizo, Julius Fridriksson, Grigori Yourganov, Argye E. Hillis, Gregory Hickok, Bratislav Misic, Chris Rorden and Leonardo Bonilha eNeuro 23 October 2017, 4 (5) ENEURO.0204-17.2017. Reproduced with permission of Society for Neuroscience.

regions that control the face, tongue, lips, palate, and vocal cords. Classically, because of the characteristics of Broca's aphasia, Broca's area has been presumed to be the neural seat of language production (but see Dronkers, 2000, *Further Reading*). Wernicke's area is located in the posterior superior temporal gyrus (STG) in the left hemisphere, occupying a portion of auditory parabelt cortex (*Figure 5.6*). It is, therefore, anatomically proximal to auditory cortex, and receives projections from A1 and belt cortex, as well as direct projections from the medial geniculate nucleus (MGN). Its function has classically been associated with the comprehension of spoken language.

The functional relevance of the connectivity of the network

Although Geschwind didn't include him in the 1970 review, a third important nineteenth-century contributor to our understanding of the brain bases of language function was the German neurologist Ludwig Lichtheim (1845–1928). Lichtheim is famous for his "house model," which diagrammatically predicted how interruptions of connections between input, semantic, and output regions would produce aphasias that would differ in subtle ways from those resulting from frank damage that was limited to Wernicke's or Broca's area (*Figure 19.2*). Remarkably, over the next century, each of the aphasias predicted by Wernicke and by Lichtheim came to be identified by subsequent generations of neurologists. To name just one example, conduction aphasia presents as fluent but **para-phasic speech**, but in the presence of *intact* language comprehension. (Wernicke's aphasia, in contrast, would present as fluent but paraphasic speech, but with severely *impaired* language comprehension.) It is most often associated with damage to the inferior parietal cortex – "AG"

in *Figure 19.1*; and corresponding to "3" in *Figure 19.2.A* – and is explained by damage to the arcuate fasciculus, a white matter pathway connecting, among others,

FIGURE 19.2 Lichtheim's (1885) "house model" of the neural substrates of language. Source: From Compston, Alastair. 2006. "On Aphasia. By L. Lichtheim, MD, Professor of Medicine in the University of Berne. Brain 1885; 7: 433–484." Brain 129 (6): 1347–1350. Oxford University Press. Public Domain.

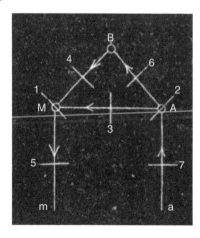

FIGURE 19.2.A "a" and "m" correspond to auditory inputs and motor outputs, respectively; "A" to Wernicke's area and "M" to Broca's. "B" is a stand-in for the neural representation of the semantic content carried by language (see *Figure 19.2.B* for more elaboration on "B"). The numerals indicate lesioning points, each known (1, 2) or (accurately) predicted (3–7) to produce a different aphasia: 1 = Broca's; 2 = Wernike's; 3 = conduction; 4 = transcortical motor; 5 = subcortical motor; 6 = transcortical sensory; 7 = subcortical sensory. Details on these can be found in Alexander (1997, *Further Reading*).

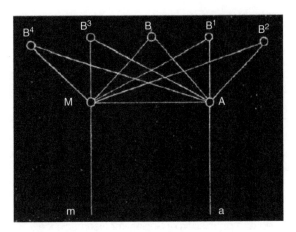

FIGURE 19.2.B Lichtheim's strikingly modern conceptualization of the distributed nature of the brain's representation of knowledge and its anatomical interfaces with Broca's and Wernicke's areas.

Wernicke's and Broca's areas. Damage to the arcuate fasciculus has the effect of leaving both Wernicke's and Broca's areas intact, but interrupting the fibers that connect them.

Thus, this Wernicke–Lichtheim model provides a framework from which we can start examining language functions more closely, beginning with speech perception, and then proceeding to grammar and production.

SPEECH PERCEPTION

Segregation of the speech signal

In *Chapter 5*, we considered in some detail the processes of the transduction of acoustic energy into a neural code, and the stages of processing of the auditory signal from brainstem to MGN to cortex that are discernible in the auditory-evoked potential (AEP). This gives us a starting point to think about speech perception and, in particular, where and how auditory signal that is related to speech is identified and extracted from the auditory stream for its presumably specialized processing as language. A straightforward way to get at this question would clearly be to contrast the processing of speech sounds vs. nonspeech sounds while measuring sound-evoked activity, and determining how the two physiological signals differ. Before just jumping in, however, there are some basic facts that we need to understand about the signal that we want to study. First, what are

the elements of spoken language? Second, at what level of organization do we want to study it? Finally, what are some fundamental physical and psychophysical properties of the speech signal itself?

The elements of language

One way to think about the organization of spoken language is as a hierarchy. At the lowest level are phonemes, the smallest unit of sound in a language that, when changed, can change the meaning of an utterance. For example, as we shall see in the next section, in US English the words "doll" and "tall" only differ by one phoneme, even though they differ by two letters. Note that, with a few exceptions, phonemes do not, by themselves, convey meaning. Rather, phonemes are assembled to create words. All human languages use many fewer phonemes than there are possible sounds that can be made with the articulatory apparatus (i.e., the diaphragm, vocal cords, tongue, jaw, lips). Although the number of phonemes differs across languages, and often differs across dialects within a language, no natural language is believed to contain more than 150. The next level in the hierarchy is *words*. Words are the smallest element of the language that have inherent meaning. The mental representation of words is referred to as the mental lexicon, and the mental lexicon of the typical adult comprises anywhere between 50,000 and 75,000 words. The meaning of a word can be changed by the addition of one or more *morphemes*: prefixes and suffixes that modify words. For example, in English, adding an "s" to the end of a regular noun makes it plural, and adding "un" to the front of one adds negation or "the opposite of."

The next level is the *clause*. A clause is constructed of words, and it is at this level of language that grammar comes into play. For example, it is the rules of grammar that determine that different ordering of the same three words can have different meanings (e.g., "Dog bites man" vs. "Man bites dog"), or no meaning (e.g., "Bites dog man"). Within the two grammatical clauses just given, the meaning was changed by changing which noun filled the role of *noun phrase* and which the role of *verb phrase*. These two clauses are also each a *sentence*.

Sentences can consist of single clauses or of a combination of two or more clauses. Indeed, it was to the infinite "combine-ability" of clauses into grammatically legal, information-conveying sentences that Chomsky (1975) was referring as "an intricate structure of specific rules and guiding principles." Collectively, these rules are referred to as "syntax," or "grammar."

Finally, there is *discourse*, which comprises any collection of sentences that can be construed as comprising a unit of communication (e.g., a paragraph, a song, a book).

If we're interested in understanding how speech is processed differently from nonspeech sounds, it makes sense to start at the most basic level, the phoneme. Studying the perception of words, for example, risks confounding neural signal associated with processing the meaning of the word vs. neural signal associated with detecting and processing a speech sound, per se. Of course, as we're about to see, limiting oneself to just studying phoneme perception doesn't mean that one's task will be easy.

The speech signal

Speech perception presents several difficult engineering problems. One of them has been summarized as the "lack of invariance" problem. That is, to understand spoken language, we need to be able to categorize sounds as corresponding to specific phonemes. However, the same phoneme can often be carried by acoustic patterns that are physically different from one another. When hearing the English words "doll" and "deal," for example, one needs to categorize the initial sound from both words as being the same: /d/. (Mistakenly categorizing one of them as /t/, for example, would lead to mistakenly hearing those words as "tall" or as "teal.") The problem of the lack of invariance, however, is that the acoustic signals when one is saying [da] vs. [di] are quite different (*Figure 19.3*). Indeed, there are no invariant cues to the identity of any phoneme! (One reason for lack of invariance of speech sounds is the phenomenon of coarticulation: the way that we shape and move our speech apparatus to utter the current phoneme is influenced by preparations already underway for the next phoneme to be spoken.)

Although lack of invariance can pose problems for the auditory system, it creates opportunities for scientists. Consider the following. A complement to the property illustrated in *Figure 19.3* is the fact there exist pairs of phonemes whose acoustic properties are very similar, yet they are perceived differently. The most studied examples are phonemes that differ only along one physical dimension, and depending on where along that dimension the sound falls, it is perceived as one or the other. A famous example from English are the phonemes [da] and [ta]. The consonants /d/ and /t/ are from a class of consonants called "stop consonants," whose utterance requires a momentary blockage of the vocal tract, typically with the tongue. The voice onset timing (VOT) of a stop consonant refers to the latency between the release of the stop and the onset of vibration in the vocal cords. [da] and [ta] differ from each other only in that [da] has a VOT of 0 ms and [ta] a VOT of approximately 70 ms. Experimentally, it has been shown that if one begins with [da] and gradually modifies the sound by increasing the VOT from 0 ms out to 70 ms [ta] in, say, 5-ms intervals, subjects do not hear a smooth transition of sounds incrementally changing from [da] to [ta]. Rather, what they hear is a series of undifferentiated "[da]"s until, at one point (typically around a VOT of 35 ms), they suddenly begin hearing a series of undifferentiated "[ta]"s. What this means is that the physical stimulus is changing continuously, but the percept has only two states. Finding the first station in the auditory system that tracks the percept, as opposed to tracking the physical signal, would be a candidate for the first area that processes speech sounds differently than nonspeech sounds.

Neural correlates of speech perception

In 2005, a team led by the French pediatrician-turned-developmental cognitive neuroscientist, Ghislaine Dehaene-Lambertz, published a study that followed a logic similar to the one just described. Specifically, it employed a mismatch paradigm, which is one of the most commonly used to study differences in the auditory/language processing system. The stimuli that Dehaene-Lambertz and colleagues (2005) presented to subjects were "synthetic sinewave analogues of speech" that could be interpreted as either whistles or as the phonemes [ba] or [da]. Without prompting, most hear them as whistles. Instead of varying VOT, which differentiates [da] vs. [ta] in English, the continuum along which [ba] vs. [da] differ in French is place of articulation — the physical location in the mouth where the tongue applies the blockage that makes the /b/ and /d/ stop consonants. Varying place of articulation influences the acoustic signal by changing the frequency of its formants (see *Figure 19.3*). The stimuli were labeled according to their phonemic identity for French speakers and according to the onset frequency of the second formant: ba975, ba1250, da1525, da1800. (Note that, to achieve such precise and replicable acoustic properties in the stimuli, it's common for speech perception research to use synthesized, rather than natural stimuli.)

Stimuli were presented in groups of four (*Figure 19.4*). In repetition trials, one of the four stimuli was repeated four times. In within-category change (WC) trials, the last stimulus differed from the previous three, but still fell on the same side of the phonemic boundary (e.g., *ba975*

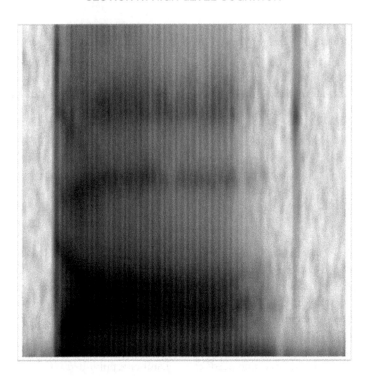

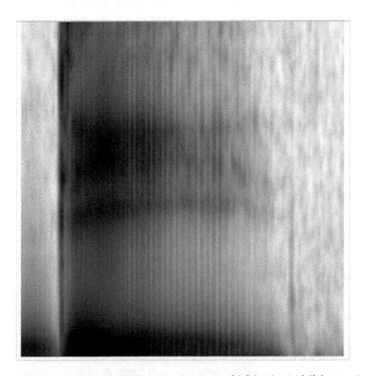

FIGURE 19.3 Spectrograms from an individual uttering the phonemes [da] (top) and [di] (bottom). The vertical axis is frequency (scale of 0–5000 Hz) and the horizontal axis is time (scale of 0–300 ms), with darker gray representing greater power. Note how the early portion of the speech signals differ, even though both are perceived as /d/. The "dark bands" of sustained, elevated oscillatory power that are visibly evident in these spectrograms are called formants. Formants are labeled from low to high frequency, with the first formant also known as the **fundamental frequency** of the tone. Source: Bradley R Postle

ba975 ba975 ba1250 or *da1800 da1800 da1800 da1525*). In across-category change (AC) trials, the last stimulus fell on the other side of the phonemic boundary (e.g., *da1525 da1525 da1525 ba1250* or *ba1250 ba1250 ba1250 da1525*). Thus, the acoustical distance between the last stimulus and the preceding ones was identical in the two types of change, the only difference being that, when the stimuli were perceived as speech sounds, the change in AC trials would be perceived as a change of phonemic category. This created the critical contrast whereby a change in the auditory percept could be dissociated from changes in the acoustic signal. This comparison was implemented experimentally by performing the experiment in two stages. In "Nonspeech mode," which always came first, the stimuli were referred to as "whistles," and subjects were instructed to indicate whether or not the fourth whistle of each trial differed from the others in pitch. Prior to the second stage – "Speech mode" – the speech-like nature of the stimuli was explained to subjects, and they were played the two most

distinctively different sounds until they could identify them as [ba] and [da]. Then the procedure was repeated.

Results from the ERP study (*Figure 19.4.A*) indicated that the identical physical change (of 275 Hz) produced significantly greater changes in the ERP for AC vs. WC changes, only in speech mode, beginning with the P1 component. Because the P1 has been localized to parabelt cortex of the STG, this suggests that speech is segregated from nonspeech at a very early stage of cortical processing. When the same experimental procedure was repeated, with different subjects, during fMRI scanning, three patterns emerged (*Figure 19.4.B*): (1) midline and lateral frontal areas, insula, and subcortical regions were sensitive to any change of stimulus; (2) the posterior portion of the superior temporal sulcus was sensitive to mode of perception (i.e., speech > whistle); and (3) supramarginal gyrus was sensitive to a change of phoneme (i.e., AC > WC).

The ERP findings from Dehaene-Lambertz et al. (2005), indicating that the earliest cortical AEPs discriminate

FIGURE 19.4 Dissociating speech perception from nonspeech perception.

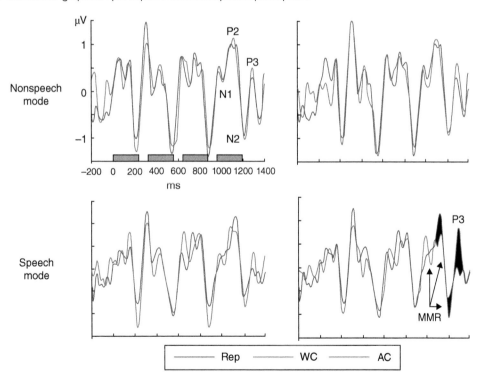

FIGURE 19.4.A ERPs from the Fz electrode for repeat trials (Rep), collapsed across stimuli, plotted with each of the two types of mismatch trials. Along the abscissa of the plot in the upper left, the timing and duration of each stimulus is illustrated with green blocks. The mismatch response (MMR) was prominent, beginning with P1, for stimuli in speech mode that produced an across-category (AC) change. Source: From Dehaene-Lambertz, Ghislaine, Christophe Pallier, Willy Serniclaes, Liliane Sprenger-Charolles, Antoinette Jobert, and Stanislas Dehaene. 2005. "Neural Correlates of Switching from Auditory to Speech Perception." NeuroImage 24 (1): 21–33. Reproduced with permission of Elsevier Ltd.

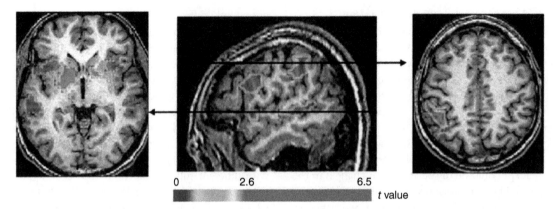

FIGURE 19.4.B Differential sensitivity of these regions to specific contrasts identified with fMRI as showing greater signal for AC vs. Rep in speech mode. Source: Dehaene-Lambertz, Pallier, Serniclaes, Sprenger-Charolles, Jobert, and Dehaene, 2005. Reproduced with permission of Elsevier.

speech signals from nonspeech ones, led them to hypothesize that the two rely on "partially distinct neural processes" (p. 29). Although they don't speculate about the origin of these processes, they do note that "Further research should determine to what extent this organization is laid down in the course of language acquisition and to what extent it is already present early in infancy" (p. 32). In the ensuing years, Dehaene-Lambertz and colleagues performed some of this "further research," with one more recent finding, using near-infrared spectroscopy (NIRS; a.k.a. "optical imaging") on preterm infants (28–32 weeks gestational age) coming down decidedly on the side of "already present" (see by Mahmoudzadeh et al., 2013, *Further Reading*).

Dual routes for speech processing

One model for speech perception that is generally consistent with the network of regions illustrated in *Figure 19.4.B*, and with the white matter tracts illustrated in *Figure 19.6*, is a "dual-route" model advocated by Hickok and Poeppel (2000) whereby "the posterior–superior temporal lobe, bilaterally, is the primary substrate for systems involved in constructing sound-based representations of heard speech" (p. 134), and processing then bifurcates into a ventral and a dorsal stream. The ventral stream is proposed to subserve auditory comprehension via connections that support "interfacing sound-based representations in auditory cortex with conceptual–semantic information [represented in] multimodal cortical fields in the vicinity of the left temporal-parietal-occipital junction" (p. 134). Note that these "multimodal cortical fields" correspond to portions of the temporal and occipital lobe areas that we studied in considerable detail in *Chapter 13, Semantic long-term memory*. Further, Hickok and Poeppel readily acknowledge that their model draws heavily on ideas first proposed by Wernicke and by Lichtheim. Thus, their ventral stream can be understood as corresponding to the pathway *A–B–M* as portrayed in *Figure 19.2.A*, and magnified in *Figure 19.2.B*. (We'll consider a modern-day depiction of this pathway in the upcoming section on grammar.)

The dorsal stream is proposed to act as an "auditory-motor interface" that, in effect, follows the course of the arcuate fasciculus, but with at least some fibers synapsing in inferior parietal cortex. The argument is that posterior temporal regions, as well as inferior parietal regions, contribute to assembling articulatory codes for speech. One consequence of this is the involvement of "dorsal stream" regions, particularly in inferior parietal cortex, in processing phonetic segments, such as stimuli of the kind featured in the study from Dehaene-Lambertz et al. (2005) that we just reviewed.

Note that although *Figure 19.4.B* does not show evidence of elevated right hemisphere activity for the "AC > Rep" contrast illustrated from the Dehaene-Lambertz et al. (2005) study, Hickok and Poeppel (2000) review several neuropsychological and neuroimaging studies that implicate right STG in at least some aspects of speech perception. Although beyond the scope of this book, a comprehensive review of data and theories relating to the lateralization of processing of speech and other acoustical signals is provided in the *Further Reading* from Zatorre and Gandour (2008). Also a *Further Reading* is an alternative dual-route model, from Scott and Johnsrude (2003).

GRAMMAR

Up until this point, we have focused on the perception of speech sounds, and referenced our earlier treatments of single-word comprehension in *Chapter 5* and *Chapter 13*. Now, let's turn to the decoding of the language signal. After all, it's one thing to understand the meaning of the individual words *dog*, *bites*, and *man*, but with this knowledge, alone, one couldn't know who did what to whom.

Genetics

Here is a partial summary of impairments of four individuals diagnosed with a developmental language disorder:

> They took a long time to name pictures of objects with which they were familiar, and tended to use approximate words, for example "glass" or "tea" for cup, and "sky" for star. Comprehension was also delayed, especially the understanding of comparatives. "The knife is longer than the pencil" was poorly understood. "The girl is chased by the horse" became "the girl is chasing the horse," and "the boy chasing the horse is fat" was interpreted as being the same as "a thin boy chasing a fat horse." Most of those affected, especially the children, could not retain three items in the correct sequence. Many were unintelligible and reluctant to offer spontaneous conversation (Hurst et al., 1990, p. 354).

The striking thing about these four is that they were all drawn from three generations of the same family (the "KE family," as Hurst and colleagues refer to them): "The affected members of this family all have the same type of speech and language difficulty, but to varying degrees of severity. The unaffected members have no speech and language difficulties and all attended normal schools" (p. 354). This was a remarkable finding because, as Hurst et al. (1990) noted,

> Few speech disorders are inherited in a simple Mendelian way, and when familial clustering does occur, both genetic and environmental factors are involved. The genetic contribution in most cases involves more than one gene (polygenic inheritance), although a dominant gene with reduced penetrance is possible. In practice, it is unusual to find full penetrance of a speech disorder occurring in a regular manner over generations. (p. 352)

Let's unpack this statement. Inspection of the KE family's pedigree (*Figure 19.5*) indicates that, beginning with the first generation on record, each child has had a 50% chance of acquiring the disorder. This is a hallmark of autosomal dominant transmission of a mutation in a single gene. That is, in one parent, one of two alleles of this single gene carries a mutation. As a consequence, because the sperm and the egg each contribute one allele from each gene, there is a 50% chance that any child produced by that couple will inherit the mutation. The mutation is dominant, meaning that it determines the transcription and translation of that gene. (For eye color, a brown allele is dominant over blue, such that if a child inherits a blue allele from Mom and a brown allele from Dad, the child's eyes will be brown.) "Penetrance" refers to likelihood that, if one does inherit the mutation, it will be expressed in that individual's phenotype; *Figure 19.5* indicates that the disorder is fully penetrant. Autosomes are all chromosomes that are not the sex chromosomes – that is, the X and the Y – and therefore an autosomal disorder can be inherited by either male or female offspring. (For a primer on inheritance of single-gene disorders, see Chial, 2008, *Further Reading*.) That's the elementary Mendelian genetics. What's most striking about this case is that it indicates that a single gene is responsible for the KE family's disorder!

Autosomal dominant disorders are uncommon. (I just looked up a list of exemplars and the only one I'm confident that I recognize is Huntington's disease.) They are almost unheard of for purely cognitive disorders. Thus, considerable excitement, and controversy, was generated by the linguist Myrna Gopnik, and subsequently others, when she proposed that the function of the mutated gene is to "affect the capacity to acquire language" (Gopnik, 1990b, p. 26). In the interpretation of some (including Pinker, 1994), this was evidence for a "grammar gene."

Is there a grammar gene?

The reason the initial summary of KE family deficits was qualified as "partial" is because the initial report from Hurst et al. (1990) also included the fact that, for example, "Articulation was also defective, and [the affected family members had] . . . a moderate to severe degree of dyspraxia. They could position the tongue and lips for simple movements, but failed when a sequence of movements was required" (p. 354). These would be considered by many to

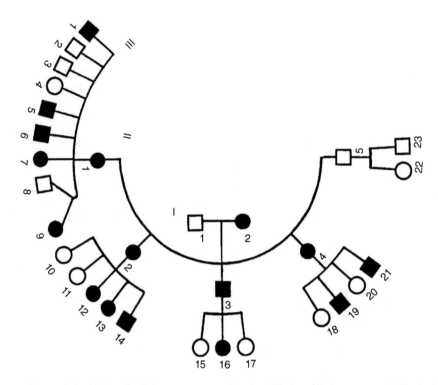

FIGURE 19.5 The pedigree of the KE family. Squares represent males; filled symbols represent affected individuals. Roman numerals identify the three generations of the family, Arabic numerals the pedigree number of an individual within a generation. Source: From Hurst, Jane A., Michael Baraitser, E. Auger, Frances Graham, and S. Norell. 1990. "An Extended Family with a Dominantly Inherited Speech Disorder." Developmental Medicine and Child Neurology 32 (4): 352–355. Reproduced with permission of John Wiley & Sons.

extend beyond grammar per se. Nonetheless, where grammar ends and where extra-grammatical, but nonetheless "language related," processes begin is not well defined.

Gopnik's original (1990a) report on the KE family prompted immediate objections to her interpretation that noted, for example, that the impairments of affected family members were much broader than portrayed, and also included "impairment in repeating phonemes, words, non-words, and sentences . . . [and in] aspects of language that are unrelated to syntax[, including] semantic errors in naming . . . [and poor] receptive vocabulary" (Vargha-Khadem and Passingham, 1990, p. 226). Gopnik's reply to these objections, in turn, was that

> What really is at stake here are not the data, but rather how to account for them. My explanation hypothesizes that the subjects are impaired because they cannot construct language rules at the necessary level of abstraction. [Fletcher (1990) and Vargha-Khadem and Passingham (1990)] prefer an explanation which hypothesizes that the subjects are impaired because they cannot hear or cannot pronounce certain

sounds, and this indirectly affects the grammar carried by these sounds (Gopnik, 1990b, p. 26).

When confronted with such a case of "he said-she said," what is one to do? The answer, of course, is that one is to perform studies that are carefully designed to assess the key questions of the debate. This additional research proceeded along two paths, one behavioral, one genetic. The behavioral, pursued by Vargha-Khadem et al. (1995), elaborated on the arguments that they had raised in their initial response, and also established that affected members had significantly lower verbal and nonverbal IQ, with the score of many being below the value "commonly considered an exclusionary criterion for classification of a subject as having a specific language impairment" (p. 932).

Perhaps the most compelling evidence against the "grammar gene" position, however, arrived with the identification of the gene itself. In 2001, geneticists Cecilia Lai and Simon Fisher, working in the laboratory of Anthony Monaco at Oxford University, reported the identification of the mutated gene as FOXP2. This was an important breakthrough, because the functions of the gene could then be studied, independent of the KE family or others carrying the mutation. From research that both preceded and followed the report of Lai et al. (2001), we know that FOXP2 acts as a regulator (primarily a repressor) of the transcription of other genes. Graham and Fisher (2013) have noted that "FoxP2 is likely to be present in all vertebrates, and is highly conserved in neural expression pattern and amino-acid coding sequence . . . Thus, ancestral versions contributed to brain development long before language appeared" (p. 44). In mice, the mutation is associated with impaired motor learning and electrophysiological abnormalities in the striatum.

The cumulative effect of the findings summarized here is that, in this "post-FOXP2 era," most would endorse the statement that a complex cognitive faculty like grammar is unlikely to be isolatable to a single gene. At a different level, another lesson that one can take away from the saga of FOXP2 is that, for some developmental disabilities, it is not uncommon for there to be disagreement about what constitutes a fundamentally linguistic impairment vs. a more general cognitive impairment. (This is also seen, for example, with so-called specific language impairment [SLI] and with dyslexia.) If we want to address the cognitive neuroscience of grammar head-on, then, let's move to a different literature.

Rules in the brain?

Grammar refers to the systematic ways by which one can combine and modify words to encode (if speaking) or decode (if listening) a precise message. One property of grammar is that it is combinatorial. For example, the phrase structure of grammar allows one to go from "The boy is fat" to "The boy chasing the horse is fat," to "The boy chasing the horse on the reduced-calorie diet is fat," with nary any uncertainty about which organism's body type is being called out. Another property of grammar is that it allows one to convey temporal context by changing verb tense: "The boy was fat" vs. "The boy is fat" vs. "The boy will be fat." These are descriptions of grammar that no one would contest as untrue. Where there is contention, however, is whether these operations – combinatorial recursion and verb inflection – depend on rules.

Pinker and Ullman (2002a) have summarized it as follows:

> The [English] past tense is of theoretical interest because it embraces two strikingly different phenomena. Regular inflection, as in walk-walked and play-played, applies predictably to thousands of verbs and is productively generalized to neologisms such as spam-spammed and mosh-moshed, even by preschool children. Irregular inflection, as in come-came and feel-felt, applies in unpredictable ways to some 180 verbs, and is seldom generalized; rather, the regular suffix is often overgeneralized by children to these irregular forms, as in holded and breaked. A simple explanation is that irregular forms must be stored in memory, whereas regular forms can be generated by a rule that suffixes -ed to the stem. (p. 456)

Indeed, they go on to argue that "rules are indispensable for explaining the past tense, and by extension, language and cognitive processes" (p. 456). Although Pinker, a cognitive psychologist/psycholinguist (we'll define this in a later section), originally made this argument on theoretical and behavioral grounds, his former student/now collaborator Michael Ullman added a neural implementation of this "Words-and-Rules" theory: whereas the past-tense forms of irregular verbs are stored in lexical memory (i.e., posterior temporoparietal cortex), past tense forms of regular verbs are computed online by a concatenation rule ("add −ed") implemented by the cortico-basal ganglia-thalamic loop that includes Broca's area. Evidence that Ullman and colleagues have marshaled in support of this model has included two demonstrations of neuropsychological double dissociations. In one, a patient with a non-fluent (for our purposes, Broca's) aphasia was shown to have a selective impairment at generating the past tense with regular verbs, whereas a patient with a fluent (for our purposes, Wernicke's) aphasia had a selective impairment with irregular verbs. In a second, patients with Parkinson's disease, which is characterized by progressive deafferentation of the neostriatum from its supply of midbrain dopamine, and thus by progressive striatal dysfunction, have selective difficulty inflecting regular verbs, whereas patients with Alzheimer's disease, associated with episodic and semantic memory loss, have selective difficulty inflecting irregular verbs.

An alternative view has been advanced by McClelland and colleagues. For example, McClelland and Patterson (2002a), in response to Pinker and Ullman (2002a), wrote that the parallel distributed processing (PDP) framework holds that "Acquisition of language and other abilities occurs via gradual adjustment of the connections among simple processing units. Characterizations of performance as 'rule-governed' are viewed as approximate descriptions of patterns of language use; no actual rules operate in the processing of language" (p. 465). Although much of their argument derives from computational modeling, McClelland and Patterson (2002a) also appeal to neural data as consistent with their perspective. For example, they note that among the findings of Vargha-Khadem et al. (1995) was the fact that affected members of the KE family (see the section *Is there a language gene?*) showed comparable levels of impaired past-tense processing of regular and irregular verbs. Further, with Bird et al. (2003), they have argued that a key difference between regular and irregular verbs in English is that "relative to most irregular past-tense forms, regular past-tense English verbs place greater demands on the phonological system, because the transformation from regular stem to past always involves the addition of extra phonemes whereas this is rarely the case for irregular verbs" (p. 503). They suggest that some studies, including the double dissociations invoked by Pinker and Ullman (2002a), may thus be explainable in terms of patients with anterior lesions having more difficulty with items with greater phonological complexity. Further, Bird et al. (2003) demonstrated in 10 nonfluent (i.e., Broca's) aphasics that "the apparent disadvantage for the production of regular past tense forms disappeared when phonological complexity was controlled" (p. 502).

"Immediate" replies to the two positions summarized here can be found at Pinker and Ullman (2002b) and McClelland and Patterson (2002b), and in many ensuing publications. The debate about whether the brain truly represents rules, or, alternatively, rule-like patterns are emergent from complex network dynamics, seems unlikely to end in a consensus agreement during the career of this author (but at this point you should know enough to be able to pick a side ;-)). In the meantime, with this context, let's consider some perspectives on the functions of Broca's area.

Broca's area

The classical view from behavioral neurology, as embodied by the Wernicke–Lichtheim model, is that Broca's area is critical for syntactic processing – that is, for the grammatical decoding of heard sentences, as well as the grammatical encoding of spoken sentences. For example, patients with Broca's aphasia have difficulty understanding sentences like "It was the squirrel that the raccoon chased," from which meaning can't be derived just from knowledge of the meaning of the individual words. Despite this fact, however, others have shown that even though these patients don't understand these sentences, they can make correct grammaticality judgments about them, suggesting that at least some grammatical knowledge may reside elsewhere in the brain. And so, what is the function of Broca's area?

The connectivity of Broca's area

Actually, before we dive into function, let's peel back the skin (get it?!?) and take a closer look at structure, specifically, at white matter tracts underlying perisylvian cortex. *Figure 19.6* is taken from a paper titled "Broca and Wernicke are dead, or moving past the classic model of language neurobiology" (just to give you a feel for how some contemporary scientists view the Wernicke–Lichtheim scheme with which we started the chapter). Indeed, this figure contrasts (in drab gray scale) the authors' characterization of the "obsolete" view of language-critical white matter tracts espoused by Geschwind (1970), with (in popping color!) the "emerging picture" of pathways contributing to language function. Certainly, in the "emerging" image, one can see that the inferior fronto-occipital

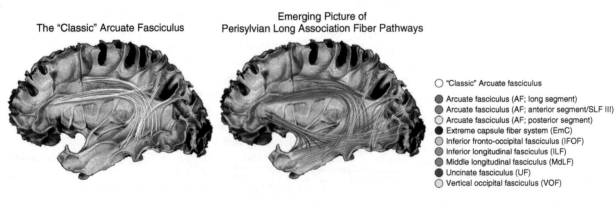

FIGURE 19.6 "Classic" vs. contemporary views of the perisylvian connectivity underlying language functions. Source: From Tremblay, P. and Dick, A. S. (2016). Broca and Wernicke are dead, or moving past the classic model of language neurobiology. Brain & Language 162: 60–71. Reproduced with permission of Elsevier Ltd.

fasciculus and the uncinate fasciculus provide a substrate for the ventral stream of dual-route models. This figure also makes clear how dorsal and ventral anatomical pathways converge in an inferior frontal region of the brain.

Functional accounts of Broca's area

Readers who have made it this far through the book won't be surprised to learn that the classical view of Broca's area as the neural seat of grammar has, over the past quarter century, been called into question by many investigators. There's not room here to detail every alternative account, but some prominent ones are verbal short-term memory (STM), cognitive control, and "unification," a hypothesized process of integrating information from phonological, lexical, syntactic, and semantic levels in the service of language comprehension and production (Hagoort, 2005; this concept will also be relevant for the upcoming section on *The electrophysiology of grammar*). Although these alternative accounts differ in terms of the extent to which each may or may not generalize to domains of cognition outside of language, what they have in common is that they propose non-syntactic functions for this region. Two responses to these non-syntactic models have argued that there may be a finer-grained functional organization of Broca's area, such that some subregions do, indeed, support

syntactic functions, and others may support non-syntactic ones. This style of reasoning is similar in spirit to one that we saw in relation to the question of whether theory-of-mind (ToM) mentalization can be localized to the right temporoparietal junction (TPJ), in *Chapter 17*. Indeed, one of the proponents of one "subregional specialization" model of Broca's area is none other than Nancy Kanwisher, whose intellectual fingerprints are also on the ToM debate, in addition, of course, to that of face processing in ventral temporal cortex. Evelina Fedorenko, John Duncan, and Nancy Kanwisher (2012) summarize what's at stake in this debate as follows:

> In 1861, Paul Broca stood up before the Anthropological Society of Paris and announced that the left frontal lobe was the seat of speech. Ever since, Broca's eponymous brain region has served as a primary battleground for one of the central debates in the science of the mind and brain: Is human cognition produced by highly specialized brain regions, each conducting a specific mental process, or instead by more general-purpose brain mechanisms, each broadly engaged in a wide range of cognitive tasks? (p. 2059)

Two "subregional specialization" models of Broca's area, defined with univariate analyses of fMRI data, are illustrated in *Figure 19.7*.

FIGURE 19.7 Two models advocating a functional compartmentalization of Broca's area.

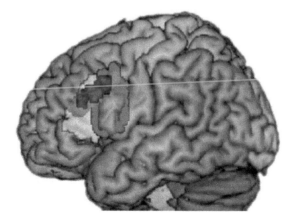

FIGURE 19.7.A fMRI evidence for subregions of Broca's area sensitive to syntactic complexity (orange) vs. working memory load (teal) of sentences. Syntactic complexity was manipulated by varying hierarchical structure, such as presence of center-embedded clauses, in sentences. Working memory was manipulated by the number of words separating a subject noun from its respective verb, and the region sensitive to this contrast is illustrated in teal. Regions in purple showed elevated connectivity with PO during more syntactically demanding sentence processing, and are interpreted as evidence for increased working-memory support for difficult sentence processing. Yellow is anatomically identified BA 45/IFGt; green is anatomically identified BA 44/IFGo. Source: From Makuuchi, Michiru, Jörg Bahlmann, Alfred Anwander, and Angela D. Friederici. 2009. "Segregating the Core Computational Faculty of Human Language from Working Memory." Proceedings of the National Academy of Sciences of the United States of America 106 (20): 8362–8367. Reproduced with permission of PNAS.

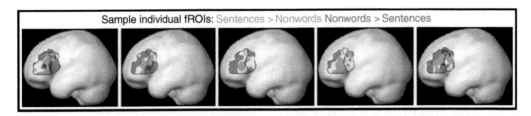

FIGURE 19.7.B Data from Broca's area of five individual subjects, with black outlines corresponding to BA 45/IFGt and BA 44/IFGo. Regions in red responded with greater fMRI signal intensity during sentence-reading followed by a single-word recognition probe, and regions in blue during nonword viewing followed by a single-nonword recognition probe. Subsequently, the sentence-sensitive regions were shown to respond only minimally to tests of mental arithmetic, spatial and verbal STM, and three tests of cognitive control, whereas the nonword-sensitive regions responded with significantly greater signal intensity to difficult vs. easy versions of many of these tasks. Source: Fedorenko, Duncan, and Kanwisher, 2012. Reproduced with permission of Elsevier.

Pyrrhic battles?

Broca's area has, indeed, been a hotly contested piece of territory in number of scientific skirmishes. Just one other example that has used martial metaphors is "The Battle for Broca's Region" (Grodzinsky and Santi, 2008), and "Broca's Region: Battles Are Not Won by Ignoring Half of the Facts" (Willems and Hagoort, 2009). So, not to be left out, I'll introduce a martial metaphor of my own: Some neuropsychological research suggests that these "Battles for Broca's area" might end up being viewed rather like the D-Day (June 6, 1944) assault by US Army Rangers on the Pointe du Hoc – after fierce fighting and heavy losses, the US soldiers took the position, only to find that the objective of their attack, large artillery guns, had been disassembled and moved days earlier.

Traditional approaches to neuropsychology, although valuable in many instances, can be susceptible to bias. One is to group patients by locus of lesion. A potential source of bias with this method is that the existence of multiple functional subregions may be obscured if many of the patients have lesions that span these subregions. (An analogous caveat about TMS was raised in *Chapter 17*'s treatment of the rTMS study featured in the section *rTMS of TPJ during moral judgments*.) A second traditional method is to select patients based on their symptoms; after identifying individuals with similar behavioral impairments, one looks for a locus of lesion overlap that might explain the syndrome. The potential bias with this approach relates to specificity: when employing a symptom-based approach, one necessarily only selects patients whose behavior manifests the symptom(s) that one is interested in. Thus, even if area X is damaged in each patient in one's sample, one cannot know whether, in recruiting patients, one may have overlooked a comparable number who also have lesions of area X, but who don't manifest the symptom(s) of interest (See also section *Experimental neuropsychology*, Chapter 3.).

A method known as voxel-based lesion-symptom mapping (VLSM) avoids these two sources of bias by allowing for the study of deficit–lesion correlations without requiring the categorization of patients, either in terms of lesion location or in terms of behavior. Rather, both can be treated as continuous variables, and the procedure evaluates the relation of tissue integrity to behavior on a voxel-by-voxel basis. Thus, it is effectively the same approach that is used in massively parallel univariate analysis of functional neuroimaging data, the difference being that there is no variation in the neural data over time.

A study by Bates and colleagues (2003) applied VSLM to 101 patients who had suffered left-hemisphere strokes and showed some degree of consequent language impairment. Behavioral measures of verbal fluency and auditory comprehension were acquired from subtests of a standard clinical assessment tool, the Western Aphasia Battery. Each patient's fluency score summarized performance on articulatory, word-finding, and sentence-production tasks, and the auditory comprehension score summarized performance on yes/no questions, single-word recognition, and enactment of one-, two- and three-part commands. Lesion maps were transformed into a common space, and each voxel then binarized for each subject as "intact" or "lesioned." (That is, tissue integrity was not treated as a continuous variable in this study, although one could do so, as is done with voxel-based morphometry [VBM] data.)

The results, illustrated in *Figure 19.8*, suggest a more prominent role for insular cortex than Broca's area for fluency and a more prominent role for middle temporal

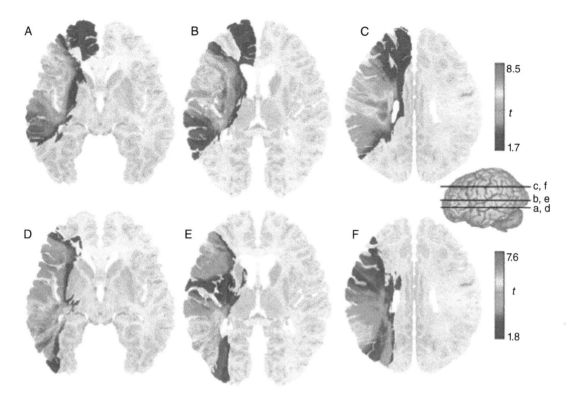

FIGURE 19.8 Voxel-based lesion-symptom maps (VLSM) from 101 left-hemisphere stroke patients. The top row illustrates the voxel-by-voxel association between lesion status and verbal fluency performance, and the lower between lesion status and auditory comprehension. The color at each voxel corresponds to its *t*-value for that contrast, just as it does for univariate functional imaging maps. Thus, warmer colors indicate regions where lesions are more strongly associated with behavioral impairment. The faintly visible gray line on each color bar indicates the Bonferroni-corrected significance cutoffs (in the teal range for fluency, and near the teal–yellow transition for auditory comprehension. These analyses suggest that lesions of Broca's area (**A**) insula (**B**) and white matter encompassing both the arcuate and superior longitudinal fasciculi (**C**) had the greatest impact on fluency, whereas lesions of the middle (**D**) and superior (**E**) temporal gyri had the greatest impact on auditory comprehension. Source: From Bates, Elizabeth, Stephen M. Wilson, Ayse Pinar Saygin, Frederic Dick, Martin I. Sereno, Robert T. Knight, and Nina F. Dronkers. 2003. "Voxel-Based Lesion-Symptom Mapping." Nature Neuroscience 6 (5): 448–450. Reproduced with permission of Nature Publishing Group.

gyrus than superior temporal gyrus for comprehension. For arguments supporting these two findings, see Dronkers (2000) and Binder (2003) in the list of *Further Readings*.

A caveat not unique to VSLM, but salient here, is that stroke damage is often correlated across regions, because one arterial branch typically feeds multiple regions. Thus, for example, it could be that the identification of the superior longitudinal fasciculus, which primarily connects posterior parietal cortex with PFC, is incidental to concurrent damage to the arcuate fasciculus, which actually does contribute importantly to production.

(Note that we also saw this approach in *Figure 7.3*, where it was used to determine regions whose damage was associated with acute vs. chronic unilateral neglect.)

The electrophysiology of grammar

Explicit in the arrows drawn onto Lichtheim's house model (*Figure 19.2.A*), as well as in many more recent accounts, is the assumption that sentence processing is carried out in a feedforward (or "bottom-up") manner: the first stage is the identification of phonemes; followed by word recognition; followed by syntactic decoding of the spoken utterance. ERP studies of sentence processing, however, have provided evidence for high levels of interactivity within the system. For example, *Figure 19.9* illustrates the results from a study by Van Berkum and colleagues (2005), led by Peter Hagoort at the Max Planck Institute for Psycholinguistics and F. C. Donders

The burglar had no trouble whatsoever to locate the secret family safe. Of course, it was situated behind a...

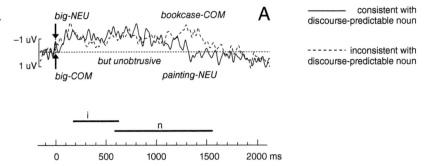

The burglar had no trouble whatsoever to locate the secret family safe. Of course, it was situated behind a...

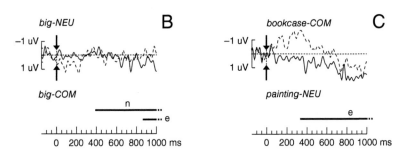

[no predictive discourse context]

Of course, it was situated behind a...

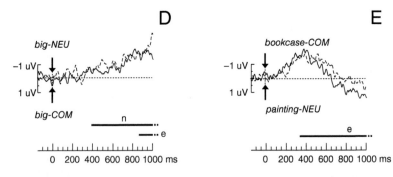

FIGURE 19.9 ERPs time-locked to the adjective that preceded the critical noun that either was (solid waveform) or was not (dotted waveform) consistent with subjects' predictions. The negative-going deviation related to the inconsistent (and, presumably, unpredicted) noun is the N400 ERP component. The data are averaged over several two-sentence stories, each with a different onset time of the inflected adjective that preceded the critical noun ("i" above the time axis), of the critical noun itself ("n" above the time axis) and of the end of the sentence ("e" above time axis), which accounts for the "smeary-ness" of the N400 in this figure. Source: From Van Berkum, J. J. A., Brown, C. M., Zwitserlood, P., Kooijman, V. and Hagoort, P. (2005). Anticipating upcoming words in discourse: Evidence from ERPs and reading times. Journal of Experimental Psychology: Learning, Memory, and Cognition 31: 443-467. Reproduced with permission of American Psychological Association.

Centre for Cognitive Neuroimaging in Nijmegen, The Netherlands. In this study, subjects listened to a series of stories that were crafted to lead subjects to predict a particular word. The key manipulation was whether or not the expected word or another grammatically and semantically valid, but unexpected, word was actually spoken. The researchers' goal was "to determine whether listeners can use their knowledge of the wider discourse to rapidly predict specific upcoming words as a sentence is unfolding" (p. 446)."

The dependent measure of interest in the study of Van Berkum and colleagues (2005) was the N400, an ERP component previously shown to be sensitive to the processing of meaning in several contexts, including spoken, written, and signed language, mathematical symbols, and videos with social content. As the name implies, it typically peaks later than ERP components associated with sensory and perceptual processing. Marta Kutas and Kara Federmeier (2011) interpret findings like these as follows: "Because, in the absence of prediction, these [critical] words constitute equally good fits to the accrued contextual information, N400 reductions when the words matched as opposed to mismatched the predicted target showed clearly that information about likely upcoming words has shaped the system in advance" (p. 634). That is, one perspective on the grammatical decoding of language is to understand it as another example of the principle of predictive coding, which we first introduced at the end of *Chapter 4* (section *Where does sensation end? Where does perception begin?*). This perspective will also carry through to the final section of this chapter, on speech production.

SPEECH PRODUCTION

The physical production of speech is an exquisitely complicated motor control problem. We have seen this, for example, in the fact that an offset of a few tens of milliseconds in the coordination of the release of the tongue from the palate relative to the contraction of the diaphragm to force air over the vocal cords can make the difference in whether one produces the word "doll" or "tall." Further, we saw with the KE family that dysfunction in the control of the articulatory apparatus can produce profound impairments in the ability to communicate. Our focus here, however, will be on the neural control of speech production. Recall that with Broca's aphasia, for example, the patients are no better able to produce language with a pen, or a keyboard, than with their voice. Thus, there's a level of language production that is abstracted from the effectors that will actually generate the output. (Having said this, however, it will be most expedient in this brief chapter to stick with spoken language.)

A psycholinguistic model of production

Within the study of language, a distinction is often made between the fields of *linguistics* and of *psycholinguistics*. Although no single definition will please everyone, it's a fair generalization to say that linguistics concerns itself primarily with computational and theoretical questions about the structures and rules of language, whereas psycholinguistics addresses how humans process and use language in everyday circumstances. As a general rule, the discipline of psycholinguistics tends to be more associated with designing and carrying out experiments. Almost all the topics that we've covered in this chapter fall under the rubric of psycholinguistics.

According to one influential model, speech production occurs via a series of stages, beginning with the formulation of a message, followed by lexical retrieval (mapping an abstract, conceptual message onto words), phonological encoding (specifying the speech sounds that comprise the words), articulatory planning (formulation of a sequential motor plan for speech output), and articulation (Indefrey and Levelt, 2004). It is likely that many of these processes recruit systems that we have already considered. For example, repetitive transcranial magnetic stimulation (rTMS) of left posterior STG, timed to interfere with the phonological encoding stage of production, does, indeed, produce errors in spoken output (Acheson, Hamidi, Binder, and Postle, 2011). The area targeted for rTMS in this study overlapped with the region of STG highlighted in *Figure 19.4.B*, and the processes disrupted by rTMS likely overlapped with those discussed earlier in the context of articulatory coding functions of the speech perception system (i.e., the "dorsal stream").

Forward models for the control of production

Decades of psycholinguistic research on speech production indicate that it, too, is a highly dynamic and interactive process. In particular, our production of speech is very sensitive to our online assessment of it. Here's a summary from Houde, Nagarajan, Sekihara, and Merzenich (2002):

Adding noise to a speaker's auditory feedback results in an elevation of voice volume . . . Delaying a speaker's auditory feedback by more than about 50 msec produces noticeable disruptions in speech . . . Shifting the spectrum of auditory feedback causes shifts in the spectrum of produced speech . . . Perturbing perceived pitch causes the subject to alter his pitch to compensate for those perturbations . . . Altering perceived formants induces compensating changes in the production of vowels (p. 1125).

These findings have given rise to the idea that speech production may involve the construction of a neural **forward model** of the planned utterance, against which the actual speech output is compared. Recall that we have already

seen the same principle at work in *Chapter 8*, in the context of the "internal models" of movements represented in the cerebellum, and their use in fine-tuning skeletomotor control. (For our purposes, the terms "forward model" and "internal model" are interchangeable.)

To test this idea, Houde and colleagues (2002) recorded the magnetoencephalogram (MEG) while subjects uttered a single vowel and heard its real-time playback, and when they passively listened to the acoustically identical signal (i.e., to a recording of themselves from the speaking session). The data illustrated in *Figure 19.10* showed a marked attenuation of the auditory-evoked response in the speaking condition, which the authors

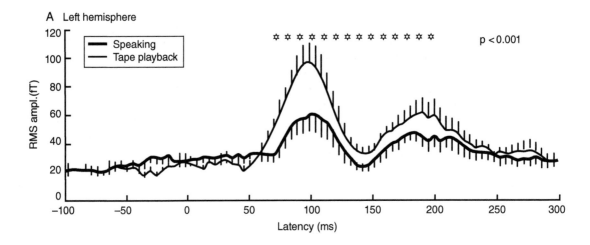

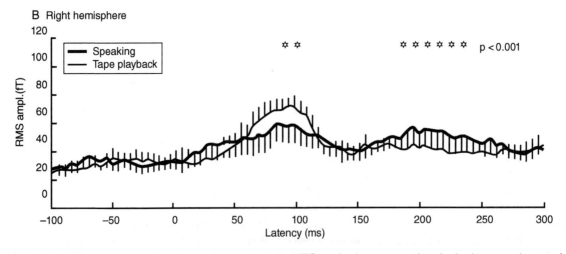

FIGURE 19.10 MEG responses to hearing one's own speech. MEG-evoked response, time-locked to sound onset, from subjects hearing their own utterance of a vowel while speaking it ("Speaking") and while passively listening to it "Tape playback"). The response is significantly attenuated in the speaking condition. Source: From Houde, John F., Srikantan S. Nagarajan, Kensuke Sekihara, and Michael M. Merzenich. 2002. "Modulation of the Auditory Cortex during Speech: An MEG Study." Journal of Cognitive Neuroscience 14 (8): 1125–1138. Reproduced with permission of MIT Press.

attributed to "a comparison between the incoming signal and an internally generated prediction of what that signal would be" (p. 1131). When the match was high, they reasoned, the "error" would be minimal, and so, too, would be the neural response. To rule out any possible confounds due to the fact that one condition involved actual production and one did not, they performed a control study that was identical to what was just described, with the exception that the auditory feedback of their voice was convolved with white noise. Thus, the auditory signal was perfectly matched to the timing of hearing one's own voice, but was a signal that the auditory system could not predict that it would hear. The result, comparable responses in both conditions, ruled out many confounding alternative explanations to the "forward model" interpretation. (Conceptually similar results have also been demonstrated with fMRI, implicating bilateral STG in processing mismatches between expected and actual auditory signals [Tourville, Reilly, and Guenther, 2008].)

Before moving on, let's take a moment to consider the potentially very profound implications of these findings and their interpretation. One construal of auditory-evoked responses is that they correspond to (they "signal," if you will) mismatches between the brain's expectation about the environment and the actual state of the environment. In relation to this section of the chapter, the studies reviewed here suggest that production is a highly interactive process that recruits many levels of neural representation of sound, meaning, and syntax.

PREDICTION

The experiment discussed in the previous section highlights the important role that prediction plays in speech production. Here we bring it full circle by illustrating research that emphasizes a role for prediction in the perception of others' speech. Starting with the rationale that the brain needs a mechanism to parse the continuous speech stream into decodable units, David Poeppel and colleagues at New York University have noted that the rate at which syllables unfold in speech falls into the delta (~1–4 Hz) and theta (~4–8 Hz) range of oscillatory frequencies, and these oscillations may provide a substrate for "pars[ing] the continuous speech signal into the necessary chunks for decoding." The scheme that they, and others, have explored is that resets of the ongoing oscillations implement an ongoing process of "sliding and resetting"

oscillatory cycles that act as temporal integration windows that optimize the initial processing of the speech signal (Poeppel, 2014, pp. 146–147). This is sometimes referred to as entrainment of activity in the auditory system to the speech signal.

To fully appreciate the proposed mechanisms, and how analyses were carried out to generate the results portrayed in *Figure 19.11*, it may be helpful to revisit section *Analysis of time-varying signals*, *Chapter 3*, and accompanying figures. For example, panel A of *Figure 19.11* uses filtering analogous to the technique that generated the "alpha-filtered gamma amplitude" from *Figure 3. 11* to generate the envelope of time-varying fluctuations in power in the MEG signal from 2 to 8 Hz, which is shown in the blue trace in panel A of *Figure 19.11*. It is evident from visual inspection that these fluctuations in the neural signal are closely following the fluctuations in energy in the concurrent speech signal (red trace). How does the brain accomplish this? In the experiment from Doelling, Arnal, Ghitza, and Poeppel (2014), from which panels A, B, and C of *Figure 19.11* are taken, the researchers used a synthesized speech signal in which they controlled whether or not transitions between syllables were accompanied by abrupt changes in the amplitude of the signal, a measure referred to as *sharpness*. (*Sharpness* is computed as the first derivative of the speech envelope over time – i.e., its rate of change.) A key finding was that decreasing the sharpness in the speech signal markedly decreased how closely the neural signal tracked the otherwise identical speech signal, and a concomitant decrease in the intelligibility of the speech. Therefore, transients in the amplitude envelope of the speech signal may provide the triggers that the auditory system uses to reset the phase of its low-frequency oscillations.

One can imagine that an entrainment-to-speech mechanism like the one proposed here would be most helpful in an environment in which there are multiple sources of acoustic input, and one needs to be able to isolate and track the signal coming from just one source – say, to follow a single conversation taking place at a crowded cocktail party. (Recall from *Chapter 5* that situations like this, which are characteristic of the vast majority of situations that we encounter every day, present a major challenge for the auditory system because the mechanical elements at the front end of the auditory system combine all incoming acoustic signals into a single complex pattern of vibration.) This is a problem formally similar to the one that confronts a visual system wanting to track the location and movements and of a single object of interest in a crowded scene;

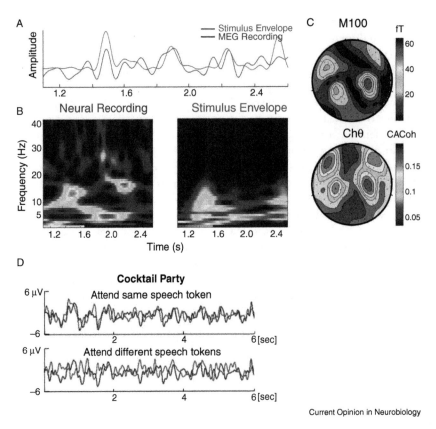

FIGURE 19.11 Synchronization of neural rhythms to incoming speech. **A.** Results from the study of Doelling and colleagues (2014) showing the close tracking of fluctuations in the speech signal ("Stimulus Envelope," red trace) by the auditory system ("MEG Recording," blue trace). **B.** Time–frequency representations (TFRs) of the data from panel **A**. **C.** Comparison of topography of the M100 with the onset of a pure tone (top) vs. the similarly timed-evoked response to the Chθs, the syllabic transitions in the synthesized speech signal (bottom). Because the M100 is known to be generated by auditory cortex, the similarity between the two validates the interpretation that it is the processing of Chθs in the auditory system that drives the synchronization illustrated in **A**. (M100 is the MEG analogue to the N1 AEP that is measured with EEG (see *Figure 5.4*). fT = femotesla (10^{-15} tesla); CACoh = cerebro-acoustic coherence, a measure of phase locking between the neural and acoustic signals. **D.** Data from one ECoG electrode from the study by Zion Golumbic et al. (2013). The top figure shows two different trials when the subject is viewing and listening to the same snippet of two people talking simultaneously, and on both trials attending to the speech being spoken by the same person. The bottom shows the same trace from one of these two trials (blue), but now compared to a trace from third trial in which the patient viewed the same snippet, but was attending to the other talker. Source: From Poeppel, D. (2014). The neuroanatomic and neurophysiological infrastructure from speech and language. Current Opinion in Neurobiology 28: 142–149. Reproduced with permission of Elsevier Ltd.

one that we know requires endogenously controlled attention. To test whether auditory entrainment could help solve what psychologists call "the cocktail party problem," a team led by Elana Zion Golumbic asked epilepsy patients with subdural grids placed within and near the Sylvian fissure to follow narrative being spoken by one of two individuals who were talking simultaneously. The data shown in panel D of *Figure 19.11* suggests that the entrainment-to-speech mechanism can, indeed, be guided by attention.

INTEGRATION

The final sections of this chapter provide the kind of synergy that a textbook author lives for! It demonstrates that the same principle of building, then updating, predictive models of the world – first introduced in *Chapter 4*'s treatment of visual perception – also apply to the production and the perception of speech. In this chapter's example, this plays out on a very fast timescale – at the speed of speech – with model updates needed each time the cadence of the talker's output changes. Like other examples that we've seen, however, this system of model building and updating is hierarchically nested. Representations of who are the talkers, what are the rules governing what I should be doing in this situation, and so on, are encoded at higher levels of representation than A1, and shifts of attention between different talkers take place much less frequently than do changes in the cadence of any one talker's output. Nonetheless, the mechanism for resetting the phase angle

of oscillation in corticothalamic circuits of the auditory system is the same regardless of whether the system is adjusting to a change within one speech train or switching to a different one. Furthermore, it's likely that a mechanism very similar to the signals from primary cortex to first-order thalamus that we covered in considerable detail in section *Information processing in primary visual cortex – interactivity, Chapter 4,* is also important for implementing the real-time "sliding and resetting" of cortical oscillations that implement the mechanism of entrainment-to-speech.

END-OF-CHAPTER QUESTIONS

1. Draw Lichtheim's house model, and indicate how lesions of three different areas would produce three kinds of aphasia.

2. What level of language processing was of interest in the study of Dehaene-Lambertz et al. (2005) that varied place of articulation of stimuli? What about in the study of Abrams et al. (2013) that compared clips of speeches vs. spectrally rotated variants of these speeches?

3. What anatomical systems comprise the dual routes of language processing? What are their proposed functions?

4. What conclusions can be drawn from the "story" of FOXP2, with regard to genetic contributions to language function?

5. There are at least four distinct accounts of the functions of Broca's area that we reviewed in this chapter. Summarize them and propose an account that you think best fits the current state of knowledge.

6. Explain the concept of predictive coding as it applies to sentence processing. What is the evidence for it?

7. Explain the proposed role of forward models in speech production, and the evidence for it.

8. Can speech perception and speech production both be interpreted within the framework of predictive coding? If so, how?

REFERENCES

Acheson, Daniel J., Massihullah Hamidi, Jeffrey R. Binder, and Bradley R. Postle. 2011. "A Common Neural Substrate for Language Production and Verbal Working Memory." *Journal of Cognitive Neuroscience* 23 (6): 1358–1367. doi: 10.1162/jocn.2010.21519.

Adolphs, Ralph, Daniel Tranel, and Antonio R. Damasio. 1998. "The Human Amygdala in Social Judgment." *Nature* 393: 470–474. doi: 10.1038/30982.

Bates, Elizabeth, Stephen M. Wilson, Ayse Pinar Saygin, Frederic Dick, Martin I. Sereno, Robert T. Knight, and Nina F. Dronkers. 2003. "Voxel-Based Lesion-Symptom Mapping." *Nature Neuroscience* 6 (5): 448–450. doi: 10.1038/nn1050.

Bird, Helen, Matthew A. Lambon Ralph, Mark S. Seidenberg, James L. McClelland, and Karalyn Patterson. 2003. "Deficits in Phonology and Past-Tense Morphology: What's the Connection?" *Journal of Memory and Language* 48 (3): 502–526. doi: 10.1016/S0749-596X(02)00538-7.

Chomsky, Noam. 1975. *Reflections on Language.* New York: Pantheon.

Compston, Alastair. 2006. "On Aphasia. By L. Lichtheim, MD, Professor of Medicine in the University of Berne. *Brain* 1885; 7: 433–484." *Brain* 129 (6): 1347–1350. doi: 10.1093/brain/awl134.

Dehaene-Lambertz, Ghislaine, Christophe Pallier, Willy Serniclaes, Liliane Sprenger-Charolles, Antoinette Jobert, and Stanislas Dehaene. 2005. "Neural Correlates of Switching from Auditory to Speech Perception." *NeuroImage* 24 (1): 21–33. doi: 10.1016/j.neuroimage.2004.09.039.

Del Gaizo, J., J. Fridriksson, G. Yourganov, A. E. Hillis, G. Hickok, B. Misic, . . . L. Bonilha. 2017. "Mapping Language Networks Using the Structural and Dynamic Brain Connectomes." *eNeuro* 4: e0204-0217.2017. doi: dx.doi.org/10.1523/ENEURO.0204-17.2017.

Doelling, K. B., L. H. Arnal, O. Ghitza, and D. Poeppel. 2014. "Acoustic Landmarks Drive Delta-Theta Oscillations to Enable Speech Comprehension by Facilitating Perceptual Parsing." *NeuroImage* 85: 761–768.

Fedorenko, Evelina, John Duncan, and Nancy Kanwisher. 2012. "Language-Selective and Domain-General Regions Lie Side by Side within Broca's Area." *Current Biology* 22 (21): 2059–2062. doi: 10.1016/j.cub.2012.09.011.

Fletcher, Paul. 1990. "Speech and Language Deficits." *Nature* 346 (6281): 226. doi: 10.1038/346226c0.

Geschwind, Norman. 1970. "The Organization of Language and the Brain: Language Disorders after Brain Damage Help in Elucidating the Neural Basis of Verbal Behavior." *Science* 170 (3961): 940–944. doi: 10.1126/science.170.3961.940.

Gopnik, Myrna. 1990a. "Feature-Blind Grammar and Dysphasia." *Nature* 344 (6268): 715. doi: 10.1038/344715a0.

Gopnik, Myrna. 1990b. "Genetic Basis of Grammar Defect." *Nature* 347 (6288): 26. doi: 10.1038/347026a0.

Graham, S. A., and S. E. Fisher. 2013. "Decoding the Genetics of Speech and Language." *Current Opinion in Neurobiology.* 23: 43–51. doi: 10.1016/j.conb.2012.11.006.

Grodzinsky, Yosef, and Andrea Santi. 2008. "The Battle for Broca's Region." *Trends in Cognitive Sciences* 12 (12): 474–480. doi: 10.1016/j.tics.2008.09.001.

Hagoort, Peter. 2005. "On Broca, Brain, and Binding: A New Framework." *Trends in Cognitive Sciences* 9 (9): 416–423. doi: 10.1016/j.tics.2005.07.004.

Hickok, Gregory, and David Poeppel. 2000. "Towards a Functional Neuroanatomy of Speech Perception." *Trends in Cognitive Sciences* 4 (4): 131–138. doi: 10.1016/S1364-6613(00)01463-7.

Houde, John F., Srikantan S. Nagarajan, Kensuke Sekihara, and Michael M. Merzenich. 2002. "Modulation of the Auditory Cortex during Speech: An MEG Study." *Journal of Cognitive Neuroscience* 14 (8):1125–1138.

Hurst, Jane A., Michael Baraitser, E. Auger, Frances Graham, and S. Norell. 1990. "An Extended Family with a Dominantly Inherited Speech Disorder." *Developmental Medicine and Child Neurology* 32 (4): 352–355. doi: 10.1111/j.1469-8749.1990.tb16948.

Indefrey, Peter, and Willem J. M. Levelt. 2004. "The Spatial and Temporal Signatures of Word Production Components." *Cognition* 92 (1–2): 101–144. doi: 10.1016/j.cognition.2002.06.001.

Kuhl, Patricia K., and James D. Miller. 1975. "Speech Perception by the Chinchilla: Voiced–Voiceless Distinction in Alveolar Plosive Consonants." *Science* 190 (4209): 69–72. doi: 10.1126/science.1166301.

Kutas, Marta, and Kara D. Federmeier. 2011. "Thirty Years and Counting: Finding Meaning in the N400 Component of the Event-Related Brain Potential (ERP)." *Annual Review of Psychology* 62: 621–647. doi: 10.1146/annurev.psych.093008.131123.

Lai, Cecilia S. L., Simon E. Fisher, Jane A. Hurst, Faraneh Vargha-Khadem, and Anthony P. Monaco. 2001. "A Forkhead-Domain Gene Is Mutated in a Severe Speech and Language Disorder." *Nature* 413 (6855):519–523. doi:10.1038/35097076.

Liberman, Alvin M. 1996. *Speech: A Special Code.* Cambridge, MA: MIT Press.

Makuuchi, Michiru, Jörg Bahlmann, Alfred Anwander, and Angela D. Friederici. 2009. "Segregating the Core Computational Faculty of Human Language from Working Memory." *Proceedings of the National Academy of Sciences of the United States of America* 106 (20): 8362-8367. doi: 10.1073/pnas.0810928106.

McClelland, James L., and Karalyn Patterson. 2002a. "Rules or Connections in Past-Tense Inflections: What Does the Evidence Rule Out?" *Trends in Cognitive Sciences* 6 (11): 465–472. doi: 10.1016/S1364-6613(02)01993-9.

McClelland, James L., and Karalyn Patterson. 2002b. "'Words *or* Rules' Cannot Exploit the Regularity in Exceptions." *Trends in Cognitive Sciences* 6 (11): 464–465. doi: 10.1016/S1364-6613(02)02012-0.

Pinker, Steven. 1994. *The Language Instinct: How the Mind Creates Language.* New York: William Morrow and Company.

Pinker, Steven, and Michael T. Ullman. 2002a. "The Past and Future of the Past Tense." *Trends in Cognitive Sciences* 6 (11): 456–463. doi: 10.1016/S1364-6613(02)01990-3.

Pinker, Steven, and Michael T. Ullman. 2002b. "Combination and Structure, Not Gradedness, Is the Issue." *Trends in Cognitive Sciences* 6 (11): 472–474. doi: 10.1016/S1364-6613(02) 02013-2.

Poeppel, D. 2014. "The Neuroanatomic and Neurophysiological Infrastructure from Speech and Language." *Current Opinion in Neurobiology* 28: 142–149.

Tremblay, P. and A. S. Dick. 2016. "Broca and Wernicke Are Dead, or Moving Past the Classic Model of Language Neurobiology." *Brain & Language* 162: 60–71.

Tourville, Jason A., Kevin J. Reilly, and Frank H. Guenther. 2008. "Neural Mechanisms Underlying Auditory Feedback Control of Speech." *NeuroImage* 39 (3): 1429–1443. doi: 10.1016/j.neuroimage.2007.09.054.

Van Berkum, J. J. A., C. M. Brown, P. Zwitserlood, V. Kooijman, and P. Hagoort. 2005. "Anticipating Upcoming Words in Discourse: Evidence from ERPs and Reading Times." *Journal of Experimental Psychology: Learning, Memory, and Cognition* 31: 443–467.

Vargha-Khadem, Faraneh, and Richard E. Passingham. 1990. "Speech and Language Deficits." *Nature* 346 (6281): 226. doi: 10.1038/346226c0.

Vargha-Khadem, Faraneh, Kate Watkins, Katie Alcock, Paul Fletcher, and Richard E. Passingham. 1995. "Praxic and Nonverbal Cognitive Deficits in a Large Family with a Genetically Transmitted Speech and Language Disorder." *Proceedings of the National Academy of Sciences of the United States of America* 92 (3): 930–933. doi: 10.1073/pnas.92.3.930.

Willems, Roel M., and Peter Hagoort. 2009. "Broca's Region: Battles Are Not Won by Ignoring Half of the Facts." *Trends in Cognitive Sciences* 13 (3): 101. doi: 10.1016/j.tics.2008.12.001.

Zion Golumbic, E. M., N. Ding, S. Bickel, P. Lakatos, C. A. Schevon, G. M. McKhann, . . . C. E. Schroeder. 2013. "Mechanisms Underlying Selective Neuronal Tracking of Attended Speech at a 'Cocktail Party'." *Neuron* 77: 980–991.

OTHER SOURCES USED

Boatman, Dana. 2002. "Neurobiological Bases of Auditory Speech Processing." In *The Handbook of Adult Language Disorders: Integrating Cognitive Neuropsychology, Neurology, and Rehabilitation*, edited by Argye E. Hillis, 269–291. New York: Psychology Press.

Carroll, John B. 1958. "Communication Theory, Linguistics, and Psycholinguistics." *Review of Educational Research* 28 (2): 79–88. doi: 10.2307/1168862.

Chomsky, Noam. 1959. "A Review of B. F. Skinner's *Verbal Behavior*." *Language* 35 (1): 26–58. doi: 10.2307/411334.

Fedorenko, Evelina, and Nancy Kanwisher. 2011. "Some Regions within Broca's Area Do Respond More Strongly to Sentences Than to Linguistically Degraded Stimuli: A Comment on Rogalsky and Hickok." *Journal of Cognitive Neuroscience* 23 (10): 2632–2635. doi: 10.1162/jocn_a_00043.

Gabrieli, John D. E. 2009. "Dyslexia: A New Synergy between Education and Cognitive Neuroscience." *Science* 325 (5938): 280–283. doi: 10.1126/science.1171999.

Goswami, Usha. 2010. "A Temporal Sampling Framework for Developmental Dyslexia." *Trends in Cognitive Sciences* 15 (1): 3–10. doi: 10.1016/j.tics.2010.10.001.

Hauser, Marc D., Noam Chomsky, and W. Tecumseh Fitch. 2002. "The Faculty of Language: What Is It, Who Has It, and How Did It Evolve?" *Science* 298 (5598): 1569–1579. doi: 10.1126/science.298.5598.1569.

Hickok, Gregory, and Corianne Rogalsky. 2011. "What Does Broca's Area Activation to Sentences Reflect?" *Journal of Cognitive Neuroscience* 23 (10): 2629–2631. doi: 10.1162/jocn_a_00044.

Pinker, Steven, and Ray Jackendoff. 2005. "The Faculty of Language: What's Special about It?" *Cognition* 95 (2): 201–236. doi: 10.1016/j.cognition.2004.08.004.

Samet, Jerry. 1999. "History of Nativism." In *The MIT Encyclopedia of the Cognitive Sciences*, edited by Robert A. Wilson and Frank C. Keil, 586–588. Cambridge, MA: MIT Press.

Serniclaes, Willy, Liliane Sprenger-Charolles, Rene Carre, and Jean-Francois Demonet. 2001. "Perceptual Discrimination of Speech Sounds in Developmental Dyslexia." *Journal of Speech, Language, and Hearing Research* 44 (2): 384–397. doi: 10.1044/1092-4388(2001/032).

Skinner, B. F. 1957. *Verbal Behavior*. Acton, MA: Copley.

Wernicke, Carl. 1874. *Der Aphasische Symptomencomplex: Eine Psychologische Studie Auf Anatomischer Basis*. Breslau: Cohn and Weigert.

FURTHER READING

Alexander, Michael P. 1997. "Aphasia: Clinical and Anatomic Issues." In *Behavioral Neurology and Neuropsychology*, edited by Todd E. Feinberg and Martha J. Farah, 133–149. New York: McGraw-Hill.
Authoritative overview of the eight classical aphasia syndromes.

Binder, Jeffrey R. 2003. "Wernicke Aphasia: A Disorder of Central Language Processing." In *Neurological Foundations of Cognitive Neuroscience*, edited by Mark D'Esposito, 175–238. Cambridge, MA: MIT Press.
Exhaustive review of the neuropsychological and neuroimaging literature, which concludes that "Wernicke's aphasia, simply put, consists of [a] combination of . . . syndromes and results from a larger lesion that encompasses both lexical-semantic and phoneme output systems" (p. 226).

Chial, Heidi. 2008. "Mendelian Genetics: Patterns of Inheritance and Single-Gene Disorders." *Nature Education* 1 (1). http://www.nature.com/scitable/topicpage/mendelian-genetics-patterns-of-inheritance-and-single-966.
Primer on inheritance of single-gene disorders.

Dehaene, Stanislas. 2009. *Reading in the Brain: The Science and Evolution of a Human Invention*. New York: Viking.
A thorough, reader-friendly introduction to the cognitive neuroscience of reading.

Dronkers, Nina F. 2000. "The Gratuitous Relationship between Broca's Aphasia and Broca's Area." *Behavioral and Brain Sciences* 23 (1): 30–31. doi: 10.1017/S0140525X00322397.
A succinct argument, thoroughly referenced, against a grammar-specific view of Broca's area.

Herholz, Sibylle C., and Robert J. Zatorre. 2012. "Musical Training as a Framework for Brain Plasticity: Behavior, Function, and Structure." *Neuron* 76 (3): 486–502. doi: 10.1016/j.neuron.2012.10.011.
A comprehensive overview of the effects of musical training on the brain.

Mahmoudzadeh, Mahdi, Ghislaine Dehaene-Lambertz, Marc Fournier, Guy Kongolo, Sabrina Goudjil, Jessica Dubois, . . . Fabrice Wallois. 2013. "Syllabic Discrimination in Premature Human Infants prior to Complete Formation of Cortical Layers." *Proceedings of the National Academy of Sciences of the United States of America* 110 (12): 4846–4851. doi: 10.1073/pnas.1212220110.
Technically impressive demonstration of speech-like functions in the brains of prematurely delivered infants whose cortices are still dramatically underdeveloped.

Salimpoor, Valorie N., Iris van den Bosch, Natasa Kovacevic, Anthony Randal McIntosh, Alain Dagher, and Robert J. Zatorre. 2013. "Interactions between the Nucleus Accumbens and Auditory Cortices Predict Music Reward Value." *Science* 340 (6129): 216–219. doi: 10.1126/science.1231059.
Integrates content from Chapter 15, Chapter 16, and Chapter 18 with "music appreciation": "aesthetic rewards arise from the interaction between mesolimbic reward circuitry and cortical networks involved in perceptual analysis and valuation."

Scott, Sophie K., and Ingrid S. Johnsrude. 2003. "The Neuroanatomical and Functional Organization of Speech Perception." *Trends in Neurosciences* 26 (2): 100–107. doi: 10.1016/S0166-2236(02)00037-1.
Presents an alternative dual-route model of perception, with a greater emphasis on homologies with the monkey auditory system.

Zatorre, Robert J., and Jackson T. Gandour. 2008. "Neural Specializations for Speech and Pitch: Moving beyond the Dichotomies." *Philosophical Transactions of the Royal Society of London, Series B: Biological Sciences* 363 (1493): 1087–1104. doi: 10.1098/rstb.2007.2161.
A detailed consideration of what factors may underlie the lateralized processing of some sounds.

CHAPTER 20
CONSCIOUSNESS

KEY THEMES

- Consciousness can be construed at many levels, and the three that we'll consider here are physiological (*What physiological conditions are necessary to support consciousness?*), content based (*What neural processes determine whether or not one is aware of a specific stimulus?*), and theoretical (*What are the neurobiological principles that make consciousness possible?*).

- The neurological syndrome of coma indicates that elevated activity in the brainstem reticular activating system is necessary for an individual to be conscious, but the persistent vegetative state/unresponsive wakefulness syndrome (UWS) indicates that elevated brainstem activity is not sufficient for consciousness.

- Lesions of the ventral pons can produce the diabolical locked-in syndrome (LIS), in which the patient is fully conscious, but only able to blink and produce vertical eye movements. LIS patients are often initially misdiagnosed as vegetative state/UWS.

- Perturbation of the brain with TMS, when coupled with EEG recording, reveals a systematic reduction of effective connectivity accompanying the loss of consciousness experienced during non-REM sleep, or while under general anesthesia.

- When visual stimuli are presented at the threshold of detection, V1 activity more closely tracks the subjective judgment of whether or not one has seen a target than the objective reality of whether or not a target was presented.

- In the monkey, detection of a target stimulus seems to depend not on the initial "feedforward sweep" of information from the retina and up through the visual hierarchy, but on reentrant, feedback signals to V1 after the feedforward sweep has been registered in higher level visual areas.

- Studies using visual masking, as well as studies of the attentional blink, have been interpreted to suggest that widespread activation of high-level cortical areas corresponding to a putative "global workspace" might be necessary for awareness of a stimulus.

- One theory of consciousness, Integrated Information Theory, has given rise to an empirical measure, the Perturbational Complexity Index (PCI), which discriminates levels of consciousness associated with neurological syndromes, level of anesthesia, and stage of sleep.

CONTENTS

THE MOST COMPLEX OBJECT IN THE UNIVERSE

Natural evolution, as brilliantly revealed by Charles Darwin (1809–1882), over these many million years gave rise to nervous systems as complex as the human brain, arguably the most complex object in the universe. And somehow, through the interactions among its 100 billion neurons, connected by trillions of synapses, emerges our conscious experience of the world and ourselves (Laureys and Tononi, 2009, p. ix).

In 2005, to mark its 125th anniversary, *Science* magazine published a feature on "What don't we know?," posing 125 questions that science has yet to satisfactorily answer. Twenty-five of these were allocated a page in the journal and, although they weren't explicitly rank ordered, one of them necessarily had to appear on the first page of the section, and one on the 25th. Coming in at number two (and thereby reconfirming that "just because it appeared in *Science* doesn't make it so") was *What is the Biological Basis of Consciousness?* Number one had something to do with the origin of the universe. What?!? I mean, come on, without consciousness, the notion of what a universe is, not to mention that it can have an origin, wouldn't exist . . .

What is it that gives you that ineffable sense of knowing that you are who you are, that you are presently reading a cognitive neuroscience textbook, that you are engrossed/tired/hungry/relaxed/stressed/whatever? These are the questions at which we will take a crack in this, our final chapter together. AND IT'S GOING TO BE AWESOME!

DIFFERENT APPROACHES TO THE PROBLEM

Even though we all experience it, there's an "unmeasurable" quality to consciousness that makes its study different from every other aspect of cognition that we have considered up to this point. There aren't obvious ways in which behavioral measures such as reaction time or selection of a response can index consciousness in the way that we can use these to study visual perception, or attention, or emotion, or sentence processing. Nor are there a priori neural signatures that everyone would agree on as being indisputably related to consciousness, short of some basal level of physiological functioning in at least some parts of the brain. But even for these, as we'll soon see, their invocation leads immediately to questions of the necessity vs. the sufficiency of activity in the systems in question. Nobody said that this would be simple.

The approach that we will take is to group cognitive neuroscience studies of consciousness into three categories: *physiological*, *content-based*, and *theoretical*. Physiological approaches to understanding consciousness will include quantitative measurement of metabolic or electrical activity of the brain during states of consciousness that can be independently validated, typically by a physician with specialized training. One area of study that has produced an enormous amount of knowledge about physiological bases of consciousness is coma and stages of recovery from coma. Two other models that are also yielding valuable information about relations between various physiological states and states of consciousness are sleep and anesthesia. Note that, for the "physiological/content-based" distinction to be useful, the former will be limited to physiological states that persist for minutes, hours, or longer. Shorter lived, momentary states, such as the instantaneous phase of an oscillation in a particular frequency band in the EEG, will be grouped under the content-based approach to consciousness. Finally, as we consider physiological bases of consciousness, it will be important to distinguish what Christoph Koch and colleagues (2016) refer to as "true neural correlates of consciousness" from "background conditions" that contribute to, and may even be necessary for, consciousness, but that aren't directly involved in determining what the contents of consciousness will be at any moment in time. "For example, global enabling factors, such as blood flow or oxygen supply to the cortex, are obviously essential for consciousness, but they do not contribute directly to its contents (p. 9604)." Although this particular example may seem uncontroversial, we'll see further along that there are other phenomena that Koch and colleagues will also classify as "background conditions," a designation with which other reasonable people will disagree.

What we'll call "content-based" approaches to consciousness are those that assess trial-by-trial awareness of specific stimuli or events and seek to infer from trials with, versus trials without, awareness of the item in question, what are the neural correlates of awareness of that item/event. One such study that we have already considered was the phosphene induction study featured all the way back in *Chapter 4, The relation between visual processing and the brain's physiological state.* Recall that the finding was that the likelihood that transcranial magnetic stimulation (TMS) of the occipital cortex would induce a phosphene was predicted by the power in the alpha band of the EEG during the 500 ms prior to the TMS pulse. This tells us something about how the instantaneous state of V1 relates to visual awareness.

Finally, there are theoretical approaches, and here, we will limit ourselves to cognitive- and systems-level neuroscience theories and models of how the brain might give rise to consciousness. Of course, it is invariably the case that tests of these theories will, themselves, be classifiable in our scheme as either "physiological" or "content-based" experiments, and so for some of this final section we'll be considering how research that we had reviewed earlier fits into theory X as opposed to theory Y.

THE PHYSIOLOGY OF CONSCIOUSNESS

Physiological states are objective, measurable phenomena, and so provide objective, measurable anchoring points to which we can remain tethered during our exploration of consciousness, and to which we can periodically return as things get progressively more abstract.

Neurological syndromes

From the syndromes of coma and the sequelae of minimally conscious state and locked-in syndrome, we can gain insight into which neuroanatomical and neurochemical systems are necessary for consciousness. From the vegetative state/unresponsive wakefulness syndrome (UWS), we can begin cataloguing physiological states that are not sufficient. (Due to the pejorative connotations of the term "vegetative" [despite its technical appropriateness], it has recently been proposed that the term "unresponsive wakefulness syndrome [UWS]" be adopted in the place of "vegetative state" [Laureys et al., 2010]. Because much of the literature cited here predates 2010, however, there will necessarily be some switching back and forth.)

Coma

Coma is a state of unarousable unconsciousness, characterized by a failure of the arousal/alerting system of the brain, the ascending reticular activating system. For practical purposes, this includes failure of eye opening to stimulation, a motor response no better than simple withdrawal type movements and a verbal response no better than simple vocalization of nonword sounds. This presupposes that the motor pathways and systems that would allow the person to respond if they were conscious are intact (Young, 2009, p. 138).

"Ascending reticular activating system" is a name given to a collection of nuclei in the brainstem that are necessary for arousal from sleep and for maintenance of wakefulness. In particular, two nuclei located in the pons (the "bridge" [from Latin] of the brainstem) deliver the neurotransmitters acetylcholine (ACh) and glutamate to targets in the thalamus via a dorsal pathway, and to the basal forebrain and posterior hypothalamus via a ventral pathway. Neurons in these structures, in turn, project diffusely throughout the brain, including cortex, releasing the excitatory/arousing neurochemicals glutamate, norepinephrine, histamine, and hypocretin/orexin. These pontine cholinergic nuclei, together with the cholinergic nuclei of the basal forebrain and the glutamatergic interlaminar nuclei of the thalamus, fire at high rates in wakefulness and during the rapid-eye-movement (REM) stage of sleep, but decrease or stop firing during non-REM stages of sleep. Thus, coma and related neurological states of compromised consciousness are often due to abnormal functioning of the brainstem. Due to the redundancy of nuclei and ascending fibers from the reticular activating system, recovery/transition from the initial state of coma often occurs within three weeks of coma onset.

At this point, we can provisionally say that elevated activity in the reticular activating system seems to be necessary for consciousness, as a background condition.

Vegetative state/UWS

The syndrome of vegetative state/UWS demonstrates, all too cruelly, that an activated state of the reticular activating system is not sufficient for consciousness. As summarized by Belgian neurologist Steven Laureys (2005):

Vegetative patients have their eyes wide open but are considered – by definition – to be unaware of themselves or their surroundings. They may grimace, cry or smile (albeit never contingent upon specific external stimuli) and move their eyes, head and limbs in a meaningless "automatic" manner. The vegetative state is often, but not always, chronic (the "persistent vegetative state"). Given proper medical care (i.e., artificial hydration and nutrition) patients can survive for many years (p. 556).

Vegetative state/UWS is the state into which patients transition from coma. How is it determined that these patients lack consciousness? The absence of consciousness is inferred from the absence of any of the following signs across repeated neurological examinations that are ideally performed at different times of the day: no meaningful response to external stimuli, including threats and noxious (i.e., painful) stimuli; no control over bladder or bowels; no sustained visual-pursuit eye movements, nor volitional fixation; and no purposeful limb withdrawal, or localization of a stimulus with a limb. By the classical terminology, the diagnosis transitions to persistent vegetative state if present for longer than one month.

Because the transition out of coma necessarily entails a recovery of at least some of the functions of the reticular activating system, UWS is often accompanied by a daily cycle of opening and closing of the eyes. Even during periods of "wakefulness," however, global brain metabolic activity in UWS rarely surpasses 50% of the level that is typical for the healthy awake state. For comparison, this 50%-of-normal level is similar to what is observed when healthy individuals are under general anesthesia or deep states of non-REM sleep.

The diagnosis of vegetative state/UWS inescapably has a dissatisfying, equivocal element because short of brain death, one can never know with absolute certainty that there isn't some residual, difficult-to-detect spark of awareness "somewhere in there." Indeed, the reality of the neurological syndromes of minimally conscious state and locked-in syndrome play into this. For me, it's analogous to (although, of course, much more consequential than) a failure to reject the null hypothesis in classical hypothesis testing – because one can never "prove" the null, one can never know with certainty that a "true" difference isn't lurking somewhere in those data, but just too subtle to detect.

Locked-in syndrome

What is like to be conscious, but paralyzed and voiceless? It's horrifying to even contemplate, right? In my prescience days, ideas that pushed me toward this career included the musing that one can lose an arm, or a leg, or both, and still be the same person. Lose just a tiny piece of brain from the wrong spot, however, or tweak its function with a pharmacological agent, and one's very essence could be dramatically altered. (And this was long before I had encountered the expression "He was no longer Gage" . . .) What if one were to take this hypothetical exercise in a different direction, losing the use of *all* of one's body, but leaving the brain intact? Unfortunately, this is not just an idle, lay-philosophical question for patients with locked-in syndrome (LIS).

LIS typically results from damage to the corticospinal and corticobulbar pathways, most often due to stroke at the level of the ventral pons. We learned about the former pathway in *Chapter 8*; the latter pathway carries fibers from the motor cortex to motor neurons controlling the muscles of the face, head, and neck (i.e., the lower cranial nerves). An authoritative definition of LIS is:

> a state in which selective supranuclear motor de-efferentation produces paralysis of all four limbs and the last cranial nerves without interfering with consciousness. The voluntary motor paralysis prevents the subjects from communicating by word or body movement. Usually, but not always, the anatomy of the responsible lesion in the brainstem is such that locked-in patients are left with the capacity to use vertical eye movements and blinking to communicate their awareness of internal and external stimuli (Plum and Posner, 1983, p. 9).

Despite the sparing of vertical eye movements and blinking in most cases of LIS, upon transition from coma, many patients with LIS go misdiagnosed as UWS for extended periods of time. One celebrated case, the Frenchman Jean-Dominique Bauby, who dictated a bestselling memoir after his stroke (1997), founded the Association for LIS (ALIS). Laureys et al. (2005) cite a survey of 44 patient-members of ALIS, which indicated that the average time between insult and diagnosis of LIS was 78 days, and for some, more than four years! Just imagine being the patient who has written that the doctors believed he was a "vegetable and treated him as such" (Laureys et al., 2005, p. 500). The same ALIS survey indicated that family members made the initial observation that their loved one might be conscious in 55% of the cases, doctors doing so in only 23% of cases. Therapists are also often credited with making the discovery. In one case a therapist was testing the gag reflex of a presumed vegetative patient, and the patient bit the therapist. As the therapist cursed and withdrew his finger, the patient managed a grin. The now-suspicious therapist asked, "How much is 2 + 2?" and the patient blinked four times.

The eloquent writing of many patients with LIS leaves no doubt about their level of consciousness. Bauby (1997) dictated his book letter by letter, blinking each time his therapist spoke the desired letter from a frequency-ordered list. Vigand (2000), the finger biter, composed his book on software controlled by eye movements tracked with an infrared camera.

Detecting awareness in the vegetative state

One of the first in the series of fMRI studies by Adrian Owen and colleagues – which has perhaps garnered more attention in the popular press than any other cognitive neuroscience research program, ever – was an fMRI study of a 23-year-old traumatic brain injury patient who had been in a persistent vegetative state for five months. First, it was observed that for spoken sentences vs. acoustically matched noise sequences, activity in posterior STG and MTG was greater for speech vs. noise. A tantalizing result, but not definitive, as it could possibly be attributable to automatic activation of auditory/language cortex spared by the accident. The second task was devised to require a volitional, internally generated response: the patient was asked, at specific times during the scan, to either "imagine yourself playing tennis," or "imagine yourself walking around your house, visiting each room." The results, illustrated in *Figure 20.1*, indicated that the patient's neural activity in response to these two sets of instructions was comparable to that of the control subjects.

Now, of course, this is an analytically simplistic example of "brain reading" relative to the sophisticated MVPA methods that we've discussed in previous chapters. Indeed, had this patient been scanned absent any comparison, it would be inferentially problematic to visually inspect activity from the scan and conclude that she was imagining tennis vs. navigating her house. However, that's not the point here. The point is that this study indicated that this patient, who was believed to be vegetative, was, in fact, able to modulate her brain activity "on command," in more or less the same way as did unequivocally conscious control subjects. Thus, one might reasonably conclude that she, too, was conscious, despite also being fully locked in, such that she was deprived of even volitional eye movements or blinks.

In a subsequent study, Monti et al. (2010) scanned 54 patients with disorders of consciousness, and found five who could willfully modulate their brain activity in this way. With one patient they took the further step of telling him that he'd be asked a series of autobiographical questions, and that he was to respond "yes" by imagining tennis, or "no" by imagining navigating through his house. In response to the question "Is your father's name Alexander?," for example, he correctly responded "yes" by producing elevated activity in predefined "tennis regions." To the question "Is your father's name Thomas?," he correctly responded "no" by producing elevated activity in navigation regions.

Recovery of consciousness from coma or UWS

In some fortunate instances, patients in a coma or UWS experience substantial recovery of function. Such cases offer unique opportunities to assess the changes in the brain that accompany recovery of conscious awareness. Of course, doing so requires having thoroughly documented the patient's condition prior to recovery, something that is not a standard procedure at most hospitals. One notable exception is the Coma Science Group, under the direction of Steven Laureys, at the Université de Liège in Belgium, which has produced a prodigious amount of groundbreaking research on the neurology of consciousness. Their work indicates that, although the UWS is associated with widespread metabolic dysfunction, measures of global metabolic activity, alone, lack both the sensitivity and the specificity that one would want in a "consciousness meter": there are cases of patients who have recovered from UWS without significant changes in brain metabolism, and there are neurologically healthy volunteers whose global brain metabolism levels are as low as those of UWS patients.

At least one intriguing case suggests that more diagnostic precision might be found in patterns of connectivity than in measures of raw metabolic activity. A patient was scanned with $H_2^{15}O$ PET two weeks after the onset of UWS, and again, at rest, four months after recovery of consciousness and "partial autonomy." While in UWS, functional connectivity between the thalamus and several

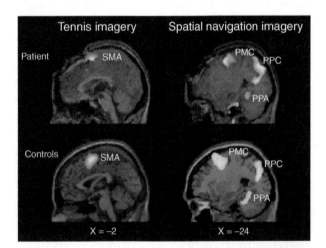

FIGURE 20.1 fMRI responses to imagery instructions, in a patient diagnosed as being in a persistent vegetative state (top row) and in the group-average of 12 control subjects. Source: Owen, Coleman, Boly, Davis, Laureys, and Pickard, 2006. Reproduced with permission of AAAS.

regions of lateral PFC (BAs 8, 9, and 10) and anterior cingulate cortex (BAs 24 and 32) was reduced relative to 18 control subjects, after clinical recovery, thalamocortical connectivity was also recovered to normal levels. Further consistent with the disconnection idea is that UWS patients (who, unfortunately, have not recovered) show "'Functional disconnections' in long-range cortico–cortical (between latero-frontal and midline-posterior areas) and cortico–thalamo–cortical (between non-specific thalamic nuclei and lateral and medial frontal cortices) pathways" (Laureys, 2005, p. 557).

Two dimensions along which consciousness can vary

This review of neurological syndromes has introduced the idea of two dissociable dimensions along which states of consciousness can vary. One, which seems to be necessary, but is clearly not sufficient, is the level of physiological arousal of the brain. This is portrayed as the horizontal axis in the diagram in *Figure 20.2*. The second, for the time being, we can label as the "level of awareness." Different clinical states that we just discussed, and physiological

states that we're about to take up in the next two subsections, can be portrayed in terms of where they fall in this 2-D space. The label of, and the idea behind, the second dimension is not satisfactory, because it's just a descriptive term. Ideally, to really understand the neural bases of consciousness, one would want to know what measurable, physical property varies along that axis, "bringing awareness along with it." Indeed, one way to construe the third section of this chapter, on theoretical approaches to consciousness, will be to evaluate each theory in terms of how well its explanatory factor would do if plugged into the label for the vertical axis of this plot. (To make this concrete, the implication of one theory will be that the physical property that varies along the vertical axis is connectivity – greater levels of connectivity within the brain, coupled with greater levels of physiological arousal, result in higher levels of awareness.)

In the next two subsections, on sleep and anesthesia, we will introduce a different technique for relating physiological state to consciousness: estimating effective connectivity from the TMS-evoked response in the EEG. Once we have established an understanding of this technique, the

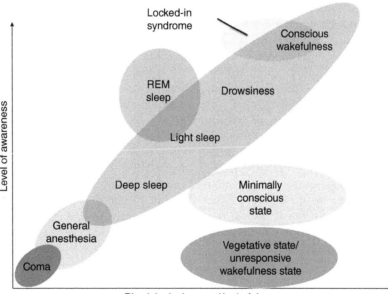

FIGURE 20.2 A depiction of neurophysiological states as they fall along two dimensions that are related to consciousness: arousal (horizontal axis) and awareness (vertical axis). The blue region corresponds to normal physiological states, pink to pathological coma, and orange to pharmacologically induced coma (a state from which patients cannot be awakened). Source: Adapted from Laureys, Steven, Frédéric Pellas, Philippe Van Eeckhout, Sofiane Ghorbel, Caroline Schnakers, Favien Perrin, … Serge Goldman. 2005. "The Locked-In Syndrome: What Is It Like to Be Conscious but Paralyzed and Voiceless?" Progress in Brain Research 150: 495–511. Elsevier.

final subsection of this section on "physiological state" will return to coma, and summarize how the recovery of consciousness is paralleled by a recovery of large-scale connectivity within the brain.

Sleep

Sleep offers another fascinating window on consciousness. As discussed previously, sleep results from an inhibition of arousing signals to thalamus and cortex from the brainstem and basal forebrain. In the absence of these signals, two phenomena come about. First, there is a marked reduction in long-range connectivity between brain areas. Second, there is a marked increase in local homogeneity of activity across populations of cortical neurons whose waking activity is typically not correlated, certainly not for extended periods of time. At the level of the EEG, the first change to occur at the transition between quiet eyes-closed wakefulness and sleep is the replacement of the prominent "high-voltage" alpha rhythm (*Figure 4.6*) with low-voltage oscillations in the theta band. Then, the EEG gradually comes to be dominated by the large-amplitude low-frequency (~1–2 Hz) "slow waves" of so-called "slow-wave sleep." High-density EEG recordings have revealed slow waves to be traveling waves that sweep across the cortical surface, typically from anterior to posterior, in a manner very reminiscent of the expanding, concentric rings produced by a single stone dropped into the proverbial flat pond: an expanding arc of neuronal depolarization, and, hence, "lock-step" elevated firing, followed by a prolonged, few-hundred milliseconds period of hyperpolarization, during which synaptic activity is reduced to almost nil. Upon falling asleep, one typically progresses through three stages of ever-deeper non-REM sleep, with only infrequent slow waves during stage N1, roughly 5 per minute in N2, and more than 10 per minute in N3. After a bout of several, if not tens of, minutes in N3 (longer at the beginning of the night, progressively shorter as the night wears on), one transitions back into N2, and then "pops" into REM.

REM sleep shares many characteristics with wakefulness, in that it entails high activity in the reticular activating system, basal forebrain, and thalamus. These, in turn, arouse the cortex via release of ACh and glutamate, just as in wakefulness. The difference between REM sleep and wakefulness is that additional neurotransmitter systems also driven by the reticular activating system during wakefulness – including norepinephrine, histamine, hypocretin, and serotonin – remain inhibited during REM. Additionally, a set of neurons in the dorsal pons applies tonic inhibition that counteracts descending cortical motor signals, thus keeping muscle tone flaccid.

REM is, of course, the stage of sleep during which we experience our most vivid dreams. Dreams can be construed as "sleep consciousness," and offer yet another window through which consciousness can be studied (see Nir and Tononi, 2010, *Further Reading*).

Studying sleep (and consciousness) with simultaneous TMS-EEG

Using TMS/EEG to investigate critical differences in the functioning of the waking and sleeping brain offers several advantages. Unlike sensory stimulation, direct cortical stimulation does not activate the reticular formation and bypasses the thalamic gate. Thus, it directly probes the ability of cortical areas to interact, unconfounded by peripheral effects (Massimini et al., 2005, p. 2229).

Marcello Massimini and colleagues, in the laboratory of Giulio Tononi at the University of Wisconsin–Madison, ran their experiments late at night, when subjects would more readily fall sleep. Subjects were fitted with an electrode cap connected to an EEG amplifier that, when triggered by the TMS stimulator, would momentarily divert the signal coming from the electrodes, such that the amplifier would not saturate in response to the TMS pulse (see Figure 3.2). (The voltage transient produced by a pulse of TMS exceeds the range within which a conventional EEG amplifier, configured to detect exquisitely small voltage fluctuations on the scalp, can detect signals. The resulting "saturation" would render the amplifier insensitive to voltage fluctuations for hundreds of milliseconds or longer, meaning that much, if not all, of the neural response evoked by the TMS pulse would not be recorded.)

Figure 20.3 illustrates the EEG response to a TMS pulse delivered to the right premotor cortex of a single subject before falling asleep (A–C) and during non-REM sleep (A′–C′). A and A′ show the traces from all 60 electrodes in the cap, all plotted on the same timeline, and color-coded red during periods when the signal deviated significantly from baseline. Comparison across the two conditions illustrates that (1) the initial TMS-evoked response was roughly twofold larger in magnitude during non-REM sleep than during wakefulness and (2) the TMS-evoked response was of considerably

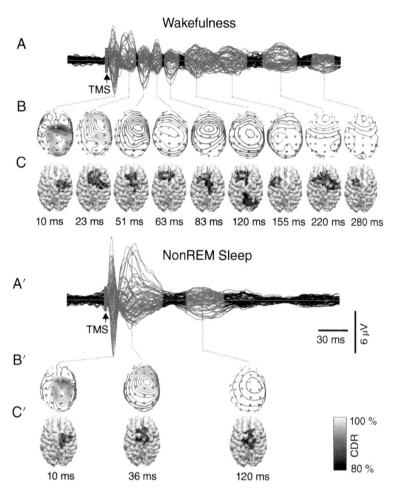

FIGURE 20.3 TMS-EEG from a single subject during wakefulness and non-REM sleep. Rows A and A′ are "butterfly plots" in which the trace from each electrode is plotted along the same time axis. B and B′ show the scalp distribution of voltages associated with the peak of each significant excursion of the signal. C and C′ show the source solutions of the maps from B and B′, displayed on the subject's MRI. CDR = current density reconstruction. Source: From Massimini, Marcello, Fabio Farrarelli, Reto Huber, Steve K. Esser, Harpreet Singh, and Giulio Tononi. 2005. "Breakdown of Cortical Effective Connectivity during Sleep." Science 309 (5744): 2228–2232. Reproduced with permission of AAAS.

longer duration during wakefulness than during non-REM sleep. The reason for the latter is evident from C and C′: whereas in wakefulness the initial response of the region directly under the coil was followed by responses in contralateral and posterior regions, then again in right PMC and then again in the left, during non-REM sleep, the response stayed local. Thus, during wakefulness, the ERP was "kept alive" longer because activity in proximal areas triggered activity in distal areas, which then reactivated the proximal area, which then reactivated the distal area. There was greater effective connectivity during wakefulness. (Recall that *functional connectivity* refers to

correlated activity between two regions, whereas *effective connectivity* refers to one region exerting a causal influence over another.)

Why was the initial response of such greater magnitude during non-REM sleep? One possibility is that, because proximal neurons are more tightly synchronized during non-REM sleep, the response to TMS in the region that was directly stimulated was also more synchronized. A second is that thalamic neurons are more likely to respond in burst mode (section *Setting the state of thalamic neurons: tonic mode vs. burst mode, Chapter 4*) during non-REM sleep, because they are in a hyperpolarized state.

The point emphasized by the authors is that the fading of consciousness as we fall asleep is accompanied by a breakdown of long-range connectivity.

Anesthesia

[M]any aspects of an EEG of anaesthesia are really consistent with patients who are in a coma. The key difference is that anaesthesia is reversible . . . If you came in for an operation and I told you I was going to put you into a coma, you would probably get up and run away! [But] the idea of being put to sleep is a euphemism . . . For your surgery you need to be in a state of general anaesthesia, not sleep. (Anesthesiologist and neuroscientist Emery Brown, quoted in Heathcoat, 2011.)

Anesthesia refers to a chemically induced, reversible, loss of sensation, and general anesthesia, of interest to us here, to a chemically induced loss of consciousness. As detailed by Brown, Lydic, and Schiff (2010, *Further Reading*), and illustrated in *Figure 20.2*, general anesthesia is, indeed, a physiological state more akin to coma than to sleep. One can be shaken awake from even the deepest sleep, but not so with anesthesia (or coma). Further, with one exception, systemically administered anesthesia causes a decline in overall brain glucose metabolism, as assessed with PET. The effects of anesthesia are dose dependent. For example, the anesthesiologist Michael Alkire (2009) describes a patient for whom a small dose of desflurane lowered global cerebral metabolism by 5–10% from baseline, rendering the patient sedated, but still able to generate a movement on request. The patient lost consciousness at a larger dose, which reduced global cerebral metabolism by 30%. (The exception is the anesthetic agent ketamine, which is "dissociative" in that it lowers awareness, and affects different brain regions in different ways, but, overall, causes an increase in global brain metabolic activity. Further, not unlike the dissociated neurological states that we have considered, including UWS, the eyes remain open during ketamine anesthesia. [At the time of the preparation of the second edition of this book, there's increasing buzz about a possible application of ketamine as an antidepressant. It's still early days, however, so we'll need to wait at least until the third edition to see whether or not this potential is realized.])

There are two additional findings from anesthesia that we will consider here, the first relating to a "breakdown of effective connectivity" reminiscent of what is seen with sleep. The second, neuroimaging evidence for why this might happen.

Studying anesthesia with TMS/EEG

A second study by Tononi's group, led by psychiatrist Fabio Ferrarelli and including anesthesiologist Robert Pearce, replicated the general procedures of the study of Massimini et al. (2005), but did so while subjects were administered the anesthetic agent midazolam. Midazolam is a benzodiazepine whose anesthetic properties derive from activating $GABA_A$ receptors, and thus imposing widespread inhibition in the brain. The effects of midazolam on the TMS-evoked response were virtually identical to those seen with non-REM sleep: the immediate response under the TMS coil was of greater magnitude, but the overall response was markedly reduced in terms of anatomical spread and temporal duration. Thus, anesthesia-induced loss of consciousness is also associated with drastic reduction of effective connectivity.

Studying anesthesia with surface EEG and deep brain EEG

Despite the considerable advantages of the TMS/EEG approach that we have described, it is limited, as are all EEG studies, by the fact that it can't directly "see inside" the brain. This problem has been addressed by a team based in Marseille, France, led by anesthesiologist Lionel Velly, that has studied the effects of general anesthesia on patients with Parkinson's disease (PD) who have electrodes chronically implanted in the subthalamic nucleus of the basal ganglia (see *Figure 8.10*). Signal from the in-dwelling electrodes provided subcortical field potential recordings, referred to as electrosubcorticography (ESCoG), that can be thought of as a "subcortical EEG." They performed two analytically independent analyses, one being a fairly conventional spectral analysis of the electrical signals, the second being a nonlinear measure based on deterministic chaos theory. "Schematically," as Velly et al. (2007) explain it, dimensional activation (D_a) estimates "the complexity of EEG and ESCoG signals: The more alert the patient is, the more complex the EEG is, and the higher the D_a is; the deeper the sleep or anesthetic state is, the simpler the EEG is, and the lower the D_a is" (pp. 204–205).

The findings from this study, illustrated in *Figure 20.4*, are striking and clear: anesthesia-induced loss of consciousness is associated with a precipitous drop in oscillatory power at frequencies above the theta range, and a comparably precipitous drop in D_a, in the cortical EEG. Measures of subcortical activity, in contrast, begin a gradual decline with anesthesia onset. This suggests that anesthesia-related loss of consciousness, and perhaps also

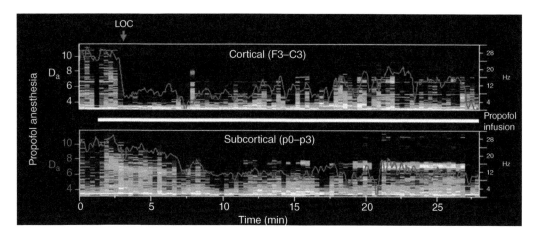

FIGURE 20.4 The time course of propofol anesthesia-induced loss of consciousness ("LOC"), as indexed by electrical activity measured from scalp EEG ("Cortical [F3–C3]") and from electrodes implanted in the subthalamic nuclei ("Subcortical [p0–p3]"). Red traces indicate level of dimensional activation (D_a) computed from the electrical signals (see text), with D_a axis on left-hand side of graphs, and green-to-white horizontal bars indicate spectral power in the EEG in 20-second time steps, with axis for the spectrograms on the right-hand side of the graphs. Source: Velly, Rey, Bruder, Gouvitsos, Witjas, Regis, Peragut, and Gouin, 2007. Reproduced with permission of Wolters Kluwer.

the commensurate loss of effective connectivity reported in the TMS/EEG study of Ferrarelli et al. (2010), is primarily due to effects in cortex.

Summary across physiological studies

Let's take a moment to consolidate what we've learned about the "physiology of consciousness." First, a factor that is necessary, but not sufficient, for conscious awareness is elevated activity in the reticular activating system, which, in turn, drives the state of arousal in thalamus and cortex. When this system is in an inactive state, whether due to neurological dysfunction (e.g., coma), sleep state, or anesthesia, one does not experience conscious awareness. The fact that elevated levels of physiological arousal are not sufficient to support consciousness, however, is seen in vegetative state/ UWS, as well as in circumstances that we have not discussed, sleepwalking and epileptic seizures. What do UWS patients, and patients under anesthesia, lack that individuals who are conscious (REM sleep, LIS, healthy wakefulness) seem to have? Complex cortical activity and substantial cortico-cortical and cortico-thalamic connectivity. In the final section of this chapter, we will consider some theoretical explanations for why these aspects of brain activity might be important for consciousness.

As a "farewell to physiology," *Figure 20.5* illustrates the evolution of the TMS/EEG response as subjects recover from coma, indicating that, just as with sleep and with anesthesia, a return to consciousness in these patients is accompanied by an increase in complex, interconnected activity.

BRAIN FUNCTIONS SUPPORTING CONSCIOUS PERCEPTION

Up until now we've considered what factors underlie the state of being conscious. Now we turn to a different question: *What determines the contents of our conscious awareness at any given moment?* As I lift my eyes from my computer screen and look out from my front patio, what neural functions underlie my becoming aware of that sign across the street, or the hawk that's perched up in the tree? It's presumably not "just" the visual processes that we considered earlier in this book, because, as we saw with the phenomenon of the attentional blink (section *Meditation as mental training in emotion regulation, Chapter 18,*), one can be presented with the same stimulus under identical external conditions and sometimes one becomes aware of it and other times one doesn't. And so minimally, we would need to specify what is different about the state of the visual system during trials when we are aware of the stimulus vs. during trials when we are not. And then there's the question of attention. Patients with hemispatial neglect are not aware of stimuli on the left side of space, so surely attention and consciousness closely linked? Undoubtedly they are, but one of the claims that we'll consider in this section is that the two can be dissociated.

Research that addresses these kinds of questions is often referred to as the study of the neural correlates of consciousness (NCC). As alluded to in the previous paragraph,

FIGURE 20.5 Tracking recovery from coma with TMS/EEG. Source: From Rosanova, Mario, Olivia Gosseries, Silvia Casarotto, Mélanie Boly, Adenauer G. Casali, Marie-Aurélie Bruno, Marcello Massimini. 2012. "Recovery of Cortical Effective Connectivity and Recovery of Consciousness in Vegetative Patients." Brain 135 (4): 1308–1320. Reproduced with permission of Oxford University Press.

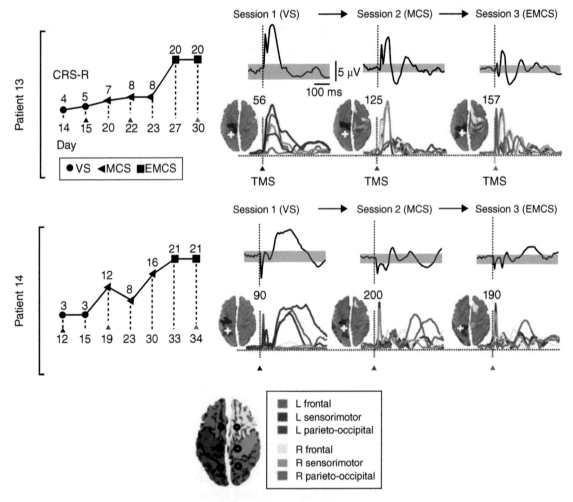

FIGURE 20.5.A Two patients whose recovery transitioned from coma to vegetative state (VS) to minimally conscious state (MCS) to emerged-from-MCS (EMCS), at which point they had recovered the ability to communicate and were classified as "recovered" by the Coma Recovery Scale-Revised (CRS-R), a standard diagnostic assessment. CRS-R score is tracked as a function of day post-trauma, with arrows indicating day of each TMS/EEG session (black = first; blue = second; red = third). At the top of the diagram for each session is the TMS-evoked response from the electrode under the coil. Lower left illustrates TMS target (white "+") and brain regions from which a TMS-evoked response occurred, color coded as per the legend. Plots to the right of each brain show the estimated TMS-evoked current from each of the six sources identified with black circles in the legend. The numbers between the two plots (e.g., 56, 125, 157 for Patient 13) indicate the number of vertices in the cortical mesh from the source solution at which a significant TMS-evoked response was detected.

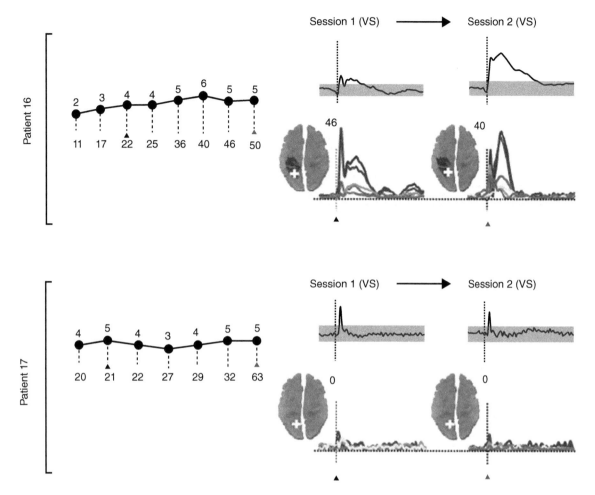

FIGURE 20.5.B Data from two patients who did not emerge from VS.

studies of the NCC often use the strategy of presenting subjects with a series of identical trials, under conditions when they may or may not be aware of the stimulus. After the experiment, trials are sorted as a function of whether the target was or was not detected, and trials from the two types are compared for differences. Although we will restrict ourselves here to the visual modality, we'll see that there are many domains of cognition that are implicated in this research.

Are we conscious of activity in early sensory cortex?

The first question we might want to address when thinking about visual awareness is: what level(s) of the system potentially *can* support it? Let's say, for example, we had

some way of determining that areas V1–V4 in the ventral stream are very machine-like, in that although the output of their processing is raw material that is needed for visual awareness, they don't actually contribute to the generation of visual awareness. If we knew this to be true, then we could focus our attention on higher level visual areas and not worry about fluctuations in signal in the earlier areas – these might relate to something like signal strength, but couldn't tell us about awareness, per se.

There are two classes of experimental procedure that have been used extensively to address this question. The one that we will consider in detail is, in a way, the simplest: it consists in presenting a single item on each trial, and comparing brain activity from trials in which the stimulus was seen vs. from trials in which it wasn't. A second experimental strategy for addressing this question takes advantage of the phenomenon of **binocular rivalry**: when a different image is presented to each of the two eyes, the percept is of seeing just one of the two, with frequent seemingly spontaneous "switches" between which of the two one is currently seeing. Because it explicitly dissociates the information being processed by the retina from the percept that we are consciously aware of, this procedure has been used extensively in studies of the NCC in both humans and monkeys. It's a rich and nuanced literature to which we can't do justice here, but that can be explored via publications by Leopold and Logothetis (1996), Crick (1996), Engel et al. (1999), and Blake and Tong (2008), all listed under *Further Reading*.

Now, back to "single-item" studies. The first that we will consider in some detail is a very simple, very elegant fMRI study by cognitive neuroscientists David Ress and David Heeger (2003). It was designed to ask, in effect, *does activity in early visual cortex correspond to the external stimulus from which photons are impacting our retinae, or does it correspond to our subjective ("internal") experience?* Subjects fixated centrally, and every two seconds a ring of "background" visual noise would appear for one second, and on a randomly determined 17% of trials a target (a sinusoidal grating, i.e., parallel lines) would be superimposed on the background (*Figure 20.6.A*). Subjects responded "yes" or "no" on every trial, and target contrast was adjusted for each subject such that they would detect it on roughly 80% of trials. Each trial could have one of four outcomes: the target is present and the subject detects it (a "hit"); the target is present and the subject does not detect it (a "miss"); the target is absent and the subject reports detecting it (a "false alarm"); or the target is absent and the subject does not detect it (a "correct rejection"). It was assumed that hits would produce a robust stimulus-evoked response in V1, and correct rejections were treated as the baseline. The questions of interest, then, were how V1 would respond on trials in which the physical display and the subject's percept were at odds. If the fMRI response to misses resembled that of hits, and the fMRI response to false alarms resembled that of correct rejections, then visual cortex activity would be congruent with the objective state of the visual world. If, on the other hand, the reverse were true, then visual cortex activity would be congruent with the subject's subjective percept. Drum roll please . . . *Figure 20.6.B* indicates that V1 activity was congruent with the percept! The implication for consciousness research, then, is that regions as early as primary sensory cortex may contribute to our conscious awareness of sensory perception.

FIGURE 20.6 Does V1 activity correspond to the external world, or to our conscious perception of it? Source: From Ress, David, and David J. Heeger. 2003. "Neural Correlates of Perception in Early Visual Cortex." Nature Neuroscience 6 (4): 414–420. Reproduced with permission of Nature Publishing Group.

A

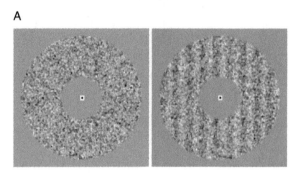

FIGURE 20.6.A A stimulus without the target vertical grating (left), and with it (right).

B

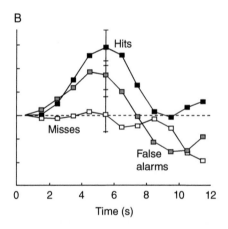

FIGURE 20.6.B fMRI responses from V1 of a single subject, time locked to stimulus onset, as a function of behavioral response. Correct-rejection trials were treated as the baseline for this analysis, meaning that the activity during misses did not differ from activity during correct rejections. (Each tick mark on the vertical axis corresponds to 0.1% signal change from baseline.)

If you stop to think about this result for a moment, it might be somewhat unsettling for your sense of agency. Is not an implication of the result from false-alarm trials that our visual system may sometimes "show us what we want to see," rather than what is really out in the world? Perhaps. But keep in mind that subjects were performing under atypically difficult conditions – low-contrast targets, continuous target presentation, speeded responses, and no feedback. Additionally, isn't it at least somewhat reassuring to encounter an effect in this, our last chapter together, that reinforces principles that we've encountered earlier in the book, in this instance the hypothesis-testing idea articulated by Sillito and colleagues all the way back in *Chapter 4*?

Although the false-alarm trials from Ress and Heeger (2003) might be spookier, for the reasons just discussed, we also need to come up with a satisfactory account for the miss trials from this study. We'll address this first, and then return to the false alarms.

What intrinsic factors influence conscious visual perception?

If one takes a deterministic view of the brain (which we do), there must have been one or more factors that differed in the momentary state of subjects' brains during miss trials vs. during hit trials from the Ress and Heeger (2003) study. One possibility that was raised at the beginning of the chapter is the spontaneous fluctuation in posterior alpha-band power. The study from Romei et al. (2008; *section The relation between visual processing and the brain's physiological state, Chapter 4*) demonstrated that the excitability of visual cortex varies systematically with spontaneous fluctuations of power in the alpha band. A second factor that influences near-threshold visual stimulus detection is the phase of low-frequency oscillations during which a stimulus is presented. Mathewson and colleagues (2011) have proposed that oscillations in the alpha band implement a "pulsed inhibition" of neuronal activity, such that the phase of the ongoing alpha-band oscillation at which a stimulus is presented can determine how effectively it is processed. Relatedly, several groups have provided evidence for periodic attentional sampling that flits between targets at a rate corresponding to the theta band. One interesting exercise would be to revisit this evidence for periodic "up" and "down" states ("open" and "closed"?) of our perceptual awareness after reading the final section of this chapter, on theories of consciousness, and to try to interpret these phenomena in the context of the theories that we're about to review.

Bottom-up and top-down influences on conscious visual perception

Now we turn to the false-alarm finding from Ress and Heeger (2003), that the V1 "response" to an absent target was almost as robust as its response to real targets, but only on the trials when subjects thought that they had seen a target. One possible explanation is that the conditions of this experiment promoted noisy retino-geniculo-striate signaling, such that on some target-absent trials there was simply more afferent input to V1 than on others. By this account, target-absent trials with randomly higher input to V1 would have produced greater V1 physiological responses, which, in turn, would trigger false-alarm behavioral responses. This explanation follows the logic that we saw in the microstimulation studies described in *Chapter 16's* treatment of perceptual decision-making. That is, from the perspective of the signal-detection theory of perceptual decision-making, perhaps there was more afferent "momentary evidence" on false-alarm trials than on correct-rejection trials. Although this explanation seems sensible when only considering false-alarm trials, it becomes more problematic when one takes into account the miss trials.

The most straightforward prediction of a "noisy feedforward" account of the Ress and Heeger (2003) data would be that one should see an orderly progression, from little-to-no V1 response on correct-rejection trials, to a small-to-intermediate level of V1 activity on false-alarm trials, to an intermediate-to-large level of V1 activity on miss trials, to a large level of V1 activity for hits. But this is not what was observed. Instead, the results seem much more consistent with an account whereby intrinsic, possibly top-down, factors suppressed momentary evidence on miss trials and boosted it on false-alarm trials. In *Chapter 4* we considered anatomical and physiological substrates for such top-down influence on V1 (and LGN) processing of an afferent motion signal. Next, we consider a study whose results suggest a physiological basis for this top-down account of perceptual awareness.

The role of reentrant signals to V1

For the experiment whose results are shown in *Figure 20.7*, the Dutch team of Supèr, Spekreijse, and Lamme (2001) recorded from V1 of monkeys performing a target-detection task with many similarities to that of Ress and Heeger (2003): each trial presented a full-screen "background" of small scratch-like lines all oriented in the same direction, with the to-be-detected "figure" defined by rotating the orientation of every line within a square-shaped region by 90°. The figure appeared in each of three locations on 26.7% of trials, the remaining 20% being catch trials with no figure. The first thing to notice from *Figure 20.7* is that the neural detection of the figure (i.e., the "modulation") lagged behind the initial, nonspecific, stimulus-evoked

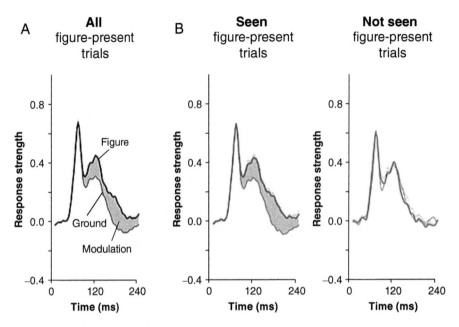

FIGURE 20.7 Normalized spiking activity ("Response strength") from V1 neurons in a monkey performing a target-detection task. Each trial began with the onset of a full-field textured "background" within which a figure (defined by differently oriented texture) was embedded in one of three locations. These plots show the response of V1 neurons when only the background fell into their receptive fields ("Ground"; light lines) and when the figure fell into their receptive fields (heavy lines). The plot in **A** presents the results from all trials when a figure was present somewhere on the screen (20% of trials, not shown here, were catch trials in which no figure was presented). The plots in **B** show the same data, but broken into hits ("Seen") and misses ("Not seen"). Source: From Supèr, Hans, Henk Spekreijse, and Victor A. F. Lamme. 2001. "Two Distinct Modes of Sensory Processing Observed in Monkey Primary Visual Cortex (V1)". Nature Neuroscience 4 (3): 304–310. Reproduced with permission of Nature Publishing Group.

ramp-up of activity by roughly 90 ms. This effect is reminiscent of the effects of object-based attention reported by Chelazzi, Duncan, Miller, and Desimone (1998; *Figure 7.15.B*), in that both show a delayed boost in activity that corresponds to the discrimination between two visual elements in the display. The effect in the Supèr, Spekreijse, and Lamme (2001) data must derive from a different source, however, because its "boost" comes much earlier than the earliest that selective attention effects are seen in V1, which is 200–250 ms after stimulus onset. (Note that the 120–175 ms latencies reported in Chelazzi et al. [1998; section *Object-based attention, Chapter 7*] were recorded from vAIT.)

The monkeys' performance was such that there were too few unambiguous false-alarm responses to analyze. What's shown in *Figure 20.7*, panel B, however, is that activity on miss (i.e., "Not seen") trials completely lacks any evidence of the figure-related modulation. (Thus, had Supèr, Spekreijse, and Lamme [2001] portrayed their data in a manner analogous to how Ress and Heeger [2003]

did theirs, by subtracting baseline [i.e., "Ground"] from target-present [i.e., "Figure"] trials, V1 activity on the miss [i.e., "Not seen"] trials from *Figure 20.7*, panel B, would also have been a flat line with a value of 0.) Supèr, Spekreijse, and Lamme (2001) suggest that the source of figure detection-related modulation of V1 was a reentrant (i.e., feedback) signal coming from a higher station in extrastriate cortex. We will also return to this idea in the final section on theories of consciousness.

Next, we turn to a study whose logic relies on extrinsic, rather than intrinsic, factors to control whether or not subjects are aware of a given stimulus.

Manipulating extrinsic factors to study conscious vs. unconscious vision

The French cognitive neuroscientist Stanislaus Dehaene has made important contributions to at least three subdisciplines within the field: numerical cognition, reading (see

Chapter 19, Further Reading), and consciousness. The study that we'll consider here combines the latter two interests. In it, Dehaene et al. (2001) explored the phenomenon that words presented very briefly and backward masked, such that they are not consciously perceived, can nonetheless produce robust repetition priming effects. (This is reminiscent of the emotional face presentation studies that we considered in *Chapter 18*.) While in the fMRI scanner, subjects viewed a serial stream of jumbled masks, each for a duration of 71 ms, with a mask periodically replaced by one or two 71-ms blank screens. Embedded in these streams were words presented for 29 ms that were either immediately preceded and followed by a blank screen ("unmasked") or immediately preceded and followed by a mask.

Behavioral data from Dehaene et al. (2001) indicated that masked words went undetected on > 99.3% of trials, whereas unmasked words were detected on > 90% of trials. Despite subjects' lack of awareness that masked words had been presented, there was evidence in both the behavioral and fMRI data that they were processed to at least the level of the mental lexicon. The behavioral evidence came in the form of repetition priming: when a masked word was followed by the 500-ms (i.e., clearly visible) presentation of a target word about which subjects had to make a "natural/manmade" judgment, responses were significantly faster when the target word was the same as had been the masked word. This was even true when the masked word was presented in an ALL CAPS typeface and the target word in lower case, meaning that it was a relatively high-level representation of the word that had been accessed, not just a low-level visual representation of what the letters physically looked like. The fMRI evidence that masked words were processed to a fairly high level came from the fact that repetition suppression (section *Repetition effects and fMRI adaptation, Chapter 13,*) was observed in left fusiform gyrus, in the "visual word form area" (labeled "word selective" in *Figure 10.5*), for 500-ms target words that were preceded by masked presentation of the same word, relative to 500-ms target words preceded by masked presentation of a different word.

As illustrated in *Figure 20.8*, the fMRI response evoked by unmasked 29-ms words was markedly stronger than that for masked 29-ms words, both in terms of signal intensity and in terms of the number of regions responding at above-threshold levels. Unmasked words evoked responses in several regions, including fusiform gyrus, parietal cortex, posterior inferior PFC/anterior insula,

anterior cingulate, precentral cortex, and supplementary motor area, bilaterally. Masked words, in contrast, evoked above-threshold responses only in left fusiform gyrus and in left precentral sulcus. Further, when signal intensity in "commonly activated" areas was compared, the response to masked words was only 8.6% as strong as the response to unmasked words in fusiform gyrus, and 5.2% as strong in left precentral sulcus. Finally, for many of the regions listed here, functional connectivity was significantly greater during the presentation of unmasked than masked words. In summarizing their findings, the authors foreshadowed one of the theories that we'll be considering in the final section of this chapter: "unmasking the words enabled the propagation of activation and the ignition of a large-scale correlated cerebral assembly" (Dehaene et al., 2001, p. 757).

ARE ATTENTION AND AWARENESS THE SAME THING?

(This is an interesting question, and one that offers a nice transition between the preceding section on empirical studies of consciousness and the upcoming one on theories of consciousness.)

My intuition certainly tells me that they are. I am aware of that to which I pay attention and, conversely, how would it be possible for me to be aware of something to which I'm not attending? The classic example is one considered earlier in the book: the feeling of the pressure of the chair that you're currently sitting in pressing against your body. The sensory information coming from your legs, etc., was no different 10 seconds ago than it is now, yet without attending to it, you weren't aware of it. But there are at least two really smart groups of people who have given this question considerable thought, and who would disagree.

Neuroscientist Victor Lamme, whose study of visual awareness in the monkey is featured in *Figure 20.7*, argues that we can be aware of much more information than we can attend to. While the capacity of the number of items we can simultaneously attend to is limited to four (or less), we can be consciously aware of much more information. One example that he points to is iconic memory. The classic demonstration is from the psychologist George Sperling (1960), in which a 4 × 3 array of letters is briefly flashed (50 ms). If subjects are asked to write down as many letters as they can remember seeing (the so-called "full report" procedure), they typically only recall three or four items. If,

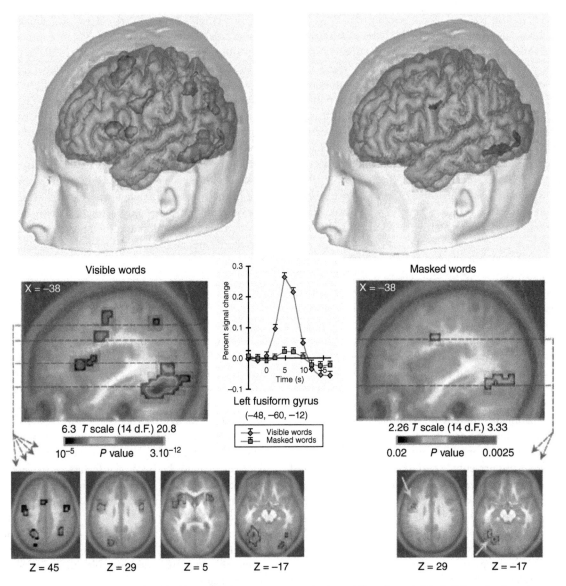

FIGURE 20.8 fMRI activity associated with words that were unmasked, and therefore visible (left-hand column), vs. words that were masked, and therefore not detected (right-hand column). The plot in the center compares the group-averaged response to visible vs. masked words in left fusiform gyrus. Source: Dehaene, Stanislas, Lionel Naccache, Laurent Cohen, Denis Le Bihan, Jean-François Mangin, Jean-Baptiste Poline, and Denis Rivière. 2001. "Cerebral Mechanisms of Word Masking and Unconscious Repetition Priming." Nature Neuroscience 4 (7): 752–758. doi: 10.1038/89551. Reproduced with permission of Nature Publishing Group

however, after the offset of the array, a high-, medium-, or low-pitched tone is played, indicating which row is to be "read out" from memory, subjects will typically remember three items from that row. Because they did not know prior to the tone which row would be cued, it must be the case that, for that brief moment, subjects had at least nine items in mind. Part of Lamme's (2003) argument is that

attentional selection is necessary for any kind of overt response, and that it's the attentional selection of items from conscious awareness into the focus of attention that is the capacity-limiting step, during which the representation of nine items shrinks to three. More generally, he argues that attentional selection is the process whereby one (or a few) mental representations that are competing to control behavior get selected for that purpose. Conscious awareness of visual representations arises through other processes (to be considered in the next section) independent of whether or not they will be selected by attention to influence behavior.

Tsuchiya and Koch (2009) also argue that these two constructs can be shown to operate independently. An example of attention without consciousness comes from a variant of the task used by Dehaene et al. (2001). In it, Naccache, Blandin, and Dehaene (2002) demonstrate that the priming for masked words (i.e., for words for which subjects have no awareness) only occurs when they are attending to the display. As an example of consciousness without attention, they point to dual-task studies in which subjects' attention is focused on a central task, yet they are able to identify each time a briefly flashed image in the periphery contains an animal, information that they became aware of without attending to it (Li, VanRullen, Koch, and Perona, 2002).

THEORIES OF CONSCIOUSNESS

It is remarkable that that most of the work in both cognitive science and the neurosciences makes no reference to consciousness (or "awareness"), especially as many would regard consciousness as the major puzzle confronting the neural view of the mind and indeed at the present time it appears deeply mysterious to many people. This attitude is partly a legacy of behaviorism and partly because most workers in these areas cannot see any useful way of approaching the problem (Crick and Koch, 1990, p. 263).

A great deal has changed in the 30 years since those words were written. Indeed, much of the early energy, the ideas, and the cheerleading driving consciousness research came from this duo, the Nobel Laureate molecular biologist, co-discoverer of the structure of DNA, and the young computational neuroscientist. Today, in contrast to 1990, there is an Association for the Scientific Study of Consciousness, and a handful of journals devoted to the topic.

In this final section, we'll summarize just a small number of neural theories of consciousness, working backward through the chapter up to this point, to highlight how theoretical and empirical work on this problem are interwoven.

Global Workspace Theory

The Global Workspace Theory was first developed by Bernard Baars, based on global workspace computer architecture with a distributed set of knowledge sources, which could cooperatively solve problems that no single

constituent could solve alone. Considerable theoretical and empirical work with this model has been carried out by Dehaene and his colleagues. In 2003 Dehaene, Sergent, and Changeux published a neural network model proposing that "the step of conscious perception, referred to as access awareness, is related to the entry of processed visual stimuli into a global brain state that links distant areas including the prefrontal cortex through reciprocal connections, and thus makes perceptual information reportable by multiple means" (p. 8520). This step of access awareness corresponds to the "ignition of a large-scale correlated cerebral assembly" triggered by the presentation of unmasked words in the Dehaene et al. (2001) study that we considered in the previous section.

A key component of the model is competition among stimulus representations for representation in the global workspace. The global workspace itself is implemented as high-level cortical areas (representing prefrontal, parietal, temporal, and cingulate) that implement "brain-scale propagation of stimulus information," which results in its accessibility for consciousness. There are several distinctive features to this model. A central tenet that differentiates it from many other models is the all-or-none property of conscious awareness of a representation:

According to the present model, the primary correlate of conscious access is a sudden self-amplifying bifurcation leading to a global brain-scale pattern of activity. This leads to several critical predictions for neuroimaging and pharmacological experiments. First, in experiments with a continuously varying and increasingly visible stimulus, we predict a sudden nonlinear transition toward a state of globally increased brain activity ("ignition"), particularly evident in prefrontal, cingulate, and parietal cortices, accompanied by a synchronous amplification of posterior perceptual activation and by thalamo-cortical brain-scale synchrony in the [gamma] range (20–100 Hz) (Dehaene, Sergent, and Changeux, 2003, p. 8525).

This "all-or-noneness" is illustrated in *Figure 20.9*, in which the simulated attentional blink effectively prevents any representation of T2 in the global workspace. Further, in simulations with the model, when long-distance top-down connections within areas C and D of the global workspace and between C and D and earlier sensory levels of the model were removed, weakened by 50%, or reduced to only short-range connections, the attentional blink no longer occurred (i.e., T2 gained access to the global workspace). An empirical demonstration that the authors point to is the finding illustrated in *Figure 20.8*: unmasked words, which were consciously perceived, evoked robust activity

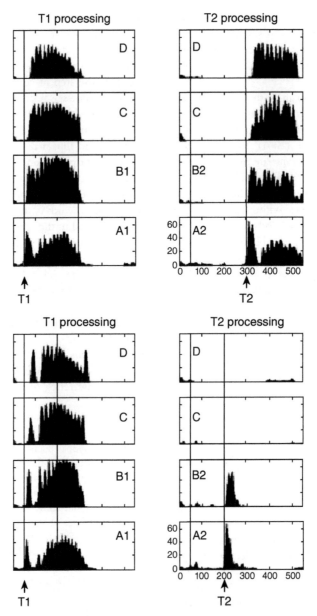

FIGURE 20.9 Simulation of two trials from the attentional blink task in a neural network model of Global Workspace Theory. In both groups, each rectangle corresponds to a level in the model, A1 and B1 being sensory channels that process the T1 stimulus; A2 and B2 being sensory channels that process the T2 stimulus. C and D are higher level areas in which the stimulus must compete for representation. Activity within each level represents firing rate. The trial illustrated in the upper panels simulates a long lag (250 ms) between the two stimuli, and as a result, T2 is presented at a time when the sustained activity evoked by T1 has decayed. T2 is therefore able to invade all cortical levels and elicits long-lasting activity comparable to T1. The trial illustrated in the lower panels simulates a "blink" trial that featured a shorter (150 ms) T1–T2 lag. Because T1 is still dominating the workspace when T2 arrives, T2 is blocked from entering, and therefore fails to evoke neural activity beyond an initial bottom-up activation in areas A and B. Source: From Dehaene, Stanislas, Claire Sergent, and Jean-Pierre, Changeux. 2003. "A Neuronal Network Model Linking Subjective Reports and Objective Physiological Data during Conscious Perception." Proceedings of the National Academy of Sciences of the United States of America 100 (14): 8520–8525. Reproduced with permission of PNAS.

in a broadly distributed array of cortical areas; masked words, in contrast, of which patients were unaware, evoked relatively weaker activity, and only in a narrow set of regions that is automatically engaged by the visual presentation of words.

Recurrent Processing Theory

Victor Lamme, whose group's empirical work is illustrated in *Figure 20.7*, has assembled a theoretical account from his analysis of the neurophysiology of visual perception:

> When a new image hits the retina, it is processed through successive levels of visual cortex, by means of feedforward connections, working at an astonishing speed. Each level takes only 10 ms of processing, so that in about 100–150 ms the whole brain "knows" about the new image before our eyes, and potential motor responses are prepared. From the very first action potentials that are fired, neurons exhibit complex tuning properties such as selectivity for motion, depth, colour or shape, and even respond selectively to faces. Thus, the feedforward sweep enables a rapid extraction of complex and meaningful features from the visual scene, and lays down potential motor responses to act on the incoming information.
>
> Are we conscious of the features extracted by the feedforward sweep? Do we see a face when a face-selective neuron becomes active? It seems not. Many studies, in both humans and monkeys, indicate that no matter what area of the brain is reached by the feedforward sweep, this in itself is not producing (reportable) conscious experience. What seems necessary for conscious experience is that neurons in visual areas engage in so-called recurrent (or re-entrant or resonant) processing . . ., where high- and low-level areas interact. This would enable the widespread exchange of information between areas processing different attributes of the visual scene, and thus support perceptual grouping. . . . In addition, when recurrent interactions span the entire sensorimotor hierarchy, or involve the frontoparietal areas, potential motor responses could modify the visual responses, which would form the neural equivalent of task set, attention, etc. (Lamme, 2006, pp. 496–497).

Thus, although the Recurrent Processing and Global Workspace models both emphasize a critical role for recurrent processing, they also differ in important ways, as illustrated in *Figure 20.10*. First, by the Recurrent Processing view, activation by the feedforward sweep is not sufficient, in and of itself, to trigger conscious awareness. For example, the initial feedforward sweep of visual information could activate parietal and frontal oculomotor centers so as to trigger an eye movement toward a target without the subject being consciously aware of the stimulus (*Figure 20.10*,

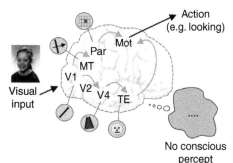

A The feedforward sweep

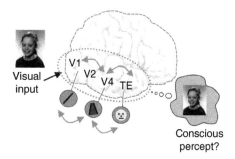

B Localized recurrent processing

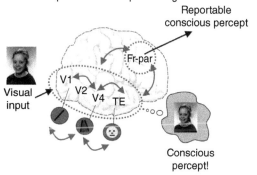

C Widespread recurrent processing

FIGURE 20.10 Three scenarios considered by Lamme's (2006) Recurrent Processing framework. In **A**, the presentation of a stimulus triggers a feedforward sweep along the ventral and dorsal processing streams of the visual system, the icons representing the features that are extracted at each level. **B** corresponds to highly interactive recurrent processing that is nonetheless restricted to the ventral stream. **C** illustrates a state of widespread recurrent processing, which is equivalent to ignition of the Global Workspace, as specified in Global Workspace Theory. Source: From Lamme, Victor A. F. 2006. "Towards a True Neural Stance on Consciousness." Trends in Cognitive Sciences 10 (11): 494–501. Reproduced with permission of Elsevier Ltd.

Panel A). Second, as illustrated in *Figure 20.10., B*, recurrent processing that is restricted to ventral-stream processing areas would be sufficient to support conscious awareness in the absence of selective attention. Examples of this type of awareness are highlighted in the arguments raised both by Lamme (2003) and Tsuchiya and Koch (2009) in the earlier section *Are attention and awareness the same thing?* Lamme has noted that the distinction between consciousness without attention (*Figure 20.10.B*) and consciousness with attention (*Figure 20.10.C*) maps onto the distinction made by philosopher Ned Block (1996) between "phenomenal awareness" and "access awareness."

An important distinction between the Recurrent Processing and Global Workspace models, therefore, is that the former allows for quantitatively graded levels of consciousness, whereas, for the latter, consciousness is an all-or-none proposition. Recurrent Processing Theory also argues, based on neurophysiological evidence, that there can be states in which we are aware of information, but this information is not accessible to overt report. This relates to Lamme's (2006) argument that when behavioral/psychological evidence (e.g., a subject's overt report or performance, or even one's subjective experience) diverges from neurophysiological evidence, the latter should be given equal weight with regard to evidence for conscious awareness of the stimulus in question. Otherwise, he reasons, our construal of consciousness reduces to the cognitive functions necessary for behavior and/or subjective experience – language, motor control, attention, and working memory – and we are no longer studying consciousness per se.

Important challenges for the Recurrent Processing model include specifying what it is about recurrent processing, per se, that gives rise to conscious awareness. Is the activation of certain layers of cortex critical, or the synchronized coactivation of two or more specific layers? Are there different neurotransmitter receptors whose activation is crucial (e.g., AMPA receptors by feedforward glutamate signaling vs. NMDA receptors by feedback)? What are the implications for consciousness of recurrent processing in brain regions whose activity seems to be inaccessible to overt awareness on the part of patients, such as in the right hemisphere of split-brain patients, or in the right hemisphere of patients suffering from unilateral neglect?

Integrated Information Theory

To convey the gist of his Integrated Information Theory (IIT), Giulio Tononi (2008) invites us to consider the following:

You are facing a blank screen that is alternately on and off, and you have been instructed to say "light" when the screen turns on and "dark" when it turns off. A photodiode – a simple light-sensitive device – has also been placed in front of the screen . . . The first problem of consciousness reduces to this: when you distinguish between the screen being on or off, you have the subjective experience of seeing light or dark. The photodiode can also distinguish between the screen being on or off, but presumably it does not have a subjective experience of light and dark. What is the key difference between you and the photodiode?

. . . the difference has to do with how much information is generated when that distinction is made. Information is classically defined as reduction of uncertainty: the more numerous the alternatives that are ruled out, the greater the reduction of uncertainty, and thus the greater the information . . . When the blank screen turns on, . . . the photodiode . . . reports "light." . . . When you see the blank screen turn on, on the other hand, the situation is quite different . . . [because] you are in fact discriminating among a much larger number of alternatives, thereby generating many more bits of information.

This is easy to see. Just imagine that, instead of turning light and dark, the screen were to turn red, then green, then blue, and then display, one after the other, every frame from every movie that was ever produced. The photodiode, inevitably, would go on signaling whether the amount of light for each frame is above or below its threshold: to a photodiode, things can only be one of two ways . . . For you, however, a light screen is different not only from a dark screen, but from a multitude of other images, so when you say "light," it really means this specific way versus countless other ways, such as a red screen, a green screen, a blue screen, this movie frame, that movie frame, and so on for every movie frame (not to mention for a sound, smell, thought, or any combination of the above). Clearly, each frame looks different to you, implying that some mechanism in your brain must be able to tell it apart from all the others. So when you say "light," whether you think about it or not (and you typically won't), you have just made a discrimination among a very large number of alternatives (pp. 217–218).

And so, according to this theory, effective connectivity is key, because the number of potential alternative states that the brain can take on determines the amount of information contained in the brain's present state. Tononi calls this measure of integrated information "phi," and quantitative models for computing the value of phi can be found in Balduzzi and Tononi (2008; 2009). Many of the facts and findings that were reviewed in the first section of this chapter can be interpreted from the perspective of Integrated Information Theory. Instead of running through these, however, we'll conclude with a set of reanalyses of data from some of these, plus other studies, that demonstrate how something so seemingly abstract and recondite as a quantitative theory of consciousness can have practical applications.

A practical application of Integrated Information Theory

A problem that we considered earlier in this chapter is determining whether patients diagnosed with UWS truly lack all conscious awareness. A multinational team co-led by Brazilian physicist-turned-neuroscientist Adenauer Casali and Belgian neuropsychologist Olivia Gosseries, together with colleagues based in Liège, Milan, and Madison, has begun to address this problem by applying the following rationale:

> Phenomenologically, each conscious experience is both differentiated – that is, it has many specific features that distinguish it from a large repertoire of other experiences – and integrated – that is, it cannot be divided into independent components. Neurophysiologically, these fundamental properties of subjective experience rely on the ability of multiple, functionally specialized areas of the thalamocortical system to interact rapidly and effectively to form an integrated whole . . . Hence, an emerging idea in theoretical neuroscience is that consciousness requires an optimal balance between functional integration and functional differentiation in thalamocortical networks – otherwise defined as brain complexity . . . This kind of complexity should be high when consciousness is present and low whenever consciousness is lost in sleep, anaesthesia or coma (Casali, Gosseries, et al., 2013, n.p.).

To assay IIT-inspired complexity in real brains, Casali Gosseries et al. (2013) developed a measure inspired by phi, called the Perturbational Complexity Index (PCI). From one perspective, it's an approach that's similar in spirit to the "dimensional activation" estimate of complexity in the EEG that was used by Velly et al. 2007 (*Figure 20.4*). However, because the PCI is derived from TMS-evoked signal in the EEG, it is (a) potentially more sensitive (because perturbation of a system with an exogenous force can sometimes reveal properties of a system that aren't evident just by observing it) and (b) not as susceptible to variability in the state of the brain at the time of measurement (because it is computed as a difference score between baseline EEG and the TMS-evoked response in the EEG). (For example, if an individual is experiencing pain or not experiencing pain at the time of EEG measurement, these measurements will be quite different. However, $[\text{TMS/EEG}_{pain} - \text{EEG}_{pain}]$ is likely to be comparable to $[\text{TMS/EEG}_{no\ pain} - \text{EEG}_{no\ pain}]$.) "[O]perationally, PCI is defined as the normalized Lempel–Ziv complexity of the spatiotemporal pattern of cortical activation triggered by a direct TMS perturbation," and those who want to dig into the details are directed to Casali Gosseries, et al. (2013).

Figure 20.11 illustrates the PCI derived from patients suffering from the neurological syndromes that we reviewed in the first section of this chapter, as well as from individuals in different states of consciousness produced by

FIGURE 20.11 The PCI applied to anesthesia, sleep, and neurological syndromes of consciousness. Source: From Casali, Adenauer G., Olivia Gosseries, Mario Rosanova, Mélanie Boly, Simone Sarasso, Karina R. Casali, Marcello Massimini. 2013. "A Theoretically Based Index of Consciousness Independent of Sensory Processing and Behavior." Science Translational Medicine 5 (198): 198ra105. Reproduced with permission of American Association for the Advancement of Science - AAAS.

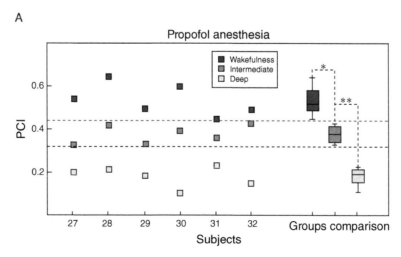

FIGURE 20.11.A PCI is sensitive to graded changes in the level of propofol anesthesia; data from six individual subjects.

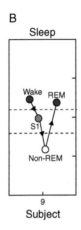

FIGURE 20.11.B PCI is sensitive to changes in sleep state, as illustrated in a single subject transitioning from wakefulness to sleep stage N1 (S1) to non-REM sleep to REM sleep.

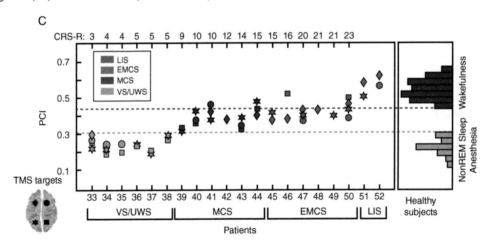

FIGURE 20.11.C PCI, estimated from TMS delivered to each of the four locations indicated on the legend, discriminates vegetative state/unresponsive wakefulness syndrome (VS/UWS) from minimally conscious state (MCS) and emerging-from-MCS (EMCS) from locked-in syndrome (LIS). For comparison, to the right, are histograms showing distribution of PCI scores from healthy individuals in non-REM sleep or under deep general anesthesia, and in wakefulness.

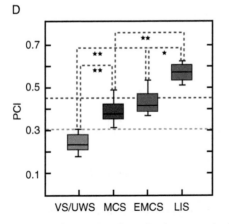

FIGURE 20.11.D Group data from the four neurological groups.

anesthesia or by sleep. Perhaps most impressively, and most promisingly, it shows the PCI to be sensitive to different levels of consciousness, as associated, for example, with pre-anesthesia wakefulness vs. intermediate levels of anesthesia vs. deep levels. This holds out the promise of a clinically practical procedure, and one far less expensive than fMRI, to objectively, quantitatively, confirm the diagnosis, for example, of persistent vegetative state/UWS.

UPDATING THE CONSCIOUSNESS GRAPH

At the conclusion of this chapter, although we haven't definitively answered the question *What is the Biological Basis of Consciousness?* (not to mention the Origin of the

Universe; nor, for that matter, the Meaning of Life . . .), we are, at least, in a position to put some more satisfactory labels on the vertical axis of our two-dimensions-of-consciousness graph (*Figure 20.2*): the level of awareness seems to depend on connectivity, particularly of lateral and feedback projections, and the complexity of cortical activity

that these connections underlie. Of course, there's so much more that we don't yet understand. But amazing discoveries about the neural bases of cognition are being made practically every day. And who knows? Perhaps, gentle reader, it will be you who one day solves the puzzle of the biological bases of consciousness!

END-OF-CHAPTER QUESTIONS

1. What factors – physiological and behavioral – differentiate coma from persistent vegetative state/unresponsive wakefulness syndrome? From locked-in syndrome?

2. Describe the role of the reticular activating system in physiological arousal. How does its activity vary in wakefulness vs. non-REM vs. REM sleep?

3. Are the effects of general anesthesia – physiologically and behaviorally – more akin to coma or to sleep? What physiological change occurs abruptly at the point of anesthesia-induced loss of consciousness?

4. How have scientists addressed the question of whether activity in early sensory cortex more closely reflects

events in the outside world or our subjective experience of events in the outside world? And what have they learned?

5. Two sets of studies have emphasized two distinct kinds of physiological response that seem to correspond to whether or not one will be consciously aware of a visually presented stimulus; summarize these.

6. What are the key tenets of Global Workspace Theory? Of Recurrent Processing Theory? Of Integrated Information Theory? Is it possible that more than one of these could be "right"?

7. What are the biological bases of consciousness?

REFERENCES

Alkire, T. Michael. 2009. "General Anaesthesia and Consciousness." In *The Neurology of Consciousness: Cognitive Neuroscience and Neuropathology*, edited by Steven Laureys and Giulio Tononi (pp. 118–134). Amsterdam: Elsevier.

Balduzzi, David, and Giulio Tononi. 2008. "Integrated Information in Discrete Dynamical Systems: Motivation and Theoretical Framework." *PLoS Computational Biology* 4 (6): e1000091. doi: 10.1371/journal.pcbi.1000091.

Balduzzi, David, and Giulio Tononi. 2009. "Qualia: The Geometry of Integrated Information." *PLoS Computational Biology* 5 (8): e1000462. doi: 10.1371/journal.pcbi.1000462.

Bauby, Jean-Dominique. 1997. *The Diving Bell and the Butterfly*. Translated by Jeremy Leggatt. New York: Knopf.

Block, Ned. 1996. "How Can We Find the Neural Correlate of Consciousness?" *Trends in Neurosciences* 19 (11): 456–459. doi: 10.1016/S0166-2236(96)20049-9.

Casali, Adenauer G., Olivia Gosseries, Mario Rosanova, Mélanie Boly, Simone Sarasso, Karina R. Casali, . . . Marcello Massimini. 2013. "A Theoretically Based Index of Consciousness Independent of Sensory Processing and Behavior." *Science Translational Medicine* 5 (198): 198ra105. doi: 10.1126/scitranslmed.3006294.

Chelazzi, Leonardo, John Duncan, Earl K. Miller, and Robert Desimone. 1998. "Responses of Neurons in Inferior Temporal Cortex during Memory-Guided Visual Search." *Journal of Neurophysiology* 80 (6): 2918–2940. http://wexler.free.fr/library/files/chelazzi%20(1998)%20responses%20of%20

neurons%20in%20inferior%20temporal%20cortex%20during%20memory-%20guided%20visual%20search.pdf.

Crick, Francis, and Christof Koch. 1990. "Towards a Neurobiological Theory of Consciousness." *Seminars in the Neurosciences* 2: 263–275. http://authors.library.caltech.edu/40352/1/148.pdf.

Dehaene, Stanislas, Lionel Naccache, Laurent Cohen, Denis Le Bihan, Jean-François Mangin, Jean-Baptiste Poline, and Denis Rivière. 2001. "Cerebral Mechanisms of Word Masking and Unconscious Repetition Priming." *Nature Neuroscience* 4 (7): 752–758. doi: 10.1038/89551.

Dehaene, Stanislas, Claire Sergent, and Jean-Pierre, Changeux. 2003. "A Neuronal Network Model Linking Subjective Reports and Objective Physiological Data during Conscious Perception." *Proceedings of the National Academy of Sciences of the United States of America* 100 (14): 8520–8525. doi: 10.1073/pnas.1332574100.

Ferrarelli, Fabio, Marcello Massimini, Simone Sarasso, Adenauer Casali, Brady A. Riedner, Giuditta Angelini, . . . Robert A. Pearce. 2010. "Breakdown in Cortical Effective Connectivity during Midazolam-Induced Loss of Consciousness." *Proceedings of the National Academy of Sciences of the United States of America* 107 (6): 2681–2686. doi: 10.1073/pnas.0913008107.

Hanks, Timothy D., Jochen Ditterich, and Michael N. Shadlen. 2006. "Microstimulation of Macaque Area LIP Affects Decision-Making in a Motion Discrimination Task." *Nature Neuroscience* 9 (5): 682–689. doi: 10.1038/nn1683.

Heathcote, Elizabeth. 2011. "Emery Brown: 'Aspects of Anaesthesia Are Consistent with Patients in a Coma.'" *The Observer*, April 9. http://www.theguardian.com/technology/2011/apr/10/anaesthesia-coma-sleep-emery-brown.

Koch, C., Massimini, M., Boly, M., & Tononi, G. 2016. Neural Correlates of Consciousness: Progress and Problems. *Nature Reviews Neuroscience* 17: 307–321.

Lamme, Victor A. F. 2003. "Why Visual Attention and Awareness Are Different." *Trends in Cognitive Sciences* 7 (1): 12–18. doi: 10.1016/S1364-6613(02)00013-X.

Lamme, Victor A. F. 2006. "Towards a True Neural Stance on Consciousness." *Trends in Cognitive Sciences* 10 (11): 494–501. doi: 10.1016/j.tics.2006.09.001.

Laureys, Steven. 2005. "The Neural Correlate of (Un)Awareness: Lessons from the Vegetative State." *Trends in Cognitive Sciences* 9 (12): 556–559. doi: 10.1016/j.tics.2005.10.010.

Laureys, Steven, Gastone G. Celesia, Francois Cohadon, Jan Lavrijsen, José León-Carrión, Walter G. Sannita, . . . European Task Force on Disorders of Consciousness. 2010. "Unresponsive Wakefulness Syndrome: A New Name for the Vegetative State or Apallic Syndrome." *BMC Medicine* 8 (1): 68. doi: 10.1186/1741-7015-8-68.

Laureys, Steven, Frédéric Pellas, Philippe Van Eeckhout, Sofiane Ghorbel, Caroline Schnakers, Favien Perrin, . . . Serge Goldman. 2005. "The Locked-In Syndrome: What Is It Like to Be Conscious but Paralyzed and Voiceless?" *Progress in Brain Research* 150: 495–511. doi: 10.1016/S0079-6123(05)50034-7.

Laureys, Steven and Giulio Tononi, eds. 2009. *The Neurology of Consciousness*. Amsterdam: Elsevier.

Li, Fei Fei, Rufin VanRullen, Christof Koch, and Pietro Perona. 2002. "Rapid Natural Scene Categorization in the Near Absence of Attention." *Proceedings of the National Academy of Sciences of the United States of America* 99 (14): 9596–9601. doi: 10.1073/pnas.092277599.

Massimini, Marcello, Fabio Farrarelli, Reto Huber, Steve K. Esser, Harpreet Singh, and Giulio Tononi. 2005. "Breakdown of Cortical Effective Connectivity during Sleep." *Science* 309 (5744): 2228–2232. doi: 10.1126/science.1117256.

Mathewson, Kyle E., Alejandro Lleras, Diane M. Beck, Monica Fabiani, Tony Ro, and Gabriele Gratton. 2011. "Pulsed Out of Awareness: EEG Alpha Oscillations Represent a Pulsed-Inhibition of Ongoing Cortical Processing." *Frontiers in Psychology* 2: 99. doi: 10.3389/fpsyg.2011.00099.

Monti, Martin M., Audrey Vanhaudenhuyse, Martin R. Coleman, Melanie Boly, John D. Pickard, Luaba Tshibanda, . . . Steven Laureys. 2010. "Willful Modulation of Brain Activity in Disorders of Consciousness." *New England Journal of Medicine* 362 (7): 579–589. doi: 10.1056/NEJMoa0905370.

Naccache, Lionel, Elise Blandin, and Stanislas Dehaene. 2002. "Unconscious Masked Priming Depends On Temporal Attention." *Psychological Science* 13 (5): 416–424. doi: 10.1111/1467-9280.00474.

Owen, Adrian M., Martin R. Coleman, Melanie Boly, Matthew H. Davis, Steven Laureys, and John D. Pickard. 2006. "Detecting Awareness in the Vegetative State." *Science* 212 (5792): 1402. doi: 10.1126/science.1130197.

Plum, Fred, and Jerome B. Posner. 1983. *The Diagnosis of Stupor and Coma*. Philadelphia, PA: F. A. Davis.

Ress, David, and David J. Heeger. 2003. "Neural Correlates of Perception in Early Visual Cortex." *Nature Neuroscience* 6 (4): 414–420. doi: 10.1038/nn1024.

Romei, Vincenzo, Verena Brodbeck, Christoph Michel, Amir Amedi, Alvaro Pascual-Leone, and Gregor Thut. 2008. "Spontaneous Fluctuations in Posterior [alpha]-Band EEG Activity Reflect Variability in Excitability of Human Visual Areas." *Cerebral Cortex* 18 (9): 2010–2018. doi:10.1093/cercor/bhm229.

Rosanova, Mario, Olivia Gosseries, Silvia Casarotto, Mélanie Boly, Adenauer G. Casali, Marie-Aurélie Bruno, . . . Marcello Massimini. 2012. "Recovery of Cortical Effective Connectivity and Recovery of Consciousness in Vegetative Patients." *Brain* 135 (4): 1308–1320. doi: 10.1093/brain/awr340 Brain 2012: 135; 1308–1320.

Soon, Chun Siong, Marcel Brass, Hans-Jochen Heinze, and John-Dylan Haynes. 2008. "Unconscious Determinants of Free Decisions in the Human Brain." *Nature Neuroscience* 11 (5): 543–545. doi: 10.1038/nn.2112.

Sperling, George. 1960. "The Information Available in Brief Visual Presentations." *Psychological Monographs* 74 (11): 1–29. doi: 10.1037/h0093759.

Supèr, Hans, Henk Spekreijse, and Victor A. F. Lamme. 2001. "Two Distinct Modes of Sensory Processing Observed in Monkey Primary Visual Cortex (V1)". *Nature Neuroscience* 4 (3): 304–310. doi: 10.1038/85170.

Tononi, Giulio. 2008. "Consciousness as Integrated Information: A Provisional Manifesto." *Biological Bulletin* 215 (3): 216–242. doi: 10.2307/25470707.

Tsuchiya, Naotsugu and Christof Koch. 2009. "The Relationship Between Consciousness and Attention." In *The Neurology of Consciousness: Cognitive Neuroscience and Neuropathology,* edited by Steven Laureys and Giulio Tononi (pp. 63–77). Amsterdam: Elsevier.

Velly, Lionel J., Marc F. Rey, Nicolas J. Bruder, François A. Gouvitsos, Tatiana Witjas, Jean Marie Regis, . . . François M. Gouin. 2007. "Differential Dynamic of Action on Cortical and Subcortical Structures of Anesthetic Agents during Induction of Anesthesia." *Anesthesiology* 107 (2): 202–212. doi: 10.1097/01.anes.0000270734.99298.b4.

Vigand, Phillipe, and Stephane Vigand. 2000. *Only the Eyes Say Yes: A Love Story*. English language edition. New York: Arcade.

Young, G. Brian. 2009. "Coma." In *The Neurology of Consciousness: Cognitive Neuroscience and Neuropathology*, edited by Steven Laureys and Giulio Tononi (pp. 137–150). Amsterdam: Elsevier.

OTHER SOURCES USED

Baars, Bernard J. 2005. "Global Workspace Theory of Consciousness: Toward a Cognitive Neuroscience of Human Experience?" *Progress in Brain Research* 150: 45–53. doi: 10.1016/S0079-6123(05)50004-9.

Boly, M., M. Massimini, N. Tsuchiya, B. R. Postle, C. Koch, , and G. Tononi. 2017. Are the Neural Correlates of Consciousness in the Front or in the Back of the Cerebral Cortex? *The Journal of Neuroscience*, 37: 9603–9613.

Gosseries, Olivia, Marie-Aurélie Bruno, Audrey Vanhaudenhuyse, Steven Laureys, and Caroline Schnakers. 2009. "Consciousness in the Locked-In Syndrome." In *The Neurology of Consciousness: Cognitive Neuroscience and Neuropathology*, edited by Steven Laureys and Giulio Tononi (pp. 191–203). Amsterdam: Elsevier.

Lau, Hakwan. 2011. "Theoretical Motivations for Investigating the Neural Correlates of Consciousness." *WIREs Cognitive Science* 2 (1): 1–7. doi: 10.1002/wcs.93.

Laureys, Steven, Maryliese E. Faymonville, André Luxen, Maurice Lamy, George Franck, and Pierre Maquet. 2000. "Restoration of Thalamocortical Connectivity after Recovery from Persistent Vegetative State." *Lancet* 355 (9217): 1790–1791. doi: 10.1016/S0140-6736(00)02271-6.

Laureys, Steven, Christian Lemaire, Pierre Maquet, Christophe Phillips, and George Franck. 1999. "Cerebral Metabolism during Vegetative State and after Recovery to Consciousness." *Journal of Neurology, Neurosurgery, and Psychiatry* 67 (1): 121–122. doi: 10.1136/jnnp.67.1.121.

Shannon, Claude E. 1948. "A Mathematical Theory of Communication." *Bell System Technical Journal* 27 (3): 379–423. doi: 10.1002/j.1538-7305.1948.tb01338.x.

Tononi, Giulio. 2009. "Sleep and Dreaming." In *The Neurology of Consciousness: Cognitive Neuroscience and Neuropathology*, edited by Steven Laureys and Giulio Tononi (pp. 89–107). Amsterdam: Elsevier.

VanRullen, Rufin, Niko A. Busch, Jan Drewes, and Julien Dubois. 2011. "Ongoing EEG Phase as a Trial-by-Trial Predictor of Perceptual and Attentional Variability." *Frontiers in Psychology* 2. doi: 10.3389/fpsyg.2011.00060.

FURTHER READING

Blake, Randolph, and Frank Tong. 2008. "Binocular Rivalry." *Scholarpedia* 3 (12): 1578. doi: 10.4249/scholarpedia.1578.
An authoritative and succinct summary of the history of the study of this intriguing phenomenon.

Brown, Emery N., Ralph Lydic, and Nicholas D. Schiff. 2010. "General Anesthesia, Sleep, and Coma." *New England Journal of Medicine* 363 (27): 2638–2650. doi: 10.1056/NEJMra0808281.
Authoritative review by prominent clinician-scientists focusing on the neural mechanisms of unconsciousness induced by selected intravenous anesthetic drugs.

Crick, Francis. 1994. *The Astonishing Hypothesis: The Scientific Search for the Soul.* New York: Scribner.
Francis Crick's influential book is among the publications that made consciousness a respectable topic for study by mainstream cognitive neuroscientists.

Crick, Francis. 1996. "Visual Perception: Rivalry and Consciousness." *Nature* 379 (6565): 485–486. doi: 10.1038/379485a0.
A succinct statement of why and how the study of binocular rivalry has important implications for understanding the neural correlates of consciousness.

Engel, Andreas K., Pascal Fries, Peter König, Michael Brecht, and Wolf Singer. 1999. "Temporal Binding, Binocular Rivalry, and Consciousness." *Consciousness and Cognition* 8 (2): 128–151. doi: 10.1006/ccog.1999.0389.
Summary of the influential view that oscillatory synchrony among anatomically distal brain areas is of central importance to conscious experience; also reviews binocular rivalry, a phenomenon that is often used in the study of visual awareness.

Laureys, Steven, Olivia Gosseries, and Giulio Tononi, eds. 2015. *The Neurology of Consciousness.* 2nd Edition. Amsterdam: Elsevier.
A volume containing chapters by a proverbial "veritable who's-who" of scientific and clinical experts on disorders of consciousness.

Leopold, David A., and Nikos K. Logothetis. 1996. "Activity Changes in Early Visual Cortex Reflect Monkeys' Percepts during Binocular Rivalry." *Nature* 379 (6565): 549–553. doi: 10.1038/379549a0.
An empirical paper that Crick (1996) declared "among the opening salvos of a concerted attack on the baffling problem of consciousness."

Nir, Yuval, and Giulio Tononi. 2010. "Dreaming and the Brain: From Phenomenology to Neurophysiology." *Trends in Cognitive Sciences* 14 (2): 88–100. doi: 10.1016/j.tics.2009.12.001.
Overview of this fascinating arena from which to study consciousness.

Seager, William. 2007. "A Brief History of the Philosophical Problem of Consciousness." In The Cambridge *Handbook of Consciousness*, edited by Philip David Zelazo, Morris Moscovitch, and Evan Thompson (pp. 9–33). New York: Cambridge University Press.
This chapter offers a nice overview for the nonphilosopher.

Tononi, Giulio. 2012. *Phi: A Voyage from the Brain to the Soul*. New York: Pantheon Books.
A literary, fantastical dream in which Galileo, alternately guided by Francis Crick, Alan Turing, and Charles Darwin, encounters a variety of neurological patients whose syndromes define the parameters of consciousness.

GLOSSARY

acoustic energy The movement of molecules of the gases that make up the air that's all around us; or, when underwater, collisions between and movement of molecules of water.

action An isolated movement, or a coordinated series of movements intended to achieve some behavioral goal.

affective neuroscience The study of the neural bases of emotions. Because disorders of emotional processing are often characteristic of, if not at the root of, many types of psychopathology (e.g., depression, addiction, generalized anxiety disorder), there is considerable overlap between affective neuroscience and clinical psychology. (*Affective* is the adjectival form of the noun *affect*, a synonym for *emotion*.)

allocentric localization The representation of space in an environment-centered coordinate system (e.g., *The cup is on the table*).

AMPA receptor Glutamate-gated sodium channel.

anatomical specificity This refers to the degree to which a particular effect on behavior is attributable to only a particular anatomical region.

anions A general term for negatively charged ions.

anomia An inability to produce the appropriate verbal label for objects (i.e., to "name" things).

anoxia Deprivation of oxygen to brain tissue.

anterior Used as a term of direction, it refers to the front (a synonym, introduced in *Chapter 2*, is "rostral").

antisaccade task A task requiring a speeded saccade in the opposite direction of a temporally unpredictable target; it requires cognitive control to overcome the prepotent tendency to saccade to unexpected appearance of stimuli in the environment.

aperture problem The small size of receptive fields at early stages of the visual system can result in ambiguities when local information present within the aperture is inconsistent with global information in the scene.

aphasia An inability to use language following brain damage.

attention The process of prioritizing the processing of a subset of the information that is available at any given moment in time.

attention deficit hyperactivity disorder A neurological condition associated with difficulty sustaining attention, difficulty controlling behavior, and, in some cases, hyperactivity.

auditory nerve The bundle of axons that carry auditory information from the cochlear nucleus to the brainstem auditory nuclei.

autism spectrum disorder A developmental disorder characterized by deficits in social communication and social interaction, and often accompanied by restricted, repetitive patterns of behavior, interests, or activities.

avoidance learning Context-dependent skeletomotor behaviors that, when engaged, enable an organism to avoid the adverse outcome predicted by a conditioned stimulus (CS).

axons The branches through which a neuron sends signals to other neurons. Most axons have many "collateral" arms splitting off from the main branch, thereby enabling a single neuron to influence hundreds, thousands, and in some cases tens of thousands of other neurons.

axon hillock Region of the cell body from which the axon emerges. The membrane of the hillock is dense with voltage-gated Na^+ channels, such that it can powerfully initiate action potentials.

axonal transport Intracellular mechanisms for transporting proteins from the cell body to synapses.

barrel cortex The region of rodent somatosensory cortex containing groupings of neurons in layer IV – "barrels" – that, together, represent somatosensory signals coming from a single whisker.

basilar membrane The membrane within the cochlea whose vibrations effect the spectral transformation of sound, which is subsequently encoded in the cochlear nucleus's place code. The base of the basilar membrane is the portion nearest the oval window, and is maximally displaced by high frequencies. The apex is the "far end" of the basilar membrane, and is maximally displaced by low frequencies.

binding problem The fact that anatomically discrete parts of the brain encode what an object is vs. where it is, yet our perception is of an object occupying (i.e., "bound to") a location.

binocular rivalry The phenomenon that when a different image is presented to each eye, one sees only one of the two at any given point of time, and the perceived image alternates, stochastically, between the two.

blood–brain barrier The layer of endothelial cells lining the inside walls of blood vessels that prevents the diffusion from capillaries of many substances carried in the blood stream.

blood oxygen level-dependent (BOLD) signal The signal most commonly detected by fMRI, measured via T2★ relaxation.

brain The organ of the body responsible for the control of all other systems of the body, and thus for the control of behavior. In technical, anatomical terms, the brain is comprised of the brainstem, cerebellum, and cerebrum (see *Chapter 2*).

brainstem auditory nuclei The many subcortical structures that process the auditory signal prior to the inferior colliculus, medial geniculate nucleus, and primary auditory cortex.

brainstem oculomotor nuclei Neurons located at the rostrocaudal level of the superior colliculus whose axons innervate the muscles that control the position and movements of the eyes (the "extraocular muscles").

cations A general term for positively charged ions.

central nervous system Comprises the brain plus the spinal cord and the motor neurons that leave the cord to innervate the muscles of the body.

cerebellum The highly folded structure that lies "below" the "back" of the cerebral cortex. (See *Figure 1.2*; correct terms of direction are introduced in *Chapter 2*.)

cerebral hemisphere The *cerebrum* is the largest part of the brain (in volume), sitting atop the brainstem. It is divided into two roughly symmetrical halves, each referred to as a hemisphere.

cerebrospinal fluid (CSF) The fluid that circulates through the ventricles (thereby inflating the brain) and also along the exterior surface of the CNS (thereby providing a medium in which the brain floats).

cerebrum Also referred to as the forebrain, the cerebrum is the anterior-most part of the brain, and largest in volume. Deriving from the embryonic telencephalon, it comprises the thalamus plus all the tissue that is rostral to it. Major elements include the five lobes of cerebral cortex, basal ganglia, hippocampus, thalamus, and all the white matter connecting these structures.

choice probability A measure derived by relating variations in neuronal activity evoked by sensory signals to variations in behavioral choices prompted by those same sensory signals.

cognition Thinking. This word derives from Greek and Latin words meaning "to know" or "to learn" and, in some translations, "to recognize."

cognitive neuroscience The studies of the neural bases (e.g., anatomical, physiological, pharmacological, genetic) of behavior that have traditionally fallen under the purview of cognitive psychology.

cognitive psychology Cognitive psychology is the study of the information processing that intervenes between sensory inputs and motoric outputs. Domains of mental activity and behavior that it covers include those covered in this book, beginning with *Chapter 4*.

conjugate eye movements Movements of the two eyes in parallel (i.e., with the same vector).

consolidation In the context of long-term memory, consolidation is the hypothesized process whereby, over time, memories become more resistant to disruption.

construct validity the extent to which there is a principled basis for asserting the existence of a psychological, or physiological, construct.

contralateral delay activity (CDA) An ERP component that scales with the amount of information being held in visual working memory, and that asymptotes at the subject's estimated capacity.

convolution (*n.*, to convolve, *v.*) The transformation of a function (e.g., a time-varying signal) that results from passing it through a filter.

coordinate transformations The conversion of spatial information from one reference frame to another, such as when re-representing a location initially encoded in eye-centered coordinates into head-centered coordinates.

corpus callosum (Latin for "tough body," presumably because, relative to other structures, it is difficult to cut through or pull apart.) The corpus callosum is made up of millions of axons that project between neurons in the two hemispheres.

cortex Refers to the layer of tissue containing cell bodies, and thus is often synonymous with "gray matter." Cortex is the Latin name for "rind," and thus the name references the outer covering of many types of fruit, such as melons or citrus fruit. The cerebrum and cerebellum each have a cortex.

cortical column A group of neurons that are stacked, as though in an imaginary tube that is oriented orthogonal to the cortical surface. Neurons within a cortical column are more strongly interconnected, and, therefore, share more functional properties, than are physically adjacent neurons that are not within the same column.

cortical magnification The phenomenon that higher-resolution regions of a sensory-transduction surface, or more finely controlled muscles, are represented by larger areas of cortex.

cotransport Process whereby more than one molecule is simultaneously carried across a cell membrane, as with reuptake.

covert attention Attention deployed toward an object or location without looking directly at it.

craniotomy Surgical procedure of removing a portion of the cranium (skull) in order to expose a portion of the underlying brain tissue.

decision variable A hypothesized representation of all the accruing information related to a decision.

declarative memory Long-term memory that entails conscious retrieval of information from one's past. Synonymous with *explicit memory*.

deep neural networks (DNNs) A class of artificial neural networks with a feedforward architecture and multiple serially connected hidden layers, which are particularly adept at classification (e.g., of visual images, or of speech sounds).

degrees of visual angle When foveating an object in the environment, the surface area of retina that that object occupies can be expressed as an angle anchored in the fovea, and whose imaginary sides project to the left and right edges of the object. The larger the object, the wider the degrees of visual angle that it occupies.

delay discounting The tendency to assign less value to a reward if its delivery will occur at some point in the future, rather than immediately.

delayed match-to-sample (DMS) A recognition memory test in which memory for a previously presented sample stimulus is assessed at test by re-presenting it along with one or more foil stimuli.

delayed recognition A recognition memory test in which memory for one or more previously presented sample stimuli are assessed at test by decision of whether or not a single probe item was presented as a sample.

dementia A progressive impairment of intellect and behavior that, over time, dramatically alters and restricts normal activities of daily living. Although typically implicating memory, dementia differs from amnesia in that it also affects at least one other domain of cognition (e.g., language, motor control, visual perception, . . .).

dendrites The branches of a neuron that receive signals from other neurons (*Figure 2.5*).

depolarize To make the membrane potential of a cell more positive. Typically, because the resting state of neurons is negative, this is equivalent to making the membrane potential less negative.

Diencephalon also commonly referred to as the thalamus. Large collection of nuclei located near the center of the cerebrum, straddling the midline and immediately rostral (see *Figure 1.2*) to the midbrain. Many thalamic nuclei serve as relay points between peripheral sensory information and the cerebral cortex.

diffusion tensor imaging A type of MRI that measures the diffusion of water to infer information about white-matter tracts.

dipole An object with a positive charge at one end and a negative charge at the other.

distributed system A system in which functions arise from the contribution of many processing elements performing simple computations (i.e., carrying out simple functions) that often bear little surface resemblance to the function in question. These elements may or may not be anatomically distant from each other.

double dissociation of function A pattern of results in which damage to area A affects performance on task X, but not on task Y; whereas damage to area B affects performance on task Y, but not on task X.

dynamics The forces and torques that produce movement or motion.

dysphasia An impairment in the ability to communicate following brain damage (compare with *aphasia*).

efference copy A copy of the motor command that is carried by a recurrent circuit.

egocentric localization The representation of space in a body-centered coordinate system (e.g., *The cup is to my right*).

electrocorticography (ECoG) Electrical recordings made from the surface of the cortex.

emergent property A property that is characteristic of a system, but that is not reducible to the properties of the system's constituent elements.

encoding In the context of memory, encoding refers to the processing of newly encountered information such that it can be stored and later retrieved.

engram The hypothetical physical representation of a memory in the brain.

ensembles From the French for "together," this term is often used to refer to a group of neurons that are "working together" as a functional unit. The analogy is to a musical ensemble, such as a string quartet.

epilepsy A chronic condition of repeating seizures.

epileptogenic Brain tissue with a low threshold for initiating seizure activity.

episodic memory Declarative memory that includes a representation of the context (e.g., place, situation, who was present) at the time that encoding of that information occurred.

equipotentiality The idea that any given piece of cortical tissue has the potential to support any brain function.

etiology The cause of a disease.

excitatory neurotransmitter A neurotransmitter whose effect on the postsynaptic membrane is to depolarize it.

explicit memory Long-term memory that entails conscious retrieval of information from one's past. Synonymous with *declarative memory*.

express saccades A very short-latency saccade, typically triggered by the abrupt appearance of an unexpected stimulus.

extinction A phenomenon often accompanying unilateral neglect, whereby a patient can be aware of a single object presented on their left, but will lose awareness of it if a second item is then presented on their right.

extracellular recording Electrical recordings made from an electrode located in extracellular space. Such recordings can be sensitive to action potentials and/or LFPs.

extrapersonal space Those portions of perceived space that are further than we can reach.

extrastriate cortex Cortex that is part of the visual system, but not a part of striate cortex (i.e., it is beyond visual cortex).

familiarity The subjective experience that accompanies correctly recognizing an item on a recognition memory test (or, in real life, maybe a face on the bus) without a memory for the context in which that item was previously experienced. Hypothesized in some models to correspond to a neural signal and/or process that is qualitatively different from recollection.

feature detector An element in a sensory processing array whose function is to signal each time it detects the presence of the feature for which it is tuned.

fMRI adaptation The experimental procedure of habituating the response of a region to repeated exposure to a stimulus (or class of stimuli), then varying the stimuli by a dimension of interest and measuring whether there is a "release from adaptation," which would indicate that the region in question is sensitive to change along the dimension being varied.

foraging decision The more-or-less-continual decision of whether to persist in one's current mode of behavior or whether to change to a different mode of behavior. Compare with *perceptual decision* and *value-based decision*.

formants Bands of sustained, elevated oscillatory power in a spectrogram of speech.

fornix The bundle of fibers leaving the hippocampus, initially in the caudal direction, and arcing dorsally and rostrally to synapse on targets in the hypothalamus and basal forebrain. It is generally understood that these fibers are important for maintaining the physiological state of the hippocampus (e.g., maintenance and regulation of hippocampal theta rhythm) rather than carrying the "informational output" of the hippocampus, which, instead, is carried by projections to entorhinal cortex.

forward model A model that predicts the consequences of an action.

Fourier transform The translation of a time-varying wave form into constituent sine and cosine waves.

fovea Region of the retina most densely packed with photoreceptors, and thus supporting highest resolution vision.

frontal lobe The rostral-most of the four lobes of the cerebral cortex.

functional connectivity A state in which the activity of two or more regions is correlated, and they are therefore presumed to be "working together." (Note that the existence of a "structural connection," need not imply that the connected areas are functionally connected in a particular context.)

functional magnetic resonance imaging (fMRI) The application of MRI that is sensitive to physiological signals (as opposed to anatomical structure).

fundamental frequency The lowest frequency formant of an utterance.

fundus The "bottom" of a sulcus, where its two banks meet.

GABA (*adj.* GABAergic) The abbreviated name of γ-aminobutyric acid, the most common inhibitory neurotransmitter in the CNS.

gamma frequency band Neural oscillations in the range of 30–80 Hz.

ganglion (*pl. -glia*) Ganglion is a synonym for *nucleus*. The *basal ganglia* are a group of nuclei located near the base of the cerebrum, and anterior and lateral to the thalamus. Their function is considered first in *Chapter 8*.

gating The process of allowing certain signals to pass through a circuit, and preventing others.

generalized anxiety disorder (GAD) A psychiatric condition characterized by incessant, excessive worry.

gestalt psychology A school of thought, originating with the study of perception, that emphasized recognition of whole objects as being more important than feature-level processes.

glial cells Glial cells (or "glia") is the designation given to cells in the brain that are not themselves neurons, but can be construed as playing supporting roles that are necessary for the normal functioning of neurons. Two types of glia mentioned in *Chapter 2* are oligodendrocytes (responsible for myelin in the CNS) and astrocytes.

glutamate (*adj.* glutamatergic) Also referred to as "glutamic acid," glutamate is the excitatory neurotransmitter released by cortical pyramidal cells, and thus the primary chemical currency for information transmission in the brain.

grand mal epileptic seizure A seizure producing a loss of consciousness and violent muscle contractions (sometimes also referred to as tonic-clonic).

gray matter Informal, but widely used, term for the cortex. Derives from fact that the cortex lacks myelin, and thus appears darker to the naked eye than does the myelin-rich white matter.

gyri (*sing.* gyrus) From the Latin for "circles," convex ridges on the surface of the cerebral cortex. Each gyrus is bounded by sulci on each side.

hidden unit layer A layer in a parallel-distributed processing (PDP) model that neither receives direct external input from, nor sends direct output to, the environment.

high-level cognition An informal, but widely used, classification for most mental functions other than sensory perception, motor control, and some aspects of memory and attention. Examples include language, problem solving, reasoning, and meta-cognition (thinking about thinking).

hippocampus A subcortical structure of the cerebrum that, in the primate, is surrounded by the cortex of the temporal lobe, and lies parallel to this lobe's long axis. The term is Latin for seahorse, of which its shape is suggestive.

homunculus A "little human" in the brain. In the context of somatosensory and motor neuroscience, the homunculus refers to the somatotopic representation of the post- and precentral gyri, respectively; in the context of psychology, the homunculus refers to a hypothetical little person inside the brain who is responsible for one's cognitive functions.

horizontal meridian The imaginary horizontal line that divides the visual field into upper and lower quadrants.

hypercolumn A hypothesized functional unit of cortical columns in V1, the totality of which represents every possible feature that could be presented to a specific V1 receptive field.

hyperpolarize To make the membrane potential of a cell more negative. (I.e., the opposite of *depolarize*.).

implicit memory The influence of past experience on present behavior that is accompanied by awareness of neither this influence, nor the past experience. Synonymous with *nondeclarative memory*.

individual differences The study of factors underlying why individuals in a sample or population differ from each other. In studies of brain–behavior relationships, for example, a between-subject correlation between levels of some neurobiological dependent measure (e.g., neuroimaging signal in Region x) and variation in a behavioral measure (e.g., RT on Task y) is taken as compelling evidence for a role for x in performance of y.

information theory A set of mathematical formalisms for understanding how messages (i.e., information) can be transmitted over a noisy channel and understood by a receiver.

inhibitory interneurons Along with pyramidal cells, the second major class of neurons in the cerebral cortex. Although there are many types of inhibitory interneuron, this book will primarily refer to the generic interneuron: a GABAergic cell with only local projections to nearby pyramidal cells.

inhibitory neurotransmission Of a neurotransmitter whose effect on the postsynaptic membrane potential is to hyperpolarize it, i.e., to make it more negative.

inhibitory surround A mechanism for increasing the signal-to-noise ratio of a neural representation by inhibiting activity corresponding to adjacent, but irrelevant, representations.

inner hair cells Cells that are attached to the basilar membrane, the vibration of which causes them to release neurotransmitter onto a neuron of the cochlear nucleus, thereby effecting the transduction of acoustic energy into a neural signal.

innervate To connect to a neuron via one or more synaptic connections.

intention When used in contradistinction to attention, intention refers to the processes required to prepare for (or, sometimes, just to contemplate) performing an action.

interaural intensity difference (IID) The phenomenon that the acoustic energy from a single sound source, if not positioned in direct alignment with the midline of the head, will be of greater intensity for one ear than the other. This provides a useful cue for sound localization.

interaural timing difference (ITD) The phenomenon that the time of arrival of the acoustic energy emanating from a single sound source, if not positioned in direct alignment with the midline of the head, will be quicker for one ear than the other. This provides a useful cue for sound localization.

intertemporal choice A situation in which an individual must choose between immediate vs. delayed reward, and often producing the phenomenon of delay discounting.

intracellular recording Electrical recordings made with a hollow-tube electrode that punctures the membrane of a cell (and forms a tight seal between membrane and electrode) so that fluctuations of voltage within a that individual neuron can be measured.

invariant The property of not responding to changes along a particular dimension.

inverse problem The fact that there is an infinite number of possible source solutions for any pattern of voltages distributed across the scalp.

isoluminant A display in which all objects, plus the background, are presented with the same luminance, and so only color information can be used to segment the scene.

kinematics The dynamics of movement or motion without consideration of the forces that impel it.

laminar "Lamina" is another word for layer; although derived from Latin, the plural form is "laminas" or "laminae." It is often used as an adjective when referring to layer-specific (i.e., "laminar") organization within cortex.

lateral inhibition When that activity of a neuron in area X inhibits the activity of other neurons in area X; i.e., the source of the inhibition is neither feedback nor feedforward, but "feed-sideways" / "feed-laterally."

lesion Used as a noun, it refers to either damage or an abnormality. It is a general term, in that it does not specify either the nature or the cause. Thus, a region of dead or otherwise compromised tissue can be referred to as a lesion regardless of whether it was caused by viral encephalitis (swelling), by anoxia (deprivation of oxygen, such as could happen as a result of a stroke), or by the experimental injection of ibotenic acid, a chemical that kills the soma of neurons, but not fibers of passage. (Note that each of these causes would produce a different pattern of damage.)

ligand A substance, typically a molecule, that binds to a specific site on a protein, thereby triggering some process. In the context of neurotransmission, the ligand is the neurotransmitter that binds with a binding site on the postsynaptic receptor.

local field potential (LFP) The oscillating electrical potential recorded from extracellular space that reflects the moment-to-moment aggregation of fluctuating voltages from all dendritic trees in the vicinity of the electrode.

localization of function The idea that a behavioral or cognitive function can arise from a discrete (i.e., "localized") anatomical region of the brain.

machine learning A branch of computer science that studies how computer algorithms can learn about structure in a set of data.

major depressive disorder (MDD) A mood state characterized by sadness, poor concentration, and loss of interest in daily activities, that can also be accompanied by fatigue, insomnia, and loss of appetite.

mass action The idea that behavioral and cognitive functions cannot be understood as arising from discrete anatomical regions of the brain.

mechanoreceptors Receptors in the skin that respond to mechanical forces: pressure, texture, flutter, vibration, and stretch.

mental lexicon The subset of semantic memory corresponding to all the words that one knows the meaning of.

motor effector The part of the body that actually carries out an action – most often used to refer to the limbs and the eyes.

motor efferent An efferent fiber is a fiber projecting away from a region (an afferent is a fiber projecting into a region). The term "motor efferents" refers to fibers leaving the cortex and synapsing in the spinal cord.

myelin The cholesterol-laden sheath that insulates axons. Myelin is formed by branches from a type of glial cell, called oligodendrocytes, that wraps the axon in several layers.

Na$^+$/K$^+$ pump Also known as "ATPase," because it's an enzyme that requires ATP for its operation. The Na$^+$/K$^+$ pump is a membrane-spanning complex that actively restores the electrochemical gradients of Na$^+$ and K$^+$ that are decreased by the mechanisms of the action potential and of neurotransmitter reuptake.

natural language Any human language commonly used for verbal or written communication that has arisen in an unpremeditated fashion. Computer-programming languages and languages used in formal logic are not natural languages.

neural code The way that the nervous system represents information via action potentials and neurochemicals.

neurodegenerative disease A disease characterized by progressive loss of brain tissue.

neuroeconomics The study of the neural bases of value-based decision making.

neurons The cells in the brain that generate electrical and chemical signals that control all other systems of the body. The typical neuron has four distinct regions: dendrites, cell body, axon, and synaptic terminals (see *Chapter 2*).

neurophysiology The study of the functioning of the brain. Two broad classes of neurophysiology are *stimulation* and *measurement*.

neuropsychology The careful analysis of the behavioral consequences resulting from damage to a particular brain structure. Neuropsychology can either be conducted by experimentally producing a surgical lesion in an animal, or by observing the effects on behavior of brain damage incurred by humans.

nigrostriatal tract The bundle of axons, originating from the substantia nigra pars compacta, responsible for the release of dopamine onto the neostriatum.

nondeclarative memory The influence of past experience on present behavior that is accompanied by awareness of neither this influence, nor the past experience. Synonymous with *implicit memory*.

nucleus (*pl. -clei*) In the context of the cell, the nucleus is the central structure containing DNA. In the context of gross anatomy, a nucleus is a subcortical group of neurons, often visible to the naked eye, that typically all perform the same function.

ontogeny The development of an organism; in the case of vertebrates, from fertilization through birth and the entirety of the lifespan.

outer hair cells Cells anchored at one end to the basilar membrane, and at the other to the cochlea; their coordinated activity can dampen the vibrations of the basilar membrane in precise areas.

overt attention Attention deployed toward an object or location by looking directly at it.

paraphasic speech Speech including unintended/inappropriate syllables, words, or phrases.

perception The identification of important features of the item or information being sensed.

perceptual decision A categorical decision about a perceived stimulus (e.g., *Is it moving up or down?, Is that tone louder or* *softer?*). In effect, *What is the objective state of this specific part of my environment?* Compare with *value-based decision* and *foraging decision*.

peripersonal space The space surrounding one's body that is within one's grasp.

peripheral nervous system The peripheral nervous system (sometimes abbreviated PNS) is made up of all neurons in the body whose cell bodies are located outside of the central nervous system. These include somatosensory (i.e., "touch") neurons whose cell bodies are in the skin and that send signals via axons that project into the spinal cord, and neurons that control autonomic functions of many organ systems of the body.

perseveration The persistent selection of a previously reinforced response (or type of response) after having received feedback that it is no longer appropriate.

phase–amplitude coupling (PAC) The periodic waxing and waning of the amplitude of a high-frequency signal (either spiking or oscillation) that is synchronized with the phase of a lower-frequency oscillating signal.

phase precession Precession is the rotation of an axis of rotation, such as the "wobble" of a spinning top. Phase precession occurs when a dot painted on the cylinder of the top is facing in a progressively different direction each time the spinning axis completes one cycle of its wobble.

phase synchronization The phase of a sine wave refers to the location within the cycle (e.g., the peak, or the trough). To synchronize the phase of two oscillators is to adjust them such that, whenever one is at a particular phase in its cycle (say, the peak) the other will always be at the same phase in its (this could also be the peak, or it could be any other phase, so long as its predictable).

phoneme The smallest unit of sound in a language that, when changed, can change the meaning of an utterance.

phototransduction The conversion of energy from the electromagnetic spectrum ("visible light") into neural signals.

phrenology The practice, initiated by Franz Josef Gall (1758–1828), of divining mental traits from patterns of convexities and concavities on the skull.

phylogeny The evolutionary history of organismal lineage.

pinwheel The actual physical layout, discovered with optical imaging, of the cortical columns that accomplish the function that had been proposed for the hypercolumn.

place code The scheme whereby the representation of frequency (pitch) in the cochlear nucleus is laid out in a linear fashion, from high to low.

place of articulation The physical location in the mouth where the tongue applies the blockage that produces stop consonants.

plasticity The property of nervous tissue that its structure and function can modify in response to a change in environmental inputs (or, indeed, to systematic changes of mental activity).

posterior Used as a term of direction, posterior refers to the back (a synonym, introduced in *Chapter 2*, is "caudal").

post-traumatic stress disorder (PTSD) A pathological condition resulting from traumatic experience. Symptoms can include re-experiencing the traumatic event (including "flashbacks" and nightmares); avoidance symptoms; and hyperarousal (includes being easily startled, and feeling tense).

power The magnitude of an oscillatory component in a signal.

power spectrum A representation of the magnitude of each frequency resulting from a Fourier (or *spectral*) transform.

predictive coding The idea that a core function of the brain is to construct, and constantly update, a model of what's likely to happen in the future.

premotor theory of attention The idea that attention may be a consequence of motor planning, covert attention being a motor plan that never gets realized as an actual movement.

primary visual cortex The part of the cerebral cortex that is responsible for the sense of sight.

principal component analysis (PCA) A statistical procedure for summarizing a data set by identifying the elements (the components) that account for the most variance.

proprioception The sense of where different parts of the body are relative to one another.

Priority map A neural representation of the stimuli that are currently the most relevant for behavior.

pyramidal cells Glutamate-releasing neurons with pyramid-shaped cell bodies that are the most common excitatory neuron in the neocortex and hippocampus.

random walk A mathematical formalization of a process that proceeds via stochastic steps.

rate code A neural code whereby information about a change in some metric property is conveyed by a change in the rate at which a neuron fires.

receiver operating characteristic (ROC) A graph plotting hits vs. false alarms as a function of criterion (often operationalized as "confidence"). Deriving from signal detection theory, this measure is intended to capture the strength of a percept (or, in memory research, of a subject's memory) independent of decision criteria.

receptive field The portion of the sensory-transduction surface that is represented by a neuron.

recollection In the context of memory retrieval, the act of remembering the context in which a fact was learned or an event occurred. Hypothesized in some models to correspond to a neural signal and/or process that is qualitatively different from familiarity.

recurrent circuit A circuit carrying feedback signals.

recurrent neural network (RNN) An artificial neural network in which interactivity between units in the hidden layer allows for the processing of information over time, making them well suited for modeling neural information processing that evolves over time, and/or for processing inputs that vary over time (e.g., speech).

region of interest (ROI) In neuroimaging, a region from which signal is extracted for analysis.

reinforcement learning The study of how a biological organism or manmade intelligent agent learns to optimize its behavior based on its past experience with gaining rewards or other desired outcomes.

resonant frequency The frequency at which oscillations are most readily induced in a system, and at which it can oscillate at the greatest amplitude.

retinal ganglion cells The neurons in the retina that send visual signals to the thalamus.

retinotopic organization, or retinotopy The property of reinstating the spatial organization of the retina in the spatial organization of visual regions of the brain, such that, for example, adjacent regions of the retina are represented by adjacent regions of visual cortex.

reuptake The active process of removing molecules of neurotransmitter from the synaptic cleft by transporting them into a cell, either the presynaptic neuron or an adjacent astrocyte.

saccade A French word for a fast eye movement that "jumps" from point A to point B, with "no interest" in what is between the two points. Saccadic eye movements differ from smooth-pursuit eye movements, in which the eye follows the movement of an object in the environment.

salience map A neural representation of the location of stimuli that have the most energy (e.g., the brightest, or the loudest).

scotoma A small area of blindness in the visual field surrounded by normal visual sensation.

seizure An incidence of abnormal brain functioning that can last from seconds to minutes, and can occur as an isolated event or as a chronic condition of repeating seizures (i.e., epilepsy). *Positive* manifestations of seizure can include the perception of flashing light or the uncontrolled, involuntary jerking of a limb; *negative* manifestations can be a transient blindness or paralysis. Seizures result from stereotypical, abnormal patterns of activity in brain cells (neurons).

semantic memory Knowledge that was acquired through learning, but for which one cannot remember the context (e.g., place, situation, who was present) at the time that encoding of that information occurred.

sensation The detection of energy in the external world via one of the sensory organs (eyes, ears, skin, nose, tongue), or of the state of an internal organ via visceral sensors.

sensorimotor circuits Neural pathways connecting the processing of *sensory* information (i.e., having to do with

vision, audition, touch, smell, or taste) with the *motor* system, which controls the movement of the body. The most basic sensorimotor circuits are reflex circuits that afford, for example, rapid withdrawal of the hand from a hot surface.

signal detection theory A framework in which decision-making factors are explicitly taken into account in tests of perception and recognition memory. Entails categorizing each response as a *hit, false alarm, miss,* or *correct rejection*.

social cognitive neuroscience One of many terms used to refer to the study of the neural bases of social interactions.

somatotopy This word derives from *soma* (Greek for *body*) and *-topy* (of or pertaining to a place). The principle of somatotopy is that adjacent parts of the body (for example, the upper lip and the nose) are represented in adjacent parts of the brain. Note that the "map" created by a somatotopically organized portion of cortex need not bear close physical resemblance to the parts of the body being represented – there are frequently distortions and discontinuities.

spectrogram A representation of how a power spectrum evolves over time.

striate cortex Another name for V1, so named for its striped appearance when viewed in just the right way.

stroke Disruption of the supply of blood to a region of the brain, either due to blockage (*occlusive stroke*) or bleeding from a vessel (*hemorrhagic stroke*).

sulci (*sing.* sulcus) Latin for "furrows," fissures on the cortical surface.

superior temporal polysensory area (STP) The region in the superior temporal gyrus of the monkey that receives inputs from visual, auditory, and somatosensory regions, and in which many neurons respond to more than one sensory modality.

synapse Name introduced by Nobel Laureate Charles Sherrington (1857–1952) for the location at which one neuron passes a signal to another cell (within the CNS, it's typically to the dendrite of another neuron; at the neuromuscular junction, it's to a muscle cell). The name is derived from the Greek for "to clasp."

synaptic cleft The gap between the presynaptic and postsynaptic terminals, across which neurotransmitter diffuses from pre- to post-.

systems neuroscience The study of how neural circuits and brain regions give rise to functions of the brain (e.g., behavior, or the regulation of autonomic functions).

thermoreceptors Receptors in the skin that are tuned to detect heat or cold.

tonotopy The organization of primary auditory cortex by frequency, which derives from the place code originating in the cochlear nucleus.

topographic organization The principle that adjacent parts of some aspect of the world outside the brain are represented in adjacent parts of the brain. Thus, somatotopy is one example of topographic organization. Other examples include the representation of the visual field, as projected onto the retina, by the visual cortex ("retinotopy"), the representation of sound in a continuous map from low to high frequencies ("tonotopy"), and the representation of geometric space ("grid cells" of the entorhinal cortex).

transduction The translation of physical energy from the environment into a neural code that represents the properties of that energy.

traveling wave A wave that propagates energy through a medium. In the case of the cochlea, it characterizes the propagation of vibrational energy from the oval window along the length of the basilar membrane.

type vs. token These two words are used to distinguish a category of stimulus (e.g., the category of faces) from a specific exemplar from within that category (e.g., my father's face).

typically developed (or "–ing") Refers to individuals who do not have a neurological or psychiatric diagnosis. In the context of studies of developmental disorders (e.g., autism spectrum disorder or fragile X syndrome), this term is often preferred to the synonym "normal," to avoid the pejorative connotations of the label "abnormal."

utilization behavior An exaggerated tendency for one's behavior to be determined by the external environment.

value-based decision Selecting between two (or more) items when faced with a choice (as, for example, when at the store, or when deciding how to spend one's Saturday evening). The purview of neuroeconomics; compare with *perceptual decision* and *foraging decision*.

ventricle A hollow chamber inside the brain through which cerebrospinal fluid (CSF) circulates. Like a chain of lakes connected by canals, the ventricles are part of a continuous system of CSF that extends to the caudal end of the spinal cord.

vertical meridian The imaginary vertical line that divides the visual field in half, thereby defining the boundary between the two hemifields.

viral encephalitis Acute inflammation of the brain due to viral infection.

visual field The full extent of what one can see while holding the eyes still, measured as degrees of angle from the point of fixation.

visual hemifield The half of the visual field represented in V1 of one hemisphere.

voxel A 3-D image element, the smallest unit of data in a PET or fMRI data set; 2-D "picture elements" having earned the nickname "pixel," these "pixels with volume" have been dubbed "voxels."

white matter Made up of billions of myelin-sheathed axons, the white matter makes up the majority of the volume of the cerebrum.

INDEX

Note: page numbers in *italic* refer to figures.